D1563756

Springer Monographs in Mathematics

Wolmer Vasconcelos

Integral Closure

Rees Algebras, Multiplicities, Algorithms

Wolmer Vasconcelos
Department of Mathematics
Rutgers University
Piscataway, NJ 08854
USA
e-mail: vasconce@math.rutgers.edu

Library of Congress Control Number: 2005923162

Mathematics Subject Classification (2000):
Primary 13D40, 13D22, 14E15
Secondary 13D45, 13P99, 14Q99

ISSN 1439-7382
ISBN-10 3-540-25540-0 Springer Berlin Heidelberg New York
ISBN-13 978-3-540-25540-6 Springer Berlin Heidelberg New York

Springer is a part of Springer Science+Business Media
springeronline.com

Printed in Germany

Typesetting: by the author using a Springer LaTeX macro package
Cover design: Erich Kirchner, Heidelberg, Germany

Printed on acid-free paper 46/sz - 5 4 3 2 1 0

This is for Aurea

Preface

There are many ways of looking at the properties of an ideal I of a Noetherian ring R, an R–module E, or even an R–algebra B and the geometric objects they represent. We single out the following aspects:

- Syzygies & Hilbert functions
- Polynomial relations
- Linkage & deformation
- Integral closure
- Primary decomposition
- Complexity of computations

These aspects often arise as outcomes of processes required to understand the deep structure of the subvariety defined by I, but also in many constructions defined on I (and similarly on E or B.) One way to look at them all is via the filtrations that initialize on I, giving rise to Rees algebras and Hilbert functions. The examination of these algebras and functions embraces all of these issues and some more.

Blowup algebras realize as rings of functions the process of blowing up a variety along a subvariety. This book will focus on Rees rings of ideals (and their generalizations), the most ubiquitous of those algebras. It is primarily aimed however at providing a survey of several recent developments (and some not so recent) on integral closure in many guises. It will be realized by looking at the numerical invariants, special divisors and attached algebras whose interplay assists in understanding the arithmetic of various algebras. The underlying techniques are fundamentally computations of cohomology on suitable algebras. It seeks out those regularity properties usually associated with Cohen–Macaulayness and normality. The first controls many of the numerical invariants of the blowup, while the latter is a required ingredient of desingularization.

The required extension of these problems, from ideals to modules, is very apparent already in the comparison of associated graded rings with Rees algebras of conormal modules. Given the centrality of finding integral closures of algebras, our subject finds its full theme: the study of integral closure of algebras, ideals and modules.

Emphasis will be placed on determining these invariants and properties from a description of the ring, ideal or module by generators and relations. Open problems and basic techniques will be stressed at the expense of individual results. Another rule of navigation used was that given the choice between two approaches to a topic, the path that seemed more constructive to the author was taken—unless the sheer beauty of the more abstract method made the choice inevitable.

A limited effort in this direction by the author [Va94b] was slightly premature, as the subject was experiencing just then an explosion of activity. Given the rate of change in the subject, these notes will also be premature, with probability of one, whenever they appear. It is worth emphasizing that while this book and [Va94b] deal with some of the same algebras, the focus here is on normalization while the other book was mainly aimed at the study of the Cohen–Macaulay property. However, these topics have grown so much that the subject deserves an exclusive treatment. We shall nevertheless touch on several of these activities.

The reader will observe an imbalance of attention given to certain topics. They are often an expression of ignorance/lack of expertise by the author, or simply that the topic is undergoing so many developments that it would be premature to stop & look around.

We are very grateful to many colleagues who over the years have shared directly (or indirectly via their publications) their ideas on these subjects with the author. Others had a very direct involvement as co-authors in some of the joint research reported here or freely took on the task of proofreading parts of the manuscript. To all of them, but particularly to Joe Brennan, Alberto Corso, Sam Huckaba, Jooyoun Hong, Claudia Polini, Aron Simis, Amelia Taylor, Bernd Ulrich and Rafael Villarreal, the author gives his heartfelt thanks. The support and care of Martin Peters and Ruth Allewelt at Springer is much appreciated. Finally, we are grateful to James Wiegold for his careful editorial work on the manuscript.

The financial support of the National Science Foundation is also gratefully acknowledged.

New Brunnswick, *Wolmer V Vasconcelos*
15 March 2005

Contents

Introduction

This book is a treatment of the notion of integral closure as it applies to ideals, modules and algebras, and associated algebras, emphasizing their relationships. Its main theme concerns the structures that arise as solutions of collections of equations of integral dependence in an algebra A,

$$\boxed{z^n + a_1 z^{n-1} + \cdots + a_n = 0.}$$

In such equations, the a_i and z are required to satisfy set-theoretic restrictions of various kinds, with the solutions assembled into algebras, ideals, or modules, each process adding a particular flavor to the subject. The study of these equations–including the search for the equations themselves–is a region of convergence of many interests in algebraic geometry, commutative algebra, number theory and computational aspects of each of these fields. The overall goal is to find and understand the equations defining these assemblages. Both challenges and opportunities arise with these issues, the former because of the difficulties current models of computation have in dealing with them, the latter in the need to develop new theoretical approaches to its understanding.

For one of these structures, $\mathbf{S}$, the analysis and/or construction of its integral closure $\overline{\mathbf{S}}$ usually passes through the study of its so–called reductions $\mathbf{S}_0$:

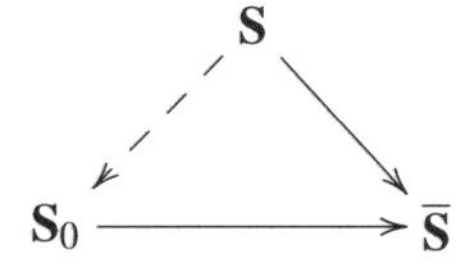

The $\mathbf{S}_0$ are structures similar to $\mathbf{S}$, with the same closure as $\mathbf{S}$, possibly providing a pathway to the closure which does not pass through $\mathbf{S}$. They are geometrically and computationally simpler than $\mathbf{S}$ and therefore provide a convenient platform from which to examine $\overline{\mathbf{S}}$, and they give rise to many structures that ultimately bear on $\mathbf{S}$.

It is not too far-fetched to make an equivalence between the study of all the $\mathbf{S}_0$ and of $\overline{\mathbf{S}}$. Rather than a source of frustration, this diversity is a mine of opportunities to examine $\mathbf{S}$, and often it is the springboard to the examination of other properties of $\mathbf{S}$ besides its closure.

An algebraic structure–a ring, an ideal or even a module–is often susceptible to smoothing processes that enhance its properties. One major process is the *integral closure* of the structure. This often enable them to support new constructions, including analytic ones. In the case of algebras, the divisors acquire a group structure, and the cohomology tends to slim down. To make this more viable, multiplicity theory–broadly seen as the assignment of measures of size to a structure–must be built up with the introduction of new families of degree functions suitable for tracking the processes through their complexity costs. The synergy between these two regions is illustrated in the diagram:

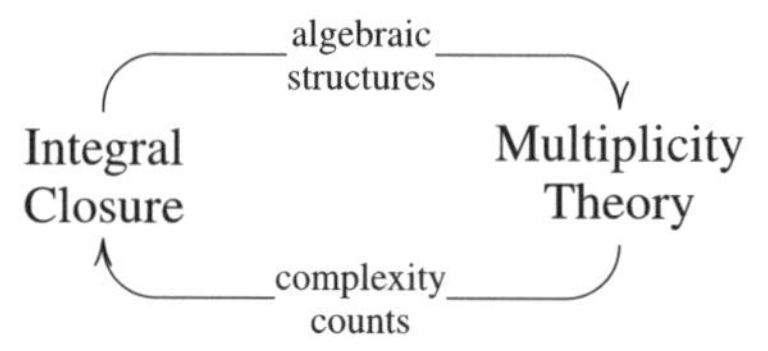

The overall goal is to describe the developments leading to this picture, and, hopefully, of setting the stage for further research.

There is an obvious organization for the several problems that arise. Without emphasizing relationships one has:

- □ *Membership Test*: $f \in \overline{\mathbf{S}}$?
- □ *Completeness Test*: $\mathbf{S} = \overline{\mathbf{S}}$?
- □ *Construction Task*: $\mathbf{S} \rightsquigarrow \overline{\mathbf{S}}$?
- □ *Complexity Cost*: $\mathrm{cx}(\mathbf{S} \rightsquigarrow \overline{\mathbf{S}})$?

None of these problems yet has an optimal solution that would permit checkoffs, certainly not for the various structures treated. In addition to these issues, predicting properties of $\overline{\mathbf{S}}$ from those of $\mathbf{S}$–such as number of generators and their degrees as the case may be–becomes rather compelling and can also be stated in terms of complexities. We shall emphasize the development of numerical signatures for algebras. These are numerical functions $\tau(\cdot)$ on algebras, usually based on their cohomology, with properties such as

$$\tau(A) = \tau(B) \Rightarrow \overline{A} = \overline{B},$$

or used to tell them apart, as in

$$\overline{A} = \overline{B} \quad \& \quad A \subsetneq B \Rightarrow \tau(A) \neq \tau(B).$$

Given an affine algebra $\mathbf{S}$ over a constructible field, the usual approach to the construction of $\overline{\mathbf{S}}$ starting from $\mathbf{S}$, or from some $\mathbf{S}_0$, involves a procedure $\mathcal{P}$ which when applied to $\mathbf{S}$ produces an extension

$$\mathbf{S} \subset \mathcal{P}(\mathbf{S}) \subset \overline{\mathbf{S}},$$

properly containing **S**, unless it is already the desired closure.

The presence of Noetherian conditions will guarantee the existence of an integer r such that $\mathcal{P}^r(\mathbf{S}) = \overline{\mathbf{S}}$. Interestingly, though, under appropriate control conditions obtained by judicious choices for $\mathcal{P}$, it is possible to estimate the order r from $\mathbf{S}_0$. More delicate issues are those regarding the number of generators of $\mathcal{P}^n(\mathbf{S})$ (and their degrees in the graded case):

- □ Determine r such that $\mathcal{P}^r(\mathbf{S}) = \overline{\mathbf{S}}$.
- □ Determine the number of (module or algebra) generators of $\overline{\mathbf{S}}$.
- □ In the graded case, determine the degrees of the (module or algebra) generators of $\overline{\mathbf{S}}$.

If $\mathcal{P}$ allows for answering the second and third problems for $\mathcal{P}(\mathbf{S})$ in terms of **S**, then we would achieve a good understanding of the whole closure process.

There has been progress along the whole front of problems outlined above. For the other algebraic structures we treat–ideals and modules–the picture is not so rosy. While each of them–to wit, the construction or understanding of the integral closure of an ideal or module **S**–can be converted into another involving an associated algebra, this process is not robust under two important criteria: The conversion process makes the complexity intractable and brings no understanding of the nature of $\overline{\mathbf{S}}$. Instead we must look for direct pathways, sensitive to the nature of the ideal or module.

In our circuitous approach to these questions–in ever-expanding circles–we look first at ideals, which turns out to be the more challenging of the endeavors. For ideals, unlike the case of algebras when platforms provided by Noether normalizations and Jacobian criteria are enabling, the approach has been indirect through the construction of associated algebras. Instead, while keeping the azimuth in view, we use an approach in which the trip itself becomes part of the goal. It will allow a treatment of several other problems in what might be fuzzily named the nonlinear theory of ideals, which we take to mean the theory of the algebras built up from processes defined on ideals. More concretely, if the syzygies of an ideal I code for its *linear* invariants, where should we look for the *nonlinear* invariants of I? Two direct partial answers are to look at the syzygies of the powers I^n of I and at the algebraic relations amongst sets of elements of I. Other invariants are the suitably-interpreted multiplicities from these algebras.

There are many ways of looking at the properties of an ideal I of a Noetherian ring R. We single out the following aspects:

- Syzygies
- Polynomial relations
- Linkage
- Integral closure of ideals, modules, and algebras
- Primary decomposition

- Complexity of computations
- Rees algebras of filtrations

One way to look at them all is through the filtrations that initialize on I, giving rise to algebras and Hilbert functions. Thus to an ideal I, we can associate several filtrations: I^n, the ordinary powers of I, $\overline{I^n}$, their integral closures, $I^{(n)}$, their symbolic powers, and if R is a ring of polynomials over a field k and $>$ is a monomial order, the filtration $\text{in}_>(I^n)$. The examination of corresponding algebras and functions embraces nearly all of these aspects, and it can also be used to capture numerical constraints on the ideals, modules and algebras.

There is another explanation for the dominant role played by the study of the integral closure of ideals. In dealing with algebras or modules, very often the most natural path to the study of integrality passes through operations involving reductions to the ideal case. Typically this occurs when the syzygetic information in projective presentations is converted into its associated determinantal ideals, but also when Noether normalizations are employed.

AGENTS

We shall describe the threads whose intermingling form the structure of the book: *Rees algebras*, *Hilbert functions* and its invariants particularly those expressible in the *cohomology of blowups*. Honestly put, our interest in these algebras is greater as vessels for many other questions in commutative and computational algebra and algebraic geometry.

The particular class of algebras that dominate our exposition are called *blowup algebras*. They arise in the process of blowing-up a variety along a subvariety: In the case of an affine variety $V(I) \subset \text{Spec}\,(R)$, the *blowing-up morphism* is the natural mapping

$$\text{Bl}_I(R) = \text{Proj}\,(R[It]) \xrightarrow{\pi} \text{Spec}\,(R),$$

where I is an ideal of the commutative ring R and $R[It]$ is the subalgebra of the polynomial ring $R[t]$ generated by the 1-forms It.

Algebras such as these appear in many other constructions in commutative algebra and algebraic geometry. They represent fibrations of a variety with fibers which are often affine spaces; a polynomial ring $R[T_1,\ldots,T_n]$ is the notorious example of such algebras. In addition to its role in Hironaka's celebrated theorem of resolution of singularities, its uses include counterexamples to Hilbert's 14th Problem, the determination of the minimal number of equations needed to define algebraic varieties, the computation of some invariants of Lie groups, and several others. An impetus for their systematic study has been the long list of beautiful Cohen-Macaulay algebras produced by the various processes. Finally, they provide a testing ground for several computational methods in commutative algebra.

A common framework for these algebras is the following. Let R be a commutative ring, $I \subset R$ an ideal, $\varphi : R \to S$ a ring homomorphism and $\theta : I \to E$ an embedding of I into an R-module:

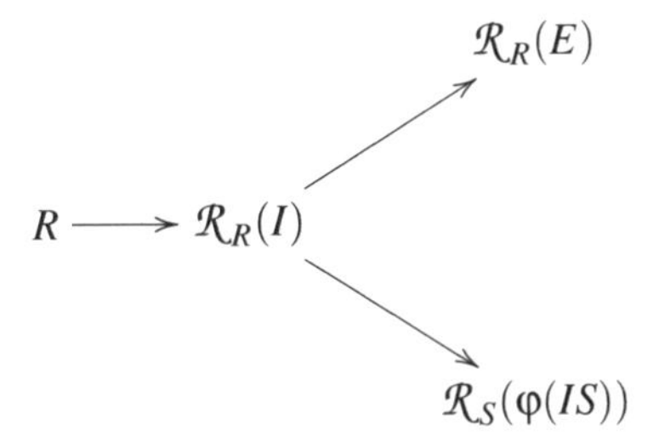

There results a series of morphisms with the expectation that some feature of R gets improved or a property of I gets exposed.

The problems discussed here arise from the following sharply framed questions. To illustrate, let I be an ideal of the Noetherian ring R.

- When is the Rees algebra $R[It]$ Cohen-Macaulay? In which way can this property be detected or expressed? What does the cohomology of $R[It]$ read in the algebra?
- What is the integral closure $\overline{I}$ of I? Are there methods, algorithms and analyses of costs to find it?
- What are the uses of $\overline{I}$ and how do they add to the understanding of I?

In principle, some of these questions have ready answers. But these are so vague that they verge on the inconsequential, as they come detached from the data that define I (we illustrate this view later). It would be significantly more valuable to have answers based on the web of relationships I that has with R (e.g., its *syzygies*), and with other related ideals (e.g., its *reductions*). These answers come with attached numbers/degrees that usually have some predictive power in uncovering properties of various varieties defined by I. Such a quest will involve generalizations of the basic issues (such as the focus on Rees algebras of modules), the preoccupation with algorithms and off-beat ways of measuring their complexity, and theoretical problems of non-classical multiplicity.

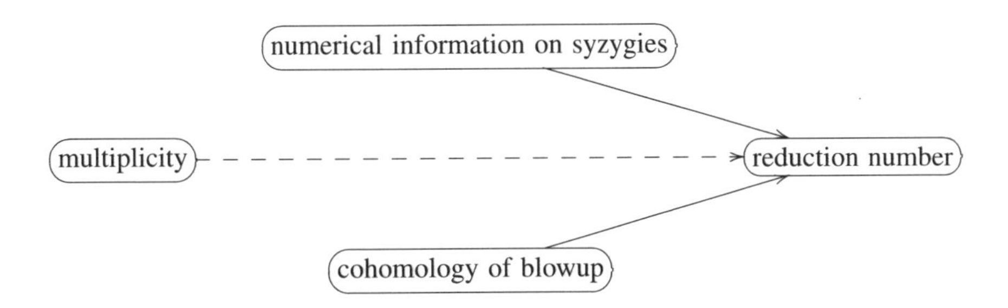

Graded Algebras and Hilbert Functions

Let $G = \bigoplus_n G_n$ be a finitely generated graded algebra over the Noetherian ring $R = G_0$. A notion that plays a pivotal role throughout is that of a *reduction subalgebra* of G. This is simply a graded R-subalgebra $H \subset G$ over which G is finite,

$$G = \sum_{i=1}^{m} H g_i.$$

The premier example of a reduction subalgebra is that of a Noether normalization of an affine graded algebra over a field k. Another is the Rees algebra $R[It]$ and the subalgebra defined by a *reduction* J of I.

Reductions provide a vehicle for studying an algebra G when they are considerably simpler in structure and the relationship between G and H can be coded conveniently. The natural notion of *minimal reductions*–such as a Noether normalization- and the corresponding degrees of the generators g_i are model examples. They are usually refined by choosing minimal reductions that yield a minimizing value for some numerical function of the degrees of the g_i. Probably the most useful of them is $\mathrm{r}(G)$, the *reduction number* of G, obtained by minimizing over all minimal reductions $\inf_H\{\max\{\deg(g_i) \mid G = \sum H g_i\}\}$.

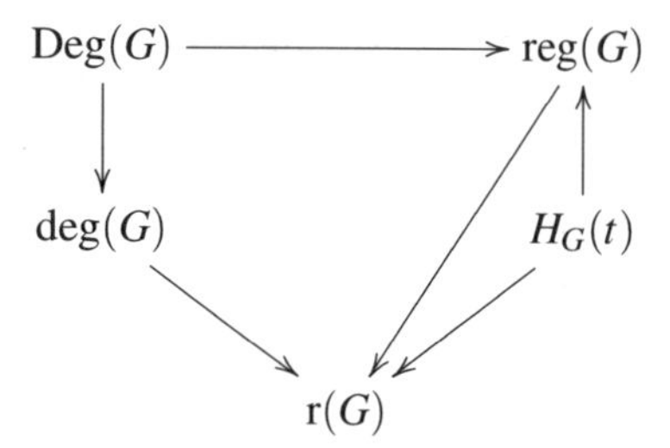

Reduction numbers are loosely related by several other measures of a graded algebra: notions of multiplicity, such as the classical multiplicity $\deg(G)$ and some extended forms of it ($\mathrm{Deg}(G)$, particularly), Castelnuovo-Mumford regularity $\mathrm{reg}(G)$, and the entire Hilbert series $H_G(t)$.

The relationships between these numbers and objects is not very direct, but are mediated by several other agents. One desirable goal would be to seek to capture these various possibilities by some low-degree polynomial inequalities. For example, to obtain expressions of the kind

$$\mathrm{r}(G) \leq f(\mathbf{h}(G)),$$

where $\mathbf{h}(G)$ is the $\mathbf{h}$-vector of G.

Filtrations, Rees Algebras and Associated Graded Rings

A *filtration* of a ring R is a family $\mathcal{F}$ of subgroups F_i of R indexed by some set S. Some of the most useful kinds are indexed by an ordered monoid S, are *multiplicative*,

$$F_i \cdot F_j \subset F_{i+j},\ i, j \in S,$$

and are either *increasing* or *decreasing*, that is $F_i \subset F_j$ if $i \leq j$ or conversely. The *Rees algebra* of the filtration $\mathcal{F}$ is the graded ring

$$R(\mathcal{F}) := \bigoplus_{i \in S} F_i,$$

with natural addition and multiplication. If the filtration is decreasing, there is another algebra attached to it, the *associated graded ring*

$$\mathrm{gr}_{\mathcal{F}}(R) := \bigoplus_{i \in S} F_i/F_{<i},$$

with $F_{<i} = \bigcup_{j<i} F_j$.

There are several issues regarding such filtrations, beginning with whether $R(\mathcal{F})$ is a Noetherian ring, in which case one wants to determine the degrees of its generators. A great deal of emphasis is placed on constructing amenable subfiltrations, $\mathcal{F}_0 \subset \mathcal{F}$, whose algebras $R(\mathcal{F}_0)$ are considerably simpler and examine its relationship to $R(\mathcal{F})$ very much in the manner Noether normalizations are used.

I-adic Filtrations

The algebras we shall study arise from special filtrations of a commutative ring, namely multiplicative decreasing $\mathbb{N}$-filtrations $\mathcal{F} = \{R_n,\ n \in \mathbb{N}\}$ of R, where each R_n is an ideal of R, and

$$R_m \cdot R_n \subset R_{m+n}.$$

Its Rees algebra can be coded as a subring of the polynomial ring

$$R(\mathcal{F}) = \sum_{n \in \mathbb{N}} R_n t^n \subset R[t].$$

In addition to the associated graded ring as above, we also have the extended Rees algebra

$$R_e(\mathcal{F}) = R(\mathcal{F})[t^{-1}] = \sum_{n \in \mathbb{Z}} R_n t^n \subset R[t, t^{-1}].$$

These representations are useful when computing Krull dimensions. Very important is the isomorphism

$$R_e(\mathcal{F})/(t^{-1}) \cong \mathrm{gr}_{\mathcal{F}}(R) = \bigoplus_{n=0}^{\infty} R_n/R_{n+1}.$$

It provides a mechanism to pass properties from $\mathrm{gr}_{\mathcal{F}}(R)$ to R itself (but we purposely leave this statement obscure for the reader to puzzle it out).

A major example is the I-adic filtration of an ideal I: $R_n = I^n$, $n \geq 0$. Its *Rees algebra*, which will be denoted by $R[It]$ or $\mathcal{R}(I)$, has its significance centered on the fact that it provides an algebraic realization for the classical notion of blowing-up a variety along a subvariety, and plays an important role in the birational study of algebraic varieties, particularly in the study of desingularization.

Symmetric Algebras

The ancestors of these rings, *symmetric algebras*, have several other interesting descendants. Given a commutative ring R and an R-module E, the symmetric algebra of E is an R-algebra $S(E)$ which together with a R-module homomorphism

$$\psi : E \to S(E)$$

solves the following universal problem:

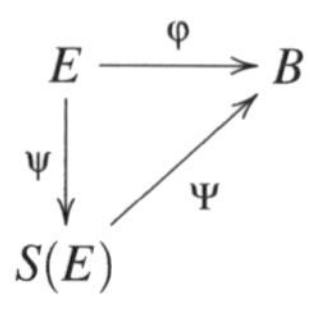

For a commutative R-algebra B and any R-module homomorphism $\varphi : E \to B$, there exists a unique R-algebra homomorphism $\Psi : S(E) \to B$ that makes the diagram commutative. Thus, if E is a free module, $S(E)$ is a polynomial ring $R[T_1,\ldots,T_n]$, one variable for each element in a given basis of E. More generally, when E is given by the presentation

$$R^m \xrightarrow{\varphi} R^n \longrightarrow E \to 0, \ \varphi = (a_{ij}),$$

its symmetric algebra is the quotient of the polynomial ring $R[T_1,\ldots,T_n]$ by the ideal $J(E)$ generated by the 1-forms

$$f_j = a_{1j}T_1 + \cdots + a_{nj}T_n, \ j = 1,\ldots,m.$$

Conversely, any quotient ring of a polynomial ring $R[T_1,\ldots,T_n]/J$, with J generated by 1-forms in the T_i, is the symmetric algebra of a module. Like the classical blowup, the morphism

$$\text{Spec}(S(E)) \to \text{Spec}(R)$$

is a fibration of $\text{Spec}(R)$ by a family of hyperplanes. The case of a vector bundle, when E is a projective module, already warrants interest.

The other algebras are derived from $S(E)$ by effecting modifications on its components, some rather mild but others brutal. To show how this comes about, consider the case of ideals. For an ideal $I \subset R$, there is a canonical surjection

$$\alpha : S(I) \to \mathcal{R}(I).$$

If, further, R is an integral domain, the kernel of α is just the R-torsion submodule of $S(I)$.

Symbolic Rees Algebra

Another filtration is that associated with the symbolic powers of the ideal I. If I is a prime ideal, its nth symbolic power is the I-primary component of I^n. (There is a more general definition if I is not prime.) Its Rees algebra

$$\mathcal{R}_s(I) = \sum_{n\geq 0} I^{(n)}t^n,$$

the *symbolic Rees algebra* of I, which also represents a blowup, inherits more readily the divisorial properties of R, but is of limited usefulness because it is not always Noetherian. The presence of Noetherianess in $\mathcal{R}_s(I)$ is loosely linked to the number of equations necessary to define set-theoretically the subvariety $V(I)$ ([Co84]). In turn, the lack of Noetherianess in certain cases has been used to construct counterexamples to Hilbert's 14th Problem.

Rees Algebra of a Module

Let E is a finitely generated R-module, what is, or what should be, the *Rees algebra* $\mathcal{R}(E)$ of E? This question has an ambiguous answer that can perhaps be used to some profit. A direct answer, intuited from the case of ideals, might be the often-used declaration,

$$\mathcal{R}(E) = S_R(E)/\mathcal{A},$$

where $S_R(E)$ is the symmetric algebra of E and $\mathcal{A}$ is the ideal of R-torsion elements (the elements of $S_R(E)$ annihilated by regular elements of R). Put otherwise, let

$$\varphi : E \to B$$

be an R-homomorphism of E into an R-algebra B. If B is torsionfree over R, the canonical extension of φ

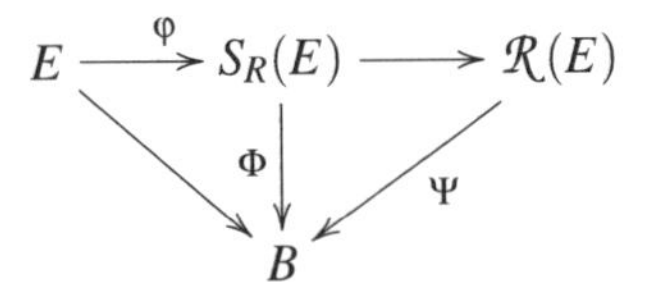

to the R-algebra homomorphism Φ restricts to the canonical homomorphism Ψ of torsionfree R-algebras. An unambiguous case is when $S_R(E) \cong \mathcal{R}(E)$, and the module is then said to be of *linear type*.

An alternative construction is the following. Suppose that E embeds into the free module $F = R^p$. In the symmetric algebra $S = S_R(F) = R[T_1, \ldots, T_p]$, the module E (viewed as the set of linear forms in S_1 defined by E), induces two multiplicative filtrations: E^n, the powers of the linear forms in the T_i, and the powers of $\widetilde{E}$, the ideal (E) generated by those forms. Two Rees algebras are obtained as

$$\mathcal{R}(E) = \bigoplus_{n\geq 0} E^n,$$
$$\mathcal{R}(\widetilde{E}) = \bigoplus_{n\geq 0} \widetilde{E}^n.$$

The first of these is what we called above the Rees algebra of E (at least when R is a domain). The second one is more orthodox however: $\mathcal{R}(\widetilde{I}) = R[It][X]$! So, which is the richer algebra? Note the diagram

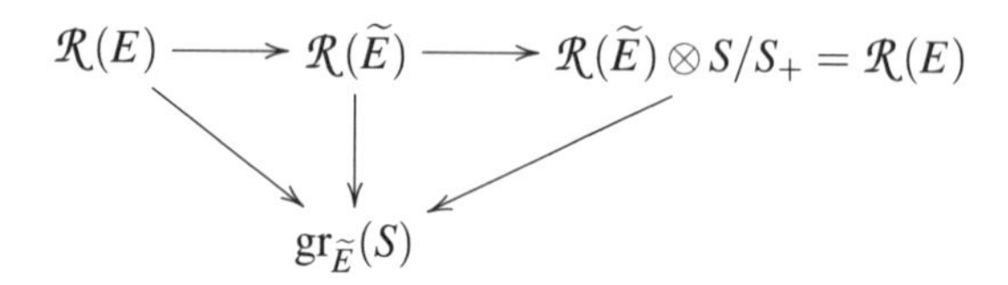

In the case of an ideal I, to obtain $\mathrm{gr}_I(R)$ we still need a reduction since $\mathrm{gr}_{\widetilde{I}}(S) = \mathrm{gr}_I(R)[X]$. Consider the case when $p = \mathrm{rank}\ (E)$. Now

$$\dim \mathcal{R}(E) = \dim \mathrm{gr}_{\widetilde{E}}(S) = \dim \mathcal{R}(\widetilde{E}) - 1 = \dim R + p.$$

In this case ($p = \mathrm{rank}\ (E)$), how is $\mathcal{R}(\widetilde{E})$ independent of the embedding?

Our definition of Rees algebras has been derived from multiplicative filtrations, the main one from the powers I^n of an ideal $I \subset R$. This emphasizes the fact that the nature of the embedding, in particular the geometry of Spec (R/I), should influence the properties of $\mathrm{Bl}_I(R)$. In efforts to extend the definition to a module E, without a preferred embedding into a free module, some attempts have been made to define the Rees algebra of E *in abstracto*. The most common approach is to set

$$\mathcal{R}(E)^t := S(E)/(R\text{-torsion}),$$

where $S(E)$ is the R-symmetric algebra of E.

In [EHU3], Eisenbud, Huneke and Ulrich proposed another notion of Rees algebra for modules. If R is a Noetherian ring and E is a finitely generated module, they define

$$\mathcal{R}(E)^f := S(E)/\bigcap \ker(f),$$

where f runs over all maps from E into free R-modules, with $\ker(f)$ denoting the kernels of the homomorphisms of symmetric algebras. Despite the apparent difficulty in using this definition, they show how to recover very directly information about the defining equations for the algebra and how to deal with issues such as reductions. At the same time, examples of ideals are given whose Rees algebras depend on the embedding.

Here is yet another candidate. Let R be a Noetherian ring and let E be a finitely generated R-module. Let

$$E \hookrightarrow E_0$$

be the embedding of E into its injective envelope. Set

$$\mathcal{R}(E)^i := \text{image } S(E) \to S(E_0).$$

The three definitions lead to different algebras, although some relationships are apparent. Thus, they are isomorphic when the total ring of fractions of R is a Gorenstein ring. They all highlight the significance of the role of the R-module structure of E.

Our attitude will be to just let all these algebras co-exist as long as they can be used to trap properties of the module. There is another significant difference between the Rees algebra of an ideal I and a submodule $E \hookrightarrow R^r$ of rank greater than 1. While the Veronese subrings of $R[It]$ are still Rees algebras of ideals, the proper Veronese subrings of $\mathcal{R}(E)$ are Rees algebras of modules no longer.

Intertwining Algebra of a Morphism

Let R be a commutative ring and consider a homogeneous homomorphism of graded R-algebras,

$$\varphi : A \mapsto B.$$

We consider an encoding of this morphism into a third algebra C. It has for benefit of converting set-theoretic properties of φ into ideal-theoretic properties of C, which often offer a more direct approach for study. A noteworthy case is that of a graded R-algebra B and one of its graded subalgebras A,

$$A \subset B.$$

A comparison of the properties of A and of B usually demands the examination of their graded components (which for simplicity we assume taken with a standard grading) and a particular role is played by the R-modules B_n/A_n. Except when B is a finite A-module, the quotient B/A lacks structure that allows various kinds of approaches. The mere embedding $A \subset B$ does not provide for much interaction between the two algebras.

To cure this problem, and put the two algebras on a more equal footing, we use a construction occurring often in the theory of Rees algebras (see [SUV1]). One of its features will be that while the modules B_n/A_n do not appear as components of Noetherian modules, we do recover all the factors of a filtration of B_n/A_n in a component of a Noetherian module.

The construction of the algebra is straightforward. Let $B = \bigoplus_{n\geq 0} B_n$ be a graded R-algebra. We begin by replacing B by an isomorphic algebra $\sum_{n\geq 0} B_n t^n$, where t is an indeterminate. Let u be a new indeterminate and set

$$C = \mathcal{R}(A,B) = B[\sum_{n\geq 0} A_n u^n],$$

which we call the *intertwining algebra of A and B*. Note that these indeterminates are used simply as degree holders of the bigraded algebra

$$C=\sum_{n\geq 0}\sum_{i=0}^{n}A_iB_{n-i}u^it^{n-i}.$$

To see the significance of this structure consider the C-module

$$T_{B/A}=C(\sum_{n\geq 1}B_nt^n)/C(\sum_{n\geq 1}A_nt^n).$$

In the case of standard graded algebras (and subalgebras), when $R=A_0=B_0$, $B_n=B_1^n$ and $A_n=A_1^n$ (a situation often achievable by taking respective Veronese subrings), we have

$$T_{B/A}=\bigoplus_{n\geq 1}T_n=\sum_{n\geq 1}\sum_{j=1}^{n}(A_{j-1}B_{n-j+1}/A_jB_{n-j})u^{j-1}t^{n-j+1},$$

which already achieves one of the desired goals by providing in the components of T_n all the factors of a filtration of B_n/A_n. In particular, as we shall have the occasion to use, if B_n/A_n is a module of finite length, then $\lambda(B_n/A_n)=\lambda(T_n)$. We shall call $T_{B/A}$ the *intertwining module of* A *and* B.

For standard graded algebras $A\subset B$, $\mathcal{R}(A,B)$ is simply the ordinary Rees algebra of the ideal A_1B of the ring B, $\mathcal{R}(A,B)=\mathcal{R}_B(A_1B)$. In other cases, such as when B is the integral closure of the standard graded algebra A, we will have to consider the general definition of the intertwining algebra.

A common thread of the algebras derived from $S(E)$ is that each is obtained by the same process of taking the ring of global sections of $\mathrm{Spec}(S(E))$ on an appropriate affine open set (see [Va94b] for details).

CONTENTS

Within the realm of these algebras, let us chart up the territory to be covered here. We single out some of the main themes.

• *Cohen-Macaulay algebras.* Let $(R,\mathfrak{m})$ be a Noetherian local ring of dimension d, let $I=(f_1,\ldots,f_n)$ be an ideal of positive height and denote by $\mathcal{R}=R[It]$ and $\mathcal{G}=\mathrm{gr}_I(R)$ its Rees algebra and associated graded ring. What makes one of these algebras Cohen-Macaulay? This question provides a certain unity of purpose and a framework for techniques that may be used in other problem areas.

There are at least three main issues regarding the Cohen-Macaulay properties of the algebras $\mathcal{R}$ and $\mathcal{G}$. First, for which classes of ideals are the conditions achieved? It occurs in more varied ways than one seems able to catalog. One of our purposes here is to examine the rich tableau on which the relationships between the Cohen-Macaulayness of $\mathcal{R}$ and of $\mathcal{G}$ are played out:

$$\begin{array}{c} \mathcal{R} \text{ is Cohen-Macaulay} \\ \Downarrow \\ \mathcal{G} \text{ is Cohen-Macaulay} \\ \Downarrow \\ \text{Proj } \mathcal{G} \text{ is Cohen-Macaulay} \\ \Updownarrow \\ \text{Proj } \mathcal{R} \text{ is Cohen-Macaulay} \end{array}$$

• *Hilbert functions.* Here one has been able to pursue a more focused approach. Finally, there are rich connections to other problems, such as the theory of Hilbert-Samuel characteristic functions. These functions–classical or extended–provide a good number of measures that are helpful in describing the relationship between ideals and their *reductions*.

• *Computational issues.* It should not be a surprise that Rees algebras play roles in several constructions away from our main setting. This may occur simply because certain ideals pop up–such as conductors in the theory of integral closure of affine algebras–and because the understanding of their limit properties are helpful. We carry out some detailed discussions in several instances.

In Rees algebra theory proper, from a computational perspective, the path to decide that a Rees algebra $\mathcal{R}$ is Cohen-Macaulay hardly ever goes by first checking that property in $\mathcal{G}$. Mixing the generators of an ideal and their relations may ask for grid lock in a computation. Nevertheless the interplay between $\mathcal{R}$ and $\mathcal{G}$ provides a rich set of guide posts in highlighting the significant numerical invariants of the algebras. These, in turn, may be addressed by more direct means.

• *Integral closure.* The study of the integral closure of ideals, modules and algebras is one of the main purposes of this book. Many of its aspects, from the construction to the analysis of its algorithms to its role on the whole landscape of Rees algebras, will be driving the exposition. More explicitly, in the particularly significant case of ideals, we shall examine the following:

□ Methods for the construction of the integral closure of certain ideals.
□ Completeness and normality testing.
□ Cohomological properties of normal ideals.
□ Exploitation of the relationship between integral closure of ideals and multiplicities.

• *Generalized Hilbert functions and their multiplicities.* Let $(R, \mathfrak{m})$ be a Noetherian local ring. A *generalized multiplicity* is a function $\text{gdeg}(\cdot)$ on finitely generated R-modules such that for any finitely generated graded module $M = \bigoplus_{n\geq 0} M_n$ over $R[x_1, \ldots, x_d]$, the series

$$H(M;t) = \sum_{n\geq 0} \text{gdeg}(M_n)t^n$$

assembles well. One rule worth requiring is Bertini's property: that generic hyperplane sections behave relatively well with respect to the numerical invariants of these functions. The practice has shown that it is important to find bounds for the values of $\text{gdeg}(M_n)$ valid for all n.

What are these functions? It turns out to be more fruitful to look at a broader range of functions, some of which do not fit the required mold above. The criterion becomes their relevance to the issue of the complexity of the algorithms.

Let $(R, \mathfrak{m})$ be a Noetherian local ring (or a Noetherian graded algebra) and let $\mathcal{M}(R)$ be the category of finitely generated R-modules (or the appropriate category of graded modules). A *degree function* is a numerical function $\mathbf{d} : \mathcal{M}(R) \mapsto \mathbb{N}$. The more interesting of them initialize on modules of finite length and have mechanisms that control how the functions behave under generic hyperplane section. Thus, for example, if L is a given module of finite length, one function may request that

$$\mathbf{d}(L) := \lambda(L),$$

or hen L is a graded module,

$$L = \bigoplus_{i \in \mathbb{Z}} L_i,$$

that

$$\mathbf{d}(L) := \sup\{i \mid L_i \neq 0\}.$$

A great deal of attention, from their construction, mutual relationships and applications, is given to these functions, particularly those that are not treated extensively in the literature. One of our aims is to provide an even broader approach to Hilbert functions that does not depend so much on the base ring being Artinian. We shall be looking at the associated multiplicities and seek ways to relate them to the reduction numbers of the algebras. Among the *multiplicities* are:

- Castelnuovo-Mumford regularity
- Classical multiplicity
- Arithmetic degree
- Extended multiplicity
- Cohomological degree
- Homological degree
- bdeg

We outline the contents more precisely. As each chapter discusses its own content, we shall be brief here. In Chapter 1, we introduce most of the main notions we shall treat throughout: the defining equations of a Rees algebra of ideals and modules and more general filtrations, the integral closure of ideals, and reduction and reductions numbers of ideals. Several major classes of ideals are exhibited.

In Chapter 2, we discuss the role of Hilbert functions in establishing bounds for reduction numbers and cohomological properties of blowups. Its main goal is to develop families of what can be described as *numerical signatures* for modules as numbers indicative of their complexity particularly with regard to the contributions of its primary components. The finer ones may pick up some of the interaction among components: multiplicity, Mumford-Castelnuovo regularity, arithmetic and geometric degrees. To these we add a discussion of the so-called DEGS (read: *big degs*), relate them to the other signatures and describe their roles in some complexity counts.

Since Hilbert functions of graded modules over rings such as $k[x_1,\ldots,x_n]$ (k a field) are so fruitful, we consider extensions to more general algebras, when k is a global ring. This is achieved by using the degree functions developed here applied to the components (much as ordinary length is used in the field case). Thus broadened, the properties of an algebra that can be read from these Hilbert functions are its *linear* properties. Those that cannot, such as their reduction numbers, are taken to be *nonlinear* properties. It is not unexpected therefore that relationships between Hilbert functions and such nonlinear invariants are expressed by inequalities.

Chapter 3 provides a discussion of the cohomology of Rees algebras and of associated graded rings that emphasizes Cohen-Macaulayness. Technically, the treatment is a subset of the study of the vanishing properties of the cohomology of such algebras. This larger picture is too broad, with too many things occurring in it worthy of their own full examination. Our more modest goal is to treat those aspects of cohomology in which reductions and their related numbers play significant roles. This phenomenon is patently visible in the 'extreme' cases of Cohen-Macaulay algebras but are also felt to be useful in any treatment of integral closure of ideals.

Chapter 4 treats the divisors of Rees algebras in their basic aspects: divisor class groups for normal algebras, the structure of the canonical ideal. Rees algebras have another divisor–which shall be called the *Serre divisor*–carrying all the information in the canonical divisor plus some. This divisor packs some of the data required to decide when a Rees algebra is Cohen-Macaulay. However, the canonical divisor is still the focus of interest, particularly deciding when it has some expected form. It also takes a look at the phenomenon of *gaps* in the degrees of the generators of the canonical modules.

Chapter 5 is an assemblage of facts on Koszul complexes and their homology modules as they pertain to Rees algebras. Given that these modules occur in the construction of other complexes–*the approximation complexes*– the emphasis is on the module-theoretic properties of the homology modules as opposed to the usual emphasis on their vanishing (which is treated alongside). In addition, we discuss some conjectural properties of annihilators of these modules related to integral closure of the defining ideal.

In a very definite sense, Chapters 6, 7 and 8 form the core of the book, as each treats one of our main themes, the integral closure of algebras, ideals and modules. Our model is Chapter 6, where for a reduced affine algebra A we focus on:

- Effective methods to compute the integral closure $\overline{A}$.
- Development and analysis of processes $\mathcal{P}(\cdot)$ that associate to an intermediate extension $A \subset B \subset \overline{A}$, another extension $B \subset \mathcal{P}(B) \subset \overline{A}$, guaranteed to be distinct from B when $B \neq \overline{A}$.
- Establishing *a priori* bounds for the 'order' of $\mathcal{P}$, that is for the integer n such that $\mathcal{P}^n(A) = \overline{A}$.

- Estimation of the number of generators (and of their degrees in the graded case) of $\overline{A}$.
- Perfunctory treatment of arithmetic algebras.
- Basic treatment of the tracking number of an algebra.

Chapter 7, dealing with the integral closure of ideals, might at first resemble a special case of the problems treated in the previous chapter. Thus, given an ideal I of the normal domain (for simplicity) A, the integral closure $\overline{I}$ of the ideal is to be found as the degree 1 component of the algebra

$$\overline{R[It]} = R + \boxed{\overline{I}t} + \overline{I^2}t^2 + \cdots,$$

so finding $\overline{I}$ goes via the construction of $\overline{R[It]}$! One would rather have direct constructions $I \rightsquigarrow \overline{I}$ taking place entirely in R. The difficulty of realizing this lies in the absence of *conductors*, such as Jacobian ideals in the algebra case: Given A by generators and relations (at least in characteristic zero), the Jacobian ideal J of A has the property

$$J \cdot \overline{A} \subset A;$$

in other words, $\overline{A} \subset A : J$. This fact lies at the base of all current algorithms to build $\overline{A}$. There is no known corresponding *annihilator* for $\overline{I}/I$. The main topics here are:

- Properties of integrally closed ideals.
- Monomial ideals, when the theory is fairly complete and efficiently implemented already.
- Computation of multiplicities.
- Normalization of ideals, which may be considered one of the main problems of the theory, particularly the examination of certain conjectures pointing to a decisive role of the Hilbert polynomial in controlling the computation.

Chapter 8 is a treatment of integral closure for modules and some algebras arising from them. There are two threads running in parallel. First, the emphasis on techniques that seek to assemble such issues with those techniques already discussed in previous chapters on algebras and ideals. There are also novel phenomena to modules with a concomitant development of appropriate methods.

The last chapter surveys a collection of methods and algorithms to process integral closure problems.

1

Numerical Invariants of a Rees Algebra

The algebras A that we are going to study are susceptible of two representation-theoretic approaches. One is via its algebra generators and relations as a homomorphic of a polynomial ring B, and the other through its module generators over one of its subrings S *of similar structure*:

$$S \hookrightarrow A \leftarrow B.$$

Let us illustrate these two views. Let R be a Noetherian ring, and let I be an ideal defined by a set of generators, $I = (f_1, \ldots, f_n)$. Consider the subring $R[f_1t, \ldots, f_nt] \subset R[t]$, t an indeterminate. It is called the *Rees* algebra of I and variously denoted by $R[It]$, $\mathcal{R}(I)$, or even $\mathcal{R}$ when the context permits.

There are two general approaches to describing properties of the Rees algebra $\mathcal{R} = R[It]$ of the ideal I. Technically they are the outside in-inside out views of $\mathcal{R}$.

- Through a presentation

$$0 \to J \longrightarrow R[T_1, \ldots, T_n] \longrightarrow \mathcal{R} \to 0, \; T_i \mapsto f_it,$$

  of the algebra as a quotient of a polynomial ring by looking at the structure of the ideal J. These equations carry (or hide, in the view of many) most of the information one might want to have about the algebra $\mathcal{R}$.
- Through the study of the reductions of the ideal I, that is subideals $J \subset I$ with the embedding $R[Jt] \subset R[It]$ imitating Noether normalization. It is a different platform from which to look at $\mathcal{R}$.

These approaches emphasize the structure of the polynomial relations amongst the elements of a generating set of I, and they tend to mimic one another. The first method consists mainly of: starting with a matrix of presentation of the ideal $I = (f_1, \ldots, f_n)$, and using elimination theory directly or via Gröbner basis computations, one seeks to describe J. The method of reductions is more supple: even partial failures in providing a full arithmetical study of $R[It]$ may yield lots of information about it.

The R-module structure of I alone gives rise to its syzygies

$$\mathbb{F}_\bullet : \qquad \cdots \longrightarrow F_2 \longrightarrow F_1 \longrightarrow F_0 \longrightarrow I \to 0;$$

they carry several attached sequences of numbers (graded Betti numbers particularly) that assemble into measures of gross size–here codimension and multiplicities are noteworthy–that are useful in comparison of ideals and their varieties. For the purpose here we view these numbers as the *linear invariants* of I.

A different set of measures arises when one considers algebras derived from *multiplicative* filtrations of R-ideals, noteworthy being the I-adic filtration that gives rise to the Rees algebra

$$\mathcal{R} = R[It] = \bigoplus_{n\geq 0} I^n.$$

These structures give rise to other numerical measures of the ideal such as reduction number, relation type, etc.; they reflect how the ideal sits inside of the ring and expose many properties of the ideal not directly related to its syzygies. They could be viewed as *nonlinear invariants* of I. In this chapter we introduce most of them and begin to examine the role they play in the arithmetical and geometric properties of the blowup. A primary focus is to understand what information about the cohomology of $\mathcal{R}$ these numbers contain.

At first glance, the invariants tend to give measures of how far an ideal is from being generated by analytically independent elements, and they reflect nuanced versions of this notion. These are best set in a local ring or a graded ring, so that often we assume that R is of this kind, with an infinite residue field.

Technically, we shall express the behavior of the ideal and of its associated algebras by several notions and measures. Some of these will be:

- Equations of Rees algebras
- Ideals of linear type
- Relation type of an ideal
- Reduction numbers
- Special fibers
- Integral closure of ideals
- Equations of integral dependence
- Intertwining algebras
- Briançon-Skoda numbers

The treatment of some of the topics is cursory, just sufficient to place them in the appropriate context, but pointers will always be given to the literature.

1.1 Equations of a Rees Algebra

Let R be a ring and E a finitely generated R-module. In this section, we discuss *presentations* of the symmetric and Rees algebras of E and their dimensions. The case of ideals will be emphasized.

The symmetric algebra $S_R(E)$, or simply $S(E)$, is the algebra

$$S_R(E) = T_R(E)/(x \otimes y - y \otimes x,\ x, y \in E),$$

where $T_R(E)$ is the *tensor algebra* of E over R. It is convenient to give it in terms of generators and relations. This arises directly from a free presentation of E, as follows. Given the first order syzygies of E,

$$0 \to L \longrightarrow R^n \longrightarrow E \to 0,$$

$$S(E) = S(R^n)/(L) = R[T_1, \ldots, T_n]/(L),$$

where (L) is the ideal generated by the linear forms $a_1T_1 + \cdots + a_nT_n$ corresponding to the syzygies $z = (a_1, \ldots, a_n)$ of E (for the chosen set of generators of E). Geometrically, Spec $(S(E))$ is a fibration of Spec (R) by a family of hyperplanes. (L) is said to be the ideal of the equations of $S(E)$.

Despite its transparency the ideal (L) encodes many interesting features that we shall explore. It is also a gateway to the equations of the Rees algebra of E. There are variations on the notion of the Rees algebra of a module. First there is the case of a module E with an embedding into a free module $E \overset{\varphi}{\hookrightarrow} R^r$: the algebra $\mathcal{R}(E)$ is the image of $S(E)$ in $S(R^r)$ induced by φ. It is significant because the properties of the embedding get rolled over into $\mathcal{R}(E)$. Thus in the case of an ideal $I \hookrightarrow R$, $\mathcal{R}(I)$ carries a great deal of information about the subvariety $V(I)$, especially of how I sits in R. The *equations* of $\mathcal{R}(E)$ are given by a (graded) presentation of

$$\mathcal{R}(E) = R[T_1, \ldots, T_n]/J, \quad J = J_1 + J_2 + \cdots,$$

where $(J_1) = (L)$.

For modules which are generically free the equations of $\mathcal{R}(E)$ are related to L in the following manner.

Proposition 1.1. *Let K be the total ring of fractions of R and $E \hookrightarrow R^r$ an R-module such that $K \otimes_R E$ is a free K-module. Then*

$$\mathcal{R}(E) \cong S(E)/\text{modulo } R\text{-torsion}.$$

Furthermore, if E has a presentation as above and f is a regular element of R such that E_f is a free R_f-module, then

$$\mathcal{R}(E) \cong R[T_1, \ldots, T_n]/J, \quad J = (L) : f^\infty := \bigcup_{n \geq 1} (L) : f^n.$$

The element f is to be found in an appropriate Fitting ideal of the module E. If

$$R^m \xrightarrow{\phi} R^n \longrightarrow E \to 0$$

is a free presentation of E, choose f to be a regular element in $\text{Fitt}(E)$, the ideal of minors of order $m - \text{rank}(E)$ of ϕ.

When E does not meet the condition of genericity, it is unclear how to get hold of the equations of $\mathcal{R}(E)$. It is an issue that will arise when one attempts to define the Rees algebra of a module E *in abstracto*, that is, irrespective of embeddings of E into free modules. In [EHU3], it is proposed to set

$$\mathcal{R}(E) := S(E)/\bigcap \ker(\varphi),$$

for all morphisms induced by mappings from E into free R-modules (not necessarily injective). Another consists in embedding E into its injective envelope E_0, and setting

$$\mathcal{R}(E) := \text{image}(S(E) \to S(E_0)).$$

They both localize (R is Noetherian) and agree with the characterization above for generically free modules, or agree with each other when the total ring of fractions K is Gorenstein.

1.1.1 The Rees Algebra of an Ideal

The first approach to studying the Rees algebra of an ideal I is focused on the degrees of a generating set for the presentation ideal J and seeks to obtain those equations from the syzygies of I. The ideal J, which we refer to as the *equations* of $R[It]$, is graded:

$$J = J_1 + J_2 + \cdots,$$

where J_1 is the R-module of linear forms $\sum a_i T_i$ such that $\sum a_i f_i = 0$. The module J_r is the module of syzygies of the r-products of the f_i.

There is a related presentation, that of the associated graded ring $\mathcal{G} = \text{gr}_I(R)$. If $\mathcal{R} = R[T_1, \ldots, T_n]/J$, then (J, I) is the presentation ideal of $\mathcal{G}$. The challenge is, from a given presentation of I,

$$R^m \xrightarrow{\varphi} R^n \longrightarrow I \to 0,$$

to describe J. In one case this is straightforward and we will return repeatedly to some of its variations).

Example 1.2. If the ideal I is generated by a regular sequence $f_1, \ldots, f_n$, the equations of $\mathcal{R} = R[It]$ are nice:

$$\mathcal{R} \cong R[T_1, \ldots, T_n]/I_2 \begin{pmatrix} T_1 & \cdots & T_n \\ f_1 & \cdots & f_n \end{pmatrix}.$$

In other words, J is generated by the Koszul relations of the f_i.

Dimension of the Rees Algebra

There are several ways to express $\dim R[It]$ beginning with the following.

Theorem 1.3. *Let R be a Noetherian ring and I an ideal of R. For each minimal prime $\wp$ of R, set $c(\wp) = 1$ if $I \not\subset \wp$, and $c(\wp) = 0$ otherwise. Then*

$$\dim R[It] = \sup\{\dim R/\wp + c(\wp)\}.$$

In particular, $\dim R[It] \leq \dim R + 1$, *and equality will hold when I is not contained in any minimal prime of R.*

Proof. There are several elementary approaches to the proof; we use that of [Va94b, Theorem 1.12.4]. We may assume that $\dim R < \infty$. Let N be the nilradical of R. Since $N[t]$ is the nilradical of $R[t]$, it follows that $N[t] \cap R[It]$ is the nilradical of $R[It]$, so that we may assume that R is reduced and the minimal primes of $R[It]$ have the form $P_i = \wp_i R[t] \cap R[It]$. In turn, this implies that to determine $\dim R[It]$ we may assume that R is a domain. When $I = 0$, $R[It] = R$. If $I \neq 0$, let V be any valuation ring of R. Since I is finitely generated, $V \cdot R[It] = V[at]$, a ring of dimension $\dim V + 1$. □

Corollary 1.4. *If R is an integral domain of Krull dimension d and $I = (a_1, \ldots, a_n)$ is a nonzero ideal with a presentation ideal,*

$$0 \to J \longrightarrow R[T_1, \ldots, T_n] \longrightarrow R[It] \to 0, \; T_i \mapsto a_i t,$$

then the ideal J has codimension $n-1$.

Equations via Elimination

A useful method of obtaining the presentation ideal of a Rees algebra uses elimination. Briefly it comes about as follows.

Let S be a ring of polynomials over a field, let $R = S/J$ and let I be an ideal of R. The equations of the Rees algebra $R[It]$ can be described as follows.

Proposition 1.5. *Let $a_1, \ldots, a_n$ be a lift in S of a set of generators of I. Let $T_1, \ldots, T_n, t$ be independent variables. Then*

$$R[It] \cong S[T_1, \ldots, T_n]/L,$$

where L is obtained as

$$L = (T_1 - a_1 t, \ldots, T_n - a_n t, J) \bigcap S[T_1, \ldots, T_n],$$

and

$$\mathrm{gr}_I(R) = S[T_1, \ldots, T_n]/(L, a_1, \ldots, a_n).$$

These eliminations are evidently very costly.

Remark 1.6. Here is a list of commands in *Macaulay2* that can be used to obtain the equations of the Rees algebra of the ideal

$$I = (a_1, \ldots, a_n) \subset R = k[x_1, \ldots, x_n].$$

For k we will take the largest indigenous finite field of *Macaulay2*:

$KK = ZZ/31991$
$R = KK[t, t_1, t_2, \ldots, t_n, x_1, x_2, \ldots, x_d, \text{MonomialOrder} \Rightarrow \text{ Eliminate } 1]$
$I1 = \text{ideal}(t_1 - t * a_1, t_2 - t * a_2, \ldots, t_n - t * a_n)$
$J = \text{selectInSubring}(1, \text{generators gb } I1);$
$L = \text{minors}(1, J)$

Degree by Degree

The difficulty in finding the equations is partly governed by the following notion.

Definition 1.7. The *relation type* of I is the smallest integer s such that $J = (J_1, \ldots, J_s)$. It is independent of the chosen set of generators.

Some of the steps of eliminations can be carried out by 'hand' as follows. We rewrite a set of generators of J_1,

$$h_j = \sum_{i=1}^{n} a_{ij} T_i, \quad j = 1, \ldots, p,$$

in matrix representation

$$J_1 = (h_1, \ldots, h_p) = [T_1, \ldots, T_n] \cdot \varphi,\ \varphi = (a_{ij}).$$

The J_s arise by elimination from these equations of the "parameters" describing the f_i (but how?). The most naive approach is the following. Let $[x_1, \ldots, x_m] \in R^m$ be determined by a set of elements generating the ideal generated by the entries of φ. We may then describe the set of the f_i by

$$J_1 = [x_1, \ldots, x_m] \cdot B(\varphi),$$

where $B(\varphi)$ is a $m \times n$ matrix whose entries are linear forms in the T_i. By Cramer's rule, the ideal generated by the $m \times m$ minors of $J(\varphi)$ are conducted into J by the x_j, consequently $I_m(B(\varphi)) \subset J$. Obviously the matrix $B(\varphi)$ depends on the generating set $x_1, \ldots, x_m$. Because of its construction, it is referred to as the *Jacobian dual* of φ ([SUV93]). It gives rise to the following notion.

Definition 1.8. The ideal I has the *expected defining equations* if $J = (J_1, I_m(B(\varphi)))$.

The ideals of $R[T_1, \ldots, T_n]$ generated by $(J_1, \ldots, J_r)$, for $r \leq s$, where s is the relation type of I, define 'approximations' or 'ancestors' of $\mathcal{R}$. The module J_1 generates the ideal of definition of the symmetric algebra $S(I)$ of I, and we have canonical surjections

$$0 \to (J_1) \longrightarrow R[T_1,\ldots,T_n] \longrightarrow S(I) \to 0,$$

$$0 \to \mathcal{A} \longrightarrow S(I) \longrightarrow R[It] \to 0,$$

where

$$J/(J_1) = \mathcal{A} = \mathcal{A}_2 + \mathcal{A}_3 + \cdots.$$

In particular the degrees of the generators of J are independent of the chosen generators of I. When $\mathcal{A} \neq 0$, some emphasis has been put on determining the first non-vanishing component $\mathcal{A}_j$ and on predicting its structure ([SUV95], [UV93]).

Definition 1.9. The ideal I is said to be of *linear type* if $J = (J_1)$, that is if $\mathcal{A} = 0$. More generally, I is said to be of *relation type* r if J can be generated by forms of degree $\leq r$. Finally, I is said to be of *co-type* $\geq s$ if $\mathcal{A}_i = 0$ for $i \leq s$.

Ideals with a Linear Presentation

A case when the theory of Jacobian duals is more transparent is that of ideals with linear presentations. For technical reasons it is preferable to consider the more general case of modules.

Let $R = k[x_1,\ldots,x_d]$ be a ring of polynomials over the field k and let

$$R^m \xrightarrow{\varphi} R^n \longrightarrow E \to 0$$

be a presentation of the module E. Suppose that all the entries of φ are linear forms of R. The equations defining the symmetric algebra of E can then be written

$$J_1 = [T_1,\ldots,T_n] \cdot \varphi = [x_1,\ldots,x_d] \cdot B(\varphi),$$

where $B(\varphi)$ is a $d \times m$ matrix of linear forms in the ring $T = k[T_1,\ldots,T_n]$. This mapping is the presentation matrix of the T-module

$$T^m \xrightarrow{B(\varphi)} T^d \longrightarrow F \to 0.$$

The 'duality' qualifier refers to the isomorphism of symmetric algebras

$$S_R(E) \cong S_T(F).$$

If any of these algebras is an integral domain, that is, E (or F) is of linear type, one has the equality

$$\dim \mathcal{R}(E) = d + \text{rank } E = n + \text{rank } F,$$

in particular

$$\text{rank } \varphi = \text{rank } B(\varphi). \tag{1.1}$$

The (transposed) Jacobian matrix of the defining ideal J of $S(E)$ is made up of two blocks

$$\left[\varphi | B(\varphi)\right]$$

which makes some calculations easy. We make some elementary observations.

Proposition 1.10. *Let E be a torsion-free module over the ring of polynomials $k[x_1,\ldots,x_d]$. Suppose that E has a linear presentation matrix φ and $S(E)$ is an integral domain. Then setting $\mathfrak{P} = (x_1,\ldots,x_d)S(E)$, we have:*

(i) *$S(E)_{\mathfrak{P}}$ is a regular local ring.*
(ii) *If k has characteristic zero and φ has rank 2 and $S(E)$ satisfies Serre's condition S_2, then $S(E)$ is a normal domain.*

Proof. (i) From the discussion above, $\mathfrak{P}$ is a prime ideal of $S(E)$ of height

$$h = \text{rank}(F) = d + \text{rank}(E) - \nu(E).$$

We must show that the minimal number of generators of the maximal ideal of $S(E)_{\mathfrak{P}}$ is h. This localization can be written

$$(x_1,\ldots,x_d)k[x_1,\ldots,x_d,T_1,\ldots,T_n]/([x_1,\ldots,x_d]\cdot B(\varphi))_{(x_1,\ldots,x_d)}.$$

Since $B(\varphi)$ has rank $d - \text{rank}(F)$, it follows that the maximal ideal of $S(E)_{\mathfrak{P}}$ has at most h generators, as desired.

(ii) By assumption the presentation ideal J of $S(E)$ has codimension 2. It suffices to show that the Jacobian ideal L has codimension at least 2 and it will provide for Serre's condition R_1.

The ideal generated by the 2×2 minors of the Jacobian matrix contains the ideals $I_2(\varphi)$ and $I_2(B(\varphi))$, both of which have codimension at least 2 since E and F are torsion-free over the corresponding polynomial rings in the x_i and T_j indeterminates. It follows that the subideal

$$(I_2(\varphi) + I_2(B(\varphi)) + J)/J \subset L$$

has codimension at least 2, as we wanted to show. □

Ideals of Linear Type

Noteworthy is the meaning of the condition $\mathcal{A} =$ nilpotent: Let R be a local ring, with maximal ideal $\mathfrak{m}$. The elements $a_1,\ldots,a_n \in R$ are said to be *analytically independent* if any homogeneous polynomial $f(X_1,\ldots,X_n)$ for which $f(a_1,\ldots,a_n) = 0$ has all of its coefficients in $\mathfrak{m}$. If $I = (a_1,\ldots,a_n)$, this means that the ring $R[It]\otimes(R/\mathfrak{m})$ is a polynomial ring in n indeterminates over $R/\mathfrak{m}$.

If I is of linear type, then I is locally generated by analytically independent elements, in particular $\nu(I_{\mathfrak{p}}) \leq \dim R_{\mathfrak{p}}$ for any prime ideal $\mathfrak{p}$ that contains I. There is no general theory describing the ideals of linear type, although many well circumscribed classes are known. If I is not of linear type, there is a beginning of a theory for ideals of quadratic type and for certain families of ideals whose equations are obtained from elimination and are concentrated in degrees 1 and another degree.

Let us formalize some of these properties in:

Definition 1.11. Let R be a Noetherian ring and I an ideal. I satisfies the condition $\mathcal{F}_1$, or G_∞, if for each prime ideal $I \subset \mathfrak{p}$, $\nu(I_\mathfrak{p}) \leq \text{height } \mathfrak{p}$. Similarly, I satisfies the condition $\mathcal{F}_0$ if for each prime ideal $I \subset \mathfrak{p}$, $\nu(I_\mathfrak{p}) \leq \text{height } \mathfrak{p} + 1$. For an R-module E of rank r, these conditions are extended by setting $\nu(E_\mathfrak{p}) \leq r - 1 + \text{height } \mathfrak{p}$ and $\nu(E_\mathfrak{p}) \leq r + \text{height } \mathfrak{p}$, respectively.

Fitting Ideals

Suppose there is a presentation for the ideal $I = (f_1, \ldots, f_n)$:

$$R^m \xrightarrow{\varphi} R^n \longrightarrow I \to 0.$$

The condition $\mathcal{F}_1$ can be recast as follows (see [Va94b, Chapter 1]).

Proposition 1.12. *Let R be an equidimensional catenary domain and I a nonzero ideal presented by a matrix φ as above. The following are equivalent:*

(a) *I satisfies the condition $\mathcal{F}_1$.*
(b) *For each $t = 1, \ldots, n-1$,* height $I_t(\varphi) \geq n - t + 1$.

Under these conditions there is a measure of control over some of the homogeneous components of the ideal J. Indeed, suppose that $(R, \mathfrak{m})$ is a local catenary domain of dimension d, height $I \geq 1$, $n = \nu(I)$, and set $S = R[T_1, \ldots, T_n]$. Then we have:

(i) height $J_1 S \leq d$, and this height can be determined precisely in terms of the heights of the determinantal ideals $I_s(\varphi)$,

$$\text{height } J_1 S = \sup\{t + \text{height } I_t(\varphi) \mid 1 \leq t \leq n-1 \quad \text{and} \quad t + \text{height } I_t(\varphi) < n\}.$$

(ii) If I satisfies the condition $\mathcal{F}_1$, then height $J_1 S = d - 1$.

d-sequences

A far-reaching extension of the notion of regular sequence is the following ([Hu80]).

Definition 1.13. Let $\mathbf{x} = \{x_1, \ldots, x_n\}$ be a sequence of elements in a ring R generating the ideal I. Then $\mathbf{x}$ is called a *d-sequence* if $(x_1, \ldots, x_i) : x_{i+1} x_k = (x_1, \ldots, x_i) : x_k$ for $i = 0, \ldots, n-1$ and $k \geq i + 1$.

The significance of this concept was recognized early ([Hu80], [Val80]):

Theorem 1.14. *Every ideal generated by a d-sequence is of linear type.*

The sequences can also be defined with respect to a module E. The natural habitat of these sequences is a family of differential graded modules whose construction we treat in Chapter 5.

Remark 1.15. A useful observation in checking whether an ideal is of linear type is the following. Let $I = (a_1, \ldots, a_n)$ be an ideal of an integral domain R and let $(J_1) \subset R[T_1, \ldots, T_n]$ be the ideal defining the symmetric algebra $S_R(I)$. Suppose that for $0 \neq x \in R$, I_x is of linear type (say, generated by a regular sequence or more generally by a d-sequence). Then I is of linear type if and only if $(J_1) : x = (J_1)$.

Expected Defining Equations and Heights

We use the Jacobian dual matrix of a presentation φ of an ideal I to sketch scenarios where the defining equations of $\mathcal{R} = R[It]$ have an expected form. In the notation above, set

$$J_1 = [x_1, \ldots, x_p] \cdot B(\varphi),$$

where $B(\varphi)$ is a $p \times m$-matrix. If R is a local ring, one can always take for the x_i a set of generators of the matrix ideal. The cases we consider depend on the nature of the matrix $B(\varphi)$. The case when $B(\varphi)$ is a skew-symmetric matrix is in many ways simpler to deal with; see [SUV93]. We discuss briefly the more general case as to what it takes to ensure the equality

$$J = ([x_1, \ldots, x_p] \cdot \varphi, I_p(B(\varphi))).$$

We assume that $(R, \mathfrak{m})$ is a regular local ring, $\dim R = d$, $n = \nu(I)$, and that the x_i form a minimal set of generators of $\mathfrak{m}$. We shall also assume that $n \geq d$.

Proposition 1.16. *If $\mathcal{R}$ has the expected defining equations the following hold:*

(i) *The ideal $I_d(B(\varphi))$ satisfies the condition* height $I_d(B(\varphi)) \geq n - d - 1$.
(ii) *The image L of $I_d(B(\varphi))$ in $R/\mathfrak{m}[T_1, \ldots, T_n]$ has height at least $n - d$.*

Proof. (i) J is an ideal of height $n-1$ of the regular ring $R[T_1, \ldots, T_n]$, while the ideal generated by J_1 has height at most d. By the Nagata-Serre theorem, we must have

$$\text{height } (J_1) + \text{height } I_d(B(\varphi)) \geq n - 1.$$

(ii) The ideal L is the kernel of the presentation

$$R/\mathfrak{m}[T_1, \ldots, T_n] \longrightarrow \text{gr}_I(R) \otimes_R R/\mathfrak{m} = F(I),$$

onto the so-called special fiber of I (see next section). Since $F(I)$ has dimension bounded by d, the kernel of the presentation must have codimension at least $n - d$. $\square$

There is a broad treatment of this question for Cohen-Macaulay ideals of codimension two and Gorenstein ideals of codimension three in [Mor96] and [MU96].

Example 1.17 (S. Morey). The following describes a Rees algebra which has the expected defining equations but is not Cohen-Macaulay. Let $R = k[x_1, x_2, x_3, x_4, x_5]$ and let

$$\varphi = \begin{bmatrix} 0 & -x_1 & -x_3 & x_2 & -x_5 & x_4 & -x_3 \\ x_1 & 0 & -x_3 & x_2 & x_1 & x_5 & -x_1 \\ x_3 & x_3 & 0 & 0 & -x_3 & x_1 & -x_4 \\ -x_2 & -x_2 & 0 & 0 & -x_4 & x_2 & 0 \\ x_5 & -x_1 & x_3 & x_4 & 0 & -x_3 & x_1 \\ -x_4 & -x_5 & -x_1 & -x_2 & x_3 & 0 & -x_2 \\ x_3 & x_1 & x_4 & 0 & -x_1 & x_2 & 0 \end{bmatrix}.$$

Let I be the ideal of 6×6 Pfaffians of φ. A calculation with Macaulay will show that $\mathcal{R} = R[It]$ has the expected equations. To check that $\mathcal{R}$ is not Cohen-Macaulay, one checks that the h-vector of its Hilbert series is not positive.

The following observation is useful as an approach to the defining equations of a Rees algebra. We make use of the notation above.

Proposition 1.18. *Let R be a Cohen-Macaulay integral domain, I a nonzero ideal and $\mathcal{R} = R[T_1,\ldots,T_n]/J$ a presentation of its Rees algebra. Let J_1 be the degree 1 component of J and L an ideal of $R[T_1,\ldots,T_n]$ with $J_1 \subset L \subset J$ such that $J = \sqrt{L}$ and let $f_1,\ldots,f_{n-1}$ be a regular sequence contained in L. If $\mathfrak{f}$ is the ideal generated by the f_i, then*

$$J = \mathfrak{f}\colon(\mathfrak{f}\colon L).$$

Proof. Passing to the field of fractions K of R gives $(J_1)K = JK$, which implies that the annihilator of the module J/L contains some nonzero elements of R. Since by hypothesis $J = \sqrt{L}$, this implies that height ann $(J/L) \geq$ height $J + 1 = n$. Consider the exact sequence of S-modules ($S = R[T_1,\ldots,T_n]$):

$$0 \to J/L \longrightarrow S/L \longrightarrow S/J \to 0.$$

Applying $\mathrm{Hom}_S(\cdot, S/\mathfrak{f})$, we have the exact sequence

$$0 \to \mathrm{Hom}_S(S/J, S/\mathfrak{f}) \longrightarrow \mathrm{Hom}_S(S/L, S/\mathfrak{f}) \longrightarrow \mathrm{Hom}_S(J/L, S/\mathfrak{f}) = 0,$$

the last assertion because $S/\mathfrak{f}$ is Cohen-Macaulay and J/L has codimension at least 1 with respect to it. Thus we have $\mathfrak{f}\colon L = \mathfrak{f} : J$ and the assertion will follow from the equality $J = \mathfrak{f}\colon(\mathfrak{f}\colon J)$, which in turn is a consequence of the fact that $J/\mathfrak{f}$ is a minimal prime ideal of the Cohen-Macaulay ring $S/\mathfrak{f}$. □

The following asymptotic statement is quite surprising (see [BF85], [JK94], [Wa98]; for a theory of ideals with relation type two, see [Rag94]):

Theorem 1.19. *Let R be a Noetherian ring and I an ideal. Then for all $n \gg 0$ the relation type of the ideal I^n is at most 2.*

1.1.2 Dimension of Symmetric and Rees Algebras of Modules

For many comparisons, we shall need dimension estimates for several algebras derived from symmetric algebras of modules. A fuller discussion is to be found in [Va94b, Chapter 1].

Let R be a commutative Noetherian ring of finite Krull dimension, and E a finitely generated R-module. For a prime ideal $\mathfrak{p}$ of R, denote by $\nu(E_{\mathfrak{p}})$ the minimal number of generators of the localization of E at $\mathfrak{p}$; in other words, it is the same as the torsion-free rank of the module $E/\mathfrak{p}E$ over the ring $R/\mathfrak{p}$. The Forster-Swan number of E is defined by

$$b(E) = \sup_{\mathfrak{p}\in\mathrm{Spec}(R)} \{\dim(R/\mathfrak{p}) + \nu(E_{\mathfrak{p}})\}.$$

The theorem of Huneke-Rossi ([HuR86, Theorem 2.6]) explains the nature of this number.

Theorem 1.20. $\dim S(E) = b(E)$.

To give the proof of this we need a few preliminary observations about dimension formulas for graded rings.

Lemma 1.21. *Let B be a Noetherian integral domain that is finitely generated over a subring A. Suppose that there exists a prime ideal Q of B such that $B = A + Q$, $A \cap Q = 0$. Then*

$$\dim(B) = \dim(A) + \text{height}(Q) = \dim(A) + \text{tr.deg.}_A(B).$$

Proof. We may assume that $\dim A$ is finite; $\dim(B) \geq \dim(A) + \text{height}(Q)$ by our assumption. On the other hand, by the standard dimension formula in [Ma86], for any prime ideal P of B, $\mathbf{p} = P \cap A$, we have

$$\text{height}(P) \leq \text{height}(\mathbf{p}) + \text{tr.deg.}_A(B) - \text{tr.deg.}_{k(\mathbf{p})} k(P).$$

The inequality

$$\dim(B) \leq \dim(A) + \text{height}(Q)$$

follows from this formula and reduction to the affine algebra obtained by localizing B at the zero ideal of A. □

There are two cases of interest here. If B is a Noetherian graded ring and A denotes its degree 0 component, then

$$\dim(B/P) = \dim(A/\mathbf{p}) + \text{tr.deg.}_{k(\mathbf{p})} k(P)$$

and

$$\dim(B_{\mathbf{p}}) = \dim(A_{\mathbf{p}}) + \text{tr.deg.}_A(B).$$

We shall need to identify the prime ideals of $S(E)$ that correspond to the extended primes when E is a free module. It is based on the observation that if R is an integral domain, then the R-torsion submodule T of a symmetric algebra $S(E)$ is a prime ideal of $S(E)$. This is clear from the embedding $S(E)/T \hookrightarrow S(E) \otimes K$, where $K =$ field of quotients of R, and the fact that the latter is a polynomial ring over K.

Let $\mathbf{p}$ be a prime ideal of R; denote by $T(\mathbf{p})$ the $R/\mathbf{p}$-torsion submodule of

$$S(E) \otimes R/\mathbf{p} = S_{R/\mathbf{p}}(E/\mathbf{p}E).$$

The torsion submodule of $S(E)$ is just $T(0)$.

Proof. By the formula above we have

$$\begin{aligned} \dim S(E)/T(\mathbf{p}) &= \dim(R/\mathbf{p}) + \text{tr.deg.}_{R/\mathbf{p}} S(E)/T(\mathbf{p}) \\ &= \dim(R/\mathbf{p}) + \nu(E_{\mathbf{p}}), \end{aligned}$$

and it follows that $\dim S(E) \geq b(E)$.

Conversely, let P be a prime of $S(E)$ and put $\mathbf{p} = P \cap R$. It is clear that $T(\mathbf{p}) \subset P$,

$$\dim S(E)/P \leq \dim S(E)/T(\mathbf{p}),$$

and $\dim S(E) \leq b(E)$, as desired. □

Corollary 1.22. *Let R be a local domain of dimension d, and E a finitely generated module that is free on the punctured spectrum of R. Then* $\dim S(E) = \sup\{\nu(E), d + \text{rank}(E)\}$.

Corollary 1.23. *Let R be a local domain of dimension d, and E a finitely generated module of rank r. Then* $\dim \mathcal{R}(E) = d + r$.

We now give an effective version of these formulas. A lower bound for $b(E)$ is $b_0(E) = \dim R + \text{rank}(E)$, the value corresponding to the generic prime ideal in the definition of $b(E)$. We show that the correction from $b_0(E)$ can be explained by how deeply the condition $\mathcal{F}_0$ is violated. Set $m_0 = \text{rank}(\varphi)$, so that $\text{rank}(E) = n - m_0$. Without loss of generality we assume that R is a local ring, $\dim R = d$. Consider the descending chain of affine closed sets:

$$V(I_{m_0}(\varphi)) \supseteq \cdots \supseteq V(I_1(\varphi)) \supseteq V(1).$$

Suppose that $P \in \text{Spec}(R)$; if $I_{m_0}(\varphi) \not\subseteq P$, then $\text{rank}(E/PE) = n - m_0$, and therefore

$$\dim(R/P) + \text{rank}(E/PE) \leq b_0(E).$$

On the other hand, if $P \in V(I_t(\varphi)) \backslash V(I_{t-1}(\varphi))$, we have $\text{rank}(E/PE) = n - t + 1$; if $\mathcal{F}_0$ holds at t, the height of P is at least $m_0 - t + 1$ and again $\dim(R/P) + \text{rank}(E/PE) \leq b_0(E)$.

Define the following integer valued function on $[1, \text{rank}(\varphi)]$:

$$d(t) = \begin{cases} m_0 - t + 1 - \text{height } I_t(\varphi) & \text{if } \mathcal{F}_0 \text{ is violated at t} \\ 0 & \text{otherwise.} \end{cases}$$

Finally, if we put $d(E) = \sup_t \{d(t)\}$, we have the following dimension formula ([SV88, Theorem 1.1.2]).

Theorem 1.24. *Let R be a quasi-unmixed domain and E a finitely generated R-module. Then*

$$b(E) = b_0(E) + d(E).$$

Proof. Assume that $\mathcal{F}_0$ fails at t, and let P be a prime as above. We have

$$\text{height}(P) \geq m_0 - t + 1 - d(t),$$

and thus

$$\dim(R/P) + \text{rank}(E/PE) \leq (d - (m_0 - t + 1 - d(t))) + (n - t + 1) = b_0(E) + d(t).$$

Conversely, choose t to be an integer where the *largest* deficit $d(t)$ occurs. Let P be a prime ideal minimal over $I_t(\varphi)$ of height exactly $m_0 - t + 1 - d(t)$. From the choice of t it follows that $P \notin V(I_{t-1}(\varphi))$, as otherwise the deficit at $t - 1$ would be higher. As R is catenary, the last displayed expression gives the desired equality. □

Corollary 1.25. *Let R be a quasi unmixed domain, and E a finitely generated R-module. Then* $\dim S(E) = \dim R + \text{rank } E$ *if and only if E satisfies* $\mathcal{F}_0$.

An useful application of these formulas occurs as follows. Let R be a Noetherian ring of finite Krull dimension and E a torsionfree R-module of rank r with an embedding $E \hookrightarrow R^r$. Set $S = S(R^r) = R[T_1, \ldots, T_r]$ and denote by (E) the ideal of S generated by the 1-forms in $E \subset RT_1 + \cdots + RT_r$; (E) is the defining ideal of the algebra $S_R(R^r/E)$.

Corollary 1.26. *Under these assumptions,* height $(E) \leq \dim R + r - b(R^r/E)$. *Equality holds if R is universally catenary.*

The Rees Algebra of the Conormal Module

Let R be a Noetherian ring and $I \subset R$ an ideal. The *conormal module* of I is I/I^2 viewed as an R/I-module. In many cases I/I^2 is generically R/I-free, when in particular it has a well-defined rank. One of these classes is that of prime ideals. Thus, if $I = \mathfrak{p} \in \text{Spec }(R)$, the $R/\mathfrak{p}$-module $\mathfrak{p}/\mathfrak{p}^2$ has as its rank the embedding dimension $\text{embdim}(R_\mathfrak{p})$ of the local ring $R_\mathfrak{p}$.

The Rees algebra of $\mathfrak{p}/\mathfrak{p}^2$ can be defined equivalently by its symmetric algebra modulo its $R/\mathfrak{p}$-torsion, or via a mapping from $\mathfrak{p}/\mathfrak{p}^2$ into a free $R/\mathfrak{p}$-module of rank $\text{embdim}(R_\mathfrak{p})$ that is an isomorphism when we localize at $\mathfrak{p}$. For example, if $(R, \mathfrak{m})$ is a local ring then $\mathcal{R}(\mathfrak{m}/\mathfrak{m}^2)$ is a polynomial ring in $\text{embdim}(\mathfrak{m})$ variables over the residue field of R.

Proposition 1.27. *Let $(R, \mathfrak{m})$ be a Noetherian local ring and $\mathfrak{p}$ a prime ideal. Then*

$$\dim \mathcal{R}(\mathfrak{p}/\mathfrak{p}^2) \leq \begin{cases} \dim R/\mathfrak{p} + \text{embdim}(R_\mathfrak{p}) \leq \text{embdim}(R) \\ \dim R/\mathfrak{p} + \text{height }(\mathfrak{p}) \leq \dim R \quad \text{(if } \mathfrak{p} \text{ is generically a CI.)} \end{cases}$$

Proof. The first part of the assertion is an application of the dimension formula in Lemma 1.21 to the algebra $\mathcal{R}(\mathfrak{p}/\mathfrak{p}^2)$. That $\dim R/\mathfrak{p} + \text{embdim}(R_\mathfrak{p}) \leq \text{embdim}(R)$ holds in any local ring is proved in [Le64, Theorem 3]. The last dimension formula is again a consequence of Lemma 1.21. □

In case $\mathfrak{p}$ is generically a complete intersection, the explanation of the last dimension estimate lies simply in the fact that $\mathcal{R}(\mathfrak{p}/\mathfrak{p}^2)$ is a homomorphic image of $G = \text{gr}_\mathfrak{p}(R)$. It will then inherit many properties of G.

Modules of Linear Type

One can extend to certain classes of modules the notion of *linear type*:

Definition 1.28. Let E be a torsionfree module over the integral domain R. Then E is said to be of *linear type* if $S_R(E)$ is an integral domain.

The dimension formulas above are useful in clarifying this notion for rings of small dimension. Let R be a Noetherian local domain of dimension $d \geq 1$; if E is a finitely generated torsionfree module of rank n, to qualify for being of linear type it must be generated by at most $\nu(E) \leq \text{rank}(E) + d - 1$ elements, according to Corollary 1.25. In particular, if $d = 1$, E must be a free R-module, while for $d = 2$ it admits a presentation

$$0 \to K \longrightarrow R^{n+1} \longrightarrow E \to 0,$$

where K is a rank one module.

Proposition 1.29. *Let R be a normal local domain and let E be a torsionfree R-module of rank n generated by $n+1$ elements. Then E is of linear type.*

Proof. In a presentation of E as above, let $(a_1, \ldots, a_{n+1})$ be a nonzero syzygy of E (for a chosen basis of R^{n+1}). Let

$$f = a_1 T_1 + \cdots + a_{n+1} T_{n+1}$$

be the corresponding polynomial in the presentation

$$S_R(E) = R[T_1, \ldots, T_{n+1}]/(K).$$

If L is the ideal of R generated by the coefficients of f, we claim that $(K) = (L^{-1}f)$, and it is a prime ideal of the polynomial ring. It is clear that since E is torsionfree, we have $L^{-1}f \subset K$, as submodules of R^{n+1}. But they are both reflexive modules that are clearly isomorphic when R is localized at any height 1 prime $\mathfrak{p}$. Now we argue that (K) is a prime ideal. Since $A = R[T_1, \ldots, T_{n+1}]$ is normal, we have $K \subset P$ for some height 1 prime of A. Note that $P \cap R = 0$ (which would imply that $(K) = P$), since otherwise the intersection would be a height one prime of R containing all the coefficients of the polynomials in $L^{-1}f$, that is, the ideal $L^{-1}L$. Since R is normal, such ideals have height at least two. □

Remark 1.30. If R is not an integral domain, let $p_1, \ldots, p_n$ be its minimal primes. It is clear from Theorem 1.20 that

$$\dim S_R(E) = \sup_{i=1}^{n} \{\dim S_{R/p_i}(E/p_iE)\}.$$

Assume that for each p_i, R/P_i is equidimensional. If we put

$$b_i(E) = b_0(E/p_iE), \quad d_i(E) = d(E/p_iE),$$

then

$$\dim S(E) = \sup_{i=1}^{n} \{b_i(E) + d_i(E)\}.$$

1.2 Rees Algebras and Reductions

The other approach to studying a Rees algebra $R[It]$ is less straightforward, but considerably more malleable than that via presentations. Let us recall the notion on which it is based.

Definition 1.31. Suppose that R is a commutative ring and J, I ideals of R with $J \subset I$. Then J is a *reduction* of I if $I^{r+1} = JI^r$ for some integer r; the least such integer is the *reduction number* of I with respect to J. It is denoted by $r_J(I)$.

Introduced by Northcott and Rees ([NR54]), this notion has played an important role in the study of blowup algebras. It can be variously expressed as we shall see but its most useful aspect is the following. J is a reduction of I when the embedding

$$R[Jt] \hookrightarrow R[It]$$

is a finite morphism of graded algebras, and $r_J(I)$ is the infimum of the top degree of any homogeneous set of generators of $R[It]$ as a module over $R[Jt]$.

There is another way to express the notion of a reduction. It is clear that an ideal I and all of its reductions have the same radical, while the converse is not true. Nevertheless, as the reader can easily verify, J is a reduction of I if and only if the ideals (Jt) and (It) have the same radical in the Rees algebra $R[It]$.

Example 1.32. (a) The simplest example of a reduction is probably the case $J = (x^2, y^2) \subset I = (x^2, y^2, xy)$, when $I^2 = JI$.

(b) The following shows the possible variation of $r_J(I)$ ([Huc87, Example 3.1]). Consider the ideal

$$I = (x^7, x^6y, x^2y^5, y^7) \subset k[x,y].$$

Then both $J = (x^7, y^7)$ and $L = (x^7, x^6y + y^7)$ are reductions, but $\mathrm{r}_J(I) = 4$ while $\mathrm{r}_L(I) = 3$.

(c) For k a field, let $f(x,y,z) \in k[x,y,z]$ be a homogeneous, irreducible polynomial of degree n, and suppose that $I = (x,y,z) \subset k[x,y,z]/(f)$. If k is infinite or f is monic in one of the variables, then I has a reduction $J = (a,b)$ with reduction number $\mathrm{r}_J(I) = n - 1$.

Ideals and Integral Extensions

Let us illustrate further the notion of reduction and reduction number with the description of an elementary process that produces reductions. Let A be a Noetherian ring with total ring of fractions K and let $A \subset B \subset K$ be a finitely generated integral extension. Clearing denominators of a set of generators of B, we can write $B = Ix^{-1}$, where $I \subset A$ is an ideal and x is a regular element of A. Since $B^2 = B$, we obtain the equality

$$I^2 = xI,$$

so I has a principal reduction, and reduction number 1 if $B \neq A$. This is a full correspondence between the isomorphism classes of such ideals and the set of finite birational extensions of A.

One can exploit this correspondence further after noticing that the depth properties of B and I are the same. For example, if A is a Gorenstein ring then B satisfies Serre's S_2 property if and only if I is divisorial. In Chapter 6 we shall return to this setting to describe an approach to the complexity of the integral closure.

1.2.1 Basic Properties of Reductions

There are several equivalent ways to describe the notion of a reduction. We recall some of them for the perspective each provides.

(a) *Integral closure*. Let L be an ideal of R. An element $z \in R$ is *integral* over L if it satisfies an equation of the form

$$z^n + a_1 z^{n-1} + \cdots + a_n = 0, \quad \text{with } a_i \in L^i.$$

Definition 1.33. The integral closure $\overline{L}$ of L is the set of the elements $z \in R$ integral over L.

Another way to frame this notion is through the Rees algebra of L. The element $z \in R$ is integral over L if the element $zt \in R[t]$ is integral over $R[Lt]$. In other words, the subalgebra $R[(L,z)t]$ of $R[t]$ is integral over $R[Lt]$. An interpretation of $\overline{L}$ arises as follows. Let $A = R[Lt]$ be the Rees algebra of the ideal L, and C the integral closure of A in the ring $R[t]$. Then C is a graded R-algebra,

$$C = R + \overline{\mathbf{L}}t + \sum_{n \geq 2} \overline{L^n} t^n.$$

In particular, $\overline{\mathbf{L}}$ is an ideal of R and

$$L \subset \overline{\mathbf{L}} \subset \sqrt{L}.$$

Proposition 1.34. *If R is a Noetherian integral domain, an element $z \in R$ is integral over the ideal L if and only if there is a nonzero element $c \in R$ such that $cz^n \in L^n$ for all $n \geq 0$.*

Proof. The condition means that as an $R[Lt]$-module, $R[(L,z)t]$ is contained in $R[Lt]c^{-1}$. □

Unlike the integral closure of algebras, which is discussed in a later chapter, the equations of integral dependence over ideals are much more specific and harder to find. These equations usually arise in the following situation. Let

$$M = (m_1, \ldots, m_n)$$

be a finitely generated faithful R-module and assume that $zM \subset L \cdot M$. Setting this inclusion as a system of equations

$$zm_i = \sum_{j=1}^{n} a_{ij} m_j, \quad a_{ij} \in L,$$

and using the determinantal trick, one gets an equation as above. Conversely, given the equation, set $M = (z,L)^{n-1}$ to have $zM \subset L \cdot M$.

These two observations highlight the difficulty one encounters while attempting to deal with one of the following four problems.

Suppose that $I = (f_1, \ldots, f_m) \subset R = k[x_1, \ldots, x_n]$, and $f \in R$. How to solve the following:

- *Membership Test*: $f \in \overline{I}$?
- *Completeness Test*: $I = \overline{I}$?
- *Construction Task*: $I \rightsquigarrow \overline{I}$?
- *Complexity Cost*: $\mathrm{cx}(I \rightsquigarrow \overline{I})$?

In the literature one does not find effective methods of dealing generally with these problems. The difficulty arises, partly, from the specialized nature of the equations the elements are to satisfy.

Theoretically, they are all reducible to the algebra problem, but at a truly high price. Ideally, all decisions and construction should take place at (or very near to) the ring R.

Proposition 1.35. *J is a reduction of I if and only if every element of I is integral over J.*

Proof. (C. Weibel) Suppose that $JI^r = I^{r+1}$ for some integer r, and set $M = I^r$ as above. For each $x \in I$, we have an element

$$z = x^n + a_1 x^{n-1} + \cdots + a_n, \quad a_i \in J^i,$$

in the annihilator of I^r that belongs to I; $z \in \mathrm{ann}\, I^r \cap I$. It follows that $z^{r+1} = z^r \cdot z = 0$.

For the converse, assemble the ideal I as above. □

Corollary 1.36. *The ideal $J \subset I$ is a reduction of I if and only if $\overline{J} = \overline{I}$.*

(b) *Integral domain.* Another property that is useful is the following observation.

Proposition 1.37. *Let R be a Noetherian ring and let J and I be ideals with $J \subset I$. Then J is a reduction of I if and only if for every minimal prime ideal $\mathfrak{p}$ of R, the ideal $J \cdot R/\mathfrak{p}$ is a reduction of $I \cdot R/\mathfrak{p}$.*

The verification is a straightforward calculation.

(c) *Valuative criterion.* Valuations are versatile tools in the examination of reductions. At the start is the following general existence result ([Kap74, Theorem 56]).

Theorem 1.38. *Let I be a proper ideal of the integral domain R. There exists a valuation ring V of R such that $IV \neq V$. If R is Noetherian, V can be chosen to be a discrete valuation ring.*

For a proof in the Noetherian case, see the discussion below. A deep way to express the notion of reductions is given in the following theorem of Zariski [ZS60, p. 350].

Theorem 1.39. *Let R be an integral domain and L an ideal. Then*

$$\overline{L} = R \cap \bigcap L \cdot V,$$

where V runs over all the valuation rings of R. If R is Noetherian, $x \in R$ is integral over L if and only if, $\varphi(x) \in \varphi(L)V$ for every homomorphism $\varphi : R \to V$, where V is a discrete valuation domain.

Proof. (*Sketch*) Let $0z$ be a nonzero element in the right hand side and consider the ring $S = R[Lz^{-1}]$. If $1 \in (Lz^{-1}S$, a direct expansion shows that $z \in \overline{L}$. If not, applying Theorem 1.38 to the pair $Iz^{-1}S, S$ gives a contradiction. The reverse inclusion is immediate. □

This is a handy theoretical tool. For instance, if J and I are ideals of an integral domain then J is a reduction of I if for some integer s, $\overline{J^s}$ is a reduction of $\overline{I^s}$. Just check at each valuation overring of R. Another version of this result, to which we shall refer later, is:

Corollary 1.40. *Let $I = (a_1, a_2, \ldots, a_n)$ be an ideal of an integral domain. For every positive integer s,*

$$\overline{I^s} = \overline{(a_1^s, a_2^s, \ldots, a_n^s)}.$$

Rees Valuations of an Ideal

The ideals of the form $R \cap LV$ for some valuation V are called *contracted ideals*. If R is a Noetherian domain, $\overline{L}$ can be represented using a finite set of contracted ideals as follows. Let $S = R[Lt, t^{-1}]$ be the extended Rees algebra of the ideal L and let A be its integral closure; A is a Krull domain. Let $\mathfrak{p}_1, \ldots, \mathfrak{p}_n$ be the minimal prime ideals of At^{-1}; the localizations $A_{\mathfrak{p}_i}$ are discrete valuation domains and thus will induce discrete valuations V_i on R. These valuations are also known as the *Rees valuations* of L. One has:

Proposition 1.41. *Let $\{V_1, \ldots, V_n\}$ be the set of Rees valuations of the ideal L. The integral closure of L in R is given by*

$$\overline{L} = R \cap (\bigcap_{i=1}^{n} LV_i).$$

Proof. To provide for flexibility later, we first describe the Rees valuations of L in another way. Set

$$S = \overline{R[Lt]} = \overline{R} + \overline{L}t + \cdots$$

for the integral closure of $R[Lt]$. Then S is a Krull domain and the ideal $\overline{L} + \overline{L^2}t + \cdots$, being isomorphic to the height 1 prime ideal S_+, is also divisorial. We can then write

$$J = \overline{L} + \overline{L^2} + \cdots + \overline{L^{m+1}}t^m + \cdots = P_1^{(e_1)} \cap \cdots \cap P_n^{(e_n)}$$

for its primary decomposition. Denote $S_i = S_{P_i}$ for the corresponding discrete valuation domains and set $V_i = K \cap S_i$, where K is the field of fractions of R.

We are now ready to prove the assertion. If $z \in R \cap (\bigcap_{i=1}^n LV_i)$, then $z \in R \cap (\bigcap_{i=1}^n LS_i)$ and therefore $z \in \overline{L}$, as desired. □

Corollary 1.42. *Let L be as above. For any positive integer r, the Rees valuations of L^r are the same as the Rees valuations of L.*

Proof. The argument is similar to the one above. Consider the ideal

$$J_r = \overline{L^r} + \overline{L^{r+1}}t + \cdots + \overline{L^{m+r}}t^m + \cdots,$$

which is isomorphic to $J_r t^r$. From the exact sequence

$$0 \to J_r t^r \longrightarrow S \longrightarrow R + \overline{L}t + \cdots + \overline{L^{r+1}}t^{r-1} \to 0$$

it will follow that $J_r t^r$ satisfies Serre's S_2-condition and therefore it will be a divisorial ideal. (See Section 4.1, where a general discussion of divisors in Rees algebras is given.) It now follows that

$$J_r = \bigcap_{i=1}^n P_i^{(re_i)} = \bigcap_{i=1}^n L^r S_i,$$

that again gives the desired decomposition for $\overline{L^r}$. □

There is another way of coding all these valuations together. Let R be an unmixed integral domain, and I a proper ideal of R. The I-adic valuation of R is the function

$$\text{If } x \neq 0, \; v_I(x) = n \text{ if } x \in I^n \setminus I^{n+1}.$$

Samuel refined this into

$$\overline{v}_I(x) = \lim_{n \to \infty} \frac{v_I(x^n)}{n}.$$

This limit always exists and is finite for $x \neq 0$. Although not always a valuation, it can be expressed in terms of the Rees valuations. A significant point is that ([Mc83])

$$x \in \overline{I} \Leftrightarrow \overline{v}_I(x) \geq 1.$$

(d) *Differential criterion.* We illustrate this with a well-known example. Other properties will be discussed in Chapter 7.

Example 1.43. An interesting process to produce ideals and some of its proper reductions is the following (see [ScS70], [Hu96, Exercise 5.1]). Suppose that $f \in R$, where R is a power series ring in the indeterminates $x_1, \ldots, x_n$ over $\mathbb{C}$,

$$J = \left(\frac{\partial f}{\partial x_1}, \ldots, \frac{\partial f}{\partial x_n} \right).$$

Then J is a reduction of $I = (J, f)$ (with some provisos, such as if f is a power series then it should not be a unit). We apply the valuative criterion of Theorem 1.39; it is enough to consider valuation rings of the form $k[[t]]$, $\mathbb{C} \subset k$, that dominate R. We can write $f = t^n g$, where g is a unit of $k[[t]]$. Using the chain rule for differentiation

$$f' = \frac{df}{dt} = \sum_{i=1}^{n} \frac{\partial f}{\partial x_i}\frac{dx_i}{dt},$$

(why does it apply here?) it becomes clear that $f \in f'k[[t]] \subset Jk[[t]]$.

(e) *Geometric interpretation.* Finally, a geometric way (from conversations with C. Weibel) to describe this notion as follows.

Proposition 1.44. *Let R be a Noetherian ring and I an ideal. Let $x_1, \ldots, x_n$ be elements of I generating the ideal J. If the x_i's are regular elements, the following conditions are equivalent:*

(a) *J is a reduction of I.*
(b) *The open sets $D_+(x_it)$ cover* Proj $(R[It])$.
(c) *The morphism* Proj $(R[It]) \to$ Proj $(R[Jt])$ *is finite.*

Proof. Let us show that (c) $\Rightarrow$ (a), the most interesting of the implications. The assumption includes the fact that each of the mappings

$$H^0(O_{\text{Proj }(R[It])}, D_+(x_it)) \to H^0(O_{\text{Proj }(R[It])}, D_+(x_it))$$

is finite. Since they are, respectively,

$$\bigcup_{n \geq 0} \frac{I^n}{x_i^n} \quad \text{and} \quad \bigcup_{n \geq 0} \frac{J^n}{x_i^n},$$

it means that for each x_i there are integers m, r such that

$$\frac{I^{m+1}}{x_i^{m+1}} \subset \frac{J^r}{x_i^r} \cdot \frac{I^m}{x_i^m},$$

that is,

$$x_{r-1} I^{m+1} \subset J^r I^m.$$

It is clear that we can arrange m, r for that inclusion to hold for all x_i. It is now immediate that

$$J^{n(r-1)} I^{m+1} \subset J^{n(r-1)+1} I^m.$$

Since the reverse inclusion is obvious we have equality between the two ideals, that is, we have an equality of the form

$$I \cdot M = J \cdot M, \quad M = J^a I^b.$$

This is the usual equational formulation of reduction since M is a faithful module. (In the case of integral domains, use Zariski's theorem.) □

(f) *Homogeneous ideals.* Let k be a field and I a homogeneous ideal of the polynomial ring $k[x_1,\ldots,x_n]$. Then $\overline{I}$, the integral closure of I, is also homogeneous. Indeed, suppose that $f(x)+g(x)$ is integral over I, where $f(x)$ and $g(x)$ are homogeneous of degrees r and s, $r>s$. Then

$$(f(x)+g(x))^m + c_1(x)(f(x)+g(x))^{m-1} + \cdots + c_m(x) = 0, \quad c_i(x) \in I^i.$$

Suppose that $0 \neq \lambda \in k$ and consider the automorphism of $k[x_1,\ldots,x_n]$ induced by multiplication of each variable by λ. Applying it to the equation above, we get

$$(f(\lambda x)+g(\lambda x))^m + c_1(\lambda x)(f(\lambda x)+g(\lambda x))^{m-1} + \cdots + c_m(\lambda x) = 0,$$

which is another equation of integral dependence over I since $c_i(\lambda x) \in I^i$, as every homogeneous ideal is invariant under the action of the automorphism. This means that $\lambda^r f(x) + \lambda^s g(x)$ is also integral over I. Since we may assume that k is infinite, sufficiently many different choices of λ allow (by the Vandermonde trick, say) that every homogeneous component is integral over I.

A similar argument applies to other orderings and even other structures (see [B6183, Chap. 5, Prop. 20]). For example, if I is a monomial ideal of $k[x_1,\ldots,x_d]$, the substitution $x_i \rightarrow t_i x_i$ leaves I invariant, and from this it will follow as above that $\overline{I}$ is a monomial ideal. An additional use refers to the integral closure of $R[It]$ itself: $\overline{R[It]} = \sum_{n\geq 0} \overline{I^n} t^n$. (Note that the integral closure of the $R[It]$ is taken in $R[t]$.)

(g) *Analytic interpretation.* As expected, the growth conditions of a numerical function f that is integral over an ideal I are controlled very tightly. The following precise statement is taken from [HRT99] (see also [LT81]).

Theorem 1.45. *Set $R = \mathbb{C}[x_1,\ldots,x_n]$, let $I \subset R$ be an ideal and suppose that $f \in R$. The following conditions are equivalent:*

(a) *f is integral over I.*
(b) *For any set of generators $(a_1,\ldots,a_r)$ of I, there exist constants $C>0$ and $m \in \mathbb{N}$ such that*

$$|f(z)| \leq C(1+|z|)^m \cdot \sup_j |a_j(z)| \quad \textit{for all } z \in \mathbb{C}^n.$$

(h) *Existence of reductions.* Let R be a Noetherian ring of Krull dimension d and let I be an ideal. A reduction J of I can be characterized by the fact that the ideal $JtR[It]$ has the same nilradical as $ItR[It]$.

While to find the integral closure of a given ideal I is a difficult task, in the other direction the problem is easily handled. To find reductions of an ideal I, possibly with fewer generators than I, we make use of the following elementary observation.

Proposition 1.46. *Let S be a Noetherian ring of Krull dimension r that contains an infinite field k. If I is an ideal generated by the elements $f_1,\ldots,f_n$, then there exist $r+1$ k-linear combinations of the f_i generating an ideal with radical $\sqrt{I}$.*

Corollary 1.47. *If R is a Noetherian ring of Krull dimension d containing an infinite field, then any ideal I contains a reduction generated by $d+2$ elements.*

Proof. We set $S = R[It]$, which has Krull dimension at most $d+1$. Choose a set $g_1,\ldots,g_n$ of generators of I and apply Proposition 1.46 to the elements $g_1t,\ldots,g_nt$ of the Rees algebra $R[It]$. □

We want to cut down the bound above to $d+1$, by proving a specialized version of Proposition 1.46, better tailored to the study of reductions.

Proposition 1.48. *Let R be a Noetherian integral domain and let $B = \sum_{n\geq 0} B_n$, $B_0 = R$, be a graded algebra of dimension d. Then there are homogeneous elements $f_1,\ldots,f_d \in B_+$ such that $B_+ = \sqrt{(f_1,\ldots,f_d)}$. Moreover, if R contains an infinite field k, then $f_1,\ldots,f_d$ can be chosen in B_1.*

Proof. Note that $d = \dim B = \dim R + \text{height}\, B_+$. If height $B_+ = 0$, then B_+ is a nilpotent ideal. We assume that $s = \text{height}\, B_+ > 0$. Choose $f_1 \in B_+$ that avoids all minimal primes of B. Some of the minimal primes of (f_1) (all of height 1 and all homogeneous) are either contained in B_+, or are of the kind $Q = \mathfrak{q} + Q_+$, $0 \neq \mathfrak{q} \subset R$. If $s > 1$, we choose f_2 homogeneous that avoids all the minimal primes of (f_1). We proceed in this manner until $f_1,\ldots,f_s$ have been chosen. The minimal primes of $(f_1,\ldots,f_s)$ are B_+, or a prime of height s such as $Q = \mathfrak{q} + Q_+$, $\mathfrak{q} \neq 0$, and $Q_+ \neq B_+$. If any such Q exists, we choose a form $f_{s+1} \in B_+$ that avoids all the Q's. Clearly the minimal primes of $(f_1,\ldots,f_s,f_{s+1})$ are B_+ and primes like Q of height $s+1$. At the d-th step of this process, if any Q shows up it would have height d and therefore would contain B_+.

The second assertion, in case the B_1 generates B and R contains an infinite field, follows by the standard prime avoidance. □

Corollary 1.49. *If R is a Noetherian integral domain of Krull dimension d containing an infinite field, then any ideal I contains a reduction generated by $d+1$ elements.*

Remark 1.50. The restriction to integral domains is easily seen to be unnecessary. For a general discussion, see [Ka94]. When R is a ring of polynomials over an infinite field, G. Lyubeznik proved that reductions always exists with $\dim R$ generators ([Ly86]).

1.2.2 Integrally Closed Ideals and Normal Ideals

Definition 1.51. Let I be an ideal of the commutative ring R.

(a) I is *integrally closed* or *complete* if $I = \overline{I}$.
(b) I is *normal* if all powers I^n are integrally closed.

There is a wide gap between these two notions. They agree however when R is a two-dimensional regular local ring, by Zariski's theory of complete ideals (see [ZS60]). More generally, according to [Li69b], this also holds for two-dimensional rings rational singularities. Furthermore, according to [Cu90], if the residue field of R is algebraically closed, then I^2 is complete for a complete primary ideal I precisely for two-dimensional rational singularities.

Example 1.52. Prime ideals are premier examples of integrally closed ideals. They are however not always normal. Let P be the defining ideal of the monomial curve (t^a, t^b, t^c), $a < b < c$, $\gcd(a,b,c) = 1$. According to [He70], P is given by the 2×2 minors of a matrix

$$\begin{bmatrix} z^{a_1} & x^{a_2} & y^{a_3} \\ y^{b_1} & z^{b_2} & x^{b_3} \end{bmatrix}.$$

The ideal is not normal (see [Va87]) precisely when one of the following occurs

(i) $\inf\{a_1, a_2, a_3\} > 1$ or $\inf\{b_1, b_2, b_3\} > 1$;
(ii) $a_2 = b_2 = 1$, and the other exponents are > 1.

The ideals corresponding to $(7,8,10)$ and $(6,7,16)$ are the 'minimal' examples of each kind. For example, if P corresponds to $(6,7,16)$ and $L = P^2 : (x,y,z)$, it is easy to verify that P^2 is minimally generated by 6 elements, while L requires 7 generators. Using the notion of $\mathfrak{m}$-full ideals discussed later in this section, it will follow that P^2 is not integrally closed.

A notion employed in examining certain classes of normal ideals is the following.

Definition 1.53. Let R be a Noetherian ring and I an ideal. Then I is *normally torsionfree* if $\operatorname{Ass} R/I = \operatorname{Ass} R/I^n$ for $n \geq 1$.

The following observation gives an illustration.

Proposition 1.54. *Let R be a regular local ring and I a reduced ideal. If I is normally torsionfree, then I is normal.*

Normal Determinantal Ideals

An important class of normal ideals arises as ideals of minors of generic matrices. We indicate some from the recent literature.

Let $X = (x_{ij})$ be a generic matrix of size $m \times n$ over a field k and R the ring of polynomials $R = k[x_{ij}]$. Let I_t be the ideal of R generated by the minors of size t of X. If $t = \max\{m,n\}$ then in any characteristic I_t is normal ([EH83]). In characteristic zero, this was later proved for all I_t ([B91]). In other characteristics the situation is very different. For instance, if $m = n = 4$ and $t = 2$ then I_t^2 is not integrally closed. In general, if char $k > \min\{t, m-t, n-t\}$ then I_t is normal ([BC98]).

Complete Ideals from Others

A direct way to obtain integrally closed ideals is to use the ideal quotient operation on a complete ideal. More precisely one has:

Proposition 1.55. *Let R be an integral domain, let $J \subset I$ be ideals and set $L = J : I$. Then $\overline{L} \subset \overline{J} : \overline{I}$. In particular if J is a complete ideal, then for any ideal I, $J : I$ is also a complete ideal.*

Proof. Let $x \in \overline{L}$. For any valuation overring V of R, $xVI \subset VL \cdot VI \subset VJ$. Thus $\bigcap xVI = x\bigcap VI = x\overline{I} \subset \bigcap VJ = \overline{J}$. □

Although this can be used to show that the minimal primary components of a complete ideals are primary, the situation is much more thorough in normal domains ([ZS60, p. 354]):

Proposition 1.56. *Let R be an integrally closed Noetherian domain. Any complete ideal of R has an irredundant primary decomposition by complete primary ideals.*

Example 1.57. The following example of C. Huneke ([Hu86b]) shows a primary ideal I in a regular local ring whose integral closure $\overline{I}$ is not primary. Let k be a field of characteristic 2 and let $\mathfrak{p}$ be the defining ideal of $k[[t^3, t^4, t^5]]$ in $k[[x,y,z]]$. The ideal $\mathfrak{p}$ is generated by $a = x^3 - yz, b = y^2 - xz, c = z^2 - x^2y$. This ideal is normal (see Example 1.52), in particular $\overline{\mathfrak{p}^2} = \mathfrak{p}^2$. Let I be the ideal $I = (a^2, b^2, c^2)$. A computation with *Macaulay* shows that I is $\mathfrak{p}$-primary. (Or, as pointed out originally, R is a regular local ring of characteristic 2, the Frobenius homomorphism is exact so that $\mathrm{Ass}(R/(a^2,b^2,c^2)) = \mathrm{Ass}(R/(a,b,c)) = \{\mathfrak{p}\}$.) Note that $I \subset \mathfrak{p}^2 \subset \overline{I}$ so that $\overline{I} = \mathfrak{p}^2 = \overline{\mathfrak{p}^2}$. But $\mathfrak{p}^2$ is not $\mathfrak{p}$-primary since it has projective dimension 2.

Tests and Tools

Testing for these properties is not easy. One approach is a recasting of the definition of integrality.

Proposition 1.58. *Let R be a Noetherian integral domain. The ideal I is integrally closed if and only if, for every ideal L,*

$$IL : L = I.$$

The difficulty lies in getting a proper test ideal. Usual choices for partial tests are the powers of I, and when $(R, \mathfrak{m})$ is a local ring, the maximal ideal.

Example 1.59. Let us illustrate the use of this formulation with a discussion of primary Gorenstein ideals; for other details we refer to [CHV98]. Let $(R, \mathfrak{m})$ be a Cohen-Macaulay local ring of dimension $d > 0$, and I an $\mathfrak{m}$-primary Gorenstein ideal. This means that $I : \mathfrak{m}$ is the minimal ideal containing I properly.

Suppose that I is integrally closed. Then for any ideal N (containing regular elements), N/IN is a faithful R/I-module, by the observation above. But we recall that every faithful module over a zero-dimensional Gorenstein ring must contain a free summand. Indeed, if $f_1, \ldots, f_n$ is a set of generators of N/IN, then we cannot have the annihilator of each f_i containing I properly, as otherwise they would all contain $I : \mathfrak{m}$, contradicting the fact that N/IN is faithful. Let then, say, f_1 have I for its annihilator. We have then an embedding

$$R/I \hookrightarrow N/IN$$

that splits since R/I is self-injective.

Suppose that $I \subset L$. Tensoring the splitting above by R/L, we get another splitting

$$R/L \hookrightarrow N/LN.$$

In particular this shows that N/LN is a faithful R/L-module, so L will be an integrally closed ideal.

When I is a perfect ideal (of codimension 3, or of arbitrary codimension with additional conditions), I must be a complete intersection of the very special kind studied in [Go87]. In particular R must be a regular local ring.

Remark 1.60. In another variation of these approaches, the valuation criterion leads to the following observation. Let J, L be two ideals of an integral domain with $\overline{J} = \overline{L}$. For any ideal I, $J \cdot I : L \subset \overline{I}$.

$\mathfrak{m}$-full Ideals

There is a test of the completeness of an ideal of a local ring $(R, \mathfrak{m})$ that is very useful in calculations. It is based on the following notion of Rees and Watanabe. We follow the treatment of [Go87].

Definition 1.61. I is called $\mathfrak{m}$-*full* if there is $x \in \mathfrak{m}$ such that $\mathfrak{m}I :_R x = I$.

Note that if R is a ring of polynomials over an infinite field, and I is a homogeneous ideal, then x can be chosen to be a linear form.

More generally, for two ideals I and L, one could define I to be L-full if there is $x \in L$ such that $LI : x = I$.

Proposition 1.62. *Let $(R, \mathfrak{m})$ be a local Noetherian ring. If the residue field of R is infinite, then nontrivial integrally closed ideals are $\mathfrak{m}$-full. More generally, nontrivial integrally closed ideals are L-full for any nonzero ideal L.*

Proof. We make use of the notion of Rees valuation in Proposition 1.41, and its notation. To simplify we assume that R is an integral domain.

Let $(V_1, \mathfrak{p}_1), \ldots, (V_n, \mathfrak{p}_n)$ be the Rees valuations associated to I. Observe that since for each i $\mathfrak{m}V_i \neq 0$, it follows that $\mathfrak{m}$ contains $\mathfrak{p}_i\mathfrak{m}V_i \cap R$ properly. Since $R/\mathfrak{m}$ is infinite, we can choose $x \in \mathfrak{m}$ that is not contained in $\bigcup_{i=1}^{n} \mathfrak{p}_i\mathfrak{m}V_i$. This means that $\mathfrak{m}V_i = xV_i$ for each i. We claim that this element will do. Suppose that $y \in \mathfrak{m}I : x$. For each V_i we have $yx \in \mathfrak{m}IV_i = xIV_i$. Thus $y \in \bigcap_{i=1}^{n} IV_i \cap R$, and therefore $y \in \overline{I} = I$.

The same argument can be used for the ideal L. □

Remark 1.63. The choice of the element x above depends only on $\overline{I}$, a fact that will be useful shortly.

Corollary 1.64. *Let $(R, \mathfrak{m})$ be a Noetherian local ring with infinite residue field. An ideal I is integrally closed if and only if it is L-full for all ideals L.*

Proof. Since I is a reduction of $\overline{I}$, there exists an integer n such that $\overline{I}^{n+1} = I \cdot \overline{I}^n$. It is enough to set $L = \overline{I}^n$. □

Example 1.65. Let $R = k[x,y]$, k a field and set $I = (x^4, x^3y^3, x^2y^4, xy^6, y^8)$. Then $\overline{I} = (x, y^2)^4 \neq I$, but $(x,y)I : y = I$, so I is $\mathfrak{m}$-full but not complete.

Definition 1.66. The $\mathfrak{m}$-*full closure* of an ideal I is the smallest $\mathfrak{m}$-full ideal L containing I.

This notion is justified by the following observation.

Proposition 1.67. *Let $(R, \mathfrak{m})$ be a Noetherian ring and I an ideal. The $\mathfrak{m}$-full closure of I exists.*

Proof. Choose x as in Remark 1.63. Since $\overline{I}$ is $\mathfrak{m}$-full,

$$I \subset L = \mathfrak{m}I : x \subset \mathfrak{m}\overline{I} : x = \overline{I}.$$

Iterating, we obtain an ideal that is $\mathfrak{m}$-full and contained in $\overline{I}$. It is also clear that such an ideal will be contained in any $\mathfrak{m}$-full ideal containing I. □

Number of Generators of Complete Ideals

$\mathfrak{m}$-full ideals have the following useful property ([Go87]). It puts a combinatorial control on the study of the integral closure of an ideal.

Proposition 1.68. *Let $(R, \mathfrak{m})$ be a Noetherian local ring and I an $\mathfrak{m}$-full ideal. Then $\nu(I) \geq \nu(L)$ for any ideal $I \subset L$ such that $\lambda(L/I) < \infty$.*

Proof. Let $x \in \mathfrak{m}$ be such that $\mathfrak{m}I : x = I$. There is an exact sequence of modules of finite length

$$0 \to I/\mathfrak{m}I \longrightarrow L/\mathfrak{m}I \stackrel{x}{\longrightarrow} L/\mathfrak{m}I \longrightarrow L/(xL + \mathfrak{m}I) \to 0.$$

Reading lengths, we have

$$\nu(L) = \lambda(L/\mathfrak{m}L) \leq \lambda(L/(xL + \mathfrak{m}I)) = \lambda(I/\mathfrak{m}I) = \nu(I),$$

as desired. □

Examples of integrally closed ideals of a simple character are the following. Let R be an integrally closed Noetherian domain. Then any divisorial ideal I is integrally closed. However, the powers of I may not always be integrally closed. For a more complicated example, suppose that $\mathfrak{p}$ is a prime ideal generated by a regular sequence; then $\mathfrak{p}$ is a normal ideal. A more delicate class is described by the following result ([Go87]).

Proposition 1.69. *Let R be a Noetherian ring and $I = (a_1, \ldots, a_g)$ a complete intersection of codimension g. Then I is integrally closed if and only if for every associated prime $\mathfrak{p}$, $R_\mathfrak{p}$ is a regular local ring of dimension g and $I_\mathfrak{p} = (b_1, \ldots, b_g)$, where at least $g-1$ of these elements are minimal generators of $\mathfrak{p}R_\mathfrak{p}$.*

Proof. Suppose that I is an integrally closed ideal and let $\mathfrak{p}$ be one of its associated primes. Localizing at $\mathfrak{p}$, we may assume that R is a local ring and that its maximal ideal is associated to I. Since I is $\mathfrak{p}$-full, we have $\nu(I : \mathfrak{p}) \leq \nu(I)$, and therefore $I : \mathfrak{p}$ must also be a complete intersection. By a theorem of Northcott (see Theorem 1.113) it follows that R is a regular local ring of dimension g. Moreover, we cannot have 2 of the b_i contained in $\mathfrak{p}^2$, as this would place I inside a $\mathfrak{p}$-primary ideal minimally generated by at least $g+1$ elements. The converse is clear. □

Corollary 1.70. *If I is an integrally closed complete intersection, then I is a normal ideal.*

Completeness Test

The following elementary observation shows some of the opportunities and difficulties in developing such tests. We recall the notion of the *socle of an ideal*. Let $(R, \mathfrak{m})$ be a local ring and I an $\mathfrak{m}$-primary ideal. The ideal $L = I : \mathfrak{m}$ is called the socle ideal of I.

Proposition 1.71. *Let $(R, \mathfrak{m})$ be a Noetherian local ring, let I be an $\mathfrak{m}$-primary ideal and let L be its socle ideal. Then I is complete if and only if no element of $L \setminus I$ is integral over I.*

Proof. If $f \in \overline{I} \setminus I$, then for some power of $\mathfrak{m}$, $\mathfrak{m}^r f$ will contain non-trivial elements in the socle of I. The converse is clear. □

In general the set $L \setminus I$ is much too large. In the case where I is a monomial ideal, we can restrict the test to the finite set of monomials that span L.

Normalization of a Blowup

A very general and geometrically significant method to find normal ideals is based on the following observations.

We first introduce the key class of rings ([Ra78]) in which to study numerical properties of reductions of ideals.

Definition 1.72. A Noetherian local ring $(R, \mathfrak{m})$ is *quasi-unmixed* if the $\mathfrak{m}$-adic completion $\widehat{R}$ is equidimensional, that is all the minimal primes of $\widehat{R}$ have the same dimension

$$\mathfrak{p} \in \operatorname{Ass} \widehat{R} \Leftrightarrow \dim \widehat{R}/\mathfrak{p} = \dim R.$$

A global Noetherian ring R will be called quasi-unmixed when its localizations have this property.

Cohen-Macaulay rings provide major examples. For our purposes, the relevant properties of these rings are given by ([Ma86], [Ra78]):

Theorem 1.73. *Let R be a Noetherian quasi-unmixed ring. Then*

(a) *For every prime ideal $\mathfrak{p}$ of R, $R/\mathfrak{p}$ is quasi-unmixed.*
(b) *For every prime ideal $\mathfrak{p}$ of R, the integral closure of $R/\mathfrak{p}$ is finite.*
(c) *The polynomial ring $R[X]$ is quasi-unmixed.*

Let R be a quasi-unmixed normal domain and I an ideal. The integral closure A of $R[It]$ is the ring

$$\sum_{n\geq 0} \overline{I_n} t^n,$$

where $\overline{I^n}$ is the integral closure of I^n. Since A is a finitely generated graded R-algebra, $A = R[I_1 t, \ldots, I_r t^r]$, we have that for each $I_n = \overline{I_n}$,

$$I_n = \sum_{i=1}^{r} I_{n-i} I_i.$$

Proposition 1.74. *Let s be the least common multiple of $\{1, 2, \ldots, r\}$. Then I_s is a normal ideal.*

The verification is left for the reader. We note that there is no claim that I_n is normal for $n \gg 0$. The algebra $R[I_s t^s]$ is a Veronese subring of A and Proj $(R[I_s t^s])$ is the integral closure of Proj $(R[It])$.

1.3 Special Fiber and Analytic Spread

The ideal I and any of its reductions J share several properties, among which is having the same radical. One of the advantages of J is that it may have a great deal fewer generators. We indicate how this may come about, with the notion of *minimal reduction.*

Definition 1.75. Let $(R, \mathfrak{m})$ be a Noetherian local ring and I an ideal. The *special fiber* of the Rees algebra $R[It]$ is the ring

$$\mathcal{F}(I) = R[It] \otimes_R R/\mathfrak{m} = \bigoplus_{s\geq 0} I^s/\mathfrak{m} I^s.$$

Its Krull dimension is called the *analytic spread* of I, and is denoted by $\ell(I)$. If $I = \mathfrak{m}$, then $\mathcal{F}(\mathfrak{m})$ is the Zariski's *tangent cone* of R. $\mathcal{F}(I)$ is also called the *fiber cone* of I.

Definition 1.76. Let $(R, \mathfrak{m})$ be a Noetherian ring and I an ideal. A reduction J is *minimal* if it does contain properly another reduction of I.

1.3.1 Special Fiber and Noether Normalization

Theorem 1.77 (Northcott-Rees). *Let $(R,\mathfrak{m})$ be a Noetherian local ring of infinite residue field. Then every ideal I admits a reduction generated by $\ell(I)$ elements. Such reductions are minimal.*

Proof. Set $k = R/\mathfrak{m}$, and let $\mathcal{F}(I)$ be the special fiber of I. Since k is infinite, we may then choose a Noether normalization of $\mathcal{F}(I)$,

$$A = k[z_1,\ldots,z_\ell] \hookrightarrow \mathcal{F}(I),$$

where $\ell = \ell(I)$, and the z_j can be chosen in degree 1. There exists then an integer n such that

$$\mathcal{F}(I)_{n+1} = (z_1,\ldots,z_\ell)\,\mathcal{F}(I)_n.$$

Let $a_1,\ldots,a_\ell \in I$ map onto $z_1,\ldots,z_\ell$ and define $J = (a_1,\ldots,a_\ell)$. The equality above can be recast as

$$I^{n+1} = JI^n + \mathfrak{m}I^{n+1},$$

which by Nakayama's Lemma gives $I^{n+1} = JI^n$, as desired.

We leave the second assertion as an exercise for the reader. □

The reduction number $\mathrm{r}_J(I)$ is expressed more precisely as follows. Let $b_1,\ldots,b_s$ be a minimal set of homogeneous module generators of $\mathcal{F}(I)$ over the algebra A:

$$\mathcal{F}(I) = \sum_{1\le i\le s} Ab_i, \ \deg(b_i) = r_i.$$

Proposition 1.78. *Let $a_1,\ldots,a_\ell$ be elements of I that are lifts of $z_1,\ldots,z_\ell$. Then $J = (a_1,\ldots,a_\ell)$ is a reduction of I and*

$$\mathrm{r}_J(I) = \sup\{\deg(b_i),\ 1\le i\le s\} \le \dim_k \mathcal{F}(I)/(z_1,\ldots,z_\ell) - \nu(I) + \ell.$$

Proof. The assertion about the generators of $\mathcal{F}(I)$ follow easily by lifting the equality $\mathcal{F}(I) = \sum_i Ab_i$ to $R[It]$ and using Nakayama's Lemma. As for the inequality for $\mathrm{r}_J(I)$, denote the Artin graded algebra $\mathcal{F}(I)/(z_1,\ldots,z_\ell)$ by F, which we express as a vector space

$$F = F_0 + F_1 + F_2 + \cdots + F_r, \quad F_r \neq 0.$$

Note that $r = \mathrm{r}_J(I)$, $\dim F_0 = 1$ and $\dim F_1 = \nu(I) - \ell$. Since F is a standard graded algebra, we have $F_n \neq 0$ for $n \le r$, which implies that

$$1 + (\nu(I) - \ell) + (r-1) \le \dim F,$$

as desired. □

Remark 1.79. If R is a local ring, the reductions of I generated by $\ell(I)$ elements are minimal and the converse holds if the residue field is infinite. The need to have infinite residue fields arise for very simple reasons. Suppose that k is a finite field, $R = k[x,y]$ and that f is the product of all linear forms in R. The ring $A = R/(f)$ admits no Noether normalization of the form $k[z] \subset A$, $\deg z = 1$. This implies that the maximal ideal $\mathfrak{m} = (x,y)A_{(x,y)}$ admits no reduction generated by one parameter.

In a few cases the integral closure of the ideal I can be obtained through the direct intervention of the integral closure of the ring:

Proposition 1.80. *Let R be a Noetherian integral domain of integral closure $\overline{R}$. Then $\overline{I} = R \cap I\overline{R}$ in the following cases:*

(a) *I is a principal ideal.*
(b) $\dim R = 1$.

Proof. (a) Suppose that $I = Rx$. If $s \in \overline{R}$ and $sx \in R$, multiplying an equation of integral dependence of s over R gives that $sx \in \overline{Rx}$. The converse uses a similar argument.

(b) We may assume that R is a local domain of infinite residue field, so that we may replace I by a principal reduction (x), and use the previous case. □

1.3.2 Explicit Reduction Numbers

The bounds obtained above for the reduction number of an ideal I via the examination of $\mathcal{F}(I)$ are rather crude. Here is another elementary approach to $\mathrm{r}(I)$.

Suppose that we have a presentation $\mathcal{F}(I) = k[x_1,\ldots,x_n]/I$, and set $\ell(I) = \ell$. Assuming that k is an infinite field, we can find linear forms $f_1,\ldots,f_\ell$ of $k[x_1,\ldots,x_n]$ such that $(I, f_1,\ldots,f_\ell)$ is $(x_1,\ldots,x_n)$-primary.

We recall that given an ideal I of the Noetherian ring R, its *index of nilpotency* is the smallest integer $\mathrm{nil}(I)$ such that

$$(\sqrt{I})^{\mathrm{nil}(I)} \subset I.$$

Proposition 1.81. *Let J be the ideal generated by the lifts to R of the forms $f_1,\ldots,f_\ell$. Then*

$$\mathrm{r}_J(I) = \mathrm{nil}(0) - 1,$$

the null ideal of $k[x_1,\ldots,x_n]/(I, f_1,\ldots,f_\ell)$.

This representation can also be used to decide whether $J = (a_1,\ldots,a_s) \subset I$ is a reduction of I. Suppose that $(R,\mathfrak{m})$ is a local ring and $\mathcal{F}(I) = k[x_1,\ldots,x_n]/I$. Let $g_1,\ldots,g_s$ be a set of 1-forms of $k[x_1,\ldots,x_n]$ that are lifts of the images of the a_i in $I/\mathfrak{m}I$.

Proposition 1.82. *J is a reduction of I if and only if the ideal $(I, g_1,\ldots,g_s)$ is $(x_1,\ldots,x_n)$-primary, or equivalently if for any term order $>$, $\mathrm{in}_>(I, g_1,\ldots,g_s)$ has finite co-length.*

Reduction Numbers and Initial Ideals

The following result of A. Conca ([Co3]) and N. V. Trung ([Tr2]) has settled the question raised in [Va98b, Conjecture 9.4.3] that sought to give another path to the estimation of reduction numbers.

Theorem 1.83. *Let I be a homogeneous ideal of $R = k[x_1, \ldots, x_n]$ and $in_>(I)$ its initial ideal with respect to some term order. Then*

$$\mathrm{r}(R/I) \leq \mathrm{r}(R/in_>(I)). \tag{1.2}$$

Remark 1.84. Examples in [Co3] and [Tr2] show that the reduction numbers of the algebras R/I and $R/in_>(I)$ can differ. In particular this shows that the reduction number of a standard graded algebra is not determined by its Hilbert function.

Cohen-Macaulay Fiber Cones

The estimates for reduction numbers start to become more significant once depth information about the fiber cone is available.

If $\mathcal{F}(I)$ is Cohen-Macaulay the reduction number $\mathrm{r}(I)$ can be read off directly from the Hilbert-Poincaré series of $\mathcal{F}(I)$:

Proposition 1.85. *If*

$$\frac{1 + h_1 t + \cdots + h_r t^r}{(1-t)^d}, \quad h_r \neq 0, \quad d = \ell(I), \quad h_1 = \nu(I) - \ell,$$

then

$$\mathrm{r}(I) = r = \mathrm{r}(\mathcal{F}(I)) = a(\mathcal{F}(I)) + \dim \mathcal{F}(I) = a(\mathcal{F}(I)) + \ell(I).$$

Here $a(\mathcal{F}(I))$ denotes the a-invariant of $\mathcal{F}(I)$ (see [BH93]). This is useful because in many cases of interest the canonical module ω of the algebra $\mathcal{F}(I)$ is known and

$$a(\mathcal{F}(I)) = -\inf\{i \mid \omega_i \neq 0\}.$$

Example 1.86. Let Δ be a simplicial complex on the set $\{x_1, \ldots, x_n\}$ and $k[\Delta]$ the corresponding face ring (see [BH93, Section 5.1]). If $k[\Delta]$ is Cohen-Macaulay of dimension d, then $\mathrm{r}(k[\Delta]) \leq d$, according to [BH93, Lemma 5.1.8].

If $\mathcal{F}(I)$ is not Cohen-Macaulay, it becomes harder to predict $\mathrm{r}(I)$, but estimating it becomes an open game. For example, if $\mathcal{F}(I)$ is a torsionfree module over one of its Noether normalizations, a condition equivalent to saying that $\mathcal{F}(I)$ satisfies the condition S_1 of Serre, then in characteristic zero (see Proposition 1.142),

$$1 + (\nu(I) - \ell) + \mathrm{r}(I) + 1 \leq \deg \mathcal{F}(I) = \sum_{j=0}^{r} h_j,$$

the multiplicity of $\mathcal{F}(I)$.

Question 1.87. There is one case in which it is straightforward to determine the reduction number of an ideal, namely when $\mathcal{F}(I)$ is Cohen-Macaulay. The number $r(I)$ can then be read off the Hilbert function of $\mathcal{F}(I)$, which obviates the need to use Noether normalization. What can be done, if we only have depth $\mathcal{F}(I) \geq \dim \mathcal{F}(I) - 1$?

Another kind of estimate for the reduction number of an ideal I arises from a presentation of $\mathcal{F}(I)$:

$$\mathcal{F}(I) \cong k[x_1,\ldots,x_n]/(f_1,\ldots,f_s),$$

where the f_i are homogeneous polynomials. We arrange things so that the f_i give a minimal set of generators of the presentation and have decreasing degree, $\deg f_i \geq \deg f_{i+1}$. If $\ell(I) = \ell$ then height $(f_1,\ldots,f_s) = r = n - \ell$. By considering elements of the form

$$g_j = f_j + \sum_{j>i} h_{ij} f_j,$$

we may assume that the first r elements in the presentation of $\mathcal{F}(I)$ form a regular sequence; changing notation, we may assume that $f_1,\ldots,f_r$ is a regular sequence. From the surjection

$$k[x_1,\ldots,x_n]/(f_1,\ldots,f_r) \longrightarrow \mathcal{F}(I)$$

of algebras of dimension $n - r = \ell$, we obtain that any Noether normalization of the first algebra maps onto a Noether normalization of $\mathcal{F}(I)$. In particular, we have

$$\mathrm{r}(k[x_1,\ldots,x_n]/(f_1,\ldots,f_r)) \geq \mathrm{r}(I).$$

Since the Hilbert series of the complete intersection is simply

$$\frac{\prod_{j=1}^{r}(1-t^{d_j})}{(1-t)^n}, \quad d_j = \deg f_j,$$

we obtain

$$\mathrm{r}(I) \leq d_1 + \cdots + d_r - r.$$

We want to argue, using very general considerations and a theorem of Mumford ([Mu66, p. 101]), for the existence of an universal function on the h-vectors that bounds the corresponding reduction. We need the notion of a *pro-polynomial function* over the ring S. This is simply a function

$$f : S^{\mathbb{N}} \longrightarrow S,$$

which is polynomial in any finite subset of arguments.

Proposition 1.88. *Given a positive integer d, there exists a pro-polynomial function f over $\mathbb{N}$ such that*

$$\mathrm{r}(A) \leq f(\mathbf{h}(A))$$

for any standard graded algebra A of embedding dimension at most d, where $\mathbf{h}(A)$ is the h-vector of A.

Proof. From a Noether normalization of A, $k[\mathbf{z}] \hookrightarrow A$, it follows easily that the Castelnuovo-Mumford regularity $\mathrm{reg}(A)$ of A (see Chapter 3) bounds the degrees of a minimum set of homogeneous module generators of A over $k[\mathbf{z}]$, and consequently it bounds $\mathrm{r}(A)$, the reduction number of the algebra.

On the other hand, according to [Mu66, p. 101], there exists a polynomial $g_d(x_1,\ldots,x_d)$ such that for any standard algebra of embedding dimension $\leq d$, we have $\mathrm{reg}(A) \leq g_d(e_0,\ldots,e_{d-1})$, where $e_0,\ldots,e_{d-1}$ are the Hilbert coefficients of A. By standard facts, the vector $\mathbf{h}(A)$ determines the Hilbert polynomial of A. □

1.3.3 Analytic Spread and Codimension

Proposition 1.89. *Let R be a local ring and let I be an ideal. The following inequalities hold*

$$\text{height}\, I \leq \ell(I) \leq \begin{cases} \nu(I) \\ \dim R \end{cases}.$$

Proof. The second inequality arises from the formula for the Krull dimension of $R[It]$ ([Val76]). The other assertion follows since every minimal prime of J is also minimal over I, which have codimension at most $\ell(I)$ by Krull principal ideal theorem. □

Example 1.90. Suppose that I is an ideal of $k[x_1,\ldots,x_n]$ generated by forms $f_1,\ldots,f_m$ of the same degree. With $\mathfrak{m} = (x_1,\ldots,x_n)$, we have

$$R[It] = k[f_1t,\ldots,f_mt] \bigoplus \mathfrak{m}R[It],$$

so that the fiber cone $\mathcal{F}(I) \cong k[f_1,\ldots,f_m]$. Suppose that each f_i is a monomial $\mathbf{x}^{v_i}$ (where $v_i = (a_{i1},\ldots,a_{in})$ is a vector of exponents). In this case, it is easy to see that

$$\ell(I) = \text{rank}\ \varphi,$$

where φ is the numerical matrix $[v_1,\ldots,v_m]$ defined by the exponent vectors. If the monomials are not of the same degree, one stills gets a surjection

$$k[f_1t,\ldots,f_mt] \longrightarrow \mathcal{F}(I) \to 0,$$

and therefore rank $\varphi \geq \ell(I)$. What is still missing are methods of finding the reduction numbers of these ideals.

When the monomials f_i are quadratic and square-free, Villarreal ([Vi90]) attaches to this set a graph (and conversely). Their interplay (see [SVV94]) is useful in determining reductions and reduction numbers. More precisely, let G be a graph whose vertices we label by the set $\{x_1,\ldots,x_n\}$ of indeterminates over a field k. The edge ideal of G is the ideal $I = I(G)$ generated by the binomials x_ix_j corresponding to the edges of G. If G is a connected graph, it is easy to show that

$$\ell(I) = \begin{cases} n-1 & \text{if } G \text{ is bipartite} \\ n & \text{otherwise.} \end{cases}$$

It is a lot more challenging to find the analytic spread of ideals generated by binomials. A tantalizing question is how the ideal of relations of semigroup algebras (e.g. toric ideals) should be dealt with in these matters. Progress on this question has already been made in [GMS99].

The notion of special fiber can also be useful in treating *normalization indices.* Let $(R,\mathfrak{m})$ be a local, normal domain and I an ideal such that its Rees algebra $A = R[It]$ has a finite integral closure $B = \sum_{n\geq 0} \overline{I^n}t^n$. Set

$$\begin{aligned} s(I) &= \inf\{n \mid B = \sum_{k\leq n} A\overline{I^k}t^k\} \\ s_0(I) &= \inf\{n \mid B = A[\overline{I}t,\ldots,\overline{I^n}t^n].\} \end{aligned}$$

This means that $s(I)$ measures the 'degree' of B as an A-module, while $s_0(I)$ measures the 'degree' of B as an A-algebra. If we set

$$G = B/(\mathfrak{m},It)B = \sum_{n\geq 0} G_n,$$

we get an Artinian local ring, and can use Nakayama's Lemma to derive simple relations between $s(I)$ and $s_0(I)$.

Proposition 1.91. *For an ideal I as above,*

$$\begin{aligned} s(I) &= \sup\{n \mid G_n \neq 0\}. \\ s_0(I) &= \inf\{n \mid G = G_0[G_1,\ldots,G_n]\}. \end{aligned}$$

Furthermore, if the index of nilpotency of G_i is r_i, then

$$s(I) \leq \sum_{i\leq s_0(I)} (r_i - 1).$$

Equimultiple Ideals

Definition 1.92. Let R be a Noetherian local ring and let I be an ideal of codimension g. If $\ell(I) = g$, the ideal I is called *equimultiple.*

There are large classes of such ideals. If $(R,\mathfrak{m})$ is a Noetherian local ring with an infinite residue field, then every $\mathfrak{m}$-primary ideal I is equimultiple.

Proposition 1.93. *Let R be a Cohen-Macaulay local ring and I an equimultiple ideal. If I is generically a complete intersection, then I is a complete intersection ideal.*

Proof. Let J be a minimal reduction of I, $J = (a_1,\ldots,a_g)$, $g = \text{height } I$. We claim that $I = J$. Since J and I have the same minimal primes, it follows that for any such prime, $\mathfrak{p}$, $J_\mathfrak{p}$ is a reduction of the complete intersection ideal $I_\mathfrak{p}$, and therefore $J_\mathfrak{p} = I_\mathfrak{p}$. As a consequence $I \subset J$, since the associated primes of J are just those primes. □

Burch's Formula

The calculation of the analytic spread $\ell(I)$ of an ideal I often results from an examination of the codimension of the ideal $\mathfrak{m}\mathcal{R}(I)$,

$$\ell(I) = \dim \mathcal{F}(I) = \dim \mathcal{R}(I) - \text{height } \mathfrak{m}\mathcal{R}(I),$$

if I has positive codimension and $\mathcal{R}(I)$ is equidimensional. Sometimes the inequality

$$\ell(I) = \mathcal{F}(I) \leq \dim \mathcal{R}(I) - \text{grade } \mathfrak{m}\mathcal{R}(I)$$

(first examined by Burch) is considered.

Theorem 1.94. *Let $(R, \mathfrak{m})$ be a Noetherian local ring and let I be an ideal of codimension at least one. Then*

$$\ell(I) \leq \dim R - \inf_n \{\text{depth } R/I^n\}.$$

Moreover, if $\text{gr}_I(R)$ is Cohen-Macaulay then equality holds.

The inequality is due to Burch ([Bu72]), while the equality comes from [EH83, Proposition 3.3]. These formulas can be slightly refined when basic facts of the so-called *asymptotic depth* are taken into account (see [Br74]).

Theorem 1.95. *Let $(R, \mathfrak{m})$ be a Noetherian local ring and I an ideal. For $n \gg 0$,* depth R/I^n *and* depth I^n/I^{n+1} *are independent of n and*

$$\text{depth } R/I^n = \text{depth } I^n/I^{n+1}.$$

The extreme values in this formula are obtained for the following class of ideals. An ideal I is *normally Cohen-Macaulay* if all powers I^n are Cohen-Macaulay. With an additional hypothesis, such ideals are very well-behaved ([CN76]):

Theorem 1.96. *Let $(R, \mathfrak{m})$ be a Cohen-Macaulay local ring with infinite residue field.*

(i) *If I is a normally Cohen-Macaulay ideal then $\ell(I) =$ height I.*
(ii) *Moreover, if I is generically a complete intersection then it is a complete intersection.*

Proof. Part (ii) is Proposition 1.93. □

Deviations of an Ideal

There are several notions of *deviation of an ideal* that essentially take from baseline complete intersections. Among those we will encounter are as in the following definition.

Definition 1.97. Let $(R, \mathfrak{m})$ be a Noetherian local ring and I an ideal.

(a) $\ell(I) -$ height I is the *analytic deviation* of I.
(b) $\nu(I) - \ell(I)$ is the *second analytic deviation* of I.

Observe that $\nu(I) -$ height I, the sum of the two deviations, is one measure of the deviation of I from being a complete intersection.

Inertial Analytic Spread of an Ideal

Let $(R,\mathfrak{m})$ be a local ring of dimension d and I an ideal. Besides the special fiber $\mathcal{F}(I)$ of $\mathcal{G} = \mathrm{gr}_I(R)$, there is another object that plays an occasional role in the study of the reductions of I: $H(I) = H^0_{\mathfrak{m}}(\mathcal{G})$, the subideal of elements of finite support of $\mathcal{G}$.

Definition 1.98. The *inertial analytic spread* of I is the dimension of the module $H(I)$. We shall denote it by $h(I) = \dim H(I)$.

The Hilbert function of the module $H(I)$—particularly through its dimension and multiplicity—plays a role in extensions of multiplicity based criteria of the integrality for ideals; see [AMa93], [FOV99], [FM01].

A first observation is that since $H(I)$ is annihilated by a power of $\mathfrak{m}$, it admits a filtration whose factors are modules over $\mathcal{G}\otimes R/\mathfrak{m} = \mathcal{F}(I)$. It follows that

$$h(I) = \dim H(I) \leq \dim \mathcal{F}(I) = \ell(I).$$

A more concrete elementary observation is the following:

Proposition 1.99. *Let $(R,\mathfrak{m})$ be a Noetherian local ring of dimension d and I an ideal. If $\ell(I) = d$, then $h(I) = d$ or $H(I) = 0$.*

Proof. Suppose that $H(I) \neq 0$. Using the Artin-Rees Lemma, we choose s to be large enough so that $H \cap \mathfrak{m}^s\mathcal{G} = 0$. This leads to the exact sequence

$$0 \to H \longrightarrow \mathcal{G}/\mathfrak{m}^s\mathcal{G} \longrightarrow \mathcal{G}'/\mathfrak{m}^s\mathcal{G}' \to 0,$$

which shows that

$$\dim H \leq \dim \mathcal{G}/\mathfrak{m}^s\mathcal{G} = \max\{\dim H, \dim \mathcal{G}'/\mathfrak{m}^s\mathcal{G}'\}.$$

But $\dim \mathcal{G}/\mathfrak{m}^s\mathcal{G} = \dim \mathcal{G}/\mathfrak{m}\mathcal{G}$, and $\dim \mathcal{G}'/\mathfrak{m}^s\mathcal{G}' < d$ since $\mathfrak{m}$ does not consist of zero divisors of $\mathcal{G}'$. This shows that we must have $h(I) = \ell(I)$, as desired. □

Let $(R,\mathfrak{m})$ be a quasi-unmixed local ring of dimension $d > 0$ and I an ideal of positive height and such that $\dim R/I > 0$. These conditions imply that the minimal primes of $\mathcal{G} = \mathrm{gr}_I(R)$ have dimension d since $\mathcal{G} = R[It, t^{-1}]/(t^{-1})$. Suppose that $\mathcal{G}$ is unmixed (for instance, suppose that I is a normal ideal). If $H(I) \neq 0$ then $\dim H(I) = \ell(I) = d$.

If $H(I) \neq 0$, it is not always the case that $\dim H(I) = \ell(I)$. Note that because of the exact sequence

$$0 \to H(I) \longrightarrow \mathcal{G} \longrightarrow \mathcal{G}' \to 0,$$

we can choose $z \in \mathfrak{m}$ which is regular on $\mathcal{G}'$. Tensoring with $R/(z)$ we get the exact sequence

$$0 \to H(I)/zH(I) \longrightarrow \mathcal{G}/z\mathcal{G} \longrightarrow \mathcal{G}'/z\mathcal{G}' \to 0,$$

where $\dim \mathcal{G}'/z\mathcal{G}' = \dim \mathcal{G}' - 1$, since z is regular on $\mathcal{G}'$, and $\dim H(I)/zH(I) = \dim H(I)$, since z is nilpotent on $H(I)$; therefore

$$\ell(I) \leq \dim \mathcal{G}/z\mathcal{G} = \sup\{\dim H(I), \dim \mathcal{G}' - 1\}.$$

In the special case that I is a non-complete intersection that is a generic complete intersection of dimension 1 (and R is Cohen-Macaulay) we have $\ell(I) = d$, so that also in this case $h(I) = d$.

Example 1.100 (Bernd Ulrich). Let $R = k[x,y,z]$ be a ring of polynomials over a field and set $I = (x^4, x^3y, x^2y^2z, xy^3, y^4)$; then I has (x,y,z) as an embedded prime. It is easy to verify that $I^n = (x,y)^{4n}$ for $n \geq 2$. This means that $H(I)$ is nonzero and concentrated in degrees 0 and 1 in $\mathrm{gr}_I(R)$, which implies that $\dim H(I) = 0$.

Let us examine now the case of a radical ideal I with $\ell(I) < d$ that is generically a complete intersection. If $H(I) \neq 0$, it follows that $H(I)$ is nilpotent since $\mathfrak{m}^s H(I) = 0$ and $\mathfrak{m}G$ has positive codimension.

Proposition 1.101. *Let $(R, \mathfrak{m})$ be a Cohen-Macaulay normal domain and I an equidimensional radical ideal that is of linear type on the punctured spectrum. If $\ell(I) < d$, then $H(I)$ is a radical ideal.*

Proof. As observed above, $H(I)$ is nilpotent. On the other hand, the quotient $G' = G/H(I)$ is the Rees algebra of the conormal module I/I^2 over the reduced ring R/I; it follows that G' is reduced as well. □

Remark 1.102. It is likely, in this case, that $\dim H(I) = \ell(I)$.

1.4 Reduction Numbers of Ideals

We introduce now another of our basic notions. We shall view it as a numerical invariant of the ideal. From the ways it is dealt with, it has a clear "nonlinear" character.

Definition 1.103. Let R be a local ring and I an ideal. The *reduction number* of I is the infimum of all $r_J(I)$ for all minimal reductions J of I. If there is no confusion, it is denoted by $\mathrm{r}(I)$.

To get minimal reductions it may be required that the residue field of R be infinite. In the next section, we show that $\sup_J \mathrm{r}_J(I)$ exists for any ideal I.

Let $(R, \mathfrak{m})$ be a local ring and I an ideal. There are several approaches to the reduction number of I. At one time or another we shall mix elements of the following ingredients:

- Multiplicities of the algebras associated to I and characteristic polynomials
- The growth of $\nu(I^n)$
- The depth of $\mathrm{gr}_I(R)$

In this section we examine relationships between an ideal and some of its reductions, particularly between their reduction numbers. To be truly useful the general methods must be grounded on a rich catalog of examples.

Determinantal Ideals

Let φ be an $m \times n$ generic matrix. Its determinantal ideals are well studied and much is known about their reduction numbers. Here is a short catalog (k is a field and $k[X]$ is the ring of polynomials in all the indeterminates that occur in X).

(a) If $m = n+1$, then $I = I_n(\varphi)$ is generated by a d-sequence ([Hu86a]).
(b) If $m = n$, then $I = I_{n-1}(\varphi)$ is of linear type but not generated by a d-sequence ([Hu86a]).
(c) If φ is the generic symmetric $n \times n$ matrix, then $I = I_{n-1}(\varphi)$ is of linear type ([Kot91]) but it is not generated by a d-sequence.
(d) If $m < n$, then $I = I_m(\varphi)$ has reduction number

$$\mathrm{r}(I) = (m-1)(n-m-1).$$

This follows from [BV88] and [BH92]: Since I is generated by forms of degree m, the special fiber $\mathcal{F}$ is the subring of $k[X]$ generated by the maximal minors of X. Thus $\mathcal{F}(I)$ is the affine cone of the variety of subspaces of dimension m of k^n, and is a Cohen-Macaulay ring of dimension $m(n-m)+1$ whose a-invariant is $-n$. It follows by Remark 1.85 that

$$\mathrm{r}(I) = \mathrm{r}(\mathcal{F}(I)) = a(\mathcal{F}(I)) + \dim \mathcal{F}(I) = -n + m(n-m) + 1 = (m-1)(n-m-1).$$

(e) If $t < m \leq n$, then $I = I_t(\varphi)$ has analytic spread $\ell(I) = mn$. However, its reduction number may depend on the characteristic of k.

Ideals of Reduction Number One

It would be useful to say that we have many ways to build ideals of low reduction numbers–say, reduction numbers one or two. Unfortunately this is not yet the case.

Links of Prime Ideals

Let us indicate a method that produces plenty of ideals of reduction number 1. Let R be a Cohen-Macaulay ring and $\mathfrak{p}$ a prime ideal of codimension g. Let J be generated by a regular sequence $\mathbf{z} = \{z_1, \ldots, z_g\}$ of g elements contained in $\mathfrak{p}$, and set $I = J : \mathfrak{p}$. Consider the following very general settings:

(L_1) $R_\mathfrak{p}$ is not a regular local ring;
(L_2) $R_\mathfrak{p}$ is a regular local ring of dimension at least 2, and two elements in the sequence $\mathbf{z}$ lie in the symbolic square $\mathfrak{p}^{(2)}$.

We then have (see [CP95], [CPV94]):

Theorem 1.104. *Let R be a Cohen-Macaulay ring, $\mathfrak{p}$ a prime ideal of codimension g, and let $\mathbf{z} = (z_1, \ldots, z_g) \subset \mathfrak{p}$ be a regular sequence. Set $J = (\mathbf{z})$ and $I = J : \mathfrak{p}$. Then I is an equimultiple ideal with reduction number one, more precisely,*

$$I^2 = JI,$$

if either condition L_1 or L_2 holds.

The proof requires the assemblage of 3 elementary techniques. We deal with them separately first.

Maximal Ideals

To establish the equality $I^2 = JI$, we examine the associated prime ideals of JI. To that end, looking at the exact sequence

$$0 \longrightarrow J/JI \longrightarrow R/JI \longrightarrow R/J \longrightarrow 0,$$

and since

$$J/JI = J/J^2 \otimes R/I = (R/I)^g,$$

we get that the associated primes of JI satisfy the condition

$$\mathrm{Ass}(R/JI) \subset \mathrm{Ass}(R/I) \cup \mathrm{Ass}(R/J).$$

Therefore all the associated primes of JI have codimension g, and since I and J are equal at any localization $R_{\mathfrak{q}}$, where $\mathfrak{q}$ is a prime of codimension g, distinct from $\mathfrak{p}$, we may finally assume that R is a local ring and $\mathfrak{p}$ its maximal ideal.

Singular Case

Following [CP95], we first deal with the case when R is not a regular local ring. We begin with the following criterion for complete intersections.

Proposition 1.105. *Let A and B be two ideals of a local ring $(R,\mathfrak{m})$ and let $\mathbf{z} = z_1,\ldots,z_n$ be a regular sequence contained both in A and in B. If $AB \subset (\mathbf{z})$ but $AB \not\subset \mathfrak{m}(\mathbf{z})$ then A and B are both generated by regular sequences of length n.*

Proof. Consider first the case $n = 1$ and write $z = z_1$, for sake of simplicity. By assumption, there exist $a \in A$ and $b \in B$ such that $ab = \alpha z$ with $\alpha \notin \mathfrak{m}$. We may even assume that $z = ab$. Choose now any x in A; since $xb \in AB \subset (z)$, we have$xb = cz$ for some $c \in R$ and so

$$xz = x(ab) = a(xb) = a(cz) = (ac)z.$$

But z is a regular element, thus $x = ac$, i.e. $x \in (a)$. Since x is an arbitrary element of A, we then conclude that $A = (a)$; similarly, $B = (b)$. Note that if the product of two elements is regular then they are both regular, and this establishes the first case.

Consider now the case $n > 1$. Since $AB \not\subset \mathfrak{m}(\mathbf{z})$, there exist $a \in A$ and $b \in B$ such that $ab = \alpha_1 z_1 + \cdots + \alpha_n z_n$ and at least one of the α_i does not belong to $\mathfrak{m}$; by reordering the z_i's we may assume $\alpha_n \notin \mathfrak{m}$. Let "$^-$" denote the homomorphic image modulo $(z_1,\ldots,z_{n-1})$: we are then in the case of two ideals $\overline{A}$ and $\overline{B}$ of a local ring $(\overline{R},\overline{\mathfrak{m}})$ both containing the regular element $\overline{z}_n$. In addition, $\overline{AB} \subset (\overline{z}_n)$ but $\overline{AB} \not\subset \mathfrak{m}(\overline{z}_n)$. Using the previous argument, we then conclude that $\overline{A} = (\overline{a})$ and $\overline{B} = (\overline{b})$, which implies that

$$A = (z_1,\ldots,z_{n-1},a) \quad \text{and} \quad B = (z_1,\ldots,z_{n-1},b),$$

as claimed. ☐

Theorem 1.106. *Let $(R,\mathfrak{m})$ be a Cohen-Macaulay local ring of dimension d, that is not a regular local ring. Set $I = J\colon \mathfrak{m}$ where $J = (z_1,\ldots,z_d)$ is a system of parameters. Then $I^2 = JI$.*

Proof. We divide the proof into two parts, first showing that $\mathfrak{m}I = \mathfrak{m}J$, and then that $I^2 = JI$.

Clearly, $\mathfrak{m}I \supseteq \mathfrak{m}J$; since $I = J\colon \mathfrak{m}$ we have that $\mathfrak{m}I \subset J$, so if $\mathfrak{m}I \not\subseteq \mathfrak{m}J$, Proposition 1.105 with $A = I$, $B = \mathfrak{m}$ and $(\mathbf{z}) = J$ would imply that $\mathfrak{m}$ is generated by a regular sequence, contradicting the assumption that R is not a regular local ring.

It will suffice to show that the product of any two elements of I is contained in JI. Suppose that $a,b \in I$; since $ab \in I^2 \subset \mathfrak{m}I \subset J$, ab can be written in the following way:

$$ab = \sum_{i=1}^{d} \alpha_i z_i. \tag{1.3}$$

The proof will be complete once it is shown that $\alpha_i \in I$ for $i = 1,\ldots,d$; to this end it is enough to prove that $\alpha_i x \in J$ for every $x \in \mathfrak{m}$.

Since R is a local ring, regular sequences permute, thus we can key on any of the coefficients, say α_d. Note that $xb \in \mathfrak{m}I = \mathfrak{m}J$, hence xb satisfies an equation of the form

$$xb = \sum_{i=1}^{d} \beta_i z_i, \qquad \beta_i \in \mathfrak{m}. \tag{1.4}$$

Multiply equation (1.3) by x and equation (1.4) by a; a quick comparison between these new equations yields the equality

$$\sum_{i=1}^{d} \alpha_i x z_i = \sum_{i=1}^{d} a\beta_i z_i,$$

which can also be written as

$$\sum_{i=1}^{d} (\alpha_i x - a\beta_i) z_i = 0. \tag{1.5}$$

Equation (1.5) modulo the ideal $J_1 = (z_1,\ldots,z_{d-1})$ gives that $(\alpha_d x - a\beta_d)z_d$ is zero modulo J_1; but z_d is regular in R/J_1, hence

$$\alpha_d x - a\beta_d \in J_1 = (z_1,\ldots,z_{d-1}) \subset J.$$

Since $a\beta_d \in I\mathfrak{m} \subset J$ we then have that $\alpha_d x \in J$ also, or equivalently $\alpha_d \in I$. □

A related class of ideals of reduction number one is taken from [CHV98]; it will not require that R be Cohen-Macaulay.

We start with the following lemma about the existence of a *dual basis*.

Lemma 1.107. *Let $(R,\mathfrak{m})$ be a Noetherian local ring with embedding dimension n at least two. Let I be an $\mathfrak{m}$-primary irreducible ideal contained in $\mathfrak{m}^2$. If $\mathfrak{m}=(x_1,\ldots,x_n)$ then there exist $y_1,\ldots,y_n$ such that*

$$x_iy_j \equiv \delta_{ij}\Delta \qquad \text{mod } I \tag{1.6}$$

for all $1 \leq i,j \leq n$, where Δ is the lift in R of the socle generator of R/I and δ_{ij} denotes the Kronecker delta.

Proof. It will be enough to find $y_j \in I\colon(x_1,\ldots,x_j^2,\ldots,x_n)$ for any j between 1 and n such that $y_j \notin I\colon(x_1,\ldots,x_n)$. If this cannot be done one must necessarily have $I\colon(x_1,\ldots,x_j^2,\ldots,x_n) = I\colon(x_1,\ldots,x_n)$ and consequently $(x_1,\ldots,x_j^2,\ldots,x_n) = (x_1,\ldots,x_n)$, as $\operatorname{Hom}(\cdot,R/I)$ is a self-dualizing functor; this is a contradiction.

The element y_j satisfies $x_iy_j \in I$ whenever $i \neq j$ and $x_jy_j \in I\colon(x_1,\ldots,x_n)$. Hence we can write $y_jx_j = a_j + g_j\Delta$ with $a_j \in I$ and $g_j \neq 0$. However, $g_j \notin (x_1,\ldots,x_n)$ as otherwise $y_jx_j \in I$. But then g_j is an invertible element, so that $g_j^{-1}y_j$ will have all the properties required in (1.6). □

Theorem 1.108. *Let $(R,\mathfrak{m})$ be a Noetherian local ring with embedding dimension n at least two. Suppose that I is an $\mathfrak{m}$-primary Gorenstein ideal contained in $\mathfrak{m}^2$. Setting $L = I\colon\mathfrak{m}$, the following conditions then hold:*

(a) *L has reduction number one with respect to I, i.e., $L^2 = IL$;*
(b) $I\mathfrak{m} = L\mathfrak{m}$*, i.e., the coefficient ideal of I and L is* $\mathfrak{m}$.

Proof. (a) As $L = (I,\Delta)$ one only need show that $\Delta^2 \in IL$. If we let $\mathfrak{m} = (x_1,\ldots,x_n)$, by Lemma 1.107 we can find $y_1,\ldots,y_n$ and $a_1,\ldots,a_n \in I$ such that $\Delta = x_iy_i + a_i$ for $1 \leq i \leq n$. As $n \geq 2$, we can write

$$\begin{aligned}\Delta^2 &= (x_1y_1+a_1)(x_2y_2+a_2) = x_1y_1x_2y_2 + x_1y_1a_2 + a_1x_2y_2 + a_1a_2\\ &= (x_1y_2)(x_2y_1) + (x_1y_1)a_2 + (x_2y_2)a_1 + a_1a_2.\end{aligned}$$

Note that each term in the last sum belongs to the ideal $I(I,\Delta) = I^2 + I\Delta = IL$.

(b) We only need to show the inclusion $L\mathfrak{m} \subseteq I\mathfrak{m}$, or better $\Delta \in I\mathfrak{m}\colon\mathfrak{m}$, as $L = (I,\Delta)$. But for any $1 \leq i \leq n$ choose $j \neq i$ and write $\Delta = x_jy_j + a_j$ with $a_j \in I$; hence

$$x_i\Delta = x_i(x_jy_j + a_j) = x_j(x_iy_j) + x_ia_j \in \mathfrak{m}I,$$

as desired. □

Syzygetic Ideals

In making comparisons between the primary ideals I^2 and JI we consider the following diagram:

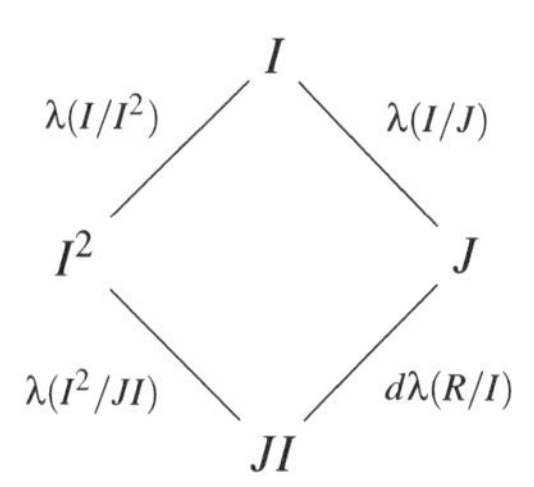

We shall establish some relationships between these lengths. The first of these is very well known ([Ab67]):

Proposition 1.109. *Let $(R,\mathfrak{m})$ be a Cohen-Macaulay local ring of dimension $d \geq 1$, let I be an $\mathfrak{m}$-primary ideal and J one of its reductions generated by a regular sequence. Then*

$$\lambda(I/I^2) \leq \lambda(R/J) + (d-1)\lambda(R/I). \tag{1.7}$$

Note that equality means that $I^2 = JI$, that is, the reduction number of I is at most 1. The term $\lambda(R/J)$ is the multiplicity of I. When $I = \mathfrak{m}$, this gives the following inequality of Abhyankar ([Ab67]):

$$\operatorname{embdim}(R) \leq e(R) + d - 1.$$

This is the initial inequality expressed by a comparison of Hilbert functions that we discuss in Chapter 2. We make a slightly different analysis of it here.

Proposition 1.110. *Let $(R,\mathfrak{m})$ be a Cohen-Macaulay local ring of Krull dimension d, let $J = (a_1,\ldots,a_d)$ be an $\mathfrak{m}$-primary ideal and $J \subset I$. Suppose that I is minimally generated by $d+r$ elements $(J,b_1,\ldots,b_r)$. Denote by H_1 the 1-dimensional Koszul homology module of I and by $\delta(I)$ the kernel of the natural surjection from the symmetric square of I onto I^2. Then*

$$\lambda(I/J) = \lambda(I^2/JI) + r\lambda(R/I) - \lambda(H_1) + \lambda(\delta(I)).$$

Proof. Consider a minimal presentation of the ideal I:

$$0 \to Z \longrightarrow R^{d+r} \longrightarrow I \to 0.$$

Tensoring with R/I we obtain the exact sequence (see [SV81b])

$$0 \longrightarrow \delta(I) \longrightarrow H_1 \longrightarrow (R/I)^{d+r} \longrightarrow I/I^2 \longrightarrow 0, \tag{1.8}$$

where $\delta(I)$ accounts for the cycles in Z whose coefficients lie in I.

This leads to the inequality $\lambda(I/I^2) - (d+r)\lambda(R/I) + \lambda(H_1) - \lambda(\delta(I)) = 0$, or equivalently

$$\lambda(I/I^2) - d\lambda(R/I) = r\lambda(R/I) - \lambda(H_1) + \lambda(\delta(I)). \tag{1.9}$$

To account for the left-hand side of (1.9), consider the following two exact sequences

$$0 \longrightarrow J/JI \longrightarrow I/JI \longrightarrow I/J \longrightarrow 0,$$

$$0 \longrightarrow I^2/JI \longrightarrow I/JI \longrightarrow I/I^2 \longrightarrow 0.$$

From the first one we get that $\lambda(I/JI) = \lambda(J/JI) + \lambda(I/J)$, while from the second one we get that $\lambda(I/JI) = \lambda(I^2/JI) + \lambda(I/I^2)$. By comparing the last two equations, we have that $\lambda(J/JI) + \lambda(I/J) = \lambda(I^2/JI) + \lambda(I/I^2)$, or equivalently

$$\lambda(I/I^2) - \lambda(J/JI) = \lambda(I/J) - \lambda(I^2/JI).$$

As we remarked earlier, J/JI is a free R/I-module of rank d, so since $\lambda(J/JI) = d\lambda(R/I)$, the last equation becomes

$$\lambda(I/I^2) - d\lambda(R/I) = \lambda(I/J) - \lambda(I^2/JI). \tag{1.10}$$

If we compare (1.10) with (1.9), we get $\lambda(I/J) - \lambda(I^2/JI) = r\lambda(R/I) - \lambda(H_1) + \lambda(\delta(I))$, or equivalently

$$\lambda(I/J) = \lambda(I^2/JI) + r\lambda(R/I) - \lambda(H_1) + \lambda(\delta(I)),$$

as desired. □

Corollary 1.111. *Under the same assumption as in* Proposition 1.110, *if R is a Gorenstein local ring and $r = 1$ then*

$$\lambda(I/J) = \lambda(I^2/JI) + \lambda(\delta(I)), \quad \text{with } \delta(I) \neq 0.$$

Proof. Since R is Gorenstein and R/I is Cohen-Macaulay, the canonical module of R/I is the last non-vanishing Koszul homology module

$$\omega_{R/I} = \operatorname{Ext}^d_R(R/I, R) = \operatorname{Hom}_{R/J}(R/I, R/J) = (J{:}I)/J = H_1,$$

as I is generated by a regular sequence plus an extra element. Recall that the canonical module of an Artinian ring is the injective envelope of its residue field and therefore $\lambda(H_1) = \lambda(R/I)$. Applying Proposition 1.110, we get

$$\lambda(I/J) = \lambda(I^2/JI) + \lambda(\delta(I)).$$

On the other hand, if $\delta(I) = 0$, from (1.8) we get the short exact sequence

$$0 \longrightarrow H_1 \longrightarrow (R/I)^{d+1} \longrightarrow I/I^2 \longrightarrow 0,$$

which splits since H_1 is injective. Therefore

$$(R/I)^{d+1} \cong H_1 \oplus I/I^2.$$

Now, $\lambda(H_1) = \lambda(R/I)$ implies that $H_1 \cong R/I$ and so $I/I^2 \cong (R/I)^d$, which implies that $\nu(I) = d$ (since $\nu(I) = \nu(I/I^2)$ by Nakayama's Lemma), contradicting the assumption. □

Remark 1.112. A similar computation would establish the following more general formula. Suppose that $(R,\mathfrak{m})$ is a Cohen-Macaulay ring of type $s \geq 2$. Then

$$\lambda(I^2/JI)+\lambda(\delta(I)) = \binom{s+1}{2}.$$

Northcott Ideals

There is an useful construction introduced by Northcott (see [Nor63]), that provides for a very explicit description of a family of links.

Let R be a Noetherian ring, $\mathbf{u} = u_1,\ldots,u_n$ a sequence of elements in R, and $\varphi = (a_{ij})$ a $n \times n$ matrix with entries in R. Denote by $\mathbf{v} = v_1,\ldots,v_n$ the sequence

$$[\mathbf{v}]^t = \varphi \cdot [\mathbf{u}]^t.$$

If grade $(\mathbf{v}) = n$, then the ideal $N = (\mathbf{v}, \det\varphi)$ (where $\det\varphi$ is the determinant of the matrix φ) is known as the *Northcott ideal* associated with the sequence $\mathbf{u}$ and the matrix φ. It has then three important properties that can be summarized in the next theorem (see [Nor63]).

Theorem 1.113. *Let* $\mathbf{v}$, $\mathbf{u}$ *and* φ *be as above. Then either* N *is the whole ring or it is a proper ideal with the following properties:*

(a) *N is a perfect ideal (i.e. its grade is equal to the projective dimension of R/N);*
(b) $(\mathbf{v})\colon N = (\mathbf{u})$;
(c) $(\mathbf{v})\colon(\mathbf{u}) = N$.

The case where $(\mathbf{u})$ is the maximal ideal of a regular local ring is particularly interesting: $\det\varphi$ will then generate the socle of the Artin algebra $R/(\mathbf{v})$.

Proof of Theorem 1.104. According to Corollary 1.111, we must show that $\delta(I) \neq 0$, as this leads to the equality $\lambda(I^2/JI) = 0$. As observed earlier, $\delta(I) = 0$ means that I is also generated by a regular sequence. This puts us in the context of Northcott ideals. Since however $\lambda(I/J) = 1$, it follows that the ideal $N = (J, \det\varphi)$ is perfect and necessarily the maximal ideal of R. In the case that R is not a regular local ring, this is a contradiction. In the other case, when at least two of the generators of J lie in the square of the maximal ideal $\mathfrak{m}$, we cannot have the equality $\mathfrak{m} = (J, \det\varphi)$. □

Content Ideals of Polynomials

Let us consider in some detail a family of examples that gives rise to ideals of all possible reduction numbers. If R is a commutative ring and $f = f(t) \in R[t]$ is a polynomial, say $f = a_0 + \cdots + a_m t^m$, the *content* of f is the R-ideal $(a_0,\ldots,a_m)$. It is denoted by $c(f)$. Given another polynomial $g(t) = b_0 + b_1 t + \cdots + b_n t^n$, the following formula–Dedekind-Mertens *formula*–holds (see [Nor59]):

$$c(fg)c(g)^m = c(f)c(g)^{m+1}. \tag{1.11}$$

Multiplying both sides of (1.11) by $c(f)^m$, we obtain

$$c(fg)[c(f)c(g)]^m = c(f)c(g)[c(f)c(g)]^m. \tag{1.12}$$

Thus (1.12) says that $J = c(fg)$ is a reduction of $I = c(f)c(g)$, and that the reduction number is at most $\min\{\deg f, \deg g\}$.

If G is a graph with vertices labelled by $x_0, \ldots, x_m$, its monomial subring $k[G]$ is the subring of $k[x_0, \ldots, x_m]$ generated by all monomials $x_i x_j$ where (x_i, x_j) is an edge of G. In general, it is difficult to find Noether normalizations of any of these two families of algebras. Following [CVV98], we consider one case that provides an explanation for (1.12).

Theorem 1.114. *Let $X = \{x_0, \ldots, x_m\}$ and $Y = \{y_0, \ldots, y_n\}$ be distinct sets of indeterminates and let*

$$f = \sum_{i=0}^{m} x_i t^i \quad \text{and} \quad g = \sum_{j=0}^{n} y_j t^j$$

be the corresponding generic polynomials over a field k. Set $R = k[X,Y]$, $I = c(f)c(g)$, and $J = c(fg)$ and suppose that $m \leq n$. Then

(a) *J is a minimal reduction of I, $\ell(I) = m+n+1$, and $r_J(I) = m$.*
(b) *The polynomials*

$$h_q = \sum_{i+j=q} x_i y_j$$

are algebraically independent and subring $k[h_q\text{'s}]$ is a Noether normalization of $k[x_i y_j\text{'s}]$.

In particular, the factor $c(f)^m$ in the content formula (1.11) *cannot be replaced by $c(f)^p$ with $p < m$.*

Proof. We note that the ideal $I = (x_i y_j\text{'s})$ is the edge ideal associated with the graph G which is the join of two discrete graphs, one with $m+1$ vertices and the other with $n+1$ vertices; G is, therefore, bipartite.

Since J is already a reduction of I by (1.12), we may assume that k is an infinite field. On the other hand, as I is generated by homogeneous polynomials of the same degree, $\mathcal{F}(I) \cong k[x_i y_j\text{'s}] = k[G]$ (see [SVV94]). Let $Q_{ij}, 0 \leq i \leq m,\ 0 \leq j \leq n$ be distinct indeterminates and map

$$\psi : k[Q_{ij}\text{'s}] \longrightarrow k[x_i y_j\text{'s}], \quad \psi(Q_{ij}) = x_i y_j.$$

We claim that the kernel of ψ is generated by the 2×2 minors of a generic $(m+1) \times (n+1)$ matrix. Indeed let $Q = (Q_{ij})$. It is clear that the ideal $I_2(Q)$ generated by the 2×2 minors of Q is contained in $\mathfrak{Q} = \ker(\psi)$. On the other hand, since the graph is bipartite, $\dim(k[G]) = m+n+1$ (see [SVV94]) and therefore

$$\text{height}(\mathfrak{Q}) = (m+1)(n+1) - (m+n+1) = mn = \text{height}(I_2(Q)),$$

the latter by the classical formula for determinantal ideals (see [BV88, Theorem 2.5]). Since they are both prime ideals, we have $I_2(Q) = \mathfrak{Q}$.

To complete the proof we note that the a-invariant of $k[Q_{ij}\text{'s}]/I_2(Q)$ is $-n-1$ according to [BH92], and therefore the reduction number of $\mathcal{F}(I)$ is $(m+n+1)-n-1=m$. □

Relation Type versus Reduction Number

A not well understood relationship is that subsisting between the relation and reduction numbers of an ideal. Suppose that R is a local ring and we have the equality

$$I^{r+1} = JI^r,$$

where $r = r(I)$, and J is a corresponding minimal reduction. By Nakayama's Lemma, we may assume that $I = (a_1, \ldots, a_n)$, $n = \nu(I)$, and that the first $\ell(I)$ of the elements a_i generate J.

Consider a presentation of $R[It]$

$$0 \to L \longrightarrow R[T_1, \ldots, T_n] \longrightarrow R[It] \to 0, \quad T_i \mapsto a_i t.$$

Reducing modulo $\mathfrak{m}$ gives a presentation of the special fiber $\mathcal{F}(I)$. In particular we have that

$$\text{relation type}(\mathcal{F}(I)) \leq \text{relation type}(R[It]).$$

In turn the reduction J gives rise to a Noether normalization $A = k[T_1, \ldots, T_{\ell(I)}]$ of $\mathcal{F}(I)$. Among the elements of L are those obtained directly from the reduction relations, that is for any monomial $\mathbf{T}^\alpha = T_1^{\alpha_1} \cdots T_n^{\alpha_n}$ of degree $r+1$ one has a polynomial

$$h_\alpha = \mathbf{T}^\alpha - \sum_{i=1}^{\ell} T_i g_{\alpha,i} \in L_{r+1}.$$

Thus for any $f = \sum_\alpha c_\alpha \mathbf{T}^\alpha \in L_{r+1}$, we have a reduction

$$f - \sum_\alpha c_\alpha h_\alpha = \sum_{i=1}^{\ell} T_i f_i.$$

The issue is what conditions on J are required so that these forms together with L_i, $i \leq r$, and the relations on J span L. This shows the importance that the minimal reductions be of linear type or even be generated by a d-sequence.

Ideals of Reduction Number Two

Unlike the situation of reduction number one, there are few known processes that builds ideals of reduction number two.

More Actors

There are several measures of 'irregularity' for ideals in local rings. The following notions will play a role in the sequel.

- The *deviation* of I is the non-negative integer $\nu(I) - \text{height } I$.
- The *analytic deviation* of I as $ad(I) = \ell(I) - \text{height } I$.
- The difference between these two numbers, $\nu(I) - \ell(I)$ is the *second deviation* of I.
- The ideals of analytic deviation zero are called *equimultiple* (see [HIO88] for a wealth of information about these ideals).

Among all measures of I that we have defined, none play a role more central than that of the reduction number $\text{r}(I)$ in deciding when the algebra $R[It]$ is Cohen-Macaulay. It should come as no surprise, therefore that this integer is the hardest one to determine. That we know, there are no explicit process to compute $rn(I)$ unless one resorts to methods that introduce many fresh variables.

A pressing question here is this. If $\text{r}(I)$ is so significant, how do the other invariants of I relate to it? There are some intriguing relationships that, however non-general, occur repeatedly. For example, a basic guess for $rn(I)$ is

$$\text{r}(I) \leq \ell(I) - \text{height } I + 1,$$

the right-hand side being a value often called the *expected reduction number* of I.

Most other relationships among the actors depend on special circumstances. Here is a special one ([Va94a]):

Theorem 1.115. *Let R be a Gorenstein local ring and I a Cohen-Macaulay ideal of codimension g that is of linear type in codimension $\leq g+1$. Then $\ell(I) \geq g+2$.*

Reductions of a Filtration

Let $\mathcal{F} = \{I_n, n \geq 0\}$ be a multiplicative, decreasing filtration of ideals of R, where $I_0 = R$ (see [OR90] and [HZ94] for a fuller discussions). Denote by

$$\mathcal{R}(\mathcal{F}) = \sum_{n \geq 0} I_n t^n$$

its Rees algebra. We shall be particularly interested in *Hilbert filtrations*. These are the filtrations $\mathcal{F}$ whose Rees algebra $\mathcal{R}(\mathcal{F})$ is finite over the Rees algebra defined by the subfiltration generated by $I = I_1$. This means that $I_{n+1} = I \cdot I_n$ for large n.

By a *reduction* of a Hilbert filtration $\mathcal{F}$ we mean a subfiltration $\mathcal{G} = \{J_n, n \geq 0\}$ such that $\mathcal{R}(\mathcal{F})$ is integral over $\mathcal{R}(\mathcal{G})$. Correspondingly, the *reduction number* $\text{r}_{\mathcal{G}}(\mathcal{F})$ of $\mathcal{F}$ over $\mathcal{G}$ is the smallest integer r such that

$$\mathcal{R}(\mathcal{F}) = \sum_{s \leq r} \mathcal{R}(\mathcal{G}) I_s t^s.$$

The typical example is that where $\mathcal{F} = \{I^n, n \geq 0\}$ and $\mathcal{F} = \{\overline{I^n}, n \geq 0\}$, where $\mathcal{R}(\mathcal{F})$ is assumed Noetherian. In this case

$$\mathrm{r}_{\mathcal{G}}(\mathcal{F}) = \inf\{r \mid \overline{I^{n+1}} = I\overline{I^n}, n \geq r\}.$$

The requirement emphasizes that we must have $\overline{I^{n+1}} = I\overline{I^n}$ for *all* $n \geq r$. This will make for a much harder task in determining this number.

Another issue concerns the notion of minimal reduction. At least, however, for filtrations with $I_n \subset \overline{I_1^n}$, the Rees algebra $R[Jt]$ will contain no proper reduction of $\mathcal{R}(\mathcal{F})$ if J is a minimal reduction of I_1.

Independence of Reduction Numbers

In Example 1.32 we illustrated the fact that the reduction numbers of an ideal I may vary with respect to the minimal reduction J. We shall visit this topic again in Chapter 3. For the moment, to give a sense of what conditions may be required to guarantee the invariance of $\mathrm{r}_J(I)$, we quote [Huc87, Theorem 2.1] (see also the note added to this result):

Theorem 1.116. *Let $(R, \mathfrak{m})$ be a Noetherian local ring with infinite residue field and let I be an equimultiple ideal of codimension $g \geq 1$. Set $\mathcal{G} = \mathrm{gr}_I(R)$ and let $\mathcal{G}_+$ be its irrelevant ideal. If* depth $\mathcal{G}_+ \geq g - 1$ *then $\mathrm{r}_J(I)$ is independent of the minimal reduction J.*

Question 1.117. Let $(R, \mathfrak{m})$ be a Noetherian local ring. There are at least two kinds of ideals for which general methods for determining reductions and their reduction numbers would be welcome:

(i) Divisorial ideals of R.
(ii) Unmixed ideals of codimension two.

For example, how to deal with the canonical module ω of R? For the second class of ideals, there is a clear need for that kind of information in view of the role such ideals play in the theory of Bourbaki sequences and therefore in the study of Rees algebras of modules in general (see Chapter 8).

1.5 Determinants and Ranks of Modules

We now develop a number of elementary techniques aimed at obtaining equations of integrality that will have several applications that include bounding reduction numbers. They are primarily based on the classical Cayley-Hamilton theorem, appropriately tailored for modules and their endomorphisms.

1.5.1 Cayley-Hamilton Theorem

Let R be a commutative Noetherian ring and let $\varphi : E \to E$ be an endomorphism of a finitely generated R-module. In the classical manner, we turn this situation into an action of $R[t]$ on E. A morphism of rings

$$\pi : R[t] \longrightarrow \mathrm{Hom}_R(E,E)$$

arises, and a *Cayley-Hamilton theorem* is a description of kernel of π in terms of some property/invariant of E. More specifically, $\ker \pi$ contains a monic polynomial and we aim at finding the degree of such elements. They will be referred to loosely as Cayley-Hamilton *polynomials*.

We are going to consider variations of this problem since it provides a handy source of equations of integral dependence, with applications to the determination of reduction numbers. Some of these cases are as follows.

- The classical Cayley-Hamilton theorem
- Torsionfree modules over normal domains
- Torsionfree modules of finite projective dimension
- Graded modules over rings of polynomials

In the first of these, the determinant proof of the usual Cayley-Hamilton theorem can be applied to show that $\ker \pi$ contains a monic polynomial of degree $\nu(E)$, the minimal number of generators of E. Indeed, if $e_1, \ldots, e_n$ is a set of generators of E, from the set of relations

$$\varphi(e_i) = t \cdot e_i = a_{1,i}e_1 + \cdots + a_{n,i}e_n, \; i = 1, \ldots, n,$$

we obtain that $\det(tI - (a_{i,j})) \cdot E = 0$, where $(a_{i,j})$ is the square matrix built on the $a_{i,j}$. Moreover, when E is a graded module and φ is a graded homomorphism, this polynomial is a homogeneous element of $R[t]$, graded so that $\deg t = \deg \varphi$.

Let us state and prove a version of the Cayley-Hamilton theorem that is directly related to reductions and reduction numbers.

Proposition 1.118. *Let R be a commutative ring and E a faithful R-module generated by n elements. Suppose that $J \subset I$ are ideals such that $IE = JE$. Then $I^n = JI^{n-1}$.*

Proof. Let $e_1, \ldots, e_n$ be a set of generators of E and $u_1, \ldots, u_n$ arbitrary elements of I. We claim that their product is contained in JI^{n-1}. Consider the following elements of E,

$$\begin{aligned} u_1e_1 &= a_{11}e_1 + a_{12}e_2 + \cdots + a_{1n}e_n \\ u_2e_2 &= a_{21}e_1 + a_{22}e_2 + \cdots + a_{2n}e_n \\ &\;\vdots \\ u_ne_n &= a_{n1}e_1 + a_{n2}e_2 + \cdots + a_{nn}e_n, \end{aligned}$$

where $a_{ij} \in J$. By the usual rules, the determinant of the matrix

$$\begin{bmatrix} a_{11}-u_1 & a_{12} & \cdots & a_{1n} \\ a_{21} & a_{22}-u_2 & \cdots & a_{2n} \\ \vdots & \vdots & \ddots & \vdots \\ a_{n1} & a_{n2} & \cdots & a_{nn}-u_n \end{bmatrix}$$

annihilates each of the e_i. It must be trivial since E is faithful. Since it has the proper format, namely

$$(-1)^n u_1 \cdot u_2 \cdots u_n + b, \quad b \in JI^{n-1},$$

the assertion is proved. □

In the next two cases, one is able to obtain lower degree polynomials. We use the notion of the *rank* $\operatorname{rank}(E)$ of a module E, in the following sense. If K is the total ring of fractions of R, and $K \otimes_R E \cong K^r$, we say that E has rank r.

Proposition 1.119. *Let A be a local normal domain and E a finitely generated torsionfree A-module. If φ is an endomorphism of E then there exists a monic polynomial $P_\varphi(t)$ of degree at most the rank r of E over A such that $P_\varphi(\varphi) = 0$.*

Proof. This goes the usual way. We turn E into an $A[t]$-module by setting $tv = \varphi(v)$ for every element $v \in E$. The annihilator L contains at least one monic polynomial of degree r, by an application of the classical Cayley-Hamilton theorem and the assumption that E is torsionfree–in other words, E contains a free A-submodule F of rank r which is invariant under φ. The characteristic polynomial of this action will be contained in L.

If $f(t) \in L$ is a nonzero polynomial of degree lower than r, the Euclidean algorithm yields an equation

$$aP_\varphi(t) = bf(t)g(t) + h(t),$$

where a, b are nonzero elements of A and $\deg h(t) < \deg f(t)$. By taking $f(t)$ of least minimal degree in L to begin with, we may assume that $h(t) = 0$.

We now take the content of the polynomials in the equation $aP_\varphi(t) = bf(t)g(t)$. By the Dedekind-Mertens formula (1.11), one obtains (with $m = \deg f(t)$) that

$$ac(g)^m = bc(f)c(g)^{m+1},$$

showing that $bc(f)c(g)$ is integral over the ideal (a). Since $a \in bc(f)c(g)$, we have the equality $(a) = bc(f)c(g)$ as A is a normal domain. But A is also a local domain, thus showing that $c(f)$ and $c(g)$ are principal ideals. Putting all this together we get an equation $P_\varphi(t) = F(t)G(t)$, with $F(t) \in L$ a monic polynomial of the same degree as $f(t)$, and therefore $L = (F(t))$. □

A simpler argument goes as follows. Let R be an integral domain with field of fractions K and φ an endomorphism of a finitely generated torsionfree R-module E. To define $\det \varphi$ as an element of R, we first extend φ to the endomorphism

$$\varphi' = K \otimes \varphi : E' = K \otimes_R E \longrightarrow K \otimes_R E,$$

and define $\det \varphi = \det \varphi'$.

Proposition 1.120. *If R is a normal domain, then* $\det \varphi' \in R$.

Proof. It will suffice to show that $\det \varphi' \in R_{\mathfrak{p}}$ for each height one prime ideal $\mathfrak{p}$ of R. (Recall that R is the intersection of these localizations.) Since $R_{\mathfrak{p}}$ is a DVR, $E_{\mathfrak{p}}$ is a free module and φ' restricted to it defines an endomorphism. Thus its determinant can be computed using a basis of $E_{\mathfrak{p}}$, since it is also a basis (over K) for E'. □

The assertions of the proposition are valid more widely. A could be replaced by a normal domain in which invertible ideals are principal. In some of our applications E will be a graded ring and A a graded normal subring; the endomorphism φ will be induced by multiplication by a homogeneous element $z \in B_1$. The relation of dependence now looks like this:

$$z^r + a_1 z^{r-1} + \cdots + a_r = 0, \qquad a_i \in A_i.$$

When A contains a field of characteristic greater than r, one has:

Proposition 1.121. *Let $A \subset B$ be a homogeneous inclusion of standard graded rings, with A a normal domain and B a finitely generated torsionfree module over A of rank r. If A contains a field of characteristic greater than r, then $B_r = A_1 B_{r-1}$.*

Proof. Let $u_1, \ldots, u_n$ be a set of generators of B_1 over A, and consider the integrality equation of

$$u = x_1 u_1 + \cdots + x_n u_n,$$

where the x_i are elements of k. By assumption, we have

$$u^r = (x_1 u_1 + \cdots + x_n u_n)^r = a_1 u^{r-1} + \cdots + a_r,$$

where $a_i \in A_1^i$. Expanding u^r we obtain

$$\sum_{\alpha} a_\alpha m_\alpha u^\alpha \in A_1 B_1^{r-1},$$

where $\alpha = (\alpha_1, \ldots, \alpha_n)$ is an exponent of total degree r, a_α is the multinomial coefficient $\binom{r}{\alpha}$, and m_α is the corresponding 'monomial' in the x_i. We must show that

$$u^\alpha \in A_1 B_1^{r-1}$$

for each α.

To prove the assertion, it suffices to show that the span of the vectors $(a_\alpha m_\alpha)$, indexed by the set of all monomials of degree r in n variables, has the dimension of the space of all such monomials. Indeed, if these vectors lie on a hyperplane

$$\sum_{\alpha} c_\alpha T_\alpha = 0,$$

we would have a homogeneous polynomial

$$f(X_1,\ldots,X_n) = \sum_{\alpha} c_\alpha m_\alpha X^\alpha$$

which vanishes on k^n. This means that all the coefficients $c_\alpha a_\alpha$ are zero, and therefore each c_α is zero since the m_α do not vanish in characteristic zero. □

The next case does not require that R be a normal domain, but assumes that E has finite projective dimension. It is taken from [Al78] (see also [Va98b, Proposition 9.3.4]). Let E be a finitely generated R-module and let

$$\varphi : E \mapsto E$$

be an endomorphism. Map a free module over E and lift φ:

$$\begin{array}{ccc} F_0 & \xrightarrow{\psi} & E \\ \downarrow{\scriptstyle\varphi} & & \downarrow{\scriptstyle\varphi} \\ F_0 & \xrightarrow[\psi]{} & E \end{array}$$

Let

$$P_{\varphi_0}(t) = \det(tI - \varphi_0) = t^n + \cdots + a_n$$

be the characteristic polynomial of φ_0, $n = \operatorname{rank}(F)$. By the usual Cayley-Hamilton theorem, we have that $P_{\varphi_0}(\varphi) = 0$. The drawback is that n, which is at least the minimal number of generators of E, may be too large. One should do much better using a trick of [Al78]. Lift φ to a mapping from a projective resolution of E into itself,

$$\begin{array}{ccccccccccc} 0 & \longrightarrow & F_s & \longrightarrow & \cdots & \longrightarrow & F_1 & \longrightarrow & F_0 & \longrightarrow & E & \longrightarrow & 0 \\ & & \downarrow{\scriptstyle\varphi_s} & & & & \downarrow{\scriptstyle\varphi_1} & & \downarrow{\scriptstyle\varphi_0} & & \downarrow{\scriptstyle\varphi} & & \\ 0 & \longrightarrow & F_s & \longrightarrow & \cdots & \longrightarrow & F_1 & \longrightarrow & F_0 & \longrightarrow & E & \longrightarrow & 0 \end{array}$$

and define

$$P_f(t) = \prod_{i=0}^{s} (P_{\varphi_i}(t))^{(-1)^i}.$$

This rational function is actually a polynomial in $R[t]$ (see [Al78]). Indeed

$$P_f(t) = \frac{a(t)}{b(t)} = \frac{t^m + a_1 t^{m-1} + \cdots + a_m}{t^n + b_1 t^{n-1} + \cdots + b_n}$$

reduces to the characteristic polynomial of f acting on the projective module $E \otimes_R K$, where K is the total ring of fractions of R, since it has finite projective dimension. It follows that

$$P_\varphi(t) = c(t) = t^e + c_1 t^{e-1} + \cdots + c_e \in K[t].$$

Multiplying out, we obtain $a(t) = b(t) \cdot c(t)$, from which it will follow that all the coefficients of $c(t)$ lie in R.

Proposition 1.122. *Let R be a commutative Noetherian ring without nontrivial idempotents and let*

$$\begin{array}{ccccccccc} 0 & \longrightarrow & E_1 & \longrightarrow & E_2 & \longrightarrow & E_3 & \longrightarrow & 0 \\ & & \downarrow{\scriptstyle \varphi_1} & & \downarrow{\scriptstyle \varphi_2} & & \downarrow{\scriptstyle \varphi_3} & & \\ 0 & \longrightarrow & E_1 & \longrightarrow & E_2 & \longrightarrow & E_3 & \longrightarrow & 0 \end{array}$$

be an exact commutative diagram of modules of finite projective dimension. Then

$$P_{\varphi_2}(t) = P_{\varphi_1}(t) \cdot P_{\varphi_3}(t).$$

Proof. This follows directly from the preceding comments. □

Remark 1.123. If E is a graded module and φ is homogeneous, then $P_\varphi(t)$ is a homogeneous polynomial and $\deg E = \deg P_\varphi(t)$.

Minimal Polynomial

Under special conditions one can obtain the minimal polynomial of certain endomorphisms instead of the characteristic polynomial. This may occur in the setting of an affine domain A and one of its Noether normalizations $R = k[z_1, \ldots, z_d]$. For $u \in A$, the kernel of the homomorphism

$$R[t] \longrightarrow A, \quad t \mapsto u,$$

is an irreducible polynomial $h_u(t)$ (appropriately homogeneous if A is a graded algebra and u is homogeneous). It is related to the characteristic polynomial $f_u(t)$ of u by an equality of the form

$$f_u(t) = h_u(t)^r.$$

The Determinant of an Endomorphism

Let R be a commutative ring and A a finite R-module. An element $a \in A$ defines an endomorphism

$$f_a : A \to A, \; f_a(x) = ax,$$

of R-modules. We seek ways to define the *determinant* of f_a relative to R. For example, if A is a free R-module, we may use the standard definition. More generally, if A has a finite free R-resolution $\mathbb{F}$,

$$0 \to F_n \longrightarrow \cdots \longrightarrow F_1 \longrightarrow F_0 \longrightarrow A \to 0,$$

as above we lift f_a to an endomorphism of $\mathbb{F}$, and 'define' $\det(f_a)$ as the alternating product

$$\det(f_a) = \prod_{i=0}^{n} \det(f_i)^{(-1)^i},$$

of ordinary determinants. If this can be done unequivocally, we might call $\det(f_a)$ the *norm* of a. We shall write

$$\mathrm{N}_{A/R}(a) = \det(f_a).$$

We note that this definition does not require that A be commutative. This also means that f_a can be replaced by an endomorphism of an R-module E. The difficulty of this definition shows up when one of the f_i is not injective. To circumvent it, we make use of the characteristic polynomial $P_{f_a}(t)$.

Definition 1.124. Let R be a commutative Noetherian ring without nontrivial idempotents and $\varphi : E \to E$ an endomorphism of a module of finite projective dimension. The *determinant* of φ is

$$\det(\varphi) = (-1)^{\deg P} P_{\varphi}(0).$$

There is a related notion:

Definition 1.125. Let R be a commutative ring with total ring of fractions K. A finitely generated R-module E is said to be of rank r if $K \otimes_R E \cong K^r$. Its *determinant* is the module

$$\det(E) = (\wedge^r E)^{**}.$$

Definition 1.126. Let A be an R-algebra as above and let $a \in A$. The *norm* of a relative to R is

$$\mathrm{N}_{A/R}(a) = P_{f_a}(0).$$

It is easy to see that the definitions agree, up to sign, when all f_i are injective and A is torsionfree as an R-module. The point is that all divisions occur in the expansion of

$$P_{f_a}(t) = \prod_{i=1}^{n} P_{f_i}(t)^{(-1)^i}.$$

In the last of our cases, R is a standard graded algebra and E is a graded module. Typically, E will be the special fiber ring $\mathcal{F}(I)$ of an ideal I, $A = \mathcal{F}(J)$ for a minimal reduction J of I, and φ is an endomorphism induced by multiplication by a homogeneous element of $\mathcal{F}(I)$. In order to obtain different bounds for the degrees of the Cayley-Hamilton polynomials we have to recall other notions of degrees of modules.

Let R be a commutative Noetherian ring and $\varphi : E \to E$ an endomorphism of a finitely generated R-module. We have already discussed a setting rich enough to define a determinant for φ. Here is a quicker approach when we assume that R is an integrally closed domain.

Definition 1.127. Suppose that R is an integral domain with field of fractions K, E a torsionfree R-module of rank r, and $\varphi : E \to E$ an R-endomorphism. Let $I(E)$ be the module $\wedge^r E$/torsion submodule and Ψ the induced mapping $\Psi : I(E) \to I(E)$. The *determinant* of φ is the unique element $\det(\varphi)$ of K that realizes this mapping.

If R is integrally closed, then $\det(\varphi) \in \mathrm{Hom}_R(I(E), I(E))$; this ring is canonically identified with R, so $\det(\varphi) \in R$.

Let K be the field of fractions of R and assume that $E \otimes_R K$ is a vector space of dimension r. Let x be an indeterminate and ψ the endomorphism of the extended module

$$\psi : E \otimes_R R[x] \to E \otimes_R R[x]$$
$$\psi = xI - \varphi \otimes 1,$$

where I is the identity endomorphism of $F = E \otimes_R R[x]$. Then F is a module of rank r, and we set

$$I(E) = \wedge^r F/\text{modulo torsion}.$$

$I(E)$ is a torsion free module of $R[x]$ and thus can be identified with an ideal of $R[x]$. Furthermore, any two such identifications, I_1 and I_2, differ by multiplication by nonzero elements of $R[x]$, that is $aI_1 = bI_2$, for suitable $a, b \in R[x]$.

The mapping ψ induces an endomorphism of $I(E)$: first consider $\wedge^r \psi$, then reduce modulo torsion. We denote the induced map by $\Psi \in \mathrm{Hom}_{R[x]}(I(E), I(E)$. Since $R[x]$ is integrally closed, there is a natural identification $\mathrm{Hom}_{R[x]}(I(E), I(E)) = R[x]$. We denote by $f_\varphi(x)$ the element of $R[x]$ corresponding to Ψ.

Proposition 1.128. *$f_\varphi(x)$ is a monic polynomial of degree r,*

$$f_\varphi(x) = a_0 x^r + a_1 x^{r-1} + \cdots + a_r, \quad a_0 = 1.$$

Proof. To prove the assertion, it suffices to pass to the K-vector space $E \otimes_R K$, when $f_\varphi(x)$ is the usual characteristic polynomial of any matrix representation of the action induced by φ. □

Corollary 1.129. *Let $\varphi : E \to E$ be an endomorphism as above. The* determinant *and the* trace *of φ are the following coefficients of $f_\varphi(x)$:*

$$\det(\varphi) = (-1)^r a_r,$$
$$\mathrm{trace}(\varphi) = -a_1.$$

Remark 1.130. We note the following property of these notions. If $\theta : R \to S$ is an extension of R and S is a torsionfree as an R-module (not necessarily R-flat), then $\det_S(\varphi \otimes 1_S) = \det_R(\varphi)S$, for any endomorphism, and similarly for the traces.

McRae Determinant

Yet another approach to obtaining equations of integrality for an endomorphism is through the notion of McRae determinants. Its fundamental application is to obtain close to a 'best fit' invertible ideal in the annihilator of a module. Briefly, it can be described as follows.

Let S be an integral domain and let E be a finitely generated torsion module of finite projective dimension. (In our applications, $S = R[t]$ and $E = A$, where R and A are as above.) If E admits a projective resolution

$$0 \to P_1 \xrightarrow{\varphi} P_0 \longrightarrow E \to 0,$$

where P_0 and P_1 are free modules (necessarily of the same rank) then the ideal generated by $\det(\varphi)$ defines the *divisor* $\underline{d}(E)$ of E. Note that $\underline{d}(E)$ is independent of the presentation; in the more general case that P_0 and P_1 are projective, $\underline{d}(E)$ is defined as the Fitting ideal of φ. Finally, if proj dim $E \geq 2$, let x be a regular element of S with $xE = 0$ and choose a presentation

$$0 \to L \longrightarrow (S/(x))^n \longrightarrow E \to 0.$$

Applying an induction hypothesis on the (smaller) projective dimension of L, one sets

$$\underline{d}(E) = x^n \underline{d}(L)^{-1}.$$

This turns out to be well-defined. For a more detailed exposition, see [Va94b, Section 1.1].

1.5.2 The Big Rank of a Module

Let R be a commutative Noetherian ring and E a finitely generated R-module. For the purpose of finding bounds on the degrees of polynomial equations satisfied by homomorphisms $\varphi : E \to E$, we introduce a notion of *rank* for E. It will agree with the usual notion of torsionfree rank in some cases, but it is usually very different.

Definition 1.131. Let $\mathcal{A} = \{\mathfrak{p}_1, \ldots, \mathfrak{p}_s\}$ be the set of associated primes of E. For each $\mathfrak{p} \in \mathcal{A}$, denote by $\text{mult}_E(\mathfrak{p})$ the length of $H^0_{\mathfrak{p}}(E_{\mathfrak{p}})$, as an $R_{\mathfrak{p}}$-module. For each chain C of elements $\mathfrak{p}_{i_1} \subset \mathfrak{p}_{i_2} \subset \cdots \subset \mathfrak{p}_{i_s}$ of $\mathcal{A}$, set

$$\text{mult}_E(C) = \sum_{j=1}^{s} \text{mult}_E(\mathfrak{p}_{i_j}).$$

The *big rank* of E over R is the integer

$$\text{bigrank}(E) = \sup\{\text{mult}_E(C) \mid \forall\, C \subset \mathcal{A}\}.$$

If R is an integral domain and E is a torsionfree R-module, *bigrank*(E) = $\text{rank}(E)$; $\text{bigrank}(R) = 1$ if and only if R is reduced.

This definition can be extended to an endomorphism $\varphi : E \to E$ by setting

$$\text{bigrank}(\varphi) = \text{bigrank}(\varphi(E)).$$

In the following we collect several elementary properties of this notion.

Proposition 1.132. *Let R be a commutative Noetherian ring and let E be a finitely generated R-module with* $\text{bigrank}(E) = r$.

(a) *For any endomorphism* φ *of a module* E *of bigrank* r,

$$\ker \varphi^s = \ker \varphi^r, \ \forall s \geq r.$$

(b) *If* $\varphi_1, \ldots, \varphi_r$ *are nilpotent commuting endomorphisms of* E, *then* $\varphi_r \cdots \varphi_1 = 0$.

Proof. (a) For any positive integer n, set $L_n = \ker \varphi^n$. We claim that it suffices to check the equality $L_r = L_{r+1}$ at the associated primes of E. Indeed, if these two modules differ, let $\mathfrak{p}$ be an associated prime of the quotient L_{r+1}/L_r. But φ^r defines an embedding of this module into $\ker \varphi \subset E$, and $\mathfrak{p}$ is also an associated prime of E.

For a chain $\mathfrak{p}_1 \subset \mathfrak{p}_2 \subset \cdots \subset \mathfrak{p}_n = \mathfrak{m}$ of associated prime ideals of E, we may localize E and φ at $\mathfrak{m}$, and the value of the bigrank $E_{\mathfrak{m}}$ cannot exceed $r = \text{bigrank}(E)$. We may thus assume that R is a local ring and that its maximal ideal $\mathfrak{m}$ is an associated prime of E. Set $E_0 = \Gamma_{\mathfrak{m}}(E)$, and note that φ induces an endomorphism φ_0 of E_0, and one of $E' = E/E_0$, which we denote by ψ.

We are now set to begin the proof by induction on the number of associated primes of E. We have already reduced to the case of a local ring whose maximal ideal is associated to E. If $\mathfrak{m}$ is the only associated prime, then $E = E_0$ is a module of length r, and the assertion is clear from the observations above. We may thus assume that $\mathcal{A}$ has other associated prime ideals besides $\mathfrak{m}$.

Set $t = \lambda(E_0)$, and note that $\text{bigrank}(E') = r - t$. Consider the commutative diagram of mappings induced by φ and some of its powers:

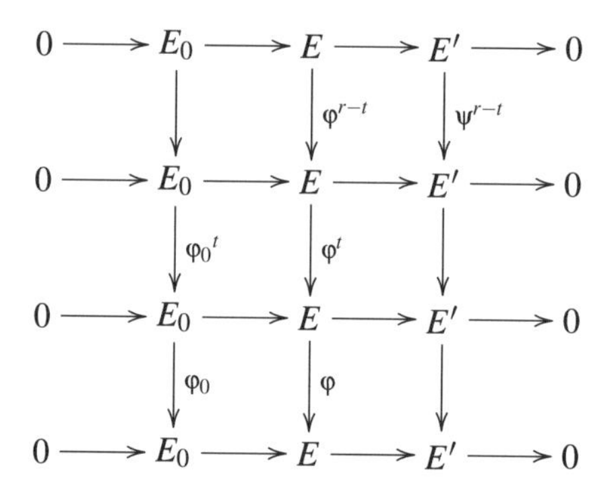

Suppose that z is an element in the kernel of the composite of the vertical mappings in the center, $\varphi^{r+1}(z) = 0$. Denote by z' its image in E'. Since $\psi^{r+1}(z') = 0$, by induction on the number of associated primes, we get $\psi^{r-t}(z') = 0$. This means that $w = \varphi^{r-t}(z) \in E_0$. As

$$\varphi_0^{t+1}(w) = \varphi^{r+1}(z) = 0$$

from the earlier case, $\varphi_0^t(w) = 0$ and therefore $\varphi^r(z) = 0$, as desired.

(b) If R contains a field of characteristic zero, (b) follows from (a) by the usual trick.

Set $L = \varphi_r \cdots \varphi_1(E)$. To show that $L = 0$, assume otherwise and let $\mathfrak{m}$ be one of its associated prime ideals. Note that $\mathfrak{m} \in \mathcal{A}$. Localizing at this ideal we may assume that R is a local ring and that its maximal ideal is an associated prime of E; note that the definition of $\text{mult}_E(\mathfrak{p})$ for those prime ideals $\mathfrak{p} \subset \mathfrak{m}$ is unaffected by the localization.

We make use of the observation that if φ and ψ are commuting, nilpotent endomorphisms of a module M, then $\ker \varphi \neq \ker \psi \cdot \varphi$ if $\varphi \neq 0$. When we apply this to a module of length $n = \lambda(M)$ it yields that the product $\varphi_n \cdots \varphi_1$ of such morphisms must vanish.

We induct on the maximal length of primes of $\mathcal{A}$, taking special account (as in the proof of (a)) of the fact that $\text{mult}_E(\mathfrak{m})$ occurs in each of the terms $\text{mult}_E(C)$. Set $t = \text{mult}_E(\mathfrak{m})$; by the induction hypothesis, localizing at any other associated prime it follows that the module $N = \varphi_{n-t}\varphi_{n-t-1} \cdots \varphi_1(E)$ has $\mathfrak{m}$ for its only possible associated prime. This means that N is contained in E_0, which is a module of length t. Since the φ_i are nilpotent and commuting and induce mappings on E_0, it is clear that $\varphi_n \cdots \varphi_{n-s+1}(E_0) = 0$, which implies our assertion. □

Corollary 1.133. *For any endomorphism φ of E,*

$$\text{bigrank}(\varphi^s) = \text{bigrank}(\varphi^r), \; s \geq r.$$

Proof. Since $\ker \varphi^r = \ker \varphi^s$, φ^{s-r} defines an isomorphism of $\varphi^r(E)$ with $\varphi^s(E)$. □

Corollary 1.134. *For an ideal I of the Noetherian ring R,*

$$\text{nil}(I) \leq \text{bigrank}(R/I).$$

Note that using the notion of Loewy length, a sharper estimate with this structure is obtained in [Va98b, Proposition 9.2.2].

1.6 Boundedness of Reduction Numbers

A general issue that arises has to do with the great number of reductions that an ideal may admit, and for which we introduce the following measure:

Definition 1.135. Let R be a Noetherian ring. The *big reduction number* of an ideal I is

$$\text{bigr}(I) = \sup\{ \text{r}_J(I) \mid J \text{ is a reduction of } I\}.$$

We are going to see that this supremum is always finite for arbitrary Noetherian rings.

1.6.1 Arithmetic Degree

Next we remind ourselves of the notion of the *arithmetic degree* of an algebra or module (see [Va98b, Chapter 9]). It can be defined for local rings or graded algebras, but we shall emphasize the latter.

Let B be a graded algebra. For each associated (homogeneous) prime ideal $\mathfrak{p}$ of B, its local contribution (to the primary decomposition of the trivial ideal of B) is the length $\text{mult}_B(\mathfrak{p})$ of the Artinian $B_\mathfrak{p}$-module

$$\Gamma_\mathfrak{p}(B_\mathfrak{p}) = H^0_\mathfrak{p}(B_\mathfrak{p}) = \bigcup_{n\geq 1}(0 :_{B_\mathfrak{p}} \mathfrak{p}^n),$$

the 0th local cohomology of $B_\mathfrak{p}$. The arithmetic degree of B is the integer

$$\text{arith-deg}(B) = \sum_{\mathfrak{p}} \text{mult}_B(\mathfrak{p})\deg(B/\mathfrak{p}),$$

where $\deg(\cdot)$ is the ordinary multiplicity function on modules over standard graded algebras, or over local rings. The justification for the notation $\text{mult}_B(\mathfrak{p})$ lies with the need to distinguish it from the ordinary multiplicity $\deg\text{mult}_B(\mathfrak{p})$ of the localization.

Rules of Computation for the Arithmetic Degree

Among various degree functions (see [Va98b, Chapter 9]) this is probably the most versatile for obtaining general estimates. Let us indicate some basic reasons. Without loss of generality we assume that R is a Gorenstein local ring (or a ring of polynomials over a field with corresponding graded modules).

One advantage of this function for obtaining general bounds lies in the fact that it can be determined in many cases when its constituent elements are not available. We recall a formula of [Va98c, Proposition 1.11].

Suppose that $S = k[x_1,\ldots,x_n]$ and $\dim M = d \leq n$. It suffices to construct graded modules M_i, $i = 1\ldots n$, such that

$$a_i(M) = \deg(M_i).$$

For each integer $i \geq 0$, set $L_i = \text{Ext}^i_S(M,S)$. By local duality ([BH93, Section 3.5]), a prime ideal $\mathfrak{p} \subset S$ of height i is associated to M if and only if $(L_i)_\mathfrak{p} \neq 0$; furthermore $\lambda((L_i)_\mathfrak{p}) = \text{mult}_M(\mathfrak{p})$.

We are set to find a path to $\text{arith-deg}(M)$. Compute for each L_i its degree $e_1(L_i)$ and codimension c_i. Then choose M_i according to the rule

$$M_i = \begin{cases} 0 & \text{if } c_i > i \\ L_i & \text{otherwise.} \end{cases}$$

Proposition 1.136. *For a graded S-module M and for each integer i denote by c_i the codimension of* $\text{Ext}^i_S(M,S)$. *Then*

$$\text{arith-deg}(M) = \sum_{i=0}^{n} \lfloor \frac{i}{c_i} \rfloor \deg(\text{Ext}^i_S(M,S)). \tag{1.13}$$

This can also be expressed as

$$\text{arith-deg}(M) = \sum_{i=0}^{n} \deg(\text{Ext}^i_S(\text{Ext}^i_S(M,S),S)). \tag{1.14}$$

Proposition 1.137. *Let R be a Cohen-Macaulay local ring of dimension d. For any exact sequence*

$$0 \to A \longrightarrow B \longrightarrow C \to 0$$

of finitely generated R-modules, one has

$$\text{arith-deg}(A) \leq \text{arith-deg}(B) \leq \text{arith-deg}(A) + \text{arith-deg}(C).$$

Proof. Since $\text{Ass}(B) \subset \text{Ass}(A) \cup \text{Ass}(C)$, it will suffice to show that for any prime $\mathfrak{p}$ which is an associated prime of either A or C, the contribution of $\mathfrak{p}$ to the arithmetic degrees satisfies the asserted inequalities.

Suppose that height $\mathfrak{p} = r$. Localizing at $\mathfrak{p}$ and taking the long exact sequence of $\text{Hom}(\cdot, R_{\mathfrak{p}})$, we have

$$\text{Ext}^{r-1}(A_{\mathfrak{p}}, R_{\mathfrak{p}}) \longrightarrow \text{Ext}^{r}(C_{\mathfrak{p}}, R_{\mathfrak{p}}) \longrightarrow \text{Ext}^{r}(B_{\mathfrak{p}}, R_{\mathfrak{p}}) \longrightarrow \text{Ext}^{r}(A_{\mathfrak{p}}, R_{\mathfrak{p}}) \to 0.$$

It follows, by duality, that if $\mathfrak{p}$ makes any contribution to $\text{arith-deg}(B)$ at all, it will bound its contribution to $\text{arith-deg}(A)$, and it is bounded by the sum of its contributions to $\text{arith-deg}(A)$ and $\text{arith-deg}(C)$. □

Hyperplane Sections

The behaviour of the arithmetic degree under hyperplane section is given by:

Theorem 1.138. *Let A be a standard graded algebra, M a finitely generated graded module and $h \in A_1$ a regular element on M. Then*

$$\text{arith-deg}(M/hM) \geq \text{arith-deg}(M).$$

For a proof, see [Va98b, Theorem 9.1.5]. A similar proof works for local rings with h taken to be a superficial element on M. Unfortunately the sense of the inequality would have to be reversed, in order for $\text{arith-deg}(\cdot)$ to be used as a scale of complexity.

Arithmetic Degree and Reduction Number

We shall now give the required Cayley-Hamilton theorem for graded modules over rings of polynomials. It will then be used as a way to bound reduction numbers.

Proposition 1.139. *Let $R = k[x_1, \ldots, x_n]$ be a ring of polynomials over an infinite field, E a finitely generated graded R-module and φ a homogeneous endomorphism of E. There exists a homogeneous Cayley-Hamilton polynomial $P_{\varphi}(t)$ of (ordinary) degree in t at most* $\text{arith-deg}(E)$.

Proof. This is similar to that of a result of [Va96]. Let $\mathfrak{p}$ be an associated prime of E that is maximal in the set $\text{Ass}(E)$, and set E_0 for the inverse image in E of $\Gamma_{\mathfrak{p}}(E_{\mathfrak{p}})$. This simply means that in the exact sequence

$$0 \to E_0 \longrightarrow E \longrightarrow E' \to 0,$$

$\mathfrak{p}$ is not an associated prime of E', so that we obtain $\text{arith-deg}(E) = \text{arith-deg}(E_0) + \text{arith-deg}(E')$. Moreover, E_0 is a graded R-module invariant under φ. If we denote by φ_0 the restriction of φ to E_0, this induces a homogeneous endomorphism φ' of E'.

We can now apply Theorem 1.119 to φ_0 using a subring of R (say, a Noether normalization of $R/\text{ann}\,(E_0)$), to obtain a Cayley-Hamilton polynomial $P_{\varphi_0}(t)$ of degree $\deg(E_0) = \text{arith-deg}(E_0)$. On the other hand, from the choice of $\mathfrak{p}$, we have that the set $\text{Ass}(E')$ is contained properly in $\text{Ass}(E)$.

Now we use induction on the number of associated primes, and finally set $P_\varphi(t) = P_{\varphi_0} \cdot P_{\varphi'}(t)$. □

Theorem 1.140. *Let B be a standard graded algebra and assume that $k[z_1,\ldots,z_d]$ is a standard Noether normalization. If k is an infinite field, then every element $f \in B_1$ satisfies a monic equation*

$$f^a + c_1 f^{a-1} + \cdots + c_a = 0, \quad c_i \in (z_1,\ldots,z_d)^i, \tag{1.15}$$

where $a \leq \text{arith-deg}(B)$. Moreover, $\text{r}(B) \leq \text{arith-deg}(B) - 1$ if $\text{char}\, k = 0$.

Remark 1.141. For a proof, see [Va98b, Theorem 9.3.4]. This result can be viewed as a Cayley-Hamilton theorem: A-endomorphisms of B satisfy monic equations over A of degree $\text{arith-deg}(B)$. The question can be raised as to the *bigrank* notion can also serve as a bound. The answer is no. If I is a singular prime of a reduced irreducible curve, the elements of I^{-1} act as endomorphisms on I but do not always satisfy a monic equation of degree 1. However, there may be two more hopeful candidates for bounds. Let φ be an endomorphism of a finitely generated module. Does φ satisfy a monic equation of degree given by any of the two numbers defined below?

(i) $\text{bigrank}(E)$ if $\text{proj. dim.}_R(E) < \infty$, or
(ii) if R is a local ring, for each associated prime $\mathfrak{p}$ of E, $\text{mult}_E(\mathfrak{p}) = \lambda(H^0_{\mathfrak{p}}(E_{\mathfrak{p}}))$, and for each chain C of elements $\mathfrak{p}_{i_1} \subset \mathfrak{p}_{i_2} \subset \cdots \subset \mathfrak{p}_{i_s}$ of associated prime ideals, set

$$\text{arith-deg}_E(C) = \sum_{j=1}^{s} \text{mult}_E(\mathfrak{p}_{i_j}) \cdot \deg(R/\mathfrak{p}_{i_j}).$$

Define now

$$\text{arith-deg}_0(E) = \sup_C \{\text{arith-deg}_C(E)\}.$$

Reduction Number and Arithmetic Degree

Returning to reduction numbers, the issue is to use the algebras $\text{gr}_I(R)$ (when I is $\mathfrak{m}$-primary) and $\mathcal{F}(I) = \text{gr}_I(R) \otimes (R/\mathfrak{m})$ (in general) in place of B. Thus in the first case, when applied to the elements of I/I^2 the equations are actually of the form

$$f^a \in (z_1,\ldots,z_d)G, \quad a \leq \text{arith-deg}(G).$$

If the characteristic of $R/\mathfrak{m}$ is zero, one can convert these equations into reduction equations

$$I^a = J \cdot I^{a-1}.$$

This approach, when applied to the special fiber $\mathcal{F}(I) = G \otimes (R/\mathfrak{m})$ of an ideal of arbitrary dimension, gives our local bound.

Proposition 1.142. *Let $(R, \mathfrak{m})$ be a Noetherian local ring, let I be an ideal and let J be one of its reductions.*[1] *Then*

(a) *If $R/\mathfrak{m}$ has characteristic zero, then*

$$\mathrm{r}_J(I) < \text{arith-deg}(\mathcal{F}(I)).$$

(b) *If $R/\mathfrak{m}$ has characteristic $p > 0$, then*

$$\mathrm{r}_J(I) \leq \nu(I) \cdot (\text{arith-deg}(\mathcal{F}(I)) - 1).$$

(c) *If I is $\mathfrak{m}$-primary and R/I contains a field, then*

$$\mathrm{r}(I) \leq \begin{cases} \text{arith-deg}(G) - 1 & \textit{if char } R/\mathfrak{m} \textit{ is } 0 \\ \nu(I)(\text{arith-deg}(G) - 1) & \textit{if char } R/\mathfrak{m} \textit{ is } p > 0. \end{cases}$$

(d) *If R is quasi-unmixed, $R/\mathfrak{m}$ has characteristic zero and G has no embedded primes, then (e.g. I is a normal $\mathfrak{m}$-primary ideal)*

$$\mathrm{r}(I) \leq e(I) - 1. \tag{1.16}$$

Proof. We may assume that the residue field of R is infinite, and then we can replace J by a minimal reduction. Now recall that minimal reductions of I correspond to graded Noether normalizations of $\mathcal{F}(I)$. All the assertions will now follow from [Va96] and the observations above. In case (d), note that the algebra G has the condition S_1 of Serre and therefore $\text{arith-deg}(G) = \deg(G) = e(I)$. □

Theorem 1.143. *Let A be a standard graded algebra over a field k that satisfies Serre's condition S_1. If $\dim A = 2$, then for any Noether normalization $R = k[x, y]$ with $x, y \in A_1$, we have $A_e = (x, y)A_{e-1}$ for $e = \deg(A)$.*

Proof. The module (actually an algebra) $B = \mathrm{Hom}_R(\mathrm{Hom}_R(A, R), R)$ can be given a natural B-module structure that is B-faithful. Since it has depth 2 over the polynomial ring R, it is a free R-module. Let $v_1, \ldots, v_e$ (where $e = \deg(A) = \deg(B)$) be a free homogeneous basis. For $u_1, \ldots, u_e \in A_1$, as in Proposition 1.118, we have equations

$$\begin{aligned} u_1 v_1 &= a_{11} v_1 + a_{12} v_2 + \cdots + a_{1e} v_e \\ u_2 v_2 &= a_{21} v_1 + a_{22} v_2 + \cdots + a_{2e} v_e \\ &\vdots \\ u_e v_e &= a_{e1} v_1 + a_{e2} v_2 + \cdots + a_{ee} v_e, \end{aligned}$$

[1] See also Exercise 2.178.

where a_{ij} are homogeneous elements of R of degree $\deg(a_{ij}) = \deg(v_i) - \deg(v_j) + 1$. The determinant of the matrix

$$\begin{bmatrix} a_{11} - u_1 & a_{12} & \cdots & a_{1e} \\ a_{21} & a_{22} - u_2 & \cdots & a_{2e} \\ \vdots & \vdots & \ddots & \vdots \\ a_{e1} & a_{e2} & \cdots & a_{ee} - u_e \end{bmatrix}$$

is a form

$$(-1)^e u_1 u_2 \cdots u_e + b, \quad b \in (x,y)A_{e-1},$$

that annihilates B, and therefore it is trivial in A. □

In section 2.7.3 we return to the discussion of the relationship between multiplicity and reduction numbers for certain higher-dimensional ideals.

1.6.2 Global Bounds of Reduction Numbers

Proposition 1.142 shows that $\text{bigr}(I)$ is finite for any ideal of a local ring. It is not clear how to use this to study the upper-semicontinuity of the $\text{bigr}(I_\mathfrak{p})$ on spec (R). We are going to settle the case of a ring of polynomials over an arbitrary field. Because the approach to a bound for $\text{bigr}(I)$ is very constructive, we use the theory of complexity of Gröbner bases as developed in [BM91], [Gi84] and [MM84].

Let $R = k[x_1, \ldots, x_n]$ be a ring of polynomials over the field k, and I an ideal generated by a set of polynomials $\{f_1, \ldots, f_m\}$. We denote by $d = d(I)$ the maximal degree of the f_i. Of course $d(I)$ will depend on the generating set. We make the following assumption.

Condition C_1 For each field k there exists a numerical function $f(n,d)$ such that for any term ordering $>$ of the polynomial ring $R = k[x_1, \ldots, x_n]$ and any ideal I with a set of generators of degree at most $d(I)$, the degree of a Gröbner basis $G_>(I)$ satisfies the inequality

$$d(G_>(I)) \leq f(n, d(I)). \tag{1.17}$$

These function play the following role in bounding reduction numbers ([Va99]):

Proposition 1.144. *Let $f(n,d)$ be a function satisfying Condition C_1. Then for any ideal $I = (f_1, \ldots, f_m)$ of the polynomial ring $R = k[x_1, \ldots, x_n]$,*

$$\text{bigr}(I) \leq \begin{cases} f(m, f(m+n+1, d(I)+1)) & \text{if } k \text{ has characteristic zero} \\ m(f(m, f(m+n+1, d(I)+1)) - 1) & \text{otherwise} \end{cases}$$

Proof. Let J be a reduction of I. It is clear that we may assume that k is algebraically closed, as $\text{bigr}(I)$ can only increase when we extend the base field. It will be enough to show that the bound described above holds at each localization at maximal ideals.

Consider first the defining ideal of the Rees algebra of I:

$$R[T_1,\ldots,T_m] \longrightarrow R[f_1t,\ldots,f_mt], \quad T_i \to f_it.$$

The kernel L of this homomorphism can be obtained as

$$L = R[T_1,\ldots,T_m]\bigcap(T_1 - f_1t,\ldots,T_m - f_mt).$$

L is obtained by elimination of the variable t, which can be realized via a computation of the Gröbner basis of the ideal

$$(T_1 - f_1t,\ldots,T_m - f_mt) \subset k[x_1,\ldots,x_n,T_1,\ldots,T_m,t].$$

According to C_1, we have

$$d(L) \leq f(m+n+1,d(I)+1)).$$

We select a set of generators of L of degree bounded by $d(L)$. Let $\mathfrak{m}$ be a maximal ideal of R. By the Nullstellensatz, $\mathfrak{m}$ is generated by linear forms in the x_i and thus the special fiber of I at $\mathfrak{m}$ is given by

$$\mathcal{F}(I) = R[It] \otimes (R/\mathfrak{m}) = k[T_1,\ldots,T_m]/L(\mathfrak{m}),$$

where $L(\mathfrak{m})$ is the evaluation of L at $\mathfrak{m}$ and is therefore generated by elements of degree at most $d(L)$.

Consider now the image K of J in the component of degree 1 of $\mathcal{F}(I)$. We must show that $K \cdot \mathcal{F}(I)_n = \mathcal{F}(I)_{n+1}$ for n greater or equal to the asserted bounds for $\operatorname{bigr}(I)$. Since k is infinite, we can find a set $\{z_1,\ldots,z_r\}$ of elements of K, where r is the analytic spread of I at the prime $\mathfrak{m}$, such that they span a Noether normalization of $\mathcal{F}(I)$. Note that the construction of these elements only involves taking linear combinations of the images of the f_i with coefficients in k. This means that we can change coordinates without increasing the degrees of the generators of the ideal of relations that corresponds to $L(\mathfrak{m})$. In other words, we may assume that we have a Noether normalization (after relabeling)

$$k[z_1,\ldots,z_r] \cong k[T_1,\ldots,T_r] \hookrightarrow \mathcal{F}(I) = k[T_1,\ldots,T_m]/L(\mathfrak{m}).$$

At this point we can proceed as in Proposition 1.142, by bounding the reduction of the algebra in terms of its arithmetic degree–once we use [BM91] to bound it with degree data–or move more crudely but more directly as follows.

Using Gröbner bases computations, we can extract for each T_i, $i > r$, a monic polynomial g_i in $L \cap k[T_1,\ldots,T_r,T_i]$ of degree at most $f(m,d(L(\mathfrak{m}))$. By simplicity we may assume that they all have this bound for their degrees. Now note that $g_r,g_{r+1},\ldots,g_m$ is a regular sequence of forms in $k[T_1,\ldots,T_m]$, so that the following surjection of finite graded algebras over $k[T_1,\ldots,T_r]$,

$$k[T_1,\ldots,T_m]/(g_{r+1},\ldots,g_m) \longrightarrow \mathcal{F}(I),$$

allows us to bound the reduction of $\mathcal{F}(I)$ with respect to $k[T_1,\ldots,T_r]$ by the reduction number of the complete intersection. This leads to the case of characteristic not zero. In characteristic zero we can proceed as in [Va96] and get the sharper bound. □

Remark 1.145. Considering that the functions $f(n,d)$ discussed in [BM91], [Gi84] and [MM84] are doubly exponential, the bound described above for $\text{bigr}(I)$ will have 4 levels of exponentiation!

The finiteness of $\text{bigr}(I)$ can now be shown ([Va99]). It is valid for global rings, not just for local rings.

Theorem 1.146. *Let I be an ideal of the Noetherian ring R. Then* $\text{bigr}(I)$ *is finite.*

Proof. First we pass from R to $R(X)$, the localization of the ring of polynomials $R[X]$ at the multiplicative set of all polynomials $f(X)$ whose coefficients generate the unit ideal of R; this ensures that all the residue fields of $R(X)$ are infinite. Note that $\text{bigr}(I) \leq \text{bigr}(IR(X))$.

To reduce to the previous proofs it suffices to start from a presentation of the Rees algebra of I. Suppose that $R[It] = R[T_1,\ldots,T_m]/L$, and denote by $d(L)$ the maximal degree in the T_i-variables of a set of generators of L. Then $\text{bigr}(I) \leq m(f(m,d(L)) - 1)$. □

Big Reduction Number and Regularity

When $(R,\mathfrak{m})$ is a local Noetherian ring with infinite residue field, it is straightforward to prove that $\text{bigr}(I) < \infty$:

Theorem 1.147. *Let $\mathcal{F}(I)$ be the special fiber cone of the ideal I. Then*

$$\text{bigr}(I) \leq \text{reg}(\mathcal{F}(I)).$$

Proof. Let $A = k[z_1,\ldots,z_\ell] \subset B = \mathcal{F}(I)$, $\ell = \ell(I)$, be a Noether normalization generated by 1-forms. The Castelnuovo-Mumford regularity of B can be expressed in several equivalent manners; intrinsically through the local cohomology of B, but also via the degree of any projective resolution of B as a module over any of its Noether normalizations (see Chapter 3, but especially [EG84]).

In our case here, for a minimal A-resolution of $B = \sum Ab_i$, the regularity $\text{reg}(B)$ is the supremum of several terms that include $\max\{\deg(b_i)\}$. □

Remark 1.148. Results of this kind (see also [Tr87]) lead to the notion of the *core of the ideal I*: $\text{core}(I) = \bigcap J$, for all reductions J of I. The existence of $\text{bigr}(I)$ shows that $\sqrt{\text{core}(I)} = \sqrt{I}$. For details on the structure of some of these ideals, see [CPU1], [HS95], [RS88]. We thus have another ideal in a chain

$$\text{core}(I) \longrightarrow I \longrightarrow \overline{I} \longrightarrow \sqrt{I},$$

of ideals with the same radical and it becomes pertinent to ask about their relationships.

Reduction Numbers of Global Rings

Let A be an affine algebra over an infinite field k. It might be of interest to consider the reduction numbers of the maximal ideals of the localizations $A_{\mathfrak{p}}$ as the means to gauge the singularities of A, given that $\mathrm{r}(\mathfrak{p}A_{\mathfrak{p}}) = 0$ if and only if $A_{\mathfrak{p}}$ is a regular local ring. Before this endeavor can be fully realized, it would be interesting to have positive answers to the following general questions:

(a) Is there a finite bound for

$$\sup\{\mathrm{r}(\mathfrak{p}A_{\mathfrak{p}}) \mid \mathfrak{p} \in \mathrm{Spec}\,(A)\}?$$

If so, what is its nature?

(b) For prime ideals $\mathfrak{p} \subset \mathfrak{q}$, does the inequality

$$\mathrm{r}(\mathfrak{p}A_{\mathfrak{p}}) \leq \mathrm{r}(\mathfrak{q}A_{\mathfrak{q}})$$

always hold?

1.7 Intertwining Algebras

Let $\varphi : A \mapsto B$ be a homogeneous homomorphism of graded algebras. We are going to create an encoding of the morphism into another algebra whose purpose is to unscramble the properties of φ into ideal-theoretic data. Although our use of it will be for the study of reductions, it will have other applications that will not be discussed here. The general treatment is given in [SUV1].

Let $(R,\mathfrak{m})$ be a local ring of dimension $d > 0$ and let J and I be two ideals of R with $J \subset I$. We have seen how the property that J is a reduction of I can be expressed by the property of a pair of algebras: "$R[It]$ is integral over $R[Jt]$." We shall now discuss how this relationship can also be found in the properties of a single algebra.

Intertwining Rees Algebra

Let R be a Noetherian ring and let A and B be homogeneous algebras with $A \subset B$. In the Introduction we defined the *intertwining algebra* $\mathcal{R}(A,B)$ *of* A *and* B ([SUV1]). It is constructed as follows. Let $B = \bigoplus_{n\geq 0} B_n$ be a graded R-algebra, which we see as $\sum_{n\geq 0} B_n t^n$, where t is an indeterminate. If u is another indeterminate, then

$$C = \mathcal{R}(A,B) = B[\sum_{n\geq 0} A_n u^n], \tag{1.18}$$

is the *intertwining algebra of* A *and* B. Of interest to us is the companion *intertwining module*

$$T_{B/A} = C(\sum_{n\geq 1} B_n t^n)/C(\sum_{n\geq 1} A_n t^n). \tag{1.19}$$

In the case of standard graded algebras $A \subset B$, $\mathcal{R}(A,B)$ is simply the ordinary Rees algebra of the ideal A_1B of the ring B, that is $\mathcal{R}(A,B) = \mathcal{R}_B(A_1B)$. When A and B are the Rees algebras of a pair of ideals J and I, with $J \subset I$, $\mathcal{R}(A,B)$ will be the Rees algebra of the module $J \oplus I$. We denote the corresponding intertwining module by $T_{B/A}$ (or even $T_{E/F}$ when A and B are the Rees algebras of a pair of modules $F \subset E$).

We begin our discussion by reading the dimension of this module.

Proposition 1.149. *Let R be a Noetherian ring of finite Krull dimension and let A and B, with $A \subset B$, be standard graded algebras such that B is a torsionfree R-module and* height ann $(B_1/A_1) > 0$. *If A is a reduction of B then* $\dim T_{B/A} \leq \dim B - 1$.

Proof. From the hypothesis on B, the ideal A_1B has positive height and therefore its Rees algebra $C = \mathcal{R}(A,B)$ has dimension $\dim C = \dim B + 1$.

Note that $L = \text{ann } T_{B/A}$ contains $(A_1t, \text{ann } (B_1/A_1))$, so that in general $\dim T_{B/A} \leq \dim C - 1 = \dim B$. When A is a reduction of B, we have $B_1t \subset \sqrt{A_1B})$. In this case $\sqrt{L} \supset (B_1t, \text{ann } (B_1/A_1))$ and therefore $\dim C/\sqrt{L} = \dim D/\text{ann } (B_1/A_1)D$, where $D = C/(B_1t)$ is isomorphic to A, and our assertion follows. □

Test Curves and Integral Closure

If $(R, \mathfrak{m})$ is a Cohen-Macaulay local ring, a theorem of [Re61] gives that two $\mathfrak{m}$-primary ideals have the same integral closure exactly when they have the same multiplicity. The following test formulation of integral closure is useful ([Sc69]).

Proposition 1.150. *Let $(R, \mathfrak{m})$ be a Cohen-Macaulay local ring of dimension $d \geq 1$, and let I and L be two $\mathfrak{m}$-primary ideals with $I \subset L$. Then $e(I) = e(L)$ if and only if for every prime ideal $\mathfrak{p}$ of dimension* 1 *we have the equality $e(IR/\mathfrak{p}) = e(LR/\mathfrak{p})$.*

Proof. If $d = 1$ the assertion follows from the additivity formula for multiplicities.

Suppose that $d \geq 2$. We have only to show that if $L = (I, z)$, with $z\mathfrak{m} \subset I$, and the conditions on the multiplicities in the rings $R/\mathfrak{p}$ are satisfied, then $e(I) = e(L)$. Let $f_1, \ldots, f_d$ be a multiplicity system for the ideal L. Each f_i can be written $f_i = g_i + a_iz$, with $g_i \in I$. For those $a_i \in \mathfrak{m}$, the f_i lie in I. Note that if a_1 and a_2 are units for two of the f_i, say f_1 and f_2, then

$$(f_1, f_2) = (f_1, g_2 - g_1), \quad f_2 = (g_2 - g_1) + a_2a_1{}^{-1}f_1.$$

This means that we can arrange things so that the first $d-1$ elements in the multiplicity system lie in I. Changing the notation, if needed, we may assume that $f_i \in I$ for $i < d$. Setting $J = (f_1, \ldots, f_{d-1})$, we have

$$\begin{aligned} e(L) &= e(L/J) = \lambda(R/(J, f_d)), \\ e(I) &\leq e(I/J), \end{aligned}$$

the last inequality because in a Cohen-Macaulay local ring, the multiplicity of an $\mathfrak{m}$-primary ideal I is the minimum of the lengths of $R/(g_1,\ldots,g_d)$, for all the systems of parameters contained in I.

We can now complete the proof. By assumption $e(I/J) = e(L/J)$, which forces the inequality $e(L) \geq e(I)$. Since the reverse inequality always holds, we have $e(I) = e(L)$, as desired. □

Let us record one case when we have more precise information about dimensions.

Corollary 1.151. *Let $(R,\mathfrak{m})$ be a Noetherian local ring of dimension $d > 0$ and let F and E be torsionfree R-modules of rank r with $F \subset E$. If F is a reduction of E, then* $\dim T \leq d+r-1$.

We identify a broad context for the converse to hold. Quite generally one has:

Proposition 1.152. *Let R be a Noetherian local quasi-unmixed ring, and let J and I be R-ideals with $J \subset I$, and write $B = \mathcal{R}(I)$. If* height $JB :_B IB \geq 2$ *then J is a reduction of I.*

Proof. We first reduce to the case where R is a complete local domain. Indeed, we may localize and complete to assume that R is a complete local equidimensional ring. We may also replace R by $R/\sqrt{0}$ since equations of integral dependence lift modulo nilpotent ideals. Next, we may adjoin a power series variable to R,J,I in order to assume that height $J > 0$, in which case B is equidimensional. Write $\mathfrak{p}_1,\ldots,\mathfrak{p}_r$ for the minimal primes of R, and $R_i = R/\mathfrak{p}_i$. Since B is equidimensional, catenary, and positively graded over a local ring, our height assumption is preserved as we pass from R to R_i. On the other hand, because $\mathcal{R}(I)$ is a subring of $\prod_{i=1}^r \mathcal{R}_{R_i}(IR_i)$, it suffices to prove that IR_i is integral over JR_i for every i. Thus we may assume that R is a complete local domain.

But then the integral closure $\overline{B}$ of B is a Noetherian catenary normal domain, positively graded over the local ring $\overline{R}$ and contained in $\overline{R}[t]$. Thus we have

$$2 \leq \text{height } JB :_B IB \leq \text{height } J\overline{B} :_{\overline{B}} I\overline{B} = \text{height } Jt\overline{B} :_{\overline{B}} It\overline{B} \leq \text{height } \overline{B}_+ :_{\overline{B}} It\overline{B}.$$

Therefore $\overline{B}_+ :_{\overline{B}} It\overline{B} = \overline{B}$ since $\overline{B}$ is a normal domain, $\overline{B}_+$ is a divisorial $\overline{B}$-ideal, and $It\overline{B}$ is a fractional $\overline{B}$-ideal. It follows that $It \subset \overline{B}$, thus proving the integrality of I over J. □

Theorem 1.153. *Let R be a quasi-unmixed local ring and let A and B be standard graded algebras with $A \subset B$ and such that B is a torsionfree R-module with* height ann $(B_1/A_1) > 0$, *and write $C = \mathcal{R}(A,B)$, $T_{B/A} = (A_1C)/(B_1C)$, and $G = \mathrm{gr}_{A_1B}(B)$. The following are equivalent:*

(i) $\dim T_{B/A} \leq \dim B - 1$;
(ii) height $(A_1C) :_C (B_1C) \geq 2$;
(iii) height $0 :_G B_1G > 0$;
(iv) *A is a reduction of B.*

Proof. We may assume $\dim B > 0$. Notice that height $A_1B > 0$, $\dim C = \dim B + 1$ and C is an equidimensional catenary positively graded algebra over a local ring. Now the equivalence of (i), (ii), (iii) is clear, (iv) implies (i) by Proposition 1.149, and (ii) implies (iv) according to Proposition 1.152 applied to the B-ideals $J = A_1B \subset B_1B = I$. □

The classical case of this result is the theorem of [Re61]. Let I and J be $\mathfrak{m}$-primary ideals and consider the exact sequence

$$0 \to T_n = I^n/J^n \longrightarrow R/J^n \longrightarrow R/I^n \to 0,$$

which gives

$$\lambda(I^n/J^n) = \lambda(R/J^n) - \lambda(R/I^n).$$

Note however that $\lambda(I^n/J^n) = \lambda(T_n)$. It follows therefore that the Hilbert polynomial of T_n has for coefficient exactly $\frac{e(J)-e(I)}{d!}$ in degree d. In particular if J is a reduction of I, since $\dim T \leq d$ we get $e(J) = e(I)$, and conversely. Thus we have:

Theorem 1.154 (Rees Theorem). *Let $(R, \mathfrak{m})$ be a quasi-unmixed local ring and J and I two $\mathfrak{m}$-primary R-ideals with $J \subset I$; then J is a reduction of I if and only if $e(J) = e(I)$.*

This was later extended to equimultiple ideals by Böger ([Bo69]). In another direction is the result of [Re85]. Suppose that I and J are arbitrary ideals such that I/J has finite length. In this case the converse of Proposition 1.151 also holds. Generalizations in this direction, that is, when the module I/J is not of finite length, will arise once we develop notions of multiplicity that are capable of reading the dimension of the module $T_{I/J}$. A treatment of this, partly based in [SUV1], will be given in Chapter 8 for the more general context of modules.

Let us record here a general property of the integral closures of Rees algebras. We state it in the case of bialgebras for simplicity.

Proposition 1.155. *Let R be an integral domain and let I, J be two ideals. The integral closure of the Rees algebra*

$$\mathcal{R} = \sum_{m,n \geq 0} I^m J^n x^m y^n$$

of $I \oplus J$ is

$$\overline{\mathcal{R}} = \sum_{m,n \geq 0} \overline{I^m J^n} x^m y^n,$$

where $\overline{I^m J^n}$ is the integral closure of $I^m J^n$.

Proof. We may assume that I and J are finitely generated. For each valuation overring V of R, we have

$$V \cdot \mathcal{R} = V[IVx, JVy] = V[ax, by], \quad a \in I, b \in J,$$

which is an integrally closed domain. It follows that

$$\overline{\mathcal{R}} \subset \bigcap_V [Ix, Jy] = \sum_{m,n>0} (\bigcap_V I^m J^n V) x^m y^n$$
$$= \sum_{m,n\geq 0} \overline{I^m J^n} x^m y^n.$$

The reverse inclusion is clear. □

Example 1.156. Let G be a graph with vertex set $\{x_1, \ldots, x_n\}$. If G is bipartite, its edge ideal I (generated by the monomials $x_i x_j$ where (x_i, x_j) is an edge of G) is normal ([SVV94]). Consequently, by the previous proposition, the Rees algebra $\mathcal{R}$ of $I \oplus I$ (or of the direct sum of any number of copies of I by the similar statement) is normal, and therefore Cohen-Macaulay by Hochster's Theorem ([Ho72]) since it is a monomial ring.

The following is another version of Proposition 1.152, but with an entirely different technique of proof.

Proposition 1.157. *Let $(R, \mathfrak{m})$ be a quasi-unmixed local ring of dimension $d > 0$ and let J and I be ideals of positive height with $J \subset I$. Let $A = R[Iu, Jv]$ be the algebra constructed above. Then J is a reduction of I if and only if the ideal $JA :_A IA$ has height at least* 2.

Proof. We have already seen that reductions leads to the asserted condition on the conductor ideal. Indeed, since A is also quasi-unmixed the ideal (J, Iu) must have height 2. To prove the converse we shall make changes of setting, some of them allowed by the assumption that R is quasi-unmixed.

We shall first assume that R is an integral domain by modding out by a minimal prime of R (which will preserve all the assumptions). Passing to the integral closure of R does not change any of the conditions either. We finally pass to the integral closure B of A. Since A is quasi-unmixed, B is finite over A. By the going-up theorem, the extension of the conductor ideal $LB = (JuA :_A Iu)B$ has height at least 2. We note that B has the following description

$$B = \sum_{i,j\geq 0} \overline{I^i J^j} u^i v^j,$$

where $(\bar{\cdot})$ denotes the integral closure of ideals function.

Consider the ideal

$$JuB = \sum_{i,j\geq 0} J\overline{I^i J^j} u^{i+1} v^j \subset \sum_{i,j\geq 0} \overline{I^i J^{j+1}} u^{i+1} v^j = Q.$$

Note that

$$Q = u(\sum_{i,j\geq 0} \overline{I^i J^j} u^i v^j)$$

$$
\begin{aligned}
&\cong v(\sum_{i,j\geq 0} \overline{I^iJ^j}u^iv^j) \\
&= \sum_{i,j\geq 0} \overline{I^iJ^j}u^iv^{j+1} = P,
\end{aligned}
$$

where $B/P \cong \sum_{i\geq 0} \overline{I^i}u^i$. This means that P is a prime ideal of B of height 1 and that Q is isomorphic to P. Since B is integrally closed, P, and Q along with it, is a module with the property S_2 of Serre. We now return to the conductor ideal L. Since

$$L\cdot Iu \subset JuB,$$

we get

$$L\cdot Iu \subset JuB \subset Q.$$

Observe that Iu consists of elements of B that are conducted into Q by an ideal of height at least 2. Since Q has S_2, $Iu \subset Q$, and in particular $I \subset \overline{J}$, as desired.

The reduction to non-domains is straightforward. For each minimal prime $\mathfrak{p}$ of R we have an equation

$$I^{r+1} + \mathfrak{p} = JI^r + \mathfrak{p},$$

from which we obtain another equality,

$$I^{r+1} = JI^r + D, \quad D \subset I^{r+1} \cap \mathfrak{p}.$$

We now can choose r large enough so that all these equalities hold for every minimal prime of R. Each equality can be raised to a power n equal to the index of nilpotency of the nil radical of R, so that we have equalities where $D \subset \mathfrak{p}^n \cap I^{n(r+1)}$. Thus if $\mathfrak{p}_1, \ldots, \mathfrak{p}_s$ are the minimal primes of R, we have (set $p = n(r+1)$)

$$I^{p+1} = JI^p + A_i, \quad A_i \subset \mathfrak{p}_i{}^n,\ i = 1, \ldots, s.$$

Finally, we multiply them together to get an equation of the form

$$I^{s(p+1)} = JI^{s(p+1)-1},$$

since $A_1 \cdot A_2 \cdots A_s = 0$. □

Corollary 1.158. *Let $(R, \mathfrak{m})$ be a quasi-unmixed local ring of dimension $d > 0$ and let J and I be ideals of positive height such that $J \subset I$ and $\lambda(I/J) < \infty$. Then for $n \gg 0$, $\lambda(I^n/J^n)$ is a polynomial function $f(n)$ of degree at most d. Furthermore, J is a reduction of I if and only if the degree of $f(n)$ is less than d.*

Proof. As observed above in the primary case, we have that $\lambda(I^n/J^n) = \lambda((IuA/JuA)_n)$, so the assertion follows from the previous theorem. □

Let C be the subalgebra of $R(E)[t]$ generated by the 1-forms in E and Ft. This is the intertwining algebra we defined earlier; it could be denoted $\mathcal{R}(F,E)$. Note that C is a subalgebra of the Rees algebra of $F \oplus E$ and coincides with it when F and E are ideals. The intertwining module defined above will be noted by $T_{E/F}$. In this case one has $\dim C = \dim R + e + 1$ and $\dim T_{E/F} \leq \dim R + e$.

Reduction and Multiplicity of the Fiber

Let $(R,\mathfrak{m})$ be a local ring of dimension $d>0$, I an ideal, and let N be an $\mathfrak{m}$-primary ideal. One can extend the notion of special fiber to include the following cases. Let $I\subset N\subset\mathfrak{m}$ and consider the algebra $G=\mathrm{gr}_I(R)$ and

$$G(I;N)=G\otimes R/N.$$

Definition 1.159. Let $(R,\mathfrak{m})$ be a local ring, $A=\bigoplus_{i\geq 0}A_i$ a finitely generated graded R-algebra and N a $\mathfrak{m}$-primary ideal. The *special fiber* of A with respect to N is the ring $A\otimes_R(R/N)$.

If I is $\mathfrak{m}$-primary, the Hilbert polynomial of $G(I;N)$ can be compared to $G=\mathrm{gr}_I(R)$ and the inequality $e_0(I)\geq f_0(I;N)$ will hold, where $f_0(I;N)$ denotes the multiplicity of $G(I;N)$. One of our purposes is to make comparisons between these multiplicities ([Va99]).

Theorem 1.160. *Let $(R,\mathfrak{m})$ be a quasi-unmixed local ring of dimension $d>0$ and let I and N be $\mathfrak{m}$-primary ideals with $I\subset N$. If $e_0(I)=f_0(I;N)$ then I is a reduction of N.*

Proof. A first point we would like to clarify is the meaning of the equality $e_0(I)=f_0(I;N)$ in general. The exact sequence of G-modules

$$0\rightarrow NG\longrightarrow G\longrightarrow G(I;N)\rightarrow 0,$$

and the additivity of Hilbert polynomials imply that the Hilbert polynomial of NG has degree less than $d=\dim R$ if $e_0(I)=f_0(I;N)$. This means that the annihilator

$$H=0:_G NG$$

is an ideal of dimension less than d. Since G is equidimensional this implies that $\mathrm{height}(H)>0$. Lifting H to L in the Rees algebra $\mathcal{R}=R[It]$ of I we have that the quotient ideal

$$L=N\mathcal{R}:_{\mathcal{R}} I\mathcal{R}$$

has height at least 2. We shall write this relation as

$$L\cdot N\mathcal{R}\subset I\mathcal{R}. \tag{1.20}$$

We first consider the case in which R is an integral domain. Let B be the integral closure of $\mathcal{R}$,

$$B=S_0+S_1t+S_2t^2+\cdots,$$

where B_n is the integral closure of I^n in the ring $B_0=R'$, the integral closure of R. We note that $\overline{I^n}=B_n\cap R$.

From the equality (1.20), we have

$$LB \cdot NB \subset IB.$$

We make two observations about this relationship. First, by the going-up theorem, LB is an ideal of height at least 2. Second, we have that

$$ItS = \sum_{n\geq 0} ItB_n t^n \subset \sum_{n\geq 1} B_n t^n = Q.$$

Note that Q is a prime ideal of height 1. We rewrite (1.20) as

$$LB \cdot Nt \subset Q,$$

where Nt consists of elements of the field of fractions of B which are conducted into the prime ideal Q by another ideal of height at least 2. Since B is integrally closed, we have $Nt \subset Q$. In particular $N \subset \overline{I}$, as desired.

The extension from the domain case to the general case is straightforward, as above. □

1.8 Briançon-Skoda Bounds

We have encountered a wide range of behavior between an ideal and its minimal reductions, and we shall see more of that in the next chapter when the multiplicities of the ideal are considered. They all tend to compare I^n with $\overline{I^n}$. This is a natural phenomenon given that (R assumed quasi-unmixed) the algebra $B = \sum_{n\geq 0} \overline{I^n} t^n$ is finite over the Rees algebra A of I,

$$B = \sum_{j=1}^{m} A u_j t^{d_j}.$$

In particular, $\overline{I^{n+s}} \subset I^n$, $\forall n$ provided that $s > \sup\{d_j\}$.

Consider the following simple example of the maximal ideal $\mathfrak{m}$ of the hypersurface ring $\mathbb{Q}[[x,y,z]]/(f)$, where $f = x^n + y^n + z^n$. Then $\mathfrak{m}$ is a normal ideal and $\mathfrak{m}^n$ is the smallest power of $\mathfrak{m}$ contained in its minimal reduction (x,y). It is a remarkable phenomenon of regular local rings, discovered by Briançon and Skoda ([BS74]), and further developed and extended by Lipman and Sathaye ([LS81]), that for such rings some uniform bounds are achieved.

Theorem 1.161 (Lipman-Sathaye). *Let R be a regular local ring containing a field and let I be an ideal of R generated by ℓ elements. Then $\overline{I^{n+\ell}} \subset I^{n+1}$ for all positive integers n.*

If $\dim R = d$ and R has an infinite residue field, the theorem implies that for any ideal I with a minimal reduction J (necessarily generated by at most d elements), one has

$$I^{n+d} \subset \overline{I^{n+d}} = \overline{J^{n+d}} \subset J^{n+1},$$

for every positive integer n.

A more precise statement that seeks to take into account the coefficients that occur in the containment relations is ([AH93, Theorem 3.3]):

Theorem 1.162. *Let $(R,\mathfrak{m})$ be a regular local ring containing a field. Let $I \subseteq R$ be any ideal with analytic spread ℓ, and let J be any reduction of I. Set $h = \text{big height}(I)$. Then for all $n \geq 0$,*

$$\overline{I^{n+\ell}} \subseteq J^{n+\ell}(J^{\ell-h})^{unm},$$

where ^{unm} is the unmixed part operation.

Applying this to a prime $\mathfrak{p}$ with $\dim R/\mathfrak{p} = 1$, and the analytic spread equal to the dimension of R gives that

$$\overline{\mathfrak{p}^d} \subseteq J\mathfrak{p}.$$

Definition 1.163. Let R be a local ring with an infinite residue field. The *Briançon-Skoda number* of the ideal I is the smallest integer $s = c(I)$ such that $\overline{I^{n+s}} \subset I^n$ for all $n \geq 0$. The *Briançon-Skoda number* of the ring R is the smallest integer $s = c(R)$ such that $\overline{I^{n+s}} \subset I^n$ for all $n \geq 0$ and all ideals.

According to Theorem 1.161, for a regular local ring R containing an infinite field, $c(R) \leq \dim R - 1$, or more precisely $c(I) \leq \ell(I) - 1$ for every ideal I. The existence of uniform values for these numbers is guaranteed in a very large number of cases. The following is an important instance ([Hu96, Theorem 11.12]).

Theorem 1.164. *Let S be a Noetherian reduced ring. If S satisfies at least one of the following conditions then there exists a positive integer k such that $\overline{I^n} \subset I^{n-k}$ for all ideals I of S.*

(i) *S is essentially of finite type over an excellent Noetherian local ring.*
(ii) *S is of characteristic p, and $S^{1/p}$ is module finite over S.*
(iii) *S is essentially of finite type over $\mathbb{Z}$.*

Note that the existence of a uniform Briançon-Skoda bound of b gives a very satisfactory solution to an issue we touched earlier (Remark 1.148): $I^b \subset \text{core}(I)$ for every ideal I.

Aberbach and Huneke ([AH96]) introduced the following gauge to measure the relationship between an ideal I and one of its reductions J:

Definition 1.165. The *coefficient ideal of a reduction* J of an ideal I is the largest ideal L such that $L \cdot I = L \cdot J$. It is denoted by $c(I,J)$.

We quote two relevant results, first [AH96, Theorem 2.7] and then [Li94b].

Theorem 1.166. *Let $(R,\mathfrak{m})$ be a regular local ring of dimension d containing a field. If I is $\mathfrak{m}$-primary and J is one of its minimal reductions, then*

$$\overline{I^{n+d-1}} \subset c(I,J)J^n, \quad n \geq 0.$$

Theorem 1.167. *Let $(R,\mathfrak{m})$ be a regular local ring essentially of finite type over a field of characteristic zero. For any ideal I and one of its minimal reduction J,*

$$\overline{I^{n+d-1}} \subset \mathrm{adj}(I^{\ell-1})J^n, \quad n \geq 0,$$

where $\ell = \ell(I)$ and $\mathrm{adj}(I^{\ell-1})$ is the adjoint ideal of $I^{\ell-1}$.

A detailed discussion, particularly when the Rees algebra of I is Cohen-Macaulay, is to be found in [Hy1].

Some special classes of ideals give rise to equalities with *coefficients* (see Theorem 7.58 and special results in Chapter 3). Let $R = k[x_1,\ldots,x_d]$ be a ring of polynomials over the field k, and let I be a monomial ideal. Then

$$\overline{I^n} = I \cdot \overline{I^{n-1}}, \quad n \geq \ell(I).$$

Remark 1.168. If R is not a regular ring, there may be other expressions of the kind

$$\overline{I^{f(n)}} \subset I^n,$$

for a linear function $f(n)$. For instance, if R is an affine domain over a field of characteristic zero and $S \subset R$ is a Noether normalization with $\mathrm{rank}_S(R) = r$, a candidate formula is

$$\overline{I^{r(n+d)}} \subset I^{n+1}, \; d = \dim R.$$

What makes this likely is the existence of a norm map $N : R \to S$ and some relationships between I and $(N(I))$ (see Chapter 9). This however would have less impact than bounding the Briançon-Skoda number by the inequality

$$c(R) \leq r \cdot d.$$

1.9 Exercises

Exercise 1.169. Let E be a finitely generated module over a Noetherian ring R, let $E \to I(E)$ be the embedding into its injective envelope and denote by A the image of the symmetric algebra of E into the symmetric algebra of $I(E)$. Show that this construction commutes with localization, and derive a formula for the Krull dimension of A.

Exercise 1.170. Let $(R,\mathfrak{m})$ be a Noetherian local ring. Prove that there is a system of parameters which is a d-sequence.

Exercise 1.171 (G. Valla). Let I be an ideal and let J be the presentation ideal of its Rees algebra. Show that $J = (J_1)$ if $(J,I) = (J_1,I)$.

Exercise 1.172. Let $I=(a_1,\ldots,a_n)$ be an ideal of the Noetherian ring R. If a_1 is a regular element, show that the equations of

$$\mathcal{R}(I)=R[T_1,\ldots,T_n]/L$$

are given by

$$L=\bigcup_{s\geq 1}(a_2T_1-a_1T_2,\ldots,a_nT_1-a_1T_n):a_1^s.$$

Exercise 1.173 (A. Micali). Let $(R,\mathfrak{m})$ be a Noetherian local ring. Show that $\mathfrak{m}$ is of linear type if and only if R is a regular local ring.

Exercise 1.174. Let R be a Noetherian ring and I an ideal generated by a d-sequence $x_1,\ldots,x_n$. If grade $I=r$, show that $x_1,\ldots,x_r$ is a regular sequence.

Exercise 1.175. Let R be a local Noetherian ring and I an ideal such that I^n/I^{n+1} is R/I-free for infinitely many n. Prove that $\ell(I)=$ height I.

Exercise 1.176 (C. Weibel). Let $(R,\mathfrak{m})$ be a Noetherian local ring with finite residue field. Show that every ideal I has a power I^q with a reduction generated by $\ell(I)$ elements.

Exercise 1.177. Let $(R,\mathfrak{m})$ be a Noetherian local ring with infinite residue field and I an ideal. If $\mathrm{r}(I)=s$, show that $\mathrm{r}(I^q)\leq\ell(I)$ for all $q\geq s$.

Exercise 1.178. Let $\mathfrak{p}$ be a prime ideal of a regular local ring R. Suppose that the ordinary and symbolic powers of $\mathfrak{p}$ coincide, that is, $\mathfrak{p}^{(n)}=\mathfrak{p}^n$ for $n\geq 1$. Prove that $\ell(\mathfrak{p})<\dim R$.

Exercise 1.179. Let R be an integrally closed domain and I an ideal. Then Proj $(R[It])$ is normal if and only if I^n is integrally closed for $n\gg 0$.

Exercise 1.180. Let $(R,\mathfrak{m})$ be a Noetherian local ring and let $J\subset I$ and $K\subset\mathfrak{m}$ be ideals. Prove that $\overline{J}=\overline{I}$ if $\overline{J+KI}=\overline{I}$.

Exercise 1.181. For some field k and sets X and Y of distinct indeterminates, let $I\subset k[X]$ and $J\subset k[Y]$ be integrally closed ideals. Prove that $(IJ)\subset k[X,Y]$ is integrally closed. (*Hint*: Consider the tensor product over k of the integral closures of the Rees algebras of the two ideals in their respective rings.)

Exercise 1.182 (K. Smith). Let $A=\bigoplus_{n\geq 0}A_n$ be a graded algebra with A_0 a reduced ring. Show that if the ideal $A_{\geq m}$ is generated by elements of degree m then it is complete.

Exercise 1.183. Devise an algorithmic approach to compute the integral closure of an ideal of a polynomial ring. (Prizes are given!)

Exercise 1.184. Let $R = k[x_1,\ldots,x_n]$ be a ring of polynomials over a field k and denote by $V_d = R^{(d)}$ the dth Veronese subring. Then V_d is generated by the monomials

$$\mathbf{x}^{\alpha} = x_1^{\alpha_1}\cdots x_n^{\alpha_n}, \quad \sum \alpha_i = d.$$

There are $\binom{n+d-1}{d}$ such monomials and for each we consider an indeterminate T_{α} and define a presentation

$$0 \to I \longrightarrow k[\mathbf{T}] \longrightarrow V_d \to 0.$$

Prove that I is generated by quadrics. Of interest is the determination of the analytic spread $\ell(I)$ and of the reduction number $\mathrm{r}(I)$ of the ideal I.

Exercise 1.185. Let R be a Noetherian local ring and I, J ideals of positive height. Prove that

$$\ell(IJ) \leq \ell(I) + \ell(J) - 1.$$

Exercise 1.186. Let I be an ideal generated by a regular sequence of n elements. Show that for every positive integer s,

$$\mathrm{r}(I^s) = \left\lceil \frac{(n-1)(s-1)}{s} \right\rceil.$$

Exercise 1.187. Let I be an ideal of a commutative ring R with a reduction $J = (a_1,\ldots,a_n)$, $I^{r+1} = JI^r$. If $r \geq n$ show that $(a_1^r,\ldots,a_n^r)$ is a reduction of I^r and the reduction number is at most n. (A treatment, with additional references, of the reduction numbers of powers of ideals can be found in [Hoa93].)

Exercise 1.188. Let R be a Noetherian local ring and I an ideal for which its symmetric and ordinary powers coincide up to order r, $S_i(I) \cong I^i$, $i \leq r$. Prove that every proper reduction J of I has reduction number $\mathrm{r}_J(I) \geq r$. In particular, for syzygetic ideals ($r = 2$), such as perfect Gorenstein ideals of codimension 3, $\mathrm{r}_J(I) \geq 2$ for any proper reduction.

Exercise 1.189. Let R be a Noetherian ring and J and I ideals with $J \subset I$. Show that J is a reduction of I if and only if there is a positive integer c such that $I^{c+n} \subset J^n$, $\forall n \geq 0$.

Exercise 1.190. Let R be a Noetherian ring and t an indeterminate. Show that $\overline{I \cdot R[t]} = \overline{I} \cdot R[t]$ for any ideal $I \subset R$.

Exercise 1.191. Let A be a standard graded algebra over an arbitrary field k. If $\dim A \leq 2$, show that $\mathrm{r}(A) < \text{arith-deg}(A)$.

Exercise 1.192. Let E be a finitely generated module such that $\mathrm{bigrank}(E) = r$, and let φ be an endomorphism of E. Does φ satisfy a monic equation over R of degree r?

Exercise 1.193 (D. Hanes). Let $(A,\mathfrak{m}) \to (B,\mathfrak{n})$ be a flat, local homomorphism of local rings. Is $\mathrm{r}(\mathfrak{m}) \leq \mathrm{r}(\mathfrak{n})$?

Exercise 1.194. Let R and S be Noetherian local rings such that S is integral over R. If I is an ideal of R prove that $\ell(I) = \ell(IS)$. If R is a quasi-unmixed local domain and S is the integral closure of R, prove that $\ell(I) = \ell(IS_{\mathfrak{n}})$ for every maximal ideal $\mathfrak{n}$ of S.

Exercise 1.195. Let E be a finitely generated module over the Noetherian ring R. Show that

$$\text{bigrank}(\text{Hom}_R(E,E)) \leq \nu(E) \cdot \text{bigrank}(E).$$

2

Hilbert Functions and Multiplicities

A standard graded algebra is a graded ring $G = \bigoplus_{n\geq 0} G_n$, finitely generated over G_0 by its elements of degree 1, $G = G_0[G_1]$. If G_0 is an Artinian local ring, the assignment $H_G(n) = \lambda(G_n)$, defines the *Hilbert function* of G. There is a great deal of interest in deriving such functions from some description of G (or more generally from mild extensions of these algebras). One of our themes here can be put rhetorically as: what does $H_G(\cdot)$ know about G? That is, which properties of G are coded into its Hilbert function? We are going to pursue several variations of this theme.

There is also interest in examining a broader class of graded algebras in which G_0 is not necessarily Artinian and we add finitely generated graded G-modules to our purview. This widening provides several additional challenges, particularly about the nature of the numerical functions extending the ordinary length function. Our aim in this chapter is to use these functions as a premier route of access to the study of G. More specifically, we seek to make comparisons between

$$\boxed{\text{numerical properties of } H_G(\cdot)} \longleftrightarrow \boxed{\text{arithmetical properties of } G}$$

via certain functions of the coefficients of $H_G(\cdot)$, the reduction number of G and the depth of G or more generally its cohomology. The function $H_G(\cdot)$ is characterized by a finite set of integers of specified size, while the properties of G one seeks to understand include its *depth, reduction number, tracking number* (discussed in Chapter 6). The second set of integers is independent of $H_G(\cdot)$, but in an uncanny way are often expressed by inequalities in terms functions of the numbers of the first series.

The Hilbert function in the ordinary case is coded by the Hilbert-Poincaré series

$$H_G(t) = \sum_{n\geq 0} H_G(n)t^n.$$

This is a rational function whose numerical elements get fully displayed in the representation

$$H_G(t) = \frac{h(t)}{(1-t)^d},$$

where $h(1) = \deg(G) \neq 0$ and $d = \dim G$ are respectively the *degree* or *multiplicity* of G and its dimension. The polynomial $h(t) = h_0 + h_1 t + \cdots + h_r t^r,\ h_r \neq 0$, and the expansion of $(1-t)^{-d}$ into a power series, lead to the polynomial

$$\sum_{i=0}^{r} h_i \binom{t+d-1-i}{d-1},$$

whose values for $n \geq r$ agree with those of the Hilbert function $H_G(n)$. This polynomial is the *Hilbert polynomial* of G , while r is known as the *postulation number* of G. It is convenient to write the Hilbert polynomial in the form

$$P_G(t) = e_0 \binom{t+d-1}{d-1} - e_1 \binom{t+d-2}{d-2} + \cdots + (-1)^{d-1} e_{d-1};$$

the integers $e_i = e_i(G)$ are the *Hilbert coefficients* of G. There are also iterated versions of these functions and we shall make use of the equality $H_G^1(n) = \sum_{i \leq n} H_G(i)$ together with the corresponding Hilbert series $H_G^1(t) = \dfrac{H_G(t)}{1-t}$ and Hilbert polynomial $P_G^1(t)$.

There is an obvious degradation of information from $H_G(t)$ to $P_G(t)$, particularly when data about the postulation number is not available. Some control over this is given by the difference between the Hilbert function of G and its Hilbert polynomial ([BH93]):

$$H_G(n) - P_G(n) = \sum_{i=0}^{d} (-1)^i \lambda(H^i_{G_+}(G)_n).$$

Despite these drawbacks, Hilbert polynomials are of interest for two major reasons: the rules used to derive them are much broader than those for the strict Hilbert functions, and the class of algebras we shall be dealing with are more tightly controlled by their Hilbert polynomials than is the case for general algebras. These observations will still apply to graded modules over G generated by finitely many elements of arbitrary degrees.

One tends to view the properties of an algebra G coded into its Hilbert function $H_G(t)$ as being *linear*. This is partly justified by the fact that there exist numerous cases of pairs of algebras G and G' with the same Hilbert functions but with different algebraic properties. A case in point is provided by Macaulay's Theorem identifying the Hilbert function of the graded rings $G = k[x_1, \ldots, x_n]/I$ and $G' = k[x_1, \ldots, x_n]/\mathrm{in}_>(I)$. There will be emphasis on obtaining comparisons of the form

$$H_G(n) \leq f(n), \quad n \geq 0,$$

where $f(t)$ is a polynomial of the same degree as $\deg P_G(t)$, defined in terms of some of the Hilbert coefficients $e_i(G)$. It will then be used to derive properties for the whole class of algebras with the same $H_G(t)$.

Let $(R, \mathfrak{m})$ be a Noetherian local ring and let I be an $\mathfrak{m}$-primary ideal. The Hilbert function of I is that of the associated graded ring

$$G = \mathrm{gr}_I(R) = \bigoplus_{n \geq 0} I^n/I^{n+1}.$$

It is a significant control of the blowup process of Spec (R) along the subvariety $V(I)$ since it codes some structure of its cohomology. A challenging problem consists in relating $H_G(t)$ directly to R and I, as $\mathrm{gr}_I(R)$ may fail to inherit some of the arithmetical properties of R.

The other kind of standard graded algebra we must examine are the special fibers of ideals, not necessarily $\mathfrak{m}$-primary. Such algebras carry the information about the reduction of I, which will be used in our treatment of Cohen-Macaulayness of $\mathcal{R}(I)$ in Chapter 3. Technically this chapter is concerned with the following general query: what information about the reduction number of an ideal I and the defining equations of its Rees algebra is carried by the various Hilbert functions attached to I?

A new development is the introduction of generalized multiplicities. It will serve several ends, to begin to deal with graded algebras over non-Artinian rings. More significant are the so-called *cohomological degrees*, a family of functions tailored to compute complexities of various kinds in commutative algebra. It coexists smoothly with Castelnuovo-Mumford regularity.

The specific topics are:

- Auxiliary structures in the form of reduction modules.
- Estimating reduction numbers from partial knowledge of the Hilbert functions.
- Studying the depth properties of $\mathrm{gr}_I(R)$ in certain critical cases.
- Extended and cohomological degrees.
- Non-classical Hilbert functions.
- Explicit formulas for multiplicities and reduction numbers.
- Bounding the number of generators of modules.

We emphasize again one of our general goals: to uncover and develop invariants of algebras which may enable the task of building and/or understanding the integral closure of algebraic structures. In this endeavor, Hilbert functions–classical or associated to extended degrees–play an important role, despite the fact that some of the most sought-after invariants (reduction numbers–studied in this chapter, and the tracking numbers introduced in Chapter 6) are independent of the Hilbert functions. Nevertheless, repeatedly, some overall relationships between them is captured in the form of inequalities.

2.1 Reduction Modules and Algebras

To set the notation for most of this chapter, we put $G = \bigoplus_{n\geq 0} G_n$ for a standard graded algebra over the Artinian graded ring G_0. If M is a finitely generated graded G-module, its Hilbert series

$$H_M(t) = \sum_n \lambda(M_n)t^n \in \mathbb{Z}[[t,t^{-1}]]$$

is a rational function

$$H_M(t) = \frac{h_M(t)}{(1-t)^d}, \quad h_M(t) \in \mathbb{Z}[t,t^{-1}],$$

where $h_M(t)$ is a Laurent polynomial. In almost all cases that we treat, M is non-negatively generated, or we force it become so by shifting degrees; thus $h_M(t) \in \mathbb{Z}[t]$. In this case, with $M \neq 0$, we may assume that $h_M(1) \neq 0$. This polynomial is then referred to as the *h-polynomial of M*.

It is very convenient to decompose $H_M(t)$ into partial fractions,

$$H_M(t) = q(t) + \sum_{i=1}^{d} \frac{a_{d-i}}{(1-t)^i},$$

where $q(t)$ is a polynomial of degree $\deg(h_M(t)) - d$, or 0 if $\deg(h_M(t)) < d$. The function coded by the coefficients of $\sum_{i=1}^d \dfrac{a_{d-i}}{(1-t)^i}$ is the *Hilbert polynomial of M*

$$P_M(n) = \sum_{i=1}^{d} a_{d-i}\binom{n+i-1}{i-1}.$$

Note that $H_M(n) = P_M(n)$, starting precisely at $n = \deg(h_M(t)) - d + 1$.

For $i = 0, \ldots, d-1$, the a_i are obtained from $h_M(t)$ as $a_i = (-1)^i \dfrac{h_M^{(i)}(1)}{i!}$. It is customary to set $e_i(M) = (-1)^i a_i$. Normalized in this way the $e_i(M)$ are called the *Hilbert coefficients of M*.

2.1.1 Structures Associated to Rees Algebras

Let R be a local ring, and I an ideal with a reduction J. To be able to use the known properties of the Rees algebra of J as a tool to obtain properties of the Rees algebra of I, it is necessary to build structures in which the two algebras intermingle.

We are going to mention three such structures, one introduced in [Va94a], another of an older vintage [VV78], and a process introduced in [RR78].

The Sally Module

Let R be a Noetherian ring and let $\{I_n, n \geq 0\}$ be a multiplicative, descending filtration of ideals. Suppose that $I_0 = R$ and that $I_{n+1} = I_1 I_n$ for $n \gg 0$. Finally suppose that $J \subset I_1$ is a reduction of I_1. This data defines the Rees algebra

$$\mathcal{R} = \sum_{n \geq 0} I_n t^n,$$

and leads to an exact sequence of finitely generated modules over $\mathcal{R}_0 = R[Jt]$, the Rees algebra of the ideal J:

$$0 \to I_1 \mathcal{R}_0 \longrightarrow \mathcal{R}_+[+1] \longrightarrow S = \bigoplus_{n \geq 1}^{\infty} I_{n+1}/I_1 J^n \to 0. \tag{2.1}$$

Definition 2.1. The *Sally module* of the filtration with respect to J is S viewed as an $R[Jt]$-module.

We are going to show how to extend the construction of the Sally module attached to an ideal I and one of its reductions J to the broader context of graded algebras.

Let R be a Noetherian ring and $B = \bigoplus_{n \geq 0} B_n$ a Noetherian graded algebra over R. Suppose that there is a standard graded subalgebra $A = \bigoplus_{n \geq 0} A_n \subset B$. We say that A is a *reduction* of B if B is finite over A. The exact sequence

$$B_1 \otimes_R A \longrightarrow B_+[+1] \longrightarrow S_{B/A} \to 0 \tag{2.2}$$

of finitely generated A-modules is the *reduction sequence of B relative to A*. Denote by $\mathrm{r}_A(B)$ the reduction number of B relative to A, that is, the smallest integer n_0 such that $B_{n+1} = A_1 B_n$, for $n \geq n_0$. This number also controls the degrees of the generators of $S_{B/A}$. Following [Vz97], one has a filtration of $S_{B/A}$ as follows If $S_{B/A} \neq 0$, then setting

$$\begin{aligned} C_1 &= S_{B/A} \\ L_1 &= A(C_1)_1 \end{aligned}$$

defines the exact sequence

$$0 \to L_1 \longrightarrow C_1 \longrightarrow C_2 \to 0.$$

If $C_2 \neq 0$, we set

$$L_2 = A(C_2)_2$$

and define C_3, and so on. The process stops at the stage $r = \mathrm{r}_A(B)$, when $C_r = 0$. We now define the A-modules

$$D_n = [B_{n+1}/A_1 B_n] \otimes_R A[-n]$$

and the exact sequences

$$0 \to K_n \longrightarrow D_n \longrightarrow L_n \to 0, \quad n = 1, \ldots, r-1.$$

They will be useful in deriving bounds for the multiplicity of special fibers of Rees algebras of ideals, and of the Rees algebras of modules in Chapter 8.

The cases of major interest are those of the I-adic filtration, when the Sally module will be denoted by $S_J(I)$. Most of the developments will occur in this case. Later, in Chapter 7, we shall study the construction for the integral closure filtration.

A motivation for this definition is the work of Sally, particularly in [Sal79], [Sal80], [Sal92], and [Sal93] in the case of $\mathfrak{m}$-primary ideals.

To be useful, this sequence requires information about $I \cdot R[Jt]$–which is readily available in many cases–and knowledge of the finer structural properties of $S_J(I)$. Its main feature is the relationship it bears with the ring $R[Jt]$, a ring that often is simpler to study than $R[It]$, such as in the case of equimultiple ideals.

Let us begin by pointing out a simple but critical property of $S_J(I)$.

Proposition 2.2. *Let $(R, \mathfrak{m})$ be a Cohen-Macaulay local ring of dimension d, I an $\mathfrak{m}$-primary ideal and J a minimal reduction of I. If $S_J(I) \neq 0$, then $S_J(I)$ only has $\mathfrak{m}R[Jt]$ for associated prime ideal, in particular its Krull dimension as an $R[Jt]$-module is d.*

Proof. We first argue that $I \cdot R[Jt]$ is a maximal Cohen-Macaulay $R[Jt]$-module. For this it suffices to consider the exact sequence

$$0 \to I \cdot R[Jt] \longrightarrow R[Jt] \longrightarrow R[Jt] \otimes_R (R/I) \to 0, \tag{2.3}$$

and observe that the module on the right is a polynomial ring in d variables over R/I, and therefore Cohen-Macaulay of dimension d. Since $R[Jt]$ is a Cohen-Macaulay ring, $I \cdot R[Jt]$ has depth $d+1$.

We may assume that $d \geq 1$, and that $S_J(I) \neq 0$. Let P, $P \subset A = R[Jt]$, be an associated prime of $S_J(I)$; it is clear that $\mathfrak{m} \subset P$. If $\mathfrak{m}A \neq P$, P has grade at least two.

Consider the homology sequence of the functor $\mathrm{Hom}_A(A/P, \cdot)$ on the sequence (2.1): we get

$$0 \to \mathrm{Hom}_A(A/P, S_J(I)) \longrightarrow \mathrm{Ext}^1_A(A/P, I \cdot R[Jt]).$$

Note that $I \cdot R[Jt]$ is a maximal Cohen-Macaulay module, hence the Ext module must vanish since P has grade two or larger. □

Proposition 2.3. *Let $(R, \mathfrak{m})$ be a Cohen-Macaulay local ring of dimension $d > 0$ and I an $\mathfrak{m}$-primary ideal. For any reduction J of I generated by a regular sequence, the Sally module $S_J(I)$ is Cohen-Macaulay if and only if $\mathrm{gr}_I(R)$ has depth at least $d-1$.*

Proof. This follows from reading depths in the defining exact sequence

$$0 \to I \cdot R[Jt] \longrightarrow I \cdot R[It] \longrightarrow S_J(I) \to 0,$$

of $S_J(I)$, where $I \cdot R[Jt]$ is a maximal Cohen-Macaulay module, together with the two exact sequences

$$0 \to I \cdot R[It] \longrightarrow R[It] \longrightarrow \mathrm{gr}_I(R) \to 0,$$

$$0 \to R[It]_+ \longrightarrow R[It] \longrightarrow R \to 0.$$

A more detailed comparison will be given in Theorem 3.25. □

Filtering the Sally Module

Let J be a minimal reduction of the ideal I. We are going to use the filtration introduced above of the Sally module $S = S_J(I)$ to capture elements of its structure.

Consider the $R[Jt]$-modules $L_1, \ldots, L_{r-1}$ defined above and for $n \geq 1$ set

$$C_n = \bigoplus_{i=n}^{\infty} I^{i+1}/J^{i-n+1}I^n.$$

Notice, in particular, that $C_1 = S$. Set L_n as the $R[Jt]$-submodule of C_n generated by the first component of C_n, $(C_n)_n = I^{n+1}/JI^n$:

$$L_n = R[Jt] \cdot I^{n+1}/JI^n = \bigoplus_{i=n}^{\infty} J^{i-n}I^{n+1}/J^{i-n+1}I^n.$$

It gives rise to a short exact sequence of $R[Jt]$-modules:

$$0 \to L_n \longrightarrow C_n \longrightarrow C_{n+1} \to 0.$$

If $r = r_J(I)$ is the reduction number of I with respect to J, then $I^{r+1} = JI^r$, $C_r = 0$, $L_r = 0$ and $L_{r-1} \cong C_{r-1}$. We obtain, therefore, the following set of exact sequences of $R[Jt]$-modules

$$\begin{array}{ccccccc} 0 \to & L_1 & \longrightarrow & S & \longrightarrow & C_2 & \to 0 \\ 0 \to & L_2 & \longrightarrow & C_2 & \longrightarrow & C_3 & \to 0 \\ & \vdots & & \vdots & & \vdots & \\ 0 \to & L_{r-2} & \longrightarrow & C_{r-2} & \longrightarrow & L_{r-1} & \to 0 \end{array} \tag{2.4}$$

We shall refer to the factors L_n as the *reduction modules of I with respect to J*, and to help recognize the structure of these factors, we shall introduce another series of $R[Jt]$-modules denoted by D_n, which we shall call the *virtual reduction modules of I with respect to J.*

The D_n are defined by extensions of R-modules:

$$D_n = I^{n+1}/JI^n[T_1, \ldots, T_d].$$

Note that D_n carries a natural structure of a $\mathrm{gr}_J(R) \otimes R/I \cong R/I[T_1, \ldots, T_d]$-module. They are finitely generated $R[Jt]$-modules, and if $D_n \neq 0$ they have depth equal to $\dim R$. We have an epimorphism θ_n: $D_n[-n] \to L_n$ of $R[Jt]$-modules, which is the identity on I^{n+1}/JI^n and sends each T_i to a generator of $R[Jt]$ (clearly, the generators of I^{n+1}/JI^n have degree zero in D_n, but degree n in L_n). If we define K_n as the kernel of θ_n, then we have the following exact sequence of B_n-modules:

$$0 \to K_n \longrightarrow D_n[-n] \xrightarrow{\theta_n} L_n \to 0. \tag{2.5}$$

Remark 2.4. The following inequalities of multiplicities result.

(a) Let s_0 be the multiplicity of $S_J(I)$ and set $r = \mathrm{r}_J(I)$. Then

$$s_0 \leq \sum_{n=1}^{r-1} \lambda(I^{n+1}/JI^n),$$

with equality if and only if $D_n = L_n$ for each n. In this case $S_J(I)$ has a filtration whose factors are the modules $I^n/JI^{n-1}[T_1,\ldots,T_d][-n+1]$. It follows that $S_J(I)$ is a Cohen-Macaulay module and its a-invariant is $r_J(I) - d - 1$.

(b) Another example of the use the Sally modules (see [Vz97]) is the following (for other approaches to this kind of problem, see [Gu95]). Let $(R,\mathfrak{m})$ be a Cohen-Macaulay local ring of dimension d and I an $\mathfrak{m}$-primary ideal with a minimal reduction J such that $I^3 = JI^2$. If the trivial submodule of I^2/JI is irreducible, then depth $\mathrm{gr}_I(R) \geq d-1$.

The following result of Huckaba ([Huc96]) explains–in the perspective of Proposition 2.3–precisely when the Sally module is Cohen-Macaulay. In view of the remarks above it yields several formulas of Marley ([Mar89]) for the Hilbert coefficients of I.

Theorem 2.5. *Let $(R,\mathfrak{m})$ be a Cohen-Macaulay local ring of dimension $d \geq 1$, I an $\mathfrak{m}$-primary ideal and let J be a reduction generated by a regular sequence. Then* depth $\mathrm{gr}_I(R) \geq d-1$ *if and only if $e_1(I) = \sum_{n=0}^{r-1} \lambda(I^{n+1}/JI^n)$ for $r = \mathrm{r}_J(I)$.*

Comparison: $e_0(I)$ versus $e_1(I)$

Let $(R,\mathfrak{m})$ be a Cohen-Macaulay local ring of dimension $d > 0$. Let I be an $\mathfrak{m}$-primary ideal and denote by $e_0(I)$ and $e_1(I)$ the first two coefficients of the Hilbert polynomial of I. The integers $e_0(I)$ and $e_1(I)$ are loosely related. When $I = \mathfrak{m}$ (see [KM82]), we have

$$e_0(\mathfrak{m}) - 1 \leq e_1(\mathfrak{m}) \leq \binom{e_0(\mathfrak{m}) - 1}{2}.$$

Using the standard change of rings techniques, the values of $e_0(I)$ and $e_1(I)$ are unaffected when we replace I by its image modulo an appropriate superficial sequence $x_1,\ldots,x_{d-1}$, turning the comparison $e_0(I) \leftrightarrow e_1(I)$ all the way to an ideal of a 1-dimensional ring.

Let R be a Cohen-Macaulay local ring of dimension one. If x is a superficial element in I, then for every non-negative integer, n

$$H_I(n) = e_0(I) - v_n,$$

where $v_n = \lambda(I^{n+1}/xI^n)$. In particular $v_0 = e_0(I) - \lambda(R/I)$, $v_1 = e_0(I) - \lambda(I/I^2)$ and if $v_n = 0$ for some n, then $v_j = 0$ for every $j \geq n$.

Proposition 2.6. *Let $(R,\mathfrak{m})$ be a Cohen-Macaulay local ring of dimension one, let I be an $\mathfrak{m}$-primary ideal. If $e_0(I) \neq e_0(\mathfrak{m})$ then*

$$e_1(I) \leq \binom{e_0(I)-2}{2}.$$

Proof. The condition $e_0(I) \neq e_0(\mathfrak{m})$ means, by the theorem of Rees (see [Re61]), that $\mathfrak{m}$ is not the integral closure of I. This implies that $I^{n+1} \neq \mathfrak{m}I^n$ for every positive integer n and therefore $\lambda(I^n/I^{n+1}) > \lambda(I^n/\mathfrak{m}I^n)$.

If x generates a minimal reduction of I, the Hilbert function of I can be written

$$H_I(n) = \lambda(I^n/I^{n+1}) = e_0(I) - \nu_n,$$

so it only reaches its stable value of $e_0(I)$ when $n = r$, the reduction number of I; that is, the smallest r for which $I^{r+1} = xI^r$. We claim that $\lambda(I^n/I^{n+1}) \geq n+2$ for all $n \leq r$. Indeed otherwise we would have $\lambda(I^n/\mathfrak{m}I^n) \leq n$, which by the main theorem of [ES76] would lead to an equality $I^n = xI^{n-1}$, contradicting the definition of r. This means that we have

$$\begin{aligned} e_1(I) &= \sum_{n=0}^{r} (e_0(I) - \lambda(I^n/I^{n+1})) \\ &\leq e_0(I) - \lambda(R/I) + \sum_{n=1}^{r} (e_0(I) - (n+2)) \\ &= e_0(I) - \lambda(R/I) + r(e_0(I)-2) - \binom{r+1}{2}. \end{aligned}$$

We recall the basic relationship between the 'universal' reduction number for the ideals of R and its multiplicity. For the $\mathfrak{m}$-primary ideal I^n,

$$\lambda(I^n/\mathfrak{m}I^n) \leq \lambda(I^n/zI^n) = e_0(\mathfrak{m}),$$

where (z) is a minimal reduction of $\mathfrak{m}$. We thus have

$$\nu(I^n) \leq \binom{n+1}{1}$$

for $n \geq e_0(\mathfrak{m})$. By Theorem 2.36, $\mathrm{r}(I) \leq e_0(\mathfrak{m}) - 1$. Since $e_0(\mathfrak{m}) \leq e_0(I)$, we have the desired inequality. □

Remark 2.7. The proof above also shows that in general, that is, without the assumption $\overline{I} \neq \mathfrak{m}$, one has $e_1(I) \leq \binom{e_0(I)-1}{2}$. A more refined examination is carried out in [RV3] (extending [E99]), proving that there are bounds of the kind

$$e_1(I) \leq \binom{e_0(I)}{2} - \binom{\nu(\tilde{I})}{2} - \lambda(R/I) + 1,$$

where $\tilde{I}$ is the Ratliff-Rush closure of I.

2.1.2 Bounding Hilbert Functions

We shall indicate the usefulness of the setting of reduction modules to establish inequalities on the coefficients of the Hilbert polynomial of a primary ideal.

Let $(R, \mathfrak{m})$ be a Noetherian local ring of Krull dimension d and I an $\mathfrak{m}$-primary ideal. For the *Hilbert function* of I we shall use the assignment

$$H_I : n \leadsto \lambda(R/I^n).$$

For $n \gg 0$ (but often not much greater than zero!) $H_I(n)$ is a polynomial $H(n)$ of degree d, the *Hilbert-Samuel polynomial of I*.

To study either of these functions, without loss of generality it is convenient to assume that the residue field of R is infinite. Here we limit ourselves mostly to Hilbert polynomials, leaving aside the rich area of *irregularities*, to wit the detailed comparison between the two functions.

We shall further assume that R is a Cohen-Macaulay ring. In such cases, if I is generated by a system of parameters, the functions are given simply by

$$H_I(n) = \lambda(R/I) \cdot \binom{n+d-1}{d}, \ \forall n.$$

This suggests that the Hilbert function of an arbitrary ideal I be approached through the Hilbert function of one of its minimal reductions J.

We look at two exact sequences as the vehicle for this comparison:

$$0 \to IJ^{n-1}/J^n \longrightarrow I^n/J^n \longrightarrow I^n/IJ^{n-1} \to 0$$

and

$$0 \to IJ^{n-1}/J^n \longrightarrow J^{n-1}/J^n \longrightarrow J^{n-1}/IJ^{n-1} \to 0.$$

The function $\lambda(I^n/J^n)$ is our center of interest since it is $\lambda(R/J^n) - \lambda(R/I^n)$, the first term being very well-behaved. Using the other sequence we have the following expression for the Hilbert function $H_I(n)$ of I:

$$\begin{aligned} H_I(n) &= \lambda(R/J^n) - \lambda(IJ^{n-1}/J^n) - \lambda(I^n/IJ^{n-1}) \\ &= \lambda(R/J^n) - \lambda(J^{n-1}/J^n) + \lambda(J^{n-1}/IJ^{n-1}) - \lambda(I^n/IJ^{n-1}) \\ &= \lambda(R/J^{n-1}) + \lambda(J^{n-1}/IJ^{n-1}) - \lambda(I^n/IJ^{n-1}), \end{aligned}$$

the first two terms of which can be collected since the ring $\bigoplus_n (J^{n-1}/IJ^{n-1})$ is a polynomial ring in d variables with coefficients in R/I. We obtain

$$H_I(n) = \lambda(R/J) \cdot \binom{n+d-2}{d} + \lambda(R/I) \cdot \binom{n+d-2}{d-1} - \lambda(I^n/IJ^{n-1}).$$

This expression turns the focus on $\lambda(I^n/IJ^{n-1})$. We note that both I^n/J^n and I^{n+1}/IJ^n are components of modules over the Rees algebra $R[Jt]$: the first comes from the quotient $R[It]/R[Jt]$, the other being a component of the Sally module $S_J(I)$. The latter has many advantages over the former, a key one being that it vanishes in cases of considerable interest.

Sometimes it will be convenient to write $H_I(\cdot)$ for the Hilbert function of I; the reader is also warned about shifts in the arguments arising from the grading of the modules. For example, the degree 1 component of $S_J(I)$ is I^2/IJ.

Proposition 2.8. *Let $(R,\mathfrak{m})$ be a Cohen-Macaulay local ring of dimension d, with infinite residue field, and let I be an $\mathfrak{m}$-primary ideal. Denote by $H_I(n) = \lambda(R/I^n)$ the Hilbert function of I, and let*

$$e_0\binom{n+d-1}{d} - e_1\binom{n+d-2}{d-1} + \cdots + (-1)^{d-1}e_{d-1}\binom{n}{1} + (-1)^d e_d \tag{2.6}$$

be its Hilbert polynomial. Suppose that J is a minimal reduction of I and let $S = S_J(I)$ be the corresponding Sally module. Then for all $n \geq 0$

$$H_I(n) = e_0\binom{n+d-1}{d} + (\lambda(R/I) - e_0)\binom{n+d-2}{d-1} - \lambda(S_{n-1}). \tag{2.7}$$

Proof. The proof is a straightforward calculation that takes into account the equality $e_0 = \lambda(R/J)$. □

We can thus write the Hilbert-Poincaré series of I as follows:

Corollary 2.9. *Let $(R,\mathfrak{m})$ be a Cohen-Macaulay local ring and I an $\mathfrak{m}$-primary ideal. Then for any minimal reduction J,*

$$\begin{aligned} H_I(t) &= \frac{\lambda(R/J)\cdot t}{(1-t)^d} + \frac{\lambda(R/I)(1-t)}{(1-t)^d} - (1-t)H_S(t) \\ &= \frac{\lambda(R/I) + (\lambda(I/J)\cdot t}{(1-t)^d} - (1-t)H_S(t). \end{aligned}$$

Corollary 2.10. *The Hilbert function of $S_J(I)$ is independent of the minimal reduction J. If $S_J(I) \neq 0$, the function $\lambda(S_n)$ has the growth of a polynomial of degree $d-1$. Further $S_J(I) = 0$ if and only if $I^2 = JI$; in this case the multiplicity of I is given by*

$$e_0(I) = \lambda(R/I^2) - d\lambda(R/I).$$

Proof. By Proposition 2.2, if $S_J(I)$ is nonzero then its Krull dimension is d. On the other hand, if $I^2 = JI$

$$e_0(I) = \lambda(R/J) = \lambda(R/I^2) - \lambda(J/JI),$$

while $J/JI \cong J/J^2 \otimes_R R/I \cong (R/I)^d$. □

Hilbert Coefficients

The following formulas can be used to establish several vanishing results on the coefficients of Hilbert functions.

Theorem 2.11. *Let $(R,\mathfrak{m})$ be a Cohen-Macaulay local ring, I an $\mathfrak{m}$-primary ideal and $S_J(I)$ the Sally module of I associated to a minimal reduction J. If $s_0, s_1, \ldots, s_{d-1}$ are the coefficients of the Hilbert polynomial of $S_J(I)$, then*

$$\begin{aligned} s_0 &= e_1 - e_0 + \lambda(R/I) \\ s_i &= e_{i+1}, \quad \text{for} \quad i \geq 1. \end{aligned}$$

Corollary 2.12. *The following hold:*

(a) $\lambda(R/I) \geq e_0 - e_1$ ([Nor60]).
(b) *If equality holds in the above, then $S_J(I) = 0$* ([Hu87], [Oo87]).

Proof. (a) This is clear since s_0, being the multiplicity of $S_J(I)$, must be non-negative.

(b) For equality to hold there must be no contribution from $S_J(I)$, which means that $S_J(I)$ is a module of Krull dimension at most $d-1$. From Proposition 2.2 this implies that $S_J(I) = 0$, that is, $I^2 = JI$. □

Another result that can be derived from the formulas above is Narita's inequality ([Nar63]): $e_2 \geq 0$. This results from an interpretation of $e_2(I)$ as $s_0(I^s)$ for some appropriate power of I ([CPVz98]). If $(R,\mathfrak{m})$ is a Cohen-Macaulay local ring of dimension 2 and I is an $\mathfrak{m}$-primary ideal, note that $e_2(I)$ puts some distance between the relation type of the high powers of I and their reduction numbers. Indeed, in Theorem 1.19 we have that the reduction type of I^n, $n \gg 0$, is at most 2, while $\mathrm{r}(I) > 1$ if $e_2(I) \neq 0$.

Proposition 2.13. *Let $(R,\mathfrak{m})$ be a Cohen-Macaulay local ring of dimension $d \geq 2$ and let I be an $\mathfrak{m}$-primary ideal. Then for all large integers q,*

$$e_2(I) = s_0(I^q) = e_1(I^q) - e_0(I^q) + \lambda(R/I^q).$$

Proof. We first consider the case for $d = 2$. Let

$$\lambda(R/I^n) = e_0\binom{n+1}{2} - e_1\binom{n}{1} + e_2, \quad n \gg 0,$$

be the Hilbert polynomial of I. Let q be an integer greater than the postulation number for this function. We have for $n \gg 0$,

$$\begin{aligned} \lambda(R/(I^q)^n) &= f_0\binom{n+1}{2} - f_1\binom{n}{1} + f_2 = \lambda(R/I^{qn}) \\ &= e_0\binom{qn+1}{2} - e_1\binom{qn}{1} + e_2. \end{aligned}$$

Comparing these two polynomials in n gives

$$\begin{aligned} f_0 &= q^2 e_0 \\ f_1 &= q e_1 + \frac{1}{2} q^2 e_0 - \frac{1}{2} q e_0 \\ f_2 &= e_2. \end{aligned}$$

By Proposition 2.11, the multiplicity $s_0(I^q)$ of the Sally module of I^q is given by

$$s_0(I^q) = f_1 - f_0 + \lambda(R/I^q),$$

which by the equalities above give $e_2(I) = s_0(I^q)$. In dimension $d > 2$, we reduce R and I modulo a superficial element in the usual manner. □

Corollary 2.14. *Suppose further that $d = 2$. Then $e_2(I) = 0$ if and only if $R[I^q t]$ is Cohen-Macaulay for $q \gg 0$.*

The assertion will follow from the discussion in Chapter 3 (see Theorem 3.39). Finally we point out a connection between the multiplicity of the Sally module and the reduction number of a primary ideal.

Corollary 2.15. *Let $(R, \mathfrak{m})$ be a Cohen-Macaulay local ring of dimension $d \geq 1$, I be an $\mathfrak{m}$-primary ideal and let J be a reduction generated by a regular sequence. If* depth $\mathrm{gr}_I(R) \geq d-1$ *then*

$$r_J(I) \leq e_1(I) - e_0(I) + \lambda(R/I) + 1.$$

Proof. The hypothesis is equivalent to saying that the Sally module $S_J(I)$ is Cohen-Macaulay. But the discussion of the structure of $S_J(I)$ shows that it has fresh generators until degree $r_J(I)$. But according to Theorem 2.5 each component I^k/JI^{k-1}, $1 \leq k \leq r_J(I)$, contributes at least 1 to the multiplicity s_0 of $S_J(I)$, from which the assertion follows. □

Remark 2.16. The bound above for $r_J(I)$ is also valid if $\dim R \leq 2$, according to [Ros0b], without the hypothesis on the depth of $\mathrm{gr}_I(R)$.

Remark 2.17. If $(R, \mathfrak{m})$ is a Noetherian local ring (with infinite residue field) and I is an $\mathfrak{m}$-primary ideal, the Sally module $S_J(I)$ associated to a minimal reduction J can still be useful in the non Cohen-Macaulay case, according to [C3]. Some of the multiplicity formulas are different and the general properties are not so sharp. This begins with the Hilbert function of $S_J(I) = \bigoplus_{n \geq 1} I^{n+1}/IJ^n$:

$$\begin{aligned} \lambda(I^{n+1}/IJ^n) &= \lambda(R/IJ^n) - \lambda(R/I^{n+1}) \\ &= \lambda(R/J^n) + \lambda(J^n/IJ^n) - \lambda(R/I^{n+1}). \end{aligned}$$

Expanding the Hilbert polynomials, we obtain the following expression for the multiplicity of $S_J(I)$:

$$\deg(S_J(I)) = e_1(I) - e_0(I) - e_1(J) + \deg(\mathrm{gr}_J(R) \otimes R/I).$$

In the Cohen-Macaulay case, $e_1(J) = 0$ and $\deg(\mathrm{gr}_J(R) \otimes R/I) = \lambda(R/I)$, while $\dim S_J(I)$ is non-negative; however, its vanishing no longer implies that $S_J(I) = 0$.

Maximal Hilbert Functions

One difficulty with the calculation of the Hilbert coefficients for a local ring $(R,\mathfrak{m})$ lies with the fact that they depend on properties of the associated ring $G = \mathrm{gr}_{\mathfrak{m}}(R)$ of R. Since the properties of G may differ considerably from those of R–in particular those related to depth conditions–the technique of hyperplane section, which is very helpful in the study of the Hilbert polynomial, is not very useful in the examination of the Hilbert series of R which ultimately controls most of the invariants we are interested in.

Theorem 2.18. *Let $(R,\mathfrak{m})$ be a Cohen-Macaulay local ring of dimension $d > 0$, and let I be an $\mathfrak{m}$-primary ideal. If $\mathrm{r}(I) > 1$ and $d \geq 2$, then for all $n \geq 0$,*

$$\lambda(I^n/I^{n+1}) \leq e(I)\binom{n+d-2}{d-1} + (\lambda(R/I)-1)\binom{n+d-2}{d-2} + \binom{n+d-3}{d-3}.$$

Equivalently, if $H_I(t)$ is the Hilbert series of $\mathrm{gr}_I(R)$ then

$$H_I(t) \leq \frac{\lambda(R/I) + (e(I) - \lambda(R/I) - 1)\cdot t + t^2}{(1-t)^d},$$

where the comparison of formal power series means coefficient-by-coefficient comparison.

Proof. The cases excluded here, $\mathrm{r}(I) \leq 1$ or $d = 1$, are easy to describe. A discussion of the case $I = \mathfrak{m}$ is given in [DGV98], a more general case is treated in [RVV1], and an enhanced version appears in [DVz0]. We shall follow the treatment of [DVz0].

According to Corollary 2.9, it suffices to know an estimate of the growth of the Hilbert function of the Sally module of the ideal I, more precisely of a lower bound for the term $(1-t)H_S(t)$. This is provided in the following calculation.

Lemma 2.19. *Let $(R,\mathfrak{m})$ be a Cohen-Macaulay local ring of dimension $d > 0$, with infinite residue field, I an $\mathfrak{m}$-primary ideal of R, $J \subset I$ a minimal reduction of I and $S = S_J(I)$ the Sally module of I with respect to J. If $S_J(I) \neq 0$ then*

$$\lambda(S_n) = \binom{n+d-2}{d-1} + f(n), \tag{2.8}$$

where $f(n)$ is non-decreasing. Moreover, if $d \geq 2$ then $\lambda(S_n)$ is strictly increasing.

Proof. Since $S \neq 0$, there exists nonzero $u \in I^2/JI = S_1$ such that $\mathfrak{m}\cdot u = 0$, and then $(R[Jt]/\mathfrak{m}\cdot R[Jt])\cdot u$ injects into S. Now

$$R[Jt]/\mathfrak{m}\cdot R[Jt] = R[Jt]\otimes_R R/\mathfrak{m} \cong R/\mathfrak{m}[T_1,\ldots,T_d] = k[T_1,\ldots,T_d],$$

thus $k[T_1,\ldots,T_d]\cdot u \twoheadrightarrow (R[Jt]/\mathfrak{m}\cdot R[Jt])\cdot u \hookrightarrow S$. We denote $k[T_1,\ldots,T_d]$ by A, and claim that $A[-1]$ embeds into S. Let L denote an ideal defining the kernel of the natural mapping

$$0 \longrightarrow L[-1] \longrightarrow A[-1] \longrightarrow S;$$

we examine the embedding $A[-1]/L[-1] \hookrightarrow S$. On the other hand, according to [Va94b, p.102], all associated primes of S have dimension d, which implies that $L = 0$. Therefore we have

$$0 \longrightarrow A[-1] \longrightarrow S \longrightarrow C \longrightarrow 0,$$

hence $\lambda(S_n) = \lambda(A[-1]_n) + \lambda(C_n)$. Since A is a d-dimensional Cohen-Macaulay $R[Jt]$-module, we have

$$\lambda(A[-1]_n) = \binom{n+d-2}{d-1},$$

and also that depth $C \geq 1$ (by the depth formula and the fact that depth $S \geq 1$). In particular there is an element h of degree 1 of $R[Jt]$ that is regular on C, giving rise to embeddings $C_n \to C_{n+1}$ induced by multiplication by h. It shows that its Hilbert function $f(n)$ is non-decreasing (and strictly increasing when $d \geq 2$), which proves the assertion. □

One can refine further the argument above by acquiring additional pieces of the Hilbert function of the Sally module. The argument (suppose throughout that $\dim R = d \geq 2$) considers an embedding

$$0 \to A[-1] \longrightarrow S \longrightarrow L_2 \to 0.$$

If s_0, the multiplicity of S, is greater than 1, then since L_2 has positive multiplicity, we can place a copy of A into L_1:

$$0 \to A[-d_2] \longrightarrow L_2 \longrightarrow L_3 \to 0.$$

A significant difference is that we do not have the fine control over d_2 that we had in the first embedding. Observe further that L_2 has positive depth, and L_3 either vanishes or will have positive depth.

We can continue in this fashion until the multiplicity of S,

$$s_0 = e_1 - e_0 + \lambda(R/I),$$

is exhausted. At the last step one is left with an exact sequence

$$0 \to A[-d_{s_0}] \longrightarrow L_{s_0} \longrightarrow L \to 0.$$

We can now assemble the Hilbert series of S:

$$H_S(t) = \frac{\sum_{k=1}^{s_0} t^{d_k}}{(1-t)^d} + H_L(t),$$

where $H_L(t)$ is the Hilbert series of a module that either is zero or has positive depth. Thus the coefficients of $H_L(t)$ are non-decreasing.

Theorem 2.20. *Let $(R,\mathfrak{m})$ be a Cohen-Macaulay local ring of dimension $d \geq 2$ and let I be an $\mathfrak{m}$-primary ideal that is not a complete intersection. The Hilbert series of I satisfies the inequality (with coefficient-wise comparison)*

$$H_I(t) \leq \frac{\lambda(R/I) + (e_0(I) - \lambda(R/I))t - (1-t)\sum_{i=1}^{s_0} t^{d_k}}{(1-t)^d},$$

where $s_0 = e_1(I) - e_0(I) + \lambda(R/I)$, $d_1 = 1$, $d_k \leq \mathrm{r}_J(I)$.

If the module L is 0, then S is Cohen-Macaulay, a condition equivalent to the requirement that depth $\mathrm{gr}_I(R)$ be at least $d-1$. The Hilbert function can then be made more precise, in terms of the lengths of modules such as I^p/JI^{p-1} (see [HM97], [Vz97]).

The Valabrega-Valla Module

Consider the following exact sequence of finitely generated $R[Jt]$-modules:

$$0 \to J \cdot R[It] \longrightarrow J \cdot R[t] \bigcap I \cdot R[It] \longrightarrow VV_J(I) = \bigoplus_{n=0}^{\infty} (J \cap I^{n+1})/JI^n \to 0.$$

Definition 2.21. The *Valabrega-Valla module* of I with respect to J is $VV_J(I)$ viewed as an R-module.

$VV_J(I)$ is actually a module over $R[Jt]$, but since it vanishes in high degrees it is more convenient to view it as an R-module. Its components appear in some filtrations of $S_J(I)$. Its significance lies in the following Cohen-Macaulay criterion in [VV78].

Theorem 2.22. *Let $(R,\mathfrak{m})$ be a Cohen-Macaulay local ring with infinite residue field, and let I be an $\mathfrak{m}$-primary ideal. Suppose that J is a minimal reduction of I. Then $\mathrm{gr}_I(R)$ is Cohen-Macaulay if and only if $VV_J(I) = 0$.*

This is, in fact, a special case of a very general assertion ([VV78, Theorem 2.3]).

Theorem 2.23. *Let R be a Noetherian ring, I an ideal of R, and $x_1,\ldots,x_n$ a regular sequence in R. Then the leading forms of the x_i form a regular sequence in $\mathrm{gr}_I(R)$ if and only if for $i = 1,\ldots,n$ and all $m \geq 1$,*

$$(x_1,\ldots,x_i) \cap I^m = \sum_{j=1}^{i} I^{m-d_j} x_j,$$

where d_j is the least integer s such that $x_j \in I^s \setminus I^{s+1}$. In particular, if $(R,\mathfrak{m})$ is a local ring and $J = (x_1,\ldots,x_n)$ is a reduction of I of reduction number $\mathrm{r}_J(I) = r$, it suffices to meet the condition for $m < r$.

The proof is straightforward enough that we leave it for the reader. This very useful criterion has been extended to more general filtrations by several authors.

The Ratliff-Rush Closure

Let R be a Noetherian ring and let I be an ideal containing nonzero divisors. For the purposes we have here, we can harmlessly assume that I can be generated by a set of regular elements. A useful condition for studying the algebra $R[It]$ is the presence of good depth conditions on $\mathrm{gr}_I(R)$, as it allows for several reductions of dimension. We shall look at a few basic reductions.

We begin by considering a filtration introduced in [RR78].

Definition 2.24. Let R be a Noetherian ring and I an ideal containing regular elements. The *Ratliff-Rush closure* of I is

$$\widetilde{I} = \bigcup_{n\geq 1} I^{n+1} :_R I^n.$$

Denote by $\mathrm{rr}(I)$ the least integer n such that $\widetilde{I} = I^{n+1} :_R I^n$.

Note that $\widetilde{I}$ is integral over I, and that $\widetilde{\widetilde{I}} = \widetilde{I}$. When the process is applied to the I-adic filtration $\{I^n,\ n \geq 1\}$, it gives rise to another multiplicative filtration $\{\widetilde{I^n},\ n \geq 1\}$, which can be organized into its Rees algebra,

$$\widetilde{\mathcal{R}}(I) = \sum_{n\geq 0} \widetilde{I^n} t^n.$$

A concise description of $\widetilde{\mathcal{R}}(I)$ is given in:

Proposition 2.25. *Let R be a Noetherian ring and let I be an ideal of grade > 0. Suppose that $\mathcal{R} = R[It]$ and set $\mathfrak{I} = (I, It)$. Then*

$$\widetilde{\mathcal{R}}(I) = \bigcup_{n\geq 0} \mathrm{Hom}_{\mathcal{R}}(\mathfrak{I}^n, \mathcal{R}).$$

One elementary but important property of this construction is as follows:

Proposition 2.26. *Let R be a Noetherian ring and I an ideal of grade > 0. Then $\widetilde{\mathcal{R}}(I) = \mathcal{R}(I)$ if and only if the irrelevant ideal $\mathrm{gr}_I(R)_+$ of $\mathrm{gr}_I(R)$ has grade > 0.*

We shall study some of its properties, beginning with:

Theorem 2.27 (Ratliff-Rush). *For every ideal I containing regular elements there exists an integer k_0 such that $\widetilde{I^k} = I^k$ for $k \geq k_0$.*

We want to see this as a statement about the cohomology of blowups. For this we recall:

Proposition 2.28. *Let S be a Noetherian ring and L an ideal containing regular elements. Then*

$$A = \bigcup_{n\geq 1} \mathrm{Hom}_S(L^n, L^n)$$

is a finite extension of S.

We note that each $\text{Hom}_S(L^n, L^n)$ can be identified with a finitely generated S-module contained in the total ring of fractions Q of S. An interpretation of A is the following. Let $X = \text{Proj}(S[Lt])$ and suppose that $L = (f_1, \ldots, f_m)$, f_i regular. Then

$$H^0(X, O_X) = \bigcap_{1 \leq i \leq m} S[Lf_i^{-1}] = \bigcup_{n \geq 1} \text{Hom}_S(L^n, L^n).$$

The finite generation of $H^0(X, O_X)$ is a special case of the finiteness theorem of Serre (see [BrS98, 20.4.6]).

Proof of Theorem 2.27. The Rees algebra $\widetilde{R}(I)$ is clearly a S-subalgebra of A, and therefore we can assume that it is generated as an S-module by its components $\widetilde{I^j} t^j$, for $j \leq r$; set $\lambda = \sup\{\text{rr}(\widetilde{I^j}) \mid j \leq r\}$. Since for $n \geq r$ we have

$$\widetilde{I^n} = \sum_{j=1}^{r} I^{n-j} \widetilde{I^j},$$

it follows that $\widetilde{I^n} = I^n$ for $n \geq 2\lambda$. □

One application is to a possible improvement of the depth in changing the ring from $\mathcal{R}(I)$ to $\widetilde{\mathcal{R}}(I)$. The proof uses some material from Chapter 3, so we delay it till then.

Theorem 2.29. *Let R be a Cohen-Macaulay local ring of dimension at least 2 and I an ideal of grade > 0. The irrelevant maximal ideal of $\widetilde{\mathcal{R}}(I)$ has depth at least 2.*

Partially Identical Hilbert Polynomials

Throughout this discussion, $(R, \mathfrak{m})$ will be an analytically equidimensional local domain of dimension d and I an ideal. To compare I to its integral closure $\overline{I}$ it is useful to track their numerical measures as expressed by appropriate Hilbert functions.

We begin with an observation derived from Theorem 2.27:

Corollary 2.30. *Let $(R, \mathfrak{m})$ be a Noetherian local ring and I an $\mathfrak{m}$-primary ideal containing regular elements. Then $\widetilde{I}$ is the largest overideal of I with the same Hilbert polynomial.*

Proof. The condition $L^n = I^n$ for $n \geq r$ necessarily implies that $L \subset \widetilde{I}$. □

Let us refine this statement. In the case of an $\mathfrak{m}$-primary ideal I, an approach was given in [Sha91] to constructing a 'canonical' sequence of ideals

$$I \subset I_0 \subset I_1 \subset \cdots \subset I_{d-1} \subset I_d = \overline{I},$$

with the following property. Let G be the associated graded ring of I and write

$$P_G(n) = \sum_{i=0}^{d} (-1)^i e_i(I) \binom{n+d-i}{d-i},$$

for its Hilbert polynomial. We give the treatment developed in [CPV3]:

Theorem 2.31. *There exists a unique largest ideal I_j with $I \subset \widetilde{I_j} \subset \overline{I}$ with the property that the corresponding Hilbert polynomial satisfies the equalities*

$$e_i(\widetilde{I_j}) = e_i(I), \quad i = 0, \ldots, d-j.$$

We begin by giving another derivation of this result and then discuss how to extend it to non-primary ideals. Denote by $A = R[It, t^{-1}]$ the extended Rees algebra of I. For each integer ℓ with $1 \leq \ell \leq d+1$, let $B^{(\ell)}$ be the subring of

$$C = R[t, t^{-1}] \cap \overline{R[It, t^{-1}]}$$

generated by

$$\{f \in C \mid \text{height } (A{:}_A f) \geq \ell, \ f \text{ homogeneous}\}.$$

This gives a filtration of subalgebras

$$A = B^{(d+2)} \subset B^{(d+1)} \subset \cdots \subset B^{(2)} \subset B^{(1)} = \sum_{n \geq 0} \overline{I^n} t^n [t^{-1}].$$

By considering only the component $A^{(\ell)}$ of $B^{(\ell)}$ in non-negative degrees (since they all have the same components in negative degrees) we have

$$\begin{aligned}
A^{(1)} &= \text{integral closure of } R[It] \text{ in } R[t] \\
A^{(2)} &= S_2\text{-ification of } R[It] \\
A^{(d+1)} &= \text{Ratliff-Rush closure of } R[It] \\
A^{(d+2)} &= \text{Rees algebra } R[It] \text{ of } I.
\end{aligned}$$

The algebras $A^{(\ell)}$, $\ell \leq d+1$, are not accessible, with the possible exception of $A^{(2)}$ for which there is a constructive pathway that starts at $R[It]$. When R is Gorenstein and J is a minimal reduction of I, one has ([NV93]):

$$A^{(2)} = \operatorname{Hom}_D(\operatorname{Hom}_D(A, D), D), \quad D = R[Jt].$$

We are now ready to relate these algebras to the ideals I_j and to prove the assertion about their Hilbert polynomials. Set I_j to be the ideal defined by the component of degree 1 of $A^{(d+1-j)}$. We have exact sequences

$$0 \to R[It, t^{-1}] \longrightarrow R[I_j t, t^{-1}] \longrightarrow M_j \to 0$$

of finitely generated modules over $R[It, t^{-1}]$, where M_j is a module of dimension at most $d+1-(d+1-j) = j$.

Tensoring this sequence by $R[It, t^{-1}]/(t^{-1})$, we obtain two exact sequences of graded modules:

$$0 \to K_j[+1] \longrightarrow M_j[+1] \xrightarrow{t^{-1}} M_j \longrightarrow C_j \to 0, \tag{2.9}$$

$$0 \to K_j \longrightarrow \mathrm{gr}_I(R) \longrightarrow \mathrm{gr}_{I_j}(R) \longrightarrow C_j \to 0. \tag{2.10}$$

Note that (2.9) was induced by multiplication by t^{-1}, an endomorphism which is nilpotent on M_j; it follows (see [Ei95, 12.1]) that the Krull dimensions of all the modules in this sequence are equal. One can further assert that the multiplicities of K_j and of C_j match. As a consequence, adding in (2.10) the Hilbert polynomials, we obtain that the Hilbert polynomials of $\mathrm{gr}_I(R)$ and of $\mathrm{gr}_{\widetilde{I_j}}(R)$ have matching coefficients not only down to degree $d-(j-1)$ (which is guaranteed by the codimension of M_j), but one position further down. □

Remark 2.32. One case of particular interest is the algebra $A^{(2)}$. For convenience, denote it by

$$S = \sum_{n \geq 0} S_n t^n,$$

the S_2-ification of $R[It]$. If $(R, \mathfrak{m})$ is a Cohen-Macaulay local ring of dimension $d > 1$ and I is $\mathfrak{m}$-primary, there is a direct explanation of all components S_n. We already mentioned that $S_1 = \widetilde{I}$, the largest ideal with the same multiplicity as I. We claim that the other components have a similar interpretation,

$$S_n = \widetilde{I^n}.$$

This holds because, for a fixed n, it is clear from the direct summand structures that the nth Veronese subring of S is the S_2-ification of the nth Veronese subring of $R[It]$. (A more refined treatment of these issues is given in [Ci1].)

Cascading Reductions

In the study of reductions $J \subset I$ and the relationship between their associated graded rings $\mathrm{gr}_J(R)$ and $\mathrm{gr}_I(R)$, there arises the issue of what role intermediate ideals L, $J \subset L \subset I$, may play.

The simplest reduction $J \subset I$ to examine is that where $I^2 = JI$. We then have that

$$I \cdot R[Jt] = I \cdot R[It],$$

and

$$0 \to I \cdot R[Jt] \longrightarrow R[Jt] \longrightarrow \mathrm{gr}_J(R) \otimes R/I \to 0. \tag{2.11}$$

If J is a regular sequence, and I is a Cohen-Macaulay ideal, it follows that $I \cdot R[It]$ is a maximal Cohen-Macaulay module. Taken into the sequences (3.4) and (3.5), this leads to the fact that $\mathrm{gr}_I(R)$ is Cohen-Macaulay.

When the reduction $r_J(I)$ is 2, the structure of $S_J(I)$ has to be taken into consideration (see [Vz97] for several cases). Let us try to bridge these reductions through simpler ones ([MV1]).

Let $(R,\mathfrak{m})$ be a Cohen-Macaulay local ring of dimension d, let I be an $\mathfrak{m}$-primary ideal, let J be a minimal reduction and suppose $r_J(I)=2$. Let us assume that the equality $I^3=JI^2$ is achieved in two steps as follows. There exists L, $J\subset L\subset I$, with $I^2=LI$ and $L^2=JL$. These conditions mean that the associated Sally modules $S_J(L)$ and $S_L(I)$ vanish, so that we have

$$L\cdot R[Jt] = L\cdot R[Lt], \tag{2.12}$$

$$I\cdot R[Lt] = I\cdot R[It]. \tag{2.13}$$

The exact sequence (2.11) applied to the pair (J,L) implies that $L\cdot R[Lt]$ is a maximal Cohen-Macaulay module. It would be interesting to have a similar assertion for $I\cdot R[It]$, but to use (2.11) we would need to have that $\mathrm{gr}_L(R)\otimes R/I$ is Cohen-Macaulay.

Consider the following commutative diagram of exact sequences:

$$\begin{array}{ccccccccc}
0\to & R[Jt] & \longrightarrow & R[Lt] & \longrightarrow & C & \to 0 \\
 & \uparrow & & \uparrow & & \uparrow & \\
0\to & L\cdot R[Jt] & \longrightarrow & L\cdot R[Lt] & \longrightarrow & S_J(L) & \to 0.
\end{array}$$

From the snake lemma there is an acyclic complex

$$0\to K_S \longrightarrow \mathrm{gr}_J(R)\otimes R/L \longrightarrow \mathrm{gr}_L(R) \longrightarrow K_C\to 0, \tag{2.14}$$

where K_S and K_C are defined via the exact sequence

$$0\to K_S\longrightarrow S_J(L)\longrightarrow C\longrightarrow K_C\to 0 \tag{2.15}$$

of natural inclusions.

Proposition 2.33. *If $r_J(L)=1$ there are exact sequences of Cohen-Macaulay modules of dimension d,*

$$0\to \mathrm{gr}_J(R)\otimes R/L\longrightarrow \mathrm{gr}_L(R)\longrightarrow C\to 0, \tag{2.16}$$

and

$$0\to C[+1]\longrightarrow \mathrm{gr}_J(R)\longrightarrow \mathrm{gr}_J(R)\otimes R/L\to 0. \tag{2.17}$$

Proof. We leave this as an exercise.

Since we are interested in $\mathrm{gr}_L(R)\otimes R/I$, the place to start is with sequence (2.16): we need to reduce it modulo R/I. Tensoring (2.17) instead with R/I over R/J, we have the exact sequence

$$\begin{aligned}
&0\to \mathrm{Tor}_1^{R/J}(\mathrm{gr}_J(R)\otimes R/L, R/I)\to C\otimes R/I[+1]\to \\
&\to \mathrm{gr}_J(R)\otimes R/I\cong \mathrm{gr}_J(R)\otimes R/L\otimes R/I\to 0,
\end{aligned}$$

from which we have

$$\begin{aligned} C\otimes R/I &\cong \mathrm{Tor}_1^{R/J}(\mathrm{gr}_J(R)\otimes R/L, R/I)[-1] \\ &\cong \mathrm{gr}_J(R)\otimes \mathrm{Tor}_1^{R/J}(R/L, R/I)[-1]. \end{aligned}$$

We thus have

Proposition 2.34. *$C\otimes R/I$ is a Cohen-Macaulay module of dimension d.*

One way to prove that $\mathrm{gr}_L(R)\otimes R/I$ is Cohen-Macaulay is by showing that

$$0\to \mathrm{gr}_J(R)\otimes R/I \longrightarrow \mathrm{gr}_L(R)\otimes R/I \longrightarrow C\otimes R/I\to 0 \tag{2.18}$$

is exact. (Tracing the Tor's carefully one is in good condition to use Theorem 2.23.) For example, if L is a Gorenstein ideal (a situation that [NV93] often forbids), the sequence (2.16) splits and therefore (2.18) will be exact. What are other more interesting cases? We point out the following ([Huc3]):

Theorem 2.35. *Let $(R,\mathfrak{m})$ be a Cohen-Macaulay local ring of dimension d and let J and I be $\mathfrak{m}$-primary ideals such that $J\subset I$ and $\lambda(I/J)=\mathrm{r}(J)=\mathrm{r}_J(I)=1$.*

(a) *If $\nu(J)\leq d+1$, then* depth $\mathrm{gr}_I(R)\geq d-1$. *Moreover, if $\nu(I)=d+1$, then $\mathrm{gr}_I(R)$ is Cohen-Macaulay.*
(b) *If $\nu(J)=d+2$ and $e_2(I)\neq 0$, then* depth $\mathrm{gr}_I(R)\geq d-1$.

2.2 Maximal Hilbert Functions

Let $(R,\mathfrak{m})$ be a local ring of dimension $d>0$ and I an ideal. In this section we consider some predictions about the reduction number of I that make use of the associated graded ring $\mathrm{gr}_I(R)$ of I in the case of an $\mathfrak{m}$-primary ideal, and of its special fiber $\mathcal{F}(I)$ in general. Because the properties of these rings are difficult to link to the ideal I directly, we shall have greater success with Cohen-Macaulay ideals.

The overall aim is to connect the reduction number of I to some polynomial $g(e_0,\ldots,e_d)$ on the Hilbert coefficients of $\mathrm{gr}_I(R)$ or $\mathcal{F}(I)$, in the form

$$\mathrm{r}(I)\leq g(e_0,\ldots,e_d).$$

By fixing the number of generators of I, the existence of such inequalities is guaranteed by a classical theorem on Castelnuovo-Mumford regularity (see [Mu66, p. 101]) asserting the existence of a polynomial $h(e_0,e_1,\ldots,e_d)$ bounding the postulation number of any ideal of G (or of $\mathcal{F}(I)$ as the case may be). In association with the Hilbert polynomial of the algebra this leads to a bound on $\mathrm{r}(I)$, however large.

2.2.1 The Eakin-Sathaye Theorem

To deal with the issues just outlined, a strategy we shall employ is the following. Given the ideal I of the local Noetherian ring $(R,\mathfrak{m})$, we shall look for estimates $f(n)$ for the Hilbert function of $\mathcal{F}(I)$ that are valid for all values of n,

$$\nu(I^n) < f(n).$$

We then appeal to the following result of Eakin and Sathaye [ES76] to solve for

$$f(r) \leq \binom{r+\ell}{\ell},$$

where ℓ is the analytic spread of I:

Theorem 2.36 (Eakin-Sathaye). *Let A be a standard graded algebra over an infinite field k. For positive integers n and ℓ, suppose that* $\dim_k A_n < \binom{n+\ell}{\ell}$. *Then there exist $z_1,\ldots,z_\ell \in A_1$ such that*

$$A_n = (z_1,\ldots,z_\ell)A_{n-1}.$$

Moreover, if $x_1,\ldots,x_p \in A_1$ are such that $(x_1,\ldots,x_p)^n = A_n$, then r generic linear combinations of $x_1,\ldots,x_p$ will define such sets.

Thus, for example, if $(R,\mathfrak{m})$ is a Noetherian local ring of dimension d with infinite residue field, I is an $\mathfrak{m}$-primary ideal and $P_I(t)$ is its Hilbert polynomial, we must solve the inequality

$$H_I(n) < \binom{n+d}{d}$$

for $n \geq h(e_0,\ldots,e_d)$, where the e_i are the Hilbert coefficients of I and h is the polynomial controlling the postulation number mentioned above. From this setup, a polynomial bound for the reduction number arises, $\mathrm{r}(I) < g(e_0,\ldots,e_d)$.

Definition 2.37. Let $(R,\mathfrak{m},k)$ be a local ring with residue field k. We say that an ideal $I \subseteq R$ satisfies condition $S(r,n)$ if it satisfies each of the following:

- $\mathfrak{m}I^n = 0$.
- $\dim_k I^n < \binom{n+r}{r}$.
- There exist $\{y_i\}_{i=1}^{\infty} \subseteq I$ and an integer p such that for all i, we have

$$I = (y_i, y_{i+1},\ldots,y_{i+p-1}).$$

- S_ω (the symmetric group on the natural numbers) acts on R as a group of automorphisms such that $\sigma(y_i) = y_{\sigma(i)}$ for all $\sigma \in S_\omega$.

Theorem 2.38. *If $I \subseteq R$ is an ideal satisfying $S(r,n)$, then*

$$(y_1,\ldots,y_r)I^{n-1} = I^n.$$

To prove the theorem, we reduce to the following lemma.

Lemma 2.39. *If $I \subseteq R$ is an ideal satisfying $S(r,n)$, then*

$$(y_1,\dots,y_{r+1})I^{n-1} = I^n \Longrightarrow (y_1,\dots,y_r)I^{n-1} = I^n.$$

Proof. Assume that the lemma is false, and let r be minimal such that the lemma does not hold. Let n be minimal for this r.

Note that for every $\sigma \in S_\omega$, $\sigma(I) = I$. Thus $y_iI^{n-1} \cong y_{\sigma(i)}I^{n-1}$ for every $\sigma \in S_\omega$. Let d be the common dimension, $d = \dim_k y_iI^{n-1}$.

Case 1. $d \geq \binom{n+r-1}{r}$. If $r = 1$, then $\dim_k y_1I^{n-1} \geq \binom{n}{1} = n$, and $\dim_k I^n < \binom{n+1}{1} = n+1$. Hence $y_1I^{n-1} = I^n$, a contradiction. If $r > 1$, let $\overline{R} = R/y_1I^{n-1}$. We claim that $\overline{R}$ and $\overline{I}$ satisfy $S(r-1,n)$:

- $\mathfrak{m}\overline{I}^n = 0$ still.
- $\dim_k \overline{I}^n = \dim_k I^n - \dim_k y_1I^{n-1} < \binom{n+r}{r} - \binom{n+r-1}{r} = \binom{n+r-1}{r-1}$.
- $\{\overline{y}_i\}_{i=2}^\infty$ suffices.
- set $G = \{\sigma \in S_\omega : \sigma(y_1) = y_1\}$. Then G fixes y_1I^{n-1}, and hence acts on $\overline{R}$ in the appropriate way.

So we may apply the lemma and conclude that

$$\begin{aligned}\overline{I}^n &= (\overline{y}_2,\dots,\overline{y}_r)\overline{I}^{n-1},\\ I^n &= (y_2,\dots,y_r)I^{n-1} + y_1I^{n-1}\\ &= (y_1,\dots,y_r)I^{n-1},\end{aligned}$$

a contradiction.

Case 2. $d < \binom{n+r-1}{r}$. Set $K = [0 : y_1]_{I^{n-1}}$, and $\overline{R} = R/K$. The map $I^{n-1} \xrightarrow{y_1} y_1I^{n-1}$ is surjective and has kernel K. Hence $\overline{I}^{n-1} \cong y_1I^{n-1}$. We claim that $\overline{R}$ and $\overline{I}$ satisfy $S(r,n-1)$:

- $\mathfrak{m}\overline{I}^{n-1} \cong \mathfrak{m}y_1I^{n-1} \subseteq \mathfrak{m}I^n = 0$.
- $\dim_k \overline{I}^{n-1} = \dim_k y_1I^{n-1} < \binom{n-1+r}{r}$.
- $\{\overline{y}_i\}_{i=2}^\infty$ suffices.
- Each $\sigma \in S_\omega$ fixes I^{n-1}, hence $G = \{\sigma \in S_\omega : \sigma(y_1) = y_1\}$ fixes K. So G acts on $\overline{R}$ in the appropriate way.

So we may apply the lemma again and conclude that

$$\overline{I}^{n-1} = (\overline{y}_2,\dots,\overline{y}_{r+1})\overline{I}^{n-1}$$

or,

$$I^{n-1} = (y_2,\dots,y_{r+1})I^{n-2} + K.$$

By hypothesis,

$$\begin{aligned}(y_1,\ldots,y_{r+1})I^{n-1} &= I^n,\\ y_1I^{n-1}+(y_2,\ldots,y_{r+1})I^{n-1} &= I^n,\\ y_1(y_2,\ldots,y_{r+1})I^{n-2}+y_1K+(y_2,\ldots,y_{r+1})I^{n-1} &= I^n,\\ (y_2,\ldots,y_{r+1})I^{n-1} &= I^n.\end{aligned}$$

The action of an appropriate σ yields the desired result. □

Proof of Theorem 2.36. We first show how the condition $S(r,n)$ is brought in. Let $\{x_1, \ldots, x_p\}$ be a set of elements in I such that $(x_1,\ldots,x_p)^n = I^n$. Let $\mathbf{U} = \{u_{j=1,i=1}^{p,\infty}\}$ be a set of distinct indeterminates over R, set $S = R[\mathbf{U}]_{\mathfrak{m}R[\mathbf{U}]}$, and let S_ω act on S by the rule $\sigma(u_{i,j}) = u_{\sigma(i),j}$. Finally, define $y_j = \sum_{i=1}^{p} u_{ij}x_i$. Note that every set $\{y_s,\ldots,y_{s+p-1}\}$ in S generates the same ideal as $(x_1,\ldots,x_p)$.

Now we pass to S and assume that $\mathfrak{m}I^n = 0$. We already have $(y_1,\ldots,y_p)I^{n-1} = I^n$, so if $p \leq r$ we are done. If $p-1 \geq r$, we have $S(r,n)$, and hence $S(p-1,n)$, so the lemma yields $(y_1,\ldots,y_{p-1})I^{n-1} = I^n$, and we proceed by induction.

It is clear how a specialization of the u_{ij} can bring us back to R. □

2.2.2 Hilbert Functions of Primary Ideals

The drawback of using Castelnuovo-Mumford regularity lies with the fact that the polynomial $h(e_0,\ldots,e_d)$ has degree that is a non-elementary function of $d = \dim R$. To avoid this, we shall want to bound $\nu(I^n)$ *for all n* by a polynomial $f(n)$. To optimize it, we must ensure that $f(n)$ has degree $d-1$ with the smallest possible multiplicity, that is, close to $\deg(\mathcal{F})$. In the Cohen-Macaulay case, we shall be able to use polynomials of multiplicity $e(I)$, and obtain quadratic bounds that involve only d and e_0. More likely, some other expressions in the e_i must play the same role, as evidenced by several noteworthy special cases.

Here we shall follow [DGV98], [RVV1] and [Va99] (see also [Va98c], [Va98b, Chapter 9]) to obtain bounds for Hilbert functions. We treat mostly the Cohen-Macaulay cases; other cases are discussed in these references.

We visit several approaches that depend on available data. As it is customary, we set $e(\mathfrak{m}) = \deg(R)$, the multiplicity of R.

Index of Nilpotency and Reduction Numbers

We first consider a case that it is too general but useful in the case of maximal ideals.

We recall that the *index of nilpotency* of an ideal I is the smallest integer s for which $(\sqrt{I})^s \subset I$. For an $\mathfrak{m}$-primary ideal I of a Cohen-Macaulay ring, one has $s \leq \lambda(R/I) \leq e(I)$.

Theorem 2.40. *Let* $(R,\mathfrak{m})$ *be a Cohen-Macaulay local ring of dimension* $d \geq 1$, *I an* $\mathfrak{m}$*-primary ideal and s the index of nilpotency of* R/I. *Then*

$$\nu(I) \leq \deg(R)\binom{s+d-2}{d-1} + \binom{s+d-2}{d-2}. \tag{2.19}$$

Proof. We may assume that the residue field of R is infinite. Let $J = (a_1,\ldots,a_d)$ be a minimal reduction of $\mathfrak{m}$. By assumption, $J^s \subset I$.

Set $J_0 = (a_1,\ldots,a_{d-1})^s$. This is a Cohen-Macaulay ideal of height $d-1$, and the multiplicity of R/J_0 is (easy exercise)

$$\deg(R/J_0) = \deg(R)\binom{s+d-2}{d-1}.$$

Consider the exact sequence

$$0 \to I/J_0 \longrightarrow R/J_0 \longrightarrow R/I \to 0.$$

We have that I/J_0 is a Cohen-Macaulay ideal of the one-dimensional Cohen-Macaulay ring R/J_0. This implies that

$$\nu(I/J_0) \leq \deg(R/J_0).$$

On the other hand, we have

$$\nu(I) \leq \nu(J_0) + \nu(I/J_0) \leq \binom{s+d-2}{d-2} + \deg(R/J_0),$$

to establish the claim. □

Corollary 2.41. *Let $(R,\mathfrak{m})$ be a Cohen-Macaulay local ring of dimension $d \geq 1$. Then for all n,*

$$H_R(n) \leq \deg(R)\binom{n+d-2}{d-1} + \binom{n+d-2}{d-2}. \tag{2.20}$$

If R/I is a Gorenstein ring, one can do better:

Theorem 2.42. *Let $(R,\mathfrak{m})$ be a Cohen-Macaulay local ring of dimension $d \geq 2$. Let I be an $\mathfrak{m}$-primary irreducible ideal and let s be the index of nilpotency of R/I. Then*

$$\nu(I) \leq 2\deg(R)\binom{s+d-3}{d-2} + \binom{s+d-3}{d-3}. \tag{2.21}$$

Proof. The case where $\dim R = 2$ follows from the main result of [BER79]. For higher dimension, we use the same reduction as above with one distinction: set $J_0 = (a_1,\ldots,a_{d-2})^s$. The ideal

$$I/J_0 \hookrightarrow R/J_0$$

is still irreducible, and R/J_0 is a Cohen-Macaulay ring of dimension 2 and multiplicity

$$\deg(R/J_0) = \deg(R)\binom{s+d-3}{d-2}.$$

Applying [BER79, Theorem 1] (see Theorem 2.118) again to I/J_0 and counting generators as above gives the stated estimate. □

Theorem 2.43. *Let $(R,\mathfrak{m})$ be a Cohen-Macaulay local ring of dimension $d \geq 1$ with infinite residue field. If R is not regular then*

$$\mathrm{r}(\mathfrak{m}) \leq d \cdot \deg(R) - 2d + 1.$$

Proof. We apply Theorem 2.40 to the powers of the ideal $\mathfrak{m}$. One has

$$\nu(\mathfrak{m}^n) \leq \deg(R)\binom{n+d-2}{d-1} + \binom{n+d-2}{d-2}.$$

According to Theorem 2.36, it suffices to find n such that

$$\nu(\mathfrak{m}^n) < \binom{n+d}{d},$$

since that will imply that $\mathrm{r}(\mathfrak{m}) \leq n-1$. To this end, choose n so that

$$\deg(R)\binom{n+d-2}{d-1} + \binom{n+d-2}{d-2} < \binom{n+d}{d},$$

which is equivalent with

$$\frac{\deg(R)n}{n+d-1} + \frac{d-1}{n+d-1} < \frac{n+d}{d}.$$

This inequality will be satisfied for $n = d \cdot \deg(R) - 2d + 2$, as desired. □

Reduction Numbers of Primary Ideals

We extend the approach of Theorem 2.40. Here we consider the Hilbert function of $\mathcal{F}(I)$, the special fiber of I. It will suffice to give estimates for the reduction number of I.

Theorem 2.44. *Let $(R,\mathfrak{m})$ be a Cohen-Macaulay local ring of dimension $d \geq 1$ and let I be an $\mathfrak{m}$-primary ideal. Then*

$$\boxed{\nu(I^n) \leq e(I)\binom{n+d-2}{d-1} + \binom{n+d-2}{d-2},} \tag{2.22}$$

where $e(I)$ is the multiplicity of I. Furthermore, if $I \subset \mathfrak{m}^s$, then

$$\boxed{\nu(I^n) \leq \frac{e(I)}{s}\binom{n+d-2}{d-1} + \binom{n+d-2}{d-2}.} \tag{2.23}$$

Proof. The argument here is similar to the previous ones except at one place. Let $J = (a_1, \ldots, a_{d-1}, a_d)$ be a minimal reduction of I. Consider the embedding

$$I^n/J_0^n \hookrightarrow R/J_0^n,$$

where $J_0 = (a_1, \ldots, a_{d-1})$. To bound the number of generators of I^n/J_0^n, we need a bound on the multiplicity of R/J_0^n. Since

$$\deg(R/J_0^n) = \deg(R/J_0)\binom{n+d-2}{d-1},$$

we need only to get hold of $\deg(R/J_0)$. It suffices to note that

$$\deg(R/J_0) = \inf\{\ \lambda(R/(J_0,x)),\ x \text{ regular on } R/J_0\ \},$$

in particular $\deg(R/J_0) \leq \lambda(R/(J_0,a_d)) = \lambda(R/J) = e(I)$. For the second assertion, note that $a_d \in \mathfrak{m}^s$, and then $\deg(R/J_0) \leq \dfrac{e(I)}{s}$. □

In a similar manner to the case $I = \mathfrak{m}$, one has:

Theorem 2.45. *Let $(R, \mathfrak{m})$ be a Cohen-Macaulay local ring of dimension $d \geq 1$ with infinite residue field. If I is an $\mathfrak{m}$-primary ideal whose multiplicity is $e(I)$, then*

$$\boxed{\mathrm{r}(I) \leq d \cdot e(I) - 2d + 1.} \tag{2.24}$$

Furthermore, if $I \subset \mathfrak{m}^s$, then

$$\boxed{\mathrm{r}(I) \leq d \cdot \frac{e(I)}{s} - 2d + 1.} \tag{2.25}$$

Remark 2.46. The argument actually shows that

$$\mathrm{bigr}(I) \leq d \cdot \frac{e(I)}{s} - 2d + 1.$$

The general formula may be strengthened when information about the depth of $\mathrm{gr}_I(R)$ is known. Let us consider one of these extensions. First we recall a procedure (see [Huc87]) when $\mathcal{G} = \mathrm{gr}_I(R)$ has depth at least 1.

Proposition 2.47. *Let $(R, \mathfrak{m})$ be a Noetherian local ring with infinite residue field and I an ideal of positive height. Let $x \in I \setminus \mathfrak{m}I$ be an element such that its initial form in $\mathcal{G}$ is a regular element. Then $\mathrm{r}(I) \leq \mathrm{r}(I/(x))$.*

Proof. The condition on x implies that $\ell(I/(x)) = \ell(I) - 1$. Let L be a minimal reduction of $I/(x)$ of reduction number $r = \mathrm{r}(I/(x))$; lifting L to R we obtain an ideal $J \subset I$ satisfying the equality

$$JI^r + xR = I^{r+1} + xR.$$

Thus for $w \in I^{r+1}$ we can write $w = z + xa$, $z \in JI^r$, and $a \in I^{r+1} : x$. Since x^*, the image of x in I/I^2, is a regular element of $\mathcal{G}$, we must have $a \in I^r$. □

Since the multiplicity behaves similarly (namely $e(I) = e(I/(x))$ if $d \geq 2$), we have:

Theorem 2.48. *Let $(R, \mathfrak{m})$ be a Cohen-Macaulay local ring of dimension $d \geq 2$ with infinite residue field. Let I be an $\mathfrak{m}$-primary ideal whose multiplicity is $e(I)$. If $I \subset \mathfrak{m}^s$ and* depth $\mathrm{gr}_I(R) \geq h \geq 1$ *for some $h < d$, then*

$$\boxed{\mathrm{r}(I) \leq \frac{(d-h)}{s} \cdot e(I) - 2(d-h) + 1.} \tag{2.26}$$

In contrast, an ideal whose associated graded ring has Serre's condition S_1 has a sharper bound for the reduction number. We formulate this in the case of normal ideals (see Proposition 1.142(d)):

Theorem 2.49. *Let $(R, \mathfrak{m})$ be a Cohen-Macaulay local ring of dimension $d \geq 2$ with residue field of characteristic zero. If I is an $\mathfrak{m}$-primary normal ideal whose multiplicity is $e(I)$, then*

$$\boxed{\mathrm{r}(I) \leq e(I) - 1.} \tag{2.27}$$

Maximal Hilbert Functions

The bound for Hilbert functions in Theorem 2.18 gives another method of estimating the reduction number of primary ideals. First, let us combine that theorem with Theorem 1.160.

Theorem 2.50. *Let $(R, \mathfrak{m})$ be a Cohen-Macaulay local ring of dimension $d > 0$ and I an $\mathfrak{m}$-primary ideal.*

(a) *For all $n \geq 0$,*

$$\lambda(I^n/\mathfrak{m}I^n) \leq e(I)\binom{n+d-2}{d-1} + \binom{n+d-2}{d-2}. \tag{2.28}$$

(b) *If* $\mathfrak{m}$ *is not the integral closure of* I, *then for all* $n \geq 0$,

$$\lambda(I^n/\mathfrak{m}I^n) \leq (e(I)-1)\binom{n+d-2}{d-1} + \lambda(R/I)\binom{n+d-2}{d-2}. \qquad (2.29)$$

The first formula is Theorem 2.44 while the second arises from a straightforward combination of Theorems 2.18 and 1.160.

Theorem 2.51. *Let* $(R,\mathfrak{m})$ *and* I *be as above. Then*

(a) *In the case of* (2.28),

$$\mathrm{r}(I) \leq d \cdot e(I) - 2d + 1.$$

(b) *In the case of* (2.29), $\mathrm{r}(I)$ *is bounded by positive integers* n *satisfying*

$$n^2 - ((e(I)-3)d+1)n - (\lambda(R/I)-1)d(d-1) > 0.$$

Proof. In the case of (2.29), by applying (2.36) it suffices to meet the requirement

$$(e(I)-1)\binom{n+d-2}{d-1} + \lambda(R/I)\binom{n+d-2}{d-2} < \binom{n+d}{d}.$$

The binomials make it easy to estimate n, and the inequality is equivalent to the quadratic condition

$$n^2 - ((e(I)-3)d+1)n - (\lambda(R/I)-1)d(d-1) > 0,$$

that can be easily solved. □

General Local Rings

For local rings that are not necessarily Cohen-Macaulay, [RVV1] provides the following bound on Hilbert functions of primary ideals. It can be used as above to fashion bounds for reduction numbers.

Theorem 2.52. *Let* $(R,\mathfrak{m})$ *be a Noetherian local ring of dimension* $d \geq 1$ *and* I *an* $\mathfrak{m}$*−primary ideal in* R. *If* $J = (x_1,\ldots,x_d)$ *is a system of parameters in* I, *then*

$$P_I(t) \leq \frac{\lambda(R/I) + \lambda(I/J)t}{(1-t)^d}.$$

We begin with some general facts about Hilbert functions:

Lemma 2.53 (B. Singh). *Let* $(R,\mathfrak{m})$ *be a Noetherian local ring and* I *an* $\mathfrak{m}$*-primary ideal in* R. *If* $x \in I$ *and* $\overline{I} = I/(x)$, *then*

$$H_I(n) = H^1_{\overline{I}}(n) - \lambda(I^{n+1} : x/I^n)$$

for every $n \geq 0$.

Proof. We set $\overline{R} = R/xR$, and then from the exact sequence

$$0 \to (I^{n+1} : x)/I^n \longrightarrow R/I^n \longrightarrow R/I^{n+1} \longrightarrow \overline{R}/\overline{I}^{n+1} \to 0$$

induced by multiplication by x, we get the desired equality. □

Proposition 2.54. *Let $(R, \mathfrak{m})$ be a Noetherian local ring and I an $\mathfrak{m}$-primary ideal in R. If $x \in I$ and $\overline{I} = I/(x)$, then $P_I(t) \leq \frac{P_{\overline{I}}(t)}{1-t}$.*

Proof. Since

$$\frac{P_{\overline{I}}(t)}{1-t} = P^1_{\overline{I}}(t) = \sum_{n \geq 0} H^1_{\overline{I}}(n)t^n,$$

the conclusion follows by Lemma 2.53. □

Proof of Theorem 2.52. We induct on d. Suppose that $d = 1$ and $J = (x)$, where x is a parameter in I. Since

$$\frac{\lambda(R/I) + \lambda(I/J)t}{1-t} = \frac{\lambda(R/I) + (\lambda(R/J) - \lambda(R/I))t}{1-t} =$$

$$\lambda(R/I) + \lambda(R/J)t + \lambda(R/J)t^2 + \cdots + \lambda(R/J)t^n + \cdots$$

and $H_I(0) = \lambda(R/I)$, we need only prove that $H_I(n) \leq \lambda(R/xR)$ for every $n \geq 1$. We remark that

$$R \supseteq I^n \supseteq I^{n+1} \supseteq xI^n,$$

$$R \supseteq xR \supseteq xI^n,$$

so that

$$\lambda(R/xR) + \lambda(xR/xI^n) = \lambda(R/I^n) + \lambda(I^n/xI^n).$$

Since $\lambda(R/I^n) \geq \lambda(xR/xI^n)$, we get

$$\lambda(R/xR) \geq \lambda(I^n/xI^n) = H_I(n) + \lambda(I^{n+1}/xI^n) \geq H_I(n).$$

Suppose that $d \geq 2$, and let x_1 be a parameter in I. We set $\overline{R} = R/x_1R$, $\overline{I} = I/x_1R$, $\overline{J} = J/x_1R$. Then $\overline{I}$ is a primary ideal in the local ring $\overline{R}$, which has $\dim \overline{R} = d - 1$. By induction we have

$$P_{\overline{I}}(t) \leq \frac{\lambda(\overline{R}/\overline{I}) + \lambda(\overline{I}/\overline{J})t}{(1-t)^{d-1}} = \frac{\lambda(R/I) + \lambda(I/J)t}{(1-t)^{d-1}}.$$

Since $\dfrac{1}{1-t} \geq 0$, we get

$$\frac{P_{\overline{I}}(t)}{1-t} \leq \frac{\lambda(R/I) + \lambda(I/J)t}{(1-t)^d}.$$

The conclusion follows since $P_I(t) \leq \dfrac{P_{\overline{I}}(t)}{1-t}$ by Proposition 2.54. □

Nearly the same treatment applies to standard graded algebras:

Proposition 2.55. *Let $(G_0, \mathfrak{m})$ be an Artinian local ring and $G = \bigoplus_{n\geq 0} G_n$ a standard graded algebra of dimension $d \geq 1$. If J is the ideal generated by a homogeneous system of parameters in G, then*

$$P_G(t) \leq \frac{\lambda(G_0) + (\lambda(G/J) - \lambda(G_0))t}{(1-t)^d}.$$

If equality holds here, then G is Cohen-Macaulay.

Proof. Set $I = G_+ = \bigoplus_{n\geq 1} G_n$, and note that I is primary for the irrelevant maximal ideal M of G, $\mathrm{gr}_I(G) \cong G$, and that J is generated by a system of parameters. Note also that the associated graded rings and lengths are not changed if G or the localization G_M are considered. Theorem 2.52 can now be applied directly. □

Remark 2.56. One application of this formula is to employ the technique we have been using to obtain estimates for the reduction number of the ideal I. In other words, to find a minimal reduction L of I and an integer r such that $I^{r+1} = LI^r$. For simplicity of notation, set $a := \lambda(R/I)$ and $b := \lambda(I/J)$. From the inequality

$$P_I(t) \leq \frac{\lambda(R/I) + \lambda(I/J)t}{(1-t)^d},$$

of Hilbert functions, we have that for each positive integer n,

$$\lambda(I^n/\mathfrak{m}I^n) \leq \lambda(I^n/I^{n+1}) \leq a\binom{n+d-1}{d-1} + b\binom{n+d-2}{d-1}.$$

According to Theorem 2.36, if for some integer n we bound the right hand side of this inequality by $\binom{n+d}{d}$, we have found a reduction L of I with reduction number less that n. This is easy to work out, since it leads to a quadratic inequality:

$$(n+d)(n+d-1) > ad(n+d-1) + bdn,$$

which will be satisfied for

$$n \leq dc - 2d + 1 + \sqrt{(a-1)(d-1)d},$$

where $c = a + b = \lambda(R/J)$.

Remark 2.57. Given a Noetherian local ring $(R, \mathfrak{m})$, then for an $\mathfrak{m}$-primary ideal I one might consider the three quantities $e = e(I)$ together with

$$\begin{aligned} e_+(I) &= \inf\{\lambda(R/J)\}, \\ e_{++}(I) &= \sup\{\lambda(R/J)\}, \end{aligned}$$

where J runs over all the minimal reductions of I. It follows from Theorem 1.146 that e_{++} is finite. It is not very clear how these integers compare. Nevertheless there are general bounds for $e_{++}(I)$. For example, if $\dim R = 1$ it is easy to see that $e_{++}(I) \leq e(I) + \lambda(H^0_{\mathfrak{m}}(R))$.

Lech's Formula

The estimates of the reduction number of a primary ideal I that we have obtained thus far all depend on the multiplicity $e(I)$ of I. However, this is a number that is very hard to get hold of if I is only described by its generators. (In Chapter 7 we shall treat numerical methods to deal with this issue.) There is a formula of C. Lech ([Le60]) that gives rise to *a priori* bounds for $e(I)$. A tight closure derivation of this bound is given in [Ha2]; in several cases, it is sharper.

Theorem 2.58. *If* $(R,\mathfrak{m})$ *is a Noetherian local ring of dimension* d *and* I *is an* $\mathfrak{m}$*-primary ideal, then*

$$\boxed{e(I) \leq d! \cdot \deg(R) \cdot \lambda(R/\overline{I}).} \tag{2.30}$$

There are some issues of concern here. The original proof is quite complicated, although rich in techniques. It would be of interest to have another rendering of it in which the criticality of the various elements of the formula are made more apparent. The original formula had $\lambda(R/I)$ instead of $\lambda(R/\overline{I})$, but the modification is valid in view of the result of Rees that I and $\overline{I}$ have the same multiplicity. On the other hand, applying this formula to the powers of I, when $\lambda(R/I^n)$ is given by the Hilbert polynomial, it becomes clear that the factorial $d!$ cannot be lowered.

Cohen-Macaulay Associated Graded Rings

To round out the overall picture of the relationship between the reduction number of a primary ideal I and the depth of its associated graded ring $\mathrm{gr}_I(R)$, we consider the following extremal case (see [Sch90, Theorem 4.4]).

Theorem 2.59. *Let* $(R,\mathfrak{m})$ *be a Cohen-Macaulay local ring with infinite residue field and* I *an* $\mathfrak{m}$*-primary ideal. If* $\mathrm{gr}_I(R)$ *is Cohen-Macaulay, then*

$$\mathrm{r}(I) \leq e(I) - \lambda(R/I). \tag{2.31}$$

Proof. We may assume that $\dim R = d \geq 1$. Let J be a minimal reduction of I. Modding out $d-1$ minimal generators of J, without loss of generality we reduce to the case $d=1$, $J=(x)$. Set $A = \mathrm{gr}_J(R) \otimes (R/I)$, $B = \mathrm{gr}_I(R)$; then $A = R/I[t]$ is an standard graded algebra of multiplicity $e(A) = \lambda(R/I)$, and it is easy to see that it embeds in B since t is regular on B by the Cohen-Macaulay hypothesis. Consider the corresponding exact sequence of A-modules

$$0 \rightarrow A \longrightarrow B \longrightarrow C \rightarrow 0.$$

Tensoring by A/tA, we get

$$0 \rightarrow \mathrm{Tor}_1^A(A/tA, C) \longrightarrow A/tA \longrightarrow B/tB \longrightarrow C/tC \rightarrow 0.$$

Since A/tA embeds in B/tB, $\mathrm{Tor}_1^A(A/tA,C)=0$, which shows that C is a Cohen-Macaulay A-module of multiplicity $e(C)=e(B)-e(A)=e(I)-\lambda(R/I)$. Since $e(C)=\lambda(C/xC)\geq \nu_{R/I}(C/xC)$, it suffices to examine the module C/xC. In degree n, $(C/xC)_n=I^n/(x^n,xI^{n-1},I^{n+1})$; if this component vanishes, the following components are also zero. It follows that $\nu_{R/I}(C/xC)\geq \mathrm{r}(I)$, as desired. □

2.3 Degree Functions

Let $(R,\mathfrak{m})$ be a Noetherian local ring (or a Noetherian graded algebra) and let $\mathcal{M}(R)$ be the category of finitely generated R-modules (or the appropriate category of graded modules). A *degree function* is a numerical function $\mathbf{d}:\mathcal{M}(R)\mapsto\mathbb{N}$. The more interesting of them initialize on modules of finite length and have mechanisms that control how the function behaves under generic hyperplane sections. Thus, for example, if L is a given module of finite length, one function may require that

$$\mathbf{d}(L):=\lambda(L),$$

or when L a graded module

$$L=\bigoplus_{i\in\mathbb{Z}}L_i,$$

the requirement is that

$$\mathbf{d}(L):=\sup\{i\mid L_i\neq 0\}.$$

In this section and in the next we study some of these functions, particularly those that are not treated extensively in the literature. One of our aims is to provide an even broader approach to Hilbert functions that does not depend so much on the base rings being Artinian. We shall be looking at the associated multiplicities and seek ways to relate them to the reduction numbers of the algebras. Among the *multiplicities* treated here and in the next section are:

- Castelnuovo-Mumford regularity
- Classical multiplicity
- Arithmetic degree
- Extended multiplicity
- Cohomological degree
- Homological degree
- bdeg

2.3.1 Classical Degrees

Castelnuovo-Mumford Regularity: $\mathrm{reg}(\cdot)$

This is one of most useful of the degree functions and has excellent treatments in [Ei95] and [EG84]. It has several equivalent formulations, one of which is the following: let $R=k[x_1,\ldots,x_n]$ be a ring of polynomials over the field k with the standard grading, and let A be a finitely generated graded R-module with a minimal graded resolution

$$0 \to F_n \longrightarrow F_{n-1} \longrightarrow \cdots \longrightarrow F_1 \longrightarrow F_0 \to 0,$$

$$F_j = \bigoplus_j R(-a_{ij}).$$

Then $\operatorname{reg}(\cdot)$ is defined by

$$\operatorname{reg}(A) = \sup\{a_{i,j} - i\}.$$

It can also be characterized by its initialization as above on modules of finite length (see [Ei95, Proposition 20.20]):

Proposition 2.60. *If h is a linear form of R whose annihilator $0 :_A h$ has finite length, then*

$$\operatorname{reg}(A) = \max\{\operatorname{reg}(0 :_A h), \operatorname{reg}(A/hA)\}.$$

As a long exercise, the reader can verify that $\operatorname{reg}(\cdot)$ is the unique degree function satisfying these rules.

Classical Multiplicity for Modules: $\deg(\cdot)$

If $(R, \mathfrak{m})$ is a Noetherian local ring, the *ordinary multiplicity* of a finitely generated R-module A is defined much as we have treated the case of R itself. For large values of n, the function

$$n \mapsto \lambda(A/\mathfrak{m}^n A),$$

is a polynomial

$$e_0(A)\binom{n+d-1}{d} - e_1(A)\binom{n+d-2}{d-1} + \cdots + (-1)^d e_d(A),$$

whose coefficients (in this normalized form) are integers encoding various properties of A. Thus its Krull dimension (if $e_0(A) \neq 0$) is d and its leading coefficient is the integer

$$\deg(A) := \lim_{n\to\infty} \frac{\lambda(A/\mathfrak{m}^n A)}{d! n^d},$$

the *multiplicity* of A.

Closely related are the Samuel multiplicities associated to $\mathfrak{m}$-primary ideals. To such an ideal I,

$$e(I;A) := \lim_{n\to\infty} \frac{\lambda(A/I^n A)}{d! n^d}.$$

These functions can also be characterized by the way they deal with generic hyperplane sections.

Proposition 2.61. *Let A be a finitely generated R-module of dimension d and let $h \in \mathfrak{m} \setminus \mathfrak{m}^2$ be such that $0 :_A h$ has finite length. Then*

$$\deg(A) = \begin{cases} \lambda(A) & \text{if } \dim A = 0 \\ \lambda(A/hA) - \lambda(0 :_A h) & \text{if } \dim A = 1 \\ \deg(A/hA) & \text{if } \dim A \geq 2. \end{cases}$$

Multiplicities are usually extracted from Hilbert functions of graded modules. The latter usually arise from constructions on other modules to which, by abuse of terminology, we attach the multiplicities. Let us consider one special case of interest.

Let $(R, \mathfrak{m})$ be a Noetherian local ring of dimension d and A a standard graded algebra over R,

$$A = A_0 + A_1 + \cdots = A_0 + A_+ = A_0[A_1] = R[T_1, \ldots, T_n]/L.$$

We shall assume, without undue restriction, that $A_0 = R$. Denote by M the irrelevant maximal ideal of A, $M = (\mathfrak{m}A_0, A_1)$.

The algebras that we have in mind are the associated graded rings $\mathrm{gr}_I(R)$ of ideals with $\dim R/I \geq 1$, and Rees algebras of modules $\mathcal{R}(E)$.

Definition 2.62. For any finitely generated A-module C, the *associated graded module* of C is

$$\mathrm{gr}_M(C) = \bigoplus_n M^n C / M^{n+1} C.$$

In the truly classical case, $(R, \mathfrak{m})$ is a local ring and C is a finitely generated R-module of Krull dimension d, and its multiplicity $\deg(C)$ is obtained, recall, from the following construction. The Hilbert function of the graded module $\mathrm{gr}_{\mathfrak{m}}(C)$ (over the graded ring $G = \mathrm{gr}_{\mathfrak{m}}(R)$) is a polynomial

$$\lambda(\mathfrak{m}^n C / \mathfrak{m}^{n+1} C) = \frac{\deg(C)}{(d-1)!} n^{d-1} + \text{lower terms} \quad n \gg 0.$$

This construction can be applied to A itself and gives rise to the identifications

$$\begin{aligned} \mathcal{G} = \mathrm{gr}_M(A) &= \bigoplus_{n \geq 0} (\mathfrak{m}, A_+)^n / (\mathfrak{m}, A_+)^{n+1} \\ &= \bigoplus_{n \geq 0} \bigoplus_{j=0}^{n} \mathfrak{m}^j A_{n-j} / \mathfrak{m}^{j+1} A_{n-j}, \end{aligned}$$

which is a bigraded $R/\mathfrak{m}$-algebra. Even though $\mathrm{gr}_M(A)$ is the ordinary tangent cone of the local ring A_M, we shall want to make use of the additional structure. If C is a graded A-module, $\mathrm{gr}_M(C)$ is a (bi-) graded module over $\mathcal{G}$. We denote the maximal irrelevant ideal of $\mathcal{G}$ by $\mathcal{M}$, $\mathcal{M} = (\mathfrak{m}/\mathfrak{m}^2 \oplus A_1/\mathfrak{m}A_1)\mathcal{G}$.

These functions must be applied with care when we deal with sequences of modules (see [BH93, Section 4.7] for details). It is a very different behavior from the case of graded modules.

Proposition 2.63. *Let $(R,\mathfrak{m})$ be a Noetherian local ring and I an $\mathfrak{m}$-primary ideal. If*

$$0 \to A \longrightarrow B \longrightarrow C \to 0$$

is an exact sequence of finitely generated R-modules of the same Krull dimension, then

$$\deg_I(B) = \deg_I(A) + \deg_I(C).$$

Proof. The proof is standard but we provide it in order to discuss an important additional point. Consider the exact sequence

$$0 \to (A \cap \mathfrak{m}^n B)/\mathfrak{m}^n A \longrightarrow A/\mathfrak{m}^n A \longrightarrow B/\mathfrak{m}^n B \longrightarrow C/\mathfrak{m}^n C \to 0,$$

and the corresponding equality of numerical functions:

$$\lambda(A/\mathfrak{m}^n A) + \lambda(C/\mathfrak{m}^n C) = \lambda(B/\mathfrak{m}^n B) + \lambda((A \cap \mathfrak{m}^n B)/\mathfrak{m}^n A).$$

For n large, the lengths of the last 3 modules are given by polynomials of the same degree, of coefficients $\deg_I(A), \deg_I(B), \deg_I(C)$, respectively. On the other hand, by the Artin-Rees lemma there is an integer $r \geq 0$ such $A \cap \mathfrak{m}^n B = \mathfrak{m}^{n-r}(A \cap \mathfrak{m}^r B)$ that for $n \geq r$. This gives an embedding

$$(A \cap \mathfrak{m}^n B)/\mathfrak{m}^n A \hookrightarrow \mathfrak{m}^{n-r} A/\mathfrak{m}^n A,$$

and the inequality

$$\lambda((A \cap \mathfrak{m}^n B)/\mathfrak{m}^n A) \leq \lambda(A/\mathfrak{m}^n A) - \lambda(A/\mathfrak{m}^{n-r} A).$$

But this difference is a polynomial in n of degree $d-1$, which is enough to establish the equality $\deg_I(B) = \deg_I(A) + \deg_I(C)$. $\square$

Needless to say, it is quite a burden to get hold of these numbers when $\dim R = d > 0$. For that it may be helpful to use the following adaptation of the notion of superficial element. The Krull dimension of $\mathcal{G}$ is $\dim A$. It is put together as

$$\dim A = \dim R + \text{height } A_+.$$

Similarly we have

$$\begin{aligned} \dim \mathcal{G} &= \dim R + \text{height } (A_1/\mathfrak{m}A_1)\,\mathcal{G} \\ &= \dim \operatorname{gr}_{\mathfrak{m}}(R) + \text{height } (A_1/\mathfrak{m}A_1)\,\mathcal{G}. \end{aligned}$$

We denote these summands by d and g, respectively, and both are assumed positive.

Proposition 2.64. *Suppose that $R/\mathfrak{m}$ is infinite and that $g > 0$. There exists $x \in A_1$ whose image $z \in A_1/\mathfrak{m}A_1$ has the following properties:*

(i) $A_1/\mathfrak{m}A_1 \subset \sqrt{0 : (0 : z)}$.

(ii) *If* $\overline{\mathcal{G}} = \mathrm{gr}_M(A/(x))$, *then* $\deg(\overline{\mathcal{G}}) = \deg(\mathcal{G})$.

Proof. Denote the set of associated primes of $\mathcal{G}$ by

$$\mathrm{Ass}(\mathcal{G}) = \{P_1, \ldots, P_m, Q_1, \ldots, Q_n\},$$

where $A_1/\mathfrak{m}A_1 \not\subset P_i$. Choose $x \in A_1$ so that its image

$$z \in A_1/\mathfrak{m}A_1 \setminus \bigcup P_i.$$

We prove that z has the asserted properties. Let $\wp$ be a minimal prime of $0 : (0 : z)$. Localizing, we get that in the local ring $\mathcal{G}_\wp$ its maximal ideal is minimal over some annihilator ideal, and therefore must it arise from one of the associated primes of $\mathcal{G}$. Since $z \in 0 : (0 : z)$, $\wp$ cannot be one of the P_i. Thus it has to be one of the Q_j, each of which contains $A_1/\mathfrak{m}A_1$, thus proving (i).

Consider the exact sequence of graded modules:

$$0 \to K = 0 : z \longrightarrow \mathcal{G}[-1] \longrightarrow \mathcal{G} \longrightarrow \widetilde{\mathcal{G}} = \mathcal{G}/(z) \to 0.$$

Since $0 : z$ is annihilated by some power of the ideal generated by $A_1/\mathfrak{m}A_1$, which has codimension $r > 0$, it follows that $0 : z$ has dimension at most d. This shows that $\deg(\mathcal{G}) = \deg(\widetilde{\mathcal{G}})$.

We must now compare the multiplicities of $\overline{\mathcal{G}}$ and $\widetilde{\mathcal{G}}$. For that we introduce some variation into the absolute case. There exists a natural exact sequence

$$0 \to L \longrightarrow \overline{\mathcal{G}} \longrightarrow \widetilde{\mathcal{G}} \to 0.$$

We shall show that $\dim K \geq \dim L$, which will be enough to establish the claim.

In terms of the ideal M, the comparable components of K and L are, respectively,

$$\begin{aligned} K_n &= (M^n \cap (M^{n+2} : x))/M^{n+1} \\ L_n &= (M^{n+1} + M^n \cap (x))/(M^{n+1} + xM^{n-1}). \end{aligned}$$

We shall show that if C is the annihilator of K as a $\mathcal{G}$-module, then some power of C will annihilate L. Let c be the Artin-Rees index of the filtration $M^n \cap (x)$: that is,

$$M^n \cap (x) = M^{n-c}(M^c \cap (x)), \quad n \geq c;$$

in particular,

$$M^n \cap (x) \subset xM^{n-c}, \quad M^n : x \subset (0 : x) + M^{n-c}.$$

We claim that $w^{c-1} \cdot L = 0$ for any $w \in C$. We shall show instead that w^{c-1} annihilates the module $M^n \cap (x)/xM^{n-1}$. For that it will suffice to show that for any $w \in C_s$ and all large m,

$$w^{c-1}(M^m \cap (x)) \subset xM^{(c-1)s+m-1}.$$

By assumption, we have that for all n,

$$w(M^n \cap (M^{n+2} : x)) \subset M^{s+n+1}.$$

In the manner of [Tr98, Lemma 4.4], we shall use this relation for $n = m - c + 2, \ldots, m$. Consider two instances,

$$w(M^{n+2s-2} \cap M^{n+2s} : x)) \subset M^{n+3s-1}$$

and

$$w(M^{n+s-3} \cap (M^{n+s-1} : x)) \subset M^{n+2s-2},$$

which, upon substituting the second relation into the first, yields

$$w(w(M^{n+s-3} \cap (M^{n+s-1} : x)) \cap (M^{n+2s} : x)) \subset M^{n+3s-1}.$$

Since $w(M^{n+s} : x) \subset M^{n+2s} : x)$, another substitution enhances the containment and we get

$$w^2(M^{n+s-3} \cap (M^{n+s} : x)) \subset M^{n+3s-1}.$$

Another iteration will give rise to

$$w^3(M^{n-4} \cap (M^n : x)) \subset M^{n+3s-1}.$$

To make the argument transparent, assume that $c = 4$. It follows that $w^3(M^n \cap (x)) \subset xM^{n-1}$ for all large n, as desired. □

In the case of modules over a local ring, this can be stated as follows:

Corollary 2.65. *Let $(R, \mathfrak{m})$ be a local ring with infinite residue field, I an $\mathfrak{m}$-primary ideal and C a finitely generated R-module of Krull dimension at least two. There exists an element $z \in I$ such that* $\dim C/zC = \dim C - 1$ *and* $\deg_I(C) = \deg_I(C/zC)$.

As for the computation of the multiplicity of general associated graded rings, the outlook is hard, the exception being the case of equimultiple ideals.

Proposition 2.66. *Let $(R, \mathfrak{m})$ be a Cohen-Macaulay local of dimension d with infinite residue field and I an equimultiple ideal of codimension g. If J is a minimal reduction of I, then* $\deg(\mathrm{gr}_I(R)) = \deg(R/J)$.

Proof. If $g = d$, the assertion follows from reading the Hilbert function of $G = \mathrm{gr}_I(R)$. Suppose then that $g < d$ and set $J = (x_1, \ldots, x_g)$. The images of these elements in G_1 form a superficial sequence, so that $\deg(G) = \deg(G')$ since $g < d = \dim G$, where G' is the associated graded ring of the ideal $I' = I/J \subset R/J = R'$. The graded ring G' has only finitely many components, so that its minimal primes are of the form $P = (\mathfrak{p}, G'_+)$, with $\mathfrak{p}$ a minimal prime of R'. We thus have

$$\begin{aligned}\deg(G') &= \sum_P \lambda(G'_P) \cdot \deg(G'/P) = \sum_P \lambda(G'_\mathfrak{p}) \cdot \deg(R'/\mathfrak{p}) \\ &= \sum_\mathfrak{p} \lambda(R'_\mathfrak{p}) \cdot \deg(R'/\mathfrak{p}) = \deg(R') = \deg(R/J).\end{aligned}$$

2.3.2 Generalized Multiplicities of Graded Modules

Let R be a Noetherian ring and denote by $\mathcal{M}(R)$ the category of finitely generated R-modules. A *degree* or *multiplicity* is a numerical function

$$\mathrm{Deg} : \mathcal{M}(R) \mapsto \mathbb{N}$$

that *assembles well* and *behaves well with respect to generic hyperplane sections.* It must also have interesting applications! Let us try to make this foggy description more concrete.

To *assemble well* we to require that $\mathrm{Deg}(\cdot)$ gives rise to a theory of Hilbert functions in the following sense. Let A be a finitely positively generated graded R-algebra and M a finitely generated graded A-module. In the wish list, the formal power series

$$H_M(t) = \sum_n \mathrm{Deg}(M_n)t^n$$

is rational and its invariants should be a reflection of the properties of M.

Thus for each prime ideal $\mathfrak{p}$ one can define the function

$$\deg_{\mathfrak{p}}(E) = \lambda(H^0_{\mathfrak{p}}(E_{\mathfrak{p}}))$$

on $\mathcal{M}(R)$ which leads to the usual Hilbert functions over the local ring $R_{\mathfrak{p}}$. Another function is

$$\text{arith-deg}(E) = \sum_{\mathfrak{p}\in \text{spec } (R)} \deg_{\mathfrak{p}}(E)\cdot \deg(R/\mathfrak{p}),$$

(this is actually a finite sum) which is the *arithmetic degree* of E.

The other requirements on $\mathrm{Deg}(\cdot)$ have to do with facilitating its calculus. This works best when $\mathrm{Deg}(\cdot)$ is additive on short exact sequences. This is the case when R is Artinian and $\mathrm{Deg}(E) = \lambda(E)$, the classical Hilbert function. There are other cases as well, which we discuss later. It is however a restrictive condition and we shall be satisfied when it is additive on certain short exact sequences. (In [Va98c] there are several applications of these degrees to the estimation of the number of generators of arbitrary ideals.)

Replacing Length by Multiplicities

Let $(R,\mathfrak{m})$ be a Noetherian local ring, $A = R[x_1,\ldots,x_d]$ a graded R-algebra and

$$M = \bigoplus_{n\geq 0} M_n$$

a finitely generated graded A-module. If R is not Artinian, it is still of interest to define numerical functions on the components of M, assemble it in the corresponding Hilbert series and look at the possible applications. Here we use one of the functions studied in [Do97].

Let $\operatorname{Deg}(\cdot)$ be one of these functions and consider the associated formal power series

$$H_M(t) = \sum_{n\geq 0} \operatorname{Deg}(M_n)t^n.$$

In view of the properties of $\operatorname{Deg}(\cdot)$, expected estimates for it can be obtained in cases of interest, which in turn can be used to bound the number of generators of the components M_n.

To illustrate, let us consider the associated graded ring $G = \operatorname{gr}_I(R)$ of an ideal I of dimension one. We have, according to Proposition 2.75,

$$\begin{aligned} H_G(t) &= \sum_{n\geq 0} \operatorname{Deg}(I^n/I^{n+1})t^n \\ &= \sum_{n\geq 0} \deg(I^n/I^{n+1})t^n + \sum_{n\geq 0} \lambda(\Gamma_{\mathfrak{m}}(I^n/I^{n+1}))t^n. \end{aligned}$$

The first formal power series will be studied next with some generality (it has a corresponding Hilbert polynomial of degree height $I-1$). The second is the ordinary Hilbert function of the ideal $\Gamma_{\mathfrak{m}}(G) \subset G$.

Proposition 2.67. *Let $(R,\mathfrak{m})$ be a Noetherian local ring, I an ideal of height $g>0$ and $G=\operatorname{gr}_I(R)$. The module $\Gamma_{\mathfrak{m}}(G)\subset G$ has dimension at most the analytic spread $\ell(I)$. If G is height unmixed then $\Gamma_{\mathfrak{m}}(G)$ wither vanishes or it has dimension $\ell(I)$.*

See the proof of Proposition 2.107.

Extended Hilbert Function for the Classical Multiplicity

Let us at least find the contribution of the multiplicity to $\operatorname{Deg}(M_n)$, that is, the formal power series

$$H_M(t) = \sum_{n\geq 0} \deg(M_n)t^n,$$

where $\deg(M_n)$ is the ordinary multiplicity of the module M_n. As in the case of Artin modules, $H_M(t)$ is a rational function. The corresponding polynomial

$$\deg(M_n) = P_M(n) = \sum_{i=0}^{r}(-1)^i E_i \binom{n+r-i}{r-i} \tag{2.32}$$

has properties similar to those of the ordinary Hilbert polynomial once its degree r has been characterized (see [Do97], who introduced this function).

Let us clarify the meaning of the degree of these polynomials. Let $(R,\mathfrak{m})$ be a Noetherian local ring and $A = R[x_1,\ldots,x_m]$ a standard graded R-algebra. Let $\mathfrak{p}_1,\ldots,\mathfrak{p}_s$ be the primes of R of maximal dimension, and M a finitely generated graded A-module. The extended degree for the multiplicity is computed as

$$n \mapsto \sum_{i=1}^{s} \lambda((M_n)_{\mathfrak{p}_i}) \deg(R/\mathfrak{p}_i),$$

an explanation for the degree lying in the growth of the Hilbert function. Let $\mathfrak{Q}$ be an associated prime of M, $\mathfrak{Q} = \mathfrak{p} + Q$, with $\mathfrak{p} \subset R$ and $Q \subset A_+$. If $\mathfrak{p}$ is one of the above primes, $A/\mathfrak{Q}$ gives a contribution to $\deg(M_n)$ equal to

$$\deg(R/\mathfrak{p})\lambda(((A_+/Q)_n)_\mathfrak{p}).$$

Note that this is the Hilbert function of $(A/\mathfrak{Q})_\mathfrak{p}$, which according to Lemma 1.21 has codimension $\dim A/\mathfrak{Q} - \dim R/\mathfrak{p}$. Note that if none of the associated primes of M have this form, the function is the zero function.

For later reference, let us collect these observations.

Proposition 2.68. *For R, A and M as above, for $n \gg 0$ the function $\deg(M_n)$ is a polynomial of degree $\max\{\dim A/\mathfrak{Q} - \dim R - 1\}$ over all associated primes $\mathfrak{Q}$ of M with $\dim R/(\mathfrak{Q} \cap R) = \dim R$.*

Here we apply this approach to the associated graded ring of an ideal I, or equivalently to the function

$$n \rightsquigarrow \deg(R/I^n).$$

This function is easy to express in terms of the multiplicities of the localizations of I at some of its minimal primes. Indeed, one has

$$\deg(R/I^n) = \sum_{\dim R/I = \dim R/\mathfrak{p}} \lambda(R_\mathfrak{p}/I_\mathfrak{p}^n)\deg(R/\mathfrak{p}).$$

It follows that this function behaves for $n \gg 0$ as a polynomial of degree

$$r = \max\{\text{height } I_\mathfrak{p} \mid \dim R/I = \dim R/\mathfrak{p}\},$$

whose leading coefficient in the binomial representation is

$$E_0(I) = \sum e_0(I_\mathfrak{p})\deg(R/\mathfrak{p}),$$

the sum extended over such primes. There are similar expressions for the $E_i(I)$ in terms of the local Hilbert coefficients $e_i(I_\mathfrak{p})$. (Soon we shall assume that R and I are well-behaved enough so that the sum may be taken over all the minimal primes of I.)

Remark 2.69. One glaring difficulty with this formula lies with the numbers $e_0(I_\mathfrak{p})$, which are hard to get hold of. On the other hand, the qualitative behavior of the $E_i(I)$ is that of the primary ideal case. For instance, if R is a Cohen-Macaulay ring and I is equidimensional and unmixed, then $E_1(I) = 0$ precisely when I is generically a complete intersection. Moreover, one has that $E_2(I) \geq 0$.

There is a special case when some of these coefficients may be found more easily. Suppose that I is an equimultiple ideal of codimension g and that $J = (a_1, \ldots, a_g)$ is one of its minimal reductions. For any minimal prime of I, one has

$$e_0(I_\mathfrak{p}) = e_0(J_\mathfrak{p}) = \lambda(R_\mathfrak{p}/J_\mathfrak{p}),$$

and this allows us to state:

Proposition 2.70. *If R is a Cohen-Macaulay local ring and I is an equimultiple ideal with a minimal reduction J, then*

$$E_0(I) = E_0(J) = \deg(R/J)$$
$$E_1(I) \leq \binom{E_0(I)-1}{2}.$$

Proof. For the second assertion, we recall the standard bound

$$e_1(I_\mathfrak{p}) \leq \binom{e_0(I_\mathfrak{p})-1}{2},$$

which, when carried into the expression for $E_1(I)$ gives the desired inequality. □

A very general bound for $E_0(I)$ arises from Lech's formula.

Proposition 2.71. *Let R be a Cohen-Macaulay local ring and I an unmixed ideal of height $g > 0$. Then*
$$E_0(I) \leq g!\deg(R/\overline{I})\deg(R).$$

Proof. We estimate $E_0(I)$ as given above using Lech's inequality: we have

$$\begin{aligned} E_0(I) &= \sum e_0(I_\mathfrak{p})\deg(R/\mathfrak{p}) \\ &\leq \sum g!\lambda(R_\mathfrak{p}/\overline{I}_\mathfrak{p})\deg(R_\mathfrak{p})\deg(R/\mathfrak{p}) \\ &= g!(\sum \lambda(R_\mathfrak{p}/\overline{I}_\mathfrak{p})\deg(R/\mathfrak{p}))\deg(R_\mathfrak{p}) \\ &\leq g!(\sum \lambda(R_\mathfrak{p}/\overline{I}_\mathfrak{p})\deg(R/\mathfrak{p}))\deg R \\ &= g!\deg(R/\overline{I})\deg R, \end{aligned}$$

where we have used the fact that $\deg R_\mathfrak{p} \leq \deg R$ ([Na62, Theorem 40.1]). □

The coefficients E_i share many properties with the ordinary Hilbert coefficients but differ in subtle ways, particularly in respect to exact sequences. In one case of interest the behavior is the expected one. One point that must be kept in sight is that for each dimension $s \leq \dim R$ there is a function $E_0(\cdot)$; therefore, while considering different modules in an exact sequence, one must be explicit about which function is being used on which module.

Proposition 2.72. *Let $S = R[x_1, \ldots, x_d]$ be a polynomial ring over the local ring R and let M be a finitely generated graded S-module. Let φ be a graded, nilpotent endomorphism of M. Then*

$$0 \neq E_0(M/\varphi(M)) = E_0(M) - E_0(\varphi(M)).$$

Proof. This follows from two observations. The associated primes of M that will contribute to $E_0(M)$ are those of maximal dimension among the primes in

$$\lim_{n\to\infty} \operatorname{Ass} M_n.$$

On the other hand, when we view M as a module over $S[t]$, where $t \cdot m = \varphi(m)$, one has the elementary equality

$$\sqrt{\text{ann }(M/tM)} = \sqrt{(t, \text{ann }(M))} = \sqrt{\text{ann }(M)}$$

of radical ideals, since t is nilpotent on M. It will follow that the minimal primes in $\lim_{n\to\infty} \text{Ass} M_n$ will be the same as those in $\lim_{n\to\infty} \text{Ass}(M/\varphi(M))_n$. □

2.4 Cohomological Degrees

Let S be a finitely generated graded algebra over a field k or a local Noetherian ring. An *extended degree*, or *cohomological degree*, is a numerical function $\text{Deg}(\cdot)$ on the category $\mathcal{M}(S)$ of finitely generated S-modules that seeks to capture the size of the module along with some of the complexity of its structure. They tend to have a homological character but are still amenable to explicit calculation by computer algebra systems. Our aim is to use the existence of these functions to get *a priori* estimates of the kind usually associated with Castelnuovo-Mumford regularity. A significant aspect of these functions lies in the fact that they can also be defined on the category of modules over local rings.

A general class of these functions was introduced in [DGV98] , while a prototype was defined earlier in [Va98a]. In his thesis ([Gu98]), T. Gunston carried out a more formal examination of such functions in order to introduce his own construction of a new cohomological degree. One of the points that must be taken care of is that of an appropriate *generic hyperplane* section. Let us recall the setting.

Throughout we suppose that the residue field k of R is infinite.

Definition 2.73. If $(R, \mathfrak{m})$ is a local ring, a *notion of genericity* on $\mathcal{M}(R)$ is a function

$$U : \{\text{isomorphism classes of } \mathcal{M}(R)\} \longrightarrow \{\text{non-empty subsets of } \mathfrak{m} \setminus \mathfrak{m}^2\}$$

subject to the following conditions for each $A \in \mathcal{M}(R)$:

(i) If $f - g \in \mathfrak{m}^2$ then $f \in U(A)$ if and only if $g \in U(A)$.
(ii) The set $\overline{U(A)} \subset \mathfrak{m}/\mathfrak{m}^2$ contains a non-empty Zariski-open subset.
(iii) If depth $A > 0$ and $f \in U(A)$, then f is regular on A.

There is a similar definition for graded modules. We shall usually switch notation, denoting the algebra by S.

Fixing a notion of genericity $U(\cdot)$ one has the following extension of the classical multiplicity.

Definition 2.74. A *cohomological degree*, or *extended multiplicity function*, is a function

$$\text{Deg}(\cdot) : \mathcal{M}(R) \mapsto \mathbb{N},$$

that satisfies the following conditions.

(i) If $L = \Gamma_{\mathfrak{m}}(A)$ is the submodule of elements of A that are annihilated by a power of the maximal ideal and $\overline{A} = A/L$, then

$$\operatorname{Deg}(A) = \operatorname{Deg}(\overline{A}) + \lambda(L), \tag{2.33}$$

where $\lambda(\cdot)$ is the ordinary length function.

(ii) (Bertini's rule) If A has positive depth and $h \in U(A)$, then

$$\operatorname{Deg}(A) \geq \operatorname{Deg}(A/hA). \tag{2.34}$$

(iii) (The calibration rule) If A is a Cohen-Macaulay module, then

$$\operatorname{Deg}(A) = \deg(A), \tag{2.35}$$

where $\deg(A)$ is the ordinary multiplicity of A.

Any such function will satisfy $\operatorname{Deg}(A) \geq \deg(A)$, with equality holding if and only if A is Cohen-Macaulay. It might be convenient to refer to the difference $\operatorname{Deg}(A) - \deg(A)$ as the Cohen-Macaulay *correction* or *deficiency* of A (for the $\operatorname{Deg}(\cdot)$ function).

In one special case this notion becomes very familiar.

Proposition 2.75. *Let R be a local Noetherian ring and M a finitely generated R-module of Krull dimension one. For any extended multiplicity function* $\operatorname{Deg}(\cdot)$,

$$\operatorname{Deg}(M) = \deg(M) + \lambda(\Gamma_{\mathfrak{m}}(M)) = \operatorname{arith-deg}(M).$$

In other words, this is just the ordinary arithmetic degree of M. In dimension at least two, however, there are more than one such function, as will be seen from our discussion. In view of how arith-deg$(\cdot)$ behaves with regard to hyperplane sections (see Theorem 1.138), in which Bertini's inequality runs in the opposite way, it follows that:

Proposition 2.76. *For any extended degree function* $\operatorname{Deg}(\cdot)$ *on a category $\mathcal{M}(R)$,*

$$\operatorname{Deg}(M) \geq \operatorname{arith-deg}(M).$$

2.4.1 Homological Degree

To establish the existence of cohomological degrees, we describe in some detail one such function introduced in [Va98a].

Definition 2.77. Let M be a finitely generated graded module over the graded algebra A and S a Gorenstein graded algebra mapping onto A, with maximal graded ideal $\mathfrak{m}$. Set $\dim S = r$, $\dim M = d$. The *homological degree* of M is the integer

$$\operatorname{hdeg}(M) = \deg(M) + \sum_{i=r-d+1}^{r} \binom{d-1}{i-r+d-1} \cdot \operatorname{hdeg}(\operatorname{Ext}_S^i(M,S)). \tag{2.36}$$

This expression becomes more compact when $\dim M = \dim S = d > 0$:

$$\operatorname{hdeg}(M) = \deg(M) + \sum_{i=1}^{d} \binom{d-1}{i-1} \cdot \operatorname{hdeg}(\operatorname{Ext}_S^i(M,S)). \tag{2.37}$$

Extended Degrees and Samuel Multiplicities

There are many variations of this definition, obtained by using Samuel's notion of multiplicity. Thus if $(R,\mathfrak{m})$ is a Noetherian local and I is an $\mathfrak{m}$-primary ideal, replacing $\deg(M)$ by Samuel's $e(I;M)$, we obtain the extended degree $\operatorname{hdeg}_I(\cdot)$. This shows that there are an infinite number of extended degrees on $\mathcal{M}(R)$ (when $\dim R \geq 1$).

The definition of hdeg can be extended to any Noetherian local ring S by setting $\operatorname{hdeg}(M) = \operatorname{hdeg}(\widehat{S} \otimes_S M)$. On other occasions, we may also assume that the residue field of S is infinite, an assumption that can be realized by replacing $(S,\mathfrak{m})$ by the local ring $S[X]_{\mathfrak{m}S[X]}$. In fact, if X is any set of indeterminates, the localization is still a Noetherian ring, so the residue field can be assumed to have any cardinality, as we shall assume in the proof.

Remark 2.78. Consider the case when $\dim M = 2$; we assume that $\dim S = 2$ also. The expression for $\operatorname{hdeg}(M)$ is now

$$\operatorname{hdeg}(M) = \deg(M) + \operatorname{hdeg}(\operatorname{Ext}_S^1(M,S)) + \operatorname{hdeg}(\operatorname{Ext}_S^2(M,S)).$$

The last summand, by duality, is the length of the submodule $\Gamma_{\mathfrak{m}}(M)$. The middle term is a module of dimension at most one, so can be described according to Proposition 2.75 by the equality

$$\operatorname{hdeg}(\operatorname{Ext}_S^1(M,S)) = \deg(\operatorname{Ext}_S^1(\operatorname{Ext}_S^1(M,S),S)) + \deg(\operatorname{Ext}_S^2(\operatorname{Ext}_S^1(M,S),S)).$$

Cohen-Macaulay Modules on the Punctured Spectrum

There are many important classes of such modules. They are characterized in a number of ways, beginning with the following. Let $(R,\mathfrak{m})$ be a Noetherian local ring of dimension d and M a finitely generated R-module of dimension d. The Cohen-Macaulayness of M can be tested using a system $\mathbf{x} = x_1, \ldots, x_d$ of parameters: the modules

$$((x_1,\ldots,x_{i-1})M :_M x_i)/(x_1,\ldots,x_{i-1})M, \quad i = 1,\ldots,d,$$

must all vanish. (In this case, one $\mathbf{x}$ suffices.) In general, the annihilator of these modules play an important role in their study. When they are $\mathfrak{m}$-primary, the local cohomology modules $H_{\mathfrak{m}}^i(M)$, $i < d$, have finite length and in particular $M_{\mathfrak{p}}$ is Cohen-Macaulay for any prime ideal $\mathfrak{p} \neq \mathfrak{m}$.

Definition 2.79. M is a Buchsbaum module if for any system of parameters $\mathbf{x}$ for M, the integer

$$\lambda(M/\mathbf{x}M) - e(\mathbf{x};M)$$

is independent of $\mathbf{x}$.

J. Stückrad and W. Vogel have developed an extensive theory of such modules ([StV86]). One of its properties is that the cohomology modules $H^i_{\mathfrak{m}}(M)$ are finite-dimensional vector spaces over $R/\mathfrak{m}$ in the range $i < \dim M$.

Example 2.80. The expression for $\operatorname{hdeg}(M)$ takes a straightforward form in three cases of interest (assume that $\dim S = \dim M = d \geq 1$). First, suppose that M is Cohen-Macaulay on the punctured spectrum of S. Then all the modules $\operatorname{Ext}^i_S(M,S)$, $i > 0$, are of finite length and we obtain

$$\begin{aligned}
\operatorname{hdeg}(M) &= \deg(M) + \sum_{i=1}^{d} \binom{d-1}{i-1} \cdot \operatorname{hdeg}(\operatorname{Ext}^i_S(M,S)) \\
&= \deg(M) + \sum_{i=1}^{d} \binom{d-1}{i-1} \cdot \lambda(\operatorname{Ext}^i_R(M,R)) \\
&= \deg(M) + \sum_{i=1}^{d} \binom{d-1}{i-1} \cdot \lambda(H^{d-i}_{\mathfrak{m}}(M)) \\
&= \deg(M) + \sum_{i=0}^{d-1} \binom{d-1}{i} \cdot \lambda(H^{i}_{\mathfrak{m}}(M)) \\
&= \deg(M) + I(M).
\end{aligned}$$

Note that the binomial term is the invariant of Buchsbaum in the theory of Stückrad-Vogel ([StV86, Chapter 1, Proposition 2.6]).

For the next example, we recall the notion of a *sequentially* Cohen-Macaulay module. This is a module M having a filtration

$$0 = M_0 \subset M_1 \subset \cdots \subset M_r = M,$$

with the property that each factor M_i/M_{i-1} is Cohen-Macaulay and

$$\dim M_i/M_{i-1} < \dim M_{i+1}/M_i, \; i = 1, \ldots, r-1.$$

If L is the leftmost nonzero submodule in such a chain, it follows easily that $N = M/L$ is also sequentially Cohen-Macaulay (L is Cohen-Macaulay by hypothesis) and we have

$$\begin{aligned}
\operatorname{Ext}^p_S(M,S) &= \operatorname{Ext}^p_S(L,S), \; p = \dim L \\
\operatorname{Ext}^i_S(M,S) &= \operatorname{Ext}^i_S(N,S), \; i < p
\end{aligned}$$

with the other Ext's vanishing. This gives the expression

$$\begin{aligned}\operatorname{hdeg}(M) &= \deg(M) + \sum_{i=1}^{d} \binom{d-1}{i-1} \cdot \deg(\operatorname{Ext}_S^i(M,S)) \\ &= \deg(M) + \sum_{i=1}^{d} \binom{d-1}{i-1} \cdot \deg(\operatorname{Ext}_S^i(M,S)\end{aligned}$$

in multiplicities of all the Ext modules. The multiplicity $\deg(\operatorname{Ext}_S^i(M,S))$ can be written as the ordinary multiplicity of one of the factors of the filtration–a fact that we leave as an exercise.

Finally, suppose that A is a graded algebra over a field k, of Krull dimension two and of positive depth. If $\widetilde{A}$ is the S_2-ification of A, consideration of the exact sequence

$$0 \to A \longrightarrow \widetilde{A} \longrightarrow \widetilde{A}/A \to 0$$

easily gives

$$\operatorname{hdeg}(A) = \deg(A) + \lambda(\widetilde{A}/A).$$

Note that $\widetilde{A}/A$ is the Hartshorne-Rao module of A.

It is important to note that $\operatorname{hdeg}(\cdot)$ is defined recursively on the dimension of the module; we refer to [Va98a] and [DGV98] for more technical aspects of these definitions. This definition will allow, in principle, for its calculation by symbolic computer packages. A similar notion can be defined for modules over Gorenstein local rings, and by completing and using the Cohen structure theorem it can be applied to any local Noetherian ring ([Va98a]).

Theorem 2.81. *The function* $\operatorname{hdeg}(\cdot)$ *is a cohomological degree.*

The proof requires a special notion of generic hyperplane sections that fits the concept of *genericity* defined earlier. Let S be a Gorenstein standard graded ring with infinite residue field and M a finitely generated graded module over S. We recall that a *superficial* element of order r for M is an element $z \in S_r$ such that $0 :_M z$ is a submodule of M of finite length.

Definition 2.82. A *special hyperplane section* of M is an element $h \in S_1$ that is superficial for all the iterated Exts

$$M_{i_1,i_2,\ldots,i_p} = \operatorname{Ext}_S^{i_1}(\operatorname{Ext}_S^{i_2}(\cdots(\operatorname{Ext}_S^{i_{p-1}}(\operatorname{Ext}_S^{i_p}(M,S),S),\cdots,S))),$$

and all sequences of integers $i_1 \geq i_2 \geq \cdots \geq i_p \geq 0$.

By local duality it follows that, up to shifts in grading, there are only finitely many such modules. Actually, it is enough to consider those sequences in which $i_1 \leq \dim S$ and $p \leq 2\dim S$, which ensures the existence of such 1-forms as h. It is clear that this property holds for generic hyperplane sections.

The following result establishes $\operatorname{hdeg}(\cdot)$ as a *bona fide* cohomological degree.

Theorem 2.83. *Let S be a standard Gorenstein graded algebra and M a finitely generated graded module of depth at least* 1. *If $h \in S$ is a generic hyperplane section on M, then*

$$\text{hdeg}(M) \geq \text{hdeg}(M/hM).$$

Proof. This will require several technical reductions. We assume that h is a regular, generic hyperplane section for M that is regular on S. We also assume that $\dim M = \dim S = d$, and derive several exact sequences from

$$0 \to M \xrightarrow{h} M \longrightarrow N \to 0. \tag{2.38}$$

For simplicity, we write $M_i = \text{Ext}_S^i(M,S)$, and $N_i = \text{Ext}_S^{i+1}(N,S)$ in the case of N. (The latter because N is a module of dimension $\dim S - 1$ and $N_i = \text{Ext}^i_{S/(h)}(N, S/(h))$.)

Using this notation, in view of the binomial coefficients in the definition of $\text{hdeg}(\cdot)$, it will be enough to show that

$$\text{hdeg}(N_i) \leq \text{hdeg}(M_i) + \text{hdeg}(M_{i+1}), \text{ for } i \geq 1.$$

The sequence (2.38) gives rise to the cohomology long sequence

$$\begin{gathered} 0 \to M_0 \longrightarrow M_0 \longrightarrow N_0 \longrightarrow M_1 \longrightarrow M_1 \longrightarrow N_1 \longrightarrow M_2 \longrightarrow \cdots \\ \cdots \longrightarrow M_{d-2} \longrightarrow M_{d-2} \longrightarrow N_{d-2} \longrightarrow M_{d-1} \longrightarrow M_{d-1} \longrightarrow N_{d-1} \to 0, \end{gathered}$$

which are broken up into shorter exact sequences as follows:

$$0 \to L_i \longrightarrow M_i \longrightarrow \widetilde{M}_i \to 0 \tag{2.39}$$

$$0 \to \widetilde{M}_i \longrightarrow M_i \longrightarrow G_i \to 0 \tag{2.40}$$

$$0 \to G_i \longrightarrow N_i \longrightarrow L_{i+1} \to 0. \tag{2.41}$$

We note that all L_i have finite length, because of the condition on h. For $i = 0$, we have the usual relation $\deg(M) = \deg(N)$. When $\widetilde{M}_i$ has finite length, then M_i, G_i and N_i have finite length, and

$$\begin{aligned} \text{hdeg}(N_i) = \lambda(N_i) &= \lambda(G_i) + \lambda(L_{i+1}) \\ &\leq \text{hdeg}(M_i) + \text{hdeg}(M_{i+1}). \end{aligned}$$

It is a similar relation that we want to establish for all other cases.

Proposition 2.84. *Let S be a Gorenstein graded algebra and let*

$$0 \to A \longrightarrow B \longrightarrow C \to 0$$

be an exact sequence of graded modules. Then

(a) *If A is a module of finite length, then*

$$\text{hdeg}(B) = \text{hdeg}(A) + \text{hdeg}(C).$$

(b) *If* C *is a module of finite length, then*

$$\operatorname{hdeg}(B) \leq \operatorname{hdeg}(A) + \operatorname{hdeg}(C).$$

(c) *Moreover, in the previous case, if* $\dim B = d$*, then*

$$\operatorname{hdeg}(A) \leq \operatorname{hdeg}(B) + (d-1)\operatorname{hdeg}(C).$$

Proof. They are all clear if B is a module of finite length so we assume that $\dim B = d \geq 1$.

(a) This is immediate since $\deg(B) = \deg(C)$ and the cohomology sequence gives

$$\begin{aligned}
\operatorname{Ext}_S^i(B,S) &= \operatorname{Ext}_S^i(C,S), \ 1 \leq i \leq d-1, \ \text{and} \\
\lambda(\operatorname{Ext}_S^d(B,S)) &= \lambda(\operatorname{Ext}_S^d(A,S)) + \lambda(\operatorname{Ext}_S^d(C,S)).
\end{aligned}$$

(b) Similarly we have

$$\operatorname{Ext}_S^i(B,S) = \operatorname{Ext}_S^i(A,S), \ 1 \leq i < d-1,$$

and the exact sequence

$$0 \to \operatorname{Ext}_S^{d-1}(B,S) \to \operatorname{Ext}_S^{d-1}(A,S) \to \operatorname{Ext}_S^d(C,S) \to \operatorname{Ext}_S^d(B,S) \to \operatorname{Ext}_S^d(A,S) \to 0. \quad (2.42)$$

If $\operatorname{Ext}_S^{d-1}(A,S)$ has finite length, then

$$\begin{aligned}
\operatorname{hdeg}(\operatorname{Ext}_S^{d-1}(B,S)) &\leq \operatorname{hdeg}(\operatorname{Ext}_S^{d-1}(A,S)) \\
\operatorname{hdeg}(\operatorname{Ext}_S^d(B,S)) &\leq \operatorname{hdeg}(\operatorname{Ext}_S^d(A,S)) + \operatorname{hdeg}(\operatorname{Ext}_S^d(C,S)).
\end{aligned}$$

Otherwise, $\dim \operatorname{Ext}_S^{d-1}(A,S) = 1$, and

$$\operatorname{hdeg}(\operatorname{Ext}_S^{d-1}(A,S)) = \deg(\operatorname{Ext}_S^{d-1}(A,S)) + \lambda(\Gamma_{\mathfrak{m}}(\operatorname{Ext}_S^{d-1}(A,S))).$$

Since we also have

$$\begin{aligned}
\deg(\operatorname{Ext}_S^{d-1}(B,S)) &= \deg(\operatorname{Ext}_S^{d-1}(A,S)), \\
\lambda(\Gamma_{\mathfrak{m}}(\operatorname{Ext}_S^{d-1}(B,S))) &\leq \lambda(\Gamma_{\mathfrak{m}}(\operatorname{Ext}_S^{d-1}(A,S))),
\end{aligned}$$

we again obtain the stated bound.

(c) In the sequence (2.42), if $\dim \operatorname{Ext}_S^{d-1}(B,S) = 0$, then

$$\lambda(\operatorname{Ext}_S^{d-1}(A,S)) \leq \lambda(\operatorname{Ext}_S^{d-1}(B,S)) + \lambda(C) \quad (2.43)$$

and also

$$\lambda(\operatorname{Ext}_S^d(A,S)) \leq \lambda(\operatorname{Ext}_S^d(B,S)).$$

When taken into the formula for $\operatorname{hdeg}(A)$, the binomial coefficient $\binom{d-1}{d-2}$ gives the desired factor for $\lambda(C)$.

On the other hand, if $\dim \operatorname{Ext}_S^{d-1}(B,S) = 1$, we also have

$$\operatorname{hdeg}(\operatorname{Ext}_S^{d-1}(A,S)) \leq \operatorname{hdeg}(\operatorname{Ext}_S^{d-1}(B,S)) + \lambda(C),$$

the dimension one case of (2.43). □

Suppose that $\dim \widetilde{M}_i \geq 1$. From Proposition 2.84(b) we have

$$\operatorname{hdeg}(N_i) \leq \operatorname{hdeg}(G_i) + \lambda(L_{i+1}). \tag{2.44}$$

We must now relate $\operatorname{hdeg}(G_i)$ to $\deg(M_i)$. Apply the functor $\Gamma_{\mathfrak{m}}(\cdot)$ to the sequence (2.40) and consider the commutative diagram

$$\begin{array}{ccccccc} 0 \rightarrow & \widetilde{M}_i & \longrightarrow & M_i & \longrightarrow & G_i & \rightarrow 0 \\ & \uparrow & & \uparrow & & \uparrow & \\ 0 \rightarrow & \Gamma_{\mathfrak{m}}(\widetilde{M}_i) & \longrightarrow & \Gamma_{\mathfrak{m}}(M_i) & \longrightarrow & \Gamma_{\mathfrak{m}}(G_i) & \end{array},$$

in which we denote by H_i the image of the natural map

$$\Gamma_{\mathfrak{m}}(M_i) \longrightarrow \Gamma_{\mathfrak{m}}(G_i).$$

Through the snake lemma, we obtain the exact sequence

$$0 \rightarrow \widetilde{M}_i/\Gamma_{\mathfrak{m}}(\widetilde{M}_i) \xrightarrow{\alpha} M_i/\Gamma_{\mathfrak{m}}(M_i) \longrightarrow G_i/H_i \rightarrow 0. \tag{2.45}$$

Furthermore, from (2.39) there is a natural isomorphism

$$\beta : M_i/\Gamma_{\mathfrak{m}}(M_i) \cong \widetilde{M}_i/\Gamma_{\mathfrak{m}}(\widetilde{M}_i),$$

while from (2.40) there is a natural injection

$$\widetilde{M}_i/\Gamma_{\mathfrak{m}}(\widetilde{M}_i) \hookrightarrow M_i/\Gamma_{\mathfrak{m}}(M_i),$$

whose composite with β is induced by multiplication by h on $M_i/\Gamma_{\mathfrak{m}}(M_i)$. We may thus replace $\widetilde{M}_i/\Gamma_{\mathfrak{m}}(\widetilde{M}_i)$ by $M_i/\Gamma_{\mathfrak{m}}(M_i)$ in (2.45) and take α as multiplication by h:

$$0 \rightarrow M_i/\Gamma_{\mathfrak{m}}(M_i) \xrightarrow{h} M_i/\Gamma_{\mathfrak{m}}(M_i) \longrightarrow G_i/H_i \rightarrow 0.$$

Observe that since

$$\operatorname{Ext}_S^j(M_i/\Gamma_{\mathfrak{m}}(M_i), S) = \operatorname{Ext}_S^j(M_i, S), \ j < \dim S,$$

h is still a regular, generic hyperplane section for $M_i/\Gamma_{\mathfrak{m}}(M_i)$. By induction on the dimension of the module, we have

$$\operatorname{hdeg}(M_i/\Gamma_{\mathfrak{m}}(M_i)) \geq \operatorname{hdeg}(G_i/H_i).$$

Now from Proposition 2.84(a), we have

$$\text{hdeg}(G_i) = \text{hdeg}(G_i/H_i) + \lambda(H_i).$$

Since these summands are bounded, by $\text{hdeg}(M_i/\Gamma_{\mathfrak{m}}(M_i))$ and $\lambda(\Gamma_{\mathfrak{m}}(M_i))$ respectively, (in fact, $\lambda(H_i) = \lambda(L_i)$), we have

$$\text{hdeg}(G_i) \leq \text{hdeg}(M_i/\Gamma_{\mathfrak{m}}(M_i)) + \lambda(\Gamma_{\mathfrak{m}}(M_i)) = \text{hdeg}(M_i),$$

the last equality by Proposition 2.84(a) again. Finally, taking this estimate into (2.44) we get

$$\begin{aligned} \text{hdeg}(N_i) &\leq \text{hdeg}(G_i) + \lambda(L_{i+1}) \\ &\leq \text{hdeg}(M_i) + \text{hdeg}(M_{i+1}), \end{aligned} \tag{2.46}$$

to establish the claim. □

Remark 2.85. That equality does not always hold is shown by the following example. Suppose that $R = k[x,y]$ and $M = (x,y)^2$. Then $\text{hdeg}(M) = 4$, but $\text{hdeg}(M/hM) = 3$ for any hyperplane section h. To get an example of a ring one takes the idealization of M.

Remark 2.86. In the definition of $\text{hdeg}(\cdot)$, we measured degrees using $\deg(\cdot)$, the Samuel multiplicity attached to the maximal ideal itself. If instead we had fixed an $\mathfrak{m}$-primary ideal I and used $\deg_I(\cdot)$ in the definition of $\text{hdeg}(\cdot)$, we would have obtained another cohomological degree, namely $\text{hdeg}_I(\cdot)$.

2.4.2 General Properties of Degs

We show how to use Deg to obtain *a priori* estimates in local algebra. We shall begin with basic comparisons with other measures such as Castelnuovo-Mumford regularity and number of generators.

Degs and Castelnuovo-Mumford Regularity

Metaphorically one can contrast $\text{reg}(\cdot)$ and $\text{Deg}(\cdot)$ as follows. While $\text{reg}(M)$ reads information about the *top* of the graded module M, $\text{Deg}(M)$ reads information about the *root* of M. The reading processes themselves have to be understood in the broad sense of collecting data from the components that occur under all hyperplane sections. While $\text{reg}(M)$ is extracted directly from the cohomology of M, $\text{Deg}(M)$ comes usually from the same cohomology, but filtered through local duality. It is not surprising therefore that they are comparable. Consider the following relationship to the Castelnuovo-Mumford regularity ([DGV98]).

Proposition 2.87. *Let M be a finitely generated R-module and x a generic hyperplane section. Then*

$$\text{Deg}(M) \geq \text{Deg}(M/xM) + \text{Deg}(x \cdot \Gamma_x(M)).$$

Proof. By the choice of x, $\Gamma_x(M) = \Gamma_{\mathfrak{m}}(M)$, which we denote by L. Consider the exact sequence

$$0 \to L \longrightarrow M \longrightarrow M_0 \to 0.$$

By the snake lemma, multiplication by x induces the exact sequence

$$0 \to {}_xL \longrightarrow {}_xM \longrightarrow {}_xM_0 \longrightarrow L/xL \longrightarrow M/xM \longrightarrow M_0/xM_0 \to 0,$$

where ${}_xM = \{m \in M \mid xm = 0\}$; note that ${}_xM_0 = 0$. We have the inequality of degrees

$$\begin{aligned}\operatorname{Deg}(M) &= \operatorname{Deg}(M_0) + \operatorname{Deg}(L)\\ &\geq \operatorname{Deg}(M_0/xM_0) + \operatorname{Deg}(L),\end{aligned}$$

while on the other hand

$$\begin{aligned}\operatorname{Deg}(M/xM) &= \operatorname{Deg}(M_0/xM_0) + \operatorname{Deg}(L/xL)\\ &= \operatorname{Deg}(M_0/xM_0) + \operatorname{Deg}(L) - \operatorname{Deg}(xL),\end{aligned}$$

from which the assertion follows. □

The following resulting a basic comparison between any $\operatorname{Deg}(\cdot)$ function and the Castelnuovo-Mumford index of regularity of a standard graded algebra.

Theorem 2.88. *Let A be a standard graded algebra over an Artin local ring with an infinite residue field k. For any function* $\operatorname{Deg}(\cdot)$,

$$\operatorname{reg}(A) < \operatorname{Deg}(A).$$

Proof. Set $d = \dim A$, and argue by induction on d. The case $d = 0$ is clear since A is a standard algebra. Suppose that $L = H^0_{\mathfrak{m}}(A)$; consider the exact sequence

$$0 \to L \longrightarrow A \longrightarrow A' \to 0.$$

Since $\operatorname{reg}(A) = \max\{\operatorname{reg}(L), \operatorname{reg}(A')\}$ (see [Ei95, Corollary 20.19(d)]), it will suffice to show that $\operatorname{Deg}(A) - 1$ bounds $\operatorname{reg}(L)$ and $\operatorname{reg}(A')$. We first consider the case where $L = 0$. This means that $A = A'$, when we can choose a generic hyperplane section h for A. Since $\operatorname{Deg}(A) \geq \operatorname{Deg}(A/hA)$ by (2.33), and $\operatorname{reg}(A) = \operatorname{reg}(A/hA)$, we are done by the induction hypothesis.

Suppose now that $d \geq 1$ and $L \neq 0$. We must show that L has no component in degrees $\operatorname{Deg}(A)$ or higher. Let h be a generic hyperplane section and consider the exact sequence

$$0 \to L_0 \longrightarrow A \stackrel{h}{\longrightarrow} A \longrightarrow \overline{A} \to 0.$$

Taking local cohomology, we have the induced exact sequence

$$0 \to L_0 \longrightarrow L \stackrel{h}{\longrightarrow} L \longrightarrow H^0_{\mathfrak{m}}(\overline{A}).$$

By induction, $H^0_{\mathfrak{m}}(\overline{A})$ has no components in degrees higher than $r = \text{Deg}(\overline{A}) - 1$. In particular for $n > r$ we must have $L_n = hL_{n-1}$. On the other hand, from Proposition 2.87, we have that $s = \lambda(hL) \leq \text{Deg}(A) - \text{Deg}(\overline{A})$. This implies that

$$L_{r+s+1} = h^{s+1}L_r = 0,$$

since the chain of submodules

$$hAL_r \supset h^2AL_r \supset \cdots \supset h^{s+1}AL_r \supset 0$$

has length at most s. Thus $\text{reg}(L) \leq r + s < \text{Deg}(A)$, as claimed. □

There is an extension to arbitrary graded modules given in [Nag3]:

Theorem 2.89. *Let A be an standard graded algebra over an infinite field and let M be a nonzero finitely generated A-module. Then for any* $\text{Deg}(\cdot)$ *function, we have*

$$\text{reg}(M) < \text{Deg}(M) + \alpha(M),$$

where $\alpha(M)$ is the maximal degree in a minimum graded generating set of M.

One application is to relate the reduction number of ideals to the $\text{Deg}(G)$ of their associated graded rings. Suppose that $(R, \mathfrak{m})$ is a Noetherian local ring and I an $\mathfrak{m}$-primary ideal. For a reduction J of I, look at $G = \text{gr}_I(R)$ as a standard graded algebra over $B = R/J[T_1, \ldots, T_n]$, where B acts through $\text{gr}_J(R)$. We have that

$$\text{reg}(G) = \text{reg}_B(G) < \text{Deg}(G),$$

while the inequality

$$\text{r}_J(I) \leq \text{reg}(G)$$

is valid for algebras of finite type over Artinian rings (see Section 3.3). This actually proves that $\text{bigr}(I) < \text{Deg}(G)$. There are other such comparisons for ideals which are not $\mathfrak{m}$-primary. This suggests support for:

Conjecture 2.90. Let $(R, \mathfrak{m})$ be a Noetherian local ring and I an R-ideal. For any Deg *function,*

$$\text{bigr}(I) < \text{Deg}(\text{gr}_I(R)).$$

Bounding the Number of Generators in terms of Degrees

We first derive a general estimate for the number of generators of a module in terms of some Deg function. It is a direct application of Proposition 2.87.

Proposition 2.91. *Let M be a finitely generated graded module over a standard graded algebra $(A, \mathfrak{m})$ and let $x_1, \ldots, x_d$ be a maximal superficial sequence for M. For $r = 1, \ldots, d$, set $M_r = M/(x_1, \ldots, x_{r-1})M$, $M^{(r)} = \Gamma_{\mathfrak{m}}(M_r)$ and $\lambda_r = \lambda(x_r M^{(r)})$. Then*

$$\nu(M) \leq \lambda(M/(x_1, \ldots, x_r)M) \leq \text{Deg}(M) - \sum_{r=1}^{d} \lambda_r.$$

In particular $\nu(M) \leq \text{Deg}(M)$.

Corollary 2.92. *Let R be a Noetherian local ring of dimension* 1 *and M a finitely generated R-module. Then every submodule N of M can be generated by*

$$\nu(N) \leq \mathrm{Deg}(R)\nu(M)$$

elements.

Proof. Let $\varphi : R^n \to M$, with $n = \nu(M)$, be a free presentation of M. The submodule $\varphi^{-1}(N)$ of R^n satisfies

$$\mathrm{Deg}(\varphi^{-1}(N)) \leq n \cdot \mathrm{Deg}(R),$$

since $\dim R = 1$. By the proposition, $\mathrm{Deg}(\varphi^{-1}(N))$ bounds the number of generators of $\varphi^{-1}(N)$, and therefore of N as well. □

We show how an extended degree function leads to estimates for the number of generators of an ideal I of a Cohen-Macaulay ring R in terms of the degrees of R/I.

Theorem 2.93. *Let $(R, \mathfrak{m})$ be a Cohen-Macaulay ring of dimension d, with an infinite residue field and let I be an ideal of codimension $g > 0$. If* depth $R/I = r$, *then*

$$\begin{aligned}\nu(I) &\leq \deg(R) + (g-1)\mathrm{Deg}(R/I) + (d-g-r)(\mathrm{Deg}(R/I) - \deg(R/I)) \\ &= \deg(R) + (g-1)\deg(R/I) + (d-r-1)(\mathrm{Deg}(R/I) - \deg(R/I)).\end{aligned} \tag{2.47}$$

Proof. Without loss of generality we may assume that R has infinite residue field. Consider the exact sequence

$$0 \to I \longrightarrow R \longrightarrow R/I \to 0.$$

If $g = d$, we have ([Val81], [Va98a]) that

$$\nu(I) \leq \deg(R) + (g-1)\deg(R/I).$$

Thus suppose that $g < d$ and let x be a generic hyperplane section for both R/I and R. First assume that $r > 0$. Reducing the sequence modulo (x) gives another sequence

$$0 \to I/xI \longrightarrow R/(x) \longrightarrow R/(I,x) \to 0, \tag{2.48}$$

where I/xI can be identified with an ideal I' of $R' = R/(x)$ with the same number of generators as I. Furthermore, by (2.33), $\deg(R'/I') = \deg(R/I)$ and $\mathrm{Deg}(R'/I') \leq \mathrm{Deg}(R/I)$. When taken in (2.47), any changes would only reinforce the inequality.

We may continue in this manner until we exhaust the depth of R/I. This means that instead of (2.48) we have the exact sequence

$$0 \to {}_x(R/I) \longrightarrow I/xI \longrightarrow R/(x) \longrightarrow R/(I,x) \to 0. \tag{2.49}$$

Note that

$${}_x(R/I) \subset L = H^0_{\mathfrak{m}}(R/I),$$

and therefore $\lambda({}_x(R/I)) \leq \mathrm{Deg}(R/I) - \deg(R/I)$. This leads to the inequality

$$\nu(I) \leq \nu(I') + (\mathrm{Deg}(R/I) - \deg(R/I)),$$

where I' is the image of I/xI in the ring $R' = R/(x)$.

On the other hand, according to Proposition 2.87, $\mathrm{Deg}(R'/I') \leq \mathrm{Deg}(R/I) - \mathrm{Deg}(xL)$. In particular we also have that $\deg(R'/I') \leq \mathrm{Deg}(R/I)$. This means that the reduction from the case $d > g$ to the case $d = g$ can be accomplished with the addition of $(d-g)(\mathrm{Deg}(R/I) - \deg(R/I))$ to the bound for $\nu(I')$. Factoring in the early reduction on r we obtain the desired estimate. □

The following is an easy observation on this technique. We assume that $(R, \mathfrak{m}, k = R/\mathfrak{m})$ is a Cohen-Macaulay local ring. For any finitely generated R-module A, we denote by $\beta_i(A)$ its i-th Betti number, and by $\mu_i(A)$ its i-th *Bass number*:

$$\begin{aligned}\beta_i(A) &= \dim_k \mathrm{Tor}_i^R(k, A)\\ \mu_i(A) &= \mathrm{Ext}_R^i(k, A).\end{aligned}$$

Theorem 2.94. *Let A be a finitely generated R-module. For any* $\mathrm{Deg}(\cdot)$ *function and any integer $i \geq 0$,*

$$\begin{aligned}\beta_i(A) &\leq \beta_i(k) \cdot \mathrm{Deg}(A),\\ \mu_i(A) &\leq \mu_i(k) \cdot \mathrm{Deg}(A).\end{aligned}$$

Proof. If L is the submodule of A of finite support, the exact sequence

$$0 \to L \longrightarrow A \longrightarrow A' \to 0$$

gives $\beta_i(A) \leq \beta_i(L) + \beta_i(A')$. For the summand $\beta_i(L)$, by induction on the length of L, one has that $\beta_i(L) \leq \lambda(L)\beta_i(k)$. For the other summand, one chooses a sufficiently generic hyperplane section (good for both A' and R if need be), and uses the exact sequences

$$0 \to \mathrm{Tor}_i^R(k, A') \longrightarrow \mathrm{Tor}_i^R(k, A'/hA') \longrightarrow \mathrm{Tor}_{i+1}^R(k, A) \to 0.$$

This gives $\beta_i(A) \leq \beta_i(A'/hA')$, so that we can induct. One can use a similar argument for $\mu_i(A)$. □

Degrees and Localization

Let $(R, \mathfrak{m})$ be a Noetherian local ring and E a finitely generated R-module. For a prime ideal $\mathfrak{p}$ and some of the *degrees* discussed, the localization $E_\mathfrak{p}$ behaves as expected, at least for geometric local rings.

Proposition 2.95. *Let R be a quasi-unmixed local ring and let* Deg *be one of the degree functions* deg, arith-deg *or* hdeg. *Then*

$$\mathrm{Deg}(E_\mathfrak{p}) \leq \mathrm{Deg}(E).$$

Proof. For any of these functions, the value of $\mathrm{Deg}(E)$ is put together from expressions of the form

$$\sum_{\mathfrak{q}} \alpha(\mathfrak{q}) \cdot \deg(R/\mathfrak{q}),$$

where $\mathfrak{q}$ is some associated prime of E (or in the case of $\mathrm{hdeg}(E)$ of some associated prime of some iterated Ext), and $\alpha(\mathfrak{q})$ is a positive integer determined by $E_{\mathfrak{q}}$. For instance,

$$\text{arith-deg}(E) = \sum \mathrm{mult}_{\mathfrak{q}}(E)\deg(R/\mathfrak{q}).$$

If $\mathfrak{q} \subset \mathfrak{p}$, for the localization $E_{\mathfrak{p}}$, the value $\alpha(\mathfrak{q})$ is unchanged while $\deg(R_{\mathfrak{p}}/\mathfrak{q}_{\mathfrak{p}}) \leq \deg(R/\mathfrak{q})$ by [Na62, Theorem 40.1]. □

The Number of Generators of $\mathrm{Hom}_R(A,B)$

We want to illustrate another use of cohomological degrees by discussing a 'problem' in Homological Algebra. Let $(R,\mathfrak{m})$ be a Cohen-Macaulay local ring with a canonical module ω, and let A and B be two finitely generated R-modules. What properties of A and of B are required to obtain an estimate of the number of generators of $\mathrm{Hom}_R(A,B)$, or of $\mathrm{Hom}_R(A,A)$? Because $\mathrm{Hom}_R(A,B)$ will depend on the interaction between A and B (e.g. the relationship between their associated prime ideals), which may be difficult to fathom, we shall instead look for uniform bounds; more precisely, we shall make comparisons between the number of generators function $\nu(\cdot)$ and the corresponding bounds after applying the functor $\mathrm{Hom}_R(A,\cdot)$ or $\mathrm{Hom}_R(\cdot,A)$.

Module Operations and Auslander Dual

Before we consider specific cases, we introduce some simplifying hypotheses and discuss the technique of the Auslander dual.

The usefulness of finding bounds for the number of generators of $\mathrm{Hom}_R(A,B)$, or for $A^{**} = \mathrm{Hom}_R(\mathrm{Hom}_R(A,R),R)$, comes in several constructions considered with emphasis on those of the integral closure of affine domains (see Chapter 6). It is the case that the modules involved are torsionfree over an algebra R that admit a Noether normalization

$$T = k[x_1,\ldots,x_d] \hookrightarrow S = T[u] \hookrightarrow R,$$

where $S \hookrightarrow R$ is a finite, birational extension. Under these conditions, if A and B are torsionfree R-modules, it is clear that

$$\mathrm{Hom}_R(A,B) = \mathrm{Hom}_S(A,B),$$

which allows us to assume that the base ring is Gorenstein. The other observation is that one may want to actually find the S_2-closure of A, the 'smallest' overmodule of A, $A \hookrightarrow \widetilde{A}$ that satisfies Serre's S_2 condition. For torsionfree modules over the algebra R, one has

$$\widetilde{A} = \mathrm{Hom}_T(\mathrm{Hom}_T(A,T),T).$$

This agrees with A^{**} if A itself satisfies the S_2 condition. From our perspective we can work over a nice ring such as T.

In some of our calculations the following construction of Auslander ([Abr69]) occurs frequently. It is such a pearl of Homological Algebra that young readers are encouraged to provide their own proofs!

Definition 2.96. Let E be a finitely generated R-module with a projective presentation

$$F_1 \stackrel{\varphi}{\longrightarrow} F_0 \longrightarrow E \to 0.$$

The *Auslander dual* of E is the module $D(E) = \text{coker}\ (\varphi^t)$,

$$0 \to E^* \longrightarrow F_0^* \stackrel{\varphi^t}{\longrightarrow} F_1^* \longrightarrow D(E) \to 0.$$

The module $D(E)$ depends on the chosen presentation but it is unique up to projective summands. In particular the values of the functors $\text{Ext}_R^i(D(E),\cdot)$ and $\text{Tor}_i^R(D(E),\cdot)$, for $i \geq 1$, are independent of the presentation. Its uses here lies in the following result (see [Abr69, Chapter 2]):

Proposition 2.97. *Let R be a Noetherian ring and E a finitely generated R-module. There are two exact sequences of functors:*

$$0 \to \text{Ext}_R^1(D(E),\cdot) \longrightarrow E \otimes_R \cdot \longrightarrow \text{Hom}_R(E^*,\cdot) \longrightarrow \text{Ext}_R^2(D(E),\cdot) \to 0 \quad (2.50)$$

$$0 \to \text{Tor}_2^R(D(E),\cdot) \longrightarrow E^* \otimes_R \cdot \longrightarrow \text{Hom}_R(E,\cdot) \longrightarrow \text{Tor}_1^R(D(E),\cdot) \to 0. \quad (2.51)$$

From these sequences, for any extended degree $\text{Deg}(\cdot)$ one has

$$\begin{aligned} \nu(E^{**}) &\leq \nu(E) + \nu(\text{Ext}_R^2(D(E),R)) \leq \nu(E) + \text{Deg}(\text{Ext}_R^2(D(E),R)) \\ \nu(\text{Hom}_R(E,E)) &\leq \nu(E^* \otimes E) + \nu(\text{Tor}_1^R(D(E),E)) \\ &\leq \nu(E^*) \cdot \nu(E) + \text{Deg}(\text{Tor}_1^R(D(E),E)). \end{aligned}$$

This requires information about E^* that can be tracked all the way up to E and as well as a great deal of control over $D(E)$.

Varied Estimation of Numbers of Generators

We begin our calculations of several examples.

Proposition 2.98. *Let A and B be R-modules of finite length. Then*

$$\lambda(\text{Hom}_R(A,B)) \leq \inf\{\nu(A) \cdot \lambda(B), \lambda(A) \cdot \lambda(s(B))\} \leq \lambda(A) \cdot \lambda(B) = \text{Deg}(A) \cdot \text{Deg}(B),$$

where $s(B)$ is the socle of B.

Proof. This just reflects some of the exactness properties of the Hom functor. □

Proposition 2.99. *Let R be a Cohen-Macaulay local ring of dimension d, and let A be a torsionfree R-module admitting an embedding*

$$0 \to A \longrightarrow R^r \longrightarrow L \to 0$$

such that L has finite length. Then

$$\mathrm{hdeg}(\mathrm{Hom}_R(A,A)) \leq \mathrm{hdeg}(A)^2.$$

Proof. We may assume that $d \geq 2$, as otherwise both A and $\mathrm{Hom}_R(A,A)$ are Cohen-Macaulay and the formula is clear. Using the definition for hdeg, we have

$$\mathrm{hdeg}(A) = r \cdot \deg(R) + (d-1)\lambda(L).$$

Applying the functor $\mathrm{Hom}_R(\cdot,A)$ to the exact sequence above, we obtain the exact sequence

$$0 \to \mathrm{Hom}_R(A,A) \longrightarrow \mathrm{Hom}_R(A,R^r) \longrightarrow \mathrm{Hom}_R(A,L).$$

With $d \geq 2$, one has $\mathrm{Hom}_R(A,R) = R^r$, so that we can write

$$0 \to \mathrm{Hom}_R(A,A) \longrightarrow R^{r^2} \longrightarrow L' \to 0,$$

where L' is a submodule of the module $\mathrm{Hom}_R(A,L)$ of finite length. We can write then

$$\begin{aligned}
\mathrm{hdeg}(\mathrm{Hom}_R(A,A)) &= r^2 \deg(R) + (d-1)\lambda(L') \\
&\leq r^2 \deg(R) + (d-1)\nu(A)\lambda(L) \\
&\leq r^2 \deg(R) + \nu(A)(\mathrm{hdeg}(A) - r^2 \deg(R)) \\
&\leq \mathrm{hdeg}(A)^2 - r^2 \deg(R)(\mathrm{hdeg}(A) - 1).
\end{aligned}$$

Remark 2.100. It is not difficult to see that this argument for obtaining quadratic bounds would apply equally when L is a Cohen-Macaulay module of dimension 1, or even to $\mathrm{Hom}_R(A,B)$, where both A and B are modules admitting embeddings into free modules whose corresponding cokernels are Cohen-Macaulay of dimension at most one.

Dimensions Two and Three

We shall give an answer to the problem in dimension two and some special cases in dimension three. We will also discuss some general reduction principles. As guidance, already observed in the two special cases above, we raise the following:

Conjecture 2.101. For every Cohen-Macaulay local ring R, there exists a polynomial $f(u,v)$ such that for any two finitely generated R-modules A, B,

$$\nu(\mathrm{Hom}_R(A,B)) \leq f(\mathrm{hdeg}(A), \mathrm{hdeg}(B)).$$

Thus far, K. Dalili has settled it in dimension up to four (and some classes of reflexive modules), and special modules of arbitrary dimensions provided the modules have finite projective dimension. Actually, it might be interesting to establish it in dimension four and on selected groups of problems, such as for $\mathrm{Hom}_R(A,A)$ and for $\mathrm{Hom}_R(\mathrm{Hom}_R(A,R),R)$. These are, after all, some of the modules that occur in our later discussion of integral closures.

To examine its validity in low dimensions, we start with several general reductions, first by taking out the elements of A and B of finite support. Write the exact sequences

$$0 \to A_0 \longrightarrow A \longrightarrow A' \to 0$$

and

$$0 \to B_0 \longrightarrow B \longrightarrow B' \to 0,$$

where $A_0 = H^0_{\mathfrak{m}}(A)$ and $B_0 = H^0_{\mathfrak{m}}(B)$.

Proposition 2.102. *If A and B are modules of dimension $d \geq 1$, then*

$$\mathrm{hdeg}(\mathrm{Hom}(A,B)) \leq \mathrm{hdeg}(\mathrm{Hom}(A',B')) + \beta_0(A)\lambda(B_0) + (d-1)\beta_1(A)\lambda(B_0).$$

Since $\beta_0(A) = \nu(A) \leq \mathrm{hdeg}(A)$, $\beta_1(A) \leq \beta_1(K)\mathrm{hdeg}(A)$ and $\lambda(B_0) \leq \mathrm{hdeg}(B)$, this achieves a reduction to more manageable modules.

Proof. We apply $\mathrm{Hom}(A,\cdot) := \mathrm{Hom}_R(A,\cdot)$ to the decomposition of B, and isolate part of the cohomology exact sequence as follows:

$$0 \to \mathrm{Hom}(A,B_0) \longrightarrow \mathrm{Hom}(A,B) \longrightarrow D \to 0,$$

$$0 \to D \longrightarrow \mathrm{Hom}(A,B') \longrightarrow E \to 0,$$

and

$$E \hookrightarrow \mathrm{Ext}^1_R(A,B_0).$$

Since $\mathrm{Hom}(A,B_0)$ is a module whose length is bounded by $\nu(A)\lambda(B_0)$, by Proposition 2.84(a) we have

$$\mathrm{hdeg}(\mathrm{Hom}(A,B)) \leq \mathrm{hdeg}(A)\lambda(B_0) + \mathrm{hdeg}(D).$$

Now, using Proposition 2.84(c), and noting that $\mathrm{Hom}(A,B') = \mathrm{Hom}(A',B')$, we have

$$\mathrm{hdeg}(D) \leq \mathrm{hdeg}(\mathrm{Hom}(A',B')) + (d-1)\lambda(E).$$

Finally, we observe that $\mathrm{Ext}^1_R(A,B_0)$ is a module of length bounded by the length of the module $\mathrm{Hom}_R(R^{\beta_1(A)},B_0)$. $\square$

The next reduction we consider is to take out the submodules of A and B of dimension $d-1$.

Proposition 2.103. *If* $\dim R = 2$ *and A and B are torsionfree R-modules, then*

$$\begin{aligned} \operatorname{hdeg}(\operatorname{Hom}_R(A,B)) &\leq \deg(A)\deg(B) + \operatorname{hdeg}(A)(\operatorname{hdeg}(B) - \deg(B)) \\ &\leq \operatorname{hdeg}(A) \cdot \operatorname{hdeg}(B). \end{aligned}$$

Proof. For the torsionfree R-module A, there is an embedding

$$0 \to A \longrightarrow \operatorname{Hom}_R(\operatorname{Hom}_R(A,\omega),\omega) = A^{**} \longrightarrow C \to 0.$$

The module A^{**} has depth two and therefore is Cohen-Macaulay, since R has dimension 2. As for the module C, it has finite length and a simple computation shows that $C = H^1_{\mathfrak{m}}(A)$. We can then assemble $\operatorname{hdeg}(A)$ as $\operatorname{hdeg}(A) = \deg(A) + \lambda(H^1_{\mathfrak{m}}(A))$.

We now apply the functor $\operatorname{Hom}_R(A,\cdot)$ to the similar sequence associated with B and obtain the exact sequence

$$0 \to \operatorname{Hom}_R(A,B) \longrightarrow \operatorname{Hom}_R(A,B^{**}) \longrightarrow \operatorname{Hom}_R(A,D).$$

We note that $\operatorname{Hom}_R(A,B^{**}) = \operatorname{Hom}_R(A^{**},B^{**}) = (\operatorname{Hom}_R(A,B))^{**}$, which shows that

$$\operatorname{hdeg}(\operatorname{Hom}_R(A,B)) \leq \deg(A)\deg(B) + \lambda(\operatorname{Hom}_R(A,D)).$$

It remains to estimate the length of $\operatorname{Hom}_R(A,D)$. Since A is generated by $\operatorname{hdeg}(A)$ elements, and D has length $\operatorname{hdeg}(B) - \deg(B)$, applying $\operatorname{Hom}_R(A,\cdot)$ to a composition series of D we obtain $\lambda(\operatorname{Hom}_R(A,D)) \leq \nu_R(A)(\operatorname{hdeg}(B) - \deg(B))$. We can now collect terms to get

$$\operatorname{hdeg}(\operatorname{Hom}_R(A,B)) \leq \deg(A)\deg(B) + \operatorname{hdeg}(A)(\operatorname{hdeg}(B) - \deg(B)),$$

as desired. □

We add another case where $\operatorname{Deg}(\cdot)$ is used to bound the number of generators of $\operatorname{Hom}_R(A,B)$. Let $R = k[x,y,z]$ be a polynomial ring over the field k, and let A, B be graded, reflexive R-modules (i.e., A and B are vector bundles on the projective plane). We again use the hdeg function.

Proposition 2.104. *For R, A and B as above,*

$$\operatorname{hdeg}(\operatorname{Hom}_R(A,B)) \leq 2 \cdot \operatorname{hdeg}(A) \cdot \operatorname{hdeg}(B).$$

Proof. Since A is reflexive, it has a projective resolution

$$0 \to R^m \longrightarrow R^n \longrightarrow A \to 0;$$

a simple calculation shows that

$$\operatorname{hdeg}(A) = \deg(A) + \lambda(\operatorname{Ext}^1_R(A,R)),$$

since $\operatorname{Ext}^1_R(A,R)$ is a module of finite length; $\operatorname{hdeg}(\operatorname{Hom}_R(A,B))$ has a similar expression.

Applying the functor $\mathrm{Hom}_R(\cdot, B)$ to the resolution yields the exact sequence

$$0 \to H = \mathrm{Hom}_R(A,B) \longrightarrow B^n \stackrel{\varphi}{\longrightarrow} B^m \longrightarrow \mathrm{Ext}_R^1(A,B) \to 0.$$

Denote by D the image of φ and consider the cohomology exact sequences (we take into account the fact that A, B and $\mathrm{Hom}_R(A,B)$ are reflexive modules and $\mathrm{Ext}_R^1(A,R)$, $\mathrm{Ext}_R^1(B,R)$ and $\mathrm{Ext}_R^1(A,B)$ are modules of finite length):

$$\begin{aligned} 0 \to \mathrm{Hom}_R(D,R) \longrightarrow \mathrm{Hom}_R(B^n,R) \longrightarrow \mathrm{Hom}_R(H,R) \longrightarrow \mathrm{Ext}_R^1(D,R) \longrightarrow \\ \longrightarrow \mathrm{Ext}_R^1(B^n,R) \longrightarrow \mathrm{Ext}_R^1(H,R) \longrightarrow \mathrm{Ext}_R^2(D,R) \to 0, \end{aligned}$$

$$\begin{aligned} \mathrm{Hom}_R(B^m,R) &\cong \mathrm{Hom}_R(D,R) \\ \mathrm{Ext}_R^1(B^m,R) &\cong \mathrm{Ext}_R^1(D,R) \\ \mathrm{Ext}_R^2(D,R) &\cong \mathrm{Ext}_R^3(\mathrm{Ext}_R^1(A,B),R). \end{aligned}$$

With these identifications, and using the exact sequence

$$\mathrm{Ext}_R^1(B,R^m) = \mathrm{Ext}_R^1(B^m,R) \longrightarrow \mathrm{Ext}_R^1(B^n,R) = \mathrm{Ext}_R^1(B,R^n) \longrightarrow \mathrm{Ext}_R^1(B,A) \to 0,$$

we obtain the exact sequence

$$0 \to \mathrm{Ext}_R^1(B,A) \longrightarrow \mathrm{Ext}_R^1(\mathrm{Hom}_R(A,B),R) \longrightarrow \mathrm{Ext}_R^3(\mathrm{Ext}_R^1(A,B),R) \to 0.$$

To complete the proof we estimate the length of the middle module in this exact sequence. For that note that $\mathrm{Ext}_R^1(B,A) \cong \mathrm{Ext}_R^1(B,R) \otimes_R A$ (where we have used the fact that B is a module of projective dimension ≤ 1), and therefore

$$\lambda(\mathrm{Ext}_R^1(B,A)) \leq \nu(A)\lambda(\mathrm{Ext}_R^1(B,A)).$$

By duality we have a similar expression for the length of $\mathrm{Ext}_R^3(\mathrm{Ext}_R^1(A,B),R)$. Assembling $\mathrm{hdeg}(\mathrm{Hom}_R(A,B))$ we have

$$\begin{aligned} \mathrm{hdeg}(\mathrm{Hom}_R(A,B)) &= \deg(\mathrm{Hom}_R(A,B)) + \lambda(\mathrm{Ext}_R^1(\mathrm{Hom}_R(A,B),R)) \\ &\leq \deg(A) \cdot \deg(B) + \nu(A)(\mathrm{hdeg}(B) - \deg(B)) \\ &+ \nu(B)(\mathrm{hdeg}(A) - \deg(A)) \\ &\leq 2 \cdot \mathrm{hdeg}(A) \cdot \mathrm{hdeg}(B), \end{aligned}$$

since $\nu(\cdot) \leq \mathrm{hdeg}(\cdot)$. □

bdeg: **The Optimal** Deg

In his thesis (which informs our discussion here), T. Gunston ([Gu98]) found a way to refine any given cohomological degree function $\mathrm{Deg}(\cdot)$ into the unique cohomological degree $\mathrm{bdeg}(\cdot)$, and he proved that it meets the Bertini condition strictly: If M has positive depth there are generic hyperplane sections such that

$$\operatorname{bdeg}(M) = \operatorname{bdeg}(M/hM).$$

He accomplishes this by defining

$$\operatorname{bdeg}(M) := \max\{\operatorname{bdeg}(M/hM) \mid h \text{ is a generic hyperplane section}\}.$$

However, it takes an intricate technical argument to establish the coherence of this definition. One drawback is that even when $\operatorname{Deg}(\cdot)$ is given by an explicit formula-such as in the case of $\operatorname{hdeg}(\cdot)$–the determination of $\operatorname{bdeg}(M)$ does not come easily. The existence of at least one $\operatorname{Deg}(\cdot)$ is used in his argument to establish the existence of the maximum in the definition above.

To see how bdeg improves on hdeg, consider a case we are familiar with, that of a module M admitting an embedding

$$0 \to M \longrightarrow R^r \longrightarrow L \to 0,$$

where L has finite length. Assuming that $\dim R = d \geq 2$, we have

$$\operatorname{hdeg}(M) = r \cdot \deg(R) + (d-1)\lambda(L).$$

On the other hand, by picking a superficial sequence appropriate for bdeg and M, and using the argument of Corollary 2.91, we obtain

$$0 \to {}_{x_1}L \longrightarrow M/x_1M \longrightarrow R^r/x_1R^r \longrightarrow L/x_1L \to 0,$$

so that

$$\operatorname{bdeg}(M) = \operatorname{bdeg}(M/x_1M) = \operatorname{bdeg}(M_1) + \lambda({}_{x_1}L) = \operatorname{bdeg}(M_1) + \lambda(L/x_1L),$$

where M_1 is the image of M in $(R/(x_1))^r$. Iterating, we eventually obtain

$$\operatorname{bdeg}(M) = r \cdot \deg(R) + \sum_{i=1}^{d-1} \lambda(L/(x_1, \ldots, x_i)L),$$

which may be considerably smaller than $\operatorname{hdeg}(M)$.

Combining this observation with Proposition 2.91, we get the following estimate for the number of generators for these modules:

Corollary 2.105. *Let M be a module as above, and let $x_1, \ldots, x_d$ be a maximal superficial sequence for M and R. For $i = 1 \ldots d$, set $M_i = M/(x_1, \ldots, x_{i-1})M$, $M^{(i)} = \Gamma_{\mathfrak{m}}(M_i)$. Then $M_1 = 0$, $M_i = L/(x_1, \ldots, x_{i-1})L$ for $i > 1$ and*

$$\begin{aligned}
\nu(M) &\leq r \cdot \deg(R) + \sum_{i=1}^{d-1} \lambda(L/(x_1, \ldots, x_i)L) - \sum_{i=1}^{d-1} \lambda(x_i(L/(x_1, \ldots, x_{i-1}L))) \\
&= r \cdot \deg(R) + \sum_{i=1}^{d-1} \lambda(L/(x_1, \ldots, x_i)L.
\end{aligned}$$

Another application of $\text{bdeg}(\cdot)$ provides a comparison with $\text{reg}(\cdot)$ in the reverse direction to Theorem 2.88. For a standard graded algebra A over a field, and a finitely generated graded A-module M, we consider the Hilbert function $H^1(M,n) = \sum_{j\leq n} \lambda(M_j)$.

Proposition 2.106. *For a graded module M as above,*

$$\text{bdeg}(M) \leq H^1(M, \text{reg}(M)).$$

Proof. Denote by L the submodule of M of finite support and define the exact sequence

$$0 \to L \longrightarrow M \longrightarrow M' \to 0.$$

We note that $\text{reg}(L) \leq \text{reg}(M)$ and $\lambda(L) = H^1(L, \text{reg}(L))$. From the additivity relations $\text{bdeg}(M) = \text{bdeg}(L) + \text{bdeg}(M')$ and $H^1(M,n) = H^1(L,n) + H^1(M',n)$, it is clear that we have only to prove the assertion for modules of positive depth since it is clear when $M = L$. Assume then $L = 0$ and let h be a generic hyperplane section such that $\text{bdeg}(M) = \text{bdeg}(M/hM)$. Since $H^1(M,n) \geq H^1(M/hM,n)$ and $\text{reg}(M) = \text{reg}(M/hM)$, the assertion follows by induction. □

Extended Hilbert Function

Let $A = \bigoplus_{i\in\mathbb{N}} A_i$ be a standard algebra over the local ring $(R,\mathfrak{m})$ and let $M = \bigoplus_{i\in\mathbb{Z}} M_i$ be a finitely generated graded A-module. For simplicity we shall assume that the residue field of R is very large–say uncountable. The *analytic spread* $\ell(M)$ of M is the Krull dimension of $M/\mathfrak{m}M$ as an $A/\mathfrak{m}A$-module.

The following result arose in a discussion with L. Doering and T. Gunston.

Proposition 2.107. *Let $(R,\mathfrak{m})$ be a Noetherian local ring, and let A and M be as above. Let $H_M(t)$ be the Hilbert function for* $\text{bdeg}(\cdot)$. *Then $H_M(t)$ is a rational function with a pole of order $\ell(M)$ at $t = 1$.*

Proof. Let

$$0 \to \Gamma_{\mathfrak{m}}(M) \longrightarrow M \longrightarrow \overline{M} \to 0 \tag{2.52}$$

be the usual reduction modulo the elements of finite support. First, by the additivity property of any $\text{Deg}(\cdot)$ function relative to such short exact sequences, we have

$$H_M(t) = H_{\Gamma_{\mathfrak{m}}(M)}(t) + H_{\overline{M}}(t).$$

It will suffice to show that both summands are rational functions and that one of them has a pole at $t = 1$ of order $\ell(M)$.

The first summand is obviously a rational function since it is the Hilbert-Poincaré series of an ordinary graded module. For the other summand, we choose a hyperplane element $h \in \mathfrak{m}$ that is generic for every component of $\overline{M}$ (it is here that one may need the uncountable condition). This means that $\mathrm{bdeg}(\overline{M}_n) = \mathrm{bdeg}(\overline{M}_n/h\overline{M}_n)$ for all n. In other words, we have that

$$H_{\overline{M}}(t) = H_{\overline{M}/h\overline{M}}(t),$$

and induction on the dimension of R shows that this formal Laurent series is rational.

Finally, we arrive at the calculation of analytic spreads. Let s be an integer large enough (via the Artin-Rees lemma) so that $\mathfrak{m}^s\Gamma_{\mathfrak{m}}(M) = 0$ and $\Gamma_{\mathfrak{m}}(M) \cap \mathfrak{m}^n M = \mathfrak{m}^{n-s}(\Gamma_{\mathfrak{m}}(M) \cap \mathfrak{m}^s M)$ for all $n \geq s$. Tensoring the sequence (2.52) by $R/\mathfrak{m}^{2s}$, we get the exact sequence

$$0 \to \Gamma_{\mathfrak{m}}(M)/\mathfrak{m}^{2s}\Gamma_{\mathfrak{m}}(M) = \Gamma_{\mathfrak{m}}(M) \longrightarrow M/\mathfrak{m}^{2s}M \longrightarrow \overline{M}/\mathfrak{m}^{2s}\overline{M} \to 0,$$

from which we read (taking into account that for any module E, $\ell(E) = \ell(E/\mathfrak{m}^n E)$ for all n)

$$\ell(M) = \max\{\ \ell(\Gamma_{\mathfrak{m}}(M)), \ell(\overline{M})\ \}.$$

A straightforward induction on the dimension of R completes the proof. □

The details of this function $H_M(t)$ can be inferred using the standard techniques of Hilbert functions. In particular, from an expression

$$H_M(t) = \frac{h(t)}{(1-t)^s} \qquad s = \ell(M) = \dim M/\mathfrak{m}M,$$

we have a (super?) Hilbert polynomial

$$H_M(n) = \varepsilon_0(M)\binom{n+s-2}{s-1} + \text{lower terms}.$$

We shall call $\varepsilon_0(M)$ the *supermultiplicity of a module.*

An immediate consequence is the kind of control $\varepsilon_0(M)$ gives over the ordinary multiplicity of the special fiber of M:

Corollary 2.108. *For a module M as above, one has $\varepsilon_0(M) \geq \deg(M/\mathfrak{m}M)$.*

Proof. This follows since $H_M(n) \geq \nu(M_n)$ for $n \gg 0$, from Proposition 2.91, since both functions are polynomials of degree $s-1$ and the multiplicities are the corresponding leading coefficients (in the binomial representation). □

Remark 2.109. One might not get that the corresponding Poincaré series for a function $\mathrm{Deg}(M_n)$ is necessarily rational.

Let us explain, in a slightly different manner, the rules for computation with bdeg. Let $(R, \mathfrak{m})$ be a Noetherian local with an infinite residue field and let E be a finitely generated R-module of dimension $d \geq 1$. The basic rule asserts that if $d \geq 1$, then

$$\mathrm{bdeg}(E) = \lambda(\Gamma_{\mathfrak{m}}(E)) + \mathrm{bdeg}(E/hE + \Gamma_{\mathfrak{m}}(E)),$$

where $h \in \mathfrak{m}$ is an appropriate generic element. If we set

$$T(E) = E/(hE + \Gamma_{\mathfrak{m}}(E)),$$

we have that $T(E)$ is a module of dimension $d-1$. The iterative formulation leads to:

Proposition 2.110. *Let $(R,\mathfrak{m})$ be a Noetherian local ring with infinite residue field. For any finitely generated R-module E,*

$$\operatorname{bdeg}(E) = \sum_{i=0}^{d} \lambda(\Gamma_{\mathfrak{m}}(T^i(E))), \quad d = \dim E.$$

In this formula, $\lambda(\Gamma_{\mathfrak{m}}(T^d(E))) = \deg E$. The formula also explains the character of Theorem 2.107.

2.5 Finiteness of Hilbert Functions

The numerical functions $h : \mathbb{N} \to \mathbb{N}$ that occur as Hilbert functions of standard graded algebras are conditioned by several well-known results, particularly a classical theorem of Macaulay (see [BH93, Theorem 4.2.10]). Out of such formulations, the number of such functions can be bounded in terms of some of its parameters.

We are going to establish the fact that the number of Hilbert functions of associated graded rings of ideals is bounded *a priori* by certain multiplicities. To obtain very sharp numbers in the Cohen-Macaulay case, it takes hard arguments by Srinivas and Trivedi ([ST97], [Tri97]). Using cohomological degrees one gets bounds for the number of Hilbert functions with a given multiplicity but that do not require any condition on the depth of the algebra. More recently, Rossi, Trung and Valla ([RTV3]) improved on all these developments by showing how to find sharp bounds for the number of Hilbert functions for all local rings $(R,\mathfrak{m})$ with given dimension and multiplicity.

We begin by restating Proposition 2.55 in the language of these functions.

Corollary 2.111. *Let $G = \bigoplus_{n\geq 0} G_n$ be a standard graded algebra over an Artinian ring and* $\operatorname{Deg}(\cdot)$ *any extended degree function defined on G. If* $\dim G = d \geq 1$*, then*

$$P_G(t) \leq \frac{\lambda(G_0) + (\operatorname{Deg}(G) - \lambda(G_0)) \cdot t}{(1-t)^d}.$$

In other words, for all $n \geq 0$,

$$H_G(n) \leq \operatorname{Deg}(G)\binom{d+n-2}{d-1} + \lambda(G_0)\binom{d+n-2}{d-2}.$$

Proof. Let J be an ideal generated by a system of parameters of degree 1 that is generic for the function $\mathrm{Deg}(\cdot)$ chosen; according to [DGV98, Proposition 2.3], $\lambda(G/J) \leq \mathrm{Deg}(G)$. Now replace $\lambda(G/J)$ by $\mathrm{Deg}(G)$ in the estimate of Proposition 2.55. □

We now establish a general bound for the coefficients of the Hilbert polynomial of a graded standard algebra solely in terms of the extended degree. The bounds are explicit but far from being strict since we are only looking for an application to the finiteness of Hilbert functions.

Theorem 2.112. *Let G be a standard graded algebra over an Artinian ring G_0. For every i with $0 \leq i \leq d$ we define recursively the integers $b_0 := 1$, $b_i := i+1+\sum_{j=0}^{i-1}(i-j+1)b_j$ and we set $e_i := e_i(G)$. Then, for $0 \leq i \leq d$, we have*

$$|e_i| \leq b_i \cdot \mathrm{Deg}(G)^{i+1}.$$

Proof. We induct on d. If $d = 0$, then $e_0(G) = \mathrm{Deg}(G)$ and $b_0 = 1$. For $d \geq 1$, we set $\overline{G} = G/hG$, where $h \in G_1$ is a generic hyperplane which is then a parameter in G. We have $\dim \overline{G} = d-1$, $e_i(G) = e_i(\overline{G})$ for $i = 0, \ldots d-1$ and $\mathrm{Deg}(\overline{G}) \leq \mathrm{Deg}(G)$.

By the induction hypothesis, for $i < d$ we have

$$|e_i| \leq b_i \cdot \mathrm{Deg}(\overline{G})^{i+1} \leq b_i \cdot \mathrm{Deg}(G)^{i+1}.$$

We recall here that the difference between the Hilbert function of G and its Hilbert polynomial is given by the following ([BH93, Theorem 4.3.5(b)]):

$$H_G(n) - \mathcal{P}_G(n) = \sum_{i=0}^{d} (-1)^i \lambda(H^i_{G_+}(G)_n). \tag{2.53}$$

We now make a key point on the vanishing of $H^i_{G_+}(G)_n$ for $n \geq 0$. If we denote by $a_i(G)$ the largest n for which this group does not vanish, we have the well-known description of the Castelnuovo-Mumford regularity of the algebra G,

$$\mathrm{reg}(G) = \sup\{a_i(G) + i \mid i \geq 0\}.$$

However, according to Theorem 2.88, $\mathrm{Deg}(G) > \mathrm{reg}(G)$ so that $H_G(n) = \mathcal{P}_G(n)$ for $n \geq \mathrm{Deg}(G)$.

This implies that $H^1_G(n) = \mathcal{P}^1_G(n)$ for $n \geq \mathrm{Deg}(G) - 1$. If we set $r := \mathrm{Deg}(G)$, we get

$$|e_d| \leq H^1_G(r) + \left| \sum_{i=0}^{d-1} (-1)^i e_i \binom{d+r-i}{d-i} \right|,$$

where now we bound $H^1_G(r)$ by using the estimate given in Corollary 2.111, namely

$$H_G^1(r) \leq r\binom{d+r-1}{d} + \lambda(G_0)\binom{d+r-1}{d-1}.$$

It follows that

$$|e_d| \leq r\binom{d+r-1}{d} + \lambda(G_0)\binom{d+r-1}{d-1} + \sum_{i=0}^{d-1} |e_i|\binom{d+r-i}{d-i}.$$

Now, as in the proof of Corollary 2.111, we have

$$\lambda(G_0) = \lambda(G/G_+) \leq \lambda(G/J) \leq r.$$

We use the inequalities

$$r \leq r^2, \quad \binom{d+r-i}{d-i} \leq r^{d-i}(d-i+1), \quad \binom{d+r-1}{d} \leq r^d$$

to get

$$|e_d| \leq r^{d+1} + r^{d+1}d + \sum_{i=0}^{d-1} b_i r^{i+1} r^{d-i}(d-i+1) = r^{d+1}b_d.$$

This gives the desired assertion. □

Corollary 2.113. *Given two positive integers A and d, there exist only a finite number of Hilbert functions associated with standard graded algebras G over Artinian rings such that* $\dim G = d$ *and* $\mathrm{Deg}(G) \leq A$.

Proof. The finiteness of the number of Hilbert functions follows from the finiteness of the possible Hilbert polynomials, after remarking that Corollary 2.111 takes care of the initial values of the Hilbert function. □

Results of Rossi, Trung and Valla

Let $(R, \mathfrak{m})$ be a Noetherian local ring of dimension $d \geq 1$ and set $G = \mathrm{gr}_{\mathfrak{m}}(R)$. To obtain sharper counts and statements on the finiteness of Hilbert functions, the authors of [RTV3] showed that it is possible to convert information of the cohomological degree of R into the Castelnuovo-Mumford regularity of G.

Theorem 2.114. *For any cohomological degree* $\mathrm{Deg}(\cdot)$, *set* $I(R) = \mathrm{Deg}(R) - \deg(R) = \mathrm{Deg}(R) - e(R)$. *The following estimate holds:*

$$\mathrm{reg}(G) \leq \begin{cases} \mathrm{Deg}(R) - 1 & \text{if } d = 1, \\ e(R)^{(d-1)!-1}[e(R)^2 + e(R)I(R) + 2I(R) - e(R)]^{(d-1)!} - I(R) & \text{if } d \geq 2. \end{cases}$$

The other main result of [RTV3] gives bounds for the Hilbert coefficients of R. Together they establish the finiteness of the Hilbert functions.

Theorem 2.115. *For any cohomological degree* $\mathrm{Deg}(\cdot)$*, the following estimate holds:*

$$e_1(R) \leq \frac{e(R)(e(R)-1)}{2} + I(R),$$
$$e_i(R) \leq e(R)^{i!-1}[e(R)^2 + e(R)I(R) + 2I(R)]^{i!} - 1 \text{ if } i \geq 2.$$

The authors also assert that their computation has a natural extension to the case of $\mathfrak{m}$-primary ideals. A challenge is the issue of whether there are formulas that apply to non-primary ideals, beginning with the case of a generic complete intersection ideal I of dimension one.

2.6 Numbers of Generators of Cohen-Macaulay Ideals

Let $(R, \mathfrak{m})$ be a Cohen-Macaulay local ring of dimension d and I an ideal of height $g > 0$. If I is a Cohen-Macaulay ideal, there are several approaches that can be used to bound the minimal number of generators of I in terms of the multiplicity data, or some other degree,

$$\nu(I) \leq f(\deg(R), \deg(R/I), e(I), g, d, \mathrm{embdim}(R), \ldots),$$

where f is a linear or quadratic polynomial.

For our purpose here, we set ourselves in the framework of three previous discussions of this issue: [BER79], [Sal76], and [Val81]. After reviewing some of these results, we first introduce some variations on them, and embark on an approach that emphasizes the computation of the lengths of chains

$$I \subset I_1 \subset I_2 \subset \cdots \subset I_n = \mathfrak{m}$$

of irreducible ideals containing I, together with an analysis of the number of generators of the subideals in a composition series of R/I containing subchains of irreducible ideals. A more intrusive use of the integral closure $\overline{I}$ of I was sought but it has yet to be realized.

2.6.1 Estimating Number of Generators with Multiplicities

Let us begin by setting the stage. Throughout, R is a local ring of dimension d with infinite residue field and multiplicity $\deg(R)$; I is a Cohen-Macaulay ideal of height g and multiplicity $\deg(R/I)$. Other elements to appear are the embedding dimension $\mathrm{embdim}(R)$ of R, the multiplicities of various localizations $R_{\mathfrak{p}}$.

We begin our discussion with a general observation, a reduction that is general but still useful.

Proposition 2.116. *If* $g < d$ *there is a regular sequence* $\mathbf{x} = x_1, \ldots, x_{d-g}$ *such that* $\nu(I) = \nu(\overline{I})$*,* $\deg(R) = \deg(\overline{R})$*,* $\deg(R/I) = \lambda(\overline{R}/\overline{I})$*, and* $\mathrm{embdim}(\overline{R}) = \mathrm{embdim}(R) - (d-g)$*, where* $\overline{R} = R/(\mathbf{x})$ *and* $\overline{I} = I\overline{R}$*.*

Proof. Since R has an infinite residue field, we can choose $x \in \mathfrak{m}$ which is a superficial element for the maximal ideals of R and of R/I. Since R and R/I are Cohen-Macaulay, x is a regular element of both rings. Thus from the exact sequence

$$0 \to I \longrightarrow R \longrightarrow R/I \to 0,$$

we obtain the following exact sequence by tensoring with $R/(x)$

$$0 \to I/xI \longrightarrow R/(x) \longrightarrow R/(I,x) \to 0.$$

This shows that $\nu(I) = \nu(I/xI) = \nu(\overline{I})$. Since the multiplicities are unchanged and the embedding dimension is decreased by one, the assertions follow by induction. □

The first method to bound $\nu(I)$ is derived from a general estimate (see 2.93):

$$\boxed{\nu(I) \leq \deg(R) + (g-1)\deg(R/I) + \underbrace{(d-r-1)(\operatorname{Deg}(R/I) - \deg(R/I))}_{\text{non-Cohen-Macaulay correction}}} \tag{2.54}$$

where $\operatorname{Deg}(\cdot)$ is any extended degree function and $r = \operatorname{depth} R/I$.

If I is a Cohen-Macaulay ideal, $\operatorname{Deg}(R/I) = \deg(R/I)$, then we get a benchmark for comparison with all other approaches (see also [Val81]):

Proposition 2.117. *If I is a Cohen-Macaulay ideal of codimension g, then*

$$\boxed{\nu(I) \leq \deg(R) + (g-1)\deg(R/I).} \tag{2.55}$$

The case $g \leq 1$ is part of a very general result valid for all Cohen-Macaulay modules ([BH93, Corollary 4.7.11]).

In some cases, instead of $\deg(R/I)$ one has other invariants of the ideal appearing in these estimates. Here is one from [BER79].

Theorem 2.118. *Let R be a Cohen-Macaulay local ring and I a Cohen-Macaulay ideal of codimension* 2. *If R/I has Cohen-Macaulay type r, then*

$$\nu(I) \leq (r+1)\deg(R).$$

Proof. The proof of [BER79] places no restriction on R, but here we assume that R contains an appropriate field so that we may complete R and assume the existence of a regular local subring S over which R is finite. Since R is Cohen-Macaulay, this means that R is a free S-module of rank $\deg(R)$. Now R/I has a minimal free resolution

$$0 \to S^p \longrightarrow S^m \longrightarrow S^n \longrightarrow R/I \to 0$$

over S, in which $n \leq \deg(R)$. The number m is the minimal number of generators of I as an S-module. Since $m = n + p$, we need to bound p. Applying the functor $\mathrm{Hom}_S(\cdot, S)$, we get that the canonical module $\omega_{R/I} = \mathrm{Ext}_S^2(R/I, S)$ of R/I can be generated minimally by p elements. But $\omega_{R/I}$ is generated by r elements as an R/I-module, and therefore it can be generated by $r \cdot \deg(R)$ elements over S, and we get our estimate. □

The next two results come from [Sal76].

Theorem 2.119. *Let $(R, \mathfrak{m})$ be a Cohen-Macaulay local ring of dimension $d > 0$, and I an $\mathfrak{m}$-primary ideal of nilpotency index t. Then*

$$\nu(I) \leq s^{d-1} \deg(R) + d - 1. \tag{2.56}$$

Proof. We use induction on d. The case $d = 1$ comes from the standard bound. Assuming that R has infinite residue field, choose a superficial element x for $\mathfrak{m}$. Pass to the Cohen-Macaulay ring $R/(x^t)$ of dimension $d - 1$. But $I/(x^t)$ is $\mathfrak{m}/(x^t)$-primary, so by induction

$$\nu(I/(x^t)) \leq t^{d-2} \deg(R/(x^t)) + d - 2.$$

Hence $\nu(I) \leq \nu(I/(x^t)) + 1 \leq t^{d-2} t \deg(R) + d - 1$. □

Theorem 2.120. *Let $(R, \mathfrak{m})$ be a Cohen-Macaulay local ring of dimension $d > 0$ and I a Cohen-Macaulay ideal of height $g > 0$. Then*

$$\nu(I) \leq \deg(R/I)^{g-1} \deg(R) + g - 1. \tag{2.57}$$

Proof. The case $g = d$ is Theorem 2.119. Suppose that $g < d$. Choose a nonzero element of $\mathfrak{m}$ that is superficial for $\mathfrak{m}$, and whose image in $\mathfrak{m}/I$ is a superficial element for $\mathfrak{m}/I$. Pass to $R/(x)$ and its Cohen-Macaulay ideal $(I, x)/(x)$ of height g. By induction,

$$\begin{aligned} \nu(I) = \nu((I,x)/(x)) &\leq \deg(R/(I,x))^{g-1} \deg(R/(x)) + g - 1 \\ &= \deg(R/I)^{g-1} \deg(R) + g - 1. \end{aligned}$$

Exploring the Koszul complex, Valla ([Val81]) derived the following different bounds:

Theorem 2.121. *Let $(R, \mathfrak{m})$ be a Cohen-Macaulay local ring of multiplicity e, and I a Cohen-Macaulay ideal of codimension g such that the multiplicity of R/I is δ. Setting $m = \min\{e, \delta\}$, we have*

(a) *If $g > 0$, then* $\nu(I) \leq e + \dfrac{\delta(g-1)^2}{g} + \dfrac{m(g-1)}{g}$.

(b) *If $g \geq 2$ and $I \subset \mathfrak{m}^2$, then* $\nu(I) \leq e + \dfrac{\delta(g-1)^2}{g} + \dfrac{\min\{m+g, \delta\}}{g} - \dbinom{g}{2}$.

The next method, which we have already made use of, has a different character. Let J_0 be an ideal generated by a regular sequence of $g-1$ elements, $J_0 \subset I$. Preferably J_0 should be part of a minimal reduction of I. We have

$$\nu(I) \leq \nu(I/J_0) + \nu(J_0) \leq \deg(R/J_0) + g - 1, \tag{2.58}$$

since I/J_0 is a maximal Cohen-Macaulay module of rank 1 over the Cohen-Macaulay ring R/J_0. The issue is to relate $\deg(R/J_0)$ to $\deg(R/I)$.

We consider two applications of the second method to the case of an ideal I that is equimultiple. The other approach is to some extent not sensitive to this additional information. First, let R be a Gorenstein ring and suppose that $J = (J_0, x)$ is a minimal reduction of I. Since x is regular modulo J_0, we have $\deg(R/J_0) \leq \deg(R/J)$, and

$$\nu(I) \leq \deg(R/J) + g - 1.$$

Consider the exact sequence

$$0 \to (J:I)/J \longrightarrow R/J \longrightarrow R/(J:I) \to 0,$$

where $(J:I)/J$ is the canonical module of R/I, so that the equality $\deg((J:I)/J) = \deg(R/I)$ yields

$$\deg(R/J) = \deg(R/I) + \deg(R/(J:I)).$$

If, for instance, I is equimultiple of reduction number 1, that is if $I^2 = JI$, we have $I \subset J:I$ and therefore $\deg(R/I) \geq \deg(R/(J:I))$, since they are both modules of dimension $d-g$. We thus obtain the estimate

$$\nu(I) \leq 2\deg(R/I) + g - 1,$$

which is, usually, for this limited class of ideals, considerably better than the bound (2.55). We extend this further to arbitrary Cohen-Macaulay rings.

Proposition 2.122. *Let R be a Cohen-Macaulay local ring of type r and I a Cohen-Macaulay ideal of height $g \geq 1$. If I is equimultiple and has reduction number ≤ 1, then*

$$\nu(I) \leq (r+1)\deg(R/I) + g - 1. \tag{2.59}$$

Proof. Taking into account the considerations above, it will suffice to show that $\deg(J:I/J) \leq r\deg(R/I)$. We may assume (passing to the completion if necessary) that R has a canonical module, say ω. The ring R/J has $\overline{\omega} = \omega/J\omega$ for its canonical module. Further $\overline{\omega}$ is a faithful R/J-module generated by r elements. In the usual manner, we can build an embedding

$$R/J \hookrightarrow \overline{\omega}^{\oplus r}.$$

Applying the functor $\mathrm{Hom}(R/I,\cdot)$, we obtain an embedding of $J:I/J$ into the direct sum of r copies of $\mathrm{Hom}(R/I,\overline{\omega})$. Since this is the canonical module of R/I, it has the same multiplicity as R/I. □

Number of Generators of Primary Ideals

Let $(R,\mathfrak{m})$ be a Cohen-Macaulay local ring and I an $\mathfrak{m}$-primary ideal. We seek bounds between the number of generators of I and other quantities attached to the ideal:

$$\nu(I) \leftrightarrow \begin{cases} d = \dim R \\ \varepsilon = \nu(\mathfrak{m}) \\ s = \lambda(R/I), \quad \text{the colength of } I \\ e(I) = \text{multiplicity of } I \\ t = \text{nil}(R/I), \quad \text{smallest } n \text{ such that } \mathfrak{m}^n \subset I \end{cases}$$

We have already seen one relation between some of these quantities (Proposition 1.109).

Proposition 2.123. *Let $(R,\mathfrak{m})$ be a Cohen-Macaulay local ring of dimension d and I an $\mathfrak{m}$-primary ideal of multiplicity $e(I)$.*

(a) *Then*

$$\nu(I) \leq e(I) + d - 1. \tag{2.60}$$

(b) *If $I \subset \mathfrak{m}^2$, then*

$$\nu(I) < e(I) \leq d! \cdot \lambda(R/\bar{I}) \deg(R). \tag{2.61}$$

Proof. We may assume that R has infinite residue field. Let J be a minimal reduction of I. From the inequalities exhibited by the diagram,

$$\begin{aligned} \nu(I) &= \lambda(I/\mathfrak{m}I) \leq \lambda(I/\mathfrak{m}J) \\ &= \lambda(I/J) + \lambda(J/\mathfrak{m}J) \\ &= \lambda(R/J) - \lambda(R/I) + d, \end{aligned}$$

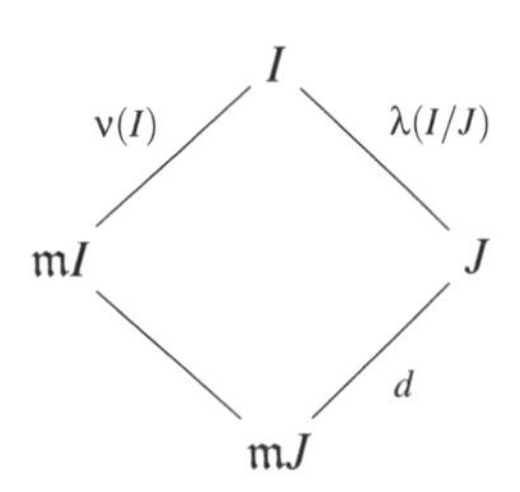

which proves the first assertion, since $e(I) = \lambda(R/J)$ as R is Cohen-Macaulay. For the other bound, just observe that with $I \subset \mathfrak{m}^2$ we have $\lambda(R/I) \geq \lambda(R/\mathfrak{m}^2) \geq \varepsilon + 1 \geq d + 1$. The final inequality arises from (2.58). $\square$

The following is a naive observation on the number of generators of an ideal of finite co-length to which we shall refer later.

Proposition 2.124. *Let $(R,\mathfrak{m})$ be a Noetherian local ring. The function*

$$I \mapsto \lambda(R/I) + \nu(I),$$

on $\mathfrak{m}$-primary ideals is non-increasing, that is, if $I \subset L$, its value on L is not larger than that on I.

The assertion is clear, the function being merely $\lambda(R/\mathfrak{m}I)$. When applied to the reductions of the integral closure of an ideal I, it attains the maximum value (when R is Cohen-Macaulay) on the minimal reductions, $e(I) + \dim R$.

2.6.2 Number of Generators and the Socle

We shall now develop new expressions for bounds on the number of generators of ideals that integrate efficiencies usually associated with Gorenstein ideals.

Proposition 2.125. *Let $(R,\mathfrak{m})$ be a Noetherian local ring and I an $\mathfrak{m}$-primary ideal.*

(a) *If $I \subset \mathfrak{m}^2$, $z \in (I : \mathfrak{m}) \setminus I$, I is irreducible and $\varepsilon \geq 2$, then*

$$\nu(I) = \nu(I,z) - 1.$$

(b) [H. Schoutens] *In general,*

$$\nu(I) \leq \nu(I,z) + \varepsilon - 1.$$

Proof. The first assertion is part of Lemma 1.107. It follows from the equality $\mathfrak{m}(I,z) = \mathfrak{m}I$.

For the other assertion, suppose that I is minimally generated by $a_1, \ldots, a_n$; after notation change of the a_i adding z may result in a set of generators for (I,z) of the form $a_1, \ldots, a_r, z$. This means (after another rewrite) that $a_i = b_i z$, for $i \geq r+1$. As z lies in the socle of I, we have $n - r \leq \varepsilon$, thus proving the assertion. □

Remark 2.126. We are going to use slightly modified versions to take into account of the possibility that I is not contained in $\mathfrak{m}^2$. Denote by ε_0 the dimension of the vector space $(I + \mathfrak{m}^2)/\mathfrak{m}^2$. (For later usage, we denote it by $\varepsilon_0(I)$.) If $\varepsilon_0 > 0$, say $\{x_1, \ldots, x_{\varepsilon_0}\} \subset I$ is part of a minimal set of generators of I, we can mod out these elements and we may still apply Proposition 2.125.a provided that $\varepsilon - \varepsilon_0 \geq 2$, and still get the equality $\nu(I) = \nu(L) - 1$. As for Proposition 2.125.b, we get the improved bound $\nu(I) \leq \nu(L) + \varepsilon - \varepsilon_0 - 1$, since $\varepsilon - \varepsilon_0$ is the embedding dimension of the actual ring.

The second assertion gives a mechanism for bounding $\nu(I)$ in terms of s and of ε. On the other hand, the first assertion allows for a possible decreased count for the number of generators. This suggests that one should go from I to $\mathfrak{m}$ by steps that meet the largest possible number of irreducible ideals. We look at this issue now.

The question reduces to the Artinian case. Let I be an ideal and consider sequences

$$I \subset I_1 \subset I_2 \subset \cdots \subset I_n = \mathfrak{m},$$

where each I_i is an irreducible ideal.

Proposition 2.127. *The supremum of the length of these sequences is at most t, the nilpotency index of R/I. Furthermore, if $A = R[x]_{(\mathfrak{m}R[x])}$ then for the ideal IA the supremum is exactly $nil(R/I) = nil(A/IA)$.*

Proof. We first assume that R is a Gorenstein ring and set $L = \operatorname{ann}(I)$. By duality, we get a chain of ideals

$$\operatorname{ann}(I) \supset \operatorname{ann}(I_1) \supset \operatorname{ann}(I_2) \supset \cdots \supset \operatorname{ann}(I_n) = \operatorname{ann}(\mathfrak{m}),$$

where $\operatorname{ann}(I_i) = Rx_i$ is principal for $i = 1, \ldots, n$. In particular we obtain that $x_n = rx_1x_2\cdots x_{n-1}$, which shows that $n-1 < t$.

Let us show that the lengths of these chains, in the ring A, attain the bound t. Since $\mathfrak{m}^t \subset I$, we have that $L\mathfrak{m}^t = 0$. Suppose we had $L\mathfrak{m}^p = 0$ for $p < t$. This would imply that $\mathfrak{m}^p \subset 0 : L = I$ by duality, contrary to the definition of t.

If L is principal, $L = (c)$, since $c\mathfrak{m}^{t-1} \neq 0$ we would build a sequence $x_1, \ldots, x_t$ of elements of $\mathfrak{m}$ with $cx_1\cdots x_t \neq 0$, which would give rise to a corresponding sequence of irreducible ideals $I_i = \operatorname{ann}(cx_1\cdots x_i)$, of the desired length.

If I is not irreducible then L is not principal, $L = (c_1, \ldots, c_m)$, pass to the ring A and consider the element

$$c = \sum_{i=1} c_i x^i.$$

It is clear that $\mathfrak{m}^t c \neq 0$. We can now build a sequence of irreducible ideals containing I, of length t.

If R is not Gorenstein, let E be the injective envelope of the residue field of R and set S to be the trivial extension of R by E,

$$S = R \times E, \quad \text{multiplication defined by } (a,b)(c,d) = (ac, ad + bc).$$

Then S is a Gorenstein local ring with maximal ideal $M = \mathfrak{m} \times E$. If we put $I' = I \times E$, then $\operatorname{nil}(S/I') = \operatorname{nil}(R/I)$ and we can obviously apply the argument above to the pair S, I' to derive the same assertion for the pair R, I. □

Let us now derive our first bound for $\nu(I)$. Since none of the measures introduced above are affected by the passage to A, we can assume the existence of chains of irreducible ideals containing I with t elements (actually $t+1$ if $t > 1$). We make a few additional observations. The approach we take is to build $\mathfrak{m}$ starting from I by adding successively socle elements along convenient chains-here we shall take chains of irreducible ideals. If anywhere in the process we get an ideal not contained in $\mathfrak{m}^2$, we can reduce the dimension of R and the embedding dimension by 1, at least for as long as $d > 0$. Finally, we remark that if $\lambda(R/I)$ is very small, then the number of generators can be easily estimated. For example, if $\lambda(R/I) = 3$, then $\lambda(\mathfrak{m}/I) = 2$, from which we get $\nu(I) \leq \varepsilon + 1$. The next case is not much more complicated, and one gets that $\nu(I) \leq \varepsilon + 3$ if $\lambda(R/I) = 4$.

The next result ([Va3b]) brings together $\nu(I)$ and the various multiplicities in the case $\lambda(R/I) > 4$. A version of part (a) is due to H. Schoutens ([Sho3]).

Theorem 2.128. *Let $(R,\mathfrak{m})$ be a Noetherian local ring of embedding dimension $\varepsilon \geq 2$, and let I be an $\mathfrak{m}$-primary ideal contained in $\mathfrak{m}^2$ of colength s.*

(a) *If $\lambda(R/I) > 4$, then*

$$\nu(I) \leq (\lambda(R/I) - 4)(\varepsilon - 1) - \frac{\varepsilon(\varepsilon - 7)}{2}.$$

(b) *If t is the nilpotency index of R/I, then*

$$\begin{aligned}\nu(I) &\leq (\lambda(R/I) - t)(\varepsilon - 1) - \frac{\varepsilon^2 - 5\varepsilon + 2}{2} - 1 \\ &\leq (s - t - \frac{\varepsilon}{2} + 2)(\varepsilon - 1).\end{aligned} \tag{2.62}$$

Proof. (a) We are going to start from I and successively add socle elements along the chain of irreducible ideals. If at some point we add to the ideal J a socle element $z \in J : \mathfrak{m}$ so that $L = (J, z)$ is also contained in the chain, we want to use Proposition 2.125 to bound $\nu(J)$ in terms of $\nu(L)$.

The argument here is a simplified version of what is needed in part (b). We choose a chain of ideals

$$I = L_0 \subset L_1 \subset \cdots \subset L_r \subset L_{r+1} \subset \mathfrak{m},$$

so that $L_{i+1} = (L_i, z_i)$, where z_i is a nonzero socle element of R/L_i; furthermore, we assume that $\lambda(R/L_{r+1}) = 4$.

Using Proposition 2.125.a and the remark that follows it, we have the following relationship between $\nu(L_i)$ and $\nu(L_{i+1})$. If $\varepsilon_0 = \varepsilon_0(L_i) > 0$, then

$$\nu(L_i) \leq \nu(L_{i+1}) + \varepsilon - \varepsilon_0 - 1.$$

This means that when we reach L_r, which satisfies $\varepsilon_0(L_r) \leq 4$ since $\lambda(\mathfrak{m}/L_r) = 4$, from the combined count

$$\nu(I) \leq \nu(L_{r+1}) + (\lambda(R/4) - 4)(\varepsilon - 1),$$

we can further take away

$$1 + 2 + \cdots + (\varepsilon - 4) = \binom{\varepsilon - 3}{2},$$

without breaking the inequality. Since $\nu(L_{r+1}) \leq \varepsilon + 3$, collecting we get the desired estimate.

(b) The assertions are not changed if we reduce R mod $\mathfrak{m}I$, so that we may assume that R is an Artinian local ring. After passage to the extension A, we may assume that we have a chain of irreducible ideals:

$$I \subset I_1 \subset I_2 \subset \cdots \subset I_t = \mathfrak{m}.$$

To apply the proposition properly, we want to distinguish the cases when J is irreducible into two subcases. An irreducible ideal $J \neq \mathfrak{m}$ is said to be of type 1 if

$$\dim \mathfrak{m}/(J+\mathfrak{m}^2) = 1;$$

otherwise J is said to be of type 2.

We now proceed with the proof. If J is not irreducible, we have

$$\nu(J) \leq \nu(L) + \varepsilon - 1,$$

which is taken into the ledger. If J is irreducible, say $J = I_{t_0}$, and of type 2, then J has a minimal set of generators of the form

$$J = (x_1, \ldots, x_m, y_1, \ldots, y_n),$$

where $x_i \in \mathfrak{m} \setminus \mathfrak{m}^2$ and $y_j \in \mathfrak{m}^2$. Furthermore, because J is of type 2, the ring $R' = R/(x_1, \ldots, x_m)$ has embedding dimension at least two and JR' is an irreducible ideal contained in the square of its maximal ideal. We can again apply Proposition 2.125.b to obtain that $\nu(J) = \nu(L) - 1$.

When J is of type 1, R/J has embedding dimension 1, and in particular it is a principal ideal ring. Thus J and all ideals containing it are generated by ε elements. Suppose this first occur at the irreducible ideal I_{t_0}, $t_0 \leq t$. This means that I_{t_0} and all ideals containing it are generated by ε elements. The number of 'socle' extensions from I up to I_{t_0} is given by $\lambda(R/I) - \lambda(R/I_{t_0})$. This means that we have

$$\nu(I) \leq (\lambda(R/I) - \lambda(R/I_{t_0}) - t_0 + 1)(\varepsilon - 1) + \nu(I_{t_0}) - t_0 - \delta,$$

where δ represents the discrepancy in using $\varepsilon - 1$ instead of $\varepsilon - \varepsilon_0 - 1$, as was done in Part (a). We claim that this contributes at least

$$\delta \geq 1 + 2 + \cdots + (\varepsilon - 2) = \frac{(\varepsilon-1)(\varepsilon-2)}{2}.$$

This is the case because, whenever we reach an irreducible ideal J, the socle element z does not affect ε_0, that is $\varepsilon_0(J) = \varepsilon_0(J, z)$, unless $\varepsilon_0(J) = \varepsilon - 1$, since socle elements of irreducible ideals in rings of embedding dimension at least two are contained in the square of the maximal ideal.

Finally, noting that $\lambda(R/I_{t_0}) = t - t_0 + 1$, we obtain

$$\begin{aligned}
\nu(I) &\leq (\lambda(R/I) - t)(\varepsilon - 1) + \varepsilon - \frac{(\varepsilon-1)(\varepsilon-2)}{2} - t_0 \\
&\leq (\lambda(R/I) - t)(\varepsilon - 1) - \frac{\varepsilon^2 - 5\varepsilon + 2}{2} - t_0.
\end{aligned}$$

Since $t_0 \geq 1$, the estimate is established after a straightforward rewrite. □

Remark 2.129. This kind of estimation is inherently uneven. If we apply it to the powers I^q of the ideal I, for $q \gg 0$ the value $\nu(I^q)$ is given by a polynomial in q of degree $d-1$, while $\lambda(R/I^q)$ grows as a polynomial in q of degree d. Given that $t(I^q)$ is bounded by a linear polynomial, the expression is asymptotically quite unbalanced.

Table 2.1. Bounds for the number of generators of $\mathfrak{m}$-primary ideals ($s = \lambda(R/I), t = \mathrm{nil}(R/I)$, $\varepsilon = \mathrm{embdim}(R) \geq 2$):

$\deg(R) + (d-1)\lambda(R/I)$	2.117
$e(I) + d - 1$	2.123(a)
$e(I) \leq d! \cdot \lambda(R/\overline{I}) \deg(R)$ if $I \subset \mathfrak{m}^2$	2.123(b)
$(s - t - \frac{\varepsilon}{2} + 2)(\varepsilon - 1)$	2.128(b)

Non-Primary Cohen-Macaulay Ideals

Let $(R, \mathfrak{m})$ be a Cohen-Macaulay local ring of dim $d \geq 1$, and let I be a Cohen-Macaulay ideal of codimension $g < d$. We seek to extend to this class of ideals some of the improvements on the number of generators for I as measured against benchmarks bounds such as (2.55).

A first step consists in examining the class of equimultiple ideals (we assume that R has infinite residue throughout).

Proposition 2.130. *Let I be an equimultiple Cohen-Macaulay ideal of codimension ≥ 1, and J one of its minimal reductions.*

(a) *If $\sqrt{I}$ is a Cohen-Macaulay ideal, then*

$$\nu(I) \leq \deg(R/J) + (g-1)\deg(R/\sqrt{I}). \tag{2.63}$$

(b) *Moreover, if $I \subset (\sqrt{I})^2$, then*

$$\nu(I) \leq \deg(R/J) - \deg(R/\sqrt{I}). \tag{2.64}$$

Proof. The proof is similar to that of Proposition 2.123. Consider the exact sequences

$$0 \to J/J\sqrt{I} \cong (R/\sqrt{I})^g \longrightarrow R/J\sqrt{I} \longrightarrow R/J \to 0,$$

$$0 \to I/J\sqrt{I} \longrightarrow R/J\sqrt{I} \longrightarrow R/I \to 0.$$

From the first sequence we get that $R/J\sqrt{I}$ is a Cohen-Macaulay module, which taken into the other sequence shows that $I/J\sqrt{I}$ is also a Cohen-Macaulay module. Thus all the modules in these sequence are Cohen-Macaulay of dimension $d-g$, so that we can compute multiplicities using the rules to compute lengths. From the second sequence we have

$$\deg(I/J\sqrt{I}) = \deg(R/J\sqrt{I}) - \deg(R/I),$$

while from the other sequence,

$$\deg(R/J\sqrt{I}) = g\deg(R/\sqrt{I}) + \deg(R/J).$$

It follows that

$$\deg(I/J\sqrt{I}) = \deg(R/J) + (g-1)\deg(R/\sqrt{I}) - \deg(\sqrt{I}/I).$$

If we drop the last summand and take into account that $\deg(I/J\sqrt{I}) \geq \nu(I/J\sqrt{I})$ (because $I/J\sqrt{I}$ is a Cohen-Macaulay module), we get Part (a) since

$$\nu(I/J\sqrt{I}) \geq \nu(I/I\sqrt{I}) = \nu(I).$$

On the other hand, if $I \subset (\sqrt{I})^2$,

$$\deg(\sqrt{I}/I) \geq \deg(\sqrt{I}/(\sqrt{I})^2) \geq g\deg(R/\sqrt{I}),$$

and the estimation above shows that

$$\deg(I/J\sqrt{I}) \leq \deg(R/J) - \deg(R/\sqrt{I}),$$

which proves Part (b). □

Remark 2.131. The requirement on $\sqrt{I}$ is strict. For example, if $I = (x^2y^2 + z^3w, x^3z + y^2z^2 + xyzw) \subset k[x,y,z,w]$, a computation with *Macaulay* will show that $\sqrt{I}$ is not Cohen-Macaulay.

The other approach to the estimation of $\nu(I)$ in the case $g < d$ is that of hyperplane sections: we can choose a regular sequence $\mathbf{z} = z_1, \ldots, z_{d-g}$ that is superficial for both rings, R/I and R, and if J is a regular sequence, for R/J as well. This has the effect that $\deg(R/I) = \lambda(R/(I,\mathbf{z}))$ and similar formulas hold for $\deg(R)$ and $\deg(R/J)$ (when J itself is Cohen-Macaulay, as when I is equimultiple). At the same time, $\nu(I) = \nu(I,\mathbf{z}) - (d-g)$. Note that if I is equimultiple, then

$$e(I,\mathbf{z}) = e(J,\mathbf{z}) = \deg(R/J).$$

Let us integrate these observations into Proposition 2.123(a):

Proposition 2.132. *For these ideals,*

$$\nu(I) \leq e(I,\mathbf{z}) + g - 1.$$

Moreover, if I is equimultiple with a minimal reduction J,

$$\nu(I) \leq \deg(R/J) + g - 1.$$

Non Cohen-Macaulay Ideals of Dimension One

Let $(R,\mathfrak{m})$ be a Cohen-Macaulay local ring of dimension $d \geq 1$ and I an ideal of dimension one. When I is Cohen-Macaulay, we have discussed a few estimates for $\nu(I)$, among which is

$$\nu(I) \leq \deg(R) + (d-2)\deg(R/I).$$

The argument leading to this does not apply when I has embedded primes, say $I = I' \cap K$, where I' is unmixed and K is $\mathfrak{m}$-primary. The nonzero module $L = I'/I$ is going to play a role in bounding $\nu(I)$.

Proposition 2.133. *Denote by* ε *the embedding dimension of* R. *Then*

$$\nu(I) \leq \begin{cases} (d-1)\lambda(L) + \deg(R) + \binom{d-1}{2}\deg(R/I) \\ (\lambda(L)-1)\varepsilon + \deg(R) + (d-2)\deg(R/I). \end{cases}$$

Proof. The first bound is simply the value of $\mathrm{hdeg}(I)$. For the other, tensor the exact sequence

$$0 \to I \longrightarrow I' \longrightarrow L \to 0$$

with the residue field $\mathfrak{k}$ of R to get the exact sequence

$$\mathrm{Tor}_1^R(L,\mathfrak{k}) \longrightarrow I \otimes_R \mathfrak{k} \longrightarrow I' \otimes_R \mathfrak{k} \longrightarrow L \otimes_R \mathfrak{k} \to 0,$$

collect $\nu(I)$ and apply to the Cohen-Macaulay ideal I' the bound $\nu(I') \leq \deg(R) + (d-2)\deg(R/I)$, since $\deg(R/I') = \deg(R/I)$. □

2.7 Multiplicities and Reduction Numbers

In this section we shall seek to extend to ideals of positive dimension the quadratic relationships between the reduction number $\mathrm{r}(I)$ and various multiplicities that were developed in [DGV98] and [Va98c] for ideals of dimension zero, and discussed early in this chapter. Figuratively, for an ideal, module or algebra A (graded in the appropriate case), where the notions of reduction number $\mathrm{r}(A)$, multiplicity $\mathrm{Deg}(A)$ and dimension d have been defined, there are usually no direct relationships between these quantities. More correctly, the relationship between $\mathrm{r}(A)$ and $\mathrm{Deg}(A)$ is mediated through various agents such as dimensions, Castelnuovo-Mumford regularity, Hilbert function, and numerical information in the syzygies associated to one of the various algebras built on I:

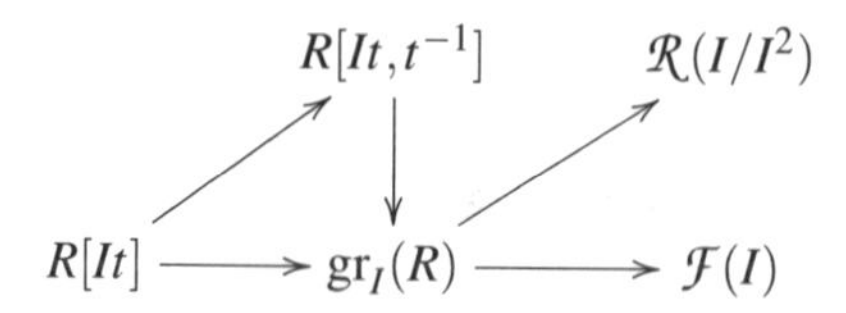

An overall aim is to capture these possibilities by inequalities of the form

$$\mathrm{r}(A) \leq f(d(A), \dim A),$$

where $d(A)$ is some 'degree' of A (such as $\deg(A)$, arith-deg(A) or even some extended multiplicity $\mathrm{Deg}(A)$), and f is a polynomial of low degree. For example, if I is an $\mathfrak{m}$-primary ideal and the residue field of R has characteristic zero, then

$$\mathrm{r}(I) < \text{arith-deg}(\mathrm{gr}_I(R)).$$

There are other bounds mediated by other agents. According to [Va98c], for a Cohen-Macaulay local ring $(R, \mathfrak{m})$ of Krull dimension d, an $\mathfrak{m}$-primary ideal I of multiplicity $e(I)$ satisfies

$$\mathrm{r}(I) \leq (d-h)e(I) - 2(d-h) + 1, \tag{2.65}$$

where h is some integer $h < d$ such that depth $\mathrm{gr}_I(R) \geq h$. The method of [Va98c] is dependent on the Cohen-Macaulayness of R. To extend to more general rings required replacing $e(I)$ in [RVV1] by the length $\lambda(R/J)$, where J is a minimal reduction of I.

The reduction number of an ideal is actually a property of its fiber cone, $\mathcal{F}(I) = \mathrm{gr}_I(R) \otimes (R/\mathfrak{m})$, and therefore a major issue is how I affects the properties of $\mathcal{F}(I)$. For example, if the residue field of R has characteristic zero,

$$\mathrm{r}(I) < \text{arith-deg}(\mathcal{F}(I)).$$

To refine this bound, in a manner that is responsive to other properties of I is hard. It is a difficult task to track the properties of I in its passage to $\mathcal{F}(I)$. A practical approach to this problem begins by attempting to bound the Hilbert function of the fiber cone of I, $\nu(I^n)$ *for all* n, by a polynomial $f(n)$ of degree $s-1$, $f(n)$ depending on the various extended multiplicities of I. We then solve for the least solution of

$$f(n) < \binom{n+s}{s},$$

which may require a convenient form for $f(n)$. Using Theorem 2.36, gives $s > \mathrm{r}(I)$. This provides for a reduction L generated by s elements and an attached bound for $\mathrm{r}_L(I)$. When I is $\mathfrak{m}$-primary, $s = \ell(I)$, and L is a minimal reduction. If however $\dim R/I \geq 1$, this approach only provides for a reduction L with fewer generators. On the other hand, it is helpful to find a minimal reduction J of L in order to use the inequality $\mathrm{r}_J(I) \leq \mathrm{r}_L(I) + \mathrm{r}_J(L)$.

We shall take two paths to these estimates. First, by extending to ideals of positive dimension the methods that have already been used on ideals of finite co-length. Then we exploit an opening provided by the consideration of the Rees algebra of the conormal module. It is a small gap, but enough to lead to several other cases of ideals of low dimension.

Another approach, one which would not require the use of extended multiplicities, would involve the Hilbert function of $\mathcal{F}(I)$. The technique of Section 2.2 would be valuable if one could obtain estimates for the whole Hilbert function.

2.7.1 The Modulo Dimension One Technique

In practice-at least in the cases it has yielded specific bounds–it is worked out by the following technique, the codimension minus one trick. Let I be an ideal of height $g < d = \dim R$, and suppose that the analytic spread of I is ℓ (we assume that the residue field of R is infinite). Let $J_0 = (a_1, \ldots, a_h)$ be part of a minimal reduction of I, with the a_i's forming a regular sequence. Now from the embedding

$$I^n/J_0^n \hookrightarrow R/J_0^n,$$

one has

$$\nu(I^n) \leq \nu(J_0^n) + \nu(I^n/J_0^n).$$

We need to bound the right-hand side by a polynomial $f(n)$ of degree at most $\ell - 1$ in order to apply the comments above. The first term is simply

$$\nu(J_0^n) = \binom{n+h-1}{h-1},$$

so it presents no difficulty since $h \leq g \leq \ell \leq d$. The other term arises from an ideal of the Cohen-Macaulay ring $S = R/J_0^n$, whose properties are easily traceable to those of R and ordinary multiplicities of R/I. The heart of the matter is to estimate the number of generators of the ideal I^n/J_0^n, using some multiplicity of S but still bounded by a polynomial of degree $\ell - 1$.

Two variants can be introduced. (i) Since one can use other kinds of multiplicities, some of which may work well for non Cohen-Macaulay rings, we do not have to choose J_0 to be a complete intersection. (ii) To achieve a minimal reduction J with the desirable reduction number, one may first use the technique to achieve an intermediate reduction L, with appropriate reduction number and much smaller number of generators; then try to handle $\mathrm{r}(L)$ (note that $\mathrm{r}(I) \leq \mathrm{r}_L(I) + \mathrm{r}(L)$).

Let us indicate how this works in practice by extending Theorem 2.44 to a much wider class of filtrations.

Proposition 2.134. *Let $(R, \mathfrak{m})$ be a Cohen-Macaulay local ring of dimension $d \geq 1$ with infinite residue field. Let $\mathfrak{F} = \{I_n, n \geq 1\}$ be a multiplicative filtration of Cohen-Macaulay ideals of dimension $d - g < d$. If $I = I_1$ is an equimultiple ideal, then for all n*

$$\nu(I_n) \leq g!\deg(R/\overline{I})\deg(R)\binom{n+g-2}{g-1} + \binom{n+g-2}{g-2}.$$

Proof. We choose J_0 as above: $J_0 = (a_1, \ldots, a_{g-1})$ is an ideal of height $g-1$ with the a_i being part of a minimal reduction of I. This already accounts for the summand

$$\nu(J_0^n) = \binom{n+g-2}{g-2}.$$

By the other assumption, I_n/J_0^n is a Cohen-Macaulay ideal of height 1 of the Cohen-Macaulay ring R/J_0^n. Thus I_n/J_0^n is a Cohen-Macaulay module, and therefore

$$\nu(I_n/J_0^n) \leq \deg(R/J_0^n) = \deg(R/J_0)\binom{n+g-2}{g-1}.$$

Let $J = (J_0, x)$ be a minimal reduction of I. Since x is regular modulo J_0, we have $\deg(R/J_0) \leq \deg(R/J)$.

What we need will be provided in the following estimate of multiplicities.

Proposition 2.135. *Let R be a Cohen-Macaulay local ring with infinite residue field, and let I be an equimultiple ideal of codimension g. If J is a minimal reduction of I, then*

$$\deg(R/J) \leq g! \deg(R/\overline{I}) \deg(R). \tag{2.66}$$

Proof. Consider how the associativity formula for multiplicities relates to R/J:

$$\deg(R/J) = \sum_{\mathfrak{p}} \lambda(R_{\mathfrak{p}}/J_{\mathfrak{p}}) \deg(R/\mathfrak{p}),$$

where $\mathfrak{p}$ runs over the minimal primes of J; since I is equimultiple, they are the same as the minimal primes of I.

We note that $\lambda(R_{\mathfrak{p}}/J_{\mathfrak{p}})$ is the Samuel multiplicity of the ideal $I_{\mathfrak{p}}$. By Theorem 2.58,

$$e(I_{\mathfrak{p}}) \leq g!\lambda(R_{\mathfrak{p}}/\overline{I}_{\mathfrak{p}}) \deg(R_{\mathfrak{p}}).$$

Noting that $\deg(R_{\mathfrak{p}}) \leq \deg(R)$, we obtain

$$\begin{aligned}
\deg(R/J) &= \sum_{\mathfrak{p}} e(I_{\mathfrak{p}}) \deg(R/\mathfrak{p}) \\
&\leq g! \sum_{\mathfrak{p}} \lambda(R_{\mathfrak{p}}/\overline{I}_{\mathfrak{p}}) \deg(R) \deg(R/\mathfrak{p}) \\
&= g!(\sum_{\mathfrak{p}} \lambda(R_{\mathfrak{p}}/\overline{I}_{\mathfrak{p}}) \deg(R/\mathfrak{p})) \cdot \deg(R) \\
&= g! \deg(R/\overline{I}) \deg(R),
\end{aligned}$$

giving the asserted bound. □

We are going to apply this result to filtrations of integral closures of equimultiple ideals of dimension 1 in Chapter 7, and to filtrations of symbolic powers of ideals of dimension 1 also in Chapter 7.

2.7.2 Special Fibers

The role of the special fibers, briefly studied in Chapter 1, will be examined here assuming various algebraic conditions on them.

Special Fibers that are Integral Domains

Let $(R,\mathfrak{m})$ be a Noetherian local ring with infinite residue field and I an ideal. Suppose that the special fiber of I, $\mathcal{F}(I)$, is a Cohen-Macaulay ring. In this case the h-vector of $\mathcal{F}(I)$ determines the reduction number of I:

$$h(\mathcal{F}(I)) = (1,h_1,\ldots,h_s), \quad h_s \neq 0 \Rightarrow \mathrm{r}(I) = s.$$

Since $\mathcal{F}(I)$ is a standard graded algebra, $h_i \neq 0$ for $i \leq s$. In terms of the multiplicity we have

$$\mathrm{r}(I) = s \leq \deg(\mathcal{F}(I)) - 1 = \sum_{i=0}^{s} h_i - 1.$$

One can do much better when $\mathcal{F}(I)$ is a Cohen-Macaulay integral domain and $R/\mathfrak{m}$ has characteristic zero, since according to [Sta91, Proposition 3.4] (see also [Val98, Theorem 2.8]) one has:

Proposition 2.136. *Suppose that A is a standard graded algebra of dimension d over a field of characteristic zero. If A is a Cohen-Macaulay integral domain, its h-vector satisfies the following property. For any two positive integers* m, $n \geq 1$ *such that* $m+n < s$, *then*

$$h_1 + \cdots + h_n \leq h_{m+1} + \cdots + h_{m+n},$$

in particular $h_i \geq h_1$ *for* $0 < i < s$.

As a consequence we have that

$$e(A) - h_0 - h_s = \sum_{i=1}^{s-1} h_i \geq (s-1)h_1.$$

Note that h_s is the Cohen-Macaulay type of A, and h_1 is its embedding codimension, so that we obtain the following bound for the reduction number of the algebra

$$\mathrm{r}(A) \leq 1 + \frac{e(A) - h_s - 1}{h_1}. \tag{2.67}$$

When we apply this to the special fiber of an $\mathfrak{m}$-primary ideal I, with $h_1 = \nu(I) - d$ and $e(I) \geq \deg(\mathcal{F}(I))$. Since $\mathcal{F}(I)$ is a homomorphic image of $\mathrm{gr}_I(R)$ (and both algebras have the same dimension) we obtain the following bound.

Corollary 2.137. *Let* $(R,\mathfrak{m})$ *be a Noetherian local ring of dimension d, with a residue field of characteristic zero and I an* $\mathfrak{m}$*-primary ideal with* $\nu(I) > d$. *If the special fiber* $\mathcal{F}(I)$ *is a Cohen-Macaulay integral domain of type t, then*

$$\boxed{\mathrm{r}(I) \leq 1 + \frac{e(I) - t - 1}{\nu(I) - d}.} \tag{2.68}$$

The difference between the approach here and that of the previous section is more pronounced when we seek to estimate the multiplicity of a Cohen-Macaulay ring of the form R/I^n for some positive integer n. The point is that the multiplicity of R/I^n, for n large, is a polynomial of degree g. On the other hand, using the second method we obtain

$$\deg(R/J_0^n) = \deg(R/J_0)\binom{n+g-1}{g-1},$$

and therefore

$$\begin{aligned}\nu(I^n) &\leq \deg(R/J_0^n) + \nu(J_0^n) \\ &\leq \deg(R/J_0)\binom{n+g-1}{g-1} + \binom{n+g-2}{g-2},\end{aligned} \tag{2.69}$$

which in the equimultiple case gives

$$\nu(I^n) \leq g!\deg(R/I)\deg(R)\binom{n+g-1}{g-1} + \binom{n+g-2}{g-2}. \tag{2.70}$$

Theorem 2.138. *Let R be a Cohen-Macaulay local ring with infinite residue field and I an ideal of height $g > 0$. If I is normally Cohen-Macaulay, then*

$$\mathrm{r}(I) \leq g \cdot g!\deg(R/I)\deg(R) - 2g + 1. \tag{2.71}$$

The proof is similar to that of Theorem 2.45.

Multiplicity of the Special Fiber

Let $(R, \mathfrak{m})$ be a Cohen-Macaulay local ring of dimension d and I an $\mathfrak{m}$-primary ideal. The surjection

$$0 \to H \longrightarrow \mathrm{gr}_I(R) \longrightarrow \mathcal{F}(I) \to 0 \tag{2.72}$$

of d-dimensional rings gives the inequality $\deg(\mathrm{gr}_I(R)) = e(I) \geq \deg(\mathcal{F}(I))$. The following observation permits the derivation of another kind of bound, one that involves the next coefficient, $e_1(I)$, of the Hilbert polynomial of I. Let $J = (a_1, \ldots, a_d)$ be a minimal reduction of I, and consider the following exact sequence of modules of finite length:

$$0 \to I^n/J^n \longrightarrow R/J^n \longrightarrow R/I^n \to 0.$$

Taking lengths, we have

$$\begin{aligned}\lambda(I^n/J^n) &= \lambda(R/J^n) - \lambda(R/I^n) \\ &= e(I)\binom{d+n-1}{d} - \lambda(R/I^n).\end{aligned}$$

For $n \gg 0$, on replacing $\lambda(R/I^n)$ by the Hilbert polynomial of I, we obtain

$$\lambda(I^n/J^n) = e_1(I)\binom{n+d-2}{d-1} + \text{lower terms},$$

from the cancelling of the term $e(I)\binom{n+d-1}{d}$ in $\lambda(R/I^n)$.

Proposition 2.139. *Let $(R,\mathfrak{m})$ be a Cohen-Macaulay local ring and I an $\mathfrak{m}$-primary ideal. Then*

$$\deg(\mathcal{F}(I)) \leq \inf\{e_0(I), e_1(I)+1\}.$$

Proof. From the estimate

$$\nu(I^n) \leq \nu(J^n) + \nu(I^n/J^n) \leq \nu(J^n) + \lambda(I^n/J^n)$$

for the number of generators of I^n, for $n \gg 0$ we obtain

$$\nu(I^n) \leq \binom{n+d-1}{d-1} + e_1(I)\binom{n+d-1}{d-1} + \text{lower terms},$$

which together with the observation in (2.72) proves the assertion. $\square$

Another bound for $\deg(\mathcal{F}(I))$ for this class of ideals is provided by the following result.

Proposition 2.140. *Let $(R,\mathfrak{m})$ be a Cohen-Macaulay local ring and I an $\mathfrak{m}$-primary ideal. Then*

$$\deg(\mathcal{F}(I)) \leq e_1(I) - e_0(I) + \lambda(R/I) + \nu(I).$$

Proof. We may assume that R has infinite residue field. Let $J \subset I$ be a minimal reduction and consider the exact sequence defining the corresponding Sally module of I:

$$0 \to I\cdot R[Jt] \longrightarrow I\cdot R[It] \longrightarrow S_J(I) \to 0.$$

Tensoring with the residue field of R, we have the exact sequence

$$I\cdot R[Jt]\otimes(R/\mathfrak{m}) \longrightarrow I\cdot R[It]\otimes(R/\mathfrak{m}) \longrightarrow S_J(I)\otimes(R/\mathfrak{m}) \to 0.$$

Observe that $I\cdot R[It]\otimes(R/\mathfrak{m})$ differs from $\mathcal{F}(I)$ in their components of degree zero only and therefore have the same multiplicity $\deg(\mathcal{F}(I))$.

We are going to estimate $\deg(\mathcal{F}(I))$ by adding bounds to the multiplicities of the ends of the exact sequence. To start with, $S_J(I)\otimes(R/\mathfrak{m})$ is a homomorphic image of $S_J(I)$ and so has multiplicity at most $\deg(S_J(I))$, which according to Theorem 2.11 is $e_1(I) - e_0(I) + \lambda(R/I)$. On the other hand $I\cdot R[Jt]\otimes(R/\mathfrak{m})$ is a homomorphic image of $I\otimes R[Jt]\otimes(R/\mathfrak{m})$, whose multiplicity is $\nu(I)$ since $R[Jt]\otimes(R/\mathfrak{m})$ is a ring of polynomials. $\square$

Several improvements are developed in [CPV3]. We describe one of them:

Theorem 2.141. *Let $(R,\mathfrak{m})$ be a Cohen-Macaulay local ring of dimension $d>0$ with infinite residue field. Let I be an $\mathfrak{m}$-primary ideal. Then the multiplicity f_0 of the special fiber ring $\mathcal{F}$ of I satisfies*

$$f_0 \leq e_1 - e_0 + \lambda(R/I) + \nu(I) - d + 1.$$

Proof. Let J be a minimal reduction of I and write $I = (J, f_1, \ldots, f_{n-d})$, where $n = \mu(I)$ and $d = \mu(J)$. We now consider the Sally module $S_J(I)$ of I with respect to J. We introduce a new defining sequence for $S_J(I)$

$$\mathcal{R}(J) \oplus \mathcal{R}(J)^{n-d}[-1] \stackrel{\varphi}{\longrightarrow} \mathcal{R}(I) \longrightarrow S_J(I)[-1] \to 0,$$

where φ is the natural inclusion on $\mathcal{R}(J)$ mapping the basis element ε_i of $\mathcal{R}(J)^{n-d}[-1]$ onto $f_i t$, for $1 \leq i \leq n-d$. Tensoring the above short exact sequence with $R/\mathfrak{m}$ yields

$$\mathcal{F}(J) \oplus \mathcal{F}(J)^{n-d}[-1] \longrightarrow \mathcal{F}(I) \longrightarrow S_J(I)[-1] \otimes R/\mathfrak{m} \to 0. \qquad (2.73)$$

As the three modules in (2.73) have the same dimension, we obtain the following multiplicity estimate:

$$f_0 = \deg(\mathcal{F}(I)) \leq \deg(S_J(I)[-1] \otimes R/\mathfrak{m}) + \deg(\mathcal{F}(J) \oplus \mathcal{F}(J)^{n-d}[-1]).$$

Since $S_J(I) \otimes R/\mathfrak{m}$ is a homomorphic image of $S_J(I)$, its multiplicity is bounded by that of $S_J(I)$, which according to Theorem 2.11 is $e_1 - e_0 + \lambda(R/I)$. On the other hand, $\mathcal{F}(J) \oplus \mathcal{F}(J)^{n-d}[-1]$ is a free $\mathcal{F}(J)$-module of rank $n-d+1$. Thus, its multiplicity is $n-d+1$, since $\mathcal{F}(J)$ is isomorphic to a ring of polynomials. □

An additional improvement that involves $\tilde{I}$, the Ratliff-Rush closure of I, arises as follows. First,

$$\deg(\mathcal{F}(I)) = \deg(\mathcal{F}(\tilde{I})),$$

since the two algebras differ in at most a finite number of components. Now observe that

$$\nu(\tilde{I}) \leq \nu(I) + \lambda(\tilde{I}/I)$$
$$\lambda(R/\tilde{I}) = \lambda(R/I) - \lambda(\tilde{I}/I).$$

Since the Hilbert coefficients are the same for the two ideals, the inequality

$$\deg(\mathcal{F}(I)) \leq e_1(I) - e_0(I) + \lambda(R/\tilde{I}) + \nu(\tilde{I}) - d + 1 \qquad (2.74)$$

sharpens Proposition 2.140.

Yet another refinement arises from the following considerations. Define the S_2-*ification of the ideal* I to be the largest ideal $I \subset I'$ such that in the embedding of Rees algebras

$$0 \to R[It] \longrightarrow R[I't] \longrightarrow C \to 0,$$

$\dim C \leq \dim R - 1$. According to Theorem 2.31, this is equivalent to asserting that I' is the largest ideal containing I with the same values for the Hilbert coefficients e_0 and e_1. It follows easily that there is an induced sequence

$$0 \to A \longrightarrow \mathcal{F}(I) \longrightarrow \mathcal{F}(I') \longrightarrow B \to 0$$

of finitely generated $\mathcal{F}(I)$-modules such that the dimensions of A and B are at most $\dim R - 1$. Thus, if we write the formula (2.74) for I' instead of for $\tilde{I}$, namely

$$\deg(\mathcal{F}(I)) \leq e_1(I) - e_0(I) + \lambda(R/I') + \nu(I') - d + 1, \tag{2.75}$$

we can only achieve gains.

Let $(R, \mathfrak{m})$ be a Noetherian local ring with infinite residue field and I an ideal with a minimal reduction J. The special fibers of these ideals will be denoted by $\mathcal{F}(I)$ and $\mathcal{F}(J)$, respectively. Our goal is to give general estimates for the multiplicity of $\mathcal{F}(I)$ (see [CPV3]). We begin with an elementary but useful formulation for the Cohen-Macaulayness of $\mathcal{F}(I)$.

Proposition 2.142. *Let $(R, \mathfrak{m})$ be a Noetherian local ring with infinite residue field and let I be an ideal with a minimal reduction J. If $r = \mathrm{r}_J(I)$ then*

$$\deg(\mathcal{F}(I)) \leq 1 + \sum_{j=1}^{r-1} \nu(I^j/JI^{j-1}),$$

and equality holds if and only if $\mathcal{F}(I)$ is Cohen-Macaulay.

Proof. Set $A = \mathcal{F}(J) = k[x_1, \ldots, x_\ell]$ and $\mathcal{F}(I) = \bigoplus_{n \geq 0} F_n$. As an A-module, $\mathcal{F}(I)$ is minimally generated by

$$\dim_k(\mathcal{F}(I)/A_+\mathcal{F}(I)) = 1 + \sum_{j=1}^{r-1} \dim_k(F_j/(x_1, \ldots, x_\ell)F_{j-1})$$

elements. But these summands are, by Nakayama's Lemma, the numbers $\nu(I^j/JI^{j-1})$. The second assertion is just the standard criterion for Cohen-Macaulayness as applied to graded A-modules. □

Theorem 2.143. *Let $(R, \mathfrak{m})$ be a Cohen-Macaulay local ring of dimension $d > 0$ and I an $\mathfrak{m}$-primary ideal with Hilbert coefficients $e_0(I)$ and $e_1(I)$. If*

$$\deg(\mathcal{F}(I)) = e_1(I) - e_0(I) + \lambda(R/I) + \nu(I) - d + 1, \tag{2.76}$$

then $\mathcal{F}(I)$ satisfies Serre's condition S_1. Furthermore, if depth $\mathrm{gr}_I(R) \geq d - 1$, *then $\mathcal{F}(I)$ is Cohen-Macaulay.*

Proof. We may assume that $\mathrm{r}(I) > 1$. We consider the sequence (2.2) associated to I,

$$\bar{I} \otimes_k \mathcal{F}(J) \stackrel{\varphi}{\longrightarrow} \mathcal{F}(I)_+[+1] \longrightarrow \overline{S_{I/J}} \rightarrow 0,$$

and place on it the requirement (2.76). Since

$$e_1(I) - e_0(I) + \lambda(R/I) = \deg(S_{I/J}) \geq \deg(\overline{S_{I/J}})$$

and

$$\mathrm{rank}(\varphi) \leq \nu(I) - d + 1,$$

it leads to equality in both of these formulas. In the case where $\deg(S_{I/J}) = \deg(\overline{S_{I/J}})$, this means that $\dim \mathfrak{m}S_{I/J} < d$, which is not possible (see 2.2) when R is Cohen-Macaulay.

The equality $\mathrm{rank}(\varphi) = \nu(I) - d + 1$ can be used as follows to conclude that image $\varphi = \mathfrak{p} \oplus \mathcal{F}(J)^{\nu(I)-d}$. Denote by $e_1, \ldots, e_\ell, \ldots, e_m$ a minimal set of generators of I, where we may assume that $e_1, \ldots, e_\ell$ is a minimal set of generators of J. The images of $e_1, \ldots, e_\ell$ in $\mathcal{F}(J)_1 = J/\mathfrak{m}J$ are denoted by $x_1, \ldots, x_\ell$; we note that $\mathcal{F}(J) = k[x_1, \ldots, x_\ell]$ and set $\mathfrak{p} = (x_1, \ldots, x_\ell)$ for its maximal ideal.

Set $L = \ker(\varphi) \subset \mathcal{F}(I)^m$. Note that the elements $e_i \otimes x_j - e_j \otimes x_i$ for $1 \leq i < j \leq \ell$ lie in L. Since they define the module Z_1 of syzygies of the maximal ideal $\mathfrak{p}$ of the polynomial ring $\mathcal{F}(J)$, we have a mapping

$$\mathfrak{p} \oplus \mathcal{F}(J)^{m-\ell} \longrightarrow \text{image } \varphi \subset \mathcal{F}(I)_+[+1].$$

If equality holds in the formula for $\deg(\mathcal{F}(I))$, the image of φ must have rank precisely $m - \ell + 1$, which means that L must be a module of rank $\ell - 1$ containing Z_1. But Z_1 is maximal among such modules, so $L = Z_1$. Packing differently, we have an exact sequence

$$0 \rightarrow \mathcal{F}(J) \oplus \mathcal{F}(J)^{m-\ell}[-1] \longrightarrow \mathcal{F}(I) \longrightarrow \overline{S_{I/J}}[-1] \rightarrow 0.$$

When depth $\mathrm{gr}_I(R) \geq d - 1$, $S_{I/J}$ is Cohen-Macaulay (see 2.3), and therefore $\mathcal{F}(I)$ will be Cohen-Macaulay as well. □

If $\mathrm{r}(I) = 1$, it seems likely that $\deg(\mathcal{F}(I)) = \nu(I) - d + 1$. This would prove a result of [CZ97, 1.1, 3.2]. One must show that $\mathrm{rank}(\varphi) = d - 1$.

The associated primes of $\mathcal{F}(I)$ can play a role in the completeness of the product $\mathfrak{m}I$. To see how this comes about, we give an observation of [HuH1] and an application in its spirit.

Proposition 2.144. *Let $(R, \mathfrak{m})$ be a normal local domain of dimension d with infinite residue field and suppose that I is a normal ideal of analytic spread d. If $\mathcal{F}(I)$ is equidimensional without embedded components, then $\mathfrak{m}I^n$ is complete for every n.*

Proof. By assumption $R[It]$ is a Krull domain and $\mathfrak{m}R[It]$ is an unmixed ideal of codimension 1, and therefore it is a divisorial ideal of a normal domain. □

Corollary 2.145. *In* (2.143), *if I is a normal ideal of the normal domain R then $\mathfrak{m}I^n$ is complete for every n.*

We can modify these bounds considerably when I is a normal ideal. Consider the exact sequence (we assume that $I \neq \mathfrak{m}$)

$$0 \to H \longrightarrow \mathrm{gr}_I(R) \longrightarrow \mathcal{F}(I) \to 0.$$

Suppose that $0 \neq u \in 0 :_H \mathfrak{m}$ and consider the ideal u generates. Observe that $\mathrm{gr}_I(R)u$ has dimension d since $\mathrm{gr}_I(R)$ is equidimensional and has no embedded primes.

This gives an embedding $\mathcal{F}(I)u \subset H$, so that

$$\deg(\mathrm{gr}_I(R)) = \deg(\mathcal{F}(I)) + \deg H \geq \deg(\mathcal{F}(I)) + \deg(\mathcal{F}(I)u).$$

Proposition 2.146. *Let $(R,\mathfrak{m})$ be a Cohen-Macaulay local ring and I a non-maximal $\mathfrak{m}$-primary normal ideal. We then have*

$$\deg(\mathcal{F}(I)) \leq \begin{cases} e_0 - 1, & \text{in general} \\ \frac{e_0}{2}, & \text{if } \mathcal{F}(I) \text{ is a domain.} \end{cases}$$

Proof. The first assertion follows because $\mathcal{F}(I)u$ is a module of dimension 1 and is therefore of multiplicity at least one. In the other case, since $\mathcal{F}(I)$ is a domain, the annihilator of $\mathcal{F}(I)u$ as a module over the special fiber is trivial, so in the inequality above we get at least two summands equal to $\deg(\mathcal{F}(I))$. □

Corollary 2.147. *Let $(R,\mathfrak{m})$ be a Noetherian local ring of dimension d, with residue field of characteristic zero, and let I be a non-maximal $\mathfrak{m}$-primary ideal with $\nu(I) > d$. If the special fiber $\mathcal{F}(I)$ is a Cohen-Macaulay integral domain of type t, then*

$$\boxed{\mathrm{r}(I) \leq 1 + \frac{e_0 - 2t - 2}{2(\nu(I) - d)}.} \tag{2.77}$$

Note that in Proposition 1.142 the estimate for the reduction number of the ideal gets cut.

When I is normally Cohen-Macaulay, bounds on $\deg(\mathcal{F}(I))$ can be derived in the same manner. If the ideal I is not normally Cohen-Macaulay, the term $\mathrm{Deg}(R/I) - \deg(R/I)$ in (2.58) must be taken into account. The simplest case is that of an ideal I of dimension one. One may proceed as follows, using the function $\mathrm{Deg}(\cdot) = \mathrm{hdeg}(\cdot)$ (here $A = R/J_0^n$ and ω_A is the canonical module of A):

$$\nu(I^n) \leq \nu(J_0^n) + \mathrm{hdeg}(I^n/J_0^n), \tag{2.78}$$

where

$$\mathrm{hdeg}(I^n/J_0^n) = \deg(I^n/J_0^n) + \mathrm{hdeg}(\mathrm{Ext}_A^1(I^n/J_0^n, \omega_A)) + \mathrm{hdeg}(\mathrm{Ext}_A^2(I^n/J_0^n, \omega_A)).$$

The last term vanishes and $\text{Ext}^1_A(I^n/J_0^n, \omega_A)$ has finite length, and therefore we may write

$$\text{hdeg}(I^n/J_0^n) = \deg(I^n/J_0^n) + \lambda(\text{Ext}^1_A(I^n/J_0^n, \omega_A)).$$

Finally an easy calculation shows that $\text{Ext}^1_A(I^n/J_0^n, \omega_A) \cong \text{Ext}^2_A(R/I^n, \omega_A)$, a module which by local duality has the same length as $H^0_{\mathfrak{m}}(R/I^n)$. To sum up, we have the estimate

$$\nu(I^n) \leq \deg(R/J_0)\binom{n+d-2}{d-2} + \binom{n+d-3}{d-3} + \lambda(H^0_{\mathfrak{m}}(R/I^n)). \tag{2.79}$$

The last term in this expression, $\lambda(H^0_{\mathfrak{m}}(R/I^n))$, is not very predictable. We shall now give the method of [BER79] that in dimension two is more suitable to our purpose.

Theorem 2.148. *Let $(R, \mathfrak{m})$ be a Cohen-Macaulay local ring of dimension two and I an ideal. If $\{x,y\}$ is a minimal reduction of $\mathfrak{m}$, then*

$$\nu(I) \leq \deg(R) + \lambda(I : (x,y)/I). \tag{2.80}$$

Proof. Let $\mathbb{K}$ be the Koszul complex associated with $\{x,y\}$. For any finitely generated R-module M, the Euler characteristic of the complex $\mathbb{K}(M) = \mathbb{K} \otimes M$ is the multiplicity $e(x,y;M)$ of M (see [BH93, Theorem 4.7.4]),

$$e(x,y;M) = \lambda(H_0(\mathbb{K}(M)) - \lambda(H_1(\mathbb{K}(M)) + \lambda(H_2(\mathbb{K}(M)).$$

Setting $M = I$, observing that $\nu(M) = \lambda(M/\mathfrak{m}M) \leq \lambda(M/(x,y)M) = \lambda(H_0(\mathbb{K}(M))$, and noting that $H_1(\mathbb{K}(I)) = H_2(\mathbb{K}(R/I)) = I : (x,y)/I$ and $H_2(\mathbb{K}(I)) = 0$, we obtain

$$e(x,y;I) \geq \nu(I) - \lambda(I : (x,y)/I).$$

From the additivity of the function $e(x,y;\cdot)$, we have $e(x,y;I) \leq \deg(R)$, giving rise to the desired inequality. □

Proposition 2.149. *Let R be a Noetherian ring and let L and J be ideals of R, with $L \subset I$. Suppose that modulo J the ideal L is a reduction of I. Then for all positive integers n, L is a reduction of I modulo J^n.*

Proof. We induct on $n \geq 1$. The assumption means that for each $x \in I$ there is a relation

$$f(x) = x^m + a_1x^{m-1} + \cdots + a_m \in J^n, \quad a_i \in L^i.$$

Squaring $f(x)$ gives another relation

$$f(x)^2 = x^{2m} + b_1x^{2m-1} + \cdots + b_{2m} \in J^{2n}, \quad b_i \in L^i.$$

Since $J^{2n} \subset J^{n+1}$, x is integral over L modulo J^{n+1}, as desired. □

We may refine the bound (2.79) as follows.

Corollary 2.150. *Let $(R,\mathfrak{m})$ be a Cohen-Macaulay local ring of dimension at least two and I an ideal of dimension one. Let J_0 be a subideal of I generated by a regular sequence of $d-2$ elements. If (x,y) is a minimal reduction of $\mathfrak{m}$ modulo J_0, then for all integers $n \geq 1$,*

$$\nu(I^n) \leq \deg(R/J_0)\binom{n+d-2}{d-2} + \binom{n+d-3}{d-3} + \lambda(I^n : (x,y)/I^n).$$

Table 2.2. Bounds for the number of generators of a Cohen-Macaulay ideal of height g (or $\mathfrak{m}$-primary in context):

$g = \text{height } I$	$\deg(R) + (g-1)\deg(R/I)$	(2.55)
$s = $ index of nilpotency of I	$s^{d-1}\deg(R) + d - 1$	(2.56)
I equimultiple & $\mathrm{r}(I) \leq 1$	$(r+1)\deg(R/I) + g - 1$	(2.59)
$d = 2$ & (x,y) is reduction of $\mathfrak{m}$	$\deg(R) + \lambda(I : (x,y)/I)$	(2.80)
I^n, $d \geq 2$	$e(I)\binom{n+d-2}{d-1} + \binom{n+d-2}{d-2}$	(2.22)
I, $d \geq 2$ & $s = $ index of nilpotency of R/I	$e(I)\binom{s+d-2}{d-1} + \binom{s+d-2}{d-2}$	(2.20)

2.7.3 Ideals of Dimension One and Two

We have examined in some detail the relationships between the reduction of an $\mathfrak{m}$-primary ideal I. Now we approach several cases of ideals of higher dimension.

Ideals of Dimension One and their Syzygies

We consider a modified approach in the case of an ideal I of dimension 1 that is generically a complete intersection. It will be based on the following simple observation (that it is a reflection of the fact that all cohomological degrees coincide in dimension one).

Proposition 2.151. *Let $(R,\mathfrak{m})$ be a Cohen-Macaulay local ring of dimension 1 and M a finitely generated R-module. Then $\nu(M) \leq \lambda(H^0_{\mathfrak{m}}(M)) + \deg(M)$.*

The point is that both summands on the right-hand side may be often easier to estimate than $\nu(M)$ itself. Let $(R,\mathfrak{m})$ be a Cohen-Macaulay local ring of dimension ≥ 1, with infinite residue field, and let I be a Cohen-Macaulay ideal of dimension 1. We suppose that I is generically a complete intersection. Let J be one of its minimal reductions and assume that J is not a complete intersection (in which case $I = J$). Then J is generated by d elements, $J = (a_1,\ldots,a_d)$, which we may assume form a d-sequence. We refer to [Va94b, Chapter 3] for details and additional references on d-sequences and how they apply to this case. One of the properties of J is that the first $d-1$ elements $a_1,\ldots,a_{d-1}$ form a regular sequence and will generate I at each of the associated primes of I.

Set $J_0 = (a_1,\ldots,a_{d-1})$. The method of [DGV98], in a formally similar situation, consisted in estimating the number of generators of I^n/J_0^n in terms of multiplicity data of R/J_0 and again making use of Theorem 2.36 to bound the reduction number of I. Let us see how this works out.

Consider the exact sequence of modules of dimension 1 (J_0 is generated by a regular sequence):

$$0 \to I^n/J_0^n \longrightarrow R/J_0^n \longrightarrow R/I^n \to 0.$$

Since I^n/J_0^n is a Cohen-Macaulay module, its number of generators is bounded by its multiplicity $\deg(I^n/J_0^n) = \deg(R/J_0^n) - \deg(R/I^n)$. This leads to

$$\begin{aligned}\nu(I^n) &\leq \nu(J_0^n) + \deg(R/J_0^n) - \deg(R/I^n)\\ &= \binom{n+d-2}{d-2} + (\deg(R/J_0) - \deg(R/I))\binom{n+d-2}{d-1}.\end{aligned}$$

Now, according to Theorem 2.36, to find a minimal reduction J with $r_J(I) \leq n-1$, it suffices to choose n such that $\nu(I^n) < \binom{n+d}{d}$. Setting $\deg(R/J_0) - \deg(R/I) = c$, it will be enough to have

$$c\binom{n+d-2}{d-1} + \binom{n+d-2}{d-2} < \binom{n+d}{d},$$

which is equivalent to

$$\frac{cn}{n+d-1} + \frac{d-1}{n+d-1} < \frac{n+d}{d},$$

an inequality that will hold for $n = d\cdot c - 2d + 2$.

Let us make an analysis of the number $c = \deg(R/J_0) - \deg(R/I) = \deg(I/J_0)$. It apparently shows the dependence of the reduction number on the choice of J and J_0. We note that $J_0 = I \cap (J_0{:}I)$, since J_0 and I have the same primary components at the associated primes of I. This leads to the equality $\deg(I/J_0) = \deg(R/(J{:}I))$ of multiplicities.

Theorem 2.152. *Let R be a Gorenstein local ring of dimension 3 and I a Cohen-Macaulay ideal of dimension 1 that is generically a complete intersection. Let $J =$*

(f,g,h) be a minimal reduction of I and denote by Z_1 the module of syzygies of these elements. Set $a = \lambda(R/c(Z_1))$ and $b = \lambda(\text{Ext}_R^1(Z_1,R))$, where $c(Z_1)$ is the Fitting ideal generated by the coefficients of the elements of Z_1 as a submodule of R^3. Then $r_J(I)$ is bounded by the larger of the roots of the quadratic equation

$$n^2 - 3(a+b-1)n + 3(a-b) + 2 = 0.$$

Proof. This approach uses the data on the reduction J in order to obtain information on another possible reduction. For that, consider the exact sequence

$$0 \to I^n/J^n \longrightarrow R/J^n \longrightarrow R/I^n \to 0.$$

The module I^n/J^n vanishes at each minimal prime of I (which are the same as those of J) so has finite length. It follows that

$$\nu(I^n) \leq \nu(J^n) + \lambda(I^n/J^n) \leq \nu(J^n) + \lambda(H^0_{\mathfrak{m}}(R/J^n)).$$

The first summand is $\binom{n+d-1}{d-1}$. Let us indicate how to estimate the other in the case of an ideal of a ring of dimension 3. Suppose that $J = (f,g,h)$ and let

$$0 \to Z_2 \otimes B_{n-2} \xrightarrow{\varphi_n} Z_1 \otimes B_{n-1} \longrightarrow B_n \longrightarrow J^n \to 0$$

be the degree n component of the $\mathcal{Z}$-complex of J (see [Va94b, Chapter 3]). Here Z_1 and Z_2 are the modules of cycles of the Koszul complex $K(a,b,c)$ and $B_n = \text{Sym}_n(R^3)$; this complex is acyclic. Break it into two short exact sequences:

$$0 \to Z_2 \otimes B_{n-2} \xrightarrow{\varphi_n} Z_1 \otimes B_{n-1} \longrightarrow L_n \to 0,$$

$$0 \to L_n \longrightarrow B_n \longrightarrow J^n \to 0.$$

We have that

$$\lambda(H^0_{\mathfrak{m}}(R/J^n)) = \lambda(\text{Ext}_R^2(J^n,R)) = \lambda(\text{Ext}_R^1(L_n,R)),$$

where we assume that R is a Gorenstein ring. We also have the exact sequence

$$\text{coker } (\varphi_n^t) \longrightarrow \text{Ext}_R^1(L_n,R) \longrightarrow \text{Ext}_R^1(Z_1 \otimes B_n,R) \to 0,$$

from which we get

$$\lambda(H^0_{\mathfrak{m}}(R/J^n)) \leq \lambda(\text{Ext}_R^1(Z_1,R))\binom{n+1}{2} + \lambda(\text{coker } (\varphi_n^t)).$$

A calculation in the $\mathcal{Z}$-complex will show that

$$\text{coker } (\varphi_n^t) = R/c(Z_1) \otimes B_{n-2}^t,$$

where $c(Z_1)$ is the Fitting ideal generated by the entries of all the elements of Z_1. Putting all these rough estimates together, we have

$$\nu(I^n) \leq \binom{n+2}{2} + a\binom{n}{2} + b\binom{n+1}{2},$$

where $a = \lambda(R/c(Z_1))$ and $b = \lambda(\mathrm{Ext}^1_R(Z_1, R))$. To make use of Theorem 2.36 it suffices to have

$$\binom{n+3}{3} > \binom{n+2}{2} + a\binom{n}{2} + b\binom{n+1}{2},$$

from which the assertion follows. □

Equimultiple Ideals of Dimension One

Let I be an ideal of dimension one as above but assume that it is equimultiple. Choose a minimal reduction $J = (a_1, \ldots, a_{d-1})$ of I. Let us discuss a reduction to the case when the ideal can be generated by d elements. Consider the sequence

$$0 \to I^n/J^n \hookrightarrow R/J^n,$$

that gives

$$\nu(I^n) \leq \nu(I^n/J^n) + \nu(J^n) \leq \deg(R/J^n) + \nu(J^n),$$

since R/J^n is Cohen-Macaulay. As we have by Proposition 2.135 that

$$\deg(R/J) \leq (d-1)!\deg(R/\bar{I})\deg(R),$$

and

$$\deg(R/J^n) \leq (d-1)!\deg(R/\bar{I})\deg(R)\binom{n+d-2}{d-1},$$

we can solve

$$(d-1)!\deg(R/\bar{I})\deg(R)\binom{n+d-2}{d-1} + \binom{n+d-2}{d-2} < \binom{n+d}{d},$$

and get a reduction L generated by d elements and satisfying

$$r = \mathrm{r}_L(I) \leq d! \cdot \deg(R/\bar{I})\deg(R) - 2d + 1.$$

What is needed now is an estimate for the reduction number of L, since if K is a minimal reduction of L with $\mathrm{r}_K(L) = s$ it follows that $\mathrm{r}_K(I) \leq r + s$.

Mixed Hilbert Functions and Multiplicities

Let $(R, \mathfrak{m})$ be a quasi-unmixed local ring of dimension d and I an ideal of codimension g that is generically a complete intersection. We denote by $\mathcal{G}$ the associated graded ring of I. The conormal module has rank g and its Rees algebra $\mathcal{R}(I/I^2)$ is isomorphic to $\mathcal{G}$ modulo its R/I-torsion. For simplicity we denote it by $\mathcal{G}'$:

$$0 \to H(I) \longrightarrow \mathcal{G} \longrightarrow \mathcal{G}' \to 0. \tag{2.81}$$

When I is an ideal of dimension 1 or R/I has isolated singularities, we have $H(I) = H^0_{\mathfrak{m}}(\mathcal{G})$. In this case, from the exact sequence in which the associated primes of $\mathcal{G}'$ do not contain those of $H(I)$, we obtain the following equality of arithmetic degrees

$$\text{arith-deg}(G) = \text{arith-deg}(H(I)) + \text{arith-deg}(\mathcal{G}').$$

Throughout this section, in order to have this description of $H(I)$ we shall assume that I/I^2 is torsionfree on the punctured spectrum. We want to make the case that the examination of both $H(I)$ and of $\mathcal{G}'$ plays a role in the determination of the reduction number of I in terms of the arithmetic degree of $\mathcal{G}$. A first observation is that if J is a minimal reduction of I, then $H(I)$ and $\mathcal{G}'$ are graded $R[Jt]$-modules and we shall seek the degrees of the equations of integral dependence of the elements of $R[It]$ acting as endomorphisms of these modules.

We shall now pay attention to some of the general properties of $H(I)$ and $\mathcal{G}'$. We note that $\dim \mathcal{G}' = \dim R = d$, and that $\dim \mathcal{G}'/\mathfrak{m}\mathcal{G}' < d$, since $\mathfrak{m}$ contains regular elements on $\mathcal{G}'$. In the theory of Rees algebras of modules, the dimension of the special fiber of $\mathcal{G}'$ is called the analytic spread of the module

$$\ell(I/I^2) = \dim \mathcal{G}'/\mathfrak{m}\mathcal{G}'.$$

As observed earlier in Proposition 1.99, since $H(I)$ is annihilated by a power of $\mathfrak{m}$, it admits a filtration whose factors are modules over $\mathcal{G} \otimes R/\mathfrak{m} = \mathcal{F}(I)$. It follows that

$$h(I) = \dim H(I) \leq \dim \mathcal{F}(I) = \ell(I).$$

A more concrete elementary observation is the following:

Corollary 2.153. *Let I be a non $\mathfrak{m}$-primary ideal such that all the associated primes of $\mathcal{G}$ have dimension d and $H(I) \neq 0$. Then $h(I) = \ell(I) = d$ and $\dim \mathcal{G}' = d$. In particular $\deg(\mathcal{G}) = \deg(H) + \deg(\mathcal{G}')$. In general we have $\text{arith-deg}(\mathcal{G}) = \text{arith-deg}(H(I)) + \text{arith-deg}(\mathcal{G}')$.*

Proof. If $H(I) \neq 0$, then $H(I)$ is a module of the same dimension as $\mathcal{G}$, since the latter is equidimensional. From Proposition 1.99, we have that $d = h(I) \leq \ell(I)$, which proves the first assertion.

Suppose now that $\dim \mathcal{G}' < d$. This means that $H(I)$ is an ideal of height at least 1 and therefore must contain regular elements of $\mathcal{G}$. On the other hand $\mathfrak{m}^s H = 0$, which implies that $\mathfrak{m}^s \subset I$, contradicting the assumption that I is not a primary ideal. The asserted addition formula follows from the ways multiplicities are computed. □

Proposition 2.154. *Let $(R, \mathfrak{m})$ be a Cohen-Macaulay local ring of dimension d and I an unmixed ideal of codimension g with $\ell(I) = d$ that is of linear type on the punctured spectrum of R. If $H(I) \neq 0$, then $\ell(I/I^2) = d - 1$.*

Proof. First, note that as $\ell(I) = d$, $\mathfrak{m}\mathcal{G}$ has codimension 0 and it is not a nilpotent ideal. On the other hand, $H(I)$ is not a nilpotent ideal either as this would yield

$\mathcal{G}_{\text{red}} = \mathcal{G}'_{\text{red}}$ and therefore $\ell(I)$ would be at most $d-1$. On the other hand, since R is Cohen-Macaulay, the ring $\mathcal{G}$ is connected in codimension one (see [Br86, 2.5]). This means that if A, B are non-nilpotent ideals with $A \cdot B = 0$, then height $(A+B) \leq 1$. Applying this to the equation $\mathfrak{m}^s H(I) = 0$, we have that height $(\mathfrak{m}^s \mathcal{G} + H(I)) \leq 1$, and therefore dim $\mathcal{G}'/\mathfrak{m}^s \mathcal{G}' \geq d-1$, as desired. □

These observations will be useful for two reasons. First, we can use the theory of Cayley-Hamilton polynomials to bound the degrees of endomorphisms on $H(I)$; second, the fact that $\ell(I/I^2)$ is usually smaller than $\ell(I)$ permits us to get a simpler reduction for the module than the ideal I itself allows.

Ideals of Dimension One

The first application of this technique is:

Proposition 2.155. *Let $(R, \mathfrak{m})$ be a Cohen-Macaulay local ring and I an ideal of dimension 1 that is generically a complete intersection. If $H = H^0_{\mathfrak{m}}(\mathcal{G}) \neq 0$ then* $\dim H = d$ *and the algebra defined by*

$$0 \to H \longrightarrow \mathcal{G} \longrightarrow \mathcal{G}' \to 0$$

satisfies $\dim \mathcal{G}' = d$, *and* $\deg(R/I) = \deg(\mathcal{G}') = \deg(\mathcal{G}) - \deg(H)$.

Proof. We only have to show that $\deg(R/I) = \deg(\mathcal{G}')$. The R/I-module

$$E = I/I^2(\text{modulo torsion})$$

admits a reduction $F \subset E$ that is a free module. Thus the Rees algebras $\mathcal{R}(F)$ and $\mathcal{R}(E)$ have the same multiplicity. Since $\mathcal{R}(F) \cong R/I[T_1, \ldots, T_{d-1}]$, we get the asserted equality of multiplicities. □

Theorem 2.156. *Let $(R, \mathfrak{m})$ be a Cohen-Macaulay local ring and I an ideal of dimension 1 that is generically a complete intersection. Suppose further that R contains a field of characteristic zero. Then*

$$\mathrm{r}(I) < \deg(\mathcal{G}).$$

Proof. We may assume that R is a complete local ring. Let J be a minimal reduction of I and view the terms of the exact sequence

$$0 \to H \longrightarrow \mathcal{G} \longrightarrow \mathcal{G}' \to 0$$

as modules over the Rees algebra $R[Jt]$. We can also view H as a module over a ring of polynomials $k[T_1, \ldots, T_d]$, which is actually the special fiber of $R[Jt]$. This happens because H is a module annihilated by a power $\mathfrak{m}^r$ of the maximal ideal and therefore is defined over the Artinian ring $R/\mathfrak{m}^r$. Since $R/\mathfrak{m}$ has characteristic zero, we can find a field of representatives of $R/\mathfrak{m}^r$.

After noting that the image of Jt in $\mathcal{G}'$ defines a reduction of the module E above, we can choose a minimal reduction F of E contained in $Jt\,\mathcal{G}'$. As R contains a field k of representatives, there is a parameter z such that the inclusion $k[[z]] \subset R/I$ makes R/I free over $k[[z]]$ of rank $\deg(\mathcal{G}') = \deg(R/I)$. After changing notation in the original T_i, let $T_1, \ldots, T_{d-1}$ be a set of generators of the module F. The ring $\mathcal{G}'$ is a finitely generated torsionfree module over $A = k[[z]][T_1, \ldots, T_{d-1}]$, of rank $\deg(R/I)$.

Finally let at be an element of It and consider the mapping it induces as an endomorphism of H and of $\mathcal{G}'$. Applying the techniques of the Cayley-Hamilton theorems of Section 1.5.1, we get polynomials $f(T)$ and $g(T)$ that are monic and graded over $R[Jt]$, of degrees $\deg(H)$ and $\deg(\mathcal{G}')$ respectively, such that $f(at)H = 0$ and $g(at)\,\mathcal{G}' = 0$. Thus $f(at)g(at)\,\mathcal{G} = 0$, which shows that $p(at) = f(at)g(at)$ is the null element of $\mathcal{G}$. By lifting to $\mathcal{R}$, this gives an equation of integral dependence of a with respect to J of degree $n = \deg(H) + \deg(\mathcal{G}')$,

$$a^n + b_1 a^{n-1} + \cdots + b_n \in (IR[It])_n = I^{n+1}, \quad b_i \in J^i.$$

In characteristic zero this implies, as remarked earlier, the desired inequality $\mathrm{r}(I) < \deg(\mathcal{G})$. □

Gorenstein Ideals of Dimension Two

Everything becomes considerably more complicated when the ideal has dimension greater than one. To simplify, we assume that R is a Gorenstein local ring, $\dim R/I = 2$, that I is a perfect Gorenstein ideal and R/I is normal.

Let us make some general observations first. Let J be a minimal reduction of I. If $\ell(I) < d$, then I is a complete intersection. Indeed, if J is generated by $d-1$ elements, then by [Va94b, Corollary 5.3.5], its associated primes have codimension at most $d-1$, which is impossible since R/I is normal.

We may then assume that I has a minimal reduction $J = (a_1, \ldots, a_d)$. In particular this implies that $H^0_{\mathfrak{m}}(\mathcal{G}) \neq 0$. The module $E = I/I^2/(\text{modulo torsion})$ has analytic spread $d-1$, so that $d-1$ elements of Jt map into a reduction F of E. Indeed a minimal reduction of E cannot be a free module F since then $\mathcal{G}'$ would be integral over a ring of polynomials over a normal ring. This would imply that E is also free, which in turn would mean that

$$I/I^2 \cong (R/I)^{d-2} \oplus L, \quad L \neq 0,$$

a condition that would say that I contains a regular sequence of $d-1$ elements. Therefore we have $\ell(E) = d-1$.

We claim that $\mathcal{G}'$ has a reduction that is a hypersurface ring over R/I,

$$\mathcal{R}(F) = R/I[T_1, \ldots, T_{d-1}]/(f), \quad f = \sum_{i=1}^{d-1} a_i T_i.$$

Let F be a minimal reduction of E. Then F is a torsionfree module with a presentation

$$0 \to K \longrightarrow (R/I)^{d-1} \longrightarrow F \to 0,$$

where K is a reflexive module of rank 1. Let $f \in (R/I)^{d-1}$ be a nonzero element of K, and let L be the ideal generated by its coordinates as an element of $(R/I)^{d-1}$. Note that $L^{-1}f \subset K$. Checking at the localizations of R/I at height one primes, it follows easily that these modules are the same.

We now make use of the fact that E is an orientable module, and therefore so will its reductions be. From the exact sequence above, $L^{-1}f$ is also an orientable module. Since it is reflexive of rank one, it must be principal.

Multiplicity of Almost Complete Intersections

Let $(R,\mathfrak{m})$ be a Cohen-Macaulay local ring and E a torsionfree R-module of rank $r > 0$ with $\ell(E) = r+1$. We assume that E is an orientable module. This will imply that any minimal reduction F has a free presentation

$$0 \to R \longrightarrow R^{r+1} \longrightarrow F \to 0.$$

We shall assume further that the Rees algebra of F is an almost complete intersection, which means that

$$\mathcal{R}(F) = R[T_1,\ldots,T_{r+1}]/(f), \quad f = a_1T_1 + \cdots + a_{r+1}T_{r+1},$$

where $(a_1,\ldots,a_{r+1})$ is an ideal of grade at least 2. From this description one has

$$\deg(\mathcal{R}(E)) = \deg(\mathcal{R}(F)) = \deg R(T_1,\ldots,T_{r+1})/(f).$$

However, we want a different expression for this multiplicity. Let us explain the goal of how to obtain a bound for the reduction number of E in terms of this multiplicity. We do this using a number of elementary observations.

Since the ring R is Cohen-Macaulay, $\mathcal{R}(F)$ is Cohen-Macaulay and the ideal generated by its 1-forms has height r, there is (we are assuming that the residue field of R is infinite) a submodule $\widetilde{F}$ of F generating a regular sequence of height r. Furthermore its generators can be chosen so that $\deg(\mathcal{R}(F)) = \deg(\mathcal{R}(F)/(\widetilde{F}))$.

Proposition 2.157. *The R-module C_n defined by the exact sequence*

$$0 \to \widetilde{F}^n \longrightarrow F^n \longrightarrow C_n \to 0$$

has multiplicity

$$\deg(C_n) = \binom{n+r-1}{r}\deg(C_1).$$

Proof. We note that the modules F^n all have projective dimension 1, so that the projective dimension of C_n is also 1. So for the purpose of putting together $\deg(C_n)$ through the usual associativity formula

$$\deg(C_n) = \sum_{\text{height } \mathfrak{p}=1} \lambda((C_n)_\mathfrak{p}) \deg(R/\mathfrak{p}),$$

we may assume that $\dim R = 1$.

We recall that we may reduce to the case of modules over discrete valuation rings, and the assertion is a simple calculation of determinants. □

Proposition 2.158. *Suppose further that* $\dim R = 2$ *and set* $a = \deg(C_1)$. *Then*

$$\mathrm{r}(E) \leq (r+1)a - 2(r+1) - 1.$$

Proof. We note that the module $E^n/\widetilde{F}^n$ is Cohen-Macaulay of dimension 1. Its multiplicity is the same as that of $F^n/\widetilde{F}^n$. This leads to the inequalities

$$\begin{aligned} \nu(E^n) &\leq \nu(E^n/\widetilde{F}^n) + \nu(\widetilde{F}^n) \\ &\leq \deg(E^n/\widetilde{F}^n) + \nu(\widetilde{F}^n) \\ &\leq a\binom{n+r-1}{r} + \binom{n+r-1}{r-1}. \end{aligned}$$

To apply Theorem 2.36 again (we may have to replace F by another minimal reduction F'), since $\ell(F) = r+1$ it will suffice to set $n = \mathrm{r}(E)+1$ such that

$$\binom{n+r+1}{r+1} > a\binom{n+r-1}{r} + \binom{n+r-1}{r-1}.$$

This is equivalent to

$$n > (r+1)a - 2(r+1),$$

which is the desired bound for the reduction number of E. □

We must now relate $a = \deg(C_1)$ to the multiplicity of $\mathcal{R}(F)$. We use the additivity formula on $\mathcal{R}(F)/(\widetilde{F})$. Let $\mathfrak{P}$ be a minimal prime of $(\widetilde{F})$, $\mathfrak{p} = R \cap \mathfrak{P}$. Note that $\mathfrak{p} \neq \mathfrak{m}$ since $\widetilde{F}$ is generated by r elements in a minimal generating set for F. In particular, C_1 is a cyclic module.

There will be two kinds of minimal primes $\mathfrak{P}$ of $(\widetilde{F})$, depending on whether $\mathfrak{p} = R \cap \mathfrak{P}$ has height 0 or 1. If $\mathfrak{p}$ is a minimal prime of R, then $\mathfrak{P} = (\mathfrak{p}, F)$. In the other case $\mathfrak{P} = (\mathfrak{p}, \widetilde{F})$, where $\mathfrak{p}$ is an associated prime of C_1.

We can now put together the multiplicity formula:

$$\begin{aligned} \deg(\mathcal{R}(F)/(\widetilde{F})) = &\sum_{\text{height } \mathfrak{p}=0} \lambda((\mathcal{R}(F)/\widetilde{F})_\mathfrak{P}) \deg(\mathcal{R}(F)/\mathfrak{P}) \\ &+ \sum_{\text{height } \mathfrak{p}=1} \lambda((\mathcal{R}(F)/\widetilde{F})_\mathfrak{P}) \deg(\mathcal{R}(F)/\mathfrak{P}). \end{aligned}$$

Since $\mathfrak{P} = (\mathfrak{p}, F)$, the first partial summation gives

$$\sum_{\text{height } \mathfrak{p}=0} \lambda((\mathcal{R}(F)/(F))_\mathfrak{p} = R_\mathfrak{p}) \deg(R/\mathfrak{p}) = \deg(R/I).$$

In the other partial summation, one has

$$\lambda((\mathcal{R}(F)/(F))_{\mathfrak{P}}) = \lambda((C_1)_{\mathfrak{p}}),$$

and the degree factor is at least 1. Altogether we obtain:

Proposition 2.159. *The following estimate holds:*

$$\deg(\mathcal{R}(F)) = \deg(\mathcal{R}(F)/(\widetilde{F})) \geq \deg(R/I) + a.$$

As a consequence, we derive the following bound for the reduction number of the Gorenstein ideal that we treated:

Theorem 2.160. *Let R be a Gorenstein local ring of dimension d and I a perfect Gorenstein ideal of dimension two such that R/I is normal. Suppose in addition that the residue field of R has characteristic zero. The following bound for the reduction number of I holds:*

$$\mathrm{r}(I) \leq (d-1)\cdot \deg(\mathcal{G}) - 4d + 5.$$

Proof. We are going to bound the reduction number of I by gluing it from the reduction number of E and the multiplicity of $H(I)$ in the exact sequence

$$0 \to H(I) \longrightarrow \mathcal{G} \longrightarrow \mathcal{G}' \to 0.$$

For each element $z \in \mathcal{G}$, we have a monic homogeneous polynomial $f(z)$ with coefficients in a minimal reduction of I (in fact, any minimal reduction) such that $f(z)H(I) = 0$. We proved earlier that f can be chosen of degree at most $e(H(I)) \leq \deg(\mathcal{G})$. On the other hand, we also proved that $E^{s+1} = FE^s$, where (recall that $r = d-2$)

$$s = (d-1)a - 2(d-1) + 1.$$

Lift F to a minimal reduction $J \subset I$. This is feasible with the assumption that R has infinite residue field and the manner in which the reductions F (or F') and J are chosen as generic linear combinations of a set of generators of I. Thus for any element $Z \in \mathcal{G}_1$, we have that $z = Z^{s+1} - g(Z) \in H(I)$, with the class of $g(Z)$ in $\mathcal{R}(E)$ lying in FE^s. Now we find an equation for z as described earlier. Altogether we have an equation of integral dependence for Z over $Jt\,\mathcal{G}$, of degree $\leq s + 1 + e(H(I))$.

Since we proved that $a \leq \deg(\mathcal{R}(F)) = \deg(\mathcal{G}) - \deg(H(I))$, we obtain a reduction number bound

$$\begin{aligned}
\mathrm{r}(I) &\leq \text{arith-deg}(H(I)) + (d-1)a - 2(d-1) + 1\\
&\leq \text{arith-deg}(H(I)) + (d-1)(\deg(\mathcal{G}) - \text{arith-deg}(H(I)) - 2d + 3\\
&\leq +(d-1)\deg(\mathcal{G}) - (d-2)\text{arith-deg}(H(I)) - 2d + 3\\
&\leq (d-1)\cdot \deg(\mathcal{G}) - 4d + 5.
\end{aligned}$$

Remark 2.161. This result is very similar to the formula in [DGV98] that considers the primary case. We note that we can drop the assumption that $\mathcal{G}$ has no embedded primes provided we use arith-deg instead of deg. This is also very similar to the formula of Theorem 2.45, that considers the primary case. More precisely, it resembles Theorem 2.48 so much that we shall formulate the following question. Let R be a Cohen-Macaulay local ring of dimension $d \geq 1$ (with infinite residue field), and I a Cohen-Macaulay ideal of codimension g, and set $h = \min\{\text{depth } \mathrm{gr}_I(R), g, d-1\}$.

Definition 2.162. The ideal I satisfies the *standard reduction formula* if

$$\boxed{\mathrm{r}(I) \leq (d-h)\cdot \deg(\mathcal{G}_I(R)) - 2(d-h) + 1.} \tag{2.82}$$

How widely is this condition met? In broad generality, we have only identified in this class $\mathfrak{m}$-primary ideals, according to Theorem 2.48, and some of the sporadic cases treated above.

Table 2.3. Reduction numbers of some classes of ideals (primary ideals of a Cohen-Macaulay local ring unless stated otherwise) of a local ring of dimension d:

depth $\mathrm{gr}_I(R) = h \leq d-1$	$\mathrm{r}(I) \leq (d-h)e(I) - 2(d-h) + 1$	(2.26)
$\mathrm{gr}_I(R)$ has S_1 & char $R/\mathfrak{m} = 0$	$\mathrm{r}(I) \leq e(I) - 1$	(1.16)
depth $\mathrm{gr}_I(R) = d$	$\mathrm{r}(I) \leq e(I) - \lambda(R/I)$	(2.31)
depth $R[It] = d+1$	$\mathrm{r}(I) \leq d-1$	(3.14)
$\mathcal{F}(I)$ C.-M. domain & char $R/\mathfrak{m} = 0$	$\mathrm{r}(I) \leq 1 + \dfrac{e(I) - t - 1}{\nu(I)}$	(2.68)
I normally C.-M. & height $I = g \geq 1$	$\mathrm{r}(I) \leq g \cdot g! \deg(R/I) \deg(R) - 2g + 1$	(2.71)

Multiplicity of the Special Fiber of Ideals of Positive Dimension

The technique introduced above can also be used to derive estimates for $\deg(\mathcal{F}(I))$, when I is not necessarily of dimension zero. We begin with the most favorable case, when I is an equimultiple Cohen-Macaulay ideal of dimension 1.

For a given minimal reduction $J \subset I$ we define the corresponding Sally module

$$S_J(I) = \bigoplus_{n \geq 2} I^n / IJ^{n-1}.$$

The exact sequence

$$0 \to J^{n-1}/IJ^{n-1} \longrightarrow R/IJ^{n-1} \longrightarrow R/J^{n-1} \to 0$$

shows that R/IJ^{n-1} is a Cohen-Macaulay module (J^{n-1}/IJ^{n-1} is a free R/I-module) and therefore in dimension 1 the embedding $I^n/IJ^{n-1} \hookrightarrow R/IJ^{n-1}$ gives that I^n/IJ^{n-1} is a Cohen-Macaulay module.

The following combines the theory of the Sally module of ideals of dimension zero and (2.32).

Proposition 2.163. *Let $(R, \mathfrak{m})$ be a Cohen-Macaulay local ring of dimension $d \geq 2$ and I an equimultiple ideal of codimension $d-1$. If $S_J(I)$ is one of its Sally modules, then*

$$\begin{aligned}\deg(S_J(I)_{n-1}) &= \deg(I^n/IJ^{n-1}) \\ &= (E_1(I) - E_0(I) + \deg(R/I))\binom{n+d-3}{d-2} + \text{lower terms.}\end{aligned}$$

We recall from Proposition 2.70 that $E_0(I) = \deg(R/J)$, but that $E_1(I)$ is only given indirectly,

$$E_1(I) = \sum e_1(I_{\mathfrak{p}}) \deg(R/\mathfrak{p}),$$

where $\mathfrak{p}$ runs over the minimal primes of I.

We can now use the approach in Proposition 2.140:

Proposition 2.164. *Let $(R, \mathfrak{m})$ be a Cohen-Macaulay local ring and I an equimultiple ideal of dimension 1. Then*

$$\begin{aligned}\deg(\mathcal{F}(I)) &\leq E_1(I) - E_0(I) + \deg(R/I) + \nu(I) \\ &\leq \binom{E_0(I)-1}{2} - E_0(I) + \deg(R/I) + \nu(I).\end{aligned}$$

Proof. From the sequence

$$0 \to I \cdot R[Jt] \longrightarrow I \cdot R[It] \longrightarrow S_J(I) \to 0,$$

tensoring with $R/\mathfrak{m}$ as in the proof of Proposition 2.140, we have only to observe that the growth of the number of generators of $S_J(I)_n$ is controlled by its multiplicity, which we read from the Proposition 2.163. □

The next case to be considered would be that of an ideal I of dimension 1 and analytic spread d. One then has the surjection $\mathrm{gr}_I(R) \to \mathcal{F}(I)$ of algebras of dimension d, so that their multiplicities can still be compared, as $\deg(\mathrm{gr}_I(R)) \geq \deg(\mathcal{F}(I))$. Unlike the dimension zero case however, $\deg(\mathrm{gr}_I(R))$ is very difficult to figure out.

2.8 Exercises

Exercise 2.165. Let $(R,\mathfrak{m})$ be a Cohen-Macaulay local ring of dimension d with infinite residue field. Let I be an ideal generated by a system of parameters, set $A = R[It]$ and let $B = \sum_{n\geq 0} I_n t^n$ be a finite A-subalgebra of $R[t]$. Define the reduction module $S_{B/A}$ (Sally module) by the exact sequence

$$0 \to I_1 A \longrightarrow B_+ \longrightarrow S_{B/A} \to 0.$$

Prove the following assertions:

1. $S_{B/A}$ is an A-module that is either 0 or has Krull dimension d.
2. The first 3 Hilbert coefficients, $e_0(B), e_1(B), e_2(B)$, of the function $\lambda(R/I^n)$ are non-negative.
3. The multiplicity of $S_{B/A}$ is $s_0(B) = e_1(B) - e_0(B) + \lambda(R/I_1)$.

Exercise 2.166. Let $(R,\mathfrak{m})$ be a Cohen-Macaulay ideal of dimension $d \geq 1$, and I an $\mathfrak{m}$-primary ideal of reduction number $\mathrm{r}(I)$. If $e_0(I)$ and $e_1(I)$ are the first Hilbert coefficients of I, show that

$$e_1(I) \leq \mathrm{r}(I) \cdot e_0(I).$$

Exercise 2.167. Let $(R,\mathfrak{m})$ be a Noetherian local ring of Krull dimension d and let I be a proper $\mathfrak{m}$-primary ideal. If $z \in \mathfrak{m}$, show that

$$e(I) - e(I,z) \leq \lambda(R/I:z) \cdot f_0(I).$$

Exercise 2.168. Let $(R,\mathfrak{m})$ be a Noetherian local ring of Krull dimension d and I a proper $\mathfrak{m}$-primary ideal. Suppose that $z \in I:\mathfrak{m} \setminus I$. Does the inequality

$$e(I) - e(I,z) \leq d!\deg(R)$$

always hold? If so, it would refine Lech's formula (2.58).

Exercise 2.169. Let I be an ideal of the local ring $(R,\mathfrak{m})$. If R has infinite residue field, show that $\mathrm{r}(I^r) \leq \ell(I)$ for all $r \gg 0$.

Exercise 2.170. Let $f_1, \ldots, f_{d+1}$ be forms in $k[x_1, \ldots, x_n]$ of degree s generating an ideal I of finite co-length. Show that

$$\mathrm{r}(I) \leq ds^{d-1} - 2d + 1.$$

Exercise 2.171. Let $I \subset R = k[x_1, \ldots, x_d]$ be an ideal of finite colength of the polynomial ring R, and let L be its initial ideal for some term order. Show that

$$e(I) \leq d!\lambda(R/\overline{L}).$$

Exercise 2.172. Let R be a ring of polynomials over a field and let

$$0 \to A \longrightarrow B \longrightarrow C \to 0$$

be an exact sequence of finitely generated graded modules (and homogeneous homomorphisms). If r is an integer such that $\dim A \leq r$ and depth $C \geq r$, show that

$$\mathrm{reg}(B) = \max\{\mathrm{reg}(A), \mathrm{reg}(C)\}.$$

Exercise 2.173. Let R be a Noetherian ring of finite Krull dimension. Is

$$\sup\{\mathrm{hdeg}(R_{\mathfrak{m}}) \mid \mathfrak{m} \in \mathrm{Spec}(R)\}$$

finite?

Exercise 2.174. (See [Va98b, Theorem 9.1.4]) Let $(R, \mathfrak{m})$ be a Noetherian local ring of dimension $d > 0$ and M a finitely generated R-module. Suppose that $x \in \mathfrak{m}$ and consider the short exact sequence induced by multiplication by x,

$$0 \to L \longrightarrow M \stackrel{x}{\longrightarrow} M \longrightarrow G \to 0.$$

Show that if L is a module of finite length, then $\ell(H^0_{\mathfrak{m}}(G)) \geq \ell(L)$. Moreover, if $d = 1$ then

$$\ell(H^0_{\mathfrak{m}}(G)) = \ell(L) + \ell(\overline{M}/x\overline{M}),$$

where $\overline{M} = M/\text{torsion}$.

Exercise 2.175 (Rossi-Trung-Valla). Let k be a field and set

$$S_n = k[[x,y,u,v]]/(u^n v^n - xy, x^3 - u^{2n}y, y^3 - xv^{2n}, u^n y^2 - x^2 v^n).$$

Show that $\mathrm{Deg}(S_n) = \deg(S_n) + n$.

Exercise 2.176. Let R be a Cohen-Macaulay local ring with canonical module ω and define the module E by the exact sequence

$$0 \to R \longrightarrow \omega^{\oplus r} \longrightarrow E \to 0,$$

where $1 \in R$ maps onto the 'vector' $(x_1, \ldots, x_r)$ determined by a generating set for ω. E is Cohen-Macaulay. What is a bound for $\nu(\mathrm{Hom}_R(E,E))$?

Exercise 2.177. Let R be a ring of polynomials over a field and let E be a graded R-module with a (graded) resolution

$$0 \to R^n \stackrel{\varphi}{\longrightarrow} R^n \longrightarrow E \to 0;$$

E is a Cohen-Macaulay module. Show that

$$\deg(E) = \deg(\det \varphi).$$

Exercise 2.178. Let $(R, \mathfrak{m})$ be a Noetherian local ring and I an ideal of analytic spread $\ell(I)$. For any extended degree function $\mathrm{Deg}(\cdot)$, prove that

$$\mathrm{bigr}(I) \leq \mathrm{Deg}(\mathcal{F}(I)) - \nu(I) + \ell(I).$$

Exercise 2.179. Let R be a Noetherian local ring and M a finitely generated R-module. If N is a submodule of M of Krull dimension at most 1, prove that for any Deg function, $\mathrm{Deg}(N) \leq \mathrm{Deg}(M)$.

Exercise 2.180. Let R be a regular local ring of dimension d and E a torsionfree R-module. Let

$$0 \to L \longrightarrow R^n \longrightarrow E \to 0$$

be a presentation of E, where $n = \text{hdeg}(E)$. Show that

$$\text{hdeg}(L) \leq \frac{d}{2}(\text{hdeg}(E) - \deg(E)).$$

Exercise 2.181. Let R be a local ring of dimension 3 and E a torsionfree R-module of depth 2. Show that

$$\nu(E^{**}) \leq 2 \cdot \text{hdeg}(E).$$

Is the factor 2 unnecessary?

Exercise 2.182. Let R be a Noetherian local ring and let

$$0 \to A \longrightarrow B \longrightarrow C \to 0$$

be an exact sequence of finitely generated R-modules. If A and B are Cohen-Macaulay modules, possibly of different dimensions, establish the various possible relationships amongst $\text{hdeg}(A)$, $\text{hdeg}(B)$ and $\text{hdeg}(C)$.

Exercise 2.183 (K. Dalili). Let R be a Noetherian local ring and

$$0 \to A \longrightarrow B \longrightarrow C \to 0$$

an exact sequence of finitely generated R-modules. Show that $\text{bdeg}(B) \leq \text{bdeg}(A) + \text{bdeg}(C)$.

Exercise 2.184. Let $(R, \mathfrak{m})$ be a Noetherian local ring and I an $\mathfrak{m}$-primary ideal. Show that Conjecture 2.90 holds for I.

Exercise 2.185. Let $(R, \mathfrak{m})$ be a Noetherian local ring and I an $\mathfrak{m}$-primary ideal. Show that

$$\deg(\mathcal{F}(I)) \geq \frac{\deg(G)}{\lambda(R/I)}.$$

Moreover, if the associated primes of $\mathcal{F}(I)$ have the same dimension, prove that equality is equivalent to he requirement that I^n/I^{n+1} be R/I-free for all $n \geq 0$.

Exercise 2.186. Let R be a Cohen-Macaulay ring (not necessarily local) and let I be an equicodimensional ideal (that is, the associated primes of I have the same codimension). Prove that the set of reductions of I that are equidimensional has minimal elements.

Exercise 2.187. Let $(R, \mathfrak{m})$ be a Noetherian local ring of dimension d, not necessarily Cohen-Macaulay, with infinite residue field, and let I be an $\mathfrak{m}$-primary ideal. Are there interesting bounds for $e_+(I)$ and $e_{++}(I)$?

Exercise 2.188. Let R be a Cohen-Macaulay local ring and $\mathfrak{p}$ a prime ideal such that $\mathfrak{p}^{(n)} = \mathfrak{p}^n$ for all n. Prove that if $\mathcal{G} = \mathrm{gr}_{\mathfrak{p}}(R)$ then $\deg(\mathcal{G}) = \deg(R/\mathfrak{p})\deg(R_{\mathfrak{p}})$.

Exercise 2.189. Let R be a Cohen-Macaulay local ring with infinite residue field and I a normally Cohen-Macaulay ideal. Prove that if J is a minimal reduction of I and $G = \mathrm{gr}_I(R)$, then $\deg(G) = \deg R/J$.

Exercise 2.190. Prove the following assertion. Let $(R, \mathfrak{m})$ be a catenary Noetherian local ring of dimension $d > 0$, let $A = R[x_1, \ldots, x_m]$ be a standard graded algebra over R and let

$$0 \to M \longrightarrow N \longrightarrow P \to 0$$

be an exact sequence of finitely graded A-modules and graded homomorphisms. Suppose that M has the condition S_2, that $\dim M = \dim N > \dim P$ and $\dim N/\mathfrak{m}N < \dim N - 1$. Then $E_0(M) = E_0(N)$ and $E_1(N) = E_1(M) + E_0(P)$.

3

Depth and Cohomology of Rees Algebras

In this chapter we explore aspects of the Cohen-Macaulay property in the interaction amongst the rings R, $\mathcal{R} = R[It]$, $\mathcal{G} = \text{gr}_I(R)$ and the geometric counterparts Proj ($\mathcal{R}$) and Proj ($\mathcal{G}$). One line we shall exploit is to look for general frameworks to allow the comparison of the depth properties of a ring R, the Rees algebra $\mathcal{R} = R[It]$ of one of its ideals, the associated graded ring $\mathcal{G} = \text{gr}_I(R)$, and for local rings R, the special fiber $\mathcal{F}(I)$. In another chapter we shall consider properties of the ideal I that can be used to build Rees algebras with the requisite depth properties.

One of the vessels for our analysis is the study of the transfer of depth properties among the following actors

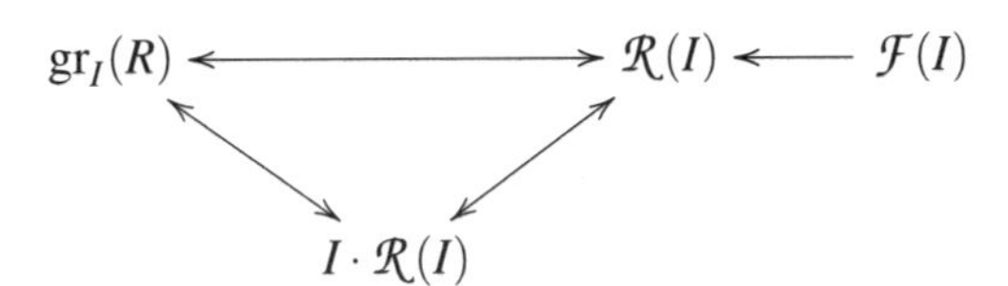

and the strong connection this carries with the reduction numbers of I at certain critical primes.

More specifically, we shall look into the following comparisons:

- Depth $\mathcal{R}$ versus depth $\mathcal{G}$
- Cohen-Macaulayness all round
- Serre's S_k-properties
- Serre's R_k-properties
- Local reduction numbers

One approach we shall pursue is that of assembling, from the vast literature on the cohomology of blowups, several results that bear directly on those properties. Since it is very hard to compute cohomology–almost any computation has an author's name attached to it–there will an emphasis on vanishing results.

3.1 Settings of Cohen-Macaulayness

Let R be a Noetherian ring and I one of its ideals. In this section we consider settings for studying general properties of Cohen-Macaulay Rees algebras. There are two frameworks at issue: one of questions, another of means.

3.1.1 Systems of Parameters and Hypersurface Sections

The technique of hyperplane section is quite useful in the study of graded rings. One of the difficulties in applying it to Rees algebras and their associated graded rings lies in finding regular sequences in these algebras. One approach is that of reduction modulo a hypersurface. The starting point is the following construction.

Given a filtration $\mathcal{F}$ of R and an ideal J, we denote by $\mathcal{F}/J$ the filtration $\{I_n \cdot R/J, n \geq 0\}$ of R/J. If $x \in I_1 \setminus I_2$, one has the following exact sequence induced by the image x^* of x in I_1/I_2:

$$0 \to 0 :_{\mathrm{gr}_{\mathcal{F}}(R)} x^* \longrightarrow \mathrm{gr}_{\mathcal{F}}(R)/(x^*) \longrightarrow \mathrm{gr}_{\mathcal{F}/(x)}(R) \to 0. \tag{3.1}$$

Consider the case of the I-adic filtration and suppose that the initial form x^* of $x \in I \setminus I^2$ in $\mathrm{gr}_I(R)$ is a regular element. We then have a connection between the two Rees algebras, $R[It]$ and $R_1[I_1t]$, where $R_1 = R/(x)$ and $I_1 = I/(x)$:

$$0 \to (x, xt) \longrightarrow R[It] \longrightarrow R_1[I_1t] \to 0. \tag{3.2}$$

System of Parameters

By a *system of parameters* of a graded algebra $G = \bigoplus_{n \geq 0} G_n$ over the local ring $R = G_0$ we mean a set of elements $\{f_1, \ldots, f_r\}$, $r = \dim G$, that generate a primary ideal for the irrelevant maximal ideal of G.

Let $(R, \mathfrak{m})$ be a Noetherian local ring of dimension $d > 0$ and I an ideal of height > 0. A system of parameters for either $\mathcal{R}(I)$ or $\mathrm{gr}_I(R)$ is very hard to get hold of from a list of generators of I. This is particularly hard when $\dim R/I > 0$. The following result provides general guidance.

Proposition 3.1. *Suppose that the residue field of R is infinite. Then*

(a) $\mathcal{R}(I)$ *has a system of parameters of the form*

$$\{x_1, x_2 + a_1t, x_3 + a_2t, \ldots, x_d + a_{d-1}t, a_dt\},$$

where $\{x_1, \ldots, x_d\}$ is a system of parameters of R and $(a_1, \ldots, a_d)$ is a reduction of I.

(b) (Valla) *If I is $\mathfrak{m}$-primary, then one can choose the a_i to define an arbitrary minimal reduction of I, and set $x_i = a_i$, $\forall i$.*

(c) *If I is an ideal of dimension 1 with a minimal reduction $J = (a_1, \ldots, a_d)$, and b is a system of parameters for R/I, then*

$$\{a_1, a_1t - a_2, a_2t - a_3, \ldots, a_{d-1}t - a_d, a_dt - b\}$$

is a system of parameters for the irrelevant maximal ideal $\mathcal{M}$ of $\mathcal{R}(I)$.

Proof. We shall only prove part (c), which is inspired by (b). Let P be a prime ideal of $\mathcal{R}(I)$ containing the sequence of elements. We claim that $P = \mathcal{M}$, for what it will suffice to show that

$$(a_1, a_2, \ldots, a_d, b, a_1t, \ldots, a_dt) \subset P.$$

We start by noting that $a_2 \cdot a_1t = a_1 \cdot a_2t \in P$ and $a_1t - a_2 \in P$ implies that a_2, a_1t both lie in P. Repeating we obtain all the listed elements belong to P. □

Definition 3.2. Let $G = \bigoplus_{n \geq 0} G_n$ be a Noetherian graded algebra. A homogeneous element $x \in G_r$ is *superficial of degree r* if $(0 :_G x)_n = 0$ for all $n \gg 0$. Those of degree 1 are simply referred to as superficial elements. (Observe that the notion can be extended to graded G-modules.) A *superficial sequence* is a set of elements $x_1, \ldots, x_k$ such that x_i is superficial for $G/(x_i, \ldots, x_{i-1})$, for $i = 1, \ldots, k$.

Note that if grade $G_+ > 0$, then superficial elements are not zero divisors. For reference, we state the basic relationship between these elements and Hilbert functions:

Proposition 3.3. *Let G be a standard graded algebra over the Artinian ring $R = G_0$, M a finitely generated graded G-module and $h \in G_1$ a superficial element for M. Then:*

(a) *The Hilbert polynomials of M and M/hM satisfy*

$$P_{M/hM}(t) = P_M(t) - P_M(t-1).$$

In particular, if $e_i(M)$ and $e_i(M/hM)$ are the coefficients of the Hilbert polynomials of M and of M/hM respectively, then $e_i(M) = e_i(M/hM)$ for $i <$ dim M.

(b) *The Hilbert series of M and M/hM satisfy the equality*

$$H_{M/hM}(t) = H_M(t) - H_M(t-1)$$

if and only if h is regular on M.

Proof. This follows immediately from the exact sequence

$$0 \to (0 :_M h)[-1] \longrightarrow M[-1] \xrightarrow{h} M \longrightarrow M/hM \to 0,$$

and the fact that $0 :_M h$ has only finitely many graded components. □

Combining this result with the exact sequence (3.1) we have:

Proposition 3.4. *Let $(R,\mathfrak{m})$ be a Noetherian local ring with infinite residue field, and having dimension $d \geq 2$. If I is an $\mathfrak{m}$-primary ideal, then there exists an element $h \in I$ such that $R' = R/(h)$ has dimension $d-1$ and the Hilbert coefficients $e_i(I)$ of $\mathrm{gr}_I(R)$ and $e_i(I')$ of $\mathrm{gr}_{I'}(R')$ satisfy the equality*

$$e_i(I) = e_i(I'), \quad i \leq d-1.$$

The construction of such elements is usually achieved via some prime avoidance device.

Proposition 3.5. *Let $G = G_0[G_1]$ be a Noetherian graded algebra over the local ring $(G_0,\mathfrak{m})$ with infinite residue field, generated by elements of degree 1. Let $P_1,\ldots,P_r$ be a family of homogeneous prime ideals that do not contain G_1. Then there exists $h \in G$ such that*

$$h \in G_1 \setminus \mathfrak{m}G_1 \cup \bigcup_{i=1}^{r} P_i.$$

Furthermore, if the P_i include all the associated primes of G that do not contain G_1, then $0\colon(0\colon hG)$ contains some power of G_+.

Proof. Denote by $C_1,\ldots,C_r$ the components of degree 1 of these homogeneous ideals. Consider the vector space $V = G_1/\mathfrak{m}G_1$ over the residue field $G_0/\mathfrak{m}$ and its subspaces $V_i = (C_i + \mathfrak{m}G_1)/\mathfrak{m}G_1$. By Nakayama's Lemma and the choices of the P_i, $V_i \neq V$, for all i. Since $G_0/\mathfrak{m}$ is infinite, it follows that

$$V \neq \bigcup_{i=1}^{r} V_i,$$

by any of the usual reasons.

For the second assertion, choose $h \in G_1$ whose image in V does not lie in any V_i. Let $\mathfrak{P}$ be a minimal prime of $0\colon(0\colon hG)$; it suffices to show that $G_+ \subset \mathfrak{P}$. (If h is a regular element, $0\colon(0\colon hG) = G$.) Note that $\mathfrak{P}$ consists of zero divisors and contains h. This means that $\mathfrak{P}$ is an associated prime of G but is distinct from any of the P_i, and therefore must contain G_+. □

Superficial Elements and Filtrations

There is a related notion of superficial element arising in the context of filtrations. Let $\mathcal{F} = \{I_n, n \geq 0\}$ be a multiplicative decreasing filtration such that the Rees algebra $\mathcal{R}(\mathcal{F})$ is Noetherian, and denote by $\mathrm{gr}_{\mathcal{F}}(R) = \bigoplus_{n\geq 0} I_n/I_{n+1}$ its associated graded ring. The filtrations we have in mind are the I-adic, the Ratliff-Rush and the integral closure $R[It]$. Although the associated graded rings are not always standard graded algebras, they are finitely generated and graded over standard algebras (at least when R is suitably restricted). Thus, for example, $x \in I$ is a superficial element for $\mathrm{gr}_{\mathcal{F}}(R)$ (or simply for $\mathcal{F}$ by abuse of terminology) if x is regular and there is an integer c

such that $[I_{n+1} :_R x] \cap I_c = I_{n-1}$ for $n > c$. Their existence is derived from the Artin-Rees Lemma. Geometrically, a superficial element x of the I-adic filtration has the property that

$$\text{Proj } \text{gr}_I(R)/(x^*) \cong \text{Proj } \text{gr}_{I/(x)}(R/(x)). \tag{3.3}$$

Remark 3.6. In the discussion of superficial elements above we treated two ways in which they interact in associated graded rings of ideals. To clarify the issue, we give some additional comments.

Let $(R, \mathfrak{m})$ be a Noetherian ring and I an ideal of positive height with associated graded ring $G = \text{gr}_I(R)$. Let $x \in I$ be a superficial of I, that is, for some integer c,

$$(I^n : x) \cap I^c = I^{n-1}, \; n > c.$$

We now compare G/xG, $x \in G_1$, and $G' = \text{gr}_{I'}(R')$, where $I' = I/(x) \subset R/(x) = R'$. The ring G/xG derives from the exact sequence of graded G-modules

$$0 \to (0 :_G x) \longrightarrow G[-1] \xrightarrow{\cdot x} G \longrightarrow G/xG \to 0.$$

By the assumption, $(0 :_G x)_n = 0$, $n \gg 0$. In turn,

$$G'_n = (I^n + (x))/(I^{n+1} + (x)).$$

Note that one has the exact sequence

$$\begin{aligned} 0 \to (I^{n+1} + I^n \cap (x))/(I^{n+1} + xI^{n-1}) &\to G_n = I^n/(I^{n+1} + xI^{n-1}) \\ &\to G'_n = (I^n + (x))/(I^{n+1} + (x)) \to 0. \end{aligned}$$

We claim that $G_n = G'_n$ for $n \gg 0$, for which it will be enough to show that $I^n \cap (x) = xI^{n-1}$, $n \gg 0$.

Consider first the case that x is a regular element of R. According to [Ma80, Proposition 11.E], by the Artin-Rees Lemma there is an integer r such that $I^n : x = I^{n-r}(I^r : x) \subset I^c$ for $n - r \geq c$. Multiplying the equality $(I^n : x) \cap I^c = I^{n-1}$ by x (since x is regular) we get $I^n \cap (x) = x(I^n : x) \subset xI^{n-1}$, as desired.

If x is not regular, its annihilator $L = 0 :_R x$ satisfies the equality $L \cap I^c = 0$ for all large c, and similarly $(0 : x^s) \cap I^c = 0$. If we set $H = \bigcup_{s \geq 1}(0 : x^s)$, and pass to the ring $S = R/H$, where the image of x is a regular element, we obtain the equality

$$I^n \cap (x) + H = xI^{n-1} + H.$$

This is actually a natural direct sum and the assertion is proved also in this case.

We next give a technique ([HM97]) that uses superficial elements to examine the depth of associated graded rings.

Proposition 3.7 (Huckaba-Marley). *Let R be a Noetherian ring and $\mathcal{F} = \{I_n, n \geq 0\}$ a multiplicative decreasing filtration such that the Rees algebra $\mathcal{R}(\mathcal{F})$ is Noetherian. Let $x \in I_1$ be a superficial element on* $\text{gr}_{\mathcal{F}}(R)$. *If* grade $\text{gr}_{\mathcal{F}/(x)}(R)_+ > 0$, *then x^* is a regular element of* $\text{gr}_{\mathcal{F}}(R)$.

Proof. Let $y \in I_t$, $t > 0$, be such that its image in $\mathrm{gr}_{\mathcal{F}/(x)}(R/(x))_t$ is a regular element. Then $(I_{n+tj} : y^t) \subset (I_n, x)$ for all n, j. Since x is superficial for $\mathcal{F}$ there exists an integer c such that $(I_{n+j} : x^j) \cap I_c = I_n$ for all $j \geq 1$ and $n \geq c$. Let n and j be arbitrary and p any integer greater than c/t. Then

$$y^p(I_{n+j} : x^j) \subset (I_{n+pt+j} : x^j) \cap I_c \subset I_{n+pt}.$$

Therefore $(I_{n+j} : x^j) \subset (I_{n+pt} : y^p) \subset (I_n, x)$. Thus $(I_{n+j} : x^j) = I_n + x(I_{n+j} : x^{j+1})$ for all n and j. Iterating this formula n times, we get that

$$(I_{n+j} : x^j) = I_n + xI_{n-1} + \cdots + x^n(I_{n+j} : x^{j+n}) = I_n.$$

Hence x^* is a regular element of $\mathrm{gr}_{\mathcal{F}}(R)$. □

Corollary 3.8. *Let $x_1, \ldots, x_k$ be a superficial sequence for $\mathcal{F}$. If*

$$\text{grade } \mathrm{gr}_{\mathcal{F}/(x_1,\ldots,x_k)}(R)_+ \geq 1,$$

then grade $\mathrm{gr}_{\mathcal{F}}(R)_+ \geq k+1$.

3.1.2 Passing Cohen-Macaulayness Around

We turn to one of the most active areas of investigation in the theory of Rees algebras. At the risk of offending friends and other sensitive listening devices, we propose:

Definition 3.9. Let R be a ring and I an ideal. We say that I is *explosively Cohen-Macaulay* if its Rees algebra $\mathcal{R} = R[It]$ is Cohen-Macaulay. Such a Rees algebra is called an *arithmetic Cohen-Macaulayfication* of R.

Unlike the ring $\mathcal{G}$, the ring $\mathcal{R}$ can be Cohen-Macaulay without R itself being Cohen-Macaulay. This terminology is slightly misleading because $\mathcal{R}$ can also be Cohen-Macaulay when I is far from sharing this property.

We derive now some connections between the Cohen-Macaulayness of R and the Rees algebra of an ideal I and associated graded ring, and some of their homomorphic images.

Proposition 3.10. *Let R be a Noetherian local ring and I an ideal of positive height. If $\mathrm{gr}_I(R)$ is Cohen-Macaulay, then R is also Cohen-Macaulay. More generally, if $\mathrm{gr}_I(R)$ has depth r, or satisfies Serre's condition S_r of Serre, R will satisfy the same conditions.*

Proof. Let $A = R[It, t^{-1}]$ be the extended Rees algebra of I. The assertion follows from the diagram

$$\begin{array}{ccc} & A = R[It,t^{-1}] : \text{ C-M} & \\ \swarrow & & \searrow \\ A/(t^{-1}-1) = R : \text{ C-M} & & A/(t^{-1}) = \mathrm{gr}_I(R) : \text{C-M} \end{array}$$

showing that A is a (flat) deformation of $\mathrm{gr}_I(R)$, where t^{-1} lies in the homogeneous maximal ideal of A. Thus the Cohen-Macaulayness of A and of $\mathrm{gr}_I(R)$ is an equivalence. In this case, R is an ordinary specialization of A. □

In addition to the basic sequences that connect R, $R[It]$ and $\mathrm{gr}_I(R)$, there are other sequences that move the depth properties of these algebras around (cf. [Hu82a, Lemma 1.1]).

Proposition 3.11. *Let R be a local ring and a an element of I be such that its image a^* in I/I^2 is a regular element of $\mathrm{gr}_I(R)$. Set $R_1 = R/(a)$ and $I_1 = I/(a)$. Then there are exact sequences*

$$0 \to (a, at) \longrightarrow R[It] \longrightarrow R_1[I_1 t] \to 0,$$

$$0 \to \mathrm{gr}_I(R)[-1] \longrightarrow R[It]/(a) \longrightarrow R_1[I_1 t] \to 0,$$

$$0 \to R \longrightarrow R[It]/(at) \longrightarrow R_1[I_1 t] \to 0,$$

$$\mathrm{gr}_{I_1}(R_1) \cong \mathrm{gr}_I(R)/(a^*).$$

Proof. The key fact, that (a, at) is the kernel of the natural surjection of Rees algebras, is a consequence of quality $0 \colon a^* = 0$ in $\mathrm{gr}_I(R)$. It will be left as an exercise for the reader. □

A framework for studying the relationship between the Cohen-Macaulayness of $\mathcal{R}$ and $\mathrm{gr}_I(R)$ are the following exact sequences (originally paired in [Hu82a]):

$$0 \to I \cdot R[It] \longrightarrow R[It] \longrightarrow \mathrm{gr}_I(R) \to 0 \tag{3.4}$$

$$0 \to It \cdot R[It] \longrightarrow R[It] \longrightarrow R \to 0, \tag{3.5}$$

with the tautological isomorphism

$$It \cdot R[It] \cong I \cdot R[It]$$

playing a pivotal role.

This leads to the result of Huneke ([Hu82a]):

Theorem 3.12. *Let R be a Cohen-Macaulay ring and I an ideal containing regular elements such that $R[It]$ is Cohen-Macaulay. Then $\mathrm{gr}_I(R)$ is Cohen-Macaulay.*

Proof. Since I contains regular elements, $\dim R[It] = \dim R + 1$. We may assume that R is a local ring. To show that the depth of $\mathrm{gr}_I(R)$ (relative to its irrelevant maximal ideal) is at least $\dim R$, we make use of the two exact sequences (3.4) and (3.5), the second of which, in this case, says that $It \cdot R[It]$ is a maximal Cohen-Macaulay module. Since it is isomorphic to $I \cdot R[It]$, the associated graded ring will be Cohen-Macaulay. □

Remark 3.13. There are other settings in which the argument above will apply. We have in mind a Rees algebra

$$A = R + F_1 t + F_2 t^2 + \cdots,$$

of a decreasing filtration of ideals. (A case in point is the integral closure of the Rees algebra of an ideal.) There is another associated graded ring,

$$G = R/F_1 + F_1/F_2 + \cdots,$$

and the Cohen-Macaulayness of G arises in the same manner from that of R and A.

Local Cohomology

The main technical tool for the examination of the arithmetic–e.g. depth, Cohen-Macaulayness–of Rees algebras is local cohomology. Here we assemble some of these techniques from several sources, particularly [BH93], [BrS98], [Gr67], [Har77], and [HIO88]. We shall emphasize aspects that take advantage of the structure of blowup algebras.

Throughout we assume that $(R, \mathfrak{m})$ is a local ring of dimension d, and that I is an ideal of positive height. Its Rees algebra $\mathcal{R} = R[It]$ has dimension $d+1$; we set $\mathfrak{M} = (\mathfrak{m}, \mathcal{R}_+)$.

Koszul and Čech Complexes

Let $A = \bigoplus_{n \geq 0} A_n$ be a finitely generated graded algebra over the Noetherian ring $R = A_0$. Let N be a finitely generated graded A-module. The cohomology of coherent sheafs over Proj (A) is expressed by the following Čech complex. Let $f_0, \ldots, f_s$ be a set of homogeneous elements of A_+ such that $A_+ \subset \sqrt{(f_0, \ldots, f_s)}$. The (limit) Koszul complex of the f_i is

$$\mathbb{K}(f_0, \ldots, f_s) = \bigotimes_{i=0}^{s} (0 \to A \longrightarrow A_{f_i} \to 0).$$

This is a complex of $\mathbb{Z}$-graded A-modules. We set $\mathbb{K}(f_0, \ldots, f_s) \otimes N = \mathbb{K}(f_0, \ldots, f_s; N)$. For a given integer n, the Čech complex of the sheaf $\mathcal{N}[n]$ is the subcomplex

$$\mathfrak{C}^i(X, \mathcal{N}[n]) = \mathbb{K}^{i+1}(f_0, \ldots, f_s; N)_n, i \geq 0,$$

of elements in degree n. Here $X = \text{Proj}\ (A)$ and $\mathcal{N}[n]$ is the sheaf associated with the module $N[n]$.

This construction defines the short exact sequence of chain complexes, where N is viewed as concentrated in dimension zero:

$$0 \to \bigoplus_n \mathfrak{C}(X, \mathcal{N}[n])[-1] \longrightarrow \mathbb{K}(f_0, \ldots, f_s; N) \longrightarrow N \to 0. \tag{3.6}$$

Since $X_{f_i} = \text{spec}\ ((A_{f_i})_0)$, the Čech complexes give rise to the cohomology of the sheafs $\mathcal{N}[n]$ on the scheme Proj (A). More precisely one has the following result.

Theorem 3.14. *Let N be a finitely generated graded A-module and denote by $\mathcal{N}$ the corresponding sheaf on $X = \text{Proj}\,(A)$. Then for all $i \geq 1$ and all integers n there exists a natural isomorphism of finitely generated R-modules*

$$H^i(X, \mathcal{N}[n]) \cong H^{i+1}_{A_+}(N[n])_n.$$

Moreover for all integers n there exists an exact sequence

$$0 \to H^0_{A_+}(N[n])_n \longrightarrow N_n \longrightarrow \Gamma(X, \mathcal{N}[n]) \longrightarrow H^1_{A_+}(N[n])_n \to 0.$$

Depth and Cohomology

The $\mathfrak{M}$-depth of $\mathcal{R}$, which affects both the arithmetic and the geometry of the blowup, is carried by the modules $H^i_{\mathfrak{M}}(\mathcal{R})$ for $i = 0, \ldots, d+1$. In turn, the analytic spread $\ell(I)$ of I and the reduction number of I are determined, as we shall see, by the modules $H^i_{\mathcal{R}_+}(\mathcal{R})$ for $i = 0, \ldots, d$.

To be useful several other factors must contribute. (i) A connection between the functors $H^i_{\mathfrak{M}}(\cdot)$ on one hand and $H^i(X, \cdot)$ on the other; (ii) explicit calculations of these various modules in many general cases of interest. As a general rule, the calculation of cohomology is very hard; some general vanishing theorems will be visited.

Let us begin with [Gr67, Lemma 1.8]; the special case here and its proof is [Br83, Lemma 3.9].

Proposition 3.15. *Let R be a commutative Noetherian ring, J an ideal, x an element of R and F a finitely generated R-module. There exists a natural exact sequence*

$$\begin{aligned} 0 \to H^0_{(J,x)}(F) \to H^0_J(F) \to H^0_{J_x}(F_x) \to H^1_{(J,x)}(F) \to \cdots \\ \cdots \to H^i_{(J,x)}(F) \to H^i_J(F) \to H^i_{J_x}(F_x) \to H^{i+1}_{(J,x)}(F) \to \cdots. \end{aligned}$$

Proof. Let

$$\mathbb{E}: \quad 0 \to F \longrightarrow E_0 \longrightarrow E_1 \longrightarrow \cdots$$

be an injective resolution of F. Each E_i is a direct sum of injective modules of the form $E(R/\mathfrak{q})$, the injective envelope of $R/\mathfrak{q}$ for some prime $\mathfrak{q}$. As a consequence, there is direct sum decomposition

$$0 \to \Gamma_{xR}(E_i) \longrightarrow E_i \longrightarrow (E_i)_x \to 0.$$

Thus for each $i \geq 0$ and any ideal I, we have the exact sequence

$$0 \to \Gamma_I(\Gamma_{xR}(E_i)) \longrightarrow \Gamma_I(E_i) \longrightarrow \Gamma_I((E_i)_x) \to 0.$$

Since $\Gamma_J \cdot \Gamma_{xR} = \Gamma_{(J,x)}$, these exact sequences give rise to the following short exact sequence of complexes:

$$0 \to \Gamma_{(J,x)}(\mathbb{E}) \longrightarrow \Gamma_J(\mathbb{E}) \longrightarrow \Gamma_J((\mathbb{E})_x) \to 0.$$

The assertion of the lemma is the long exact cohomology sequence that results. □

As in the original source ([Gr67]), one can use more general combinations of the functors $\Gamma_J(\cdot)$. For example, instead of $\Gamma_{xR}(\cdot)$, we apply $\Gamma_{\mathfrak{m}}(\cdot)$ and then compose with $\Gamma_{\mathcal{R}_+}(\cdot)$ to obtain another exact sequence of chain complexes:

$$0 \to \Gamma_{\mathfrak{M}}(\mathbb{E}) \longrightarrow \Gamma_{\mathcal{R}_+}(\mathbb{E}) \longrightarrow \Gamma_{\mathcal{R}_+}((\mathbb{E})') \to 0.$$

Local Duality

To make amenable the 'huge' modules that occur in local cohomology one appeals to Grothendieck's local duality. It can be found in the sources mentioned above (see in particular [BH93, Section 3.6]).

Since we shall deal almost exclusively with graded algebras over local rings. For the purpose of stating some results in local cohomology, we refer to a ring such as $S = \bigoplus_{n\geq 0} S_n$, with $(S_0, \mathfrak{m}_0)$ local, as local instead of *local, as is the usage. If E is the injective envelope of $S_0/\mathfrak{m}_0$ as an S_0-module, the corresponding Matlis dual of a graded S-module will be $M^\vee = {}^*\mathrm{Hom}_{S_0}(M,E)$. Its grading is given by $(M^\vee)_n = \mathrm{Hom}_{S_0}(M_{-n},E)$.

Theorem 3.16 (Local duality). *Let $(S,\mathfrak{M})$ be a Cohen-Macaulay, complete local ring of dimension d. Then*

(a) *S has a canonical module ω_S, and $\omega_S \cong ({}^*H^d_{\mathfrak{M}}(S))^\vee$.*
(b) *For all finite graded S-modules M and all integers i there exist natural homogeneous isomorphisms*

$$({}^*H^i_{\mathfrak{M}}(M))^\vee \cong {}^*\mathrm{Ext}^{d-i}_S(M,\omega_S).$$

Theorem 3.17. *Let $(A,\mathfrak{m})$ be a Gorenstein local ring of dimension s and M a finitely generated A-module of dimension d. For an integer $k \geq 1$ the following conditions are equivalent:*

(i) *M satisfies Serre's condition S_k.*
(ii) *The natural map $\tau_M : M \to \mathrm{Ext}^{s-d}_A(\mathrm{Ext}^{s-d}_A(M,A),A)$ is an isomorphism, and $H^n_{\mathfrak{m}}(M) = 0$ for all $d-k+2 \leq n < d$.*

Corollary 3.18. *Suppose moreover that M satisfies Serre's condition S_2. Then for an integer $k \geq 2$ the following conditions are equivalent:*

(i) *$\mathrm{Ext}^{s-d}_A(M,A)$ satisfies Serre's condition S_k.*
(ii) *$H^n_{\mathfrak{m}}(M) = 0$ for all $d-k+2 \leq n < d$.*

Sancho de Salas Sequence

Let $(R, \mathfrak{m})$ be a Noetherian local ring and let $\mathcal{R} = R[It]$ be the Rees algebra of an ideal of positive height. In the sequence (3.6), set $A = \mathcal{R}$, let J be an ideal of R and apply to it the functor $\Gamma_{J\mathcal{R}}(\cdot)$. Taking the hypercohomology (see [We94]) of the sequence of complexes, one obtains ([SS87] (we follow [Li94a] also):

Proposition 3.19. *Suppose that* $\mathfrak{M} = (J, \mathcal{R}_+)$ *and* $E = \text{Proj}\ (\mathcal{R} \otimes R/J)$. *For any finitely generated graded A-module N there exists a long exact sequence*

$$\cdots \longrightarrow H^i_{\mathfrak{M}}(N) \longrightarrow \bigoplus_{n\in\mathbb{Z}} H^i_{\mathfrak{m}}(N_n) \longrightarrow \bigoplus_{n\in\mathbb{Z}} H^i_E(\mathcal{N}[n]) \longrightarrow H^{i+1}_{\mathfrak{M}}(N) \longrightarrow \cdots.$$

3.2 Cohen-Macaulayness of Proj ($\mathcal{R}$) and Cohomology

Let R be a Noetherian ring, I an ideal and set $X = \text{Proj}\ (R[It])$. We recall that X is said to be Cohen-Macaulay if each of its local rings at the homogeneous prime ideals that do not contain (It) is Cohen-Macaulay. A general question asks about the relationship between the Cohen-Macaulay property of $R[It]$ and that of X.

Let us consider first the cruder properties of X vis-a-vis $R[It]$. The following elementary observation is useful here. About terminology: For a local ring the punctured spectrum has the usual meaning, the set of non-maximal prime ideals.

Proposition 3.20. *Let R be a Noetherian local ring and I an ideal of positive grade. Suppose that R is Cohen-Macaulay. Then* Proj $(\text{gr}_I(R))$ *is Cohen-Macaulay if and only if* Proj ($\mathcal{R}$) *is Cohen-Macaulay.*

Proof. We may assume that $I = (f_1, \ldots, f_n)$, with each f_i an R-regular element. Set $\mathcal{R} = R[f_1t, \ldots, f_nt]$; it suffices to prove that $\mathcal{R}_{f_it}$ is Cohen-Macaulay for each f_i. Note that f_i is a regular element of $\mathcal{R}_{f_it}$ and there is the canonical isomorphism

$$\mathcal{R}_{f_it}/(f_i)_{f_it} \cong \text{gr}_I(R)_{f_it}.$$

The assertion follows since the last ring is Cohen-Macaulay. □

Remark 3.21. These results establish the implications for the Cohen-Macaulay property,

$$\mathcal{R} \Rightarrow \mathcal{G} \Rightarrow \text{Proj}\ (\mathcal{G}) \Leftrightarrow \text{Proj}\ (\mathcal{R}).$$

None of the reverse implications hold without restrictions. The simplest situation to consider is that of a local ring $(R, \mathfrak{m})$ with infinite residue field. Let I be a primary ideal and suppose that a is an element of I generating a minimal reduction. It follows that a, at is a system of parameters of the irrelevant maximal ideal of $R[It]$, which is obviously a regular sequence only if $I = (a)$. To find examples, it suffices to consider the irrelevant ideal $I = A_+$ of a 1-dimensional graded ring A, so that $\text{gr}_I(A) = A$.

We discuss the significance of the properties of a graded algebra of the form $A = S/L = R[x_1,\ldots,x_n]/L$. Then A is Cohen-Macaulay and Proj (A) is Cohen-Macaulay; see [Har77, Theorem 7.6] for a geometric discussion. For simplicity we assume that R is a Cohen-Macaulay local ring with a canonical module ω_R; note that $\omega_S = \omega_R \otimes_R S[-n]$ is the canonical module of S. Set $d = \dim A$, $p = \dim S = \dim R + n$. Define the following graded modules:

$$\Omega_i = \operatorname{Ext}_S^{p-d+i}(A, \omega_S), \qquad i = 1, \ldots, d.$$

Of these modules, Ω_d has the simplest interpretation. It is a module of finite length whose associated prime ideals are the maximal homogeneous primes that are associated to A. In particular Ω_d is a finitely generated R-module.

Thus, for instance, if R is an Artinian ring, the Ω_i have a natural interpretation using Serre's duality theorem: If $E(k)$ is the injective envelope of $R/\mathfrak{m}$ as an R-module, and $X = \text{Proj }(A)$, then

$$\Omega_i \cong \bigoplus_{m \in \mathbb{Z}} \operatorname{Hom}_R(H^{d-1-i}(X, \omega_X[-m]), E(k)).$$

Proposition 3.22. *The following assertions hold:*

(a) *A is Cohen-Macaulay if and only if $\Omega_i = 0$ for $i = 1, \ldots, d$.*
(b) Proj (A) *is Cohen-Macaulay if and only if Ω_i is a finitely generated R-module for $i = 1, \ldots, d-1, d$.*

Proof. Part (a) is the usual way the Gorenstein ring S recognizes its Cohen-Macaulay modules, while part (b) is the application of part (a) to the localizations S_f, where the f are 1-forms of S. □

Cohen-Macaulayness of Veronese Subrings

Let $A = \bigoplus_{n \in \mathbb{Z}} A_n$ be a graded algebra and s a fixed integer. The subring $A^{(s)} = \bigoplus_{n \in \mathbb{Z}} A_{sn}$ is the sth Veronese subring of A. To a given graded A-module M there is a similar definition of the $A^{(s)}$-module $M^{(s)}$. Their constructions are functorial and give rise to natural isomorphisms from Proj (A) to Proj $(A^{(s)})$.

One of its cohomological properties is the following (see [HIO88, 47.5]):

Proposition 3.23. *Let A be a Noetherian graded ring, I a homogeneous ideal and M a graded A-module. Then for all $i \geq 0$, $H^i_{I^{(s)}}(M^{(s)}) \cong H^i_I(M)^{(s)}$ as graded $A^{(s)}$-modules.*

Corollary 3.24. *Let R be a Cohen-Macaulay local ring with a canonical module ω_R, let $A = S/L$ be a graded algebra, define the modules Ω_i as above and let s be a positive integer. Suppose that* Proj (A) *is Cohen-Macaulay. Then for all $s \gg 0$, $A^{(s)}$ is Cohen-Macaulay if and only if $[\Omega_i]_0 = 0$ for $i = 1, \ldots, d$.*

Proof. We may assume that R is a complete local ring. Denote by $\mathfrak{M}$ the maximal homogeneous ideal of A. By hypothesis and Proposition 3.22, the modules Ω_i, for $i \geq 1$, have components in finitely many degrees only. If the integer s avoids all such degrees then by local duality and Proposition 3.23,

$$H^i_{\mathfrak{M}^{(s)}}(A^{(s)})_n = 0, \quad \forall n, i < d,$$

as desired. □

The Depth of $\mathcal{G}$ versus the Depth of $\mathcal{R}$

In a situation where $\mathrm{gr}_I(R)$ is not Cohen-Macaulay, the comparison between the depth properties of $R[It]$ and of $\mathrm{gr}_I(R)$ are straightforward according to the result of [HM94].

Theorem 3.25. *Let R be a Noetherian local ring with* depth $R \geq d$ *and I an ideal. If* depth $\mathrm{gr}_I(R) < d$*, then* depth $R[It] =$ depth $\mathrm{gr}_I(R) + 1$.

Proof. We compute all depths with respect to the homogeneous maximal ideal $P = (\mathfrak{m}, ItR[It])$ of $R[It]$. For simplicity put $Q = ItR[It]$ and $Q[+1] = IR[It]$.

We are going to determine the depth of $R[It]$ by examining the exact sequences of local cohomology modules derived from the sequences (3.4) and (3.5). For simplicity of notation we denote $H^j_P(\cdot)$ by $H^j(\cdot)$. Since depth $R = d$ and depth $\mathcal{G} = r < d$, we have

$$\begin{aligned} H^j(R) &= 0, \ j < d \\ H^j(\mathcal{G}) &= 0, \ j < r. \end{aligned}$$

The portions of the cohomology sequences that we are interested in are:

$$\begin{aligned} &H^{j-1}(R) \longrightarrow H^j(Q) \longrightarrow H^j(\mathcal{R}) \longrightarrow H^j(R) \\ &H^{j-1}(\mathcal{G}) \longrightarrow H^j(Q[+1]) \longrightarrow H^j(\mathcal{R}) \longrightarrow H^j(\mathcal{G}). \end{aligned}$$

Taking into these two sequences the conditions above yield the exact sequences

$$H^j(Q) \cong H^j(\mathcal{R}), \ j < d \tag{3.7}$$

$$0 \longrightarrow H^j(Q[+1]) \longrightarrow H^j(\mathcal{R}) \longrightarrow H^j(\mathcal{G}), \ j \leq r, \tag{3.8}$$

from which we claim that $H^j(\mathcal{R}) = 0$ for $j \leq r$. This will suffice to prove the assertion. Denote the modules $H^j(Q)$ and $H^j(\mathcal{R})$ respectively by M_j and N_j. The sequences (3.7) and (3.8) then give rise to graded isomorphisms (of degree zero),

$$\begin{aligned} M_j[+1] &\cong N_j, \ j < r \\ M_j &\cong N_j, \ j < d \end{aligned}$$

and to the monomorphism

$$0 \longrightarrow M_r[+1] \xrightarrow{\varphi} N_r.$$

Since $N_r \cong M_r[+1] \cong M_r$ as (ungraded) modules, φ is a monomorphism of isomorphic Artinian modules and therefore must be an isomorphism. This means that we have isomorphisms of Artinian graded modules,

$$M_j \cong N_j \cong M_j[+1], \ j \leq r,$$

with mappings of degree zero. As the graded components of these modules are zero in all sufficiently high degrees, the modules must vanish. □

3.2.1 Castelnuovo-Mumford Regularity and a-invariants

We introduce a major control of the cohomology of a graded module as defined by Goto and Watanabe ([GW78]). We follow the exposition of [BH93] and [Tr98].

Let $R = \bigoplus_{n\geq 0} R_n = R_0[R_1]$ be a finitely generated graded algebra over the Noetherian ring R_0. For any graded R-module F, define

$$\alpha(F) = \begin{cases} \sup\{n \mid F_n \neq 0\} & \text{if } F \neq 0, \\ -\infty & \text{if } F = 0. \end{cases}$$

Let $R = \bigoplus_{n\geq 0} R_n$ be a standard graded ring of Krull dimension d with irrelevant maximal ideal $M = (\mathfrak{m}, R_+)$ (note that $(R_0, \mathfrak{m})$ is a local ring), and let E be a finitely generated graded R-module. Some local cohomology modules $H^i_J(E)$ of E often give rise to graded modules with $\alpha(H^i_J(E)) < \infty$. This occurs, for instance, when $J = R_+$ or $J = M$. This fact gives rise to several numerical measures of the cohomology of E.

Definition 3.26. For any finitely generated graded R-module F, and for each integer $i \geq 0$, the integer

$$a_i(F) = \alpha(H^i_M(F))$$

is the ith *a-invariant* of F.

We shall also make use of the following complementary notion.

Definition 3.27. For any finitely generated graded R-module F, and for each integer $i \geq 0$, the integer

$$\underline{a}_i(F) = \alpha(H^i_{R_+}(F))$$

is the ith *$\underline{a}$-invariant* of F.

By abuse of terminology, if F has Krull dimension d, we shall refer to $a_d(F)$ as simply the a-invariant of F. In the case $F = R$, if ω_R is the canonical module of R, by local duality it follows that

$$a(R) = -\inf\{\, i \mid (\omega_R/M\cdot\omega_R)_i \neq 0 \,\}. \tag{3.9}$$

The $\underline{a}_i$-invariants are usually assembled into the *Castelnuovo-Mumford regularity* of F

$$\operatorname{reg}(F) = \sup\{\underline{a}_i(F) + i \mid i \geq 0\}.$$

Example 3.28. For several general examples, we direct the reader to [BH93, Section 3.6]. If R is Cohen-Macaulay and I is an ideal of positive height such that $\mathcal{R}(I)$ is Cohen-Macaulay, it is easy to see that $a(\mathcal{R}(I)) = -1$; ultimately this is a consequence of the fact that for a field k the canonical ideal of $k[t]$ is $tk[t]$. More detailed comments will be found when we treat canonical modules in the next chapter.

The a- and $\underline{a}$-invariants are closely related in several cases of interest. Suppose that $(R,\mathfrak{m})$ is a Cohen-Macaulay local ring of dimension $d > 0$, and I an ideal of dimension one that is generically a complete intersection. Set $\mathcal{G} = \mathrm{gr}_I(R)$, choose $x \in \mathfrak{m} \setminus I$ so that (I,x) is $\mathfrak{m}$-primary (in particular I_x is a complete intersection by hypothesis). We let $\mathcal{M}$ be the maximal homogeneous ideal of $\mathcal{G}$ and set $\mathcal{G}' = \mathcal{G}_x$. Note that $\mathcal{G}'$ is a ring of polynomials so all of its invariants are known. Let us apply Proposition 3.15 where $J = \mathcal{G}_+$. We have exact sequences

$$H^i_{\mathcal{G}_+}(\mathcal{G}) \cong H^i_{\mathcal{M}}(\mathcal{G}), \quad i < d-1$$

$$0 \to H^{d-1}_{\mathcal{M}}(\mathcal{G}) \longrightarrow H^{d-1}_{\mathcal{G}_+}(\mathcal{G}) \longrightarrow H^{d-1}_{\mathcal{G}'_+}(\mathcal{G}') \longrightarrow H^d_{\mathcal{M}}(\mathcal{G}) \longrightarrow H^d_{\mathcal{G}_+}(\mathcal{G}) \to 0.$$

Therefore

$$\begin{aligned} a_i(\mathcal{G}) &= \underline{a}_i(\mathcal{G}), \quad i < d-1, \\ a_{d-1}(\mathcal{G}) &\leq \underline{a}_{d-1}(\mathcal{G}) \leq \max\{a_{d-1}(\mathcal{G}), d-1,\} \\ \max\{d-1, \underline{a}_d(\mathcal{G})\} \geq a_d(\mathcal{G}) &\geq \underline{a}_d(\mathcal{G}). \end{aligned}$$

The Criterion of Ikeda and Trung

The relationship between the Cohen-Macaulayness of $R[It]$ and $\mathrm{gr}_I(R)$ was shown by Ikeda and Trung ([IT89]) to depend on the degrees of the minimal generators of the canonical module of $\mathrm{gr}_I(R)$. It provides for a very broad setting in which to look at $\mathcal{R}$ vis-a-vis $\mathcal{G}$.

If $\mathcal{G} = \mathrm{gr}_I(R)$ is Cohen-Macaulay, it is not always the case that $\mathcal{R}$ will be Cohen-Macaulay. The next result by Ikeda and Trung [IT89] is central to our understanding of the relationship between the depth properties of $\mathcal{R}$ and $\mathcal{G}$ in this critical case. Its proof is a model of clarity.

Theorem 3.29 (Ikeda-Trung). *Let $(R,\mathfrak{m})$ be a Noetherian local ring of dimension $d > 0$, I an ideal of positive height, and set $\mathcal{R} = R[It]$. The following equivalence holds:*

$$\mathcal{R} \text{ is Cohen-Macaulay} \iff \begin{cases} [H^i_{\mathcal{M}}(\mathcal{G})]_n = 0 \text{ for } i < d, \quad n \neq 1, \\ a(\mathcal{G}) < 0. \end{cases} \tag{3.10}$$

Proof. Suppose that $\mathcal{G}$ is Cohen-Macaulay and $a(\mathcal{G}) < 0$. For $j = d$, the exact sequences (3.7) and (3.7) become

$$\begin{aligned} &0 \longrightarrow H^d(Q) \longrightarrow H^d(\mathcal{R}) \longrightarrow H^d(R) \\ &0 \longrightarrow H^d(Q[+1]) \longrightarrow H^d(\mathcal{R}) \longrightarrow H^d(\mathcal{G}). \end{aligned}$$

Because these graded modules vanish in high degrees, taken together the sequences imply that $H^d(\mathcal{Q})_i = H^d(\mathcal{R})_i = 0$ for $i > 0$. However, since $a(\mathcal{G}) < 0$, it follows that $H^d(\mathcal{R})_0 = 0$ as well. We are then in a position to argue as in Theorem 3.25.

Conversely, if $\mathcal{R}$ is Cohen-Macaulay then $\mathcal{G}$ is Cohen-Macaulay by Theorem 3.12. To show that $a(\mathcal{G}) < 0$, consider the exact sequence

$$0 \to H^d(\mathcal{G}) \longrightarrow H^{d+1}(I \cdot \mathcal{R}) \longrightarrow H^{d+1}(\mathcal{R}) \to 0,$$

and observe that $I \cdot \mathcal{R}$ is Cohen-Macaulay and one a-invariant that must be negative (we come back more fully to this point in Theorem 4.40). □

We shall have several opportunities to apply this theorem when R is Cohen-Macaulay:

Theorem 3.30. *Let $(R, \mathfrak{m})$ be a Cohen-Macaulay local ring of dimension d, I an ideal of positive height, and set $\mathcal{R} = R[It]$. The following equivalence holds:*

$$\mathcal{R} \text{ is Cohen-Macaulay} \iff \begin{cases} \mathcal{G} \text{ is Cohen-Macaulay} \\ a(\mathcal{G}) < 0. \end{cases} \tag{3.11}$$

Remark 3.31. In continuation to Remark 3.13, we add that the same assertions of Theorem 3.30 will apply in the case of the Rees algebra of a decreasing filtration, such as that of the integral closure of the I-adic filtration.

Theorem 3.32. *Let $(R, \mathfrak{m})$ be a Cohen-Macaulay local ring of dimension d, let $\{I_n, n \geq 1\}$ be a decreasing multiplicative filtration of ideals of positive height such that the Rees algebra $B = \sum_{n \geq 0} I_n t^n$ is integral over $A = \mathcal{R}(I) = R[It]$. The following equivalence holds:*

$$B \text{ is Cohen-Macaulay} \iff \begin{cases} \mathcal{G} = \sum_{n \geq} I_n / I_{n+1} \text{ is Cohen-Macaulay} \\ a(\mathcal{G}) < 0. \end{cases} \tag{3.12}$$

Proof. Let us walk quickly over the proof of the forward direction of the assertion. The exact sequences

$$0 \to B_+ \longrightarrow B \longrightarrow R \to 0$$
$$0 \to B_+[-1] \longrightarrow B \longrightarrow \mathcal{G} \to 0$$

show that if B is Cohen-Macaulay then $\mathcal{G}$ is also Cohen-Macaulay.

To exploit them further, let J be a minimal reduction of I. Then B and $\mathcal{G}$ are finitely generated modules over the Rees algebra $C = \mathcal{R}(J)$ of J. If we denote by M the maximal homogeneous ideal of C, then the local cohomology of B, $\mathcal{G}$ and R relative to M will provide us with the required numerical data controlling $a(\mathcal{G})$. Since R may be taken to be a complete local ring, we can assume that there is a (graded) Gorenstein ring D of dimension $d+1$ equipped with a finite R-morphism $D \to B$. Note that $D[-1]$ is a canonical module for D. Applying $\mathrm{Hom}_D(\cdot, D[-1])$ to the sequences of Cohen-Macaulay D-modules above, we obtain the exact sequence

$$0 \to \mathrm{Hom}_D(B_+[-1], D[-1]) \longrightarrow \mathrm{Hom}_D(B, D[-1]) \longrightarrow \mathrm{Ext}^1_D(\mathcal{G}, D[-1]) \to 0.$$

This sequence also gives that $\mathrm{Ext}^1_D(\mathcal{G}, D[-1])$, the canonical module of $\mathcal{G}$, is generated in degrees > 0. Since $\mathcal{G}$ is Cohen-Macaulay, this means that $a(\mathcal{G}) < 0$. □

Cohen-Macaulay Rees Algebras

This section would not be complete without mentioning one the highlights of theory, the beautiful theorem of Kawasaki ([Kaw2]) on *Arithmetic Cohen-Macaulayfication*:

Theorem 3.33 (Kawasaki Theorem). *Let R be a Noetherian local ring of positive dimension. Then the following statements are equivalent:*

(i) *R has an arithmetic Cohen-Macaulayfication;*
(ii) *R is unmixed and all formal fibers of R are Cohen-Macaulay.*

3.2.2 Vanishing of Cohomology

In this section we offer some calculations of the cohomology of Proj $(R[It])$ with the aim of detecting the vanishing of $H^i_{\mathcal{M}}(\mathcal{R})$ for $i \leq \dim R$. These groups play a significant role in predicting the geometric properties of Proj $(\mathcal{R})$, but are very hard to determine explicitly in greater generality.

Ideals of Linear Type

We first treat an interesting general property of ideals of linear type discovered by Huckaba and Marley ([HH99]). Here is their approach to the calculation of $a_d(\mathcal{R})$, with some variations.

Proposition 3.34. *Let* $(R, \mathfrak{m})$ *be a Cohen-Macaulay local ring of dimension* $d > 0$, *let I be an ideal positive height, and set* $\mathcal{R} = R[It]$. *Then* $a_d(\mathcal{R}) < 0$ *in the following cases:*

(i) *R is Gorenstein and I is of linear type;*
(ii) *R is an integral domain with a canonical module and I is of linear type;*
(iii) *R is a Gorenstein integral domain and I is an ideal such that* $\nu(I_{\mathfrak{p}}) \leq \text{height } \mathfrak{p} + 1$ *for each prime ideal* $\mathfrak{p}$, *and the canonical module of* $\mathcal{R}$ *satisfies Serre's condition* S_3.

Proof. (i) Let $(a_1, \ldots, a_n)$ be a set of generators of I, and $\mathcal{R} = R[T_1, \ldots, T_n]/J$ a presentation. By hypothesis J is generated by forms of degree 1. Since we may assume that the residue field of R is infinite and J is an ideal of height $n-1$, there are $n-1$ forms $f_1, \ldots, f_{n-1}$ of J of degree 1 generating a regular sequence. Set $A = S/F$, where $S = R[T_1, \ldots, T_n]$, $F = (f_1, \ldots, f_{n-1})$, and consider the presentation $\mathcal{R} = A/L$, where $L = J/F$. Note that since $\omega_S = S[-n]$ is the canonical module of S, $\omega_A = S/F[-1] = A[-1]$.

Let M be the maximal homogeneous ideal of $\mathcal{R}$. We show that $H^d_M(\mathcal{R})_n = 0$ for $n \geq 0$. By local duality (see [BH93, Section 3.6]), if E is the graded injective envelope of $R/\mathfrak{m}$ and if for a graded A-module N we set $N^\vee = \text{Hom}_A(N, E)$, we have $(H^d_M(\mathcal{R}))^\vee = \text{Ext}^1_A(\mathcal{R}, A[-1])$. Thus the isomorphisms

$$(H^d_M(\mathcal{R})_n)^\vee = (H^d_M(\mathcal{R}))^\vee_{-n} = \text{Ext}^1_A(\mathcal{R}, A[-1])_n$$

mean that it suffices to show that $\mathrm{Ext}^1_A(\mathcal{R},A[-1])_n = 0$ for $n \leq 0$.

Observe that from the exact sequence

$$0 \to L \longrightarrow A \longrightarrow \mathcal{R} \to 0$$

and the fact that $\mathrm{Hom}_A(A,A[-1])_n = 0$ for $n < 0$, we have

$$\mathrm{Hom}_A(L,A[-1])_n \cong \mathrm{Ext}^1_A(\mathcal{R},A[-1])_n \quad \text{for } n \leq 0.$$

Suppose that $f \in \mathrm{Hom}_S(L,A[-1])_n$ and let L_1 be the set of homogeneous elements of L of degree 1. Then $f(L_1) \subset A[-1]_{n+1} = A_n$. If $n < 0$ then $A_n = 0$ and $f(L_1) = 0$. Since L is generated by L_1, in this case $f = 0$. Therefore it suffices to establish the case $n = 0$.

At this point we remark that if R is assumed to be just Cohen-Macaulay and ω_R is its canonical module (which we may assume after completing R), then the argument above already shows that $a_d(\mathcal{R}) \leq 0$. It is to ascertain the case $n = 0$ that we must consider additional restrictions on R.

In case (i), we claim that $L = 0 : (0 : L)$. Both ideals are unmixed of height zero, so it suffices to check equality at the localizations of A at prime ideals $\wp$ of height 0. Since $A_\wp$ is a Gorenstein Artinian ring, the equality holds for any of its ideals. Now observe that $(0 : L)f(L) = f((0 : L)L) = f(0) = 0$, and therefore $f(L) \subset 0 : (0 : L) = L$. In particular $f(L_1) \subset A_0 \cap L = 0$, as desired.

(ii) This case is very similar. First assume that R is complete. Let $K \subset R$ be a canonical ideal of R. Then the canonical module of A is $\omega_A = (KS/KFS)[-1]$. Notice that $(\omega_A)_n = 0$ for $n \leq 0$, and that $(\omega_A)_1 = K$.

Note that L is a prime ideal, and, since A is a Cohen-Macaulay ring, we must have $L = 0 : (0 : L)$, as the right hand side consists of zero divisors and therefore cannot properly contain the minimal prime L of the Cohen-Macaulay ring A. Now when we consider the module $\mathrm{Hom}_A(L, \omega_A)$, as above $f(L_1)$ must live in degree 1 of ω_A, so it is a subideal H of K such that $(0 : L)H = 0$. In the original ring S this equality means that

$$(F : J)H \subset KFS \subset FS,$$

and therefore $H \subset F : (F : J) = J$. But this is a contradiction, since J contains no elements of degree 0.

If R is not complete, we start with the equality $L = 0 : (0 : L)$ and then complete. Now $\widehat{K} \subset \widehat{R}$ is a canonical ideal and we can proceed as above.

(iii) The condition on the local minimal number of generators of I means that in a presentation $S = R[T_1, \ldots, T_n] \to \mathcal{R}$ the component of degree 1 of the kernel has height $n-1$. The ring A is defined as above, $A = S/(f_1, \ldots, f_{n-1})$ and $A[-1]$ is its canonical module. As in case (i), we must show that the module $\mathrm{Hom}_A(L,A)$ has no elements in negative degrees.

Observe that $0 :_A L \not\subset L$. Indeed, localizing at the nonzero elements of R, the ideal $F = (f_1, \ldots, f_{n-1})$ becomes a regular sequence generated by linear forms over a field, and therefore it is a prime ideal of height $n-1$. This shows that $F_J = J_J$, which implies the assertion. Furthermore, since L is a prime ideal, the ideal $L + 0 :_A L$ has height greater than 0 and thus contains regular elements since A is a Cohen-Macaulay ring; this implies that $L + 0_A : L = L \oplus 0 :_A L$. We further note that the image of $0 :_A L$ in $A/L = \mathcal{R}$ is a canonical ideal of $\mathcal{R}$.

For any homomorphism $f \in \mathrm{Hom}_A(L, A)$ (respectively $f \in \mathrm{Hom}_A(0 :_A L, A)$), the equality $f((0 :_A L)L) = (0 :_A L)f(L) = 0$ shows that $f(L) \subset 0 :_A (0 :_A L) = L$, that is $f \in \mathrm{Hom}_A(L, L)$ (respectively $f \in \mathrm{Hom}_A(0 :_A L, 0 :_A L)$). Consider the exact sequence

$$0 \to L \oplus 0 :_A L \longrightarrow A \longrightarrow A/(L \oplus 0 :_A L) = (A/L)/((L \oplus 0 :_A L)/L) = \mathcal{R}/\omega_{\mathcal{R}} \to 0.$$

Applying $\mathrm{Hom}_A(\cdot, A)$, we have the short exact sequence

$$0 \to \mathrm{Hom}_A(A, A) \to \mathrm{Hom}_A(L, L) \oplus \mathrm{Hom}_A(\omega_{\mathcal{R}}, \omega_{\mathcal{R}}) \to \mathrm{Ext}^1_A(\mathcal{R}/\omega_{\mathcal{R}}, A) \to 0. \quad (3.13)$$

We recall that $\mathrm{Hom}_A(\omega_{\mathcal{R}}, \omega_{\mathcal{R}}) = \widetilde{\mathcal{R}}$, the S_2-ification of $\mathcal{R}$ (see [NV93] for a discussion), and that $\omega_{\mathcal{R}} = \omega_{\widetilde{\mathcal{R}}}$. Moreover, there is a canonical isomorphism

$$\mathrm{Ext}^1_A(\mathcal{R}/\omega_{\mathcal{R}}, A) = \mathrm{Ext}^1_A(\widetilde{\mathcal{R}}/\omega_{\mathcal{R}}, A),$$

which is the canonical module of both rings $\mathcal{R}/\omega_{\mathcal{R}} \hookrightarrow \widetilde{\mathcal{R}}/\omega_{\mathcal{R}}$. We now use an argument that goes back to C. Peskine: that under the condition that $\omega_{\mathcal{R}}$ has S_3, the ring $\widetilde{\mathcal{R}}/\omega_{\mathcal{R}}$ is quasi-Gorenstein. Indeed from the sequences

$$0 \to \omega_{\mathcal{R}} \longrightarrow \widetilde{\mathcal{R}} \longrightarrow \widetilde{\mathcal{R}}/\omega_{\mathcal{R}} \to 0,$$

$$0 \to \mathrm{Hom}_A(\widetilde{\mathcal{R}}, A) \longrightarrow \mathrm{Hom}_A(\omega_{\mathcal{R}}, A) \longrightarrow \mathrm{Ext}^1_A(\widetilde{\mathcal{R}}/\omega_{\mathcal{R}}, A),$$

and the identifications $\mathrm{Hom}_A(\widetilde{\mathcal{R}}, A) = \omega_{\mathcal{R}}$, $\mathrm{Hom}_A(\omega_{\mathcal{R}}, A) = \widetilde{\mathcal{R}}$, we have a natural embedding

$$\widetilde{\mathcal{R}}/\omega_{\mathcal{R}} \hookrightarrow \mathrm{Ext}^1_A(\widetilde{\mathcal{R}}/\omega_{\mathcal{R}}, A).$$

This inclusion is an isomorphism whenever $\widetilde{\mathcal{R}}$ is Cohen-Macaulay. It is thus an isomorphism at each localization at height at most 1 in the support of $\widetilde{\mathcal{R}}$. Since $\omega_{\mathcal{R}}$ has the condition S_3 by hypothesis, the cokernel has positive depth or vanishes. It follows easily that it must be zero.

It follows that the modules at the two ends of the exact sequence (3.13) have no elements in negative degrees, and therefore neither does $\mathrm{Hom}_A(L, L)$. □

We note that this is a calculation at the edge of a very general result on local duality (see Corollary 3.18).

Equimultiple Ideals

Theorem 3.35. *Let* $(R,\mathfrak{m})$ *be a Cohen-Macaulay local ring of dimension* $d > 1$ *and* I *an* $\mathfrak{m}$*-primary ideal. Suppose that* $\operatorname{depth} \operatorname{gr}_I(R) = d-1$ *and* $\operatorname{r}(I) < d$*. Then* $a_d(R[It]) < 0$.

Proof. We make use of the special nature of the Sally module of the ideal I. We may assume that the residue field of R is infinite and let J be a minimal reduction. Consider the Sally module $S_J(I)$:

$$0 \to IR[Jt] \longrightarrow IR[It] \longrightarrow S_J(I) \to 0.$$

By Theorem 3.25, depth $\mathcal{R}(I) = d$, and therefore depth $IR[It] \geq d$ from (3.4). It follows that $S_J(I)$ is a Cohen-Macaulay module (over $R[Jt]$). Taking local cohomology in the sequence above with respect to the homogeneous maximal ideal $\mathcal{N}$ of $R[Jt]$, we have the exact sequence

$$0 \to H^d_{\mathcal{N}}(IR[It]) \longrightarrow H^d_{\mathcal{N}}(S_J(I)) \longrightarrow H^{d+1}_{\mathcal{N}}(IR[Jt]) \longrightarrow H^{d+1}_{\mathcal{N}}(IR[It]) \to 0.$$

On the other hand, we have the exact sequence

$$0 \to IR[Jt] \longrightarrow R[Jt] \longrightarrow B = R/I[T_1,\ldots,T_d] \to 0$$

of Cohen-Macaulay modules. From the cohomology exact sequence

$$0 \to H^d_{\mathcal{N}}(B) \longrightarrow H^{d+1}_{\mathcal{N}}(IR[Jt]) \longrightarrow H^{d+1}_{\mathcal{N}}(R[Jt]) \to 0,$$

we have that $H^{d+1}_{\mathcal{N}}(IR[Jt])_n = 0$ for $n \geq 0$.

Now we appeal to the fact that the Sally module $S_J(I)$, being Cohen-Macaulay, admits a filtration whose factors are the modules

$$I^i/JI^{i-1}[T_1,\ldots,T_d][-i+1], \qquad i \leq r = \operatorname{r}(I).$$

As a consequence, $H^d_{\mathcal{N}}(S_J(I))$ admits a similar filtration of cohomology modules and we obtain that the a-invariant of $S_J(I)$ is at most $r-d-1$, which shows that $a_d(IR[It]) < -1$, as we are assuming that $r < d$. The cohomology sequence of the Sally module also yields the isomorphism

$$H^{d+1}_{\mathcal{N}}(IR[It])_{-1} \cong H^{d+1}_{\mathcal{N}}(IR[Jt])_{-1} \cong H^{d+1}_{\mathcal{N}}(R[Jt])_{-1} \cong H^d_{\mathfrak{m}}(R).$$

Finally, we consider the cohomology of the sequence

$$0 \to tIR[It] = IR[It][-1] \longrightarrow R[It] \longrightarrow R \to 0.$$

We obtain in degree 0 the exact sequence

$$0 = H^d_{\mathcal{N}}(IR[It])_{-1} \longrightarrow H^d_{\mathcal{N}}(R[It])_0 \longrightarrow H^d_{\mathfrak{m}}(R) \longrightarrow H^{d+1}_{\mathcal{N}}(IR[It])_1 \to 0,$$

from which we get $H^d_{\mathcal{N}}(R[It])_0 = 0$, since as observed above $H^{d+1}_{\mathcal{N}}(IR[It])_{-1} = H^d_{\mathfrak{m}}(R)$. □

Corollary 3.36. *Suppose further that* Proj $(\mathcal{R}(I))$ *is Cohen-Macaulay. Then some Veronese subring of* $\mathcal{R}(I)$ *is Cohen-Macaulay.*

Links of Primes

In this extended example we calculate the cohomology of the Rees algebra of a direct link of a prime ideal. We shall use Theorem 1.104 and other constructions from [CP95].

Let $(R, \mathfrak{m})$ be a Cohen-Macaulay local ring of dimension d and $\mathfrak{p}$ a prime ideal of height $g \geq 2$. Choose a complete intersection ideal $J \subset \mathfrak{p}$ of height g and set $I = J : \mathfrak{p}$. We assume that the conditions of Theorem 1.104 are in place, so that $I^2 = JI$. In particular this will occur if $R_{\mathfrak{p}}$ is not a regular local ring. Finally set $\mathcal{R} = R[It]$.

If R is a Gorenstein ring and $\mathfrak{p}$ is a Cohen-Macaulay ideal, then I is a Cohen-Macaulay ideal and the algebra $\mathcal{R}$ is Cohen-Macaulay. To make the following calculation more interesting we shall not assume that $\mathfrak{p}$ is necessarily Cohen-Macaulay.

Theorem 3.37. *For all integers $i, n \geq 0$, $H^i_{\mathfrak{M}}(\mathcal{R})_n = 0$.*

Proof. Set $\mathcal{R}_0 = R[Jt]$ and observe that $I\mathcal{R}_0 = I\mathcal{R}$. Consider the exact sequences

$$0 \to I\mathcal{R}_0 \longrightarrow \mathcal{R}_0 \longrightarrow G' = R/I[T_1, \ldots, T_g] \to 0$$

and

$$0 \to I\mathcal{R}_0[-1] \longrightarrow \mathcal{R} \longrightarrow R \to 0.$$

We may take local cohomology with respect to the maximal homogeneous ideal of $\mathcal{R}_0$, which we still denote by $\mathfrak{M} = (\mathfrak{m}, Jt)$. Since $\mathcal{R}_0$ and R are Cohen-Macaulay, for $i \leq d$ we have

$$\begin{aligned} H^i_{\mathfrak{M}}(I\mathcal{R}_0) \cong H^{i-1}_{\mathfrak{M}}(\mathcal{G}') &\cong H^{i-1-g}_{\mathfrak{m}}(R/I) \otimes H^g_{(T)}(R/I[T_1, \ldots, T_g]) \\ &\cong H^{i-1-g}_{\mathfrak{m}}(R/I) \otimes R/I[T_1^{-1}, \ldots, T_g^{-1}][g]. \end{aligned}$$

Since $g \geq 2$, it follows from the isomorphism $H^i_{\mathfrak{M}}(I\mathcal{R}_0[-1]) \cong H^i_{\mathfrak{M}}(\mathcal{R})$ that $H^i_{\mathfrak{M}}(\mathcal{R})_n = 0$ for $n \geq 0$ and $i \leq d-1$.

Another simple inspection shows that $H^{d+1}_{\mathfrak{M}}(I\mathcal{R}[-1])_0 = H^d_{\mathfrak{m}}(R)$, from which the remaining assertions follow. □

3.3 Reduction Number and Cohen-Macaulayness

Let $A = A_0 + A_1 + \cdots$ be a Noetherian ring and F a graded A-module. For a given ideal $\mathfrak{p} = \mathfrak{p}_0 + A_+$, $\mathfrak{p}_0 \subset A_0$ of A we study briefly the relation between the local cohomology of F with regard to $\mathfrak{p}$ and to A_+. From the point of view of Rees algebras, this is justified because the vanishing of one cohomology reflects depth and the other reflects reduction number.

Local Reduction Numbers

We now give a characterization of the Cohen-Macaulayness of a Rees algebra $R[It]$ that involves the reduction numbers of the localizations $I_{\mathfrak{p}}$. In fact, only finitely many come into play. We discuss the theorem of [JK95] (see also [AHT95], [SUV95]).

Local Analytic Spread of an Ideal

Let I be an ideal of the Noetherian ring R. The function

$$\mathfrak{p} \in \text{Spec}\,(R) \mapsto \ell(I_{\mathfrak{p}})$$

gives information about the associated primes of I^n and also about other ideals related to these powers. This will be explored in more detail in Proposition 4.11.

Proposition 3.38. *Suppose that R is universally catenary and I an ideal of positive height. For any prime ideal $I \subset \mathfrak{p}$ we have $\ell(I_{\mathfrak{p}}) \leq \text{height}\,\mathfrak{p}$. The set $\mathcal{P}$ of prime ideals $\mathfrak{p}$ such that $\ell(I_{\mathfrak{p}}) = \text{height}\,\mathfrak{p}$ is finite. More precisely,*

$$\mathcal{P} = \{\mathfrak{p} = P \cap R \mid P \in \text{Min}(IR[It])\}.$$

Proof. When the equality $\ell(I_{\mathfrak{p}}) = \text{height}\,\mathfrak{p}$ holds we have $\dim R_{\mathfrak{p}}[I_{\mathfrak{p}}t] = \dim R_{\mathfrak{p}} + 1$, so that height $\mathfrak{p}R[It] = 1$, and conversely. This means that $\mathfrak{p}R[It]$ is contained in some minimal prime of $IR[It]$. □

Theorem 3.39. *Let R be a Cohen-Macaulay local ring and I an ideal of positive height. The following equivalence holds:*

$$\mathcal{R} \text{ is Cohen-Macaulay} \iff \begin{cases} \mathcal{G} \text{ is Cohen-Macaulay and} \\ \text{for each prime ideal } \mathfrak{p} \text{ such that} \\ \ell(I_{\mathfrak{p}}) = \text{height}\,\mathfrak{p},\ r(I_{\mathfrak{p}}) < \text{height}\,\mathfrak{p}. \end{cases} \tag{3.14}$$

The point of the argument is to see how the assertions impact on the invariant $a(\mathcal{G})$. To this end, we are going to relate $a(\mathcal{G})$ to the cohomology of Proj $\mathcal{G}$. We first consider a broad technique from [JK95] on how to approach this.

Let S be a positively graded Noetherian ring of dimension d with $S_0 = R$ local, M its irrelevant maximal ideal, and N any homogeneous ideal containing S_+. We denote by $\mathfrak{m}$ the maximal ideal of R. For $\mathfrak{p} \in \text{Spec}\,(R)$, we set $S_{\mathfrak{p}} = S \otimes_R R_{\mathfrak{p}}$ and $d(\mathfrak{p}) = \dim S_{\mathfrak{p}}$, and let P be the irrelevant maximal ideal of $S_{\mathfrak{p}}$.

Proposition 3.40. *Let S be a positively graded Noetherian ring of dimension d with $S_0 = R$ local, C a finitely generated graded S-module, and r a positive integer.*

(a) *If $H^i_P(C_{\mathfrak{p}})$ is concentrated in degrees $\leq r$ for every $\mathfrak{p} \in \text{Spec}\,(R)$, then $H_N{}^i(C)$ is concentrated in degrees $\leq r$ for every homogeneous ideal N containing S_+.*
(b) *If $H_{S_{\mathfrak{p}_+}}{}^{d(\mathfrak{p})}(C_{\mathfrak{p}})$ is concentrated in degrees $\leq r$ for every $\mathfrak{p} \in \text{Spec}\,(R)$, then $H^d_N(C)$ is concentrated in degrees $\leq r$ for every homogeneous ideal N containing S_+.*

Proof. (a) We induct on $\dim S/N$, the assertion being obvious if $\dim S/I = 0$. So suppose that $\dim S/N > 0$ and choose $x \in \mathfrak{m}$ such that $\dim S/(N,x) < \dim S/N$. Then by the induction hypothesis, $H^i_{(N,x)}(C)$ is concentrated in degrees $\leq r$. Furthermore,

$H^i_{N_x}(C_x)$ is concentrated in degrees $\leq r$, by an induction hypothesis, since local cohomology commutes with localization. Now a graded version of Proposition 3.15 implies that $H^i_N(C)$ is concentrated in degrees $\leq r$ as well.

(b) Write $I = N_0$ and induct on $\nu(I)$. If $\nu(I) = 0$, then $H^d_N(C) = H^d_{S_{\mathfrak{m}+}}(C_{\mathfrak{m}})$, and we are done. If $\nu(I) > 0$, write $I = (I', x)$ with $\nu(I') < \nu(I)$, and set $N' = (I', S_+)$. Now using that local cohomology commutes with localization and that $\dim S_x \leq d-1$, we deduce from the induction hypothesis that $H^{d-1}_{N'_x}(C_x)$, as well as $H^d_{N'}(C)$, are concentrated in degrees $\leq r$. Thus again by Proposition 3.15, $H^d_{(N',x)}(C) = H^d_N(C)$ is concentrated in degrees $\leq r$. □

Proof of Theorem 3.39. Denote by N the irrelevant ideal of $\mathcal{G}$, $N = \mathcal{G}_+$, and let $M = (\mathfrak{m}, N)$ be the maximal irrelevant ideal of $\mathcal{G}$. We argue by induction on $\dim R = d$ and on $\dim R/I$ that $H^d_M(\mathcal{G})$ vanishes in non-negative degrees if and only if $H^d_N(\mathcal{G})$ does so. If $\dim R/I = 0$, then I is an $\mathfrak{m}$-primary ideal and $\sqrt{N} = M$, and the functors Γ_M and Γ_N are equivalent. Suppose then that $\dim R/I > 0$, and choose $x \in \mathfrak{m}$ such that height $(N,x) =$ height $N+1$.

By Proposition 3.15, we have the exact sequence

$$H^{d-1}_{N_x}(\mathcal{G}) \longrightarrow H^d_{(N,x)}(\mathcal{G}) \longrightarrow H^d_N(\mathcal{G}) \longrightarrow H^d_{N_x}(\mathcal{G}) \to 0,$$

where $H^d_{N_x}(\mathcal{G}) = 0$ since $\dim R_x < \dim R = d$, and $H^{d-1}_{N_x}(\mathcal{G})$ is zero in non-negative degrees by induction. If (N,x) is not M-primary, we replace N by (N,x) and find an element $y \in \mathfrak{m}$ for which height $(N,x,y) =$ height $N+2$.

If $J = (a_1, \ldots, a_d)$ is a reduction of I, then $\sqrt{(a_1^*, \ldots, a_d^*)} = \sqrt{N = \mathcal{G}_+}$, so that in computing the local cohomology modules $H^i_N(\mathcal{G})$ we may use the ideal $(a_1^*, \ldots, a_d^*)$. This means that may assume that $\ell(I) = d$ as in the assertion of the theorem.

We are now ready for the proof of the theorem. Again we may assume that the assertions are valid on the punctured spectrum of R. Suppose that $\mathcal{R}$ is Cohen-Macaulay but $\mathrm{r}(I) \geq d$. Set $J = (a_1, \ldots, a_d)$, and denote by $\mathbb{K}$ the Koszul complex defined by the 1-forms $a_1^*, \ldots, a_d^*$ of $\mathcal{G}$:

$$0 \to K_d = G[-d] \longrightarrow K_{d-1} = G[-d+1]^d \longrightarrow \cdots \longrightarrow K_1 \longrightarrow K_0 \to 0.$$

Under these conditions a calculation of Trung ([Tr94, Proposition 3.2]) establishes:

Proposition 3.41. *Let $(R, \mathfrak{m})$ be a local ring of dimension d and I an ideal. Suppose that J is a reduction of I generated by d elements. If $r_J(I) = r$, then*

$$a_d(\mathcal{G}_+, \mathcal{G}) + d \leq r \leq \max\{a_i(\mathcal{G}_+, \mathcal{G}) + i \mid i = 0 \ldots d\}.$$

To continue with our proof, this proposition implies that $a(\mathcal{G}) < 0$ if $r < d$. The converse makes use of the transfer of the vanishing of cohomology sketched above (see Proposition 3.40). □

Example 3.42. Let $R = k[x_1,\ldots,x_d]/(f)$ be a hypersurface ring, where f is a form of degree $\geq d \geq 3$. The maximal ideal $\mathfrak{m}$ of R has reduction number $\deg f - 1$. Suppose that R has isolated singularities, in particular R is a normal domain. Since $\mathrm{gr}_{\mathfrak{m}}(R) \cong R$, $\mathrm{gr}_{\mathfrak{m}}(R)$ is Cohen-Macaulay; but, since $\mathrm{r}(\mathfrak{m}) = \deg f - 1 > d - 2 = \dim R - 1$, the Rees algebra $\mathcal{R}(\mathfrak{m})$ is normal but not Cohen-Macaulay, by Theorem 3.39.

Rules of Computation

We derive now, directly from the construction of the Čech complex, several rules that are useful in dealing with the cohomology of Rees algebras. We also derive Proposition 3.41 in a manner to suit some applications later.

For convenience, we introduce a minor notational change to the setting above. We assume that $\{f_1,\ldots,f_s\}$ is a set of forms in A of degree 1 with $A_+ \subset \sqrt{(f_1,\ldots,f_s)}$.

Proposition 3.43. *Let N be a finitely generated graded A-module. Then*

(a) $H^i_{A_+}(N) = 0$ *for* $i > s$.
(b) *If* $N_j = 0$ *for* $j \gg 0$, *then* $H^0_{A_+}(N) = N$; *and* $H^i_{A_+}(N) = 0$ *for* $i \geq 1$.

Proof. The proofs of both parts are readouts of the Čech complex of N,

$$0 \to N \to \bigoplus_i N_{f_i} \to \bigoplus_{i<j} N_{f_i f_j} \to \cdots \to N_{f_1\cdots f_s} \to 0;$$

this is clear in case (a), and in case (b) since all localizations of N vanish. □

The next rule is useful in induction arguments.

Proposition 3.44. *Suppose that the module $L = 0 :_N f_1$ vanishes in all high degrees. Then there is an exact sequence*

$$\begin{aligned} 0 \to L \to H^0_{A_+}(N)[-1] \xrightarrow{f} H^0_{A_+}(N) \to H^0_{A_+}(N/f_1N) \to H^1_{A_+}(N)[-1] \to H^1_{A_+}(N) \\ \to H^1_{A_+}(N/f_1N) \to \cdots \to H^{s-1}_{A_+}(N/f_1N) \to H^s_{A_+}(N)[-1] \to H^s_{A_+}(N) \to 0. \end{aligned}$$

Proof. Set $N' = N/L$ and consider the cohomology sequences of the two natural short exact sequences

$$0 \to L \longrightarrow N \longrightarrow N' \to 0$$

and

$$0 \to N' \longrightarrow N \longrightarrow N/f_1N \to 0.$$

From the rules above, the first of these givers the exact sequence

$$0 \to L \longrightarrow H^0_{A_+}(N) \longrightarrow H^0_{A_+}(N') \to 0$$

and the isomorphisms $H^i_{A_+}(N) \cong H^i_{A_+}(N')$ for $i \geq 1$. The other sequence gives

$$0 \to H^0_{A_+}(N')[-1] \to H^0_{A_+}(N) \to H^0_{A_+}(N/f_1N) \to H^1_{A_+}(N')[-1] \to H^1_{A_+}(N)$$
$$\to H^1_{A_+}(N/f_1N) \to \cdots \to H^s_{A_+}(N) \to H^s_{A_+}(N/f_1N) \to 0.$$

Assembled, they give the desired long exact sequence. □

This exact sequence is useful in comparing the $\underline{a}$-invariants of N and N/fN. Several such examinations were carried out by N. V. Trung ([Tr87], [Tr94], [Tr98], [Tr99]). We shall recall some of them, particularly those that relate to the reduction number of an associated graded ring. First however we state two direct consequences. For simplicity of notation, we set $\underline{a}_i(N) = \underline{a}_i(A_+, N)$.

Proposition 3.45. *If N and f are as above, then $\alpha(L) = \underline{a}_0(N)$, and for $i \geq 1$*

$$\underline{a}_{i+1} + 1 \leq \underline{a}_i(N/fN) \leq \max\{\underline{a}_i, \underline{a}_{i+1} + 1\}.$$

We shall now assume that A_0 is a local ring with maximal ideal $\mathfrak{m}$, of infinite residue field. The *analytic spread* $\ell(N)$ of N is the Krull dimension of $N/\mathfrak{m}N$. We may also assume (by replacing A by an appropriate homomorphic image) that this dimension is the Krull dimension of $A/\mathfrak{m}A$, the ordinary analytic spread of A, which we set to $s = \ell(N)$. By abuse of language, we refer to the reduction of (A_+) obtained in this manner as reductions of N.

If we choose a minimal reduction J for the ideal (A_+), $J = (f_1, \ldots, f_s)$, with the property that $\mathbf{f} = \{f_1, f_2, \ldots, f_s\}$ is a *superficial sequence*, or a *filter-regular sequence*: for each $i \geq 1$, f_i is superficial relative to $N/(f_1, \ldots, f_{i-1})N$. This means that for each i,

$$\underline{a}_i(\mathbf{f}) = \alpha(((f_1, \ldots, f_{i-1})N :_N f_i)/(f_1, \ldots, f_{i-1})N)$$

is well defined. We put

$$\alpha(\mathbf{f}) = \max\{\underline{a}_i(\mathbf{f}) \mid i = 1 \ldots s\}.$$

If we set $J = (f_1, \ldots, f_s)$, we define the *reduction number of N relative to J* by

$$\mathrm{r}_J(N) = \alpha(N/JN).$$

It is simply the smallest integer n for which $N_{m+1} = JN_m$ for $m \geq n$.

Proposition 3.46. *With N and $\mathbf{f}$ as above,*

$$\alpha(\mathbf{f}) = \max\{\underline{a}_i(N) + i \mid i = 0 \ldots s-1\}.$$

Proof. This follows from Proposition 3.45 and the equality $\alpha(\mathbf{f}) = \max\{\alpha(L), \alpha(\mathbf{f}')\}$, where $\mathbf{f}'$ is the image of $\{f_2, \ldots, f_s\}$ in $A/(f_1)$. □

Theorem 3.47. *Let A be a standard graded algebra over a local ring with infinite residue field and N a finitely generated graded A-module. If $s = \ell(N)$ and J is a minimal reduction for N, then*

$$\underline{a}_s(N) + s \leq \mathrm{r}_J(N) \leq \max\{\underline{a}_i(N) + i \mid i = 0 \ldots s\}.$$

Proof. To apply Proposition 3.45, it will suffice to obtain from a minimal reduction of N another minimal set of generators forming a filter-regular sequence. This is always possible since we have assumed that the residue field of the base ring is infinite. □

Reduction Number of Good Filtrations

We have, on several occasions, mentioned the construction of Rees algebras associated with filtrations different from adic ones. An important case is the integral closure filtration associated to an ideal. Let R be a quasi-unmixed integral domain and I a nonzero ideal. Denote by

$$A = \sum_{n\geq 0} \overline{I^n} t^n$$

the Rees algebra attached to $\{\overline{I^n}\}$. Note that while A may not be a standard graded algebra, it is finitely generated over $R[It]$. We may thus apply the theory of Castelnuovo-Mumford regularity to A in order to deduce results about the reduction number of the filtration. We state one of these extensions (see also [HZ94]).

We assume that $(R,\mathfrak{m})$ is a local ring and set $\ell(A) = \dim A/\mathfrak{m}A$ for the analytic spread of A (which is equal to the analytic spread of I). If J is a minimal reduction of I, then on setting $B = R[Jt]$, we can apply the previous results to the pair (B,A):

Theorem 3.48. *Let $(R,\mathfrak{m})$ be a Cohen-Macaulay local ring and $\{I_n, \neq 0, I_0 = R\}$ a multiplicative filtration such that the Rees algebra $A = \sum_{n\geq 0} I_n t^n$ is Cohen-Macaulay and finite over $R[I_1 t]$. Suppose that* height $I_1 \geq 1$, *and let J be a minimal reduction of I_1. Then*

$$I_{n+1} = JI_n = I_1 I_n, \quad n \geq \ell(I_1) - 1,$$

and in particular, A is generated over $R[It]$ by forms of degree at most $\ell - 1 = \ell(I_1) - 1$,

$$\sum_{n\geq 0} I_n t^n = R[It, \ldots, I_{\ell-1} t^{\ell-1}].$$

Proof. At the outset, we may complete R and assume that B has a canonical module. We set $\mathcal{G} = \sum_{n\geq 0} I_n/I_{n+1}$, and can view A and $\mathcal{G}$ as finitely generated modules over $B = R[Jt]$. We set M for the maximal homogeneous ideal of B.

Taking local cohomology of the exact sequences

$$0 \rightarrow A_+ \longrightarrow A \longrightarrow R \rightarrow 0$$

and

$$0 \rightarrow A_+[+1] \longrightarrow A \longrightarrow \mathcal{G} \rightarrow 0$$

of Cohen-Macaulay B-modules, we obtain the exact sequences of the top cohomology modules (the lower-dimensional modules all vanish):

$$0 \rightarrow H^d_M(R) \longrightarrow H^{d+1}_M(A_+) \longrightarrow H^{d+1}_M(A) \rightarrow 0,$$

$$0 \rightarrow H^d_M(\mathcal{G}) \longrightarrow H^{d+1}_M(A_+[+1]) \longrightarrow H^{d+1}_M(A) \rightarrow 0.$$

Since A is Cohen-Macaulay its a-invariant is determined entirely by the degrees of the generators of its canonical module. Since height $I_1 > 0$, localizing at the total ring of fractions of R, we get that $\alpha(A) = -1$. The other sequence then yields that $a(\mathcal{G}) < 0$.

From the fact that the local cohomology of the Cohen-Macaulay algebra A is always concentrated in negative degrees, it follows (see Proposition 3.40) that the groups $H^i_{B_+}(\mathcal{G})$ are also concentrated in negative degrees. As a consequence, the Castelnuovo-Mumford regularity of $\mathcal{G}$ is at most $\ell(I) - 1$, and in particular $\mathcal{G}$ is generated over B, more precisely over $\mathrm{gr}_J(R)$, by elements of degree at most $\ell(I) - 1$.

To complete the proof, we need a general observation. Suppose that we have a filtration $\{I_n\}$ finite over the I-adic filtration ($I = I_1$). Suppose that the reduction number of $\mathcal{G}$ over $\mathrm{gr}_J(R)$ is s: this means that

$$I_{n+1}/I_{n+2} = J(I_n/I_{n+1}), \quad n \geq s,$$

and therefore, for the same range,

$$I_{n+1} = JI_n + I_{n+2}.$$

On iterating, we get

$$I_{n+1} = JI_n + I_{m+1},$$

but since $I_{m+1} = JI_m$ for $m \gg 0$, from Nakayama's Lemma we obtain

$$I_{n+1} = JI_n, \quad n \geq s.$$

Reduction Number and the Cohen-Macaulayness of $I\mathcal{R}(I)$

Let I be an ideal of the Cohen-Macaulay local ring $(R, \mathfrak{m})$. It is to be expected that $I\mathcal{R}(I)$ will play a significant role in our study of the Cohen-Macaulayness of $\mathrm{gr}_I(R)$ and $\mathcal{R}(I)$. If $\mathcal{R}(I)$ is Cohen-Macaulay, then from the exact sequences we examined earlier, e.g.

$$0 \rightarrow I\mathcal{R}(I) \longrightarrow \mathcal{R}(I) \longrightarrow \mathrm{gr}_I(R) \rightarrow 0,$$

one has that $I\mathcal{R}(I)$ is a maximal Cohen-Macaulay module over $\mathcal{R}(I)$. It would be of interest to examine the extent of a converse.

That a full converse does not hold is easy to see. Consider a one-dimensional Cohen-Macaulay local ring $(R, \mathfrak{m})$ and let I be an $\mathfrak{m}$-primary ideal of reduction number one. For a minimal reduction (x), the equality $I^2 = xI$ will show that $IR[xt] = IR[It]$, from which it follows that $IR[It]$ is a maximal Cohen-Macaulay module whereas $R[It]$ itself is not Cohen-Macaulay.

It follows easily that if $I\mathcal{R}(I)$ is a maximal Cohen-Macaulay module of dimension $1 + \dim R$, then $\mathrm{gr}_I(R)$ is Cohen-Macaulay and depth $\mathcal{R}(I) \geq \dim R$.

Theorem 3.49. *Let $(R, \mathfrak{m})$ be a Cohen-Macaulay local ring of dimension $d \geq 1$ and I an $\mathfrak{m}$-primary ideal. Let J be a reduction of I generated by a regular sequence. If $I\mathcal{R}(I)$ is a maximal Cohen-Macaulay module over $\mathcal{R}(I)$ then $\mathrm{r}_J(I) \leq d$.*

Proof. Let J be the minimal reduction of I, and denote by $\mathcal{R}_0$ and $\mathcal{R}$ the Rees algebras of J and of I, respectively. Consider the corresponding Sally module,

$$0 \to I \cdot \mathcal{R}_0 \longrightarrow I \cdot \mathcal{R} \longrightarrow S = S_J(I) \to 0.$$

We suppose that $\mathrm{r}_J(I) = r > 1$. Note that $S_J(I)$ is then a Cohen-Macaulay $\mathcal{R}_0$-module of dimension d. According to Remark 2.4, $S_J(I)$ admits a filtration by extended modules from R whose factors are $I^n/JI^{n-1}[T_1,\ldots,T_d][-n+1]$, for $n = 2,\ldots,r$. An easy calculation shows that the a-invariant of $S_J(I)$ is given by $a(S_J(I)) = -d+r-1$.

Consider the exact sequence of local cohomology of the sequence above with respect to the irrelevant maximal ideal of $\mathcal{R}_0$. Assuming that $I\mathcal{R}(I)$ is a maximal Cohen-Macaulay module over $\mathcal{R}$ (and therefore over $\mathcal{R}_0$ as well), we have

$$H^d(I\mathcal{R}) = 0 \longrightarrow H^d(S_J(I)) \longrightarrow H^{d+1}(I\mathcal{R}_0) \longrightarrow H^{d+1}(I\mathcal{R}) \to 0.$$

On the other hand, from the exact sequence

$$0 \to I\mathcal{R}_0 \longrightarrow \mathcal{R}_0 \longrightarrow R/I[T_1,\ldots,T_d] \to 0$$

we easily obtain that $a(I\mathcal{R}_0) = -1$. The embedding of $H^d(S_J(I))$ into $H^{d+1}(I\mathcal{R}_0)$ then implies that $-d+r-1 \leq -1$, as desired. □

3.4 S_k-Conditions on Rees Algebras

We start with a comparison of the conditions (S_k) in the two algebras $\mathcal{R}(I)$ and $\mathrm{gr}_I(R)$. For that we recall Serre's conditions (S_k).

Definition 3.50. *Let R be a Noetherian ring, E a finitely generated R-module, and k a non-negative integer. Then E satisfies the condition (S_k) if for every prime ideal $\mathfrak{p}$ of R*

$$\mathfrak{p}\text{-depth } E \geq \inf\{k, \text{height } \mathfrak{p}\}.$$

Equivalently, if for every prime ideal $\mathfrak{p}$ of R,

$$\text{depth } E_{\mathfrak{p}} \geq \inf\{k, \text{height } \mathfrak{p}\}.$$

Thus the ring R satisfies (S_1) if it has no embedded primes. Further, R has (S_2) if it has no embedded prime and $\text{height}(\mathfrak{p}) = 1$ for all $\mathfrak{p} \in \mathrm{Ass}(R/xR)$ for every regular element $x \in R$.

3.4.1 Detecting (S_k)

Let R be a ring that is a homomorphic image of a Gorenstein ring S. A method to test for this condition is the following:

Proposition 3.51. *Suppose that R is equidimensional and the surjection $S \twoheadrightarrow R$ has codimension g. Then R satisfies S_k if and only if*

$$\text{height}(\text{ann}(\text{Ext}_S^i(R,S))) \geq k+i, \quad i > g.$$

For ease of reference we quote the following well-known result.

Lemma 3.52. *Let R be a non-negatively graded Noetherian ring such that R_0 is local, and denote by M its maximal irrelevant ideal. Let*

$$0 \to A \longrightarrow B \longrightarrow C \to 0$$

be an exact sequence of finitely generated R-modules where the maps are all homogeneous. Then, with depths taken relative to M, one of the following holds:

(a) depth $A \geq$ depth $B =$ depth C,
(b) depth $B \geq$ depth $A =$ depth $C+1$,
(c) depth $C >$ depth $A =$ depth B.

Theorem 3.53. *Let R be a Noetherian ring with (S_{k+1}) and I an ideal of positive height such that $R_\mathfrak{p}(I_\mathfrak{p})$ is Cohen-Macaulay for any prime ideal $\mathfrak{p}$ of height $\leq k$. The following are equivalent:*

(a) *$R[It]$ has (S_{k+1});*
(b) *$\text{gr}_I(R)$ has (S_k).*

Proof. (a) $\Rightarrow$ (b) Let P be a prime ideal of $\mathcal{G} = \text{gr}_I R$ and Q the lift of P in $\mathcal{R} = R[It]$. We may assume that $(R,\mathfrak{m})$ is a local ring and $\mathfrak{m} \subset Q$. There are two cases to consider. If $It \not\subset Q$, we may assume that $xt \notin Q$ for some regular element $x \in I$. In this case the isomorphism

$$\mathcal{R}_{xt}/(x) \cong \mathcal{G}_{xt}$$

shows that $\mathcal{R}_{xt}$ and $\mathcal{G}_{xt}$ satisfy, for any prime such as Q,

$$\text{depth } \mathcal{R}_Q = \text{depth } \mathcal{G}_P + 1.$$

If $It \subset Q$, then Q is the maximal homogeneous ideal of $\mathcal{R}$ and Theorem 3.25 takes over.

(b) $\Rightarrow$ (a). Let Q be a prime ideal of $\mathcal{R}$. If $I \not\subset Q$, then $\mathcal{R}_Q$ is a localization of $R[t]$, which inherits (S_{k+1}) from R. If $I \subset Q$, we may assume as above that $It \subset Q$ and finally, after localizing at $Q \cap R$, we may assume that R is a local ring and Q is the maximal homogeneous ideal of $R[It]$.

There are two cases to consider. If $\dim R \leq k+1$, then by hypothesis $R[It]$ is Cohen-Macaulay. If $\dim R > k+1$, an application of Theorem 3.25 shows that depth $R[It] \geq k+1$, as desired. □

The following two corollaries supply the requirement on the Cohen-Macaulayness of the $R_\mathfrak{p}(I_\mathfrak{p})$ for prime ideals of codimension at most k.

Corollary 3.54. *Let R be a Noetherian ring with (S_{k+1}) and I an ideal of R such that* $\text{height}\, I \geq k+1$. *Then $R[It]$ has (S_{k+1}) if and only if $\text{gr}_I(R)$ has (S_k).*

Corollary 3.55. *Let R be a Noetherian ring with (S_{k+1}) and I an ideal of R such that* $\text{r}(I_\mathfrak{p}) < \text{height}\, \mathfrak{p}$ *for all prime ideals of height $\leq k$. Then $R[It]$ has (S_{k+1}) if and only if $\text{gr}_I(R)$ has (S_k).*

Corollary 3.56. *Let $(R,\mathfrak{m})$ be a Cohen-Macaulay local ring of dimension $d > 0$, let I be an ideal of positive height and set $\mathcal{R} = R[It]$. The following equivalence holds:*

$$\mathcal{R} \text{ has } S_d \iff \begin{cases} \text{depth}\ \mathcal{R} \geq d \\ \text{Proj}\ (\mathcal{R}) \text{ is Cohen-Macaulay.} \end{cases} \tag{3.15}$$

Proof. We shall use the criterion of Proposition 3.51. We may assume that R is complete (guess why?) and that there is an a homogeneous surjection $S \to \mathcal{R}$, where S is a Gorenstein ring of dimension $d+1$. Suppose that the right-hand side of the assertion holds. We must show that

$$\text{height}\ (\text{ann}\ (\text{Ext}^i_S(\mathcal{R},S))) \geq i+d, \quad i \geq 1,$$

which is clear from the previous corollaries. □

3.4.2 R_k-Conditions on Rees Algebras

Let R be a Noetherian ring and I an ideal. If the Rees algebra $\mathcal{R} = R[It]$ is integrally closed, its divisor class group is intricately connected to the divisor class of R. Indeed, if I is a prime ideal of finite projective dimension such that the associated graded ring $\text{gr}_I(R)$ is normal, then R/I and $\text{gr}_I(R)$ have isomorphic divisor class groups. This fact ([Hon3]) is a consequence of two results of Johnson & Ulrich ([JU99]) and Huneke ([Hu82a]) that circumscribe very explicitly the Serre condition R_k for G.

Let us begin with a discussion of these results.

Proposition 3.57. *Let $(R,\mathfrak{m})$ be a Noetherian local ring and I an ideal such that* $\text{height}\, I > 0$ *and $I \subset \mathfrak{m}^2$. If $\text{gr}_I(R)$ satisfies R_k on $\text{Proj}(G) \cap V(\mathfrak{m}G)$, then $\ell(I) \leq$* $\dim R - k - 1$.

Proof. Write $d = \dim R$. Since $\dim G/\mathfrak{m}G = \ell(I) \geq \text{height}\, I > 0$, there exists a homogeneous prime ideal $\mathfrak{q}$ of G with $\mathfrak{m}G \subset \mathfrak{q}$ and $\dim G_\mathfrak{q} \leq d - \ell(I) < d$. Thus $\mathfrak{q} \in \text{Proj}(G) \cap V(\mathfrak{m}G)$ and $\dim G_\mathfrak{q} \leq d - \ell(I)$. If the asserted inequality does not hold, then $d - \ell(I) \leq k$ and hence $G_\mathfrak{q}$ is regular, say of dimension s.

Let Q be the preimage of $\mathfrak{q}$ in $\mathcal{R} = R[It]$, and let $x_1,\ldots,x_s$ be elements of $Q\mathcal{R}_Q$ whose images in $G_\mathfrak{q}$ form a regular system of parameters. Then $Q\mathcal{R}_Q = (I,x_1,\ldots,x_s)\mathcal{R}_Q$. But $I \subset \mathfrak{m}^2 \subset Q^2\mathcal{R}_Q$, and therefore $Q\mathcal{R}_Q = (x_1,\ldots,x_s)\mathcal{R}_Q$ by Nakayama's Lemma. But then $I\mathcal{R}_Q = 0$ and hence I is contained in a minimal prime of $\mathcal{R}$, which is impossible since height $I > 0$ always implies that height $I\mathcal{R} > 0$. □

Theorem 3.58. *Let R be a Noetherian ring, I an ideal of finite projective dimension, and write $G = \mathrm{gr}_I(R)$. If G satisfies R_k on Proj(G), then $\ell(I_\mathfrak{p}) \leq \max\{\text{height}\,\mathfrak{p}, \dim R_\mathfrak{p} - k - 1\}$ for every $\mathfrak{p} \in V(I)$.*

Proof. We may localize and assume that $\mathfrak{p} = \mathfrak{m}$ is the maximal ideal of R. We show by induction on $g = \text{grade}\, I$ that $\ell(I) \leq \{\text{height}\, I, \dim R - k - 1\}$.

If $g = 0$, then $I = 0$ since proj. $\dim._R R/I < \infty$. So suppose that $g > 0$. By Proposition 3.57, we may assume that $I \not\subset \mathfrak{m}^2$. Choose generators $f_1, \ldots, f_n$ of I, let $Z_1, \ldots, Z_n$ be variables and write $\widetilde{R} = R(Z_1, \ldots, Z_n)$, $\widetilde{\mathfrak{m}} = \mathfrak{m}\widetilde{R}$, $\widetilde{I} = I\widetilde{R}$, and $\widetilde{G} = G \otimes_R \widetilde{R}$. Further, set $x = \sum_{i=1}^n Z_i f_i$, $x' = x + \widetilde{I}^2 \in \widetilde{G}_1$, and $\overline{I} = \widetilde{I}/(x) \subset \overline{R} = \widetilde{R}/(x)$. Since $I \not\subset \mathfrak{m}^2$ and $g > 0$, it follows that x is not in $\widetilde{\mathfrak{m}}^2$ and is $\widetilde{R}$-regular, being a generic element of I. Thus, proj. $\dim._{\widetilde{R}} \overline{R}/\overline{I} < \infty$ by [Na62, 27.5]. On the other hand, x' is a generic element for $\widetilde{G}_+$. Therefore, x is a superficial element of $\widetilde{I}$, and $\widetilde{G}/(x')$ still satisfies R_k on $\mathrm{Proj}(\widetilde{G}/(x'))$ since for every $Q \in \mathrm{Proj}(\widetilde{G}/(x'))$, $(\widetilde{G}/(x'))_Q$ is the localization of a polynomial ring over G. Hence by equality (3.3), $\mathrm{gr}_{\overline{I}}(\overline{R})$ satisfies $R_k$on $\mathrm{Proj}(\mathrm{gr}_{\overline{I}}(\overline{R}))$. Now the induction hypothesis yields $\ell(\overline{I}) \leq \max\{\text{height}\,\overline{I}, \dim \overline{R} - k - 1\} = \max\{\text{height}\, I, \dim R - k - 1\} - 1$.

It remains to establish the inequality $\ell(I) - 1 \leq \ell(\overline{I})$. Indeed, writing $K = \widetilde{R}/\widetilde{\mathfrak{m}}$ and using the convention $\dim \emptyset = -1$, one concludes from equality (3.3) that

$$\begin{aligned}\ell(\overline{I}) &= \dim \mathrm{Proj}(\mathrm{gr}_{\overline{I}}(\overline{R})) + 1 = \dim \mathrm{Proj}(\widetilde{G}/(x') \otimes_{\widetilde{R}} K) + 1 \\ &= \dim(\widetilde{G} \otimes_{\widetilde{R}} K)/(x') \geq \ell(I) - 1,\end{aligned}$$

to establish the claim. □

Remark 3.59. In the special case where $\dim R/I \leq 2$, I is locally generated by a regular sequence, and therefore G is the symmetric algebra of the conormal module, $G = S_{R/I}(I/I^2)$. Indeed, since $\ell(I_\mathfrak{p}) \leq \text{height}\, I$, I is locally equimultiple (after a harmless extension to provide for infinite residue fields), and since it is generically a complete intersection, by [CN76] it must be a complete intersection.

Theorem 3.60. *Let R be a Cohen-Macaulay ring and I an R-ideal of finite projective dimension that is a complete intersection locally at each of its minimal primes. Then $\mathrm{gr}_I(R)$ satisfies R_k if and only if R/I satisfies R_k and $\ell(I_\mathfrak{p}) \leq \max\{\text{height}\, I_\mathfrak{p}, \dim R_\mathfrak{p} - k - 1\}$ for every $\mathfrak{p} \in V(I)$.*

3.5 Exercises

Exercise 3.61. Let $(R, \mathfrak{m})$ be a Noetherian local ring and I an ideal of R. Show that if $\mathrm{gr}_I(R)$ is Cohen-Macaulay then R is also Cohen-Macaulay.

Exercise 3.62. Let R be a Cohen-Macaulay local ring of dimension at least 2 and let I be a Cohen-Macaulay equimultiple ideal. Show that if $\mathrm{r}(I) = 1$ then $\mathcal{R}(I)$ is Cohen-Macaulay.

Exercise 3.63. Let R be a Cohen-Macaulay local ring and let I be an ideal of positive height such that $G = \mathrm{gr}_I(R)$ is Cohen-Macaulay. Prove that there is a ring $S = R[x_1,\ldots,x_n]$ of polynomials such that the Rees algebra of the ideal $L = (I, x_1,\ldots,x_n)$ is Cohen-Macaulay.

Exercise 3.64. Let R be a local ring and I an ideal of positive height. Show that for $n \gg 0$ all Veronese subrings $\mathcal{R}(I^n)$ have the same depth.

Exercise 3.65. Let R be a quasi-unmixed integrally closed domain and I an ideal. Prove the following implications:

$$R[It] \text{ is normal } \Longrightarrow \text{Proj } (R[It]) \text{ is normal } \Longleftrightarrow I^n \text{ is normal for all } n \gg 0.$$

(This behavior is very different from the Cohen-Macaulay condition.)

4

Divisors of a Rees Algebra

The divisors of a commutative Noetherian ring A are the distinguished rank one A-modules that satisfy Serre's condition S_2. It is a notion based on that of unmixed height 1 ideals of a normal domain. The rationale for studying divisors is that they encode many properties of the algebra that would otherwise stay hidden. Noteworthy is their use is obtaining presentations of the algebra in terms of more standard algebras. In the case of Rees algebras, several divisors stand out and our aim is to study their role in the arithmetic of the algebra.

The approach that we shall follow here is to define the ds of A through the intervention of a finite homomorphism $\varphi : S \to A$, with S admitting a canonical module ω_S. *Grosso modo*, the divisors of A are those A-modules arising in the following manner. Suppose that $\dim A = d$, $\dim S = n$, and let L be an A-module of rank 1. We shall call

$$\omega_L = \mathrm{Ext}_S^{n-d}(L, \omega_S)$$

the *canonical module* of L. The set of divisors is made up of all ω_L, more precisely, of their isomorphism classes $\mathrm{Div}(A)$.

A distinctive divisor of A is its canonical module

$$\omega_A = \mathrm{Ext}_S^{n-d}(A, \omega_S).$$

Another divisor is A itself when it already satisfies property S_2. It is a consequence of the theory of the canonical module that $\mathrm{Div}(A)$ is independent of the homomorphism φ (see [BH93]).

A Rees algebra $A = R[It]$ that satisfies the condition S_2 admits several other remarkable divisors besides A: (i) the *canonical module* ω_A, a centerpiece of the cohomology theory of A; (ii) the *exceptional divisor* $IR[It]$; and (iii), its dual,

$$\mathcal{D}(I) = \mathrm{Ext}_S^{n-d}(IR[It], \omega_S),$$

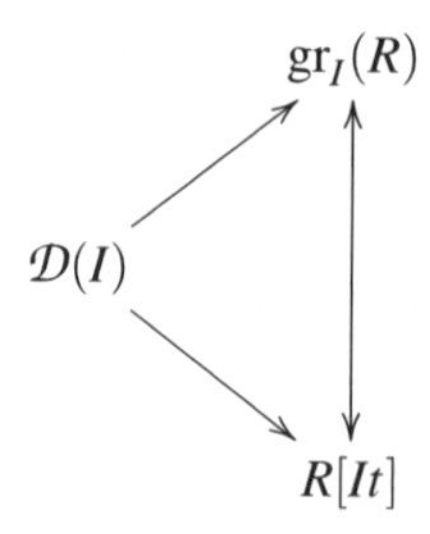

which will play an even more critical role than ω_A. It will be labeled the *fundamental divisor*, and one of our aims is to explore its role in the arithmetical study of the Rees algebra $\mathcal{R} = R[It]$. One can recover $\omega_{R[It]}$ immediately from $\mathcal{D}(I)$. It turns out that $\mathcal{D}(I)$ codes for the situation when the Cohen-Macaulayness of $\mathrm{gr}_I(R)$ implies the same property in $R[It]$. This is not totally unexpected since when $R[It]$ is normal, $\mathcal{D}(I)$ is obtained in the operation

$$\mathcal{D}(I) = \mathrm{Hom}_{\mathcal{R}}(IR[It], \omega_{\mathcal{R}})$$

linking the canonical and the exceptional divisors.

We shall emphasize the study of graded divisors for the following reason that we illustrate with the canonical module of $\mathcal{R}$. Suppose that R is a Noetherian local ring and I an ideal of positive height. We shall assume that R has a canonical module ω_R. (As a consequence both $\mathcal{R} = R[It]$ and $\mathcal{G} = \mathrm{gr}_I(R)$ also have canonical modules.) If $f \in I$ is a regular element of R, then f will also be a regular element on $\omega_{\mathcal{R}}$, so that

$$(\omega_{\mathcal{R}})_f = \omega_{R_f[It]} = \omega_{R_f[t]} = t\omega_{R_f}[t].$$

From this one has that the a-invariant of $\mathcal{R}$ is -1. Thus the graded module $\omega_{\mathcal{R}}$ (and $\mathcal{D}(I)$ likewise) has a special representation

$$\omega_{\mathcal{R}} = D_1t + D_2t^2 + \cdots + D_nt^n + \cdots,$$

where D_i are fractionary ideals of R.

There are divisors arising from a different process. Suppose that R is a quasi-unmixed integral domain, and denote by B the integral closure of $A = R[It]$. Then any divisor of B is also a divisor of A. Noteworthy is $\sum_{n\geq 1} \overline{I^n}t^n$.

A basic property of divisors, shared by many other graded divisors in general is the behavior of its components. For a divisor D

$$D = D_1t + \cdots + D_nt^n + \cdots,$$

the components D_i define a decreasing filtration $D_1 \supset D_2 \supset \cdots$. This peculiar property will be exploited systematically. It gives rise to an operation on the set $\mathrm{Divh}(\mathcal{R})$ of such divisors,

$$\mathcal{S} : \mathrm{Divh}(\mathcal{R}) \mapsto \mathrm{Divh}(\mathcal{R})$$

$$\mathcal{S}(L) = \mathcal{S}(\sum_n D_n t^n) = \sum_n D_{n+1} t^n,$$

and a corresponding construction of a graded module over $\mathrm{gr}_I(R)$, namely

$$\mathsf{G}(L) = L/\mathcal{S}(L).$$

We shall call $\mathcal{S}$ the *shifting* operator of $\mathrm{Divh}(\mathcal{R})$, and one of our goals is to determine conditions for a divisor L to be an *end*, that is, not to lie in the image of $\mathcal{S}$. This is connected to the notion of *prolongation* of a divisor, meaning from a given D what conditions are required to solve the equation $\mathcal{S}(L) = D$, and what is the structure of the set of solutions.

One of the results will show that under very general conditions a Cohen-Macaulay divisor L can hardly be in the image of $\mathcal{S}^n$ for n large (determined by $\dim R$). Under some specific conditions, such as the normality of $R[It]$, the existence and uniqueness (up to isomorphism) of prolongations is established.

Applying the construction to the module $\mathcal{D}(I)$ yields $\mathsf{G}(I)$. The structures of $\mathsf{G}(\omega_{\mathcal{R}})$ and of $\mathsf{G}(I)$ as $\mathrm{gr}_I(R)$-modules will be used to describe very explicitly in many cases of interest the canonical modules of $\mathcal{R}$ and $\mathcal{G}$, but also will shed light on when the Cohen-Macaulayness of $\mathcal{G}$ forces $\mathcal{R}$ to be Cohen-Macaulay.

Another goal is to identify Cohen-Macaulay divisors and to examine their relationship to the reduction number of the ideal. Additionally, we are interested in finding the divisors that carry information about the Briançon-Skoda number of I.

4.1 Divisors of an Algebra

Let R be a Noetherian ring and I be an ideal, $I = (f_1, \ldots, f_n)$. We assume that R has finite Krull dimension, $\dim R = d$, that its total ring of quotients is Artinian, and that height $I > 0$. If R has a canonical module ω_R, we can use for the presentation morphism the standard one, $S = R[T_1, \ldots, T_n] \to R[It]$, when $\omega_S = \omega_R S[-n]$.

Reflexive Modules

We provide, for ease of reference, a brief primer on reflexive modules. It can be viewed as a small fragment of the general duality theory associated with the canonical module (see [BH93, Chapter 3]).

Let R be a commutative ring and let E and F be R-modules. There is a pairing

$$E \times \mathrm{Hom}_R(E, F) \longrightarrow F, \quad (v, \varphi) \to \varphi(v),$$

that induces a natural mapping

$$\pi : E \longrightarrow \mathrm{Hom}_R(\mathrm{Hom}_R(E, F), F),$$

and we consider some conditions for it to be an isomorphism. The module E is then said to be F-reflexive (or simply *reflexive* when $F = R$). The cases of special interest to us are those with $F = R$ and $F = C = \omega_R$, the canonical module of R (if it exists).

The features of the module E that we want to emphasize are the properties of the localizations $E_{\mathfrak{p}}$ for prime ideals $\mathfrak{p}$ of low codimension or grade.

Proposition 4.1. *Let R be a Noetherian ring and E a submodule of a finitely generated free R-module. Then E is reflexive if and only if for every prime ideal $\mathfrak{p}$ such that* depth $R_{\mathfrak{p}} \leq 1$, *the localization $E_{\mathfrak{p}}$ is $R_{\mathfrak{p}}$-reflexive and for every prime with* depth $R_{\mathfrak{p}} \geq 2$ *the localization $E_{\mathfrak{p}}$ has depth at least* 2.

Proof. In view of the elementary fact that $\text{Ass}(\text{Hom}_R(E,R)) = \text{Supp}(E) \cap \text{Ass}(R)$, the forward implication is clear. Conversely, suppose that the localization assumptions hold. We consider the natural sequence

$$0 \to K \longrightarrow E \stackrel{\pi}{\longrightarrow} \text{Hom}_R(\text{Hom}_R(E,R),R) = E^{**} \longrightarrow L \to 0,$$

where K and L are the kernel and cokernel of π. Since the associated primes of E are found in $\text{Ass}(R)$ and by assumption E is reflexive in depth 0, it follows that $K = 0$. We are left with an exact sequence

$$0 \to E \longrightarrow E^{**} \longrightarrow L \to 0,$$

where we note E^{**}, being a dual, satisfies all the depth conditions assumed for E. If $L \neq 0$, it follows from the assumptions that the associated primes of L have grade at least two, which is contradicted by the depth lemma (Lemma 3.52). □

One formulation of this result that uses a canonical ideal C in place of R is the following:

Proposition 4.2. *Let R be a Noetherian domain that admits a canonical ideal C. A finitely generated torsionfree R-module E is C-reflexive if for every prime ideal $\mathfrak{p}$ of codimension at least two the localization $E_{\mathfrak{p}}$ has depth at least two.*

There are other variations leading to slightly different formulations. In one of these we assume that R is Gorenstein in codimension one. They all share the proof above in its main thrust.

The Semigroup of Divisors

To obtain divisors we must supply fodder to the process outlined above. When A is the ring of polynomials $R[t]$, some are simply extensions of divisorial ideals J of R, that is, $JR[t]$. However if A is a Rees algebra $R[It]$, there are constructions like $J(1,t)^m \subset R[t]$, or more generally $(J_1t, \ldots, J_mt^m)$.

Let us begin with the formal definition of divisors.

Definition 4.3. Let A be a Noetherian ring. An A-module L is a *divisor* if it satisfies the following conditions:

(i) L satisfies Serre's condition S_2;
(ii) If K is the total ring of fractions of A, then $K \otimes_A L \cong K$.

We begin by listing some elementary properties of divisors as defined here. We emphasize that a module L over the Noetherian ring A has rank 1 if it is torsionfree and $L \otimes_A K \cong K$, where K is the total ring of quotients of A.

Proposition 4.4. *Let S be a Cohen-Macaulay ring of dimension n with a canonical module ω_S. Let A be a Noetherian ring of dimension d with an Artinian total ring of quotients that admits a presentation $S \to A$ as above. The following hold:*

(i) *For any fractionary ideal L the module*

$$\omega_L = \operatorname{Ext}_S^{n-d}(L, \omega_S)$$

satisfies condition S_2.

(ii) *There is a natural isomorphism*

$$D \to \operatorname{Ext}_S^{n-d}(\operatorname{Ext}_S^{n-d}(D, \omega_S), \omega_S)$$

on modules $D = \omega_L$.

(iii) *If D_1 and D_2 are divisors, the operation*

$$D_1 \circ D_2 = \operatorname{Ext}_S^{n-d}(\operatorname{Ext}_S^{n-d}(D_1 D_2, \omega_S), \omega_S)$$

defines a monoid structure on the elements of $\mathrm{Div}(A)$ that are principal in codimension one.

(iv) *If A satisfies condition S_2 and is Gorenstein in codimension one, then*

$$\omega_L \cong \operatorname{Hom}_A(\operatorname{Hom}_A(L, A), A)$$

is an isomorphism for every rank one module L.

(v) *If A is an integrally closed domain, then $L \circ \operatorname{Hom}_A(L, A) \cong A$ for every divisor L. As a consequence, the isomorphism classes of divisors form a group, the so-called divisor class group of A.*

(vi) *If A is an integral domain with a canonical module ω_A, the smallest rational extension $\widetilde{A}$ of A that has the property S_2 is called its S_2-ification,*

$$0 \to A \longrightarrow \widetilde{A} \longrightarrow H \to 0.$$

In the graded case H is the so-called Hartshorne-Rao module of A. It can be obtained as

$$\widetilde{A} = \operatorname{Hom}_A(\omega_A, \omega_A) = \mathcal{D}(\omega_A).$$

(vii) *A and $\widetilde{A}$ have the same sets of divisors, $\mathrm{Div}(A) = \mathrm{Div}(\widetilde{A})$.*

Homogeneous Divisors and the Shifting Operation

The homogeneous divisors of $\mathcal{R} = R[It]$ are those of the form

$$D = \sum_{n \geq r} D_n t^n.$$

The role of the components D_n is an important feature of D. We begin with an observation which will be used in several constructions. Note also its extensions to more general filtrations such as the integral closure filtration, the Ratliff-Rush filtration and some symbolic powers.

There are other divisors of a Rees algebra $R[It]$ with the property that for every nonzero divisor $f \in I$, $L_f \cong J_f R[t]$ for some ideal $J \subset R$–that is L_f is 'extended' from R. This is, for instance, the case for all divisors if R is integrally closed. There are also examples of homogeneous ideals in more general cases–and we shall consider one later.

Definition 4.5. For each graded divisorial ideal

$$L = \sum_{n \geq r} L_n t^n, \quad L_r \neq 0$$

of $R[It]$, the ideal

$$\mathcal{S}(L) = \sum_{n \geq r} L_{n+1} t^n$$

is the *shifting* of L.

The following shows that the shifting operation is well-behaved in all integral domains.

Proposition 4.6. *Let R be a Noetherian integral domain with a canonical module and I a nonzero ideal. Then $\mathcal{S}(L) \subset L$ for every homogeneous divisor L of $R[It]$.*

Proof. It will suffice to show that the conductor ideal $L : \mathcal{S}(L)$ has codimension at least 2. From the construction, It is contained in the conductor $L : \mathcal{S}(L)$. On the other hand, passing to the field of fractions K of R, we obtain $KL = K\mathcal{S}(L) = K[t]t^r$, which shows that the conductor contains a nonzero element x of R. This means that it contains (It, x), which is an ideal of height at least two as (It) is a prime ideal. Since L has condition S_2, $\mathcal{S}(L) \subset L$. □

Note that in view of the exact sequence

$$0 \to \mathcal{S}(L[-1]) \longrightarrow L \longrightarrow L_r t^r \to 0, \quad r = \inf\{n \mid L_n \neq 0\},$$

$\mathcal{S}(L)$ is also a divisor. It guarantees that the graded module

$$0 \to \mathcal{S}(L) \longrightarrow L \longrightarrow \mathrm{G}(L) \to 0$$

over $\mathrm{gr}_I(R)$ has positive depth.

Prolongation of a Divisor

The process of shifting can run backwards under some conditions. Let

$$L = L_1 t + L_2 t^2 + \cdots$$

be a divisor. By a *prolongation* we mean another divisor

$$D = L_0 t + L_1 t^2 + \cdots,$$

that is, $\mathcal{S}(D) = L$. As a prerequisite, we must have that $I \cdot L_0 \subset L_1$ but more conditions are to be met. We consider a case when prolongation can be achieved.

Proposition 4.7. *Let R be a Cohen-Macaulay domain with canonical ideal ω_R, and let I be an ideal of height at least 2. Assume that the divisor*

$$L = L_1 t + L_2 t^2 + \cdots$$

of $R[It]$ satisfies the condition $I\omega_R \subset L_1 \subset \omega_R$. Then the ideal D defined by

$$0 \to Lt \longrightarrow D \longrightarrow \omega_R t \to 0,$$

is a prolongation of L.

Proof. As before we are still assuming that locally R is a homomorphic image of a Gorenstein ring. We may assume that R is local of dimension d and write $A = R[It]$ as a homomorphic image of a graded Gorenstein domain S of dimension $d+1$.

The condition S_2 for an ideal such as D is equivalent to the requirement

$$\text{height ann } (\text{Ext}^i_S(D, \omega_S)) \geq 2 + i, \quad i > 0.$$

Consider the usual Ext sequence

$$\begin{aligned} 0 &\to \text{Hom}(\omega_R t, \omega_S) = 0 \to \text{Hom}(D, \omega_S) \to \text{Hom}(Lt, \omega_S) \stackrel{\varphi}{\to} \text{Ext}^1(\omega_R t, \omega_S) \\ &= Rt^{-1} \to \text{Ext}^1(D, \omega_S) \to \text{Ext}^1(Lt, \omega_S) \to \text{Ext}^2(\omega_R t, \omega_S) = 0, \end{aligned}$$

and the isomorphisms

$$\text{Ext}^i(D, \omega_S) \cong \text{Ext}^i(Lt, \omega_S) \quad i > 1.$$

For $i > 1$ the isomorphism above says that height ann $\text{Ext}^i(D, \omega_S) \geq 2 + i$. For $i = 1$, note that Rt^{-1} is already annihilated by It. What is needed is for the cokernel of the map φ to have an annihilator, as R-module, of height at least 2. Then its annihilator over S will have codimension at least 3, and $\text{Ext}^1(D, \omega_S)$, sitting as it is in the middle of a short sequence of modules with annihilators of codimension at least 3 has the same property.

We can test the cokernel by localizing at height 1 primes of R, when $R[It]$ localizes into $R_\mathfrak{p}[t]$, a polynomial ring. But in this case $D_\mathfrak{p} \cong R_\mathfrak{p}[t]$, so that $\text{Ext}^1(D, \omega_S)_\mathfrak{p} = 0$ and therefore the cokernel of $\varphi_\mathfrak{p}$ will vanish as well. □

Question 4.8. An obvious issue is whether every divisor has a proper prolongation. Another point is whether among the prolongations of a divisor there exists a canonical (minimal) one. The argument above suggests that if L_0 and L_0' are the new components of prolongations of L, then $L_0 \cap L_0'$ may also work.

Let us work out a more general approach to prolongations. Suppose that $L = \sum_{n\geq 1} L_n t^n$ is a divisor of $R[It]$ that we seek to prolong into $D = \sum_{n\geq 1} L_{n-1} t^n$. There are at least two requirements on L_0:

$$\begin{aligned} L_1 &\subset L_0 \\ I \cdot L_0 &\subset L_1, \end{aligned}$$

plus some divisoriality condition on L_0 yet to be fully determined. There is, however, at least one situation when it leads to a unique solution for L_0.

Proposition 4.9. *Suppose that* height $I \geq 2$ *and* L_1 *is an ideal with Serre's condition* S_3*. Then* $L_0 = L_1$ *and* D *is a prolongation of* L.

Proof. Consider the exact sequence

$$0 \to I \longrightarrow R \longrightarrow R/I \to 0,$$

and apply $\mathrm{Hom}_R(\cdot, L_1)$:

$$0 \to \mathrm{Hom}_R(R/I, L_1) = 0 \longrightarrow \mathrm{Hom}_R(R, L_1) \longrightarrow \mathrm{Hom}_R(I, L_1) \longrightarrow \mathrm{Ext}^1_R(R/I, L_1).$$

Since I has height at least two and L_1 satisfies condition S_2, I contains two elements forming a regular sequence on L_1; thus $\mathrm{Ext}^1_R(R/I, L_1) = 0$. This means that $L_1 = \mathrm{Hom}_R(I, L_1)$. Now note that $L_0 \subset \mathrm{Hom}_R(I, L_0)$, which with the other containment shows that $L_0 = L_1$.

One can now repeat the proof of Proposition 4.7. Using the sequence

$$\mathrm{Ext}^i_S(L_1, \omega_S) \longrightarrow \mathrm{Ext}^i_S(D, \omega_S) \longrightarrow \mathrm{Ext}^i_S(L, \omega_S)$$

one sees that the codimension of the module in the middle must be at least $i+2$. (This is the place where we have used the condition S_3 on L_1.) □

The process of shifting cannot be indefinitely applied without changing key properties of the divisor. Consider for instance the case of a divisor

$$L = Rt + L_2 t^2 + \cdots + L_n t^n + \cdots, \quad L_{n+1} \subset L_n.$$

The exact sequence

$$0 \to L_{n\geq 2} \longrightarrow L \longrightarrow Rt \to 0$$

shows that if R and L are both Cohen-Macaulay then both $\mathcal{S}(L)$ and $\mathrm{G}(L)$ will also be Cohen-Macaulay, by the same argument as that employed in (3.4) and (3.5). The following shows that some restriction must be placed on iterated applications of shifting. For simplicity we use the fact that the early components are isomorphic to R, whereas using Cohen-Macaulay ideals of height one would serve the same purpose.

Proposition 4.10. *Let R be a Cohen-Macaulay ring and I an ideal of height at least 2. Suppose that*

$$L = Rt + \cdots + Rt^r + L_{r+1}t^{r+1} + \cdots$$

is a Cohen-Macaulay divisor. Then $r \leq$ height I.

Proof. We may assume that $(R, \mathfrak{m})$ is a local ring of dimension $d \geq 2$ and that I is $\mathfrak{m}$-primary. Let $J = (a_1, \ldots, a_d)$ be a minimal reduction of I. According to Proposition 3.1, the elements $a_1, a_2 - a_1 t, \ldots, a_d - a_{d-1}t, a_d t$ form a system of parameters for $R[It]$ and therefore will form a regular sequence on the Cohen-Macaulay module L. (More properly on L_M, where M is the maximal homogeneous ideal of $R[It]$.)

Observe the result of the action of this system of parameters on $Rt + \cdots + Rt^r$: after reduction modulo a_1, multiplication by $a_2 - a_1 t$ on $R/(a_1)t + \cdots + R/(a_1)t^{r-1}$ has the same effect as multiplication by a_2 only, so that

$$L/(a_1, a_2 - a_1 t)L = R/(a_1, a_2)t + \cdots + R/(a_1, a_2)t^{r-1} + \text{higher degree}.$$

Repeating the argument up to the element $a_d - a_{d-1}t$ gives a module

$$L/(a_1, \ldots, a_d - a_{d-1}t)L = R/(a_1, \ldots, a_d)t + \cdots + R/(a_1, \ldots, a_d)t^{r-d+1} + \text{higher degree}.$$

It is clear that $a_d t$ is not regular on this module when $r > d$. $\square$

Dimension One

Let R be an integral domain of Krull dimension 1 and I a nonzero ideal. Set $\widetilde{R} = \bigcup_{n \geq 1} \text{Hom}_R(I^n, I^n)$ and $\widetilde{I} = I\widetilde{R}$. Note that if ω_R is the canonical module of R, then $\omega_{\widetilde{R}} = \omega_R$. The module $\widetilde{I}$ is a locally free ideal of $\widetilde{R}$, and the Rees algebra $\widetilde{\mathcal{R}} = \widetilde{R}[\widetilde{I}t]$ is the S_2-ification of $\mathcal{R} = R[It]$ (see [NV93]). The canonical and fundamental divisors of $\mathcal{R}$ are now easy to describe: $\omega_{\mathcal{R}} = \omega_R t\widetilde{R}$, and $\mathcal{D}(I) \cong \omega_{\mathcal{R}}$.

The other divisors are more difficult to describe, but the following observation is useful. For simplicity suppose that R is a local ring and that $J = (a)$ is a minimal reduction of I. The homogeneous divisors of $\mathcal{R}$ are the fractionary ideals of $\widetilde{\mathcal{R}} = \widetilde{R}[at]$ on which $\{a, at\}$ is a regular sequence. Let $J = \sum_{n \geq 1} J_n t^n$ be such a divisor. We claim that $J = J_1 t\widetilde{R}$. Using the shifting operator repeatedly it will suffice to show that $J_2 = aJ_1$. Suppose that $bt^2 \in J$; since $J_2 \subset J_1$, we have the following relation with coefficients in J:

$$at \cdot bt = a \cdot bt^2.$$

Thus $bt^2 = at \cdot ct$, with $ct \in J_1 t$.

Associated Primes of the Components of a Divisor

Let R be a Noetherian integral domain with the property S_2 and let I be a nonzero ideal. We examine the role of the local analytic spread of I on the associated primes of the components of any divisor

$$D = D_1 t + \cdots + D_n t^n + \cdots, \; D_n \subset R$$

of $A = R[It]$.

Proposition 4.11. *The associated prime ideals of codimension greater than one of each component are contained in the set of prime ideals* $\mathcal{P} = \{\mathfrak{p} \mid \ell(I_{\mathfrak{p}}) = \text{height } \mathfrak{p}\}$.

In Proposition 3.38, we have already identified such primes as the restriction to R of the minimal primes of $IR[It]$.

Proof. Let $D \subset R[It]$ be a divisor and D_n one of its components. Let $I \subset \mathfrak{p}$ be an associated prime of D_n and suppose that $\mathfrak{p} \notin \mathcal{P}$. Localizing at $\mathfrak{p}$, according to the proof of Proposition 3.38, we get that height $\mathfrak{p}R_{\mathfrak{p}}[I_{\mathfrak{p}}t] \geq 2$, which means that there is a regular sequence of two elements in $\mathfrak{p}$ that gives a regular sequence on D and therefore on each of its component. But this is clearly not possible from the exact sequence

$$0 \to D_n \longrightarrow R \longrightarrow R/D_n \to 0.$$

On the other hand, if $I \not\subset \mathfrak{p}$, then $A_{\mathfrak{p}}$ is a ring of polynomials and height $\mathfrak{p} = 1$ as otherwise $\mathfrak{p}R_{\mathfrak{p}}[t]$ would contain a regular sequence of two elements. $\square$

This leads to the following pretty result of Ratliff:

Corollary 4.12. *Let R be a quasi-unmixed integral domain, and let $I = (a_1, \ldots, a_q)$ be an ideal of codimension q. For each integer n, the associated primes of $\overline{I^n}$ are the minimal primes of I.*

Proof. First note that if R is integrally closed, we just take $D = \sum_{n \geq 1} \overline{I^n} t^n$ as divisor. In general, if A is the integral closure of R, then $\overline{I^n} = R \cap \overline{AI^n}$. From the embedding

$$R/\overline{I^n} \hookrightarrow A/\overline{AI^n}$$

we have that the associated primes of $\overline{I^n}$ must be contained in the associated primes of the A-ideal $\overline{AI^n}$. $\square$

4.2 Divisor Class Group

The divisors of a Rees algebra $\mathcal{R} = R[It]$ are easier to control when R is an integrally closed domain and I is a normal ideal. We shall assume that these conditions hold in this section and give a general description of the divisor class group $\mathrm{Cl}(\mathcal{R})$.

There are two approaches to the calculation of $\mathrm{Cl}(\mathcal{R})$. One can apply to $\mathcal{R}$ general localization properties that isolate distinguished batches of divisors (see [Fo73]), or make use of specific properties of Rees algebras. The second method will also be useful for certain algebras of symbolic powers and for the integral closure of ordinary Rees algebras. Our discussion is a merge of several sources: [HSV87], [HSV91], [HV85], [Vil89], [Vi88], and especially [ST88].

Theorem 4.13. *Let R be an integrally closed Noetherian domain and I a normal ideal of height at least two. There is an exact sequence*

$$0 \to H \longrightarrow \mathsf{Cl}(\mathcal{R}) \stackrel{\varphi}{\longrightarrow} \mathsf{Cl}(R) \to 0$$

of divisor class groups, where H is the free abelian group generated by the classes $[P]$ defined by the minimal primes of $I\mathcal{R}$. In particular, if R is a factorial domain and I is a normal ideal of height ≥ 1 then $\mathsf{Cl}(\mathcal{R})$ is a free abelian group of rank equal to the number of Rees valuations of I.

Proof. Since $\mathcal{R}$ is a graded ring to construct $\mathsf{Cl}(\mathcal{R})$, it will suffice to consider homogeneous divisors. Let $P = \sum_{n\geq 0} P_n t^n$ be a height one prime of $\mathcal{R}$. Define a function from this set of divisors into the group $\mathsf{Cl}(R)$ by putting

$$\varphi(P) = [P_0].$$

This map clearly defines a homomorphism of divisor class groups, which we denote by the same symbol.

The following properties are easy to verify. (i) For each divisorial prime ideal $\mathfrak{p}$ of R, $T(\mathfrak{p}) = \mathcal{R} \cap \mathfrak{p}\mathcal{R}_{\mathfrak{p}}$ is a divisorial prime of $\mathcal{R}$ and $\varphi([T(\mathfrak{p})]) = [\mathfrak{p}]$ (and therefore φ is surjective). (ii) Another set of valuations for $\mathcal{R}$ are restrictions of valuations of $K[t]$, where K is the field of fractions of R; their classes in $\mathsf{Cl}(\mathcal{R})$ are clearly linear combinations of some $[T(\mathfrak{p})]$. (iii) If for a homogeneous prime P, the height of P_0 is at least two and $I \not\subset P_0$, then localizing at P_0 yields that P is not divisorial. This means that the kernel of φ must indeed by given by certain combinations of $[Q_1], \ldots, [Q_s]$, where $\{Q_1, \ldots, Q_s\}$ is the set of associated primes of $I\mathcal{R}$ (note that this ideal is unmixed).

Suppose there is a relation among the $[Q_i]$. We write it as

$$a_1[Q_1] + \cdots + a_r[Q_r] = a_{r+1}[Q_{r+1}] + \cdots + a_s[Q_s],$$

where the a_i are non-negative integers. For each Q_i denote by $\mathfrak{q}_i$ its component of degree 0. The $\mathfrak{q}_i$ are prime ideals of height at least two. Converted to a relation in $\mathcal{R}$, this means that there exist homogeneous elements $f = at^m$, $g = bt^n$ of $\mathcal{R}$ such that the ideals

$$f \cdot \prod_{i=1}^{r} Q_i^{a_i} \quad \text{and} \quad g \cdot \prod_{i=r+1}^{s} Q_i^{a_i}$$

have the same value at each valuation of $\mathcal{R}$. In particular, for a valuation $v_{\mathfrak{p}}$ defined by $T(\mathfrak{p})$ or arising from a valuation of $K[t]$,

$$v_{\mathfrak{p}}(f) = v_{\mathfrak{p}}(at^m) = v_{\mathfrak{p}}(bt^n) = v_{\mathfrak{p}}(g).$$

In particular $a = b$, $m = n$, so we can cancel f and g, which leads to a contradiction as $\prod_{i=1}^{r} Q_i^{a_i}$ has a nonzero value at the valuation defined by Q_1. □

Remark 4.14. The same description of divisor class groups applies equally to other filtrations. For example, if R is a normal domain, I is an ideal of height at least two and the integral closure $\overline{\mathcal{R}}$ of $\mathcal{R}$ is Noetherian, then $\mathsf{Cl}(\overline{\mathcal{R}})$ has a presentation as above. Similar comments apply to Noetherian algebras of symbolic powers.

A very general framework for studying divisor class groups of blowup algebras was given by Simis and Trung in [ST88]. Following the suggestions of an anonymous referee according to [Hon3a], we provide a proof (a near translate) of [ST88, Theorem 2.1] in a format suitable to our applications.

Theorem 4.15. *Let $A \hookrightarrow S$ be two Krull domains. Assume that the following conditions are satisfied:*

(a) *$S \otimes_A K$ is a polynomial ring over K, where K denotes the field of fractions of A.*
(b) *For every height 1 prime ideal $\mathfrak{q}$ of A, the ideal $\mathfrak{q}S_{\mathfrak{q}}$ is a prime ideal of $S_{\mathfrak{q}}$.*

Then there exists a canonical surjective homomorphism $\varphi : \mathsf{Cl}(S) \to \mathsf{Cl}(A)$ whose kernel is the free abelian group generated by the divisors of the form $[\mathfrak{Q}]$, where $\mathfrak{Q}$ is a height 1 prime ideal of S with height $(\mathfrak{Q} \cap A) > 1$.

Proof. Set $S \otimes_A K = K[t_1, \ldots, t_n] = K[\mathbf{t}]$, and denote by $K(\mathbf{t})$ its field of quotients. For a prime divisor $\mathfrak{P}$ of the Krull domain (say) S, we denote by $v_{\mathfrak{P}}(\cdot)$ the corresponding valuation. There are 3 distinguished sets of prime divisors and corresponding valuations of S. First, those $\mathfrak{P}$ such that height $(\mathfrak{P} \cap A) = 1$: according to condition (b), the valuations are essentially those of A. Second, those $\mathfrak{P}$ such that height $(\mathfrak{P} \cap A) = 0$: according to condition (a), the valuations are essentially those of $K(\mathbf{t})$. Lastly, those $\mathfrak{P}$ such that height $(\mathfrak{P} \cap A) > 1$.

In the group $\mathcal{D}(S)$ of divisors, consider the (free abelian) subgroup $\mathcal{D}^A(S)$ generated by the prime divisors (height 1 prime ideals) $\mathfrak{Q}$ of S such that height $\mathfrak{Q} \cap A \neq 0$. Condition (a) implies that the image of $\mathcal{D}^A(S)$ in $\mathsf{Cl}(S)$ is surjective, and therefore

$$\mathcal{D}^A(S)/(\mathcal{D}^A(S) \cap \mathrm{Prin}(S)) \cong \mathcal{D}(S)/\mathrm{Prin}(S) = \mathsf{Cl}(S).$$

To construct the homomorphism $\varphi : \mathsf{Cl}(S) \to \mathsf{Cl}(A)$, we first define it at the level of the divisor groups $\phi : \mathcal{D}(S) \to \mathcal{D}(A)$ by setting $\phi([\mathfrak{Q}]) = [\mathfrak{q}]$ if $\mathfrak{q} = \mathfrak{Q} \cap A$ has height 1; and 0 otherwise. Thus with $a_i \in \mathbb{Z}$ the divisor $\sum_i a_i[\mathfrak{Q}_i]$, is mapped to $\sum_i a_i \phi([\mathfrak{Q}_i]) \in \mathcal{D}(A)$.

Condition (b) ensures that ϕ is a surjection and remains so when restricted to $\mathcal{D}^A(S)$, and we denote it by ϕ again. It is this restriction that will be the focus of our argument. We first examine the principal divisors lying in $\mathcal{D}^A(S)$. Suppose that $f \in K(\mathbf{t})$, and denote its divisor by

$$[f]_S = \sum_i v_{\mathfrak{Q}_i}(f)[\mathfrak{Q}_i] + \sum_j v_{\mathfrak{P}_j}(f)[\mathfrak{P}_j],$$

where $[\mathfrak{Q}_i]$ lies in $\mathcal{D}^A(S)$ and $[\mathfrak{P}_j]$ does not, and note that

$$\phi([f]_S) = \sum_i{}' v_{\mathfrak{Q}_i}(f)[\mathfrak{Q}_i],$$

for those $\mathfrak{Q}_i$ such that height $\mathfrak{Q}_i \cap A = 1$. For $[f]_S \in \mathcal{D}^A(S)$ we must have $v_{\mathfrak{P}}(f) = 0$ for $[\mathfrak{P}] \notin \mathcal{D}^A(S)$, which means that $v_{\mathfrak{P}}(f) = 0$ for all $\mathfrak{P}$ with $\mathfrak{P} \cap A = 0$, that is, the prime valuations of the polynomial ring $K[\mathbf{t}]$, forcing $f \in K$. Appealing to condition (b) again we have that the image of $[f]_S$ in $\mathcal{D}(A)$ is the divisor $[f]_A$. We denote by φ the induced homomorphism (surjection) $\varphi \colon \mathsf{Cl}(S) \to \mathsf{Cl}(A)$.

We claim that

$$H = \ker(\mathsf{Cl}(S) \to \mathsf{Cl}(A)) = \langle [\mathfrak{Q}] \in \mathcal{D}(S) \mid \text{height } (Q \cap A) \geq 2 \rangle (\text{mod principal divisors}).$$

If $D \in \mathcal{D}^A(S)$ is such that $\varphi(D) = [f]_A$ with $f \in K$, then by Step (ii),

$$D - fS \in \langle \mathfrak{Q} \in \mathcal{D}(S) \mid \text{height } (\mathfrak{Q} \cap A) \geq 2 \rangle.$$

Finally, we verify that H is free on these generators. Otherwise, for such prime divisors $\mathfrak{Q}_1, \ldots, \mathfrak{Q}_n$ of S, a relation means that there exist integers a_i such that

$$\sum_i a_i [\mathfrak{Q}_i] = [f]_S,$$

for some $f \in K(\mathbf{t})$. This would mean that $f \in K$ and also $v_{\mathfrak{P}}(f) = 0$ for all prime divisors such that height $\mathfrak{P} \cap A = 1$. But by condition (b) these give all the valuations of A, thus showing that f is a unit of A. Thus each $a_i = 0$, and H is indeed a free group on the specified generators. □

Our main vehicle for applications is the following result ([Hon3a]).

Theorem 4.16. *Let A be a Noetherian integrally closed domain, let E be a finitely generated torsionfree A-module and S the integral closure of the Rees algebra of E. There exists an exact sequence*

$$0 \to H \longrightarrow \mathsf{Cl}(S) \longrightarrow \mathsf{Cl}(A) \to 0$$

of divisor class groups, where H is a free abelian group of finite rank.

Proof. We begin with two observations about how changes of rings affect Theorem 4.15. Firstly, if x is a fresh indeterminate, we can replace A and S respectively by $A[x]$ and $S[x]$, without changing any of the groups. This is the case because (say) the prime divisors $\mathfrak{M}$ of $S[x]$ are either of the form $\mathfrak{Q}S[x]$ for a prime divisor $\mathfrak{Q}$ of S, or satisfy $\mathfrak{M} \cap S = 0$. The latter leave H unchanged.

Secondly, if h is a nonzero element of A, in the corresponding sequence

$$0 \to H_0 \longrightarrow \mathsf{Cl}(S_f) \longrightarrow \mathsf{Cl}(A_f) \to 0$$

of divisor class groups, H_0 is the image of H under the natural mapping $\mathsf{Cl}(S) \to \mathsf{Cl}(S_f)$; this follows from the construction above.

Let us now establish the claim that H has finite rank. Since E is a torsionfree R-module, $E_{\mathfrak{p}}$ is $A_{\mathfrak{p}}$-free for every height 1 prime $\mathfrak{p}$ of A. Since the free locus of a module is an open set of Spec (A), say $D(I)$, we must have that height $I \geq 2$. Add an indeterminate x to both A and S, choose $a, b \in I$ forming a regular sequence and set $h = a + bx$. This is possible since the grade of I is at least 2. On the other hand, by Nagata's Lemma h is a prime element of A, and S_h is the symmetric algebra of the projective A_h-module E_h. This implies that there is no prime divisor $\mathfrak{Q}$ of S_h such height $(\mathfrak{Q} \cap A_h) \geq 2$. This shows that the mapping $\mathsf{Cl}(S_f) \to \mathsf{Cl}(A_h)$ is an isomorphism. For such h, $\ker(\mathsf{Cl}(S) \to \mathsf{Cl}(S_f))$ is a subgroup of finite rank generated by the divisor classes of height 1 primes of S that contain h. Since this subgroup is naturally identified with H, we establish the claim. □

Rees Valuations of Determinantal Ideals

Let $X = (x_{ij})$ be a generic matrix of size $m \times n$ over a field k and R the ring of polynomials $R = k[x_{ij}]$. Let I_t be the ideal of R generated by the minors of size t of X. In Chapter 1, Section 2, we listed some classes of normal determinantal ideals. In the case where $t = \min\{m,n\}$, $I = I_t$ is normal and $I\mathcal{R}(I)$ is a prime ideal in any characteristic. It follows that $\mathsf{Cl}(\mathcal{R}(I)) \cong \mathbb{Z}$.

Assume that $t < \inf\{m,n\}$. In the cases when $\mathcal{R}(I)$ is normal (e.g. in characteristic zero), to find the rank of $\mathsf{Cl}(\mathcal{R}(I))$ we must find the minimal primes of $I\mathcal{R}(I)$. They are the primes that give rise to the Rees valuations of I. There are t of them, according to the detailed description of [BC1], so $\mathsf{CL}(\mathcal{R}(I)) \cong \mathbb{Z}^t$.

Rank of the Divisor Class Group

To identify the set $\{Q_1, \ldots, Q_s\}$ of minimal primes of $I\mathcal{R}$ is usually very difficult. We consider a few cases of interest.

(i) Suppose that I is an unmixed normal ideal of height one. The inclusion $R \hookrightarrow \mathcal{R}$ defines an embedding

$$\mathsf{Cl}(R) \longrightarrow \mathsf{Cl}(\mathcal{R}), \quad [\mathfrak{p}] \mapsto [T(\mathfrak{p})]$$

of divisor class groups that is an isomorphism.

(ii) Let R be a factorial domain and I a prime ideal that is generically a complete intersection and is such that the associated graded ring $\mathrm{gr}_I(R)$ is an integral domain (in this case $\mathcal{R}$ is automatically normal: see the comment below). In this case, $\mathsf{Cl}(R) = 0$ and $\mathsf{Cl}(\mathcal{R}) = \mathbb{Z}[I\mathcal{R}]$.

Ideals of Linear Type

Let I be a (normal) ideal of linear type, that is, $\mathcal{R}$ is the symmetric algebra $S_R(I)$ (see [Va94b, p. 138]). Let

$$R^m \xrightarrow{\psi} R^n \longrightarrow I \to 0$$

be a presentation of I. One can determine the minimal primes of $I\mathcal{R}$ using the matrix ψ. For a prime ideal $\mathfrak{p} \subset R$ there is an associated prime ideal

$$T(\mathfrak{p}) = \ker(S_R(I) \longrightarrow S_{R/\mathfrak{p}R}(I/\mathfrak{p}I)_{\mathfrak{p}}).$$

Assume that R is universally catenary in order to avoid dwelling on technicalities. In this case, if in addition $S(I)$ is equidimensional, one sees that

$$\text{height } T(\mathfrak{p}) = 1 \text{ if and only if } \nu(I_{\mathfrak{p}}) = \text{height } \mathfrak{p}.$$

We also observe that unless $\mathfrak{p}$ is a height one prime itself, I is not free if $T(\mathfrak{p})$ is to have height one.

Proposition 4.17. *Let R be a universally catenary Noetherian ring and let I be an ideal such that $S(I)$ is a domain. Then the set*

$$\{T(\mathfrak{p}) \mid \text{height } \mathfrak{p} \geq 2 \textit{ and } \text{height } T(\mathfrak{p}) = 1\}$$

is finite. More precisely, this set is in bijection with the set

$$\{\mathfrak{p} \subset R \mid I_{\mathfrak{p}} \textit{ not free}, \mathfrak{p} \in \text{Min}(R/I_t(\psi)) \textit{ and } \text{height } \mathfrak{p} = \text{rank}(\psi) - t + 2\},$$

where $1 \leq t \leq \text{rank}(\psi)$.

Proof. Given a prime $\mathfrak{p} \subset R$, set $t = n - \nu(I_{\mathfrak{p}})$. If height $\mathfrak{p} \geq 2$ and height $T(\mathfrak{p}) = 1$ then, since $\text{rank}(\psi) = n - 1$, by the preceding remarks we have

$$\text{height } \mathfrak{p} = n - 1 - t + 2 \text{ and } \mathfrak{p} \supset I_t(\psi) \setminus I_{t-1}(\psi).$$

On the other hand, since $S(I)$ is a domain, $\nu(I_{\mathfrak{p}}) \leq \text{height } \mathfrak{p}$ for every nonzero prime ideal. Therefore, one has height $I_t(\psi) \geq \text{rank}(\psi) - t + 2$. It follows that $\mathfrak{p} \in \text{Min}(R/I_t(\psi))$. The converse is similar. □

Reducedness and Normality

There are very few criteria of normality for Rees algebra other than general Jacobian tests, which do not take into account the nature of a Rees algebra.

We begin with the following observation:

Proposition 4.18. *Let R be a Noetherian domain and let t be a nonzero element such that (t) is a radical ideal. If $R[t^{-1}]$ is normal then R is normal.*

Proof. Let $P_1, \ldots, P_s$ be the minimal primes of (t). Every localization R_{P_i} is a discrete valuation domain. It is easy to verify that

$$R = R_{P_1} \cap \cdots \cap R_{P_s} \cap R[t^{-1}],$$

and therefore R is integrally closed as well. □

Proposition 4.19. *Let R be a Noetherian normal domain and I an ideal such that $\text{gr}_I(R)$ is a reduced ring. Then $\mathcal{R} = R[It]$ is integrally closed.*

Proof. Consider the extended Rees algebra, $\mathcal{R}_e = R[It, t^{-1}]$. Since $\mathcal{G} = \mathcal{R}_e/(t^{-1})$ is reduced it follows that $\mathcal{R}_e$ satisfies the S_2 and (R_1) conditions of Serre. Thus $\mathcal{R}_e$ is normal and $\mathcal{R}$ is also integrally closed since $\mathcal{R} = \mathcal{R}_e \cap R[t]$. □

Corollary 4.20. *Let R be a factorial regular ring and I a height unmixed radical ideal of codimension two. If the symbolic power algebra $\mathcal{S} = \mathcal{R}_s = \sum_{n\geq 0} I^{(n)}t^n$ is Noetherian, then $\mathcal{S}$ is quasi-Gorenstein.*

Proof. By the proposition, $\mathcal{S}$ is a Krull domain whose divisor class group is freely generated by the classes of the minimal primes $Q_1, \ldots, Q_s$ of $I\mathcal{S}$,

$$\mathrm{Cl}(\mathcal{S}) = \mathbb{Z}[Q_1] \oplus \cdots \oplus \mathbb{Z}[Q_s].$$

If $\omega_{\mathcal{S}}$ is the canonical module of $\mathcal{S}$, let

$$[\omega_{\mathcal{S}}] = \sum_{i=1}^{s} a_i [Q_i]$$

denote its divisor class. For each Q_i, on localizing at $\mathfrak{p}_i = Q_i \cap R$ we obtain that $\mathcal{S}_{\mathfrak{p}_i}$ is a complete intersection, being the Rees algebra of the maximal ideal of the two-dimensional regular local ring $R_{\mathfrak{p}_i}$. Since the divisor class group of $\mathcal{S}_{\mathfrak{p}}$ is $\mathbb{Z}[Q_i]$, it follows that $a_i = 0$. This actually shows that $\omega_{\mathcal{S}} = t\mathcal{S}$. □

Under some additional conditions on the ideal I one obtains a very explicit description of the divisor class group of Rees algebras whose associated graded are reduced ([HSV89]).

Theorem 4.21. *Let $(R, \mathfrak{m})$ be a quasi-unmixed local ring and I an ideal of finite projective dimension. If $\mathcal{G} = \mathrm{gr}_I(R)$ is a reduced ring then $\mathcal{G}$ is a torsionfree R/I-algebra. Moreover, if I is a prime ideal then $\mathcal{G}$ is a domain.*

Proof. We may assume that the residue field of R is infinite. Let P be a minimal prime of $I\mathcal{R}$; then $P\mathcal{R}_P = I\mathcal{R}_P$. We want to relate these primes to the primes of I.

Since $\mathcal{G}$ is reduced, I is a radical ideal. Let $\{\mathfrak{p}_1, \ldots, \mathfrak{p}_s\}$ be the set of minimal primes of I. For a prime $\mathfrak{p} = \mathfrak{p}_i$, set

$$T(\mathfrak{p}) = \text{kernel } (\mathcal{R} \longrightarrow (\mathcal{R}/\mathfrak{p}\mathcal{R})_{\mathfrak{p}}).$$

Since $I_{\mathfrak{p}} = \mathfrak{p}R_{\mathfrak{p}}$ has finite projective dimension, $R_{\mathfrak{p}}$ is a regular local ring and therefore $T(\mathfrak{p})$ will be a prime ideal.

The claim is equivalent to saying that any minimal prime of $I\mathcal{R}$ is one of the $T(\mathfrak{p}_i)$, i.e., that $P \cap R = \mathfrak{p}_i$ for some i. We suppose otherwise, set $\mathfrak{q} = P \cap R$, localize at $\mathfrak{q}$, change notation and assume that $P \cap R = \mathfrak{m}$ is the maximal ideal of R. Two observations arise: (i) $I \not\subset \mathfrak{m}^2$ and

$$\text{(ii)} \quad \dim R = \dim \mathcal{R}/P \leq \dim \mathcal{R}/\mathfrak{m}\mathcal{R} = \ell(I) \leq \dim R.$$

By Theorem 3.60, (i) and (ii) are untenable. □

Example 4.22. Suppose that $R = k[[x,y,z]]/(y^2 - xz)$, k a field, and let $\mathfrak{p} = (y,x)R$. Since R is a domain and $(y) : x = (y) : x^2$, it follows that y,x is a d-sequence so that $\mathrm{gr}_{\mathfrak{p}}(R) \cong S_{R/\mathfrak{p}}(\mathfrak{p}/\mathfrak{p}^2)$, the symmetric algebra of $\mathfrak{p}/\mathfrak{p}^2$. We then obtain that $\mathrm{gr}_{\mathfrak{p}}(R) = k[z,u,v]/(uv)$, which is reduced but not a domain. Note however that $\mathrm{proj}\ \dim_R \mathfrak{p} = \infty$.

Corollary 4.23. *Let R be a quasi-unmixed normal domain and I an ideal such that $\mathcal{G} = \mathrm{gr}_I(R)$ is reduced. Denote by $\{\mathfrak{p}_1, \ldots, \mathfrak{p}_s\}$ the minimal primes of I and by $T(\mathfrak{p}_i)$ the corresponding minimal primes of $I\mathcal{R}$. Then $\mathcal{R} = R[It]$ is a Krull domain whose divisor class group has a presentation*

$$0 \to H \longrightarrow \mathrm{Cl}(\mathcal{R}) \stackrel{\varphi}{\longrightarrow} \mathrm{Cl}(R) \to 0,$$

where H is freely generated by the $[T(\mathfrak{p}_i)]$.

Divisor Class under Shifting and Prolongation

Our purpose is to use the divisor class group of a normal Rees algebra to establish the existence and uniqueness of the prolongation of a divisor. More precisely, we provide in this case an *explicit prolongation* for any divisor.

Theorem 4.24. *Let R be an integrally closed Noetherian domain and let I be a normal ideal of height at least 2. Every divisor L of $R[It]$ admits a prolongation D, unique up to isomorphism.*

Proof. We first prove uniqueness. Suppose that $L = \sum_{n\geq 1} L_n t^n$, with $L_1 \neq 0$, is a divisor and let $D = \sum_{n\geq 1} D_n t^n$ be one of its prolongations. From the sequence

$$0 \to L[-1] \longrightarrow D \longrightarrow D_1 t \to 0,$$

we observe that the annihilator of D_1 is exactly the divisorial prime ideal $\mathfrak{P} = (It)$, as $\mathrm{ann}\ (D/L[-1])$ cannot contain properly $\mathfrak{P}$ since $L[-t]$ satisfies Serre's condition S_2. We are going to consider how the relation

$$\mathfrak{P} \cdot L[-1] \subset D$$

is reflected in the divisor class group of $R[It]$. We claim that

$$[\mathfrak{P}] + [L[-1]] = [\mathfrak{P}] + [L] = [D].$$

It suffices to observe that if v is the valuation associated to $\mathfrak{P}$, we have $v(L[-1]) = v(D) - 1$. At the valuations $v_{\mathfrak{Q}}$ associated with the height one primes distinct from $\mathfrak{P}$, we have $v_{\mathfrak{Q}}(L[-1]) = v_{\mathfrak{Q}}(D)$. This show that the divisor class of D is $[L] + [\mathfrak{P}]$, and thus D is well determined by L.

To prove existence, let A be a (graded) divisorial ideal of divisor class $[A] = [L[-1]] + [\mathfrak{P}]$. For example, set

$$A = \mathrm{Hom}_{R[It]}(\mathfrak{P}, L[-1]).$$

It is easy to see that A is a divisorial ideal of the form $A = \sum_{n\geq 1} A_n t^n$. Consider the component of degree $n \geq 2$ of the quotient $A/L[-1]$, A_n/L_{n-1}. This is a torsion R-module, so its annihilator contains $\mathfrak{P}$ and some nonzero element of R. But this would be an ideal of height at least 2, which is not possible since L has condition S_1. $\square$

Divisor Class Groups of Associated Graded Rings

Let R be a Noetherian ring and I an ideal. If the Rees algebra $\mathcal{R} = R[It]$ of I is integrally closed, its divisor class group is intricately connected to the divisor class of R. Indeed, if I is a prime ideal of finite projective dimension such that the associated graded ring $G = \mathrm{gr}_I(R)$ is normal, then R/I and G have isomorphic divisor class groups. This fact ([Hon3a]) is a consequence of results of Johnson and Ulrich ([JU99]) and Huneke ([Hu82a]) that circumscribe very explicitly the Serre's condition R_k for G.

Proposition 4.25. *Let R be a Cohen-Macaulay ring and I a prime ideal of finite projective dimension. Suppose that $G = \mathrm{gr}_I(R)$ is a normal ring. Suppose that $\mathfrak{p} \in V(I)$ has codimension at least* height $I+2$. *Then* height $\mathfrak{p}G \geq 2$.

Proof. We begin by pointing out that perfect prime ideals are generically complete intersections. Consequently, if $\ell(I_\mathfrak{p}) = \text{height}\, I$ for some prime ideal $I \subset \mathfrak{p}$, then by [CN76], $I_\mathfrak{p}$ is a complete intersection.

Suppose that $\mathfrak{p}G$ is contained in a prime ideal P of height 1. Setting $\mathfrak{m}$ for the inverse image of P in R and localizing, we may assume that $(R, \mathfrak{m})$ is a local ring and that $\mathfrak{p}G \subset \mathfrak{m}G \subset P$. Note that if I is a complete intersection, G is a polynomial ring over R/I and height $\mathfrak{p} = \mathfrak{p}G \geq 2$. In all cases, if height $\mathfrak{m}G = 1$, then $\ell(I) = \dim R - 1$, which by Theorem 3.60 would mean that height $I = d-1$, thus contradicting the choice of $\mathfrak{p}$. □

Theorem 4.26. *Let R be a Noetherian ring and I a perfect prime ideal. If $G = \mathrm{gr}_I(R)$ is integrally closed, the mapping*

$$\mathrm{Cl}(G) \mapsto \mathrm{Cl}(R/I)$$

of divisor class groups defined in Theorem 4.15 *is a group isomorphism. In particular, if G is normal and R/I is factorial, then G is factorial.*

Proof. According to Proposition 4.25, there are no prime divisor $\mathfrak{Q}$ of G such that height $(\mathfrak{Q} \cap R/I) \geq 2$. Thus the subgroup H in Theorem 4.15 is trivial. □

4.3 The Expected Canonical Module

There is a need for standard models for canonical modules of Rees algebras. One that surfaced early is given by the following example.

Example 4.27. If the ideal I is generated by a regular sequence $f_1, \ldots, f_g$, $g \geq 2$, the equations of $\mathcal{R} = R[It]$ are nice:

$$\mathcal{R} \cong R[T_1, \ldots, T_g]/I_2 \begin{pmatrix} T_1 & \cdots & T_g \\ f_1 & \cdots & f_g \end{pmatrix}.$$

In other words, the defining equations of the Rees algebra are generated by the Koszul relations of the f_i.

Knowing this description of $\mathcal{R}$ leads immediately to its canonical module ([B82])

$$\omega_{\mathcal{R}} = \omega_R t(1,t)^{g-2}\mathcal{R} = \bigoplus_{n\geq 1} I^{n-g+1}\omega_R t^n.$$

There are however many other instances of Rees algebras whose canonical modules have this form. It warrants the following:

Definition 4.28. Let R be a Noetherian local ring with a canonical module ω_R and let I be an ideal of positive grade. The canonical module of $\mathcal{R} = R[It]$ is said to have the *expected form* if

$$\omega_{\mathcal{R}} = \omega_R(1,t)^b t\mathcal{R}$$

for some integer b.

We extend this slightly to accommodate the following

Definition 4.29. Let R be a Noetherian local ring with a canonical module ω_R and I an ideal of positive grade. A divisor L of $\mathcal{R} = R[It]$ has the *expected form* if

$$L \cong J(1,t)^b\mathcal{R}$$

for some ideal J of R and for some integer b.

The connection with the canonical module of $\mathrm{gr}_I(R)$ is provided in the following result ([HSV87], [Za92]).

Theorem 4.30. *Let $(R,\mathfrak{m})$ be a Noetherian local ring with a canonical module $\omega = \omega_R$, and I an ideal of positive grade. Assume that $\mathcal{R} = R[It]$ is Cohen-Macaulay and set $a = -a(\mathcal{G})$ for $\mathcal{G} = \mathrm{gr}_I(R)$. Then the following are equivalent:*

(a) $\omega_{\mathcal{G}} \cong \mathrm{gr}_I(\omega_R)[-a]$.
(b) $\omega_{\mathcal{R}} \cong \omega_R(1,t)^{a-2}\mathcal{R}[-1]$.

Proof. Denote by $\omega_{\mathcal{R}}$ the canonical module of $\mathcal{R}$. Applying $\mathrm{Hom}_{\mathcal{R}}(\cdot,\omega_{\mathcal{R}})$ to the sequences

$$0 \to I\mathcal{R} \longrightarrow \mathcal{R} \longrightarrow \mathcal{G} \to 0$$

and

$$0 \to It\mathcal{R} \longrightarrow \mathcal{R} \longrightarrow R \to 0,$$

we obtain

$$0 \to \omega_{\mathcal{R}} \longrightarrow \omega_{I\mathcal{R}} \longrightarrow \omega_{\mathcal{G}} \to 0$$

and

$$0 \to \omega_{\mathcal{R}} \longrightarrow \omega_{It\mathcal{R}} \longrightarrow \omega_R \to 0.$$

Suppose that (b) holds. From the last sequence we get that $\omega_{It\mathcal{R}} = \omega_R(1,t)^{a-1}\mathcal{R}$. Feeding this into the previous sequence, given that $\omega_{I\mathcal{R}} = t\omega_{It\mathcal{R}}$, we obtain that the canonical module of $\mathcal{G}$ is as expected.

For the converse it is convenient to express these exact sequences of canonical modules in the following manner:

$$0 \to \omega_{\mathcal{R}} = \sum_{n\geq 1} D_n t^n \longrightarrow \omega_{I\mathcal{R}} = \sum_{n\geq 1} E_n t^n \longrightarrow \omega_{\mathcal{G}} = \sum_{n\geq a} F_n t^n \to 0$$

and

$$0 \to \omega_{\mathcal{R}} = \sum_{n\geq 1} D_n t^n \longrightarrow \omega_{It\mathcal{R}} = \sum_{n\geq 1} E_n t^{n-1} \longrightarrow \omega \to 0.$$

From which we get (noting that $a > 0$),

$$\begin{aligned} E_1 &= \omega \\ E_n &= D_{n-1} \qquad n \geq 2 \\ E_n &= D_n \qquad 1 \leq n < a \\ E_n/D_n &= F_n \qquad n \geq a. \end{aligned}$$

It follows that $D_n = \omega$ for $n < a$. For $n = a$, we have

$$E_a/D_a = D_{a-1}/D_a = \omega/D_a \cong \omega/I\omega.$$

On the other hand, the inclusion $ID_n \subset D_{n+1}$ for all n gives epimorphisms

$$\omega/I\omega \twoheadrightarrow \omega/D_a \twoheadrightarrow \omega/I\omega.$$

It follows that $D_a = I\omega$. The remainder of the argument is similar. □

Remark 4.31. For an ideal I in the situation of this theorem, one has $a(\mathrm{gr}_I(R)) = a(\mathrm{gr}_{I_\mathfrak{p}}(R_\mathfrak{p}))$ for every minimal prime $\mathfrak{p}$ of I. Thus if I is a generic complete intersection, $a(\mathrm{gr}_I(R)) = -\mathrm{height}\, I$. More generally, $a(\mathrm{gr}_I(R)) = \mathrm{r}(I_\mathfrak{p}) - \mathrm{height}\, I$.

Gorenstein Algebras

The arrangement of the proof above can be used in the proof of the following result of Ikeda ([Ik86]).

Theorem 4.32. *Let $(R,\mathfrak{m})$ be a Noetherian local ring and I an ideal of grade at least 2. If $R[It]$ is Cohen-Macaulay, the following conditions are equivalent:*

(a) *$R[It]$ is Gorenstein.*
(b) *$\omega_R \cong R$ and $\omega_{\mathrm{gr}_I(R)} \cong \mathrm{gr}_I(R)[-2]$.*

Proof. (a) $\Rightarrow$ (b) Setting $\mathcal{R} = R[It]$ and $\omega_{\mathcal{R}} = \sum_{n\geq 1} D_n t^n$, we have

$$\mathcal{D}(I) = \omega_R t + \sum_{n\geq 1} D_n t^{n+1}.$$

The embedding

$$0 \to \mathsf{S}(\mathcal{D}(I)) = \omega_{\mathcal{R}} \longrightarrow \mathcal{D}(I) \longrightarrow \mathsf{G}(\mathcal{D}(I)) = \omega_{\mathcal{G}} \to 0$$

shows that if $\omega_{\mathcal{R}} \cong \mathcal{R}$ then $R = D_1 \subset \omega_R$ and $I \cdot \omega_R \subset R$; it follows that $D_1 = R = \omega_R$ since ω_R lies in the total ring of quotients of R and grade $I \geq 2$. As $D_n = I^{n-1}$ for $n \geq 2$, the canonical module of $\mathcal{G}$ is as asserted.

(b) $\Rightarrow$ (a). $\omega_{\mathcal{R}} = \omega_R t(1,t)^{a-2}\mathcal{R} \cong \mathcal{R}$ by the hypothesis and Theorem 4.30. □

Symbolic Powers

Let R be a factorial regular domain and I a height unmixed radical ideal of codimension $g \geq 2$ whose symbolic power algebra $\mathcal{S} = \sum_{n\geq 0} I^{(n)} t^n$ is Noetherian. With the same argument used in Corollary 4.20, one can describe the canonical module of $\mathcal{S}$.

Theorem 4.33. *Under these conditions,*

$$\omega_{\mathcal{S}} = t(1,t)^{g-2}\mathcal{S}.$$

Moreover, if $\mathcal{S}$ is Cohen-Macaulay the associated graded ring of the symbolic filtration is quasi-Gorenstein.

Proof. We leave the details of the proof to the reader. As in Corollary 4.20 one gets a similar description for the divisor class group of $\mathcal{S}$ with the coefficients a_i of that proof calculated as in Example 4.27. □

Gaps in Divisors

There is another property that has been observed in certain divisors of Rees algebras that require more understanding. It is related to the issue of gaps in the generating sequence of a module. Let $(R, \mathfrak{m})$ be a local ring, A a homogeneous graded R-algebra and E a nonzero finitely generated graded A-module. The module $\overline{E} = E/(\mathfrak{m}, A_+)E$ is a finite R-module,

$$\overline{E} = \overline{E}_{\alpha_1} \oplus \cdots \oplus \overline{E}_{\alpha_n}, \quad \overline{E}_{\alpha_i} \neq 0, \quad \alpha_1 < \alpha_2 < \cdots < \alpha_n.$$

The α's form the *generating sequence* of E, and the module is said to have *gaps* if the sequence is not contiguous.

Question 4.34. Let $(R, \mathfrak{m})$ be a Cohen-Macaulay local ring and B the Rees algebra of the ideal I. If B is 'nice', does ω_B have gaps?

We are going to give a discussion of the special case of an $\mathfrak{m}$-primary ideal I. To further simplify, we will assume that R is Gorenstein with infinite residue field.

Set $\dim R = d > 1$, let J be a minimal reduction, and $A = R[Jt]$, $B = R[It]$. Set $r = \mathrm{r}_J(I)$ ($r \leq d-1$ when B is Cohen-Macaulay). We consider the reduction sequence

$$0 \to I \cdot A \longrightarrow B_+[+1] \longrightarrow S_{I/J} \to 0, \tag{4.1}$$

and the sequence

$$0 \to I \cdot A \longrightarrow A \longrightarrow R/I[T_1, \ldots, T_d] \to 0.$$

We now derive the canonical modules associated with these Cohen-Macaulay modules:

$$0 \to \omega_{B_+}[-1] \longrightarrow \omega_{IA} \longrightarrow \omega_{S_{I/J}} \to 0, \tag{4.2}$$

$$0 \to \omega_A \longrightarrow \omega_{IA} \to \omega_{R/I}[T_1, \ldots, T_d][-d] \to 0. \tag{4.3}$$

On the other hand, from the exact sequence

$$0 \to B_+ \longrightarrow B \longrightarrow R \to 0,$$

we have

$$0 \to \omega_B \longrightarrow \omega_{B_+} \longrightarrow \omega_R \to 0.$$

It is simpler to start with the sequence (4.3). The canonical module of A being simply $(1,t)^{d-2}tA$ (in particular it has no gaps), it allows us to describe the generators of ω_{IA},

$$\omega_{IA} = Rt + \cdots + Rt^{d-1} + (J : I)t^d + \cdots.$$

Thus ω_{IA} also has no gaps (as an A-module) in its generating sequence.

To carry this information into (4.1), we recall that when depth $\mathrm{gr}_I(R) \geq d-1$, the Sally module $S_{I/J}$ admits a filtration by modules of the form $(I^{j+1}/JI^j)[T_1, \ldots, T_d][-j]$, $1 \leq j \leq r-1$. As a consequence, $\omega_{S_{I/J}}$ admits a filtration whose factors are

$$(I^{j+1}/JI^j)^\vee[T_1, \ldots, T_d][-d+j], \; 1 \leq j \leq r-1.$$

Let us write

$$\begin{aligned} \omega_{B_+}[-1] &= Rt + \omega_2 t^2 + \cdots + \omega_{d-r}t^{d-r} + \omega_{d-r+1}t^{d-r+1} + \cdots \\ &+ \omega_{d-1}t^{d-1} + \omega_d t^d + \omega_{d+1}t^{d+1} + \cdots. \end{aligned}$$

Since

$$\omega_{B_+}[-1] = Rt + t\omega_B,$$

we have only to recover ω_i from (4.2). We first observe that

$$\omega_2 = \cdots = \omega_{d-r} = R,$$

from which point on we have exact sequences

$$\begin{array}{c}
0 \to \omega_{d-r+1} \longrightarrow R \longrightarrow (I^r/JI^{r-1})^\vee \to 0 \\
0 \to \omega_{d-r+2} \longrightarrow R \longrightarrow (I^{r-1}/JI^{r-2})^\vee \to 0 \\
\vdots \\
0 \to \omega_{d-1} \longrightarrow R \longrightarrow (I^2/JI)^\vee \to 0 \\
0 \to \omega_d \longrightarrow J:I \longrightarrow Jt\cdot(I^2/JI)^\vee = (I^3/IJ^2)^\vee \to 0 \\
0 \to \omega_{d+1} \longrightarrow J(J:I) \longrightarrow J^2t^2\cdot(I^2/JI)^\vee = (I^4/IJ^3)^\vee \to 0 \\
\vdots
\end{array}$$

Let us illustrate with the case $r = 1$. Then $S_{I/J} = 0$ and $\omega_{B_+}[-1] = \omega_{IA}$, which has no gaps as remarked above.

Suppose that $r = 2$. If $\omega_{d-1} = I$, that is there are no fresh generators in degree $d-1$, we should find no fresh generators thereafter. Let us argue the next degree directly. By assumption, $R/I \cong (I^2/JI)^\vee$, and we claim that $(J:I)/I^2 \cong (I^3/IJ^2)^\vee$. Note that by duality the first isomorphism means that $(J:I)/I \cong I^2/JI$. From the diagram

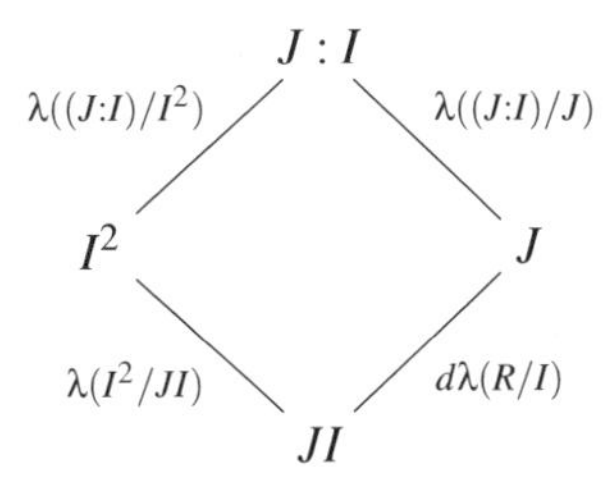

and since $\lambda(I^2/JI) = \lambda(R/I) = \lambda((J:I)/J)$, we have $\lambda((J:I)/I^2) = d\lambda(R/I) = \lambda(I^3/IJ^2)$, as this module is the corresponding component of

$$S_{I/J} = (I^2/JI)[T_1,\ldots,T_d][-d].$$

This shows that $\omega_{d-1} = I^2$.

To show that $\omega_d = I^3$, we must examine the diagram

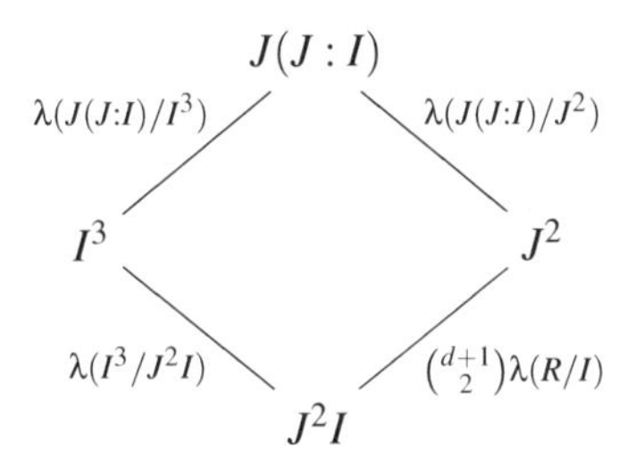

in which, from the Sally module, we have $\lambda(I^3/J^2I) = \binom{d+1}{2}\lambda(R/I)$. We must show that this is also the length of $J(J:I)/J^2$. But this follows from the isomorphism

$(J:I)/J \cong I^2/IJ$ and the induced $A = R[Jt]$-morphism $(J:I)tA/A_+ \to S_{I/J}$, which we claim is an isomorphism. Indeed, in each degree n it is obviously a surjection. Furthermore, $(S_{I/J})_n = (I^2/JI) \otimes_R R[T_1, \ldots, T_d]_{n-1}$, while $J^{n-1}(J:I)/J^n$ is a homomorphic image of $(J:I)/J \otimes_R R[T_1, \ldots, T_d]_{n-1}$.

The hypothesis $\omega_{d-2} = I$ seems very strong, as it leads to ω_B having the expected form.

Conjecture 4.35. Let $(R, \mathfrak{m})$ be a Cohen-Macaulay local ring and I an ideal such that the Rees algebra $B = R[It]$ is Cohen-Macaulay. Then ω_B has no gaps.

Remark 4.36. Several other issues are being without examination, among them the connection between the Hilbert coefficients and the canonical module of $R[It]$ or $\mathrm{gr}_I(R)$; see [HyJ2].

4.4 The Fundamental Divisor

Our main purpose is to discuss a divisorial ideal associated with a Rees algebra and to sketch out some of its applications ([MV1]). It helps to explain old puzzles while at the same time providing quite direct proofs of earlier results. The reader will note that it is a mirror image of local cohomology modules of Rees algebras. However, its Noetherian character permits a control of computation that is not always possible with Artinian modules. The properties of this divisor seem to be related to some sheaf cohomology results of Grauert-Riemenschneider ([GRi70]), but we have not worked out this similarity (if there is one).

Let $(R, \mathfrak{m})$ be a Cohen-Macaulay local ring of dimension d with a canonical module ω. Let $I = (f_1, \ldots, f_n)$ be an ideal of positive height. Fix a presentation $B = R[T_1, \ldots, T_n] \to \mathcal{R}$ of $\mathcal{R}$. We set $\omega_B = \omega \otimes_R B[-n]$ as the canonical module of the polynomial ring B. The canonical module of $\mathcal{R}$ is the module

$$\omega_{\mathcal{R}} = \mathrm{Ext}_B^{n-d+1}(\mathcal{R}, \omega_B),$$

while the *fundamental divisor* of $\mathcal{R}$ is the module

$$\mathcal{D}(I) = \mathrm{Hom}_{\mathcal{R}}(IR[It], \omega_{\mathcal{R}}).$$

It can be written out as

$$\mathcal{D}(I) = D_1 t + D_2 t^2 + D_3 t^3 + \cdots,$$

where each D_i is an R-submodule of K, the total ring of fractions of R. We fix this representation of $\mathcal{D}(I)$ from a given projective resolution of $I \cdot \mathcal{R}$ and the computation of the cohomology.

This divisorial ideal carries more information than the canonical module $\omega_{\mathcal{R}}$. Indeed, we can view $\mathcal{D}(I)$ as made up of two parts,

$$\mathcal{D}(I) = \underbrace{D_1 t}_{\text{(i)}} + \underbrace{D_2 t^2 + D_3 t^3 + \cdots}_{\text{(ii)}}.$$

Then D_1 (or $D_1 t$ when the powers of t are used as tags) will be called the leading part of $\mathcal{D}(I)$, while (ii) is called its canonical part in agreement with the next observation.

A consequence of the proof of Theorem 4.30 is the following relationship between the canonical and fundamental divisors of a Rees algebra.

Proposition 4.37. *Let R be a Noetherian ring with a canonical module and let I be an ideal of positive grade. Then the graded components of $\omega_{\mathcal{R}}$ and of $\mathcal{D}(I)$ are related in the following manner:*

$$\omega_{\mathcal{R}} \cong \sum_{n\geq 1} D_{n+1} t^n,$$
$$\mathcal{D}(I) \cong D_1 t + \sum_{n\geq 2} D_n t^n,$$

that is, $\mathcal{D}(I)$ is a prolongation of $\omega_{\mathcal{R}}$.

Definition 4.38. Let R be a Noetherian local ring with a canonical module ω_R and I an ideal of positive grade. The fundamental divisor of $\mathcal{R} = R[It]$ is said to have the *expected form* if $D_1 \cong \omega_R$.

Let us phrase the condition $D_1 \cong \omega_R$ in terms of cohomology. We assume that R is a Cohen-Macaulay local ring of dimension $d > 0$. From the exact sequence

$$0 \to IR[It][-1] \longrightarrow \mathcal{R} \longrightarrow R \to 0$$

we have

$$0 \to H^d_{\mathcal{M}}(I\mathcal{R})_{-1} \stackrel{\psi}{\longrightarrow} H^d_{\mathcal{M}}(\mathcal{R})_0 \stackrel{\varphi}{\longrightarrow} H^d_{\mathfrak{m}}(R) \stackrel{\theta}{\longrightarrow} H^{d+1}_{\mathcal{M}}(I\mathcal{R})_{-1} \longrightarrow H^{d+1}_{\mathcal{M}}(\mathcal{R})_0 = 0.$$

Proposition 4.39. *Let D be the fundamental divisor of $\mathcal{R}$. Then $D_1 \cong \omega_R$ if and only if $\varphi = 0$.*

One issue is what is the most approachable mapping to examine. Of course the most productive move thus far has been to prove directly that $H^d_{\mathcal{M}}(\mathcal{R})_0 = 0$; we have followed this in several calculations in Chapter 3. Now we must look at the mapping φ in closer detail.

$\mathcal{D}(I)$ and the Cohen-Macaulayness of $R[It]$

The next result (see [MV1]) recasts aspects of the characterization of the Cohen-Macaulay property of $\mathcal{R}$ in terms other than the vanishing of local cohomology.

Theorem 4.40 (Arithmetical criterion). *Let $(R,\mathfrak{m})$ be a Cohen-Macaulay local ring with a canonical module ω and I an ideal of positive height. The following equivalence holds:*

$$\mathcal{R} \text{ is Cohen-Macaulay} \Longleftrightarrow \begin{cases} \mathcal{G} \text{ is Cohen-Macaulay and} \\ D_1 \cong \omega. \end{cases} \tag{4.4}$$

Before we give a proof we consider the case of 1-dimensional rings, when the assertions are stronger.

Theorem 4.41. *Let $(R,\mathfrak{m})$ be a 1-dimensional Cohen-Macaulay local ring with a canonical module ω and let I be an $\mathfrak{m}$-primary ideal. The following equivalence holds:*

$$\mathcal{R} \text{ is Cohen-Macaulay} \Longleftrightarrow D_1 \cong \omega. \tag{4.5}$$

Proof. We prove that if $D_1 \cong \omega$, then I is a principal ideal. The other assertions have been established before or will be proved in the full theorem.

We may assume that the residue field of R is infinite. Let (a) be a reduction of I and suppose that $I^{r+1} = aI^r$, where r is the reduction number of I relative to (a). We claim that $r = 0$. To this end, consider $R[at]$, whose canonical module is $at\omega R[at]$. Let

$$\mathcal{D}(I) = D_1t + D_2t^2 + \cdots \cong \operatorname{Hom}_{R[at]}(I \cdot R[It], at\omega R[at]).$$

Then D_1 is defined by the relations

$$\begin{aligned} D_1 \cdot I &\subset a\omega \\ D_1 \cdot I^2 &\subset a^2\omega \\ &\vdots \\ D_1 \cdot I^r &\subset a^r\omega. \end{aligned}$$

The descending chain

$$a^r\omega : I^r \subset \cdots \subset a\omega : I$$

of fractional ideals of R implies that

$$D_1 = \lambda\omega = a^r\omega : I^r,$$

where λ is some element in the total ring of fractions of R. This equality means that

$$\omega = \operatorname{Hom}_R(I^r a^{-r}\lambda, \omega),$$

and therefore that $I^r a^{-r}\lambda \cong R$, since $\operatorname{Hom}_R(\cdot, \omega)$ is self-dualizing on the fractional ideals of R. This means that I^r is a principal ideal and I will also be principal, as R is a local ring. □

Proof of Theorem 4.40. We consider the long exact sequences of graded B-modules that result from applying the functor $\operatorname{Hom}_B(\cdot, \omega_B)$ to the sequences (3.4) and (3.5). We have

$$0 \longrightarrow \mathrm{Ext}_B^{n-1}(\mathcal{R}, \omega_B) \longrightarrow \mathrm{Ext}_B^{n-1}(It \cdot \mathcal{R}, \omega_B) \longrightarrow \mathrm{Ext}_B^{n}(R, \omega_B) = \omega$$
$$\cdots \longrightarrow \mathrm{Ext}_B^{i}(\mathcal{R}, \omega_B) \xrightarrow{\psi_i} \mathrm{Ext}_B^{i}(It \cdot \mathcal{R}, \omega_B) \longrightarrow 0, \ i \geq n.$$

and

$$0 \longrightarrow \mathrm{Ext}_B^{n-1}(\mathcal{R}, \omega_B) \longrightarrow \mathrm{Ext}_B^{n-1}(I \cdot \mathcal{R}, \omega_B) \longrightarrow \mathrm{Ext}_B^{n}(\mathcal{G}, \omega_B) = \omega_{\mathcal{G}}$$
$$\cdots \longrightarrow \mathrm{Ext}_B^{i}(\mathcal{R}, \omega_B) \xrightarrow{\theta_i} \mathrm{Ext}_B^{i}(I \cdot \mathcal{R}, \omega_B) \longrightarrow 0, \ i \geq n.$$

In the first of these sequences, in degree 0 we have the injection

$$0 \to D_1 \xrightarrow{\varphi} \omega \longrightarrow \Box \to 0$$

that is fixed and that we are going to exploit repeatedly. Suppose that $\mathcal{G}$ is Cohen-Macaulay and $D_1 \cong \omega$. In this case, φ is an injection of modules with the S_2 property that is an isomorphism in codimension 1. Thus φ is an isomorphism, and this implies that the mappings ψ_i are (graded) isomorphisms for all $i > n$. In the other sequences meanwhile, the mappings θ_i are surjections for all $i \geq n$. However, since $\mathrm{Ext}_B^i(I \cdot \mathcal{R}, \omega_B) \cong \mathrm{Ext}_B^i(It \cdot \mathcal{R}, \omega_B)$, as ungraded modules, the θ_i must be isomorphisms because they are surjections of isomorphic Noetherian modules. This implies that $\mathrm{Ext}_B^i(I \cdot \mathcal{R}, \omega_B) \cong \mathrm{Ext}_B^i(It \cdot \mathcal{R}, \omega_B)$ as graded modules, which is a contradiction since one is obtained from the other by a non-trivial shift in the grading.

Conversely, if $\mathcal{R}$ is Cohen-Macaulay, from (3.4) we have that $\mathcal{G}$ is Cohen-Macaulay, and since $\mathrm{Ext}_B^n(\mathcal{R}, \omega_B) = 0$, we have an isomorphism $D_1 \cong \omega$. □

Divisorial Component versus Expected Form

We consider two ways in which divisorial properties of the components of $\mathcal{D}(I)$ imply that $D_1 \sim \omega$, that is, the fundamental divisor of I has the expected form.

Proposition 4.42. *Let R be a Cohen-Macaulay local ring with a canonical ideal ω and let I be an ideal of height ≥ 2. If $\omega_{\mathcal{R}} = \sum_{n \geq 1} L_n t^n$ has a component L_n that is a divisorial ideal of R, then the fundamental divisor of I has the expected form.*

Proof. By Proposition 4.9, we have that once some $L_n = K$ is divisorial, all the previous components of any prolongation are equal to K. This gives us that $D_1 = K$. In addition we have a homomorphism $\sigma : K \to \omega$ that is clearly an isomorphism of divisorial ideals of R in codimension one. But this is all that is needed to identify K and ω. □

Proposition 4.43. *Let $(R, \mathfrak{m})$ be a Cohen-Macaulay local ring with a canonical ideal and infinite residue field, and let I be an ideal of height $g \geq 2$. If $D_1 = D_2$ then the fundamental divisor of I is of the expected form.*

Proof. Let $a \in I$ be a regular element and choose $b \in I$ satisfying the following two requirements. (i) b is regular on $R/(a)$, and (ii) b is a minimal generator of I and its initial form $b^* \in \mathcal{G}_1$ does not belong to any minimal prime of $\mathcal{G}$. We claim that the ideal $(a, bt)\mathcal{R}$ has height 2. If P is a prime ideal of height 1 containing a and bt, it cannot contain I, since the image b^* of bt in $\mathcal{R}/I\mathcal{R} = \mathcal{G}$ does not lie in any minimal prime of $\mathcal{G}$. This shows that $\mathcal{R}_P$ is a localization of a polynomial ring $R_c[t]$, and in this case (a, bt) obviously has height 2. We then have that a, bt is a regular sequence on the $\mathcal{R}$-module $\mathcal{D}(I)$ that has the property S_2. As in the proof of Theorem 4.40, if D_1 is not isomorphic to ω, we may assume that in the natural sequence

$$0 \to D_1 \xrightarrow{\varphi} \omega \longrightarrow \square \to 0$$

cokernel φ is a nonzero module of finite length. If $D_1 = D_2$, then since a is regular on $\mathcal{D}(I)/bt \cdot \mathcal{D}(I)$, this implies that a is regular on $D_2/bD_1 = D_1/bD_1$. But this is a contradiction since D_1 has depth 1. $\square$

Veronese Subrings

A simple application of Theorem 4.40 is to show that a common device, passing from a graded algebra to one of its Veronese subrings in order to possibly enhance Cohen-Macaulayness, will not be helpful in the setting of ideals with associated graded rings which are already Cohen-Macaulay. (C. Huneke has informed me that J. Lipman has also observed this.)

Let $\mathcal{R} = R[It]$ be the Rees algebra of an ideal I, let $q \geq 1$ be a positive integer and set

$$\mathcal{R}_0 = \sum_{j \geq 0} I^{jq} t^{jq}$$

for the qth Veronese subring of $\mathcal{R}$. Our purpose here is to prove:

Theorem 4.44. *Let R be a Cohen-Macaulay ring and I an ideal of positive height such that the associated graded ring $\mathcal{G} = \mathrm{gr}_I(R)$ is Cohen-Macaulay. Then $\mathcal{R}$ is Cohen-Macaulay if and only if every Veronese subring $\mathcal{R}_0$ is Cohen-Macaulay.*

Proof. Most of the assertions are clear, following from the fact that as an $\mathcal{R}_0$-module, $\mathcal{R}$ is finitely generated and contains $\mathcal{R}_0$ as a summand. As for the hypotheses, if $\mathcal{G}$ is Cohen-Macaulay, the extended Rees algebra $A = R[It, t^{-1}]$ will also be Cohen-Macaulay, and the ring $A/(t^{-q})$ with it. Since the associated graded ring $\mathcal{G}_0$ of I^q is a direct summand of the latter, $\mathcal{G}_0$ is Cohen-Macaulay.

It will suffice to show that the fundamental divisors of

$$\mathcal{D}(I) = D_1 t + D_2 t^2 + \cdots$$
$$\mathcal{D}(I^q) = L_q t^q + L_{2q} t^{2q} + \cdots$$

$\mathcal{R}$ and $\mathcal{R}_0$ relative to the respective algebras, satisfy $D_1 \cong L_q$.

Let ω_0 denote the canonical module of $\mathcal{R}_0$. We calculate $\mathcal{D}(I)$ as

$$\begin{aligned}\mathcal{D}(I) &\cong \mathrm{Hom}_{\mathcal{R}_0}(I \cdot \mathcal{R}, \omega_0) \\ &= \bigoplus_{s=1}^{q} \mathrm{Hom}_{\mathcal{R}_0}(t^{s-1}I^s \cdot \mathcal{R}_0, \omega_0) \\ &\cong \bigoplus_{s=1}^{q} \mathrm{Hom}_{\mathcal{R}_0}(I^s \cdot \mathcal{R}_0, \omega_0)[s-1].\end{aligned}$$

The degrees have been kept track of, permitting us to match the components of degree 1, respectively D_1 on the left and L_q on the right. The remaining assertion will then follow from Theorem 4.40. □

Symbolic Powers

Ideals whose ordinary and symbolic powers coincide provide a clear path to the fundamental divisor.

Proposition 4.45. *Let $(R, \mathfrak{m})$ be a Cohen-Macaulay local ring with a canonical module ω, and I an ideal that is generically a complete intersection. Suppose that $\ell(I_\mathfrak{p}) <$ height $\mathfrak{p}$ for each prime ideal $\mathfrak{p} \supset I$, with height$(\mathfrak{p}/I) \geq 1$. Then $D_1 \cong \omega$.*

Proof. We claim that the mapping labeled φ above is an isomorphism:

$$0 \to D_1 \xrightarrow{\varphi} \omega \longrightarrow C \to 0.$$

We must show that $C = 0$. By induction on the dimension of R, we may assume that C is a module of finite length. If $\mathfrak{m}$ is a minimal prime of I, this ideal is a complete intersection. Suppose then that I is not $\mathfrak{m}$-primary. By assumption $\ell(I) <$ height $\mathfrak{m}$, so that height $\mathfrak{m}\mathcal{R} \geq 2$. We may thus find $a, b \in \mathfrak{m}$ so that height $(a,b)\mathcal{R} = 2$. Since $\mathcal{D}(I)$ is an S_2-module over $\mathcal{R}$, it follows that a, b must be a regular sequence on $\mathcal{D}(I)$. In particular, a, b is a regular sequence on D_1, which is clearly impossible if C is a nonzero module of finite length. □

The following is an application to the symbolic powers of a prime ideal (see [Va94a] for the Gorenstein case, and [AHT95] for the general case).

Corollary 4.46. *Let R be a Cohen-Macaulay ring and $\mathfrak{p}$ a prime ideal of positive height such that $R_\mathfrak{p}$ is a regular local ring. Suppose that $\mathfrak{p}^{(n)} = \mathfrak{p}^n$ for $n \geq 1$. Then $R[\mathfrak{p}t]$ is Cohen-Macaulay if and only if $\mathrm{gr}_\mathfrak{p}(R)$ is Cohen-Macaulay.*

Proof. The condition on the equality of the ordinary and symbolic powers of $\mathfrak{p}$ implies the condition on the local analytic spread of $\mathfrak{p}$. In turn, this condition is preserved after we localize R at any prime ideal and complete. □

For these ideals one can weaken the hypothesis that $\mathcal{G}$ be Cohen-Macaulay in a number of ways. Here is a result from [MNV95]:

Theorem 4.47. *Let R be a Gorenstein local ring of dimension d and I an unmixed ideal of codimension $g \geq 1$ that is generically a complete intersection and is such that $I^{(n)} = I^n$ for $n \geq 1$. Then $\mathcal{R}$ is Cohen-Macaulay if and only if $\mathcal{G}$ satisfies (S_r) for $r = \lceil \frac{d+1}{2} \rceil$.*

A first step in the proof consists in the following calculation ([MNV95]):

Proposition 4.48. *Let R be a Gorenstein local ring and I an unmixed ideal of codimension $g \geq 1$ that is generically a complete intersection and is such that $I^{(n)} = I^n$ for $n \geq 1$. Then the canonical module of $\mathcal{R} = R[It]$ has the expected form, that is, $\omega_{\mathcal{R}} \cong (t(1,t)^{g-2})\mathcal{R}$.*

Question 4.49. Which toric prime ideals $\mathfrak{p}$ have the property that $\mathfrak{p}^{(n)} = \mathfrak{p}^n$ for $n \geq 1$? Particularly interesting are those of codimension 2 and dimension 4.

Equimultiple Ideals

One landmark result in the relationship between $\mathcal{R}$ and $\mathrm{gr}_I(R)$ was discovered by Goto-Shimoda [GS79] (later extended in [GHO84]).

Theorem 4.50 (Goto-Shimoda). *Let $(R, \mathfrak{m})$ be a Cohen-Macaulay ring of dimension $d > 1$ with infinite residue field, and I an equimultiple ideal of codimension $g \geq 2$. Then*

$$\mathcal{R} \text{ is Cohen-Macaulay} \iff \begin{cases} \mathcal{G} \text{ is Cohen-Macaulay and} \\ r(I) < g. \end{cases} \tag{4.6}$$

Proof. We may assume that R is a complete local ring, and therefore that there is a canonical module ω. Let J be a minimal reduction of I. Since J is generated by a regular sequence, the Rees algebra $\mathcal{R}_0 = R[Jt]$ is determinantal and its canonical module is (see Example 4.27)

$$\omega_0 = \omega \cdot t(1,t)^{g-2} = \omega \cdot t + \cdots + \omega \cdot t^{g-1} + J\omega \cdot t^g + \cdots.$$

We can calculate $\mathcal{D}(I)$ as

$$\mathcal{D}(I) = \mathrm{Hom}_{\mathcal{R}_0}(I\mathcal{R}, \omega_0) = D_1 t + D_2 t^2 + \cdots,$$

where D_1 must satisfy the equations

$$\begin{aligned} I \cdot D_1 &\subset \omega \\ &\vdots \\ I^{g-1} \cdot D_1 &\subset \omega \\ I^g \cdot D_1 &\subset J \cdot \omega \\ &\vdots \end{aligned}$$

Note that since ω has S_2 and height $I > 1$, D_1 can be identified with a subideal of ω and coincides with ω in codimension 1.

Suppose that $\mathcal{R}$ is Cohen-Macaulay, so that $D_1 \cong \omega$. But $D_1 \subset \omega$ and both fractionary ideals are S_2 and thus they must coincide since they are equal in codimension 1. From the equation $I^g \cdot \omega \subset J \cdot \omega$, it follows that I^g is contained in the annihilator of $\omega/J\cdot\omega$. But this is the canonical module of R/J, and therefore $I^g \subset J$. Since $\mathcal{G}$ is Cohen-Macaulay, by Theorem 2.23 we must have $I^g = J \cdot I^{g-1}$.

For the converse, the equations give that $D_1 = \omega$, so we may simply apply Theorem 4.40. □

We are going to reinforce a one-way connection between the reduction number of an $\mathfrak{m}$-primary ideal I and $D_1(I)$ in another case.

Proposition 4.51. *Let $(R,\mathfrak{m})$ be a Cohen-Macaulay local ring of dimension $d \geq 2$ with a canonical module ω_R, and let I be an $\mathfrak{m}$-primary ideal of reduction number* $\mathrm{r}(I) < d$. *If* depth $\mathrm{gr}_I(R) \geq d-1$, *then $D_1(I) \cong \omega_R$.*

Proof. Let J be a minimal reduction such that $I^{r+1} = JI^r$. By hypothesis $r < d$. Consider the associated Sally module,

$$0 \to I\mathcal{R}_0 \longrightarrow I\mathcal{R} \longrightarrow S \to 0, \tag{4.7}$$

with $\mathcal{R}_0 = R[Jt]$, $\mathcal{R} = R[It]$. As we have observed in Proposition 2.3, the condition depth $g_I(R) \geq d-1$ means that S is a Cohen-Macaulay module. In particular, according to [Vz97] (see Remark 2.4.a), S will then admit a filtration whose factors are the $\mathcal{R}_0$-modules

$$I^n/JI^{n-1}[T_1,\ldots,T_d][-n+1], \quad n \leq r.$$

Let B be a presenting Gorenstein ring for $\mathcal{R}_0$, that is, a Gorenstein ring of the same dimension as $\mathcal{R}_0$ together with a finite homomorphism $\varphi : B \to \mathcal{R}_0$. This just means that we can define the divisors of $\mathcal{R}_0$ and of $\mathcal{R}$ using B. Dualizing the exact sequence above with B gives the exact sequence

$$0 \to \omega_{I\mathcal{R}} \longrightarrow \omega_{I\mathcal{R}_0} \longrightarrow \omega_S \longrightarrow \mathrm{Ext}^1_B(I\mathcal{R}, B) \to 0. \tag{4.8}$$

We make two observations about some terms of the sequence. First, from the exact sequence

$$0 \to I\mathcal{R}_0 \longrightarrow \mathcal{R}_0 \longrightarrow R/I[T_1,\ldots,T_d] \to 0,$$

and the fact that $\omega_{\mathcal{R}_0}$ has the expected form $\omega_R(1,t)^{d-2}t\mathcal{R}_0$, we get that $\omega_{I\mathcal{R}_0}$ is generated in degree ≥ 1. On the other hand, given the factors of the filtration of S, it follows that ω_S is a Cohen-Macaulay module admitting a filtration whose factors are the duals

$$(I^n/JI^{n-1})^{\vee}[T_1,\ldots,T_d][-d+n-1],$$

where $(\cdot)^{\vee}$ means the Matlis dual functor. When these are considered in (4.8), in degree $n \leq r < d$, we get that the components of degree 1 of the terms on the left coincide, that is $D_1(I) = \omega_R$, as desired. □

Regular Local Rings

The following is a rather surprising property discovered by Lipman ([Li94a]; see also [Ul94b]).

Theorem 4.52. *If $(R,\mathfrak{m})$ is a regular local ring, then for any nonzero ideal I the fundamental divisor of $R[It]$ has the expected form.*

If I is an $\mathfrak{m}$-primary ideal, this statement follows from a basic form of the theorem of Briançon-Skoda (see a discussion of this theorem and its role in the Cohen-Macaulayness of Rees algebras in [AH93], [AHT95]). One of its consequences is ([Li94a]):

Corollary 4.53. *Let R be a regular local ring and I an ideal. Then $\mathcal{R} = R[It]$ is Cohen-Macaulay if and only if $\mathcal{G} = \mathrm{gr}_I(R)$ is Cohen-Macaulay.*

Proof of Theorem 4.52. We argue as suggested in [Hy1]. Let $(R,\mathfrak{m})$ be a regular local ring such that $\dim R = d$ and I a nonzero ideal. Write $\mathcal{R} = R[It]$, $\mathfrak{M} = (\mathfrak{m}, It)$, $X = \mathrm{Proj}\ (\mathcal{R})$ and $E = X \times_R R/\mathfrak{m}$. The Sancho de Salas sequence (3.6), in degree 0, is

$$\cdots \longrightarrow [H^d_{\mathfrak{M}}(\mathcal{R})]_0 \longrightarrow H^d_{\mathfrak{m}}(R) \stackrel{\varphi}{\longrightarrow} H^d_E(X, \mathcal{O}_X) \longrightarrow [H^{d+1}_{\mathfrak{M}}(\mathcal{R})]_0 \rightarrow 0.$$

Note that $[H^{d+1}_{\mathfrak{M}}(\mathcal{R})]_0 = 0$.

On the other hand, according to [Li69b] and [LT81], for any regular local ring,

$$H^d_{\mathfrak{m}}(R) \cong H^d_E(X, \mathcal{O}_X),$$

from which it follows easily that φ is an isomorphism.

Finally, after noting that $\mathcal{O}_X = I\mathcal{O}_X[-1]$, another application of the Sancho de Salas sequence to the module $I \cdot \mathcal{R}$ in degree -1 yields the isomorphism

$$H^d_E(X, \mathcal{O}_X) \cong [H^{d+1}_{\mathfrak{M}}(I\mathcal{R})]_1.$$

In other words, we have that $D_1 \cong R$, as desired. □

4.5 Cohen-Macaulay Divisors and Reduction Numbers

We want to study the relationships between the algebra $R[It]$ having one of the distinguished divisors we have examined thus far–Serre, canonical and fundamental–and the reduction number of I. We assume that R is a Cohen-Macaulay local ring of dimension $d \geq 1$ and that I is an ideal of positive codimension.

Exceptional Divisor

In this case the relationship is straightforward, according to the following result.

Proposition 4.54. *Let $(R,\mathfrak{m})$ be a Cohen-Macaulay local ring of dimension $d>0$ and I an ideal of positive height. If $IR[It]$ is Cohen-Macaulay then $\mathrm{r}(I)\leq \ell(I)$.*

Proof. Set $\mathcal{R}=R[It]$, $\mathcal{G}=\mathrm{gr}_I(R)$, and $\mathfrak{M}=(\mathfrak{m},\mathcal{R}_+)$. We consider the cohomology exact sequences related to (3.4) and (3.5):

$$H^d_{\mathfrak{M}}(\mathcal{R}_+)=0\longrightarrow H^d_{\mathfrak{M}}(\mathcal{R})\longrightarrow H^d_{\mathfrak{M}}(R)\longrightarrow H^{d+1}_{\mathfrak{M}}(\mathcal{R}_+)$$

and

$$H^d_{\mathfrak{M}}(I\mathcal{R})=0\longrightarrow H^d_{\mathfrak{M}}(\mathcal{R})\longrightarrow H^d_{\mathfrak{M}}(\mathcal{G})\longrightarrow H^{d+1}_{\mathfrak{M}}(I\mathcal{R}).$$

Since $I\mathcal{R}$ is Cohen-Macaulay we can find its a-invariant localizing at the total ring of fractions of R, $a(I\mathcal{R})=a(K[t])=-1$. The first of the sequences above says that $H^d_{\mathfrak{M}}(\mathcal{R})$ is concentrated in degree 0, and therefore from the second sequence we obtain $a(\mathcal{G})\leq 0$. Since $\mathcal{G}$ is Cohen-Macaulay, for any minimal reduction J one has $a(\mathcal{G})\geq \mathrm{r}_J(I)-\ell(I)$, to complete the proof. □

Canonical Divisor

The basic listing of the properties of divisors (Proposition 4.4) gives a crude interpretation of the Cohen-Macaulayness of the canonical module.

Proposition 4.55. *Let $(R,\mathfrak{m})$ be a Cohen-Macaulay local ring of dimension $d>0$ and I an ideal of positive height. The canonical module $\omega_{\mathcal{R}}$ of $\mathcal{R}$ is Cohen-Macaulay if and only if the S_2-ification $\widetilde{\mathcal{R}}$ of $\mathcal{R}$ is Cohen-Macaulay.*

Let $\mathcal{D}(I)$ be the fundamental divisor of $\mathcal{R}$. Suppose that $\mathcal{D}(I)$ is Cohen-Macaulay. According to Proposition 4.37, $\omega_{\mathcal{R}}$ is the first shifting of $\mathcal{D}(I)$,

$$0\to\omega_{\mathcal{R}}\longrightarrow \mathcal{D}(I)[+1]\longrightarrow D_1\to 0,$$

and therefore $\omega_{\mathcal{R}}$ is Cohen-Macaulay if and only if D_1 is Cohen-Macaulay. Thus when $\mathcal{D}(I)$ has the expected form, that is $D_1\cong\omega_R$, $\omega_{\mathcal{R}}$ will be Cohen-Macaulay. This will always be the case if R is a regular local ring.

4.6 Exercises

Exercise 4.56. Let $A=R[It]$ be a Rees algebra as above, not necessarily satisfying the condition S_2. Describe the components of the divisor associated to $IR[It]$.

Exercise 4.57. Let I be an ideal of the Cohen-Macaulay ring R and x an indeterminate over R. Study the relationship between $\mathcal{D}(I)$ and $\mathcal{D}(I,x)$.

Exercise 4.58. Let R be a Cohen-Macaulay local ring with a canonical ideal and I an ideal of positive height. Show that the fundamental divisor of $R[It]$ is the unique prolongation of the canonical ideal of $R[It]$.

Exercise 4.59. Suppose that the Rees algebra R[It] of the ideal I is normal. Compare the divisor class groups of $R[It]$ and of $R[I^m t]$ for some integer m.

Exercise 4.60. Let $R = k[x_1,\ldots,x_n]$ be the ring of polynomials over the field k. Let I be a homogeneous ideal (for the standard grading) of codimension c and let $S = k[y_1,\ldots,y_d]$, $d = n - c$, be a standard Noether normalization of $A = R/I$. Show by a *direct* calculation that the two graded modules $\mathrm{Ext}^c_R(A, R[-n])$ and $\mathrm{Hom}_S(A, S[-d])$ are isomorphic.

5

Koszul Homology

There are many chain complexes associated with Rees algebras whose (co-)homologies carry information on the arithmetic of the algebras. We begin by recalling some of these. Let R be a Noetherian ring and $I = (a_1, \ldots, a_n)$ one of its ideals. Two of the basic complexes built on I are its projective resolutions, particularly the minimal free resolution when R is a local ring, or I is a homogeneous ideal of a graded ring, and the Koszul complexes:

$$\mathbb{P}: \qquad \cdots \longrightarrow F_m \stackrel{\varphi_m}{\longrightarrow} \cdots \longrightarrow F_1 \stackrel{\varphi_1}{\longrightarrow} F_0 \longrightarrow I \to 0,$$

$$\mathbb{K}: \qquad 0 \to K_n \longrightarrow \cdots \longrightarrow K_1 \longrightarrow K_0 \longrightarrow R/I \to 0.$$

These complexes carry a great deal of information about the relationship between I and R. Another class of complexes, such as the *de Rham* complex, is derived from the differential structures carried by R and R/I. Additional flexibility is obtained with complexes similar to $\mathbb{P}$ but with components that are Cohen-Macaulay modules instead of projective modules, or even more general components.

Early in Chapter 1, we explored the relationship between the defining equations of $\mathcal{R} = R[It]$ and the syzygies of I. It had a limited range of possibilities as it involved mostly the first order syzygies of the ideal I,

$$0 \to Z_1(I) \longrightarrow R^n \longrightarrow I \to 0.$$

Part of the difficulty is due to the fact that $\mathbb{P}$ is essentially a collection of matrices as components that do not intermingle more closely, as would be the case if the resolution carried an algebra structure. Instead the complex $\mathbb{K}$ is a differential graded algebra and its cycles $Z(\mathbb{K})$ inherit a subalgebra structure. The boundaries $B(\mathbb{K})$ themselves define an ideal of $Z(\mathbb{B})$. It is the *Koszul homology algebra* $H(\mathbb{K}) = Z(\mathbb{K})/B(\mathbb{K})$ that will be used to feed information from $\mathbb{F}$ into the Rees algebra $\mathcal{R}$.

There are two other complexes derived from Koszul complexes that have applications to the Rees algebras of certain ideals I. The more useful one is a differential graded algebra of the form

$$\mathcal{M}(I;R):\quad 0\to H_n(\mathbb{K})\otimes S[-n]\to\cdots\to H_1(\mathbb{K})\otimes S[-1]\to H_0(\mathbb{K})\otimes S\to \mathrm{gr}_I(R)\to 0.$$

Here $\mathbb{K}$ is a Koszul complex on a set of generators $(a_1,\ldots,a_n)$ of I, and S is a ring of polynomials $R[T_1,\ldots,T_n]$. The task is to find conditions for $\mathcal{M}(I;R)$ to be acyclic. Once that is done, it becomes clear how to derive numerical information about $\mathrm{gr}_I(R)$ in the same manner as one does with projective resolutions.

For this approach to succeed, it will depend on the depth properties of the Koszul homology modules. More concretely, we examine the depths of $H_i(\mathbb{K})$ and how it may vary when I is changed along one of its linkage classes. It turns out that linkage theory, and more generally residual intersection theory, gives a good framework leading to many cases of interest. The other outlet consists in seeking how one can pass this kind of depth information between I and some of its reductions.

Our overall goal in studying the Koszul homology of an ideal I is to examine its impact on the cohomology of the Rees algebra $\mathcal{R}(I)$, but also on more other ideals definable from I, such as its reductions, its integral closure $\overline{I}$ and the radical $\sqrt{I}$:

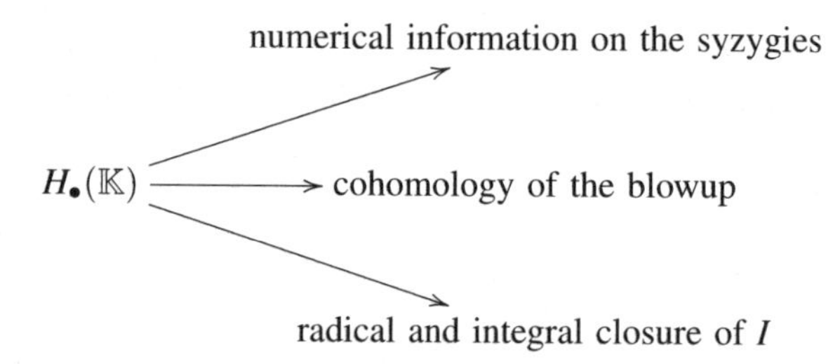

The highlights of this chapter are the following. Section 5.1 is a quick discussion of Koszul complexes, such as that found in texts on commutative algebra or homological algebra, and which are required for the study of projective dimensions over local rings and local cohomology. The subject also encompasses treatments of the theory of classical multiplicity, but they will not be carried out here. Section 5.2 is a foray into an area full of puzzles, the module structure of the Koszul homology modules. A good number of questions have been raised about the nature of these modules, from their projective dimensions to the structure of their annihilators to their relationship to radicals. The overall theme is dominated by the following set of issues. Let R be a local ring and I an ideal I with a minimal presentation

$$R^m \xrightarrow{\ \varphi\ } R^n \longrightarrow I \to 0.$$

Under restrictions of diverse nature, we will seek relationships between I and the annihilator of the Koszul homology modules $H_i(\mathbb{K}(I))$, $I : I_1(\varphi)$, the integral closure $\overline{I}$ of I and its radical $\sqrt{I}$.

Section 3 treats aspects of linkage theory required to provide some control over the depth of the Koszul homology modules. They are in the next section in our discussion of the approximation complexes. The last section examines the Koszul homology of an ideal vis-a-vis that of one of its minimal reductions. It is a theory with many applications of the Cohen-Macaulayness of Rees algebras.

5.1 Koszul Complexes of Ideals and Modules

We give here a discussion of what is one of the most useful complexes in commutative algebra. Directly it permits the introduction of various measures of size for ideals and modules. It is an ingredient in several other more elaborate complexes.

Let E be an R-module and denote by $\bigwedge E$ the exterior algebra of E. Given an element $\varphi \in \operatorname{Hom}_R(E,R)$, one defines a mapping ∂ on $\bigwedge E$ given in degree r by the rule

$$\partial(e_1 \wedge \cdots \wedge e_r) = \sum_{i=1}^{r} (-1)^{i-1} \varphi(e_i)(e_1 \wedge \cdots \wedge \widehat{e_i} \wedge \cdots \wedge e_r).$$

Then ∂ sends $\wedge^r E$ to $\wedge^{r-1} E$, and it is easy to see that $\partial^2 = 0$. We shall refer to the complex

$$\mathbb{K} = \mathbb{K}(E,\varphi) = (\bigwedge E, \partial)$$

as the *Koszul complex* associated to E and φ. For an R-module M, we can attach coefficients to $\mathbb{K}(E,\varphi)$ by forming the chain complex $\mathbb{K}(E,\varphi;M) = \mathbb{K}(E,\varphi) \otimes_R M$.

A consequence of the definition of ∂ is that, if ω and ω' are homogeneous elements of $\bigwedge E$ of degrees p and q, respectively, then

$$\partial(\omega \wedge \omega') = \partial(\omega) \wedge \omega' + (-1)^p \omega \wedge \partial(\omega'). \tag{5.1}$$

This implies that the cycles $Z(\mathbb{K})$ form a subalgebra of $\mathbb{K}$, and that the boundaries $B(\mathbb{K})$ form a two-sided ideal of $Z(\mathbb{K})$. Thus the homology $H(\mathbb{K})$ of the complex, inherits an R-algebra structure.

Proposition 5.1. *$H(\mathbb{K}(E,\varphi;M))$ is annihilated by the ideal $\varphi(E)$.*

Proof. If $M \cong R$, it suffices to note that if $e \in E$ and $\omega \in Z_r(\mathbb{K})$, then from (5.1) $\partial(e \wedge \omega) = \varphi(e)\omega$. The same argument holds when coefficients are attached. □

Since $H_0(\mathbb{K}(E,\varphi;M)) = M/\varphi(E)M$, the main problem of the elementary theory of these complexes is to find criteria for the vanishing of the higher homology modules.

In a variation, one defines a chain complex structure on $\bigwedge E$ as follows. Choose $z \in E$ and set

$$(\bigwedge E, d_z): \quad 0 \to R \longrightarrow E \longrightarrow \wedge^2 E \longrightarrow \cdots, \quad d_z(w) = z \wedge w.$$

In general, this complex is very different from the chain complex above because certain dualities are not there. Partly for this reason, and the need to attach coefficients, the most satisfying setting is the case when E is a free R-module, $E \cong R^n = Re_1 \oplus \cdots \oplus Re_n$. In view of Proposition 5.1, such complexes are more interesting when $\varphi(R^n)$ is a proper ideal of R. It is convenient to consider the elements $x_i = \varphi(e_i)$ and view $\mathbb{K}(R^n,\varphi)$ as the (graded) tensor product of n Koszul complexes associated with maps of the kind $R \xrightarrow{x} R$. That is, if we denote such a complex by $\mathbb{K}(x)$, we have

$$\mathbb{K}(R^n, \varphi) = \mathbb{K}(x_1) \otimes \cdots \otimes \mathbb{K}(x_n).$$

We shall denote such complex by $\mathbb{K}(x_1, \ldots, x_n)$ or $\mathbb{K}(\mathbf{x})$, with $\mathbf{x} = \{x_1, \ldots, x_n\}$. For a module M, the complex $\mathbb{K}(\mathbf{x}) \otimes_R M$ is denoted $\mathbb{K}(\mathbf{x}; M)$.

Koszul complexes come in many guises, each appropriate for a definite usage. In a basic setting, $(R, \mathfrak{m})$ is a Noetherian local and I is an ideal. If $\mathbf{x} = \{x_1, \ldots, x_n\}$ is a minimal set of generators of $\mathfrak{m}$, the homology $H_*(R)$ of the Koszul complex $\mathbb{K}(R/I; \mathbf{x})$ plays an important role in the study of the structure of R/I. Partly this is due to the fact that for this construction, R may be assumed to be a regular local ring and the homology algebra is a finite-dimensional associative algebra over the residue field of R. For a discussion of some properties of $H_*(R)$ we refer the reader to [BH93, Section 2.3].

Our aim being to examine the properties of the ideal I, the Koszul complex we examine most is $\mathbb{K}(x_1, \ldots, x_n)$, where the x_i form a system of generators of I. The homology algebra $H_*(\mathbf{x}; R)$ is now an associative R/I-algebra whose properties should reflect how the ideal I sits in R.

Naturally, there are many other uses for the homology of Koszul complexes, of particular interest to us being the result of Serre expressing the multiplicity of a finitely generated module M with respect to an ideal of definition I as the *Euler characteristic* of a Koszul complex. This is material that has excellent presentations in [Ser65] and [BH93].

To allow us to focus on certain objects attached to the Koszul homology, we shall quickly review some of its general properties.

Vanishing of Koszul Homology and Depth

We begin our exploration of the homology of these complexes by examining its vanishing, more precisely the *grade-sensitivity of Koszul complexes*. The vanishing of the homology modules of $\mathbb{K}(\mathbf{x}; M)$ has the following module-theoretic significance.

Proposition 5.2. *Let R be a Noetherian ring, and $\mathbf{x} = \{x_1, \ldots, x_n\}$ a sequence of elements generating the ideal I. Let M be a finitely generated R-module with $M \neq IM$, let $\mathbb{K}(\mathbf{x}; M)$ be the corresponding Koszul complex and q the largest integer for which $H_q(\mathbb{K}(\mathbf{x}; M)) \neq 0$. Then all maximal M-regular sequences in I have length $n - q$.*

Proof. Note that $0 \leq q \leq n$ since $H_0(\mathbb{K}(\mathbf{x}; M)) = M/IM$ and the complex $\mathbb{K}(\mathbf{x}; M)$ has length n.

We use descending induction on q. From the definition of $\mathbb{K}(\mathbf{x}; M)$, $H_n(\mathbb{K}(\mathbf{x}; M))$ consists of the elements of M that are annihilated by I. If this module is nonzero we are done. If not, that is $q < n$, the ideal I is not contained in any associated prime of M and therefore there is $a \in I$ that is a regular element on M. Consider the short exact sequence induced by multiplication by a,

$$0 \to M \xrightarrow{a} M \longrightarrow M/aM \to 0.$$

Tensoring it with the complex of free modules $\mathbb{K}(\mathbf{x})$, we get the following exact sequence of Koszul complexes:

$$0 \to \mathbb{K}(\mathbf{x};M) \xrightarrow{a} \mathbb{K}(\mathbf{x};M) \longrightarrow \mathbb{K}(\mathbf{x};M/aM) \to 0.$$

In homology we get the long exact sequence

$$H_{q+1}(\mathbb{K}(\mathbf{x};M)) \to H_{q+1}(\mathbb{K}(\mathbf{x};M/aM)) \to H_q(\mathbb{K}(\mathbf{x};M)) \xrightarrow{a} H_q(\mathbb{K}(\mathbf{x};M)).$$

From the definition of q, we obtain $H_i(\mathbb{K}(\mathbf{x};M/aM)) = 0$ for $i > q+1$. On the other hand, by Proposition 5.1 $H_q(\mathbb{K}(\mathbf{x};M))$ is annihilated by I, and thus $aH_q(\mathbb{K}(\mathbf{x};M)) = 0$. Taken together we have

$$H_{q+1}(\mathbb{K}(\mathbf{x};M/aM)) \cong H_q(\mathbb{K}(\mathbf{x};M)),$$

from which an easy induction suffices to complete the proof. □

The last equality in the proof also gives the next result.

Corollary 5.3. *If* $\mathbf{a} = a_1,\ldots,a_{n-q}$ *is a maximal regular sequence on* M *contained in* I*, then*

$$H_q(\mathbb{K}(\mathbf{x};M)) = (\mathbf{a}M\colon I)/\mathbf{a}M.$$

One can recast the homology of the Koszul complex $\mathbb{K}(\mathbf{x};M)$ in a somewhat different way. Given the sequence $\mathbf{x} = \{x_1,\ldots,x_n\}$ and the R-module M, set $S = R[X_1,\ldots,X_n] = R[\mathbf{X}]$, a polynomial ring with an indeterminate for each element x_i. Define M as an S-module by the rule $X_i m = x_i m$, and $X_i r = x_i r$ for $r \in R$. The Koszul complex $\mathbb{K}(\mathbf{X};S)$ is a projective resolution of R as an S-module, which with the natural identifications give

$$H_i(\mathbb{K}(\mathbf{x};M)) \cong H_i(\mathbb{K}(\mathbf{X};M)) \cong \operatorname{Tor}_i^S(R,M).$$

Koszul Homology and the Canonical Module

An useful reformulation of Corollary 5.3 will be the focus of our interest.

Proposition 5.4. *If* $\mathbf{a} = a_1,\ldots,a_{n-q}$ *is a maximal regular sequence on* M *and on* R *contained in* I*, then*

$$H_q(\mathbb{K}(\mathbf{x};M)) \cong \operatorname{Ext}_R^{n-q}(R/I,M).$$

In the case that R is a Cohen-Macaulay ring with a *canonical module* ω_R, and I is an ideal of codimension g, the last non-vanishing homology module of $\mathbb{K}(I;\omega_R)$ is $\operatorname{Ext}_R^g(R/I,\omega_R)$, the canonical module of R/I.

Depth of a Module

The following is one of the most basic notions of Commutative Algebra. When paired with *dimension* it provides the environment for most other concepts in the area.

Definition 5.5. Let I be an ideal of a Noetherian ring R, and let M be a finitely generated R-module. The *I-depth* of M is the length of a maximal regular sequence on M contained in I. If R is a local ring and I is the maximal ideal (in which case the condition $M/IM \neq 0$ is automatically satisfied, by Nakayama's Lemma), the I-depth of M is called the depth of M. If $M = R$, the I-depth of R is called the *grade* of I, and denoted by grade I. If $(R, \mathfrak{m})$ is a local ring, the $\mathfrak{m}$-depth of M will be denoted by depth M and it is called simply the *depth* of M.

Heuristically, grade I is a measure of the number of independent 'indeterminates' that may be found in I.

Proposition 5.6. *Let I be an ideal contained in the Jacobson radical of the Noetherian ring R, and let*

$$0 \to E \longrightarrow F \longrightarrow G \to 0$$

be an exact sequence of finitely generated R-modules. Then

$$\begin{aligned}
&\textit{If } I\text{-depth } F < I\text{-depth } G, \textit{ then } I\text{-depth } E = I\text{-depth } F;\\
&\textit{If } I\text{-depth } F > I\text{-depth } G, \textit{ then } I\text{-depth } E = I\text{-depth } G + 1;\\
&\textit{If } I\text{-depth } F = I\text{-depth } G, \textit{ then } I\text{-depth } E \geq I\text{-depth } G.
\end{aligned}$$

Proof. Let $\mathbb{K}(\mathbf{x})$ be the Koszul complex on a set $\mathbf{x}$ of generators of I. Tensoring the exact sequence of modules with $\mathbb{K}(\mathbf{x})$ gives the following exact sequence of chain complexes,

$$0 \to \mathbb{K}(\mathbf{x}; E) \longrightarrow \mathbb{K}(\mathbf{x}; F) \longrightarrow \mathbb{K}(\mathbf{x}; G) \to 0.$$

The assertions will follow from a scan of the long homology exact sequence and the interpretation of depth given in the previous proposition. □

The next result is the basis for several inductive arguments with ordinary Koszul complexes.

Proposition 5.7. *Let $\mathbb{C}$ be a chain complex and $\mathbb{F} = \{F_1, F_0\}$ a chain complex of free modules concentrated in degrees 1 and 0. Then for each integer $q \geq 0$ there is an exact sequence*

$$0 \to H_0(H_q(\mathbb{C}) \otimes \mathbb{F}) \longrightarrow H_q(\mathbb{C} \otimes \mathbb{F}) \longrightarrow H_1(H_{q-1}(\mathbb{C}) \otimes \mathbb{F}) \to 0.$$

Proof. Construct the exact sequence of chain complexes

$$0 \to \widehat{\mathbb{F}_0} \stackrel{f}{\longrightarrow} \mathbb{F} \stackrel{g}{\longrightarrow} \widehat{\mathbb{F}_1} \to 0,$$

whose components are

$$(\widehat{\mathbb{F}_0})_0 = F_0$$
$$(\widehat{\mathbb{F}_0})_1 = 0$$
$$(\widehat{\mathbb{F}_1})_0 = 0$$
$$(\widehat{\mathbb{F}_1})_1 = F_1,$$

where f and g are the obvious injection and surjection mappings. Tensoring with $\mathbb{C}$ and writing the homology exact sequence, we get

$$H_{q+1}(\mathbb{C}\otimes\widehat{\mathbb{F}_1}) \xrightarrow{\partial} H_q(\mathbb{C}\otimes\widehat{\mathbb{F}_0}) \to H_q(\mathbb{C}\otimes\mathbb{F}) \to H_q(\mathbb{C}\otimes\widehat{\mathbb{F}_1}) \xrightarrow{\partial} H_{q-1}(\mathbb{C}\otimes\widehat{\mathbb{F}_0}),$$

where the connecting homomorphism ∂, is up to sign, the differentiation of $\mathbb{F}$ tensored with $H_q(\mathbb{C})$. Noting that $H_{q+1}(\mathbb{C}\otimes\widehat{\mathbb{F}_1}) = H_q(\mathbb{C})\otimes F_1$, and $H_q(\mathbb{C}\otimes\widehat{\mathbb{F}_0}) = H_q(\mathbb{C})\otimes F_0$, we obtain the desired exact sequence. □

Corollary 5.8. *Let R be a Noetherian ring, $\mathbf{x} = x_1,\ldots,x_n$ a sequence generating the ideal I, and M a finitely generated R-module. If y be an element of R, then*

$$(I,y)\text{-depth } M \le 1 + I\text{-depth } M.$$

Moreover, if I-depth $M = n-q$ then y is regular on $H_q(\mathbb{K}(\mathbf{x};M))$ if and only if the equality holds.

Example 5.9. One of the more curious applications of Koszul complexes is the following proof of Hilbert's Syzygy Theorem. Suppose that $R = k[x_1,\ldots,x_n]$ where k is a field, and let E be an R-module. Set $S = R\otimes_k R \cong k[y_1,\ldots,y_n;x_1,\ldots,x_n]$, which we view as being equipped with an R-module structure in two manners, via left and right multiplications. The linear forms of $z_1 = y_1 - x_1,\ldots,z_n = y_n - x_n$ of S form an S-regular sequence, and $S/(z_1,\ldots,z_n) \cong R$. It follows that the Koszul complex $\mathbb{K}(z;S)$ gives a S-projective resolution of R,

$$0 \to \wedge^n S^n \longrightarrow \wedge^{n-1} S^n \longrightarrow \cdots \longrightarrow \wedge^1 S^n \longrightarrow S \longrightarrow R \to 0.$$

Viewing this as a complex of R-modules with either left or right multiplication structures, it splits off completely. Making use of the right R-structure on S, on tensoring with E we obtain the exact sequence

$$0 \to \wedge^n S^n \otimes_R E \to \wedge^{n-1} S^n \otimes_R E \to \cdots \to \wedge^1 S^n \otimes_R E \to S\otimes_R E \to R\otimes_R E = E \to 0.$$

For the component in degree i we have

$$\wedge^i S^n \otimes_R E \cong (S\otimes_R E)^{\binom{n}{i}} \cong (R\otimes_k R\otimes_R E)^{\binom{n}{i}} \cong (R\otimes_k E)^{\binom{n}{i}},$$

which is a projective module over R. We thus obtain a R-projective resolution of E of length n.

Rigidity of the Koszul Complex

One of the remarkable properties of these complexes is given in the next result.

Theorem 5.10. *Let* $\mathbf{x} = \{x_1, \ldots, x_n\}$ *be a sequence of elements contained in the Jacobson radical of* R*, and let* M *be a finitely generated module. If* $H_q(\mathbb{K}(\mathbf{x};M)) = 0$*, then* $H_i(\mathbb{K}(\mathbf{x};M)) = 0$ *for* $i \geq q$*.*

Proof. Set $\mathbf{y} = \{x_1, \ldots, x_{n-1}\}$ and $a = x_n$. In Proposition 5.7 set $\mathbb{C} = \mathbb{K}(\mathbf{y};M)$, $\mathbb{F} = \mathbb{K}(a)$, so that $\mathbb{C} \otimes \mathbb{F} = \mathbb{K}(\mathbf{x};M)$. For each $i \geq 0$, we have the exact sequence

$$0 \to H_0(H_i(\mathbb{K}(\mathbf{y};M)) \otimes \mathbb{K}(a)) \to H_i(\mathbb{K}(\mathbf{x};M)) \to H_1(H_{i-1}(\mathbb{K}(\mathbf{y};M)) \otimes \mathbb{K}(a)) \to 0.$$

If $H_q(\mathbb{K}(\mathbf{x};M)) = 0$, then

$$H_0(H_q(\mathbb{K}(\mathbf{y};M)) \otimes \mathbb{K}(a)) = H_q(\mathbb{K}(\mathbf{y};M)/aH_q\mathbb{K}(\mathbf{y};M)) = 0,$$

which by Nakayama's Lemma implies that $H_q(\mathbb{K}(\mathbf{y};M)) = 0$. Inducting on n, we get

$$H_i(\mathbb{K}(\mathbf{y};M)) = 0 \quad \text{for } i \geq q.$$

Taking this into the exact sequence gives that $H_i(\mathbb{K}(\mathbf{x};M)) = 0$ for $i \geq q$. $\square$

Remark 5.11. Proposition 5.7 also controls the behavior of the homology of a Koszul complex when the generators are changed. For instance, suppose that $\mathbf{x} = x_1, \ldots, x_n$ is a generating set of the ideal I and consider the extended sequence $\mathbf{x}' = \{\mathbf{x}, y\}$, where y is an element of I. Then

$$H_q(\mathbb{K}(\mathbf{x}';M)) = H_q(\mathbb{K}(\mathbf{x};M)) \bigoplus H_{q-1}(\mathbb{K}(\mathbf{x};M)), \quad q \geq 0.$$

More precisely, if $\mathbf{x}$ is a sequence and $\mathbf{y} = \{y_1, \ldots, y_r\}$ is another sequence contained in the ideal $(\mathbf{x})$, then for any R-module M,

$$H_\bullet(\mathbb{K}(\mathbf{x},\mathbf{y};M)) = H_\bullet(\mathbb{K}(\mathbf{x};M)) \bigotimes \bigwedge R^r,$$

where $\bigwedge R^r$ is the (graded) exterior algebra of R^r.

Codimension Two

We give the full description of the Koszul homology of an ideal I of codimension two with a projective resolution ([AH80]):

$$0 \to R^{n-1} \xrightarrow{\varphi} R^n \longrightarrow I \to 0.$$

We may identify the image of φ with the module Z_1 of cycles in the Koszul complex $\mathbb{K}$ on the corresponding set of n generators.

Proposition 5.12. *For each integer i, the natural map*

$$\bigwedge^i Z_1 \longrightarrow Z_i$$

is an isomorphism. In particular, for $0 \leq i \leq n-2$ the modules $H_i(\mathbb{K})$ have, as R-modules, projective dimension two.

Proof. Since they are both modules with Serre's condition S_2, it suffices to check the property at the localization $R_\mathfrak{p}$ of depth ≤ 1. But in this case it is clear, as $IR_\mathfrak{p} = R_\mathfrak{p}$.
□

Corollary 5.13. *For a perfect ideal I of codimension two, its Koszul homology algebra $H(\mathbb{K})$ is generated by $H_1(\mathbb{K})$.*

Auslander-Buchsbaum Equality

The following is a major control element in the homology of local rings.

Theorem 5.14 (Auslander-Buchsbaum). *Let $(R,\mathfrak{m})$ be a Noetherian local ring. For any nonzero finitely generated module E of finite projective dimension, the following equality holds:*

$$\text{proj. dim. } E + \text{depth } E = \text{depth } R.$$

Proof. Assume that proj. dim. $E = r$, and that the equality holds for all modules of smaller projective dimension. We may assume that $r \geq 1$. This assumption implies that depth $R \geq 1$. Indeed, let

$$0 \to F_r \xrightarrow{\varphi_r} F_{r-1} \longrightarrow \cdots \longrightarrow F_1 \xrightarrow{\varphi_1} F_0 \longrightarrow E \to 0,$$

be a minimal free resolution of E, that is, the entries of the 'matrices' φ_i lie in $\mathfrak{m}$. In particular, since φ_r is injective,

$$0 = \varphi_r((\text{ann } \mathfrak{m})F_r) = (\text{ann } \mathfrak{m})\varphi_r(F_r),$$

which implies that ann $\mathfrak{m} = 0$, as desired.

Let $L = \varphi_1(F_1)$, and consider the short exact sequence

$$0 \to L \longrightarrow F_0 \longrightarrow E \to 0.$$

Note that proj. dim. $L =$ proj. dim. $E - 1$, so that the asserted equality holds for L by induction, depth $R = (r-1) +$ depth L.

Let us show the exact reverse change in the depths of the modules E and L, depth $L =$ depth $E + 1$. Let $\mathbb{K}$ be the Koszul complex on a set $\mathbf{x} = \{x_1, \ldots, x_n\}$ of generators of $\mathfrak{m}$, and set $s =$ depth R. If we tensor the short exact sequence above by $\mathbb{K}$, we obtain the short exact sequences of Koszul complexes,

$$0 \to \mathbb{K}(L) \longrightarrow \mathbb{K}(F_0) \longrightarrow \mathbb{K}(E) \to 0.$$

Consider the homology exact sequence

$$H_{n-(s-r)}(\mathbb{K}(F_0)) \longrightarrow H_{n-(s-r)}(\mathbb{K}(E)) \longrightarrow H_{n-(s-r+1)}(\mathbb{K}(L)) \longrightarrow H_{n-(s-r+1)}(\mathbb{K}(F_0)).$$

A simple inspection shows that $H_{n-(s-r)}(\mathbb{K}(E))$ is nonzero while all higher homology modules of $\mathbb{K}(E)$ vanish. Thus depth $E = s - r$, which completes the proof. □

Koszul Homology as Tor

Let R be a commutative ring and $\mathbf{a} = \{a_1, \ldots, a_n\}$ a sequence of elements in R, and denote by $\mathbb{K}(\mathbf{a})$ the corresponding Koszul complex. We can interpret the homology of this complex (and of related complexes) as derived functors as follows. Suppose that $S = R[x_1, \ldots, x_n]$, one variable x_i to each element a_i. For a given R-module E, define an action of S on E by setting $x_i e = a_i e$, $e \in E$. The sequence $\mathbf{z} = \{x_1, \ldots, x_n\}$ is a regular sequence in S so that its Koszul complex $\mathbb{T}$ is a projective S-resolution of $S/(\mathbf{z}) \cong R$. Natural identifications yield:

Proposition 5.15. *For any R-module E with the S-module structure defined above,*

$$\mathbb{T} \otimes_S E \cong \mathbb{K}(\mathbf{a}; E).$$

In particular, for $i \geq 0$

$$\operatorname{Tor}_i^S(R, E) \cong H_i(\mathbb{K}(\mathbf{a}; E)).$$

For an application, suppose that in the sequence above, $a_1, \ldots, a_r$, for $r \leq n$, is a regular sequence on E.

Corollary 5.16. *Set $\mathbf{a}' = \{a_{r+1}, \ldots, a_n\}$ and $E' = E/(a_1, \ldots, a_r)E$. Then for $i \geq 0$,*

$$H_i(\mathbb{K}(\mathbf{a}; E)) = H_i(\mathbb{K}(\mathbf{a}'; E')).$$

Proof. We compute $\operatorname{Tor}_i^S(R, E)$ in two steps by considering a projective S-resolution $\mathbb{L}$ of E. We set also $S' = S/(x_1, \ldots, x_r)$. Consider the fact that

$$\mathbb{L} \otimes_S R = (\mathbb{L} \otimes_S S/(x_1, \ldots, x_r)) \otimes_S S/(x_{r+1}, \ldots, x_n).$$

The first term in this expression,

$$\mathbb{L} \otimes_S S/(x_1, \ldots, x_r),$$

is a projective S'-resolution of the module E', since $x_1, \ldots, x_r$ is a regular sequence on E. Furthermore,

$$\mathbb{L} \otimes_S R = (\mathbb{L} \otimes_S S/(x_1, \ldots, x_r)) \otimes_S S/(x_{r+1}, \ldots, x_n) \cong (\mathbb{L} \otimes_S S') \otimes_{S'} R,$$

we can compute the desired Tor's as $\operatorname{Tor}_i^{S'}(E', R)$, which in turn, by the proposition above, are given by the homology of the complex $\mathbb{K}(\mathbf{a}'; E')$. □

5.2 Module Structure of Koszul Homology

While the vanishing of the homology of a Koszul complex $\mathbb{K}(\mathbf{x}; M)$ is easy to track, the module-theoretic properties of its homology, with the exception of the ends, is difficult to fathom. For instance, just trying to see whether a prime is associated to some $H_i(\mathbb{K}(\mathbf{x}; M))$ can be very hard.

In this section we examine a patchwork of module-theoretic features of Koszul homology:

- Dimension
- Associated primes and depth
- Projective dimension
- Annihilators
- Linkage
- Variations of Koszul complexes

Let us clarify some of the more accessible properties at least, beginning with their Krull dimension.

Proposition 5.17. *Let R be a Cohen-Macaulay local ring. If I is an unmixed ideal then all nonzero homology modules of $\mathbb{K}(I;R)$ have Krull dimension equal to $\dim R/I$.*

Proof. Let $\mathfrak{p}$ be an associated prime of $H_0(\mathbb{K}) = R/I$, where $\mathbb{K} = \mathbb{K}(I;R)$ is the Koszul complex on a minimal set of generators of I. (From the remark above, we may consider this case only.) If $\mathbf{z} = z_1, \ldots, z_g$ is a maximal regular sequence in I, then $H_{n-g}(\mathbb{K}) = (\mathbf{z}) : I/(\mathbf{z})$ is the last non-vanishing homology module. This module is a submodule of $R/(\mathbf{z})$, which is equidimensional and therefore so is all of its submodules. By rigidity we cannot have any intermediate $H_i(\mathbb{K})$ equal to 0, and that these modules have the same dimension as $H_{n-g}(\mathbb{K})$ follows by localization. □

While the information carried by Koszul homology has a well-known role in the classical theory of multiplicity, its module theoretic properties (depth particularly) play a significant role in the study of certain Rees algebras of ideals of positive dimension. This will be the subject of the next two sections.

Syzygetic Ideals

Let R be a local ring and I an ideal minimally generated by the sequence $\mathbf{x} = \{x_1, \ldots, x_n\}$. Let $\mathbb{K}$ be the corresponding Koszul complex and denote by $\mathbb{Z}$ its cycles. The subalgebra $\mathbb{Z}' = \mathbb{Z} \cap I\mathbb{K}$ consists of the cycles with coefficients in I.

Definition 5.18. The modules $H_i'(I) = H_i(\mathbb{Z}'/\mathbb{B})$ are the *syzygetic homology modules* of the ideal I. The ideal is *i-syzygetic* if $H_i'(I) = 0$. When this holds for $i = 1$, I is said to be a *syzygetic* ideal.

The following is a straightforward interpretation of $H_1'(I)$ ([SV81b]):

Proposition 5.19. *For any ideal I there is an exact sequence*

$$0 \to H_1'(I) \longrightarrow S_2(I) \longrightarrow I^2 \to 0$$

induced by the multiplication $I \times I \to I^2$. In particular, if I is syzygetic then $S_2(I) = I^2$. As a consequence, a syzygetic ideal does not admit proper reductions of reduction number one.

There is another way to describe $H_1'(I)$ that will be useful later. If I is generated by n elements, with a presentation

$$R^m \xrightarrow{\varphi} R^n \longrightarrow I \to 0$$

we get an exact sequence

$$0 \to H_1'(I) \longrightarrow H_1(I) \longrightarrow (R/I)^n \longrightarrow I/I^2 \to 0.$$

Homological Rigidity of the Koszul Homology Modules

We are now going to examine some-module theoretic properties of $H_\bullet(\mathbb{K})$ that depend on the algebra structure of $\mathbb{K}$. Part of the discussion will involve a quick mention of conjectures of long standing.

Gulliksen Theorem

One of the strangest properties of $H_1(\mathbb{K})$ is the following result of Gulliksen ([GL67, Theorem 1.4.9]). Let $(R, \mathfrak{m})$ be a Noetherian local ring and I an ideal generated by elements $x_1, \ldots, x_n$. We shall assume that $n = \nu(I)$ is the minimal number of generators of I, and denote by $\mathbb{K}$ the corresponding Koszul complex.

If I is generated by a regular sequence, then $\mathbb{K}$ is a free resolution of R/I, otherwise $H_1(\mathbb{K}) \neq 0$. In this case, $\mathbb{K}$ can be extended by the so-called Tate resolution of R/I. This is a graded algebra $\mathbb{T} = \bigoplus_{i \geq 0} T_i$, whose components are finitely generated free R-modules, and a differential $d : T_i \to T_{i-1}$, providing for a R-free resolution of R/I. Roughly, it is obtained as a union of graded subalgebras of divided powers,

$$\mathbb{B}_{s+1} = \mathbb{B}_s[U_{s,1}, \ldots, U_{s,m_s}],$$

where B_s is acyclic up to degree $s-1$, and the $U_{s,i}$ are variables of degree $s+1$ chosen so that the differential of $\mathbb{B}_{s+1}$ in degree $s+1$ maps onto the cycles of $\mathbb{B}_s$ of degree s.

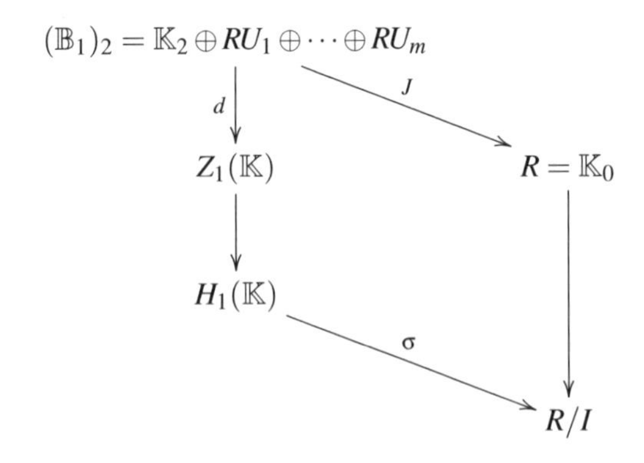

Gulliksen observed how certain derivations of degree -2 commuting with the differential of $\mathbb{T}$ arise. That is, certain derivations defined on $\mathbb{B}_1$ (set $\mathbb{B}_0 = \mathbb{K}$), can be extended to $\mathbb{T}$. An important case for us occurs when $H_1(\mathbb{K}) = R/I \oplus C = (R/I)z \oplus C$. Define $d(U_1)$ so that its image in $H_1(\mathbb{K})$ maps to z, σ is the coordinate map into that component and J is chosen to make the diagram commutative. Then J will define a derivation of $\mathbb{B}_1$ that commutes with its differential,

$$d(U_1) \in \mathfrak{m}\mathbb{K}_1, \quad J \cdot d = d \cdot J.$$

More remarkably, J can be extended all the way up to a derivation of $\mathbb{T}$, commuting with its differential. (In the original construction, Gulliksen had $H_1(\mathbb{K})$ R/I-free, but actually needed only one free direct summand.)

We are going to make use of this to obtain several results about the homological rigidity of $H_1(\mathbb{K})$.

Theorem 5.20. *Let $(R, \mathfrak{m})$ be a Noetherian local ring and I an ideal of finite projective dimension. If I is minimally generated by $x_1, \ldots, x_n$ and $\mathbb{K}$ is the associated Koszul complex, then $H_1(\mathbb{K})$ cannot have a direct summand isomorphic to R/I.*

Proof. (We follow the end of the original proof.) Consider the complex $\mathbb{T} \otimes R/\mathfrak{m}$. Since $Z_1(\mathbb{K}) \subset \mathfrak{m}\mathbb{K}_1$, $U_1 \otimes 1$ is a cycle of $\mathbb{T} \otimes R/\mathfrak{m}$; in fact, for any positive integer i, $U_1^{(i)} \otimes 1 \in Z(\mathbb{T} \otimes R/\mathfrak{m})$ as

$$d(U_1^{(i)} \otimes 1) = (U_1^{(i-1)} \otimes 1)d(U_1 \otimes 1) = 0.$$

Since $\mathbb{T}$ is a projective R-resolution of R/I and R/I has finite projective dimension, there exists an integer r such that

$$U_1^{(r)} \otimes 1 \in B(\mathbb{T} \otimes R/\mathfrak{m}).$$

However, since $J^r(U_1^{(r)}) = 1$, and $J^r \otimes 1$ commutes with the differential of $\mathbb{T} \otimes R/\mathfrak{m}$ we have

$$1 \otimes 1 = (J^r \otimes 1)(U_1^{(r)} \otimes 1) \in B_0(\mathbb{T} \otimes R/\mathfrak{m}) = 0,$$

which is a contradiction. □

Question 5.21. Let R be a Noetherian local ring. It might be worthwhile to see whether this can be extended to other Koszul homology modules. More precisely, if I is an ideal of codimension r minimally generated by $n > r$ elements and of finite projective dimension, then for the corresponding Koszul complex $\mathbb{K}$, can $H_i(\mathbb{K})$ contain a summand isomorphic to R/I for $1 \leq i \leq n - r - 1$?

A first application of Theorem 5.20 is to a result of [Ku74], which rules out certain forms of Gorenstein, or more generally of quasi-Gorenstein, ideals. We recall that an ideal I of a local ring R is said to *quasi-Gorenstein* if the canonical module of R/I is isomorphic to R/I.

Corollary 5.22. *Let R be a Gorenstein local ring and I an ideal of finite projective dimension. If I is an almost complete intersection, then the canonical module of R/I cannot be isomorphic to R/I.*

Proof. If $I = (x_1, \ldots, x_g, x_{g+1})$, $g = \text{height } I$, then $H_1(\mathbb{K})$ is the canonical module of R/I. □

Corollary 5.23. *Let R be a Noetherian local ring and I an ideal of finite projective dimension. As an R/I-module, the conormal module I/I^2 cannot have projective dimension* 1.

Proof. Suppose that I is minimally generated by n elements. We have the exact sequence

$$H_1(\mathbb{K}) \longrightarrow (R/I)^n \longrightarrow I/I^2 \to 0,$$

from which we obtain a free summand for $H_1(\mathbb{K})$. □

This provides additional support for a conjecture of [Va78].

Conjecture 5.24. Let R be a Noetherian local ring and I an ideal of finite projective dimension. Then

$$\text{proj. dim. }_{R/I} I/I^2 = 0 \text{ or } \infty.$$

In [AH94], this is settled affirmatively in the case of homogeneous ideals of graded rings. In any event, this condition on the projective dimension of I/I^2 will always imply that R/I is quasi-Gorenstein, that is, the last non-vanishing Koszul homology module is isomorphic to R/I ([He81]).

The same issue can occur indirectly as in the following:

Theorem 5.25. *Let R be a Gorenstein local ring and I a Cohen-Macaulay ideal of height g of finite projective dimension that is a complete intersection in codimension $g+1$. Denote by H_i the Koszul homology modules on a minimal generating set of n elements. If $n > g$, one cannot have that H_1 is a reflexive R/I-module and H_{n-g-1} contains a summand isomorphic to H_{n-g}.*

Proof. Making use of the fact that the homology of the Koszul complex is an R/I-algebra, there is a natural pairing

$$H_1 \times H_{n-g-1} \longrightarrow H_{n-g}$$

that leads to a mapping

$$H_1 \longrightarrow \text{Hom}(H_{n-g-1}, H_{n-g}),$$

whose image contains a copy of S/I since H_{n-g} is the canonical module of R/I. By assumption H_1 is a reflexive module, so that the mapping of reflexive modules is an isomorphism in codimension one (in the ring R/I), and consequently it is an isomorphism, and H_1 will contain a copy of R/I. Now one uses Theorem 5.20 to get a contradiction to the assumption that $n > g$. □

A Result of G. Levin

Let R be a local ring and $I = (a_1, \ldots, a_n)$ an ideal minimally generated by n elements. Let

$$R^m \xrightarrow{\varphi} R^n \longrightarrow I \to 0$$

be the corresponding presentation. The Fitting ideal $I_1(\varphi)$, that generated by the entries of φ, is called the *content* of I.

Several years ago, G. Levin observed the following property of $I : J$.

Theorem 5.26. *Let R be a local ring, I an ideal of finite projective dimension that is not a complete intersection, and $J = I_1(\varphi)$ the content of I. If $r > \mathrm{pd}_R R/I$ then $(I : J)^r \subset I$.*

The proof depends on the construction in Theorem 5.20 on the extension of derivations. If A is a differential graded R-algebra, a derivation j of A is an R- linear mapping $A \mapsto A$ of degree w such that, for $x \in A_p$, $y \in A_q$,

$$j(xy) = (-1)^{wq} j(x)y + xj(y).$$

Suppose that z is a cycle in A_p and set $B = \{A\langle S\rangle; dS = z\}$, the algebra obtained by adjoining to A a variable to kill z. Gulliksen's Lemma says that a derivation j of A of negative degree $-w$ can be extended to B if and only if $j(d(A_{p+1})) \subset d(A_{p-w+1})$.

In particular, if A is an *acyclic closure* of the homomorphism $R \to R/I$ (to be exact, begin with the Koszul complex over R on a set of generators of I and adjoin a sequence of variables $S_1, S_2, \ldots$ to kill all homology in degrees > 0), then an R-linear map $j : A_p \mapsto R$ can be extended to a derivation of A of degree $-p$ if and only if j takes the p-boundaries to 0-boundaries, i.e. elements of I, but this is precisely the condition that the composite $A_p \to R \to R/I$ be a cocycle in $\mathrm{Hom}_R(A_p, R/I)$.

This leads to the following:

Proposition 5.27. *Let $x_1, \ldots, x_s$ be elements in $\mathrm{Ext}^1_R(R/I, R/I)$ and $y_1, \ldots, y_s$ elements of $\mathrm{Tor}^R_1(R/I, R/I)$. Then*

$$(x_1 \cdots x_s)(y_1 \cdots y_s) = \pm \det(M),$$

where M is the $s \times s$ matrix whose i, j-th entry is $x_i y_j$. (Here the product of the x_i is the Yoneda product while the product of the y_i is the usual algebra product in $\mathrm{Tor}^R(R/I, R/I)$.)

Proof. The usual way to compute the action of $\mathrm{Ext}^1_R(R/I, R/I)$ on $\mathrm{Tor}^R_1(R/I, R/I)$ is to take a representative cocycle in $\mathrm{Hom}_R(A_1, R/I)$, lift it to a map ξ of degree -1 from $A \mapsto A$ commuting with the differential, and then apply $\xi \otimes 1$ to a cycle in $A_1 \otimes_R (R/I)$ representing a class in $\mathrm{Tor}^R_1(R/I, R/I)$. However, in this situation, we can take ξ to be a derivation so multiplication by any x_i satisfies the derivation rule. The assertion follows. □

Corollary 5.28. *Let $f_1,\ldots,f_r$ be any r cocycles in $\mathrm{Hom}_R(A_1,R/I)$, and $T_1,\ldots,T_n$ a basis for A_1 with $dT_i = t_i$. (This means that for any relation $\sum_{i=1}^r a_i t_i = 0 \in R$, it is also true that $\sum_{i=1}^n a_i f_j(T_i) = 0 \in R/I$.) Let $b_{ij} \in R$ represent the elements $f_i(T_j) \in R/I$ and let M be the $r \times n$ matrix with these entries. Then all subdeterminants of M of order $> \mathrm{pd}_R R/I$ are elements of I.*

Proof. Since each dT_i lies in I, each $T_i \otimes 1$ is a 1-cycle in $A \otimes (R/I)$. Now apply the previous proposition. □

Proof of Theorem 5.26. If $b_1,\ldots,b_r \in (I:J)$ the maps

$$f_i(T_j) = \begin{cases} b_i, & j = i \\ 0, & j \neq i \end{cases}$$

when reduced mod R/I become cocycles in $\mathrm{Hom}_R(A_1,R/I)$. The matrix M is then 'diagonal' and the subdeterminant is $\pm b_1 \cdots b_r \in I$ since $\mathrm{Tor}_r^R(R/I,R/I) = 0$. □

Annihilation of Koszul Homology

Annihilators of (co)homology modules, such as $\mathrm{Ext}_R^r(R/I,M)$, have an established role in uncovering properties of the primary components of I. This was used in [EHV92] to study primary decomposition in rings of polynomials. More surprisingly though, other objects related to I seem to appear from such constructions. In this regard, a very puzzling question was uncovered by A. Corso and others (see [CHKV3]). After extensive computer experiments, Corso raised the following possibility:

Conjecture 5.29. Let R be a Gorenstein local ring and I an unmixed, equidimensional ideal. If $\mathbb{K}$ is the Koszul complex on any set of generators of I, then in the non-vanishing range,

$$\mathrm{ann}\ (H_i(\mathbb{K})) \subset \bar{I}.$$

The need to circumscribe consideration to unmixed ideals is very clear, for elementary reasons. On the other hand, additional experiments have shown that the integral closure of I also occurs when R is not necessarily Gorenstein. There is not a lot of intuitive reason for these expectations. In general,

$$I \subset \mathrm{ann}\ (H_i(\mathbb{K})) \subset \sqrt{I},$$

and often the Koszul homology modules are R/I-faithful, for instance in the case of perfect codimension two ideals (see Proposition 5.12), or (in the case of the conjecture) for all ideals, the top Koszul homology module, since it gives the canonical module of R/I.

We thank B. Ulrich for the following result.

Theorem 5.30. *Let $(R,\mathfrak{m})$ be a Cohen-Macaulay local ring, I an $\mathfrak{m}$-primary ideal, and let J contained in I be a complete intersection. Then $J:(J:I) \subset \bar{I}$.*

Proof. We may assume that height J = height I. We may also assume that R has a canonical module ω. We first prove:

Lemma 5.31. *Let A be an Artinian local ring with canonical module ω and I an ideal. Then* $0 :_\omega (0 :_A I) = I\omega$.

Proof. Note that $0 :_\omega (0 :_A I) = \omega_{R/0:I}$. To show that $I\omega = \omega_{R/0:I}$, note that the socle of $I\omega$ is 1-dimensional as it is contained in the socle of ω. Hence we only need to show that $I\omega$ is faithful over $R/0 : I$. If $x \in \text{ann}\,_R I\omega$, then $xI\omega = 0$, hence $xI = 0$, hence $x \in 0 : I$. □

Returning to the proof of the theorem, it suffices to show that $(J : (J : I))\omega \subset I\omega$. But $(J : (J : I))\omega \subset J\omega :_\omega (J :_R I)$. So it suffices to show that $J\omega :_\omega (J :_R I) \subset I\omega$. Replacing R and ω by R/J and $\omega_{R/J} = \omega/J\omega$, we have to show that $0 :_\omega (0 :_R I) \subset I\omega$, but this holds by the lemma. □

Corollary 5.32. *Let $(R, \mathfrak{m})$ be a Cohen-Macaulay local ring of dimension d and I an $\mathfrak{m}$-primary ideal. If $\nu(I) = n$ then* $\text{ann}\,(H_{n-d}(\mathbb{K})) \subset \bar{I}$.

Variations on Koszul Complexes

Let R be a Noetherian ring and denote by $\mathbb{T}$ the exterior algebra of the free module R^n. We can define on $\mathbb{T}$ structures that mimic Koszul complexes. Let $w \in \mathbb{T}_r$ be a form of degree r with the property that $w \wedge w = 0$, for example

$$w = v_1 \wedge v_2 \wedge \cdots \wedge v_r,\ v_i \in R^n.$$

Left multiplication by w induces a (co-)chain structure on $\mathbb{T}$, which is actually a direct sum of r complexes. To obtain a more integrated homology, we define a cohomology module $H_*(w) = H_*(w; R^n)$ by setting (we switch notation to lower indices):

$$\begin{aligned} Z_s(w) &= \{z \in \mathbb{T}_s \mid w \wedge z = 0\} \\ B_s(w) &= \sum_{i=1}^{r} v_i \wedge \mathbb{T}_{s-1} \\ H_s(w) &= Z_s(w)/B_s(w). \end{aligned}$$

When $r = 0$, set all $Z_s(w) = 0$.

When $r = 1$, this is the homology of one of the Koszul complexes discussed. If $e_1, \ldots, e_n$ is a basis of R^n and $v_1, \ldots, v_r$ is defined by a $n \times r$ matrix φ with entries in R, the grade of the determinantal ideal $I = I_r(\varphi)$ should play a role in the vanishing of the 'cohomology' module $H_*(w)$. Note that $H_0(w) = 0 : I$. That this is indeed the case is set up in the following ([Sa76]):

Theorem 5.33. *Let R be a Noetherian ring and suppose that $H_*(w)$ is defined as above. Then*

(i) *there exists an integer* $m \geq 0$ *such that*

$$I^m \cdot H_s(w) = 0 \mid s = 0, \ldots, n.$$

(ii) *if* grade $I = q$, *then*

$$H_s(w) = 0 \mid 0 \leq s < q.$$

Proof. (i). We observe that if $v_1, \ldots, v_r$ are vectors in a basis of R^n, the assertion is clear (and $I = R$). We note also that localization preserves definitions and the assertions here. Suppose that $\mathfrak{p}$ is a prime ideal that does not contain I. Localizing at $\mathfrak{p}$ we get that some maximal minor of φ is a unit in the local ring $R_{\mathfrak{p}}$, which will imply that the v_i can be extended into a basis of the free modules. This means that the annihilator of each $H_s(w)$ is not contained in $\mathfrak{p}$ either, from which the assertion follows.

(ii). The proof (following [Sa76]) is by double induction on the pair (q, r). The case $r = 1$ is the classical Koszul complex and the case $q = 0$ was commented on above.

Suppose that grade $I \geq 1$ and choose $x \in I$ to be a regular element of R contained in the annihilator of $H_*(w)$. We denote by $\bar{y}$ the image of an element $y \in \mathbb{T}_p$ in $\overline{\mathbb{T}_p} = \wedge^p(R/xR)$. For $z \in Z_s$, one has

$$xz = \sum_{i=1}^{r} v_i \wedge z_i, \; z_i \in \mathbb{T}_{s-1}. \tag{5.2}$$

Since $\sum_{i=1}^{r} \overline{v_i} \wedge \overline{z_i} = 0$, for any $1 \leq j \leq r$, one has

$$\overline{z_j} \wedge \overline{w} = 0,$$

and the ideal $I_r(\overline{\varphi})$ has grade $q - 1 \geq 0$. By the induction hypothesis for $(q-1, r)$, there exist $z_{ji} \in \mathbb{T}_{p-2}$, for $i, j = 1, \ldots, r$, such that

$$\overline{z_j} = \sum_{i=1}^{r} \overline{v_i} \wedge \overline{z_{ji}}, \; j = 1, \ldots, r.$$

Lifting this relation to $\mathbb{T}_{s-1}$, we find $\zeta_j \in \mathbb{T}_{s-1}$, $j = 1, \ldots, r$, such that

$$z_j = \sum_{i=1}^{r} v_i \wedge z_{ji} + x\zeta_j, \; j = 1, \ldots, r.$$

Replacing z_j in (5.2), we obtain

$$x(z - \sum_{j=1}^{r} v_j \wedge \zeta_j) = \sum_{i,j=1}^{r} v_i \wedge v_j \wedge z_{ji}.$$

Multiplying this expression by $w' = v_2 \wedge v_3 \wedge \cdots \wedge v_r$, and taking into account that x is a nonzero divisor, gives the equality

$$w' \wedge (z - \sum_{i=1}^{r} v_i \wedge \zeta_i) = 0.$$

Note that the ideal I' defined by the elements $v_2, \ldots, v_r$ contains I, and therefore grade $I' \geq q > s$. By the induction hypothesis for $(q, r-1)$, we find $y_j \in \mathbb{T}_{s-1}$, $j = 2, \ldots, r$, with

$$z - \sum_{i=1}^{r} v_i \wedge \zeta_i = \sum_{j=2}^{r} v_j \wedge y_j,$$

to complete the proof. □

5.3 Linkage and Residual Intersections

In order to provide a source of interesting ideals for the construction of the approximation complexes in the next section, we must visit briefly certain aspects of linkage theory and one of its generalizations. The goal is to obtain depth properties of the homology of Koszul complexes of certain ideals that are shared across whole classes of associated ideals.

We first need to frame the notion of sliding depth in a natural setting. This is the program of residual intersection initiated by Artin and Nagata (see [AN72]) and carried out to full development by Huneke and Ulrich in several papers. A derivative source arises from minimal reductions (see [Ul94a], [Va94a], [Va94b, Chapter 5] for more details).

We begin by recalling the basic notion of linkage theory in a manner that will be appropriate for us.

Definition 5.34. Let $(R, \mathfrak{m})$ be a Cohen-Macaulay local ring, an let I, J be two ideals.

(a) I and J are *linked* if there exists a regular sequence $\mathbf{x} = x_1, \ldots, x_g$ in $I \cap J$ such that $I = (\mathbf{x}) : J$ and $J = (\mathbf{x}) : I$.
(b) I and J are *geometrically linked* if I and J are unmixed ideals of grade g without common components and $I \cap J = (\mathbf{x})$ for some regular sequence of length g.

It is convenient to denote this relation by $I \sim_{(\mathbf{x})} J$, or simply $I \sim J$. One way to generate linked ideals is through the following construction:

Proposition 5.35. *Let $(R, \mathfrak{m})$ be a Gorenstein local ring, I an unmixed ideal of grade g, and $\mathbf{x} = x_1, \ldots, x_g$ a regular sequence contained in I. The ideal $J = (\mathbf{x}) : I$ is linked to I; more precisely, $I = (\mathbf{x}) : J$.*

This notion converts to a relation as follows: two ideals, I and J, are said to be in the same *linkage class* if there exists a sequence of ideals connecting I and J:

$$I = K_0 \sim K_1 \sim \cdots \sim K_n = J.$$

The notion is refined by distinguishing ideals that are linked by an *odd* or an *even* number of direct links. An ideal I that lies in the linkage class of a complete intersection is called a *licci ideal*.

We are going to require an extended notion of linkage, obtained by replacing the regular sequences by some slightly more general ideals.

Definition 5.36. Let R be a Noetherian ring, I an ideal and s an integer such that $s \geq \text{height } I$.

(a) An *s-residual intersection* of I is an ideal J such that height $J \geq s$ and $J = \mathfrak{a}:I$ for some s-generated ideal $\mathfrak{a} \subset I$.
(b) A *geometric s-residual intersection* is an s-residual intersection J of I such that height $(I+J) \geq s+1$.

Remark 5.37. These definitions mean the following. Suppose $\mathfrak{a} = (a_1, \ldots, a_s) \subset I$, $J = \mathfrak{a}:I$. Then J is an s-residual intersection of I if for all prime ideals $\mathfrak{p}$ with $\dim R_{\mathfrak{p}} \leq s-1$ we have $I_{\mathfrak{p}} = \mathfrak{a}_{\mathfrak{p}}$. A geometric s-residual intersection requires in addition that for all $\mathfrak{p} \in V(I)$ with $\dim R_{\mathfrak{p}} = s$, the equality $\mathfrak{a}_{\mathfrak{p}} = I_{\mathfrak{p}}$ also holds. The case where $s = \text{height } I$ is the notion of linkage.

Remark 5.38. We shall be particularly interested in residual intersections that arise in the following fashion. Let I be an ideal of R of height g and let $\mathbf{x} = \{x_1, \ldots, x_s\}$ be a sequence of elements of I satisfying the following requirements:.

(1) height $(\mathbf{x}):I \geq s \geq g$.
(2) For all primes $\mathfrak{p} \supset I$ with height $\mathfrak{p} \leq s$, one has
   (i) $(\mathbf{x})_{\mathfrak{p}} = I_{\mathfrak{p}}$;
   (ii) $\nu((\mathbf{x})_{\mathfrak{p}}) \leq \text{height } \mathfrak{p}$.

These sequences have the following additional properties:

(a) height $(\mathbf{x}) = \text{height } I$;
(b) $\nu((\mathbf{x})_{\mathfrak{p}}) \leq \text{height } \mathfrak{p}$ for all primes $(\mathbf{x}) \subset \mathfrak{p}$.

To prove (a), let $\mathfrak{p}$ be a minimal prime of $(\mathbf{x})$. Suppose that $I \not\subset \mathfrak{p}$; then $((\mathbf{x}):I)_{\mathfrak{p}} = (\mathbf{x})_{\mathfrak{p}}$. It will follow from (1) that height $\mathfrak{p} \geq s \geq \text{height } I$.

To verify (b), if height $\mathfrak{p} \geq s$, the assertion is trivial; meanwhile, if height $\mathfrak{p} < s$, the proof of (a) shows that $\mathfrak{p} \supset I$ and (2) applies.

Definition 5.39. The ideal I is said to be *residually Cohen-Macaulay* if for any sequence $\mathbf{x} \subset I$ with the properties (1) and (2) of the previous remark, the following hold:

(a) $R/((\mathbf{x}):I)$ is Cohen-Macaulay of dimension $d-s$;
(b) $((\mathbf{x}):I) \cap I = (\mathbf{x})$;
(c) height $((\mathbf{x}):I + I) > \text{height } (\mathbf{x}):I$.

Ideals with Sliding Depth and Strongly Cohen-Macaulay Ideals

The following class of ideals plays an important role in the theory of Rees algebras, particularly in the construction of Cohen-Macaulay algebras.

Definition 5.40. Let R be a Noetherian local ring of dimension d and I an ideal. Let $\mathbb{K}$ be the Koszul complex on the set $\mathbf{a} = \{a_1, \ldots, a_n\}$ of generators of I and k a positive integer. Then I has the *sliding depth* condition SD_k if the homology modules of $\mathbb{K}$ satisfy

$$\operatorname{depth} H_i(\mathbb{K}) \geq d - n + i + k, \text{ for all } i.$$

Example 5.41. Let G be a graph whose vertices are labelled by the indeterminates $\{x_1, \ldots, x_n\}$, and denote by I the ideal generated by the monomials $x_i x_j$ defined by the edges of G. According to [Va94b, Theorem 4.4.9], I has sliding depth for $k = 0$.

The unspecified *sliding depth* is the case $k = 0$. There is a more stringent condition on the depth of the Koszul homology modules ([Hu82b], [Hu83]):

Definition 5.42. I is *strongly Cohen-Macaulay* if the Koszul homology modules of I with respect to one (and then to any) generating set are Cohen-Macaulay.

Example 5.43. Let I be the ideal generated by the minors of order 2 of the generic symmetric matrix

$$\begin{bmatrix} x_1 & x_2 & x_3 \\ x_2 & x_4 & x_5 \\ x_3 & x_5 & x_6 \end{bmatrix}.$$

A calculation with *Macaulay* shows that

$$\operatorname{depth} H_i(\mathbb{K}) = \begin{cases} 3, & i = 0, 2, 3 \\ 1, & i = 1. \end{cases}$$

Thus I satisfies sliding depth but is not strongly Cohen-Macaulay.

One can express these two conditions in terms of the depths of the modules of cycles of the Koszul complex (see [Va94b, Proposition 3.3.11]):

Proposition 5.44. *Let $(R, \mathfrak{m})$ be a Cohen-Macaulay local ring of dimension d, let I be an ideal of height g generated by sequence of n elements and denote by Z_i the modules of cycles of the corresponding Koszul complex. Then*

$$\operatorname{depth} Z_i \geq \begin{cases} \min\{d, d - n + i + 1\}, & \text{for } SD \text{ if } g \geq 1 \\ \min\{d, d - g + 2\}, & \text{for } SCM. \end{cases}$$

Although distinct notions, there are significant connections between them, as illustrated in the following case.

Theorem 5.45. *Let R be a Gorenstein local ring and I a Cohen-Macaulay ideal. If I has sliding depth and*

$$\nu(I_\mathfrak{p}) \leq \max\{\text{height } I, \text{height } \mathfrak{p} - 1\}$$

for every prime ideal $\mathfrak{p} \supset I$, then I is strongly Cohen-Macaulay.

Proof. See [Va94b, Theorem 3.3.17].

Licci Ideals

The broadest class of examples of strongly Cohen-Macaulay ideals arises from a theorem of Huneke ([Hu82b]).

Theorem 5.46. *Let R be a Cohen-Macaulay local ring and suppose that L is an ideal of grade n. Let $\mathbf{x} = x_1, \ldots, x_n$ and $\mathbf{y} = y_1, \ldots, y_n$ be two regular sequences in L and set $I = (\mathbf{x}) : L$ and $J = (\mathbf{y}) : L$. If I has sliding depth then J has also sliding depth.*

Corollary 5.47. *Let I be a Cohen-Macaulay ideal in the linkage class of a complete intersection. Then I is strongly Cohen-Macaulay.*

This is particularly important when applied to major classes of licci ideals:

Corollary 5.48. *Let R be a regular local ring and I be either a Cohen-Macaulay ideal of codimension two or a Gorenstein ideal of codimension three. Then I is strongly Cohen-Macaulay.*

The next two statements spell out the significance of these rather technical definitions (see [HVV85], [Va94b, Chapter 4]).

Theorem 5.49. *Let R be a Cohen-Macaulay local ring and I an ideal. If I has sliding depth then it is a residually Cohen-Macaulay ideal.*

Theorem 5.50. *Let R be a Cohen-Macaulay local ring and I an ideal satisfying $\nu(I_\mathfrak{p}) \leq \dim R_\mathfrak{p}$ for every prime ideal in $V(I)$. The following conditions are equivalent:*

(a) *I satisfies the sliding depth condition.*
(b) *I is residually Cohen-Macaulay.*
(c) *I can be generated by a d-sequence $\{x_1, \ldots, x_n\}$ such that*

$$(x_1, \ldots, x_{i+1})/(x_1, \ldots, x_i)$$

is a Cohen-Macaulay module of dimension $d - i$ for $i = 0, \ldots, n-1$.

Remark 5.51. A fuller examination of these results is given in [Va94b, Chapter 4]; it includes a discussion of relationships between the depth of some symmetric powers of the canonical module of R/I and the depth of $H_1(R)$.

5.4 Approximation Complexes

We introduce a family of extensions of Koszul complexes that lead to many classes of ideals whose Rees algebras are Cohen-Macaulay ([HSV81] has an extensive discussion). They are intimately related to d-sequences in the same manner that Koszul complexes are related to regular sequences.

Construction of the Complexes

Consider a double Koszul complex in the following sense. Let F and G be R-modules and suppose there are two mappings:

$$\begin{array}{ccc} F & \xrightarrow{\psi} & G \\ {\scriptstyle\varphi}\downarrow & & \\ R & & \end{array}$$

Let $\wedge(F)$ and $S(G)$ denote the exterior algebra of F and the symmetric algebra of G, respectively. The algebra $\wedge F \otimes S(G)$ is a double complex with differentials d_φ and d_ψ:

$$d_\varphi = \partial \colon \wedge^r(F) \otimes S_t(G) \longrightarrow \wedge^{r-1}(F) \otimes S_t(G),$$

$$\partial(e_1 \wedge \cdots \wedge e_r \otimes g) = \sum (-1)^{i-1} \varphi(e_i)(e_1 \wedge \cdots \wedge \widehat{e_i} \wedge \cdots \wedge e_r) \otimes g,$$

which is the Koszul complex associated to the mapping φ and coefficients in $S(G)$, while

$$d_\psi = \partial' \colon \wedge^r(F) \otimes S_t(G) \longrightarrow \wedge^{r-1}(F) \otimes S_{t+1}(G),$$

$$\partial\prime(e_1 \wedge \cdots \wedge e_r \otimes g) = \sum (-1)^{i-1} (e_1 \wedge \cdots \wedge \widehat{e_i} \wedge \cdots \wedge e_r) \otimes \psi(e_i) \cdot g,$$

defines the Koszul mapping of ψ. It is easy to verify that

$$\partial \cdot \partial' + \partial' \cdot \partial = 0.$$

This skew-commutativity is nice for the following reason. Each differential will then induce on the cycles of the other a differential graded structure. The resulting double complex will be denoted by $\mathcal{L} = \mathcal{L}(\varphi, \psi)$. Of significance here is the case where $\varphi \colon F = R^n \to R$ is the mapping associated with a sequence $\mathbf{x} = \{x_1, \ldots, x_n\}$ of elements of R.

Definition 5.52. The complex $\mathcal{L} = \mathcal{L}(\mathbf{x}) = \mathcal{L}(\varphi, \text{ identity})$, will be called the *double Koszul complex* of $\mathbf{x}$.

The complex $\mathcal{L}(\partial)$ is the Koszul complex associated to the sequence $\mathbf{x}$ in the polynomial ring $S = S(R^n) = R[T_1, \ldots, T_n]$. $\mathcal{L}(\partial')$, on the other hand, is also an ordinary Koszul complex but constructed over the sequence $\mathbf{T} = T_1, \ldots, T_n$. In particular $\mathcal{L}(\partial')$ is acyclic. Its graded strands,

$$\mathcal{L}(\partial') = \sum \mathcal{L}_t,$$

are complexes of R-modules,

$$\mathcal{L}_t = \sum_{r+s=t} \wedge^r(R^n) \otimes S_s(R^n),$$

and the $\mathcal{L}_t$ are exact for $t > 0$.

Attaching coefficients from an R-module E extends the construction to the complex

$$\mathcal{L}(\mathbf{x},E) = E \otimes \wedge(R^n) \otimes S(R^n).$$

If we denote by $Z_\bullet = Z_\bullet(\mathbf{x};E)$ the module of cycles of $K(\mathbf{x};E)$ and by $H_\bullet = H_\bullet(\mathbf{x};E)$ its homology, the (skew-) commutativity of ∂ and ∂' yield several new complexes, among which we single out those defined as follows.

Definition 5.53. Let I be an ideal generated by the sequence $\mathbf{x} = x_1,\ldots,x_n$. The *approximation complexes* of $\mathbf{x}$ are the following chain complexes of $R[T_1,\ldots,T_n]$-modules:

$$\begin{aligned} \mathcal{Z}_\bullet &= \mathcal{Z}_\bullet(\mathbf{x};E) = \{Z_\bullet \otimes S, \partial'\} \\ \mathcal{M}_\bullet &= \mathcal{M}_\bullet(\mathbf{x};E) = \{H_\bullet \otimes S, \partial'\}, \end{aligned}$$

called, respectively, the *$\mathcal{Z}$-complex* and the *$\mathcal{M}$-complex* of I with coefficients in E.

The ambiguity in the definition is taken care of by the fact that the homology of the complexes are independent of the chosen generating set for the ideal. More precisely, these complexes are somewhat unlike Koszul complexes in the following aspect:

Proposition 5.54. *The homology modules $H_*(\mathcal{M}_\bullet(\mathbf{x};E))$ are independent of the generating set $\mathbf{x}$ of the ideal I.*

This and other properties of these complexes, particularly a thorough discussion of their acyclicity, are treated in detail in [HSV81] and [HSV84]. The $\mathcal{M}$-complex, as well as the $\mathcal{Z}$-complex, are graded complexes over the polynomial ring $S = R[T_1,\ldots,T_n]$. The rth homogeneous component $\mathcal{M}_r$ of $\mathcal{M}_\bullet$ is a complex of finitely generated R-modules:

$$0 \to H_n \otimes S_{r-n} \longrightarrow \cdots \longrightarrow H_1 \otimes S_{r-1} \longrightarrow H_0 \otimes S_r \to 0.$$

For certain uses however we shall view them as defined over S:

$$0 \to H_n \otimes S[-n] \longrightarrow \cdots \longrightarrow H_1 \otimes S[-1] \longrightarrow H_0 \otimes S \to 0. \tag{5.3}$$

One key aspect of these complexes lies in the fact that $H_0(\mathcal{M}(\mathbf{x};E))$ maps onto $\mathrm{gr}_{(\mathbf{x})}(E)$, the associated graded module of E with respect to $(\mathbf{x})$, often allowing us to predict arithmetical properties of $\mathrm{gr}_{(\mathbf{x})}(E)$ and of its torsion free version. More specifically, when $E = R$,

$$\begin{aligned} H_0(\mathcal{Z}(\mathbf{x})) &= S(I) \\ H_0(\mathcal{M}(\mathbf{x})) &= S(I/I^2), \end{aligned}$$

the symmetric algebras of I and its conormal module I/I^2.

In view of the natural surjections $S(I) \to R[It]$ and $S(I/I^2) \to \mathrm{gr}_I(R)$, the acyclicity of the complexes will throw considerable light on the Rees algebra and the associated graded ring of the ideal I. Given the near impossibility of finding projective resolutions for the symmetric powers of I and of I/I^2, the approximation complexes, with its very explicit components, provide an entry to estimating the depths of $R[It]$ and $\mathrm{gr}_I(R)$.

Ideals Generated by d-sequences and Acyclicity

The relationship between d-sequences and the acyclicity of the complexes $\mathcal{M}_\bullet$ and $\mathcal{Z}_\bullet$ are discussed in detail in [HSV81, Theorem 12.9].

Let us examine in detail the approximation complexes $\mathcal{M}_\bullet$ in the case of an *almost complete intersection* ideal $I = (J, x)$ of height g (J is an ideal generated by a regular sequence of g elements). The approximation complex $\mathcal{M}_\bullet(I)$ is

$$0 \to H_1 \otimes S \xrightarrow{\varphi} H_0 \otimes S \to 0,$$

with $S = R[T_1, \ldots, T_{g+1}]$. Let us examine its acyclicity, that is whether φ is injective. If $K = \ker \varphi$, we examine its associated primes (as an R-module). Since $H_1 \cong \operatorname{Ext}^g(R/I, R)$, it follows easily that $\operatorname{Ass}(H_1) \subset \operatorname{Ass}(H_0) = \operatorname{Ass}(R/I)$. In particular $\mathcal{M}_\bullet(I)$ is acyclic if I is generated by a regular sequence at each of its associated primes. This is obviously burdensome to verify. Let us consider a special case:

Proposition 5.55. *Let R be a Cohen-Macaulay ring and I an almost complete intersection of codimension g. If I is a complete intersection at its associated primes of codimension g, then the approximation complex $\mathcal{M}(I)$ is acyclic. In particular, I is an ideal of linear type.*

Proof. It suffices to observe that all the associated primes of H_1 are of codimension g. Indeed, we have $H_1 = 0 :_{R/J} x \subset R/J$, and therefore its associated primes are some of the associated primes $\mathfrak{p}$ of R/I of codimension g. By hypothesis, $I_\mathfrak{p}$ is a complete intersection and $K_\mathfrak{p} = 0$. □

If the approximation complex $\mathcal{M}(I)$ is not acyclic, a first focus is on $H_0(\mathcal{M}(I))_1$, the kernel of the natural map

$$0 \to \delta(I) \longrightarrow H_1(I) \to H_0(I) \otimes S_1 = (R/I)^n \longrightarrow I/I^2 \to 0, \tag{5.4}$$

which can also be described by the exact sequence

$$0 \to \delta(I) \longrightarrow S_2(I) \longrightarrow I^2 \to 0.$$

$\delta(I)$ is discussed in detail in [SV81b] (or [Va94b, p. 30]).

To illustrate, let us go through the calculation of $\delta(I)$ in the special case of an almost complete intersection of finite colength. Let $(R, \mathfrak{m})$ be a Gorenstein local ring of dimension d and $I = (a_1, \ldots, a_{d+1})$ an $\mathfrak{m}$-primary ideal. We may assume that any proper subset of the a_i form a regular sequence. Let us note in (5.4) the nature of the mapping from $H_1(I)$ into $(R/I)^{d+1}$. This mapping arises simply by taking the class $[z]$ of a syzygy $(z_1, \ldots, z_{d+1}) \in R^{d+1}$ of the a_i modulo I. Now $H_1(I) = \omega$ is the canonical module of $A = R/I$, so that dualizing the sequence with respect to ω, we get the exact sequence

$$\omega^{d+1} \xrightarrow{\varphi} A \longrightarrow \operatorname{Hom}_A(\delta(I), \omega) = \delta(I)^\vee \to 0.$$

To determine the image of φ, consider an element of ω^{d+1}, say $\mathbf{Y} = (y, 0, \ldots, 0)$ for simplicity. Its image in $\mathrm{Hom}_A(\mathrm{Hom}_1(I), \omega)$ acts on the class $[z] = [(z_1, \ldots, z_{d+1})]$ by the rule $\varphi(\mathbf{Y})([z]) = z_1[y]$. Now observe that y is a syzygy $(y_1, \ldots, y_{d+1})$ and that the syzygy

$$z_1 \cdot y - y_1 \cdot z$$

has the form $(0, u_2, \ldots, u_{d+1})$, and therefore it is a trivial syzygy in $a_2, \ldots, a_{d+1}$, from the selection of the a_i. This proves the following.

Proposition 5.56. *Let $I_1(Z_1)$ be the Fitting ideal of all coefficients of the syzygies of any arbitrary set of $d+1$ generators of I. Then $\delta(I)$ is the canonical module of $R/I_1(Z_1)$.*

Proper Sequences

In Section 1.1, we mentioned the role of *d-sequences* in the examination of symmetric algebras. There is a companion intermediate notion more directly related to the Koszul homology of an ideal.

Definition 5.57. A sequence $x_1, \ldots, x_n$ of elements of the ring R is a *proper sequence* if

$$x_{i+1} \cdot H_j(x_1, \ldots, x_i; R) = 0, \quad i = 0, \ldots, n-1, j > 0,$$

where $H_j(x_1, \ldots, x_i; R)$ denotes the Koszul homology associated with the subsequence $x_1, \ldots, x_i$. (Actually it is enough to assume that $j = 1$.)

Thus any element $x \in R$ defines a proper sequence, while for it to define a d-sequence requires that $0 : x = 0 : x^2$. In general we have

$$\text{regular sequence} \Longrightarrow d\text{-sequence} \Longrightarrow \text{proper sequence}.$$

One way in which they are similar is by saying that proper sequences are d-sequences for the symmetric algebra of I. We refer the reader to [HSV81] for an elaboration.

Theorem 5.58. *Let $(R, \mathfrak{m})$ be a local ring with infinite residue field. Let I be an ideal of R. The following conditions are equivalent:*

(a) *$\mathcal{M}(I)$ is acyclic.*
(b) *I is generated by a d-sequence.*

Similarly

(a) *$\mathcal{Z}(I)$ is acyclic.*
(b) *I is generated by a proper sequence.*

The $\mathcal{Z}$-complex of a Module

There are several other approximation complexes (see [HSV81] and [Va94b]), of which we point out the following. We denote by $Z^*(\mathbf{x};E)$ the cycles of the Koszul complex $\mathbb{K}(\mathbf{x};E)$ contained in $I\cdot\mathbb{K}(\mathbf{x};E)$, and set $H^*(\mathbf{x};E)=Z(\mathbf{x};E)/Z^*(\mathbf{x};E)$.

Definition 5.59. The $\mathcal{M}^*$-complex of E relative to $\mathbf{x}$ is

$$\mathcal{M}^*(\mathbf{x};E)=\{H^*\otimes S,\partial'\}.$$

In the complexes constructed above the role of modules is fundamentally that of coefficients. There is a natural extension where the role is intrinsic.

Let R be a Noetherian ring and E a finitely generated module. The following notation will be used: $S=S(E)$ and $S_+=S(E)_+=\sum_{t\geq 1}S_t(E)$. We also single out a generating set of E, that is a surjection $\varphi\colon F=R^n\longrightarrow E$; the ring $S(R^n)=R[T_1,\ldots,T_n]$ will be denoted by B.

Definition 5.60. The complex $\mathcal{Z}(E)=\mathcal{M}^*(S_+;S)$ is the $\mathcal{Z}$-complex of the module E.

In other words, $\mathcal{Z}(E)$ is a complex of finitely generated B-modules whose ith component is

$$\mathcal{Z}(E)_i=H_i(S_+;S)_i\otimes B[-i],$$

where $H_i(S_+;S)_i$ denotes the ith graded part of the Koszul homology of S with respect to a set of linear generators of S_+, that is, of E itself. Note that while the total Koszul homology may turn out to be cumbersome, its ith graded part has a natural description

$$H_i(S_+;S)_i=\ker(\wedge^i F\overset{\partial}{\longrightarrow}\wedge^{i-1}F\otimes E),$$

$$\partial(a_1\wedge\cdots\wedge a_i)=\sum(-1)^j(a_1\wedge\cdots\widehat{a_j}\cdots\wedge a_i)\otimes\varphi(a_j).$$

Note that $H_1(S_+;S)_1=\ker\varphi=$ first syzygy module of E, which explains the notation $\mathcal{Z}(E)$. If we write $H_i(S_+;S)_i=Z_i(E)=Z_i$, the complex $\mathcal{Z}(E)$ has the form

$$\mathcal{Z}(E):0\to Z_n\otimes B[-n]\to\cdots\to Z_1\otimes B[-1]\to B\to H_0(\mathcal{Z}(E))=S(E)\to 0.$$

The differential of $\mathcal{Z}(E)$ is that induced by ∂'. An interesting point is its length; if E has a rank, say rank $E=e$, then $Z_i=0$ for $i>n-e$. Furthermore, its homology is independent of the chosen mapping φ.

Acyclicity: Summary of Results

Before we assemble results on these complexes, we recall a condition on the Fitting ideals of a module.

Definition 5.61. Let φ be a matrix with entries in R defining the R-module E. For an integer k, the matrix φ (or E) satisfies the condition $\mathcal{F}_k$ if:

$$\text{height } I_t(\varphi) \geq \text{rank}(\varphi) - t + 1 + k, \quad 1 \leq t \leq \text{rank}(\varphi).$$

Remark 5.62. If E is an ideal, then $\mathcal{F}_1$ is the condition G_∞ of [AN72]. There are constraints as to which $\mathcal{F}_k$ condition a module may support; for instance, for an ideal the condition $\mathcal{F}_2$ would contradict Krull's principal ideal theorem. More discriminating is the condition G_s of [AN72] requiring $\mathcal{F}_1$ but only in codimension at most $s-1$.

The following statement encapsulates some of the most important aspects of these complexes. We first record

Theorem 5.63. *Let R be a local ring with infinite residue field and E a finitely generated R-module.*

1. *The following are equivalent:*
   (a) *$\mathcal{Z}(E)$ is acyclic.*
   (b) *S_+ is generated by a d-sequence of linear forms of $S(E)$.*
2. *If $\mathcal{Z}(E)$ is acyclic, then the Betti numbers of $S(E)$ as a module over $B = S(R^n)$ are given by*

$$\beta_i^B(S(E)) = \sum_j \beta_j^R(Z_{i-j}(E)).$$

3. *If R is Cohen-Macaulay and E has rank e, the following conditions are equivalent:*
   (a) *$\mathcal{Z}(E)$ is acyclic and $S(E)$ is Cohen-Macaulay.*
   (b) *E satisfies $\mathcal{F}_0$ and*

$$\text{depth } Z_i(E) \geq d - n + i + e, \; i \geq 0.$$

4. *Moreover, if R is Cohen-Macaulay with canonical module ω_R then*
   (a)

$$\omega_S/S_+\omega_S = \bigoplus_{i=0}^{\ell} \text{Ext}_R^{\ell-i}(Z_i(E), \omega_R)[-i].$$

   (b) *$S(E)$ is Gorenstein if and only if $\text{Hom}_R(Z_{n-e}(E), \omega_R) = R$ and*

$$\text{depth } Z_i(E) \geq d - n + i + e + 1, \; i \leq n - e - 1.$$

We apply this to the conormal module I/I^2 of an ideal I.

Corollary 5.64. *Let R be a Cohen-Macaulay ring and I an ideal generated by a d-sequence. The following conditions are equivalent:*

(a) *The Rees algebra $\mathcal{R}$ is Cohen-Macaulay.*
(b) *The associated graded ring $\mathcal{G}$ is Cohen-Macaulay.*
(c) *I has sliding depth.*

Let $(R,\mathfrak{m})$ be a Cohen-Macaulay local ring and I an ideal of height g generated by $x_1,\ldots,x_{g+1}$. The sliding depth condition for I means that depth $R/I \geq d-g-1$. A ready source of Cohen-Macaulay Rees algebras is (see [Br82], [Hu81, Theorem 3.1], [Va94b, Corollary 3.3.22]).

Corollary 5.65. *Let R be a Cohen-Macaulay local ring and I an almost complete intersection. If I is a generic complete intersection, then I is of linear type. In this case, $R[It]$ is Cohen-Macaulay if and only if* depth $R/I \geq \dim R/I - 1$.

Corollary 5.66. *Let R be a Gorenstein ring and I an ideal generated by a d-sequence. The following conditions are equivalent:*

(a) *The associated graded ring $\mathcal{G}$ is Gorenstein.*
(b) *I is strongly Cohen-Macaulay.*

The approximation complex is also useful for deriving arithmetical properties of the associated graded ring of these ideals. Suppose that I is a prime ideal and we want to see when $\mathcal{G} = \mathrm{gr}_I(R)$ is an integral domain.

Theorem 5.67. *Let R be a Cohen-Macaulay local ring and let I be a prime ideal generated by a d-sequence. If I satisfies sliding depth, then $\mathcal{G}$ is an integral domain if and only if $\nu(I_\mathfrak{p}) \leq$ height $\mathfrak{p} - 1$ for every proper prime ideal $I \subset \mathfrak{p}$.*

In this case, if R is Gorenstein then I is strongly Cohen-Macaulay.

Proof. Since I is generically a complete intersection (all ideals generated by d-sequences have this property), $\mathcal{G}$ is an integral domain if and only if each of its components $\mathcal{G}_n$ is a torsionfree R/I-module. We look at the depth of $\mathcal{G}_n$ at each of the localizations of R. We test then at the maximal ideal itself.

Set height $I = g$, $\nu(I) = q+g$. We have that $q+g \leq \dim R$. Consider the component of degree n of the approximation complex

$$0 \to H_q \otimes S[-q]_n \longrightarrow \cdots \longrightarrow H_1 \otimes S[-1]_n \longrightarrow H_0 \otimes S_n \longrightarrow \mathcal{G}_n \to 0.$$

If $\nu(I) < \dim R$, then since I has sliding depth, by using the depth lemma along the exact sequence we get that depth $\mathcal{G}_n \geq 1$. This condition holding locally implies that $\mathcal{G}$ is an integral domain.

The converse does not entirely require that I have sliding depth: If $\mathcal{G}$ is a domain and $I \neq \mathfrak{m}$, then the special fiber $\mathcal{F}(I) = \mathcal{G}/\mathfrak{m}\mathcal{G}$ must have Krull dimension $< \dim R$.

The last assertion follows from Theorem 5.45. □

Fundamental Divisor

Theorem 5.68. *Let R be a Cohen-Macaulay local ring with a canonical ideal ω and I an ideal of height $g \geq 2$ generated by a d-sequence. Then the first component of the fundamental divisor of I has the expected form.*

Proof. From (4.6) we have the exact sequence

$$0 \to (D_2 t + D_3 t^2 + \cdots) \longrightarrow (D_1 t + D_2 t^2 + \cdots) \longrightarrow \omega_{\mathcal{G}}, \tag{5.5}$$

showing that D_1/D_2 embeds in the degree 1 component of $\omega_{\mathcal{G}}$.

In order to apply Proposition 4.43, it will be enough to show that ω_G is generated by elements of degree $g \geq 2$ or higher. For this purpose we use the acyclicity of the approximation complex (5.3). Suppose that $\nu(I) = n = q + g$, so that the approximation complex of I provides a complex over $\mathcal{G}$:

$$0 \to H_q \otimes S[-q] \longrightarrow \cdots \longrightarrow H_1 \otimes S[-1] \longrightarrow H_0 \otimes S \longrightarrow \mathcal{G} \to 0. \tag{5.6}$$

In this complex, we may assume that S is actually a polynomial ring $S = A[T_1, \ldots, T_n]$, where A is a Cohen-Macaulay ring of dimension $d - g$, by simply taking $A = R$ modulo a regular sequence of g elements contained in I. Since the canonical module of S is $\omega_S = \omega_A \otimes S[-n]$, we can express $\omega_{\mathcal{G}}$ as

$$\omega_{\mathcal{G}} = \mathrm{Ext}_S^q(\mathcal{G}, \omega_S).$$

Applying $\mathrm{Hom}_S(\cdot, \omega_S)$ to the complex (5.6), it is easy to see that the module $\mathrm{Ext}_S^q(\mathcal{G}, \omega_S)$ is contained in a short exact sequence of modules derived entirely from submodules of the module

$$\mathrm{Ext}_S^i(H_j \otimes_A S[-j], \omega_S) = \mathrm{Ext}_S^i(H_j \otimes_A S[-j], \omega_A \otimes S[-n]) = \mathrm{Ext}_A^i(H_j, \omega_A) \otimes_A S[-n+j],$$

all of which nonzero submodules are generated by elements of degree at least $n - j \geq g \geq 2$. □

5.5 Ideals with Good Reductions

Let R be a Cohen-Macaulay local ring, I an ideal and J a minimal reduction of I. If the reduction number $r_J(I)$ is low–in practice, meaning 1 or 2–it is often possible to determine the fundamental divisor of I from that of J. We examine two cases in this section, and a process under which they arise.

We explore briefly a merging of the techniques of the approximation complexes–which tend to work best for ideals without proper reductions–with the general method of minimal reductions, but this will require special ideals.

Koszul Homology of Reductions

Theorem 5.69. *Let R be a Cohen-Macaulay local ring of dimension d, I an ideal, J a reduction of I with $\nu(J) = s \leq d$, and assume that I satisfies $\mathcal{F}_1$ locally in codimension $\leq s - 1$. If I has sliding depth then J has sliding depth, and in particular J is of linear type and $R[Jt]$ is Cohen-Macaulay.*

Proof. We may assume that $g = \text{height}\, J < s \leq d$. Further, notice that $I_{\mathfrak{p}} = J_{\mathfrak{p}}$ for all $\mathfrak{p} \in \text{Spec}(R)$ with $\dim R_{\mathfrak{p}} \leq s-1$.

At this juncture, it is no restriction to assume that the residue field of R is infinite. Now let $a_1, \ldots, a_s$ be a generating sequence of J, and for $g \leq i \leq s-1$ write $L_i = (a_1, \ldots, a_i)$ and $K_i = L_i\colon J$. Since J satisfies $\mathcal{F}_1$, we may choose $a_1, \ldots, a_s$ in such a way that $J_{\mathfrak{p}} = (L_i)_{\mathfrak{p}}$ for all $\mathfrak{p} \in \text{Spec}(R)$ with $\dim R_{\mathfrak{p}} \leq i-1$ and all $\mathfrak{p} \in V(J)$ with $\dim R_i \leq i$. In other words, height $K_i \geq i$ and $\text{height}(J + K_i) \geq i+1$. We claim that for $g \leq i \leq s-1$, R/K_i is Cohen-Macaulay of dimension $d-i$ and $J \cap K_i = L_i$. Once this is shown, the theorem will follow from Theorem 5.50.

To prove the claim, notice that since $i \leq s-1$, we have $I_{\mathfrak{p}} = J_{\mathfrak{p}} = (L_i)_{\mathfrak{p}}$ for all $\mathfrak{p} \in \text{Spec}(R)$ with $\dim R_{\mathfrak{p}} \leq i-1$ and all $\mathfrak{p} \in V(I)$ with $\dim R_{\mathfrak{p}} \leq i$. Thus height $L_i\colon I \geq i$ and $\text{height}(I + (L_i\colon I)) \geq i+1$. However, $\nu(I_{\mathfrak{p}}) \leq \dim R_{\mathfrak{p}}$ for all $\mathfrak{p} \in V(I)$ with $\dim R_{\mathfrak{p}} \leq i \leq s-1$, and I is assumed to have sliding depth. In this situation, Theorem 5.49 implies that $R/L_i\colon I$ is a Cohen-Macaulay ideal of dimension $d-i$ and $I \cap (L_i\colon I) = L_i$.

Now it suffices to prove that $L_i\colon I = K_i$. The inclusion $L_i\colon I \subset L_i\colon J = K_i$ being trivial, we only need to establish the asserted equality at every associated prime $\mathfrak{p}$ of $L_i\colon I$. Since the latter ideal is Cohen-Macaulay of height i, we know that $\dim R_{\mathfrak{p}} = i \leq s-1$. Thus $I_{\mathfrak{p}} = J_{\mathfrak{p}}$, and $(L_i\colon I)_{\mathfrak{p}} = (K_i)_{\mathfrak{p}}$.

Finally, that $R[Jt]$ is Cohen-Macaulay follows from Theorem 5.63. □

Corollary 5.70. *Let I and J be as in* Theorem 5.69. *Let $\mathbb{K}$ be the Koszul complex on a minimal set of generators of J. Then I annihilates $H_i(\mathbb{K})$ for $i > 0$.*

Proof. Since $I \cdot H_i(\mathbb{K}) \hookrightarrow H_i(\mathbb{K})$, and this is a module of depth $\geq d-s+i$, it suffices to check the prime ideals $\mathfrak{p}$ of codimension at most $s-i \leq s-1$. But in this range we have $I_{\mathfrak{p}} = J_{\mathfrak{p}}$. □

Corollary 5.71. *Let R be a Cohen-Macaulay local ring of dimension d with infinite residue field and I an ideal with sliding depth. If I is of linear type in codimension h, then*

$$\ell(I) \geq \inf\{h+1, \nu(I)\}.$$

The Top Koszul Homology of Reductions

We begin to derive properties of the Koszul homology of a reduction.

Proposition 5.72. *Let I be a height unmixed ideal which is generically a complete intersection, and let J be a reduction of I. Then R/I and R/J have the same canonical module and the same top Koszul homology module.*

Proof. Suppose that height $I = g$, and consider the sequence

$$0 \to I/J \longrightarrow R/J \longrightarrow R/I \to 0.$$

Now I/J is a module annihilated by a power of I, whose support contains no minimal prime of I; therefore it has dimension at most $d-g-1$, $d = \dim R$. In the cohomology sequence

$$\operatorname{Ext}_R^{g-1}(I/J,\omega_R) \to \operatorname{Ext}_R^g(R/I,\omega_R) \to \operatorname{Ext}_R^g(R/J,\omega_R) \to \operatorname{Ext}_R^g(I/J,\omega_R),$$

the modules at the ends vanish. The other assertion has a similar proof. □

The next result reveals several obstructions for a Cohen-Macaulay ideal to have analytic deviation 1. It is the strand of an idea traced to [Ul92, Proposition 4].

Theorem 5.73. *Let R be a Gorenstein local ring of dimension d with infinite residue field. Let I be an ideal of codimension g that is generically a complete intersection and has analytic spread $\ell(I) = g+1$. Let J be a proper minimal reduction of I.*

(a) *If I is Cohen-Macaulay then* depth $R/J = d-g-1$.
(b) *If R/I has Serre's condition S_2 then J has no associated prime ideal of codimension $g+2$.*

Proof. (a) Assume that $J = (a_1,\ldots,a_{g+1})$, and consider the acyclic portion of the Koszul complex $\mathbb{K}$ built on the a_i:

$$0 \to K_{g+1} \longrightarrow K_g \longrightarrow \cdots \longrightarrow K_2 \longrightarrow B_1 \to 0.$$

From Proposition 5.72 we have that $H_1(\mathbb{K})$, the canonical module of R/J, is isomorphic to the canonical module of R/I, and therefore is Cohen-Macaulay.

If Z_1 denotes the module of 1-cycles of $\mathbb{K}$, the assertion is that depth $Z_1 = d-g+1$. Let

$$0 \to B_1 \longrightarrow Z_1 \longrightarrow H_1(\mathbb{K}) \to 0$$

be the sequence that defines $H_1(\mathbb{K})$. Since B_1 has projective dimension $g-1$, it has depth $d-g+1$ and thus the depth of Z_1 is at least $d-g$, since $H_1(\mathbb{K})$ is Cohen-Macaulay of dimension $d-g$. Into the cohomology exact sequence

$$\operatorname{Ext}_R^{g-1}(B_1,R) \longrightarrow \operatorname{Ext}_R^g(H_1(\mathbb{K}),R) \longrightarrow \operatorname{Ext}_R^g(Z_1,R) \longrightarrow \operatorname{Ext}_R^g(B_1,R)$$

we feed the facts that

$$\begin{cases} \operatorname{Ext}_R^{g-1}(B_1,R) = R/J \\ \operatorname{Ext}_R^g(B_1,R) = 0 \\ \operatorname{Ext}_R^g(H_1(\mathbb{K}),R) = R/I, \end{cases}$$

the first two from the Koszul complex, and the last from the local duality theorem since R/I is Cohen-Macaulay:

$$R/J \longrightarrow \operatorname{Ext}_R^g(H_1(\mathbb{K}),R) = R/I \longrightarrow \operatorname{Ext}_R^g(Z_1,R) \to 0.$$

We claim that $\operatorname{Ext}_R^g(Z_1,R)$ vanishes. By induction on the dimension of R, we may assume that this module has finite length and $\dim R/I \geq 2$. If the module is different from zero, from the sequence it must have the form $R/(I,a)$ for some a. By Krull's theorem this is not possible.

Finally, to argue that the equality of depth holds, note that if depth $R/J = d-g$, then I and J would necessarily have the same associated prime ideals, and therefore would be equal since they agree at such primes.

(b) The equality $\mathrm{Ext}_R^g(H_1(\mathbb{K}),R) = R/I$ still holds, and the module $\mathrm{Ext}_R^g(Z_1,R)$ will vanish. This means that $\mathrm{Ext}_R^{g+2}(R/J,R)=0$, which is a strong form of the assertion. □

In the following corollaries, I is still assumed to be generically a complete intersection and not generated by analytically independent elements (note that $\ell(I) > g$).

Corollary 5.74. *Let I be a Cohen-Macaulay ideal of codimension g that is of linear type in codimension $\leq g+1$. Then $\ell(I) \geq g+2$.*

Corollary 5.75. *Let I be a Cohen-Macaulay prime ideal all of whose powers I^m are unmixed. Then $g+2 \leq \ell(I) \leq \dim R - 1$. In particular, if $\dim R/I = 2$ no such prime exists, and if $\dim R/I = 3$ then $\ell(I) = g+2$.*

Proof. The upper bound is given by Theorem 1.94, while the equality $\ell(I) = g+1$ is ruled out by [CN76] and the previous corollary. □

The following is a special case of [HH93, Theorem 2.1].

Corollary 5.76. *Let I be a Cohen-Macaulay ideal of analytic deviation* 1 *and reduction number* 1. *Then $\mathrm{gr}_I(R)$ is Cohen-Macaulay.*

Proof. Let J be a minimal reduction of I of reduction number 1. The previous result says that J is an ideal with sliding depth, so we have by Corollary 5.65 that $R[Jt]$ is Cohen-Macaulay. On the other hand, the approximation complex of J,

$$0 \to H_1(\mathbb{K}) \otimes B[-1] \to R/J \otimes B \to S(J/J^2) = R[Jt]/(J \cdot R[Jt]) \to 0, \qquad (5.7)$$

where $B = R[T_1, \ldots T_{g+1}]$, splits as a complex of R-modules at the minimal primes of I. Since $H_1(\mathbb{K})$ is already a R/I-module of rank 1, on tensoring the sequence with R/I we obtain an exact sequence

$$0 \to H_1(\mathbb{K}) \otimes B[-1] \to R/I \otimes B \to R[Jt]/(I \cdot R[Jt]) \to 0 \qquad (5.8)$$

of R/I-modules, from which it is easy to see that $I \cdot R[Jt]$ is a maximal Cohen-Macaulay module. We can then argue as in Theorem 2.22 since $I^2 = JI$ by hypothesis. □

An immediate consequence for ideals of reduction number one is:

Corollary 5.77. *Let R, I and J be as above, and assume that* depth $R/I \geq \dim R - s$. *Then $I \cdot R[Jt]$ is a maximal Cohen-Macaulay module. If the reduction number $r_J(I)$ is* 1, *then $\mathrm{gr}_I(R)$ is Cohen-Macaulay.*

We make some observations about what is required for the equality $r_J(I)=1$ to hold.

Theorem 5.78. *Let I and J be ideals as in* Theorem 5.69, *and suppose that every associated prime ideal of I has codimension at most s. If the equality $I_{\mathfrak{p}}^2=(JI)_{\mathfrak{p}}$ holds for each prime ideal $\mathfrak{p}$ of codimension s, then $I^2=JI$.*

Proof. It will be enough to show that the associated prime ideals of JI have codimension at most s. From the proofs of Corollaries 5.70 and 5.77 we use the exact sequence

$$0\rightarrow H_1(J)\longrightarrow (R/I)^{\ell}\longrightarrow J/JI\rightarrow 0,$$

which will be combined with the sequence

$$0\rightarrow J/JI\longrightarrow R/JI\longrightarrow R/J\rightarrow 0.$$

Since depth $H_1(J)\geq d-s+1$, it follows from the first sequence and the condition that $\dim R_{\mathfrak{p}}\leq s$ for every associated prime $\mathfrak{p}$ of I, that every prime in $\mathrm{Ass}(J/JI)$ has codimension at most s. The claim now follows from the second sequence, since the associated prime ideals of R/J have codimension at most s.

Alternatively, we can argue as follows to establish the vanishing of the Sally module $S_J(I)$. In the exact sequence (2.1), as in Proposition 2.2, if $I\cdot R[Jt]$ is a maximal Cohen-Macaulay module (and therefore an unmixed ideal of codimension one), then $S_J(I)$ either vanishes or has Krull dimension d. By induction on $\dim R$ we may assume that the equality $I^2=JI$ holds on the punctured spectrum of the local ring $(R,\mathfrak{m})$. This means that $S_J(I)$ is annihilated by some power of $\mathfrak{m}$, so that the dimension of $S_J(I)$ is at most $\nu(J)=s<d$, which is a contradiction unless $r_J(I)=1$. □

These methods come in full fruition in the following result ([Va94a]).

Theorem 5.79. *Let R be a Cohen-Macaulay local ring and let I and J be ideals as in* Theorem 5.69. *Suppose that* depth $R/I\geq \dim R-s$. *If I has codimension at least two and $r_J(I)=1$ then $R[It]$ is a Cohen-Macaulay algebra.*

Remark 5.80. We consider an application to fundamental divisors. Let I be an ideal with sliding depth that satisfies $\mathcal{F}_1$, and suppose that N is an ideal containing I such that $N^2=IN$. The canonical module

$$\omega_{\mathcal{R}}=C_1t+C_2t^2+\cdots$$

has the property that $C_1\cong\omega$. This implies that in the representation of

$$\mathcal{D}(I)=D_1t+D_2t^2+\cdots,$$

we also have $D_1\cong\omega$. Indeed, from the sequence (3.4), we have the exact sequence

$$0\rightarrow\omega_{\mathcal{R}}\longrightarrow L\longrightarrow\omega_{\mathcal{G}};$$

since I annihilates $\mathcal{G}$, we have $I \cdot D_1 \subset C_1$, which means that $D_1 \subset C_1$, and therefore $D_1 \cong \omega$.

Consider now the ideal N and use the same notation for the components of its fundamental divisor. From the sequence

$$0 \to N \cdot R[It] \longrightarrow R[It] \longrightarrow U \to 0$$

and the corresponding sequence of 'canonical modules'

$$0 \to \omega_{R[It]} \longrightarrow \omega_{N \cdot R[It]} \longrightarrow \omega_U$$

we obtain that $C_1 = D_1$, the first component of the canonical module of $N \cdot R[It]$, since D_1 is conducted into C_1 by an ideal of height at least two. On the other hand, we have $N \cdot R[It] = N \cdot R[Nt]$, by the hypothesis on the reduction number.

Symbolic Powers

We make an application of these techniques to the comparison of the ordinary and symbolic powers of a prime ideal.

Theorem 5.81. *Let I and J be ideals as in* Theorem 5.69 *with $r_J(I) = 1$. Suppose that I is a prime ideal and* depth $R/I \geq d - s$. *If for each prime ideal $\mathfrak{p}$ of codimension at most s the powers $I_{\mathfrak{p}}^m$ are primary ideals, then the ideals I^m are also primary.*

Proof. This follows from the exact sequences (2.16) and (2.17). Together they say that the associated prime ideals of the conormal modules I^m/I^{m+1} have codimension at most s. This suffices, along with the hypothesis, to ensure that each such module is a torsion-free R/I-module. □

Here is an application observed by B. Ulrich, who proved it by other means in the case of ideals lying in the linkage class of a complete intersection (see also [Ul94a] for other developments).

Theorem 5.82. *Let R be a Cohen-Macaulay local ring and I an ideal.*

(a) *Suppose that for all integers m, the module I^m/I^{m+1} has depth at least r. Then $\ell(I) \leq \dim R - r$.*
(b) *In addition, suppose that R is a regular local ring and assume that I is a Cohen-Macaulay ideal that satisfies sliding depth and $\mathcal{F}_1$. If $r \geq 1$, then I is strongly Cohen-Macaulay and R/I is locally a complete intersection in codimension r.*

Proof. (a) Follows from the fact that (cf. [Br74])

$$\inf_m\{\text{depth } R/I^m\} = \inf_m\{\text{depth } I^m/I^{m+1}\}.$$

(b) That such ideals are strongly Cohen-Macaulay is Theorem 5.45; the other assertion follows from [SV81a, p. 356]. □

Theorem 5.83. *Let* $(R,\mathfrak{m})$ *be a Gorenstein local ring and* I *a Cohen-Macaulay ideal satisfying sliding depth which is generically a complete intersection. Suppose that for all integers* m *the conormal module* I^m/I^{m+1} *is torsionfree as an* R/I*-module. Then* I *satisfies* $\mathcal{F}_1$*, it is strongly Cohen-Macaulay (and in particular its Rees algebra* $R[It]$ *is Cohen-Macaulay), and is generated by at most* $\dim R - 1$ *elements.*

Proof. The main point is to show that for each prime ideal $\mathfrak{p}$ which is not a minimal prime of I, the localization $I_{\mathfrak{p}}$ is generated by at most $\dim R_{\mathfrak{p}} - 1$ elements. We may assume that the assertion holds for the punctured spectrum of R, and thus that I is generated by analytically independent elements on the punctured spectrum of R.

We may also assume that the residue field of R is infinite. Let J be a minimal reduction of I; by Theorem 5.82.a, $\ell(I) \leq \dim R - 1$, so that $\nu(J) \leq \dim R - 1$. By Theorem 5.69 however, J satisfies sliding depth so that its associated primes have codimension at most $\nu(J)$ and therefore $\mathfrak{m}$ cannot be one of them. By the induction hypothesis this implies that $I = J$.

Now we use Theorem 5.45 to get that I is strongly Cohen-Macaulay. The assertion about the Rees algebra will then follow from Theorem 5.63(3). □

5.6 Exercises

Exercise 5.84. Let $(R,\mathfrak{m})$ be a Noetherian local ring and I an ideal. Show that if $S_i(I) \simeq I^i$ for $i \leq r$ and I is not of linear type, then $\mathrm{r}(I) \geq r$.

Exercise 5.85. Let $(R,\mathfrak{m})$ be a Noetherian local and I an $\mathfrak{m}$-primary ideal generated by a d-sequence. Show that depth $R =$ depth $\mathrm{gr}_I(R)$.

Exercise 5.86. Let R be a Noetherian ring and I an ideal generated by a d–sequence and of finite projective dimension. Prove (5.24) for I. (This is more of an open problem.)

6

Integral Closure of Algebras

The main theme of this chapter will be consideration of a reduced affine algebra A over a field k and its relationship to its integral closure $\overline{A}$. Setting up processes to construct $\overline{A}$, analyzing their complexities and making predictions about properties and descriptions of $\overline{A}$ constitute a significant area of questions, with individualized issues but also with an array of related connections. Those links are the main focus of our discussion.

We develop frameworks that allow for discussions of various complexities associated with the construction of $\overline{A}$:

$$A = k[x_1, \ldots, x_n]/I \overset{\mathbb{P}}{\mapsto} \overline{A} = k[y_1, \ldots, y_m]/J,$$

where $\mathbb{P}$ is some algorithm. The process is often characterized by iterations of a basic procedure $\mathcal{P}$ outputting integral, rational extensions of the affine ring A terminating at its integral closure $\overline{A}$:

- $A = A_0 \mapsto A_1 \mapsto A_2 \mapsto \cdots \mapsto A_n = \overline{A}.$

Short of a direct description of the integral closure of A by a single operation-a situation achieved in some cases-the construction of $\overline{A}$ is usually achieved by a 'smoothing' procedure: an operation $\mathcal{P}$ on affine rings with the properties

- $A \subset \mathcal{P}(A) \subset \overline{A}$;
- if $A \neq \overline{A}$ then $A \neq \mathcal{P}(A)$.

The general style of the operation $\mathcal{P}$ is of the form

$$A \rightsquigarrow I(A) \rightsquigarrow \mathcal{P}(A) = \mathrm{Hom}_A(I(A), I(A)),$$

where $I(A)$ is an ideal somewhat related to the conductor of A. The ring

$$\mathcal{P}(A) = \mathrm{Hom}_A(I(A), I(A))$$

is called the *idealizer of* $I(A)$. Two examples of such methods, using Jacobian ideals, are featured in [Va91b] and [Jo98], respectively. Let J be the Jacobian ideal of A. The corresponding smoothing operations are the following.

(i) $\mathcal{P}(A) = \mathrm{Hom}_A(J^{-1}, J^{-1})$
(ii) $\mathcal{P}(A) = \mathrm{Hom}_A(\sqrt{J}, \sqrt{J})$

The 'order' of the construction is the smallest integer n such that $\mathcal{P}^n(A) = \overline{A}$. The 'cost' of the computation $C(\overline{A})$ however will consist of $\sum_{i=1}^n c(i)$, where $c(i)$ is the complexity of the operation $\mathcal{P}$ on the data set represented by $\mathcal{P}^{i-1}(A)$.

Obviously, in a Gröbner basis setting, different iterations of $\mathcal{P}$ may carry non-comparable costs. This holds true particularly if each iteration uses its own local variables.

Almost irresponsibly, one could define the astronomical complexity of a smoothing operation by the order of $\mathcal{P}$:

$$\boxed{n = C_{\mathcal{P}}(\overline{A}).}$$

It is not yet clear what significance is carried by this number. The fact is however, that $\mathcal{P}$ usually acts not on the full set $B(A)$ of integral birational extensions of A,

$$\mathcal{P} : B(A) \mapsto B(A),$$

but also on much smaller subsets of extensions (containing $\overline{A}$)

$$\mathcal{P} : B_0(A) \mapsto B_0(A).$$

We shall discuss the following issues:

- Descriptions of several such $\mathcal{P}$.
- How long these chains might be?
- How long is the description of $\overline{A}$ in terms of a description of A?

The first section is a compiling of several resources that appear in the actual construction of integral closures of rings. To describe the results of the next sections, we introduce some notation and terminology. The two most basic invariants of an affine ring A (assumed reduced, equidimensional throughout) are its dimension $\dim A = d$ and its 'degree', the rank of A with respect to one of its Noether normalizations $k[x_1, \ldots, x_d]$. If k has characteristic zero and A is a standard graded algebra, we can find an embedding

$$S = k[x_1, \ldots, x_{d+1}]/(f) \hookrightarrow A$$

with f monic in x_{d+1}, where $\deg f = \deg(A)$. In the non graded case, choose f of least possible total degree. Observe that

$$\overline{S} = \overline{A},$$

so we could always assume that A is a hypersurface ring. Often, however, we may want to use information on A that is not shared by S.

In Section 2 we shall describe an approach to the issue of complexity by developing a setting for analyzing the efficiency of algorithms that compute the integral closure of affine rings. It gives quadratic (cubic in the non-homogeneous case) multiplicity-based but dimension-independent bounds for the number of passes any construction of a broad class will make, thereby leading to the notion of astronomical complexity.

Unlike the treatment of Section 5, which seeks to estimate the complexity of $\overline{A}$ in terms of the number of generators that will be required to give a presentation of $\overline{A}$ (i.e. its *embedding dimension*), here we are going to estimate the number of passes by any method that progressively builds $\overline{A}$ by taking larger integral extensions.

Here we will argue that while the finite generation of $\overline{A}$ as an A-module guarantees that chains of integral extensions,

$$A = A_0 \subsetneq A_1 \subsetneq A_2 \subsetneq \cdots \subsetneq A_n \subset \overline{A},$$

are stationary, there are no bounds for n if the Krull dimension of A is at least 2. We show that if the extensions A_i are taken satisfying Serre's condition S_2 then n can be bound by data essentially contained in the Jacobian ideal of A (in characteristic zero at least).

A preferred approach to the computation of the integral closure should have the following general properties:

- All calculations are carried out in the same ring of polynomials. If the ring A is not Gorenstein, it will require one Noether normalization.
- It uses the Jacobian ideal of A, or of an appropriate hypersurface subring, only theoretically to control the length of the chains of extensions.
- There is an explicit quadratic bound (cubic in the non-homogeneous case) on the multiplicity for the number of passes the basic operation has to be carried out. In particular, and surprisingly, the bound is independent of the dimension.
- It can make use of known properties of the ring A.

It will thus differ from the algorithms proposed in either [Jo98] or [Va91b], by the fact that it does not require changes in the rings for each of its basic cycles of computation. The primitive operation itself is based on elementary facts of the theory of Rees algebras. To enable the calculation we shall introduce the notion of a *proper construction* and discuss instances of it.

Key to our discussion here is the elementary, but somehow surprising, observation that the set $\mathcal{S}_2(A)$ of extensions with the condition S_2 that lie between an equidimensional, reduced affine algebra A and its integral closure $\overline{A}$ satisfy the *ascending* chain condition inherited from the finiteness of $\overline{A}$ over A, but the *descending* condition as well.

The explanation requires the presence of a Gorenstein subalgebra S of A, over which A is birational and integral. If A is not Gorenstein, S is obtained from a Noether normalization of A and the theorem of the primitive element. One shows (Theorem 6.57) that there is an inclusion-reversing one-one correspondence between $\mathcal{S}_2(A)$ and a subset of the set $\mathcal{D}(\mathfrak{c})$ of divisorial ideals of S that contain the conductor $\mathfrak{c}$ of $\overline{A}$ relative to S. Since $\mathfrak{c}$ is not accessible one uses the Jacobian ideal as an approximation in order to determine the maximal length of chains of elements of $\mathcal{S}_2(A)$. In Corollary 6.59, it is shown that if A is a standard graded algebra over a field of characteristic zero, and of multiplicity e, then any chain in $\mathcal{S}_2(A)$ has at most $(e-1)^2$ elements. In the non-homogeneous case of a hypersurface ring defined by an equation of degree e, the best bound we know is $e(e-1)^2$.

Next we describe a process of creating chains in $\mathcal{S}_2(A)$. Actually the focus is on processes that produce shorter chains in $\mathcal{S}_2(A)$ such that its elements are amenable to computation. A standard construction in the cohomology of blowups provides chains of length at most $\left\lceil \frac{(e-1)^2}{2} \right\rceil$ (correspondingly, $\frac{e(e-1)^2}{2}$, in the non-homogeneous case).

Section 4 is an abstract treatment of the length of divisorial extensions between a graded domain and its integral closure, according to [DV3]. It does not improve significantly on the bounds already discussed for characteristic zero nevertheless in arbitrary characteristics it imposes absolute bounds.

In section 5, we shall discuss, following [UV3], the problem of estimating the number of generators that the integral closure $\overline{A}$ of an affine domain A may require. This number, and the degrees of the generators in the graded case, are major measures of costs of the computation. We discuss current developments on this question for various kinds of algebras, particularly algebras with a small singular locus. At worst, these estimates have the double exponential shape of Gröbner bases computations.

6.1 Normalization Toolbox

We shall assemble in this section basic resources that intervene in any discussion of the integral closure of an algebra. Most can be found in one of [BH93], [Ku86], [Ma86], or [Va98b], but for convenience they will be re-assembled here.

6.1.1 Noether Normalization

Let R be a commutative ring and B a finitely generated R-algebra, $B = R[x_1, \ldots, x_d]$. The expression *Noether normalization* usually refers to the search-as effectively as possible-of more amenable finitely generated R-subalgebras $A \subset B$ over which B is finite. This allows for looking at B as a finitely generated A-module and therefore applying to it methods from homological algebra or even from linear algebra.

When R is a field, two such results are: (i) the classical *Noether normalization lemma*, that asserts when it is possible to choose A to be a ring of polynomials, or (ii) how to choose A to be a hypersurface ring over which B is birational. We review these results since their constructive steps are very useful in our discussion of the integral closure of affine rings. The situation is not so amenable when R has Krull dimension ≥ 1.

Affine Rings

Let $B = k[x_1, \ldots, x_n]$ be a finitely generated algebra over a field k and assume that the x_i are algebraically dependent. Our goal is to find a new set of generators $y_1, \ldots, y_n$ for B such that

$$k[y_2,\ldots,y_n] \hookrightarrow B = k[y_1,\ldots,y_n]$$

is an integral extension.

Let $k[X_1,\ldots,X_n]$ be the ring of polynomials over k in n variables; to say that the x_i are algebraically dependent means that the map

$$\pi\colon k[X_1,\ldots,X_n] \to B, \quad X_i \mapsto x_i$$

has non-trivial kernel, call it I. Assume that f is a nonzero polynomial in I,

$$f(X_1,\ldots,X_n) = \sum_{\alpha} a_\alpha X_1^{\alpha_1} X_2^{\alpha_2} \cdots X_n^{\alpha_n},$$

where $0 \neq a_\alpha \in k$ and all the multi-indices $\alpha = (\alpha_1,\ldots,\alpha_n)$ are distinct. Our goal will be fulfilled if we can change the X_i into a new set of variables, the Y_i, such that f can be written as a monic (up to a scalar multiple) polynomial in Y_1 and with coefficients in the remaining variables, i.e.

$$f = aY_1^m + b_{m-1}Y_1^{m-1} + \cdots + b_1Y_1 + b_0, \tag{6.1}$$

where $0 \neq a \in k$ and $b_i \in k[Y_2,\ldots,Y_n]$.

We are going to consider two changes of variables that work for our purposes: the first one, a clever idea of Nagata, does not assume anything about k; the second one assumes k to be infinite and has certain efficiencies attached to it.

The first change of variables replaces the X_i by Y_i given by

$$Y_1 = X_1,\; Y_i = X_i - X_1{}^{p^{i-1}} \text{for } i \geq 2,$$

where p is some integer yet to be chosen. If we rewrite f using the Y_i instead of the X_i, it becomes

$$f = \sum_{\alpha} a_\alpha Y_1^{\alpha_1} (Y_2 + Y_1^p)^{\alpha_2} \cdots (Y_n + Y_1^{p^{n-1}})^{\alpha_n}. \tag{6.2}$$

Expanding each term of this sum, there will be only one term pure in Y_1, namely

$$a_\alpha Y_1^{\alpha_1 + \alpha_2 p + \cdots + \alpha_n p^{n-1}}.$$

Furthermore, from each term in (6.2) we are going to get one and only one such power of Y_1. Such monomials have higher degree in Y_1 than any other monomial in which Y_1 occurs. If we choose $p > \sup\{\alpha_i \mid a_\alpha \neq 0\}$, then the exponents $\alpha_1 + \alpha_2 p + \cdots + \alpha_n p^{n-1}$ are distinct since they have different p-adic expansions. This provides for the required equation.

If k is an infinite field, we consider another change of variables that preserves degrees. It will have the form

$$Y_1 = X_1,\; Y_i = X_i - c_iX_1 \text{ for } i \geq 2,$$

where the c_i are to be properly chosen. Using this change of variables in the polynomial f, we obtain

$$f = \sum_{\alpha} a_\alpha Y_1^{\alpha_1} (Y_2 + c_2 Y_1)^{\alpha_2} \cdots (Y_n + c_n Y_1)^{\alpha_n}. \qquad (6.3)$$

We want to make choices of the c_i in such a way that when we expand (6.3) we achieve the same goal as before, i.e. a form like that in (6.1). For that, it is enough to work on the homogeneous component f_d of f of highest degree, in other words, we can deal with f_d alone. But

$$f_d(Y_1, \ldots, Y_n) = h_0(1, c_2, \ldots, c_n) Y_1^d + h_1 Y_1^{d-1} + \cdots + h_d,$$

where h_i are homogeneous polynomials in $k[Y_2, \ldots, Y_n]$, with $\deg h_i = i$, and we can view $h_0(1, c_2, \ldots, c_n)$ as a nontrivial polynomial function in the c_i. Since k is infinite, we can choose the c_i, so that $0 \neq h_0(1, c_2, \ldots, c_n) \in k$.

Theorem 6.1 (Noether Normalization). *Let k be a field and $B = k[x_1, \ldots, x_n]$ a finitely generated k-algebra; then there exist algebraically independent elements $z_1, \ldots, z_d$ of B such that B is integral over the polynomial ring $A = k[z_1, \ldots, z_d]$.*

Proof. We may assume that the x_i are algebraically dependent. From the preceding, we can find $y_1, \ldots, y_n$ in B such that

$$k[y_2, \ldots, y_n] \hookrightarrow k[y_1, \ldots, y_n] = B$$

is an integral extension, and if necessary we iterate. □

Corollary 6.2. *Let k be a field and $\psi : A \mapsto B$ a k-homomorphism of finitely generated k-algebras. If $\mathfrak{P}$ is a maximal ideal of B then $\mathfrak{p} = \psi^{-1}(\mathfrak{P})$ is a maximal ideal of A.*

Proof. Consider the embedding

$$A/\mathfrak{p} \hookrightarrow B/\mathfrak{P}$$

of k-algebras, where by the preceding $B/\mathfrak{P}$ is a finite dimensional k-algebra. It follows that the integral domain $A/\mathfrak{p}$ is also a finite dimensional k-vector space and therefore must be a field. □

Corollary 6.3. *Let k be a field of characteristic zero and B a finitely generated k-algebra. If B is an integral domain there exists a Noether normalization A of B and an element $z \in B$ such that the subring $H = A[z]$ of B and B have the same torsionfree ranks as A-modules.*

Proof. After choosing A as above, it suffices to apply the theorem of the primitive element to the fields of fractions of A and B, followed possibly by some clearing of denominators in A. □

Noether Normalization Module

In nonzero characteristics, when the theorem of the primitive element may fail to apply, the following result is often useful.

Proposition 6.4. *Let $A = R[y_1, \ldots, y_n]$ be an integral domain finite over the subring R, and set* $\mathrm{rank}_R(A) = e$. *Let K be the field of fractions of R and define the field extensions*

$$F_0 = K, \quad F_i = K[y_1, \ldots, y_i], \quad i = 1 \ldots n.$$

Then the module

$$E = \sum R y_1^{j_1} \cdots y_n^{j_n}, \quad 0 \leq j_i < [F_i : F_{i-1}]$$

is R-free of rank e.

Proof. As the rank satisfies the equality $e = \prod_{i=1}^{n} r_i$, where $r_i = [F_i : F_{i-1}]$, it follows that e is the number of 'monomials' $y_1^{j_1} \cdots y_n^{j_n}$. Their linear independence over R is a simple verification. □

Degree of a Prime Ideal

The following is a notion of degree for ideals that are not necessarily graded. Let $\mathfrak{p}$ be a prime ideal of the polynomial ring $A = k[x_1, \ldots, x_n]$. We can use Noether normalizations to define the *degree* of $\mathfrak{p}$ or of $A/\mathfrak{p}$.

(a) If $\mathfrak{p}$ is a homogeneous ideal, its degree $\deg(\mathfrak{p})$ is the multiplicity of the graded algebra $A/\mathfrak{p}$. This means that if

$$C = k[z_1, \ldots, z_d] \hookrightarrow A/\mathfrak{p}$$

is a Noether normalization in which the z_i are forms of degree 1, then $\deg(\mathfrak{p})$ is the torsionfree rank of $A/\mathfrak{p}$ as a C-module.

(b) If $\mathfrak{p}$ is not homogeneous, $\deg(\mathfrak{p})$ is the minimum rank of $A/\mathfrak{p}$ relative to all possible Noether normalizations.

Remark 6.5. If B is a finitely generated standard graded algebra over the field k,

$$B = k \oplus B_1 \oplus \cdots = k[B_1],$$

the Noether normalization process can be described differently. The argument is that establishing the Hilbert polynomial associated with B. More precisely, if B has Krull dimension 0, then $A = k$ is a Noether normalization. If $\dim B \geq 1$, one begins by choosing a homogeneous element $z \in B_+$ which is not contained in any of the minimal prime ideals of B. When k is infinite, the usual prime avoidance finds $z \in B_1$. If k is finite, however, one argues differently. If $P_1, \ldots, P_n$ are the minimal primes of B, we must find a form h which is not contained in any P_i. Suppose that the best one can do is to find a form $f \notin \bigcup_{i \leq s} P_i$, for s maximum $< n$. This means that $f \in P_{s+1}$. If $g \in \bigcap_{i \leq P_i} \setminus P_{s+1}$ is a form (easy to show), then $h = f^a + g^b$, where $a = \deg g$, and $b = \deg f$ is a form not contained in a larger subset of the associated primes, contradicting the definition of s.

Example 6.6 (Stanley). Let k be a field and set $B = k[x_1,\ldots,x_n]/I$, where I is generated by monomials. If $\dim B = d$ and $z_1,\ldots,z_d$ are the first d elementary symmetric functions

$$z_r = \sum x_{i_1}x_{i_2}\cdots x_{i_r},$$

then the subring $k[z_1,\ldots,x_d]$ is a Noether normalization of B. We leave the proof as an exercise for the reader.

The Ideal of Leading Coefficients of an Ideal of Polynomials

The following *simulacrum* of an initial ideal is useful in theoretical arguments and in Noether normalization.

Theorem 6.7 (Suslin). *Let S be a Noetherian ring and I an ideal of the polynomial ring $S[x]$. Denote by $c(I)$ the ideal of S generated by the leading coefficients of all the elements of I. Then*

$$\text{height } c(I) \geq \text{height } I.$$

Proof. Let $P_1,\ldots,P_n$ be the minimal primes of I. From the equality $\cap P_i = \sqrt{I}$ we have that for some integer N,

$$(\prod_{i=1}^{n} P_i)^N \subset I.$$

Now for any two ideals, I,J one has $c(I)\cdot c(J) \subset c(IJ)$ since, if a is the leading coefficient of $f \in I$ and b is that of $g \in J$ then either $ab = 0$ or ab is the leading coefficient of fg. Therefore if $c(I) \subset \mathfrak{q}$ for a prime $\mathfrak{q} \subset S$ we have $c(P_i) \subset \mathfrak{q}$ for some i. This means that if the assertion of the theorem is true for prime ideals, then

$$\text{height } \mathfrak{q} \geq \text{height } c(P_i) \geq \text{height } P_i \geq \text{height } I$$

and we are done since height $c(I) = \inf\{\text{height } \mathfrak{q}\}$.

Assume then that $I = P$ is prime and suppose that $\mathfrak{p} = P \cap S$. If $P = \mathfrak{p}[x]$ then $c(P) = \mathfrak{p}$ and the statement follows from the Krull principal ideal theorem. If $P \neq \mathfrak{p}[x]$ then $c(P) \neq \mathfrak{p}$; indeed otherwise for

$$f = a_n x^n + \cdots + a_0 \in P,$$

we have $a_n \in \mathfrak{p} \subset P$ so $f - a_n x^n \in P$ and $a_{n-1} \in \mathfrak{p}$ by an easy induction. It follows that

$$\text{height } c(P) > \text{height } \mathfrak{p} = \text{height } P - 1,$$

to establish the claim. □

Integral Closure of Affine Domains

Proposition 6.8. *Let A be an integrally closed domain with field of fractions K and F a finite separable field extension of K. The integral closure B of A in F is contained in a finitely generated A-module. In particular, if A is Noetherian then B is also Noetherian. Furthermore, if A is an affine algebra over a field k, then B is an affine algebra over k.*

Proof. Let $y_1,\dots,y_m$ be a basis of F/K; by premultiplying the elements of a basis by elements of A we may assume that $y_i \in B$. Denote by Tr the trace function of F/K; by assumption Tr is a non-degenerate quadratic form on the K-vector space F. Denote by $z_1,\dots,z_m$ a basis dual to the y_i,

$$\mathrm{Tr}(y_i z_j) = \delta_{ij}.$$

Suppose that $z = \sum_j a_j z_j$, $a_j \in K$. If z is integral over A then each $zy_i \in B$ and its trace,

$$\mathrm{Tr}(zy_i) = a_i \in K,$$

are also integral over A. It follows that $B \subset \sum_j A z_j$. □

Remark 6.9. In this situation, there exists a basis of F/K of the form $y_i = y^i$, $y \in B$, $i \leq m$. If $f(t)$ is the minimal polynomial of y and D is its discriminant, the argument above can be rearranged to show that $D \cdot B \subset \sum_i A y^i$.

Theorem 6.10. *The integral closure of an affine domain over a field k is an affine domain over k.*

Proof. We follow the proof in [Ser65]. Let A be an affine domain over k with field of fractions K, denote by F a finite field extension of K, and let B be the integral closure of A in F. By Noether normalization, A is integral over a subring $A_0 = k[x_1,\dots,x_n]$ generated by algebraically independent elements. From the transitivity of integrality, B is the integral closure of A_0 in F, so we may take $A_0 = A$. We claim that B is a finitely generated A-module. Without loss of generality, we can assume that the extension F/K is a normal extension.

Let E be the subfield of elements of F that are purely inseparable over K. If char $K = 0$, we simply apply Proposition 6.8; otherwise it will suffice to prove that the integral closure of A in E is as asserted. Since $E = k(z_1,\dots,z_m)$ is a finite extension of $k(x_1,\dots,x_n)$, there exists an exponent $q = p^s$ such that $z^q \in k(x_1,\dots,x_n)$, $\forall z \in E$. Let k' be the finite extension of k obtained by adjoining to k the qth roots of all the coefficients that occur when z_i^q is written as a rational function of the x_j; note that $E \subset E' = k_0(x_1^{\frac{1}{q}},\dots,x_n^{\frac{1}{q}})$. But the polynomial ring $B' = k_0[x_1^{\frac{1}{q}},\dots,x_n^{\frac{1}{q}}]$ is clearly the integral closure of A in E', and B is contained in it. □

For certain coefficient rings a similar result holds. For example, if R is an integral domain for which the integral closure of R in a finite extension of its field of fractions is R-finite, then affine algebras over R have the same property. Typical examples are that of $\mathbb{Z}$ and complete local domains. The study of this phenomenon, due mostly to Nagata (and also to Rees) is well understood (see treatments in [Ma80], [Na62]).

Graded Algebras

One refinement of the nature of the integral closure is contained in the next result (a proof can be found in [B6183, Chap. V, Proposition 21]).

Theorem 6.11. *Let $B = k[y_1,\dots,y_n]$ be a $\mathbb{Z}$-graded affine domain over the field k. The integral closure of B is an affine domain with the same grading.*

Krull-Serre Normality Criterion

Throughout we shall assume that A is a reduced Noetherian ring, and denote by K its total ring of fractions. For each prime ideal $\mathfrak{p}$, the localization $A_{\mathfrak{p}}$ can be naturally identified with a subring of K so that intersections of such subrings may be considered. The ring A itself may be represented as

$$A = \bigcap_{\mathfrak{p}} A_{\mathfrak{p}}, \quad \text{grade } \mathfrak{p}A_{\mathfrak{p}} \leq 1.$$

It is then easy to describe, using these localizations, when A is integrally closed:

Proposition 6.12 (Krull-Serre). *Let A be a reduced Noetherian ring. The following conditions are equivalent:*

(a) *A is integrally closed.*
(b) *For each prime ideal $\mathfrak{p}$ of A associated to a principal ideal, $A_{\mathfrak{p}}$ is a discrete valuation domain.*
(c) *A satisfies the following conditions.*
  (i) *The condition S_2: For each prime ideal $\mathfrak{p}$ associated to a principal ideal, $\dim A_{\mathfrak{p}} \leq 1$.*
  (ii) *The condition R_1: For each prime ideal $\mathfrak{p}$ of codimension 1, $A_{\mathfrak{p}}$ is a discrete valuation domain.*

Remark 6.13. If A is an integral domain with a finite integral closure $\overline{A}$ and f is a nonzero element in ann $(\overline{A}/A)$, these conditions can be simply stated as: (i) (f) has no embedded primes, and (ii) for each minimal prime $\mathfrak{p}$ of (f) $R_{\mathfrak{p}}$ is a discrete valuation ring. We leave the proof as an exercise.

Proposition 6.14. *Let A be a Noetherian local ring or an affine graded algebra over a field k. Then A is a normal domain if and only if the following conditions hold: (i) A satisfies the condition S_2 and (ii) $A_{\mathfrak{p}}$ is a field or a DVR for every prime ideal $\mathfrak{p}$ of codimension at most* 1.

The only issue is to show that A is an integral domain. Under the hypotheses, this will follow from the Abhyankar-Hartshorne lemma:

Lemma 6.15 (Abhyankar-Hartshorne). *Let R be a Noetherian ring with no idempotents other than 0 or 1, and let I and J be nonzero ideals of R such that $I \cdot J = 0$. Then* grade $(I+J) \leq 1$.

Proof. We may assume that $I \cap J \subset (0:J) \cap (0:I) = 0$, as otherwise for $0 \neq x \in (0:J) \cap (0:I)$ we have $x(I+J) = 0$. Now by the connectedness of $\operatorname{Spec}(R)$,

$$I + J \subset [(0:J) + (0:I)] \neq R.$$

Hence localizing at a prime containing $0:J + 0:I$, we preserve our assumptions $I \neq 0$, $J \neq 0$, $I \cap J = 0$, and $I + J \neq R$. Thus we may from now on assume that R is local.

Suppose that grade $(I+J) > 0$. Let $x = a+b$, $a \in I$, $b \in J$, be a nonzero divisor; it is clear that $a \neq 0$, $b \neq 0$. Moreover, $a \notin R(a+b)$, for an equality $a = r(a+b)$ yields $(1-r)a = rb$, which is a contradiction, whether r is a unit or not. Since $(I+J)a \subset R(a+b)$ it follows that grade $(I+J) = 1$. □

Corollary 6.16. *Let A be an indecomposable Noetherian ring. If A satisfies the S_2 condition and $A_\mathfrak{p}$ is an integral domain for each prime of codimension at most one, then A is an integral domain.*

Remark 6.17. This is particularly useful showing that certain ideals are prime. For example, if I is a homogeneous ideal of a polynomial ring and $A = k[x_1,\ldots,x_n]/I$, char $k = 0$, then A is a normal domain if the Jacobian ideal has grade ≥ 2.

6.1.2 Canonical Module and S_2-ification

We are going to use a notion of "canonical ideal" of a reduced algebra A that is appropriate for our treatment of normality. A general reference is [BH93, Chapter 4]; we shall also make use of [Va98b, Chapter 6].

Definition 6.18. Let A be a reduced affine algebra. A *weak* canonical ideal of A is an ideal L of A with the following properties:

(i) L satisfies Serre's condition S_2.
(ii) For any ideal I that satisfies Serre's condition S_2, the canonical mapping

$$I \mapsto \mathrm{Hom}_A(\mathrm{Hom}_A(I,L),L)$$

is an isomorphism.

There are almost always many choices for L, but once fixed we shall denote it by ω_A, or simply ω if no confusion can arise. Here are some of its properties:

Proposition 6.19. *Let A be a reduced affine algebra with a canonical ideal ω.*

(i) *For any ideal I,*

$$\mathrm{Hom}_A(\mathrm{Hom}_A(I,\omega_A),\omega_A))$$

is the smallest ideal containing I that satisfies condition S_2.
(ii) *A satisfies condition S_2 if and only if*

$$A = \mathrm{Hom}_A(\omega,\omega).$$

Definition 6.20. If A is a Noetherian ring with total ring of fractions K, the smallest finite extension $A \to A' \subset K$ with the S_2 property is called the S_2-ification of A.

An extension with this property is unique (prove it!). It may not always exist, even when A is an integral domain. However, in most of the rings we discuss its existence is guaranteed simply from the theory of the canonical module. The extension

$$A \subset \mathrm{Hom}_A(\omega,\omega) \subset \overline{A}$$

is the *S_2-ification* of A.

Sometimes one can avoid the direct use of canonical modules, as in the following:

Proposition 6.21. *Let R be a Noetherian domain and A a torsionfree finite R-algebra. If $R_{\mathfrak{p}}$ is a Gorenstein ring for every prime ideal of height 1, then $A^{**} = \mathrm{Hom}_R(\mathrm{Hom}_R(A,R),R)$ is the S_2-ification of A.*

Proof. Consider the natural embedding

$$0 \to A \longrightarrow A^{**} \longrightarrow C \to 0.$$

For each prime $\mathfrak{p} \subset R$ of height 1, $A_{\mathfrak{p}}$ is a reflexive module since it is torsionfree over the Gorenstein ring $R_{\mathfrak{p}}$. This shows that C has codimension at least two. We have

$$A' = \bigcap A_{\mathfrak{p}} \subset \bigcap A_{\mathfrak{p}}^{**},$$

taken over all such primes of R. The second module is just A^{**}, from elementary properties of dual modules. Since every $A_{\mathfrak{p}}$ contains A^{**}, we have $A' = A^{**}$. □

Two lazy ways to find an integral extension of a ring with the property S_2, which do not necessarily involve canonical modules, arise in the following manner. The proofs are left as exercises.

Proposition 6.22. *Let R be a Noetherian domain with the condition S_2 and A an integral, rational extension of R. Let $\mathfrak{c} = \mathrm{Hom}_R(A,R)$ be the conductor ideal of A over R. Then $B = \mathrm{Hom}_R(\mathfrak{c},\mathfrak{c})$ is an integral, rational extension of A with the condition S_2.*

Proposition 6.23. *Let R be a Noetherian domain with the condition S_2 and let A be an integral domain that contains R and is R-finite. Then*

$$B = \mathrm{Hom}_A(\mathrm{Hom}_R(A,R),\mathrm{Hom}_R(A,R))$$

is an integral, rational extension of A with the condition S_2.

Remark 6.24. One important case this result applies is to Rees algebras of equimultiple ideals. Suppose that R is a Gorenstein ring, I an equimultiple ideal and J a complete intersection that is a reduction of I If we set $S = R[Jt]$ and $A = R[It]$, then S is a Cohen-Macaulay ring which is Gorenstein in codimension 1 (proof left to the reader) and $\mathrm{Hom}_S(\mathrm{Hom}_S(A,S),S)$ is the S_2-ification of A.

Computation of the Canonical Ideal

We describe a procedure (see [Va98b, Chapter 6]) to represent the canonical module of an affine domain A as one of its ideals. This facilitates the computation of the S_2-ification of A.

Suppose that

$$A = R/P = k[x_1,\ldots,x_n]/(f_1,\ldots,f_m) \tag{6.4}$$

where P is a prime ideal of codimension g. Choose a regular sequence

$$\mathbf{z} = \{z_1, \ldots, z_g\} \subset P.$$

The canonical module ω_A has the following representation:

$$\omega_A = \operatorname{Ext}_R^g(R/P, R) \cong ((\mathbf{z}) : P)/(\mathbf{z}).$$

Theorem 6.25. *Let P and $(\mathbf{z})$ be as above. There exists $a \in (\mathbf{z}) : P \setminus (\mathbf{z})$ and $b \in (a, \mathbf{z}) : ((\mathbf{z}) : P) \setminus P$ such that if we set $L = (b((\mathbf{z}) : P) + (\mathbf{z})) : a$, then*

$$\omega_A = ((\mathbf{z}) : P)/(\mathbf{z}) \cong L/P \subset R/P. \tag{6.5}$$

Proof. It is clear that a can be so selected. To show the existence of b we localize at P. In the regular local ring R_P, PR_P is generated by a regular sequence $\mathbf{y} = \{y_1, \ldots, y_g\}$ so that we can write

$$\mathbf{z} = \mathbf{y} \cdot \varphi$$

for some $g \times g$ matrix φ. By Theorem 1.113,

$$((\mathbf{z}) : P)_P = (\mathbf{z}, \det \varphi)$$

and the image of $\det \varphi$ in $(R/(\mathbf{z}))_P$ generates its socle.

On the other hand, ω_A is a torsionfree A-module, so that the image of a in $((\mathbf{z}) : P/(\mathbf{z}))_P$ does not vanish. This means that $(a, \mathbf{z})_P$ must contain $(\mathbf{z}, \det \varphi)$, which establishes (6.5).

Let L be defined as above. For $r \in (\mathbf{z}) : P$ we can write

$$rb = ta \bmod (\mathbf{z}).$$

Another such representation

$$rb = sa \bmod (\mathbf{z})$$

would lead to an inclusion

$$(t - s)a \in (\mathbf{z}).$$

But a already conducts P into $(\mathbf{z})$, so that $t - s$ must be contained in P, since the associated prime ideals of $(\mathbf{z})$ have codimension g. This defines a mapping

$$(\mathbf{z}) : P/(\mathbf{z}) \longrightarrow L/P,$$

in which the (non-trivial) class of a is mapped to the (non-trivial) class of $b \in R/P$. This completes the proof, as both are torsionfree R/P-modules of rank one. □

Remark 6.26. It is useful to keep in mind that once the canonical module ω_A of an algebra has been determined by a process like that above, we have $\omega_B = \operatorname{Hom}_A(B, \omega_A)$ for a finite extension $A \subset B$.

6.2 Conductors and Affine Algebras

Let A be a reduced Noetherian ring and denote by $\overline{A}$ its integral closure in the total ring of fractions K. The *conductor* of A is the ideal

$$\mathfrak{c}(A) = \text{ann}\,(\overline{A}/A) = \text{Hom}_A(\overline{A}, A).$$

We note that

$$\overline{A} = \text{Hom}_A(\mathfrak{c}(A), \mathfrak{c}(A)),$$

since $\mathfrak{c}(A)$ is an ideal of $\overline{A}$. This shows that there is an equivalence of access $\mathfrak{c}(A) \Leftrightarrow \overline{A}$.

Since access to $\mathfrak{c}(A)$ is usually only found once $\overline{A}$ has been determined, one may have to appeal to what was flippantly termed in [Va98b, Chapter 6] *semi-conductors* of A: nonzero subideals of $\mathfrak{c}(A)$. Let us denote one such nonzero ideal as C and assume that it contains regular elements.

Let us give a scenario for a description of $\overline{A}$:

Proposition 6.27. *If* height $(C) \geq 2$, *then* $\overline{A} = C^{-1}$.

Proof. First, observe that $C \not\subset \mathfrak{p}$ for each prime ideal $\mathfrak{p}$ of A of codimension at most one and therefore $A_{\mathfrak{p}} = \overline{A}_{\mathfrak{p}}$. This implies that

$$A_{\mathfrak{p}} = \bigcap_{\mathfrak{q}} \overline{A}_{\mathfrak{q}},$$

where $\mathfrak{q}$ runs over the primes of $\overline{A}$ lying over $\mathfrak{p}$, which in turn shows that

$$\overline{A} = \bigcap_{\mathfrak{p}} A_{\mathfrak{p}}.$$

We can now prove our assertion. Since $(C^{-1})_{\mathfrak{p}} = A_{\mathfrak{p}}$, C^{-1} is contained in all these intersections and therefore is contained in $\overline{A}$. On the other hand, for any regular element $b \in \overline{A}$, from $Cb \subset A$ we get $C^{-1}b^{-1} \supset A$, that is, $b \in C^{-1}$. Finally, we observe that $\overline{A}$ is generated by regular elements as an A-module. $\square$

6.2.1 The Jacobian Ideal

One of the most useful of such (conductor) ideals is the Jacobian ideal of an affine algebra A. Suppose that

$$A = k[x_1, \ldots, x_n]/I,$$

where I is an unmixed ideal of codimension g. If $I = (f_1, \ldots, f_m)$, the *Jacobian ideal* $J(A)$ is the image in A of the ideal generated by the $g \times g$ minors of the Jacobian matrix

$$\frac{\partial(f_1, \ldots, f_m)}{\partial(x_1, \ldots, x_n)}.$$

There are relative versions one, of which we are going to use.

The relationship between Jacobian ideals and conductors was pointed out in [No50] and strengthened in [LS81].

Theorem 6.28. *Let A be an affine domain and S one of its Noether normalizations. Suppose that the field of fractions of A is a separable extension of the field of fractions of S. Let $A = S[x_1,\ldots,x_n]/P$ be a presentation of A and let J_0 denote the Jacobian ideal of A relative to the variables $x_1,\ldots,x_n$. If $\overline{A}$ is the integral closure of A, then $J_0 \cdot \overline{A} \subset A$. In particular*

$$A = A : J_0$$

if and only if A is integrally closed.

Theorem 6.29. *Let A be an affine domain over a field of characteristic zero. If $J(A)$ and $\mathfrak{c}(A)$ denote the Jacobian ideal and the conductor of A, then*

$$\boxed{J(A) \subset \mathfrak{c}(A).} \tag{6.6}$$

Algebras with the Condition R_1 and One-Step Normalization

There are very few general 'explicit' descriptions of the integral closure of an algebra. One that comes close to meeting this requirement occurs from the computation of a very special module of syzygies.

Theorem 6.30. *Let A be a reduced affine algebra that satisfies Serre's condition R_1. There exist two elements f and g in the conductor ideal $\mathfrak{c}(A)$ such that*

$$\boxed{L = \{(a,b) \in A \times A \mid af - bg = 0\} \rightsquigarrow \overline{A} = \{a/g \mid (a,b) \in L\} = Af^{-1} \textstyle\bigcap Ag^{-1}.}$$

Proof. By assumption, the conductor ideal $\mathfrak{c}(A)$ has codimension at least 2; choose $f, g \in \mathfrak{c}(A)$, both regular elements of A, generating an ideal of codimension 2 (like in the converse of Krull's principal ideal theorem). We note that the module defined by the right-hand side of (6.7) is

$$A :_K (f, g).$$

Since f and g are contained in the conductor of A, we have

$$\overline{A} \subset A :_K (f, g).$$

To prove the reverse containment, we use the fact that

$$\overline{A} = \bigcap_{\mathfrak{p}} \overline{A}_{\mathfrak{p}}, \quad \mathfrak{p} \text{ prime ideal of } A \text{ of codimension } 1.$$

Note that each of these localizations is a semi-local ring whose maximal ideals correspond to the prime ideals of $\overline{A}$ lying above $\mathfrak{p}$. □

Remark 6.31. If A is a reduced equidimensional affine algebra over a field of characteristic zero with the condition R_1 then the elements f, g can be chosen in the Jacobian ideal $J(A)$.

Corollary 6.32. *If A is an algebra as above, satisfying Serre's condition R_1, then $\overline{A}$ is the S_2-ification of A, that is, the smallest extension $A \subset B \subset K$ that satisfies the condition S_2.*

This formulation permits other representations for $\overline{A}$. If the Jacobian ideal $J(A)$ has codimension at least 2, and $S \subset A$ is a hypersurface ring over which A is integral, then

$$\overline{A} = \text{Hom}_S(\text{Hom}_S(A, S), S).$$

Localization and Normalization

We expound a role that localization plays in the construction of the normalization of certain rings. The setting will be that of an affine domain A over a field k with a Noether normalization

$$R = k[x_1, \ldots, x_d] \hookrightarrow A.$$

To start with, consider the following recasting of the Krull-Serre criterion:

Proposition 6.33. *Let A be a Noetherian domain and f, g a regular sequence of R. If $\overline{A}$ is the integral closure of A, then*

$$\overline{A} = \overline{A}_f \cap \overline{A}_g.$$

The assertion holds for arbitrary Noetherian domains, although we shall use it only for affine domains. When f and g are taken as forming a regular sequence in R, we still have the equality $\overline{A} = \overline{A}_f \cap \overline{A}_g$. Since $\overline{A_f} = \overline{A}_f$ (and similarly for g), there are R-submodules $C = (c_1, \ldots, c_r)$ and $D = (d_1, \ldots, d_s)$ such that $C_f = \overline{A}_f$ and $D_f = \overline{A}_g$.

Proposition 6.34. *If f, g is a regular sequence in R, then setting B for ideal generated by $c_1, \ldots, c_r, d_1, \ldots, d_s$, one has a natural isomorphism*

$$\boxed{B^{**} = \text{Hom}_R(\text{Hom}_R(B, R), R) \cong \overline{A}.} \tag{6.7}$$

Proof. Consider the inclusion $B \subset \overline{A}$. Since $\overline{A}$ satisfies condition S_2, it will also satisfy S_2 relative to the subring R. This means that the bidual of B will be contained in $\overline{A}$, $B^{**} \hookrightarrow \overline{A}$. To prove they are equal, it will suffice to show $B^{**}_{\mathfrak{p}} = \overline{A}_{\mathfrak{p}}$ for each prime ideal $\mathfrak{p} \subset R$ of codimension 1. But from our choices of f and g, $B_{\mathfrak{p}} = \overline{A}_{\mathfrak{p}}$. □

A special case is when f belongs to the conductor of A, since $A_f = \overline{A}_f$ already, so that we simply take $C = A$. For simplicity we set $B = A(f, g)$, and the special case by $A(1, g)$. As an application, let us consider a reduction technique that converts the problem of finding the integral closure of a standard graded algebra into another involving finding the integral closure of a lower-dimensional affine domain (but not graded) and the computation of duals.

Proposition 6.35. *Suppose as above that f lies in the conductor of A and that g is a form of degree 1 (in particular, $d \geq 2$). The integral closure of A is obtained as the R-bidual of a set of generators of the integral closure of $A/(g-1)$.*

Proof. The localization A_f has a natural identification with the ring $A_g = S[T, T^{-1}]$, where S is the set of fractions in A_g of degree 0 and T is an indeterminate. Further, as it is well-known (see [Ei95, Exercise 2.17]), $S \cong A/(g-1)$. □

6.2.2 R_1-ification

We consider two constructions considered in [Va91b] and [Jo98] (using [GRe84, p. 127]), respectively; our presentation is also informed by [Mat0]. Its tests provide proactive normality criteria.

Theorem 6.36. *Let A be a reduced equidimensional affine ring over a field of characteristic zero, and denote by J its Jacobian ideal. Then A is integrally closed in the following two cases:*

(i) $A = \operatorname{Hom}_A(J^{-1}, J^{-1})$;
(ii) $A = \operatorname{Hom}_A(\sqrt{J}, \sqrt{J})$.

Proof. Let $\mathfrak{p}$ be a prime associated to a principal ideal (f) generated by a regular element, and consider $A_{\mathfrak{p}}$. Both conditions (i) and (ii) are maintained by localizing. We will then use the symbol A for $A_{\mathfrak{p}}$.

(i) The first idealizer ring condition can be rewritten as $A = (J \cdot J^{-1})^{-1}$. This means that the ideal $J \cdot J^{-1}$ is A, and therefore J is a principal ideal. By [Li69a] $J = A$, and A is a discrete valuation domain by the Jacobian criterion.

(ii) Suppose that $A \neq \overline{A}$. According to Theorem 6.28, J is contained in the conductor ideal of A. Let $\mathfrak{p}$ be an associated prime of $\overline{A}/A$. Choose $x \in \overline{A} \setminus A$ such that $A : x = \mathfrak{p}$. Since $J \subset \mathfrak{p}$, we also have $\sqrt{J} \subset \mathfrak{p}$. Let

$$x^n + a_{n-1}x^{n-1} + \cdots + a_0 = 0$$

be an equation of integral dependence of x over A. For any $y \in \sqrt{J}$, multiplying the equation by y^n we get that $(xy)^n \in \sqrt{J}$. Since $xy \in A$, we obtain $xy \in \sqrt{J}$, and therefore $x \in A$ by the hypothesis. □

Unlike the operation defined earlier, each application of $\mathcal{P}$ takes place in a different ring, as the new Jacobian ideal has to be assembled from a presentation of the algebra $\mathcal{P}(B)$.

Remark 6.37. Observe that the normality criterion expressed in the equality

$$\mathrm{Hom}_A(J^{-1},J^{-1}) = A$$

does not require A to satisfy Serre's condition S_2 (as stated originally in [Va91b]). We note that one way to configure the endomorphism ring is simply as follows. Let $x \in J$ be a regular element, and denote by $(x) :_A J$ the ideal quotient in A. Since $J^{-1} = ((x) :_R J)x^{-1}$,

$$\mathrm{Hom}_A(J^{-1},J^{-1}) = \mathrm{Hom}_A((x) : J, (x) : J).$$

Example 6.38. Let us illustrate with one example how the two methods differ in the presence the condition R_1. Let A be an affine domain over a field of characteristic zero and J its Jacobian ideal. Suppose that height $J \geq 2$ (the R_1 condition). As we have discussed,

$$\overline{A} = J^{-1} \subset \mathrm{Hom}_A(J^{-1},J^{-1}) \subset \overline{A}.$$

Consider now the following example. Let n be a positive integer, and set

$$A = \mathbb{R}[x,y] + (x,y)^n\mathbb{C}[x,y],$$

whose integral closure is the ring of polynomials $\mathbb{C}[x,y]$. Note that $A_x = \mathbb{C}[x,y]_x$ and $A_y = \mathbb{C}[x,y]_y$. This means that $\sqrt{J}$ is the maximal ideal

$$M = (x,y)\mathbb{R}[x,y] + (x,y)^n\mathbb{C}[x,y].$$

It is clear that

$$\mathrm{Hom}_A(M,M) = \mathbb{R}[x,y] + (x,y)^{n-1}\mathbb{C}[x,y].$$

It will take precisely n passes of the operation to produce the integral closure. In particular, neither the dimension ($d = 2$), nor the multiplicity ($e = 1$) play any role.

Remark 6.39. It is worth noting that the first of these Jacobian basic operations has the property that if A has condition S_2, then all $\mathcal{P}^i(A)$ also have S_2. This will put the same limits on the order of $\mathcal{P}$ that were developed in this section.

In the case of (i), if $J = J_0 \cap L$, where grade $J_0 = 1$ and grade $L \geq 2$, one has $J^{-1} = J_0^{-1}$. Indeed, if $x \in J^{-1}$, since $J_0Lx \subset A$ we have $J_0x \subset A$ since L has grade at least two. It follows that $(JJ^{-1})^{-1} = (J_0J_0^{-1})^{-1}$, since they are both equal to $\mathrm{Hom}_A(J^{-1},J^{-1})$.

In the case of (ii) it would be interesting to find bounds for the 'order' of the basic operation. There a minor modification of (ii) that makes it amenable to the bounds discussed. Let us assume that the algebra A satisfies the condition S_2 and let J be its Jacobian ideal (assume that the characteristic is either zero or does not place any obstruction). The radical of J is made up (possibly) of

$$\sqrt{J} = K \cap L,$$

where K has codimension one and L has codimension at least two (if $\sqrt{J} = L$, A is normal by the Jacobian criterion). Observe that

$$\mathrm{Hom}_A(K \cap L, K \cap L) \subset \mathrm{Hom}_A(K,K).$$

Indeed, for any element $f \in \mathrm{Hom}(K \cap L, K \cap L)$, we have

$$L \cdot f \cdot K = f \cdot (K \cdot L) \subset f(K \cap L) \subset K \cap L \subset K.$$

Since K is a divisorial ideal and L has grade at least two, $f \cdot K \subset K$, as desired (we have used the fact that A has condition S_2, which it passes to all its divisorial ideals).

There is a payoff: the computation of $\sqrt{J}$ often requires the computation of K as a preliminary step. The observation above states that $\mathrm{Hom}_A(K,K)$ is a possibly larger ring than $\mathrm{Hom}_A(\sqrt{J},\sqrt{J})$, and requires fewer steps.

Let us indicate the calculation of K for a hypersurface $A = k[x_1,\ldots,x_{d+1}]/(f)$, in the case of characteristic greater than $\deg f$ (we are assuming that f is a square-free polynomial). Let g be a partial derivative of f for which $\gcd(f,g) = 1$. Let

$$J = (f, \frac{\partial f}{\partial x_1}, \ldots, \frac{\partial f}{\partial x_{d+1}}),$$

and let J_0 be the Jacobian ideal of (f,g). Then K is given (see [Va98b, Theorem 5.4.3]) by the image in A of the ideal of $k[x_1,\ldots,x_{d+1}]$, namely

$$((f,g) : J_0) : (((f,g) : J_0) : J).$$

As an issue there remains what other idealizer construction based on the Jacobian ideal J can be used as normality criteria? A standard construction on a nonzero ideal I is the limit

$$\mathcal{P}(I) = \bigcup_{n\geq 1} \mathrm{Hom}_A(I^n, I^n) \subset \overline{A}.$$

An equality $\mathcal{P}(I) = A$ will always imply that I is principal in codimension 1, so that when $I = J$ it will imply that A satisfies condition R_1.

A simpler question that has not been resolved is whether $\mathrm{Hom}_A(J,J) = A$ means that A is normal. It will be the case if A is Gorenstein (e.g. A is a plane curve), but not many other cases are known.

Presentation of $\mathrm{Hom}_A(J,J)$

The Jacobian algorithms require presentations of $\mathrm{Hom}_A(J,J)$ as A-algebras, where J is some ideal obtained from the Jacobian ideal of the algebra A. More precisely, the passage from J_n to J_{n+1} requires a set of generators and relations for $A_{n+1} = \mathrm{Hom}(J_n,J_n)$. When $w_1,\ldots,w_s$ is a set of generators of A_{n+1} as an A-module, a set of generators and relations of A_{n+1} as a k-algebra can be obtained from the following observation ([Ca84]):

$$A_{n+1} = k[x_1,\ldots,x_m,W_1,\ldots,W_s]/(I,H_n,K_n),$$

where H_n are the generators for the linear syzygies of the w_i over A, that is, the expressions

$$a_1W_1 + \cdots + a_sW_s,$$

and K_n is the set of quadratic relations

$$H_{ij} = W_iW_j - \sum_{k=1}^{s} a_{ijk}W_k,$$

that express the algebra structure of A_{n+1}.

6.2.3 The Integral Closure of Subrings Defined by Graphs

Let G be a graph with vertices $x_1, \ldots, x_n$ and $R = k[x_1, \ldots, x_n]$ a polynomial ring over a field k. The *edge subring* of G is the k-subalgebra

$$k[G] = k[\{x_ix_j \mid x_i \text{ is adjacent to } x_j\}] \subset R,$$

and the *edge ideal* of G, denoted by $I(G)$, is the ideal of R generated by the set of x_ix_j such that x_i is adjacent to x_j.

We will give combinatorial descriptions of the integral closure of $k[G]$ and $\mathcal{R}(I(G))$, where $\mathcal{R}(I(G))$ is the Rees algebra of $I(G)$. The itinerary provided here was written by R. Villarreal, to whom we are grateful.

The next result ([CVV98]) is essential to understand the normality of those two algebras.

Theorem 6.40. *Let G be a graph and $I = I(G)$ its edge ideal. The following conditions are equivalent:*

(i) *G is bipartite.*
(ii) *I is normally torsionfree.*

Corollary 6.41. *Let G be a graph and $I = I(G)$ its edge ideal. If G is bipartite, then the Rees algebra $\mathcal{R}(I)$ is a normal domain.*

Proof. Note that $I(G)$ is a radical ideal that is generically a complete intersection and use [Vi1, Theorem 3.3.31]. □

Example 6.42. As an application, we give an extensive class of integrally closed ideals. Let $\mathbf{X}$ be a finite set of indeterminates over a field k and let $I_1, I_2, \ldots, I_n$ be ideals of $R = k[\mathbf{X}]$, not necessarily distinct, generated by indeterminates. We claim that the product

$$I = I_1 \cdot I_2 \cdots I_n$$

is integrally closed. Let $y_1, y_2, \ldots, y_n$ be new indeterminates and define the ideal

$$I = \sum_{i=1}^{n} I_iy_jS \subset S = R[y_1, \ldots, y_n].$$

According to Corollary 6.41, I is a normal S-ideal since the graph it defines is bipartite. As an R-module,

$$I^n = Iy_1 \cdots y_n \oplus L,$$

where L is a summand no term of which involves the monomial $Y = y_1 \cdots y_n$. Clearly $\overline{I}Y$ is integral over I^n, and therefore $\overline{I} = I$.

Corollary 6.43. *If G is a bipartite graph and k is a field, then the edge subring $k[G]$ is a normal domain.*

Proof. This follows at once from Corollary 6.41 and by descent of normality [Vi1, Proposition 7.4.1]. □

For use below, let $\mathcal{A} = \{v_1, \ldots, v_q\}$ be the set of sum, s $e_i + e_j$ such that x_i is adjacent to x_j, where e_i is the *ith* unit vector in $\mathbb{R}^n$, and set $f_i = x^{v_i}$. One has the following description of the integral closure of $k[G]$

$$\overline{k[G]} = k[\{x^a \mid a \in \mathbb{R}_+\mathcal{A} \cap \mathbb{Z}\mathcal{A}\}],$$

where $\mathbb{R}_+\mathcal{A}$ is the cone in $\mathbb{R}^n$ generated by $\mathcal{A}$ and $\mathbb{Z}\mathcal{A}$ is the subgroup of $\mathbb{Z}^n$ generated by $\mathcal{A}$. See [Vi1, Theorem 7.2.28].

Definition 6.44. A *bow tie* of a graph G is an induced subgraph w of G consisting of two edge disjoint odd cycles

$$Z_1 = \{z_0, z_1, \ldots, z_r = z_0\} \text{ and } Z_2 = \{z_s, z_{s+1}, \ldots, z_t = z_s\}$$

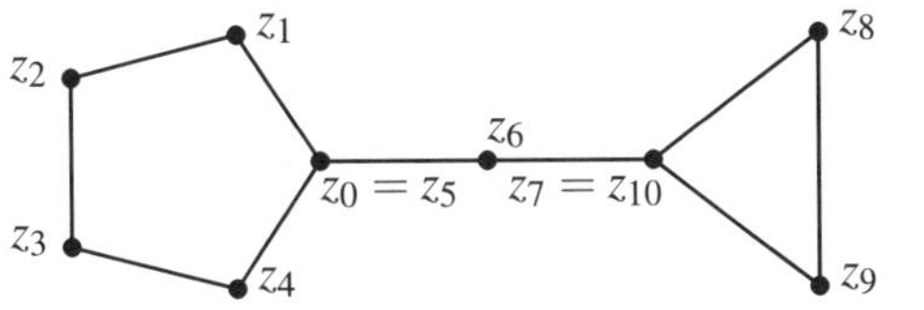

joined by a path $\{z_r, \ldots, z_s\}$. In this case we set $M_w = z_1 \cdots z_r z_{s+1} \cdots z_t$.

If w is a bow tie of a graph G, as above, then M_w is in the integral closure of $k[G]$. Indeed if $f_i = z_{i-1}z_i$, then

$$z_1^2 \cdots z_r^2 = f_1 \cdots f_r, \quad z_s^2 \cdots z_{t-1}^2 = f_{s+1} \cdots f_t, \tag{6.8}$$

which together with the identities

$$M_w = \prod_{i \text{ odd}} f_i \prod_{\substack{i \text{ even} \\ r < i \leq s}} f_i^{-1} \quad \text{and} \quad M_w^2 = f_1 \cdots f_r f_{s+1} \cdots f_t$$

gives $M_w \in \overline{k[G]}$.

We can now provide the promised description of the integral closure of $k[G]$ ([SVV98]).

Theorem 6.45. *Let G be a graph and let k be a field. Then the integral closure $\overline{k[G]}$ of $k[G]$ is generated as a k-algebra by the set*

$$\mathcal{B} = \{f_1, \ldots, f_q\} \cup \{M_w \mid w \text{ is a bow tie}\},$$

where $f_1, \ldots, f_q$ denote the monomials defining the edges of G.

As an immediate consequence of Theorem 6.45 one has the following full characterization of conditions under which $k[G]$ is normal.

Corollary 6.46. *Let G be a connected graph. Then $k[G]$ is normal if and only if, for any two edge-disjoint odd cycles Z_1 and Z_2, either Z_1 and Z_2 have a common vertex, or Z_1 and Z_2 are connected by an edge.*

Corollary 6.47. *Let G be a connected graph and $I = I(G)$ its edge ideal. Then $k[G]$ is normal if and only if the Rees algebra $\mathcal{R}(I)$ is normal.*

Proof. Let $C(G)$ be the cone over the graph G; it is obtained by adding a new vertex t to G and joining every vertex of G to t. According to [Vi1, Proposition 8.2.15] there is a isomorphism

$$\mathcal{R}(I(G)) \cong k[C(G)].$$

If M_w be a bow tie of $C(G)$ with edge disjoint cycles Z_1 and Z_2 joined by the path P, it is enough to verify that $M_w \in k[C(G)]$. If $t \notin Z_1 \cup Z_2 \cup P$, then w is a bow tie of G and $M_w \in k[G]$. Assume that $t \in Z_1 \cup Z_2$, say $t \in Z_1$. If $Z_1 \cap Z_2 \neq \emptyset$, then $M_w \in k[C(G)]$. On the other hand, if $Z_1 \cap Z_2 = \emptyset$ then $M_w \in k[C(G)]$ because in this case Z_1 and Z_2 are joined by the edge $\{t, z\}$, where z is any vertex in Z_2. It remains to consider the case where $t \notin Z_1 \cup Z_2$ and $t \in P$ since G is connected there is a path in G joining Z_1 with Z_2. Therefore $M_w = M_{w_1}$ for some bow tie w_1 of G and $M_w \in k[G]$.

Conversely, if $\mathcal{R}(I(G))$ is normal, then by [Vi1, Proposition 7.4.1] we obtain that $k[G]$ is normal. □

In the corollary above the hypothesis that G is connected is essential. To see this, consider the graph G consisting of two disjoint triangles, when $k[G]$ is a polynomial ring and $\mathcal{R}(I)$ is not normal.

Proposition 6.48. *Let G be a graph and $G_1, \ldots, G_r$ the connected components of G. Then $\mathcal{R}(I(G))$ is normal if and only if one of the following holds:*

(i) *G is bipartite,*
(ii) *exactly one of the components, say G_1, is non bipartite and $K[G_1]$ is a normal domain.*

Proof. One may proceed as in the proof of Corollary 6.47. □

Integral Closure of Rees Algebras

Let G be a graph with vertices $x_1, \ldots, x_n$, and $I = I(G)$ its edge ideal. Consider the endomorphism ϕ of the field $k(x_1, \ldots, x_n, t)$ defined by $x_i \mapsto x_i t$, $t \mapsto t^{-2}$. It induces an isomorphism

$$R[It] \overset{\phi}{\longmapsto} k[C(G)], \quad x_i \mapsto t x_i, \quad x_i x_j t \mapsto x_i x_j,$$

where $R = k[x_1, \ldots, x_n]$ is a polynomial ring over k and $C(G)$ is the cone over G obtained by adding a new variable t.

To describe the integral closure of $R[It]$ we describe first the integral closure of $k[C(G)]$. If $Z_1 = \{x_1, x_2, \ldots, x_r = x_1\}$ and $Z_2 = \{z_1, z_2, \ldots, z_s = z_1\}$ are two edge disjoint odd cycles in $C(G)$, then one has

$$M_w = x_1 \cdots x_r z_1 \cdots z_s$$

for some bow tie w of $C(G)$; this follows readily because $C(G)$ is a connected graph. In particular M_w is in the integral closure of $k[C(G)]$. Observe that if t occurs in Z_1 or Z_2, then $M_w \in k[C(G)]$ because in this case either Z_1 and Z_2 meet at a point, or Z_1 and Z_2 are joined by an edge in $C(G)$. Thus in order to compute the integral closure of $k[C(G)]$, the only bow ties that matter are those defining the following set:

$$\mathcal{B} = \{M_w \mid w \text{ is a bow tie in } C(G) \text{ such that } t \notin \operatorname{supp}(M_w)\}.$$

Therefore by Theorem 6.45 we have proved:

Proposition 6.49. $\overline{k[C(G)]} = k[C(G)][\mathcal{B}]$.

To obtain a description of the integral closure of the Rees algebra of I note that using ϕ together with the equality

$$M_w = \frac{(tz_1) \prod_{i=1}^{\frac{r+1}{2}} (x_{2i-1}x_{2i}) \prod_{i=1}^{\frac{s-1}{2}} (z_{2i}z_{2i+1})}{tx_1},$$

where $x_{r+1} = x_1$, we see that M_w is mapped back to the Rees algebra in the element

$$M_w t^{\frac{r+s}{2}} = x_1 x_2 \cdots x_r z_1 z_2 \cdots z_s t^{\frac{r+s}{2}}.$$

If $\mathcal{B}'$ is the set of all monomials of this form, then one obtains:

Proposition 6.50. $\overline{R[It]} = R[It][\mathcal{B}']$.

If the monomials do not have the underlying structure of a graph, one may attempt to solve both problems using NORMALIZ, a program developed by W. Bruns and R. Koch ([BK98]).

6.3 Divisorial Extensions of an Affine Algebra

We treat here an approach of [Va0] to an analysis of the construction of the integral closure $\overline{A}$ of an affine ring A. It will apply to an examination of the complexity of some of the existing algorithms ([BV1], [Jo98], [Va91b]). These algorithms put their trust blindly in the Noetherian condition, without any *a priori* numerical certificate of termination. We remedy this for any algorithm that uses a particular class of extensions termed *divisorial*. In addition, we analyze in detail an approach to the computation of $\overline{A}$ that is theoretically distinct from the current methods.

6.3.1 Divisorial Extensions of Gorenstein Rings

Throughout we shall assume that A is a reduced affine ring and $\overline{A}$ is its integral closure.

Definition 6.51. An integral extension B of A is *divisorial* if $A \subset B \subset \overline{A}$ and B satisfies Serre's S_2 condition. The set of divisorial extensions of A will be denoted by $\mathcal{S}_2(A)$.

Let $A = k[x_1,\ldots,x_n]/I$ be a reduced equidimensional affine algebra over a field k of characteristic zero, let $R = k[x_1,\ldots,x_d] \subset A$ be a Noether normalization and $S = k[x_1,\ldots,x_d,x_{d+1}]/(f)$ a hypersurface ring such that the extension $S \subset A$ is birational. Denote by J the Jacobian ideal of S, that is the image in S of the ideal generated by the partial derivatives of the polynomial f.

Since $S \subset A \subset \overline{S} = \overline{A}$, we have from [No50] that J is contained in the conductor of $\overline{S}$. To fix the terminology, we denote the annihilator of the S-module A/S by $\mathfrak{c}(A/S)$. Note the identification $\mathfrak{c}(A/S) = \mathrm{Hom}_S(A,S)$.

We want to benefit from the fact that S is a Gorenstein ring, in particular that its divisorial ideals have a rich structure. Let us recall some of these. Denote by K the total ring of fractions of S. A finitely generated submodule L of K is said to be *divisorial* if it is faithful and the canonical mapping

$$L \mapsto \mathrm{Hom}_S(\mathrm{Hom}_S(L,S),S)$$

is an isomorphism. Since S is Gorenstein, for a proper ideal $L \subset S$ this simply means that all the primary components of L have codimension 1. We sum up some of these properties in (see [BH93] for general properties of Gorenstein rings and [Va98b, Section 6.3] for specific details on Serre's condition S_2):

Proposition 6.52. *Let S be a hypersurface ring as above (or more generally a Gorenstein ring). Then*

(a) *A finitely generated faithful submodule of K is divisorial if and only if it satisfies condition S_2.*
(b) *Let $A \subset B$ be finite birational extensions of S such that A and B have condition S_2. Then $A = B$ if and only if $\mathfrak{c}(A/S) = \mathfrak{c}(B/S)$.*
(c) *If A is an algebra such that $S \subset A \subset \overline{S}$, then*

$$\mathrm{Hom}_S(\mathrm{Hom}_S(A,S),S) = \mathrm{Hom}_S(\mathfrak{c}(A/S),S)$$

is the S_2-closure of A.

Remark 6.53. We shall also set $A^{-1} = \mathrm{Hom}_S(A,S)$, and denote the S_2-closure of an algebra A by $C(A) = (A^{-1})^{-1}$. Note that one has the equality of conductors $\mathfrak{c}(A/S) = \mathfrak{c}(C(A)/S)$. Actually these same properties of the duality theory over Gorenstein rings hold if S is S_2 and is Gorenstein in codimension at most one.

We now define our two notions and begin exploring their relationship.

Definition 6.54. Let I be an ideal containing regular elements of a Noetherian ring S. The *divisorial degree* of I is the integer

$$\deg_0(I) = \sum_{\text{height } \mathfrak{p}=1} \lambda((S/I)_{\mathfrak{p}}).$$

Remark 6.55. We observe that the constituents of $\deg_0(I)$ occur in the expression of the multiplicity $\deg(R/I)$ when I is an ideal of codimension 1 (R a graded or local ring):

$$\deg(R/I) = \sum_{\dim /\mathfrak{p}=\dim R-1} \lambda((R/I)_{\mathfrak{p}}) \cdot \deg(R/\mathfrak{p}).$$

A significant difference between $\deg_0(I)$ and $\deg(R/I)$ lies in the fact that while the former is almost always smaller, the latter is more amenable to calculation since it can also be expressed in other ways.

Definition 6.56. A *proper operation* for the purpose of computing the integral closure of S is a method such that whenever inclusions $S \subset A \subsetneq \overline{S}$ are given, the operation produces a divisorial extension B such that $A \subsetneq B \subset \overline{S}$.

The next result highlights the fact that most extensions used in the computation of the integral closure satisfy the descending chain condition.

Theorem 6.57. *Let A be a Gorenstein ring with a finite integral closure $\overline{A}$. Let $\mathcal{S}_2(A)$ be the set of extensions $A \subset B \subset \overline{A}$ that satisfy condition S_2 and denote by $\mathfrak{c}$ the conductor of $\overline{A}$ over A. There is an inclusion-reversing one-one correspondence between that elements of $\mathcal{S}_2(A)$ and a subset of divisorial ideals of S containing $\mathfrak{c}$. In particular $\mathcal{S}_2(A)$ satisfies the descending chain condition.*

Proof. Any ascending chain of divisorial extensions

$$A \subsetneq A_1 \subsetneq A_2 \subsetneq \cdots \subsetneq A_n \subset \overline{A}$$

gives rise to a descending chain of divisorial ideals

$$\mathfrak{c}(A_1/A) \supset \mathfrak{c}(A_2/A) \supset \cdots \supset \mathfrak{c}(A_n/A)$$

of the same length, by Proposition 6.52(b). But each of these divisorial ideals contains $\mathfrak{c}$, which gives the assertion. □

We collect these properties: let $\mathfrak{p}_1, \ldots, \mathfrak{p}_n$ be the associated primes of $\mathfrak{c}$ and set $U = \bigcup_{i=1}^n \mathfrak{p}_i$. There is an embedding of partially ordered sets,

$$\mathcal{S}_2(A)' \hookrightarrow \text{Ideals}(S_U/\mathfrak{c}_U),$$

where $\mathcal{S}_2(A)'$ is $\mathcal{S}_2(A)$ with the order reversed. Note that $S_U/\mathfrak{c}_U$ is an Artinian ring.

Of course the conductor ideal $\mathfrak{c}$ is usually not known in advance, or the ring A is not always Gorenstein. When A is a reduced equidimensional affine algebra over a field of large characteristic we may replace it by a hypersurface subring S with integral closure $\overline{A}$. On the other hand, by [No50] (see more general results in [Ku86], [LS81]), the Jacobian ideal J of S is contained in the conductor of S. We get the following less tight but more explicit rephrasing of Theorem 6.57.

Theorem 6.58. *Let S be a reduced hypersurface ring,*

$$S = k[x_1, \ldots, x_{d+1}]/(f)$$

over a field of characteristic zero, and J its Jacobian ideal. Then the integral closure of S can be obtained by carrying out at most $\deg_0(J)$ *proper operations on S.*

In characteristic zero, the relationship between the Jacobian ideal J of S and the conductor $\mathfrak{c}$ of $\overline{S}$ is difficult to express in detail. In any event the two ideals have the same associated primes of codimension one, a condition we can write as the equality $\sqrt{\mathfrak{c}} = \sqrt{(J^{-1})^{-1}}$ of radical ideals.

Corollary 6.59. *Let k be a field of characteristic zero and A a standard graded domain over k of dimension d and multiplicity e. Let S be a hypersurface subring of A such that $S \subset A$ is finite and birational. Then the integral closure of A can be obtained after $(e-1)^2$ proper operations on S.*

Proof. Set $S = k[x_1, \ldots, x_{d+1}]/(f)$, where f is a form of degree $e = \deg(A)$. By Euler's formula, $f \in L = \left(\frac{\partial f}{\partial x_1}, \ldots, \frac{\partial f}{\partial x_{d+1}}\right)$. Let then g, h be forms of degree $e-1$ in L forming a regular sequence in $T = k[x_1, \ldots, x_{d+1}]$. Clearly we have that $\deg(g,h)S \geq \deg(J)$. On the other hand, we have the following estimation of ordinary multiplicities:

$$\begin{aligned}
(e-1)^2 = \deg(T/(g,h)) &= \sum_{\text{height } \mathfrak{P}=2} \lambda(T/(g,h)_{\mathfrak{P}}) \deg(T/\mathfrak{P}) \\
&\geq \sum_{\text{height } \mathfrak{P}=2} \lambda(T/(g,h)_{\mathfrak{P}}) \\
&\geq \sum_{\text{height } \mathfrak{P}=2} \lambda(T/(f,g,h)_{\mathfrak{P}}) \\
&= \deg((g,h)S) \\
&\geq \deg(\mathfrak{c}),
\end{aligned}$$

as required. □

Note that this is a pessimistic bound that would be cut in half by the simple hypothesis that no minimal prime of J is monomial. One does not need the base ring to be a hypersurface ring; the Gorenstein condition will do. We illustrate with the following:

Corollary 6.60. *Let A be a reduced equidimensional Gorenstein algebra over a field of characteristic zero. Let J be the Jacobian ideal of S and L the corresponding divisorial ideal, $L = (J^{-1})^{-1}$. If L is a radical ideal then $\overline{A}$ is the only proper divisorial extension of A.*

Proof. Let B and C with $A \subset B \subset C$ be divisorial extensions of A, and let $K \subset I$ be the conductors of the extensions C and B, respectively. We shall show that $K = I$. If

$\mathfrak{p}$ be a prime of codimension one that it is not associated to I, then $B_{\mathfrak{p}}$ is integrally closed and $B_{\mathfrak{p}} = C_{\mathfrak{p}}$. Thus I and K have the same primary components since they both contain L. $\square$

Let us highlight the boundedness of chains of divisorial subalgebras.

Corollary 6.61. *Let A be a reduced equidimensional standard graded algebra over a field of characteristic zero, and set $e = \deg(A)$. Then any sequence*

$$A = A_1 \subset A_2 \subset \cdots \subset A_n \subset \overline{A}$$

of finite extensions of A with Serre's property S_2 has length at most $(e-1)^2$.

6.3.2 Non-Homogeneous Algebras

We shall now treat affine algebras that are not homogeneous. Suppose that A is a reduced equidimensional algebra over a field of characteristic zero, of dimension d. Let

$$S = k[x_1, \ldots, x_d, x_{d+1}]/(f) \hookrightarrow A$$

be a hypersurface ring over which A is finite and birational. The degree of the polynomial f will play the role of the multiplicity of A. Of course, we may choose f of as small degree as possible.

Our aim is to find estimates for the length of chains of algebras

$$S = A_0 \subset A_1 \subset \cdots \subset A_q = \overline{A}$$

satisfying the condition S_2, between S and its integral closure $\overline{A}$. The argument we have used required the length estimates for the length of the total ring of fractions of $S/\mathfrak{c}$, where $\mathfrak{c}$ is the conductor ideal of S, ann $(\overline{A}/S)$. Actually, it only needs estimates for the length of the total ring of fractions of $(S/\mathfrak{c})_{\mathfrak{m}}$, where $\mathfrak{m}$ ranges over the maximal ideals of S.

In the homogeneous case, we have found it convenient to estimate these lengths in terms of the multiplicities of $(S/\mathfrak{c})_{\mathfrak{m}}$; we shall do likewise here.

A first point to be made is the observation that we may replace k by $K \cong S/\mathfrak{m}$ and $\mathfrak{m}$ by a maximal ideal $\mathfrak{M}$ of $K \otimes_k A$ lying over it. In other words, we can replace R by a faithfully flat (local) extension R'. The conditions are all preserved in that $S' = K \otimes_k S$ is reduced, $\overline{S'} = K \otimes_k \overline{A}$, the conductor of S extends to the conductor of S', and chains of extensions with the S_2 conditions give rise to similar extensions of K-algebras. Furthermore the length of the total ring of fractions of $R/\mathfrak{c}$ is bounded by the length of the total ring of fractions of $R'/\mathfrak{c}'$.

What this all means is that we may assume that $\mathfrak{m}$ is a rational point of the hypersurface $f = 0$. We may change the coordinates so that $\mathfrak{m}$ corresponds to the actual origin.

Proposition 6.62. *Let $A = k[x_1, \ldots, x_d]$ be the ring of polynomials over the infinite field k and let f, g be polynomials in A vanishing at the origin. Suppose that f, g is a regular sequence and $\deg f = m \leq n = \deg g$. Then the multiplicity of the local ring $(A/(f,g))_{(x_1,\ldots,x_n)}$ is at most nm^2.*

Proof. Write f as the sum of its homogeneous components,

$$f = f_m + f_{m-1} + \cdots + f_r,$$

and similarly for g,

$$g = g_n + g_{n-1} + \cdots + g_s.$$

We first discuss the route the argument will take. Suppose that g_s is not a multiple of f_r. We denote by R the localization of A at the origin, and its maximal ideal by $\mathfrak{m}$. We observe that $A/(f_r)$ is the associated graded ring of $R/(f)$, and the image of g_s is the initial form g. Thus the associated graded ring of $R/(f,g)$ is a homomorphic image of $A/(f_r, g_s)$. If f_r and g_s are relatively prime polynomials, it will follow that the multiplicity of $R/(f,g)$ will be bounded by $r \cdot s$,

$$\deg R/(f,g) \leq r \cdot s.$$

We are going to ensure that these conditions on f and g are realized for f and another element h of the ideal (f,g). After a linear homogeneous change of variables (as k is infinite), we may assume that each non-vanishing component of f and of g has unit coefficient in the variable x_d. To that end it suffices to use the usual procedure on the product of all nonzero components of f and g. At this point we may assume that f and g are monic.

Rewrite now

$$\begin{aligned} f &= x_d^m + a_{m-1}x_d^{m-1} + \cdots + a_0, \\ g &= x_d^n + b_{n-1}x_d^{n-1} + \cdots + b_0 \end{aligned}$$

with the a_i, b_j in $k[x_1, \ldots, x_{d-1}]$. Consider the resultant of these two polynomials with respect to x_d:

$$\operatorname{Res}(f,g) = \det \begin{bmatrix} 1 & a_{m-1} & a_{m-2} & \cdots & a_0 & & & \\ & 1 & a_{m-1} & \cdots & a_1 & a_0 & & \\ \vdots & \vdots & \vdots & \vdots & \vdots & \vdots & \vdots & \vdots \\ & & & 1 & a_{m-1} & a_{m-2} & \cdots & a_0 \\ 1 & b_{n-1} & b_{n-2} & \cdots & b_0 & & & \\ & 1 & b_{n-1} & \cdots & b_1 & b_0 & & \\ \vdots & \vdots & \vdots & \vdots & \vdots & \vdots & \vdots & \vdots \\ & & & 1 & b_{n-1} & b_{n-2} & \cdots & a_0 \end{bmatrix}.$$

We recall that $h = \operatorname{Res}(f,g)$ lies in the ideal (f,g). Scanning the rows of the matrix above (n rows of entries of degree at most m, and m rows of entries of degree at most n), it follows that $\deg h \leq 2mn$. A closer examination of the distribution of the degrees shows that $\deg h \leq mn$ (see [Wal62, Theorem 10.9]). If h_p is the initial form of h, then clearly h_p and f_r are relatively prime since the latter is monic in x_d, while h_p lacks any term in x_d.

Assembling the estimates, one has

$$\deg R/(f,g) \leq \deg R/(f,h) \leq r \cdot p \leq m \cdot mn = nm^2,$$

as claimed. □

Corollary 6.63. *Let $S = k[x_1, \ldots, x_{d+1}]/(f)$ be a reduced hypersurface ring over a field of characteristic zero, with $\deg f = e$. Then any chain of algebras between S and its integral closure, satisfying the condition S_2, has length at most $e(e-1)^2$.*

Remark 6.64. One must be careful to use this notion of *multiplicity* in these estimates instead of the more classical one. Consider the case of the graded ring $S = k[x^2, x^{2n+1}]$. Its ordinary multiplicity at the origin is 2, while the multiplicity derived from the Noether normalization is $e = n+1$. This is the value to be used in the estimation since there are chains of extensions of length $n+1$ between S and its integral closure $k[x]$. (The author thanks W. Heinzer for this observation.)

A Rees Algebra Approach to the Integral Closure

We now introduce and analyze in detail one proper operation which does not use Jacobian ideals so extensively. It is part of another methodology of finding integral closures. Thus far, our approach to passing from one extension to another has involved the addition of a batch of new variables. More precisely, given a setting as above (using the same notation)

$$S \hookrightarrow A, \quad S \neq A,$$

and using a presentation of A to obtain its Jacobian ideal, one defines

$$B = \mathrm{Hom}_S(\tilde{J}, \tilde{J}),$$

where $\tilde{J}$ is a divisorial ideal constructed from J. The net result is that iteration of this construction quickly leads to too many variables. It might be advisable to carry out most of the construction in the original ring or even in S. Let us describe such an approach. We shall refer to it as a *modified* proper operation.

Proposition 6.65. *There is at least one modified proper operation whose iteration leads to the integral closure of S.*

Proof. We assume that we have $S \subsetneq A \subset \overline{S}$, and now we introduce a construction of a divisorial algebra B that enlarges A whenever $A \neq \overline{S}$. Set $I = \mathfrak{c}(A/S)$. Note that $I = \mathrm{Hom}_S(A,S)$ is the canonical ideal of A, and two possibilities may occur:

(a) $I \cong A$. In this case A is quasi-Gorenstein.
(b) $I \ncong A$.

(a) In this case the method we are going to use for case (b) would come to a halt. To avoid that, we apply the Jacobian method to A to get (if $A \neq \overline{A}$, when the whole process would be halted anyway) an extension $A \neq B$. Observe that $B = \mathrm{Hom}_S(\tilde{J}, \tilde{J})$

is computed in S and $I = \mathfrak{c}(B)$ as well. In other words, we can discard the variables used to find J. In case again $\mathfrak{c}(B) \cong B$ we have no choice but repeat the previous step.

(b) A natural choice is:

$$B = \bigcup_{n \geq 1} \mathrm{Hom}_S(I^n, I^n).$$

In other words, if $R[It]$ is the Rees algebra of the ideal I and $X = \mathrm{Proj}(R[It])$, then $B = H^0(X, O_X)$. To show that B is properly larger than A it suffices to exhibit this at one of the associated prime ideals of I. Let $\mathfrak{p}$ be a minimal prime of I for which $S_\mathfrak{p} \neq A_\mathfrak{p}$. If $I_\mathfrak{p}$ is principal we have that $I_\mathfrak{p} = tS_\mathfrak{p}$ and therefore $S_\mathfrak{p} = A_\mathfrak{p}$, which is a contradiction. This means that the ideal $I_\mathfrak{p}$ has a minimal reduction of reduction number at least one, that is, $I_\mathfrak{p}^{r+1} = uI_\mathfrak{p}^r$ for some $r \geq 1$. This equality shows that

$$I_\mathfrak{p}^r u^{-r} \subset B_\mathfrak{p},$$

which gives the desired contradiction since $I_\mathfrak{p}^r u^{-r}$ contains $A_\mathfrak{p}$ properly as $I_\mathfrak{p}$ is not principal.

We observe that a bound for the integer r can be found as well. Let $g \in J$ be a form of degree $e-1$ which together with f form a regular sequence in $R = k[x_1, \ldots, x_{d+1}]$. According to Theorem 2.45, the reduction number of the ideal $I_\mathfrak{p}$ is bounded by the Hilbert-Samuel multiplicity $e(I_\mathfrak{p})$ of $I_\mathfrak{p}$ minus one. In a manner similar to our earlier argument,

$$e(I_\mathfrak{p}) \leq e((g_\mathfrak{p})) \leq e(e-1).$$

By taking divisorial closures, we have

$$C(B) = C(\mathrm{Hom}_S(I^r, I^r))$$

for $r < e(e-1)$, as desired. □

Suppose that there is no halt in a sequence of computations. The proper operation above then leads to a faster approach to the integral closure according to the following observation:

Proposition 6.66. *Let $S \subset A$ be a divisorial extension and set $I = \mathfrak{c}(A/S)$. If $B \neq A$ is the divisorial closure of $\bigcup_{n \geq 1} \mathrm{Hom}_S(I^n, I^n)$, then*

$$\deg(\mathfrak{c}(A/S)) \geq \deg(\mathfrak{c}(B/S)) + 2.$$

In particular, for any standard graded algebra of multiplicity e, the number of terms in any chain of algebras obtained in this manner will have at most $\left\lceil \frac{(e-1)^2}{2} \right\rceil$ divisorial extensions.

Proof. Suppose that the degrees of the conductors of A and B differ by 1. This means that the two algebras agree at all localizations of S at height 1 primes, except at

$R = S_{\mathfrak{p}}$, and that $\lambda((B/A)_{\mathfrak{p}}) = 1$. We shall show this is a contradiction to the given choice of how B is built.

We localize S, A, B, I at $\mathfrak{p}$ but keep simpler notation $S = S_{\mathfrak{p}}$, etc. Let (u) be a minimal reduction of I, $I^{r+1} = uI^r$. We know that $r \geq 1$ since I is not a principal ideal, as $A \neq B$. We note that $A \subset Iu^{-1} \subset B$. Since B/A is a simple S-module, we have that either $A = Iu^{-1}$ or $B = Iu^{-1}$. We can readily rule out the first possibility. The other leads to the equality

$$Iu^{-1} \cdot Iu^{-1} = Iu^{-1},$$

since B is an algebra. But this implies that

$$Iu^{-1} \cdot I \subset I \subset S,$$

in other words, that $Iu^{-1} \subset \operatorname{Hom}_S(I, S) = A$. □

Remark 6.67. When A is a non-homogeneous algebra, the cubic estimate of Corollary 6.63 must be considered.

6.4 Tracking Number of an Algebra

We now give a more conceptual explanation of the boundedness of the chains of the previous section based on another family of divisors attached to the extensions ([DV3]). Its added usefulness will include extensions to all characteristics.

Let E be a finitely generated graded module over the polynomial ring $R = k[x_1, \ldots, x_d]$. If $\dim E = d$, denote by $\det_R(E)$ the determinantal divisor of E: if E has multiplicity e,

$$\det(E) = (\wedge^e E)^{**} \cong R[-\delta].$$

Definition 6.68. The integer δ will be called the *tracking number* of E: $\delta = \operatorname{tn}(E)$.

The terminology *tracking number* (or *twist*) refers to the use of integers as locators, or tags, for modules and algebras in partially ordered sets. A forerunner of this use was made in the previous section, when divisorial ideals were employed to bound chains of algebras with Serre's property S_2.

There are immediate generalizations of this notion to more general gradings. Here we have in mind just applications to graded algebras that admit a standard Noether normalization, that is, the so-called *semistandard* graded algebras. In a brief discussion we treat abstract tracking numbers.

6.4.1 Chern Coefficients

We first develop the basic properties of this notion. It may help to begin with these examples:

Example 6.69. If A is a homogeneous domain over a field k, R a homogeneous Noether normalization and S a hypersurface ring over which A is birational,

$$R \subset S = R[t]/(f(t)) \subset A$$

($f(t)$ is a homogeneous polynomial of degree e), we have $(\wedge^e S)^{**} = R[-\binom{e}{2}]$. As a consequence $\mathrm{tn}(A) \leq \binom{e}{2}$.

Another illustrative example is that of the fractionary ideal $I = (x^2/y, y)$. This I is positively generated and $I^{**} = R(1/y)$: $\mathrm{tn}(I) = -1$.

The following observation shows the use of tracking numbers to locate the members of certain chains of modules.

Proposition 6.70. *If $E \subset F$ are modules with the same multiplicity that satisfy condition S_2, then $\mathrm{tn}(E) \geq \mathrm{tn}(F)$ with equality only if $E = F$.*

Proof. Consider the exact sequence

$$0 \to E \stackrel{\varphi}{\longrightarrow} F \longrightarrow C \to 0,$$

and we shall key on the annihilator of C (the *conductor* of F into E). Localization at a height 1 prime $\mathfrak{p}$ gives a free resolution of $C_{\mathfrak{p}}$ and the equality

$$\det(F) = \det(\varphi_{\mathfrak{p}}) \cdot \det(E).$$

Therefore $\det(E) = \det(F)$ if and only if C has codimension at least two, which is not possible if $C \neq 0$ since it has depth at least 1. □

Corollary 6.71. *If the modules in the strictly increasing chain*

$$E_0 \subset E_1 \subset \cdots \subset E_n$$

have the same multiplicity and satisfy condition S_2, then $n \leq \mathrm{tn}(E_0) - \mathrm{tn}(E_n)$.

Remark 6.72. The inequality $\mathrm{tn}(E_0) - \mathrm{tn}(E_n) \geq n$ is rarely an equality, for the obvious reasons. Let $E \subset F$ be distinct graded modules of the same multiplicity satisfying S_2. Then $\det(E) = f \cdot \det(F)$ where f is a homogeneous polynomial of degree given by $\mathrm{tn}(E) - \mathrm{tn}(F)$. If

$$f = f_1^{a_1} \cdots f_r^{a_r}$$

is a primary decomposition of f, a tighter measure of the "spread" between E and F is the integer

$$a_1 + \cdots + a_r,$$

rather than

$$\mathrm{tn}(E) - \mathrm{tn}(F) = a_1 \cdot \deg(f_1) \cdots a_r \cdot \deg(f_r).$$

Proposition 6.73. *If the complex*

$$0 \to A \xrightarrow{\varphi} B \xrightarrow{\psi} C \to 0$$

of finitely generated graded R-modules is an exact sequence of free modules in every localization $R_{\mathfrak{p}}$ at height one primes, then $\operatorname{tn}(B) = \operatorname{tn}(A) + \operatorname{tn}(C)$.

Proof. We break up the complex into simpler exact complexes:

$$0 \to \ker(\varphi) \longrightarrow A \longrightarrow A' = \operatorname{image}(\varphi) \to 0$$

$$0 \to A' \longrightarrow \ker(\psi) \longrightarrow \ker(\psi)/A' \to 0$$

$$0 \to B' = \operatorname{image}(\psi) \longrightarrow C \longrightarrow C/B' \to 0$$

and

$$0 \to \ker(\psi) \longrightarrow B \longrightarrow B' \to 0.$$

We note that $\operatorname{codim}\ker(\varphi) \geq 1$, $\operatorname{codim} C/B' \geq 2$, $\operatorname{codim}\ker(\psi)/A' \geq 2$ by hypothesis, so that we have the equality of determinantal divisors:

$$\det(A) = \det(A') = \det(\ker(\psi)), \text{ and } \det(C) = \det(B').$$

What this all means is that we may assume that the given complex is exact.

Suppose that $r = \operatorname{rank}(A)$ and $\operatorname{rank}(C) = s$ and set $n = r + s$. Consider the pair $\wedge^r A$, $\wedge^s C$. For $v_1, \ldots, v_r \in A$ and $u_1, \ldots, u_s \in C$, choose w_i in B such that $\psi(w_i) = u_i$ and consider

$$v_1 \wedge \cdots \wedge v_r \wedge w_1 \wedge \cdots \wedge w_s \in \wedge^n B.$$

Different choices for w_i would produce elements in $\wedge^n B$ that differ from the above by terms that contain at least $r+1$ factors of the form

$$v_1 \wedge \cdots \wedge v_r \wedge v_{r+1} \wedge \cdots,$$

with $v_i \in A$. Such products are torsion elements in $\wedge^n B$. This implies that modulo torsion we have a well-defined pairing:

$$[\wedge^r A/\text{torsion}] \otimes_R [\wedge^s C/\text{torsion}] \longrightarrow [\wedge^n B/\text{torsion}].$$

When localized at primes $\mathfrak{p}$ of codimension at most 1, the complex becomes an exact complex of projective $R_{\mathfrak{p}}$-modules and the pairing is an isomorphism. Upon taking biduals and the $\circ$ divisorial composition, we obtain the asserted isomorphism. □

Corollary 6.74. *Let*

$$0 \to A_1 \longrightarrow A_2 \longrightarrow \cdots \longrightarrow A_n \to 0$$

be a complex of graded R-modules and homogeneous homomorphisms that is an exact complex of free modules in codimension 1. *Then*

$$\sum_{i=1}^{n} (-1)^i \operatorname{tn}(A_i) = 0.$$

Proposition 6.75. *Suppose that* $R = k[x_1, \ldots, x_d]$ *and let*

$$0 \to A \longrightarrow B \longrightarrow C \longrightarrow D \to 0,$$

be an exact sequence of graded R-modules and homogeneous homomorphisms. If $\dim B = \dim C = d$, $\operatorname{codim} A \geq 1$ *and* $\operatorname{codim} D \geq 2$, *then* $\operatorname{tn}(B) = \operatorname{tn}(C)$.

Corollary 6.76. *If E is a graded R-module of dimension d, then*

$$\operatorname{tn}(E) = \operatorname{tn}(E/\text{mod torsion}) = \operatorname{tn}(E^{**}).$$

Let A be a homogeneous algebra defined over a field k that admits a Noether normalization $R = k[x_1, \ldots, x_d]$; then clearly $\operatorname{tn}_R(A) = \operatorname{tn}_{R'}(A')$, where K is a field extension of k, $R' = K \otimes_k R$ and $A' = K \otimes_k A$. Partly for this reason, we can always define the tracking number of an algebra by first enlarging the ground field. Having done that and chosen a Noether normalization R that is a standard graded algebra, it will follow that $\operatorname{tn}_R(A)$ is independent of R: the R-torsion submodule A_0 of A is actually an ideal of A whose definition is independent of R.

Calculation Rules

We shall now derive several rules to facilitate the computation of tracking numbers.

Proposition 6.77. *Let E be a finitely generated graded module over the polynomial ring* $R = k[x_1, \ldots, x_d]$. *If E is torsionfree over R, then* $\operatorname{tn}(E) = e_1(E)$, *the first Chern number of E.*

Proof. Let

$$0 \to \oplus_j R[-\beta_{d,j}] \to \cdots \to \oplus_j R[-\beta_{1,j}] \to \oplus_j R[-\beta_{0,j}] \to E \to 0$$

be a (graded) free resolution of E. The integer

$$e_1(E) = \sum_{i,j} (-1)^i \beta_{i,j}$$

is (see [BH93, Proposition 4.1.9]) the next to the leading Hilbert coefficient of E. It is also the integer that one gets by taking the alternating product of the determinants in the free graded resolution (see Corollary 6.74). □

Example 6.78. Suppose that $A = k[(x,y,z)^2]$ a subalgebra of the ring of polynomials $k[x,y,z]$. Setting the weight of the indeterminates to $\frac{1}{2}$, A becomes a standard graded algebra with $R = k[x^2, y^2, z^2]$ as a Noether normalization. A calculation easily shows that the Hilbert series of A is

$$H_A(t) = \frac{h_A(t)}{(1-t)^3} = \frac{1+3t}{(1-t)^3}.$$

Thus $\deg(A) = h_A(1) = 4$, $\operatorname{tn}(A) = e_1(A) = h'_A(1) = 3$.

In general, the connection between the tracking number and the first Hilbert coefficient has to be 'adjusted' in the following manner.

Proposition 6.79. *Let E be a finitely generated graded module over $R = k[x_1, \ldots, x_d]$. If $\dim E = d$ and E_0 is its torsion submodule,*

$$0 \to E_0 \longrightarrow E \longrightarrow E' \to 0,$$

then

$$\mathrm{tn}(E) = \mathrm{tn}(E') = e_1(E') = e_1(E) + \hat{e}_0(E_0),$$

where $\hat{e}_0(E)$ is the multiplicity of E_0 if $\dim E_0 = d-1$, and 0 otherwise.

Proof. Denote by $H_A(t)$ the Hilbert series of an R-module A (see [BH93, Chap. 4]) and write

$$H_A(t) = \frac{h_A(t)}{(1-t)^d},$$

if $\dim A = d$. For the exact sequence defining E', we have

$$h_E(t) = h_{E'}(t) + (1-t)^r h_{E_0}(t),$$

where $r = 1$ if $\dim E_0 = d-1$, and $r \geq 2$ otherwise. Since

$$e_1(E) = h'_E(1) = h'_{E'}(1) + r(1-t)^{r-1}|_{t=1} h_{E_0}(1),$$

the assertion follows. □

Remark 6.80. This suggests a reformulation of the notion of tracking number. By using exclusively the Hilbert function, the definition could be extended to all finite modules over a graded algebra.

Corollary 6.81. *Let E and F be graded R-modules of dimension d. Then*

$$\mathrm{tn}(E \otimes_R F) = \deg(E) \cdot \mathrm{tn}(F) + \deg(F) \cdot \mathrm{tn}(E).$$

Proof. By Corollary 6.76, we may assume that E and F are torsionfree modules. Let $\mathbb{P}$ and $\mathbb{Q}$ be minimal projective resolutions of E and F respectively. The complex $\mathbb{P} \otimes_R \mathbb{Q}$ is acyclic in codimension 1, by the assumption on E and F. We can then use Corollary 6.74,

$$\mathrm{tn}(E \otimes_R F) = \sum_{k \geq 0} (-1)^k \mathrm{tn}(\oplus_{i+j=k} \mathbb{P}_i \otimes_R \mathbb{Q}_j).$$

Expanding gives the desired formula. □

Very similar to Proposition 6.79 is:

Proposition 6.82. *Let $E \subset F$ be graded torsionfree R-modules of the same multiplicity. If E is reflexive, then*

$$\mathrm{tn}(E) = \mathrm{tn}(F) + \deg(F/E).$$

Simplicial Complexes

Theorem 6.83. *Let Δ be a simplicial complex on the vertex set $V = \{x_1,\ldots,x_n\}$, and denote by $k[\Delta]$ the corresponding Stanley-Reisner ring. If* $\dim k[\Delta] = d$*, then*

$$\operatorname{tn}(k[\Delta]) = d f_{d-1} - f_{d-2} + f'_{d-2},$$

where f_i denotes the number of faces of dimension i, and f'_{d-2} denotes the number of maximal faces of dimension $d-2$.

Proof. Set $k[\Delta] = S/I_\Delta$, and consider the decomposition $I_\Delta = I_1 \cap I_2$, where I_1 is the intersection of the primary components of dimension d and I_2 of the remaining components. The exact sequence

$$0 \to I_1/I_\Delta \longrightarrow S/I_\Delta \longrightarrow S/I_1 \to 0$$

gives, according to Proposition 6.79,

$$\operatorname{tn}(k[\Delta]) = e_1(k[\Delta]) + \hat{e}_0(I_1/I_\Delta).$$

From the Hilbert function of $k[\Delta]$ ([BH93, Lemma 5.1.8]), we have that $e_1 = d f_{d-1} - f_{d-2}$, while if I_1/I_Δ is a module of dimension $d-1$, its multiplicity is the number of maximal faces of dimension $d-2$. □

Theorem 6.84. *Let $S = k[x_1,\ldots,x_n]$ be a ring of polynomials and $A = S/I$ a graded algebra. For a monomial ordering $>$, denote by $I' = in_>(I)$ the initial ideal associated with I and set $B = S/I'$. Then* $\operatorname{tn}(B) \geq \operatorname{tn}(A)$.

Proof. Let J be the component of I of maximal dimension and consider the exact sequence

$$0 \to J/I \longrightarrow S/I \longrightarrow S/J \to 0.$$

Now $\dim J/I < \dim A$, and therefore $\operatorname{tn}(A) = \operatorname{tn}(S/J) = e_1(S/J)$. Denote by J' the corresponding initial ideal of J, and consider the sequence

$$0 \to J'/I' \longrightarrow S/I' \longrightarrow S/J' \to 0.$$

Noting that S/I and S/J have the same multiplicity, as do S/I' and S/J' by Macaulay's theorem, we find that $\dim J'/I' < \dim A$. This means that

$$\operatorname{tn}(S/I') = \operatorname{tn}(S/J') = e_1(S/J') + \hat{e}_0(J'/I') = e_1(S/J) + \hat{e}_0(J'/I') = \operatorname{tn}(A) + \hat{e}_0(J'/I').$$

Example 6.85. Set $A = k[x,y,z,w]/(x^3 - yzw, x^2y - zw^2)$. The Hilbert series of this (Cohen-Macaulay) algebra is

$$H_A(t) = \frac{h_A(t)}{(1-t)^2} = \frac{(1+t+t^2)^2}{(1-t)^2},$$

so that

$$\mathrm{tn}(A) = e_1(A) = h'_A(1) = 18.$$

Consider now the algebra $B = k[x,y,z,w]/J$, where J is the initial ideal of I for the Deglex order. A calculation with *Macaulay2* gives

$$J = (x^2y, x^3, xzw^2, xy^3zw, y^5zw).$$

By Macaulay's Theorem, B has the same Hilbert function as A. An examination of the components of B gives an exact sequence

$$0 \to B_0 \longrightarrow B \longrightarrow B' \to 0,$$

where B_0 is the ideal of elements with support in codimension 1. By Corollary 6.76,

$$\mathrm{tn}(B) = \mathrm{tn}(B') = e_1(B').$$

At same time one has the following equality of h-polynomials,

$$h_B(t) = h_{B'}(t) + (1-t)h_{B_0}(t),$$

and therefore

$$e_1(B') = e_1(B) + e_0(B_0).$$

A final calculation of multiplicities gives $e_0(B_0) = 5$, and

$$\mathrm{tn}(B) = 18 + 5 = 23.$$

The example shows that $\mathrm{tn}(A)$ is independent of the Hilbert function of the algebra.

6.4.2 Bounding Tracking Numbers

We now describe how the technique of generic hyperplane sections leads to bounds of various kinds. We are going to assume that the algebras are defined over infinite fields.

One of the important properties of the tracking number is that it will not change under hyperplane sections as long as the dimension of the ring is at least 3. So one can answer questions about the tracking number just by studying the 2-dimensional case. The idea here is that tracking number is more or less the same material as e_1 and hence cutting by a superficial element will not change it unless the dimension drops below 2.

Proposition 6.86. *Let E be a finitely generated graded module of dimension d over $R = k[x_1,\ldots,x_d]$ with $d > 2$. Then for a general element h of degree one, $R' = R/(h)$ is also a polynomial ring, and $\mathrm{tn}_R(E) = \mathrm{tn}_{R'}(E')$, where $E' = E/hE$.*

Proof. First we shall prove the statement for a torsionfree module E. Consider the exact sequence

$$0 \to E \longrightarrow E^{**} \longrightarrow C \to 0.$$

Note that C has codimension at least 2 since after localization at any height 1 prime E and E^{**} are equal. Now for a linear form h in R that is a superficial element for C we can tensor the above exact sequence with $R/(h)$ to get the complex

$$\mathrm{Tor}_1(C,R/(h)) \longrightarrow E/hE \longrightarrow E^{**}/hE^{**} \longrightarrow C/hC \to 0.$$

Now as an R-module, C/hC has codimension at least 3, so as an $R' = R/(h)$ module it has codimension at least 2. Also as $\mathrm{Tor}_1(C,R/(h))$ has codimension at least 2 as an R-module, it is a torsion $R/(h)$-module. Hence we have $\mathrm{tn}_{R'}(E/hE) = \mathrm{tn}_{R'}(E^{**}/hE^{**})$. But E^{**} is a torsion free $R/(h)$-module, so

$$\mathrm{tn}_{R'}(E/hE) = e_1(E^{**}/hE^{**}) = e_1(E^{**}) = \mathrm{tn}_R(E^{**}) = \mathrm{tn}_R(E).$$

To prove the statement for a general R-module E, we consider the short exact sequence

$$0 \to E_0 \longrightarrow E \longrightarrow E' \to 0,$$

where E_0 is the torsion submodule of E. But E' is torsionfree, so by the first case we know that $\mathrm{tn}_{R/(h)}(E'/hE') = \mathrm{tn}_R(E')$ for a general linear element h of R.

Now if in addition we restrict ourselves to those h that are superficial for E and E_0, we can tensor the above exact sequence with $R/(h)$ and get

$$0 = \mathrm{Tor}_1(E',R/(h)) \longrightarrow E_0/hE_0 \longrightarrow E/hE \longrightarrow E'/hE' \to 0,$$

but since E_0/hE_0 is a torsion R'-module, where $R' = R/(h)$, we have $\mathrm{tn}_{R'}(E/hE) = \mathrm{tn}_{R'}(E'/hE') = \mathrm{tn}_R(E') = \mathrm{tn}_R(E)$. □

We shall now derive the first of our general bounds for $\mathrm{tn}(E)$ in terms of the Castelnuovo-Mumford regularity $\mathrm{reg}(E)$ of the module. For terminology and basic properties of the $\mathrm{reg}(\cdot)$ function, we shall use [Ei95, Section 20.5].

Theorem 6.87. *Suppose that $R = k[x_1,\ldots,x_d]$ and that E a finitely generated graded R-module of dimension d. Then*

$$\mathrm{tn}(E) \leq \deg(E) \cdot \mathrm{reg}(E).$$

Proof. The assertion is clear if $d = 0$. For $d \geq 1$, if E_0 denotes the submodule of E consisting of the elements with finite support, then $\deg(E) = \deg(E/E_0)$, $\mathrm{tn}(E) = \mathrm{tn}(E/E_0)$ and $\mathrm{reg}(E/E_0) \leq \mathrm{reg}(E)$, the latter according to [Ei95, Corollary 20.19(d)]. From this reduction, the assertion is also clear if $d = 1$.

If $d \geq 3$, we use a hyperplane section h so that $\mathrm{tn}_R(E) = \mathrm{tn}_{R/(h)}(E/hE)$ according to Proposition 6.86, and $\mathrm{reg}(E/hE) \leq \mathrm{reg}(E)$ according to [Ei95, Proposition 20.20]. (Of course, $\deg(E) = \deg(E/hE)$.)

With these reductions, we may assume that $d = 2$ and that depth $E > 0$. Denote by E_0 the torsion submodule of E and consider the exact sequence

$$0 \to E_0 \longrightarrow E \longrightarrow E' \to 0.$$

Noting that either E_0 is zero or depth $E_0 > 0$, on taking local cohomology with respect to the maximal ideal $\mathfrak{m} = (x_1, x_2)$, we have the exact sequence

$$0 \to H^1_{\mathfrak{m}}(E_0) \to H^1_{\mathfrak{m}}(E) \to H^1_{\mathfrak{m}}(E') \to H^2_{\mathfrak{m}}(E_0) = 0 \to H^2_{\mathfrak{m}}(E) \to H^2_{\mathfrak{m}}(E') \to 0,$$

from which we get

$$\mathrm{reg}(E) = \max\{\mathrm{reg}(E_0), \mathrm{reg}(E')\}.$$

This provides the final reduction to $d = 2$ and E torsionfree. Let

$$0 \to \bigoplus_{j=1}^{s} R[-b_j] \longrightarrow \bigoplus_{i=1}^{r} R[-a_i] \longrightarrow E \longrightarrow 0$$

be a minimal projective resolution of E. From Corollary 6.74, we have

$$\mathrm{tn}(E) = \sum_{i=1}^{r} a_i - \sum_{j=1}^{s} b_j.$$

Reducing this complex modulo a hyperplane section h, we get a minimal free resolution for the graded module E/hE over the PID $R/(h)$. By the basic theorem for modules over such rings, after basis change we may assume that

$$b_j = a_j + c_j, \quad c_j > 0, \quad j = 1 \ldots s.$$

Noting that $\alpha = \mathrm{reg}(E) = \max\{a_i, b_j - 1 \mid i = 1 \ldots r, j = 1 \ldots s\} \geq 0$ and $\deg(E) = r - s$, we have

$$\begin{aligned} \deg(E)\mathrm{reg}(E) - \mathrm{tn}(E) &= (r-s)\alpha - \sum_{i=1}^{r} a_i + \sum_{j=1}^{s} (a_j + c_j) \\ &= \sum_{i=s+1}^{r} (\alpha - a_i) + \sum_{j=1}^{s} c_j \\ &\geq 0, \end{aligned}$$

as desired. □

Positivity of Tracking Numbers

We shall now prove our main result, a somewhat surprising positivity result for a reduced homogeneous algebra A. Since such algebras already admit a general upper bound for $\mathrm{tn}(A)$ in terms of its multiplicity, together these statements are useful in the construction of integral closures by all algorithms that use intermediate extensions that satisfy condition S_2.

Theorem 6.88. *Let A be a reduced non-negatively graded algebra that is finite over a standard graded Noether normalization R. Then $\mathrm{tn}(A) \geq 0$. Moreover, if A is an integral domain and k is algebraically closed, then $\mathrm{tn}(A) \geq \deg(A) - 1$.*

Proof. Let $A = S/I$, $S = k[x_1,\ldots,x_n]$, be a graded presentation of A. From our earlier discussion, we may assume that I is height-unmixed (as otherwise the lower dimensional components gives rise to the torsion part of A, which is dropped in the calculation of $\mathrm{tn}(E)$ anyway).

Let $I = P_1 \cap \cdots \cap P_r$ be the primary decomposition of I, and define the natural exact sequence

$$0 \to S/I \longrightarrow S/P_1 \times \cdots \times S/P_r \longrightarrow C \to 0,$$

from which a calculation with Hilbert coefficients gives

$$\mathrm{tn}(A) = \sum_{i=1}^{r} \mathrm{tn}(S/P_i) + \hat{e}_0(C).$$

This shows that it suffices to assume that A is a domain.

Let $\overline{A}$ denote the integral closure of A. Note that $\overline{A}$ is also a non-negatively graded algebra and that the same Noether normalization R can be used. Since $\mathrm{tn}(A) \geq \mathrm{tn}(\overline{A})$, we may assume that A is integrally closed.

Since the cases $\dim A \leq 1$ are trivial, we may assume $\dim A = d \geq 2$. The case $d = 2$ is also clear since A is then Cohen-Macaulay. Assume then that $d > 2$. We are going to change the base field using rational extensions of the form $k(t)$, which do not affect the integral closure condition. (Of course we may assume that the base field is infinite.)

If h_1 and h_2 are linearly independent hyperplane sections in R they define a regular sequence in A, since the algebra is normal and therefore satisfies condition S_2. Effecting a change of ring of the type $k \to k(t)$ gives a hyperplane section $h_1 - t \cdot h_2 \in R(t)$, which is a prime element in A, according to Nagata's trick ([Fo73, Lemma 14.1]). Clearly we can choose h_1 and h_2 so that $h_1 - t \cdot h_2$ is a generic hyperplane section, for the purpose of applying Proposition 6.86 to A. This completes the reduction to domains in dimension $d - 1$.

The last assertion follows in the reduction to the case where $\dim R$ is 2, as $\mathrm{tn}(A) \geq \mathrm{tn}(\overline{A})$ and $\deg(A) = \deg(\overline{A})$. The algebra $\overline{A}$ is Cohen–Macaulay so that its h–polynomial $h(t)$ has only nonnegative coefficients and it follows easily that $h'(1) \geq h(1) - 1$, since $h(0) = 1$ as k is algebraically closed. □

We single out:

Corollary 6.89. *Let A be a semistandard graded domain over an algebraically closed field. If A has S_2 and* $\mathrm{tn}(A) = \deg(A) - 1$, *then A is normal.*

One application is to the study of constructions of the integral closure of an affine domain in arbitrary characteristics.

Theorem 6.90. *Let A be a semistandard graded domain over a field k and let $\overline{A}$ be its integral closure. Then any chain*

$$A \subset A_1 \subset \cdots \subset A_n = \overline{A}$$

of distinct subalgebras satisfying Serre's condition S_2 has length at most $\binom{e}{2}$, where $e = \deg(A)$. *If k is algebraically closed, such a chain has length at most $\binom{e-1}{2}$.*

Proof. It will suffice, according to Corollary 6.71, to show that $0 \leq \mathrm{tn}(A) \leq \binom{e}{2}$. The non-negativity having been established in Theorem 6.88, we now prove the upper bound.

If k is a field of characteristic zero, then by the theorem of the primitive element A contains a hypersurface ring $S = R[t]/(f(t))$, where R is a ring of polynomials $R = k[z_1, \ldots, z_d]$, $\deg(z_i) = 1$, and $f(t)$ is a homogeneous polynomial in t of degree e. As $\mathrm{tn}(A) \leq \mathrm{tn}(S) = \binom{e}{2}$, the assertion holds in this case. (We also observe that when $\dim A \geq 3$, any hyperplane section, say h, used to reduce the dimension that were employed in the proof of Theorem 6.88, could be chosen so that the image of S in $\overline{A}/h\overline{A}$ would be $S/(h)$, and therefore we would maintain the same upper bound.)

To complete the proof in other characteristics we resort to the following construction developed in Proposition 6.4. If $A = R[y_1, \ldots, y_n]$, let

$$E = \sum R y_1^{j_1} \cdots y_n^{j_n}, \quad 0 \leq j_i < r_i = [F_i : F_{i-1}]$$

be the R-module of rank e constructed there. As the rank satisfies the equality $e = \prod_{i=1}^{n} r_i$, e is the number of 'monomials' $y_1^{j_1} \cdots y_n^{j_n}$. Their linear independence over R is a simple verification. Note also that there are monomials of all degrees between 0 and $(r_1 - 1, \ldots, r_n - 1)$. Thus according to Proposition 6.77, the bound for $\mathrm{tn}(E)$ is obvious, with equality holding only when E is a hypersurface ring over R. (The precise value for $\mathrm{tn}(E)$ could be derived from Corollary 6.81.)

For the case of an algebraically closed field k the last assertion follows from Theorem 6.88: $\mathrm{tn}(A) \geq e - 1$, so that the length of the chains is at most $\mathrm{tn}(E) - \mathrm{tn}(\overline{A}) \leq \binom{e}{2} - (e-1) = \binom{e-1}{2}$. □

Remark 6.91. When the operations used to create the chain of divisorial extensions are those described by Proposition 6.66, the last bound further reduces to

$$\left\lceil \frac{(e-1)(e-2)}{4} \right\rceil.$$

There is another way in which the length of the divisorial chains may shorten. Let $A \subset B \subset \overline{A}$ be divisorial extensions of a graded algebra of dimension d as above and let $R = k[z_1, \ldots, z_d]$ be a Noether normalization. The canonical module of B is the graded module $\omega_B = \mathrm{Hom}_R(B, R[-d])$. We recall that B is *quasi-Gorenstein* if $\omega_B \simeq B[-a]$. In terms of their tracking numbers this means that

$$\begin{aligned} \mathrm{tn}(\omega_B) &= -\mathrm{tn}(B) + d \cdot \deg(B) \\ &= \mathrm{tn}(B) + a \cdot \deg(B), \end{aligned}$$

and therefore

$$\mathrm{tn}(B) = \deg(B) \cdot \text{half-integer}.$$

A consequence is that if more than one quasi-Gorenstein extension occur in a same divisorial chain, they must lie fairly far apart.

Abstract Tracking Numbers

If R is an integrally closed local ring, it is unclear how to construct tracking numbers for its R-modules. While the determinant of an R–module E (here $\dim E = \dim R$) can be formed,

$$\det(E) = (\wedge^e E)^{**},$$

there does not seem to be a natural way to attach a degree to it.

In the case of an R–algebra A of finite integral closure $\overline{A}$, there is an *ad hoc* solution for the set of submodules of $\overline{A}$ of rank $r = \operatorname{rank}_R(A)$. The construction proceeds as follows. Let F be a normalizing free R–module,

$$F = R^r = Re_1 \oplus \cdots \oplus Re_r \subset A \subset \overline{A}.$$

There exists f with $0 \neq f \in R$ such that $f \cdot \overline{A} \subset F$. For the R–submodule E of $\overline{A}$,

$$\det(fE) \subset R(e_1 \wedge \cdots \wedge e_r) = R\varepsilon.$$

This means that

$$\det(fE) = I.R\varepsilon,$$

where I is a divisorial ideal of R, so If^{-r} is also a divisorial ideal with a primary decomposition

$$If^{-r} = \bigcap p_i^{(r_i)}.$$

Definition 6.92. The tracking number of E (offset by F) is the integer

$$\operatorname{tn}(E) = \sum_i r_i \cdot \deg(R/p_i) + \operatorname{tn}(R\varepsilon).$$

The value $\operatorname{tn}(E)$ is defined up to an offset but it is independent of f. It will have several of the properties of the tracking number defined for graded modules and can play the same role in the comparison of the lengths of chains of subalgebras lying between A and $\overline{A}$. When R is more general one can define $\operatorname{tn}(E)$ as the supremum of its local values.

6.5 Embedding Dimension of the Integral Closure

Let k be a field and A a reduced ring which is a finitely generated k-algebra. The integral closure $\overline{A}$ is also an affine algebra over k,

$$A = k[x_1, \ldots, x_n]/I \hookrightarrow \overline{A} = k[y_1, \ldots, y_m]/J.$$

The least n among all such presentations of A is the *embedding dimension* of A, $\operatorname{embdim}(A)$. This section will be focused on the number of indeterminates needed to present $\overline{A}$. If A is $\mathbb{N}$-graded, $\overline{A}$ is similarly graded and there will be presentations where the y_i are homogeneous elements. The degree of the presentation of $\overline{A}$ is the maximum of the $\deg(y_i)$. We shall denote by $\text{embdeg}(\overline{A})$ the minimum achieved among all presentations; we call that integer the *embedding degree* of $\overline{A}$. (For the moment we shall blur the fact that a short presentation-that is, one yielding $\text{embdim}(\overline{A})$-may not correspond to the presentation giving rise to $\text{embdeg}(\overline{A})$.)

These numbers are major measures of the complexity of computing $\overline{A}$. One aim here is to give estimates for the embedding dimension of $\overline{A}$, and in the graded case to $\text{embdeg}(\overline{A})$, under some restrictions. Whenever possible one should seek to control the embedding dimension of all intermediate rings that occur in the construction of $\overline{A}$. A major concern is to make use of known properties of A, such as information about its singular locus, or expected geometric properties of $\overline{A}$, in particular those regarding its depth.

Following [UV3], where the topic is treated in greater detail and depth, we give here a geometric approach to this problem; complete proofs are going to be found in it. A direct approach would be to attempt to bound degree data on the presentation ideal J in terms of n and I. We shall follow a different path, as we shall assume that we possess information about A with a geometric content, such as the dimension of A and its multiplicity, and on certain instances some fine data on its singular locus. Our results will then be expressed by bounding either $\text{embdim}(B)$ or $\text{embdeg}(B)$ in terms of that data.

We shall assume that A is reduced and equidimensional. One source of difficulty lies in that the ring $B = \overline{A}$ may also be the integral closure of other rings. This fuzziness is at the same time a path to dealing with this problem in some cases of interest. Assume that k is an infinite field, so that after a possible change of variables, the subring generated by the images of the first d variables of the x_i is a Noether normalization of A,

$$T = k[x_1, \ldots, x_d] \hookrightarrow A.$$

If A is reduced and equidimensional, the rank of A over T is the ordinary torsionfree rank over A as a T-module. Denote by $\deg(A)$ the least torsionfree rank of A over T for all possible Noether normalizations; this number is equal to the ordinary multiplicity $\deg(A)$ of A, when A is a graded algebra. If k is perfect, there exists an element $u \in A$ satisfying an equation $f(u) = 0$ where $f(t)$ is a monic, irreducible polynomial in $T[t]$, $\deg f(t) = \deg(A)$. This will imply that $S = T[t]/(f(t))$ is a subring of A such that B is also the integral closure of S. This shows that the only general numerical invariants we can really use for $\overline{A}$ are its dimension and multiplicity, and to a lesser extent the Jacobian ideal of S.

Our overall aim is to derive elementary functions $\beta(d,e)$ and $\delta(d,e)$, polynomial in e for fixed d, such that for any standard graded equidimensional reduced algebra A of dimension $d = \dim A$ and multiplicity $e = \deg(A)$,

$$\begin{aligned} \text{embdim}(\overline{A}) &\leq \beta(\dim A, \deg(A)) \\ \text{embdeg}(\overline{A}) &\leq \delta(\dim A, \deg(A)). \end{aligned}$$

The existence of such functions, albeit not in any explicit form, and therefore without the link to complexity, has been established in [DK84, Theorem 3.1] in a model-theoretic formulation grounded on the explicit construction of integral closures of [Sei75] and [Sto68]. The bounds given here, beyond their effective character, seek to derive formulas for $\beta(d,e)$ and $\delta(d,e)$ that are sensitive to additional information that is known about A, such as the case when the dimension is at most 3 or the singular locus is small. To exercise this kind of control one must, however, work in very strict characteristics, usually zero.

6.5.1 Cohen-Macaulay Integral Closure

Suppose that A is an affine algebra and $T = k[x_1,\ldots,x_d]$ is one of its Noether normalizations. The rank of A over T is the dimension of the K–vector space $A\otimes_T K$, where K is the field of fractions of T. This dimension may vary with the choice of the Noether normalization. When A is a standard graded algebra and the x_i are homogeneous of degree 1, this rank is equal to the multiplicity $\deg(A)$ of A as provided by its Hilbert function. It is thus independent of the choice of such a Noether normalization. In the non-graded case, by abuse of terminology, we let $\deg(A)$ denote the infimum of the ranks of A over its various Noether normalizations, a terminology that is consistent in the case of standard graded algebras over infinite fields (see also [SUV1, p. 251]). If S is a ring we write $\nu_S(\cdot)$ for the minimal number of generators of S-modules.

Theorem 6.93. *Let A be a reduced and equidimensional affine algebra over a field k with $e = \deg(A)$. Let $A\subset B$ be a finite and birational ring extension, and assume that B is Cohen–Macaulay.*

(a) $\nu_A(B)\leq e$ *and* $\operatorname{embdim}(B)\leq e+d-1$.
(b) *If k is a perfect field, A is a standard graded k–algebra, and $A\subset B$ is an extension of graded rings, then the graded A-module B is generated in degrees at most $\max\{0,e-2\}$; in particular* $\operatorname{embdeg}(B)\leq \max\{1,e-2\}$.

Proof. Let T be a Noether normalization of A with $\operatorname{rank}_T A = \deg(A) = e$. Being a Cohen–Macaulay ring, B is also a maximal Cohen–Macaulay T-module. Hence the T-module B is free, necessarily of rank e.

(a) The first assertion is obvious now and the second one follows because 1 is locally basic for the T-module B, forcing B/T to be free of rank $e-1$.

(b) We may assume that k is an infinite field. There exists a standard graded subring $S = T[t]/(f(t))$ of A with $f(t)$ a monic polynomial in t of degree e; see for instance Proposition 6.107. Notice that S is a graded free T-module of rank e generated in degrees at most $e-1$. Consider the following exact sequence of graded T-modules:

$$0\rightarrow S\longrightarrow B\longrightarrow C\rightarrow 0.$$

Since B is free, S is a syzygy-module of C. Thus a syzygy-module of C is generated in degrees at most $e-1$. But C has rank zero and hence does not have a nontrivial free summand. It follows that the graded T-module C is generated in degrees at most $e-2$. Hence as a graded S-module, B is generated in degrees at most $\max\{0,e-2\}$, and our assertions follow. □

The Cohen-Macaulay assumption in Theorem 6.93 is satisfied for $B=\overline{A}$ if $d \leq 2$ or $\deg(A) \leq 2$, and by a theorem of Hochster ([Ho72, Theorem 1]) if the algebra A is generated by monomials in a polynomial ring over k. For non-birational extensions, however, it can easily fail to hold. Here is a partial explanation:

Proposition 6.94. *Let A be an affine normal domain over a field of characteristic zero. If A is not Cohen-Macaulay then any finite extension B that is a domain is not Cohen-Macaulay either.*

Proof. Let K and L, $K \subset L$, be the quotient fields of A and B, respectively. The trace map $\mathrm{tr}_{L/K}$ induces an A–linear map

$$\frac{1}{s}\mathrm{tr}_{L/K} : B \mapsto A,$$

where s is the relative degree $[L:K]$ of the field extension. It provides a splitting of A-modules

$$B = A \oplus M.$$

On the other hand, for every prime ideal $\mathfrak{p}$ of A we have the standard inequality for the depth of direct summands,

$$\mathrm{depth}_{A_{\mathfrak{p}}} B_{\mathfrak{p}} \leq \mathrm{depth}\, A_{\mathfrak{p}},$$

and then B cannot be a Cohen-Macaulay ring. □

An example of such rings is built as follows. Let k be a field of characteristic zero, set $R = k[x,y,z]/(f)$, say $f = x^3+y^3+z^3$, and let A be the Rees algebra of $(x,y,z)R$. It is easy to see that A is normal, but not Cohen-Macaulay (as the reduction number of $(x,y,z) = \dim R = 2$).

Dimension 3^{++}

Since we have considered the case $\dim A \leq 2$ above, we focus here on the conditions that are always present when $\dim A \geq 3$ and $\deg(A) \geq 3$. Because of its dependence on Jacobian ideals most of our assertions only be valid only over fields of characteristic zero.

Let $A = k[x_1,\ldots,x_n]/I$ be a reduced equidimensional affine algebra over a field k of characteristic zero, $n = \mathrm{embdim}(A)$, of dimension $\dim A = d$. We call $\mathrm{ecodim}(A) = n-d$ the *embedding codimension* of A. We denote by $\Omega_k(A)$ the module of Kähler differentials of A/k. By $J(A)$ we denote the Jacobian ideal $\mathrm{Fitt}_d(\Omega_k(A))$ of A. Recall that $V(J(A)) = \mathrm{Sing}(A)$ if k is perfect and A is equidimensional.

Lemma 6.95. *Let k be a field of characteristic zero, S a reduced and equidimensional standard graded k-algebra of dimension d and multiplicity e, and assume that* $\mathrm{embdim}(S) \leq d+1$. *Let $S \subset B$ be a finite and birational extension of graded rings.*

(a) *The S-module B/S satisfies*

$$\deg(B/S) \leq (e-1)^2.$$

If $e \geq 3$ then

$$\deg(B/S) \leq (e-1)^2 - 1.$$

If $e \geq 4$ or if B is S_2 but not Cohen-Macaulay, then

$$\deg(B/S) \leq (e-1)^2 - 2.$$

(b) *If B is Cohen-Macaulay, then*

$$\deg(B/S) \leq \binom{e}{2}.$$

If moreover k is algebraically closed and S is a domain, then

$$\deg(B/S) \leq \binom{e-1}{2}.$$

Proof. (a) We may assume that $S \neq B$. As S satisfies S_2 and the extension $S \subset B$ is finite and birational, it follows that the S-module B/S is of codimension one. Thus, since S is Gorenstein,

$$\deg(B/S) = \deg(\mathrm{Ext}^1_S(B/S, S)).$$

Applying $\mathrm{Hom}_S(\cdot, S)$ to the exact sequence

$$0 \to S \longrightarrow B \longrightarrow B/S \to 0$$

yields an exact sequence

$$0 \to S/S :_S B \longrightarrow \mathrm{Ext}^1_S(B/S, S) \longrightarrow \mathrm{Ext}^1_S(B, S) \to 0.$$

Since B is torsionfree over the Gorenstein ring S, the S-module $\mathrm{Ext}^1_S(B, S)$ has codimension at least 2. Thus

$$\deg(\mathrm{Ext}^1_S(B/S, S)) = \deg(S/S :_S B).$$

On the other hand, write $S = R/(f)$ with $R = k[x_1, \ldots, x_n]$ a polynomial ring and f a form of degree e. Consider the R–ideal $J = \left(\frac{\partial f}{\partial x_1}, \ldots, \frac{\partial f}{\partial x_n}\right)$. Then $f \in J$ since $\mathrm{char}\, k = 0$, and therefore $J/(f) = J(S)$. In particular J is an R–ideal of height at least 2. As J is generated by forms of degree $e-1$, there exist homogeneous polynomials g, h in J of degree $e-1$ so that g, h and f, h are regular sequences. Write K for the preimage of $S :_S B$ in R, which is an R–ideal of height 2. Then $J(S) \subset \mathfrak{c}(S) = S :_S \overline{S}$ by [LS81, Theorem 2]. Hence $J(S) \subset S :_S B$, which implies that $J \subset K$. Consequently, the regular sequences g, h and f, h are contained in the height 2 ideal K. Therefore

$$\begin{aligned}\deg(S/S :_S B) = \deg(R/K) &= \deg(R/(g,h)) - \deg(R/(g,h) :_R K) \\ &= (e-1)^2 - \deg(R/(g,h) :_R K).\end{aligned}$$

Suppose that $\deg(R/(g,h) :_R K)$ is 0 or 1, respectively. In this case the ideal $(g,h) :_R K$ is the unit ideal or is generated by linear forms, hence the double link $(f,h) :_R K$ contains a linear or a quadratic form, respectively. On the other hand,

$$((f,h) :_R K)/(f) = hS :_S (S :_S B) \subset h\overline{S}.$$

As the nonzero homogeneous elements of $h\overline{S}$ have degrees at least $e-1$, it follows that $e \leq 2$ or $e \leq 3$, respectively. Furthermore, if B is S_2 then $hS :_S (S :_S B) = hB$, hence $B[-(e-1)] \cong ((f,h) :_R K)/(f)$. But the latter is a Cohen-Macaulay S-module because the R-ideal $(g,h) :_R K$ is perfect. Thus B is a Cohen-Macaulay ring.

(b) We use the same notation as in the proof of Theorem 6.93(b). In particular we consider the exact sequence of graded T-modules,

$$0 \to S \xrightarrow{\phi} B \longrightarrow C \to 0.$$

Since S and B are graded free T-modules of rank e, it follows that $\deg_T(C) = \deg(T/(\det(\phi))$, which is equal to the degree of the form $\det(\phi)$. Hence

$$\deg_S(B/S) = \deg_T(C) = \deg\det(\phi).$$

However, $S \cong \oplus_{i=0}^{e-1} T(-i)$ and $B \cong \oplus_{i=0}^{e-1} T(-a_i)$ with $a_i \geq 0$. Hence the degree of $\det(\phi)$ is at most

$$1+2+\cdots+(e-1) = \binom{e}{2}$$

in the first case.

If k is algebraically closed and S is a domain, then only one a_i is zero so that the degree of $\det(\phi)$ is bounded by

$$(1-1)+(2-1)+\cdots+(e-1-1) = \binom{e-1}{2},$$

as asserted. □

Theorem 6.96. *Let k be a field of characteristic zero and A a reduced and equidimensional standard graded k–algebra of dimension d and multiplicity e. If $A \subset B$ is a finite and birational extension of graded rings with* depth $_A B \geq d-1$ *then*

$$\nu_A(B) \leq (e-1)^2 + 1$$

and

$$\mathrm{embdim}(B) \leq (e-1)^2 + d + 1.$$

Proof. There exists a homogeneous subalgebra S of A such that $\text{embdim}(S) \leq d+1$ and the extension $S \subset A$ is finite and birational; see for instance Proposition 6.107. Notice that $\deg(S) = \deg(A) = e$. The S-module B/S is Cohen-Macaulay. Therefore $\nu_S(B/S) \leq \deg(B/S)$ and the assertions follow from Lemma 6.95(a). $\square$

Remark 6.97. Theorem 6.93(a) and Lemma 6.95 show that the estimates of Theorem 6.96 can be sharpened under suitable additional assumptions. Indeed, if $e \geq 3$ then

$$\nu_A(B) \leq (e-1)^2 \quad \text{and} \quad \text{embdim}(B) \leq (e-1)^2 + d.$$

If $e \geq 4$ or if B satisfies S_2 and $e \geq 3$, then

$$\nu_A(B) \leq (e-1)^2 - 1 \quad \text{and} \quad \text{embdim}(B) \leq (e-1)^2 + d - 1.$$

Corollary 6.98. *Let k be a field of characteristic zero and A a reduced and equidimensional standard graded k–algebra of dimension 3 and multiplicity $e \geq 3$. The integral closure $B = \overline{A}$ satisfies the inequality*

$$\text{embdim}(B) \leq (e-1)^2 + 2.$$

Remark 6.99. If in the setting of Theorem 6.96 and its proof, T is a homogeneous Noether normalization of A, then $\nu_T(B) \leq e(e-1)^2 + e$, as can be seen by applying the theorem with $A = S$. This bound is not strictly module-theoretic: the algebra structure of B really matters. The assertion fails if B is merely a finite T-module, even a graded reflexive T-module, with $\text{depth}\,_T B \geq d-1$. For example, let T be a polynomial ring in d variables over an infinite field and I a homogeneous perfect ideal of T of height 2 that is generically a complete intersection but has a large number of generators (such ideals exist whenever $d \geq 3$). There is an exact sequence of the form

$$0 \to T[-a] \longrightarrow E \longrightarrow I \to 0,$$

with E a graded reflexive T-module. Now indeed $\text{depth}\, E = d-1$ and $\deg(E) = 2$, whereas $\nu_T(E) \gg 0$.

Non-Homogeneous Algebras

We now formulate a version of the estimates of the previous section valid for general affine algebras. We shall deal with the non-homogeneous version of Lemma 6.95 and Theorem 6.96.

Proposition 6.100. *Let $R = k[x_1, \ldots, x_{d+1}]$ be a polynomial ring over a perfect field k, let f, g be an R–regular sequence and $\mathfrak{p}$ a prime ideal of R.*

(a) *If the degrees of f and g in the variables $x_1, \ldots, x_{d+1}$ are m and n, then*

$$\deg((R/(f,g))_{\mathfrak{p}}) \leq mn.$$

(b) *If the degrees of f and g in the variables $x_1,\dots,x_d$ are m' and n', the degrees of f and g in x_{d+1} are m'' and n'', and g is monic in x_{d+1}, then*

$$\deg((R/(f,g))_{\mathfrak{p}}) \le (m'n''+m''n')n''.$$

Proof. We may assume that $\mathfrak{p}$ is a maximal ideal containing f,g, and after passing to the algebraic closure of k we may suppose that $\mathfrak{p}=(x_1,\dots,x_{d+1})$.

To prove (a), let z be a new variable, $\widetilde{R}=R[z]$, $\widetilde{\mathfrak{p}}=\mathfrak{p}\widetilde{R}$, and $\widetilde{f},\widetilde{g}$ the homogenizations of f and g with respect to z. Notice that $\widetilde{f},\widetilde{g}$ are homogeneous of degrees m,n and form a regular sequence in $\widetilde{R}$. Thus $\deg(\widetilde{R}/(\widetilde{f},\widetilde{g}))=mn$. On the other hand, $\widetilde{\mathfrak{p}}$ is contained in the homogeneous maximal ideal of $\widetilde{R}$ and $(\widetilde{R}/(\widetilde{f},\widetilde{g}))_{\widetilde{\mathfrak{p}}}\cong(\widetilde{R}/(\widetilde{f},\widetilde{g}))_{(\widetilde{\mathfrak{p}})}(z)$. Since $(\widetilde{R}/(\widetilde{f},\widetilde{g}))_{(\widetilde{\mathfrak{p}})}\cong(R/(f,g))_{\mathfrak{p}}$, we conclude that $\deg((R/(f,g))_{\mathfrak{p}})\le mn$.

To prove (b), write $T=k[x_1,\dots,x_d]$, $\mathfrak{q}=\mathfrak{p}\cap T$, and let h be the resultant of the polynomials f,g with respect to the variable x_{d+1}. Now $h\neq 0$ since f and g form a regular sequence, $h\in(f,g)\cap T$, and h has degree at most $m'n''+m''n'$ in the variables $x_1,\dots,x_d$. Thus $\deg((T/(h))_{\mathfrak{q}})\le m'n''+m''n'$. Furthermore, $(R/(f,g))_{\mathfrak{q}}$ is a module over $(T/(h))_{\mathfrak{q}}$ whose dimension is maximal and whose number of generators is at most n''. Thus $\deg(R/(f,g))_{\mathfrak{q}}\le(m'n''+m''n')n''$. Finally, the multiplicity of the ring $(R/(f,g))_{\mathfrak{p}}$ cannot exceed that of the module $(R/(f,g))_{\mathfrak{q}}$. □

Lemma 6.101. *Let k be a field of characteristic zero, let $R=k[x_1,\dots,x_{d+1}]$ be a polynomial ring and f a nonzero squarefree polynomial, and write $S=R/(f)$. Let $S\subset B$ be a finite and birational extension of rings and $\mathfrak{p}$ a prime ideal of R.*

(a) *If the degree of f in the variables $x_1,\dots,x_{d+1}$ is e, then*

$$\deg(B/S)_{\mathfrak{p}}\le e(e-1).$$

(b) *If the degree of f in the variables $x_1,\dots,x_d$ is e' and f is monic in x_{d+1} of degree e'', then*

$$\deg(B/S)_{\mathfrak{p}}\le e'(e''-1)(2e''-1).$$

Proof. In part (a) we may assume that $e\ge 1$ and apply a linear change of variables to suppose that f is monic in x_{d+1}. Thus in either case, the homomorphism $k[x_1,\dots,x_d]\to S$ is finite and generically smooth. Therefore f and its derivative $f'=\frac{\partial f}{\partial x_{d+1}}$ generate the unit ideal or form a regular sequence in R. Applying Proposition 6.100(a) we conclude that $\deg((R/(f,f'))_{\mathfrak{p}})\le e(e-1)$ in part (a), and from Proposition 6.100(b) we obtain that

$$\deg((R/(f,f'))_{\mathfrak{p}})\le(e'(e''-1)+e''e')(e''-1)=e'(e''-1)(2e''-1)$$

in the setting of (b).

Now one proceeds as in the proof of Lemma 6.95(a). □

Theorem 6.102. *Let k be a field of characteristic zero and A a reduced and equidimensional affine k–algebra of dimension d. Let S be a k–subalgebra of A with* $\operatorname{embdim}(S) \leq d+1$ *such that the extension $S \subset A$ is finite and birational, and write* $S = k[x_1,\ldots,x_{d+1}]/(f)$. *Further let $A \subset B$ be a finite and birational extension of rings with $S \neq B$ satisfying* $\operatorname{depth}_{A_\mathfrak{p}} B_\mathfrak{p} \geq d-1$ *for every maximal ideal $\mathfrak{p}$ of A.*

(a) *If the degree of f in the variables $x_1,\ldots,x_{d+1}$ is e, then*

$$\nu_A(B) \leq e(e-1)+d-1$$

and

$$\operatorname{embdim}(B) \leq e(e-1)+2d-1.$$

(b) *If the degree of f in the variables $x_1,\ldots,x_d$ is e' and f is monic in x_{d+1} of degree e'', then*

$$\operatorname{embdim}(B) \leq e'(e''-1)(2e''-1)+2d-1.$$

Proof. In either case we may assume that f is monic in x_{d+1}. Using Lemma 6.101 one proceeds as in the proof of Theorem 6.96 to obtain estimates for the local numbers of generators $\nu_{S_\mathfrak{p}}((B/S)_\mathfrak{p})$ with $\mathfrak{p}$ any prime ideal of S. Also notice that locally in codimension one in $T = k[x_1,\ldots,x_d]$, the T-module B/T is free of rank at most $e-1$ for (a) and of rank $e''-1$ for (b). Now the appropriate global bounds for $\nu_S(B/S)$ follow from [Sw67]. □

Remark 6.103. Let A be an affine k–algebra and write $A = R/I$ with R a polynomial ring over k. For z a new variable, let $\widetilde{I}$ denote the homogenization of I in $\widetilde{R} = R[z]$ and set $\widetilde{A} = \widetilde{R}/\widetilde{I}$. The integer e occurring in the estimate of Theorem 6.102(a) satisfies the inequality $e \geq \deg(A)$ and for a suitable choice of S, the equality $e = \deg(\widetilde{A})$. Unlike Theorems 6.93(a) and 6.96, the assertion of the present theorem is no longer true with $\deg(A)$ in place of e, even if A is a positively graded k–algebra. In fact it is impossible to estimate the embedding dimension of B only in terms of the dimension and the degree of A, unless A is standard graded or B is Cohen-Macaulay.

For example, let T be a polynomial ring in $d \geq 2$ variables over a field k, $u \in T$ a quadratic form that is not a square (which exists), and let $I \neq 0$ be a homogeneous T–ideal of projective dimension at most 1. Write t for the image of the indeterminate z in the ring $C = T[z]/(z^2-u)$, and set $B = T \oplus It \subset C$. Now C is a standard graded domain over k and B is a positively graded k–subalgebra of C (with integral closure C if $\operatorname{char} k \neq 2$ and u is squarefree). The assumptions of Theorem 6.102(a) are satisfied for $A = B$ (if $\operatorname{char} k = 0$). Notice that $\deg(A) = 2$, whereas $\operatorname{embdim}(B) = n+d$ with $n = \nu(I)$, which can be arbitrarily large. We compare this to the prediction of Theorem 6.102 if I has height 2 and a linear presentation matrix (and $\operatorname{char} k = 0$). In this case one can choose $e = e' = 2n$ and $e'' = 2$. Therefore parts (a) and (b) of the theorem give the estimates $\operatorname{embdim}(B) \leq 2n(2n-1)+2d-1$ and $\operatorname{embdim}(B) \leq 6n+2d-1$, respectively. Also note the significance of Serre's condition S_2 for B: it forces I to be a divisorial ideal and hence implies that $\operatorname{embdim}(B) = d+1$, contrasting with the example of Remark 6.99.

6.5.2 Small Singularities

Let A be a reduced equidimensional affine algebra over a perfect field k, let J be its Jacobian ideal and B its integral closure. We treat the case of $\dim A = d \geq 4$ and assume that the singular locus of A is suitably small.

Suppose that height $J \geq 2$, a condition equivalent to $A_\wp = B_\wp$ for each prime ideal of A of height 1. This means that B is the S_2-ification of A, and one may describe B as the outcome of a single operation in the total ring of quotients of A. According to Theorem 6.30,

$$B = A : J.$$

The issue is to estimate how many generators this process requires. One of our results, for $d = 4$, will do precisely this. For higher dimensions we shall require a condition of the order R_{d-3} (see later on for more precise statements). Before we can get to this we need to examine various 'approximations' of B by some tractable subrings much in the manner that the hypersurface subring S was used in Section 3. They will be aimed at converting information about the degrees of the generators of a 'thick' portion of the Jacobian ideal J in terms of the multiplicity of A.

Proposition 6.104. *Let k be an infinite perfect field, A a finitely generated k-algebra of dimension d, and $\mathfrak{q}$ a prime ideal of A. If* embdim $(A_\mathfrak{q}) \leq g$ *for some $g \geq 1$, then there exists a k-subalgebra $S = k[x_1, \ldots, x_{d+g}] \subset A$ such that the extension $S \subset A$ is finite and*

$$S_{\mathfrak{q}\cap S} = A_{\mathfrak{q}\cap S} = A_\mathfrak{q}.$$

If in addition A is standard graded, then $x_1, \ldots, x_{d+g}$ can be chosen to be linear forms, and if A is reduced and equidimensional then the extension $S \subset A$ can be chosen to be birational.

The assertion about finiteness is obvious. The claim about birationality follows from the above equalities applied to he minimal primes of A.

Set $\mathfrak{q} = \mathfrak{p} \cap S$. To establish the equalities $S_\mathfrak{q} = A_\mathfrak{q} = A_\mathfrak{p}$, write $k(\mathfrak{p})$ for the residue field of $\mathfrak{p}$ and consider the exact sequence

$$0 \to \mathbb{D} \longrightarrow R = k(\mathfrak{p}) \otimes_k A \longrightarrow k(\mathfrak{p}) \to 0 \tag{6.9}$$

induced by the multiplication map $A \otimes_k A \to A$. Notice that $\mathbb{D} = \sum_{i=1}^n R(\overline{y_i} \otimes 1 - 1 \otimes y_i)$ is a maximal ideal of R. Since

$$\nu(\mathbb{D}_\mathbb{D}) \leq \nu(\Omega_k(A_\mathfrak{p})) \leq \text{embdim}(A_\mathfrak{p}) + \dim A/\mathfrak{p} \leq d + g,$$

one has $\mathbb{D}_\mathbb{D} = \sum_{i=1}^{d+g} R_\mathbb{D}(\overline{x_i} \otimes 1 - 1 \otimes x_i)$. On the other hand, as $\dim R = d < d + g$, we have

$$\mathbb{D} = \sqrt{\sum_{i=1}^{d+g} R(\overline{x_i} \otimes 1 - 1 \otimes x_i)};$$

to see this, recall that every ideal in a d–dimensional ring containing an infinite field k is generated up to radical by $d+1$ general k–linear combinations of its generators. It follows that $\mathbb{D} = \sum_{i=1}^{d+g} R(\overline{x_i} \otimes 1 - 1 \otimes x_i)$. Thus by (6.9), $k(\mathfrak{p}) \otimes_S A \cong k(\mathfrak{p})$. Comparing numbers of generators over the ring $S_\mathfrak{q}$ we conclude that $\nu_{S_\mathfrak{q}}(A_\mathfrak{q}) = 1$, hence $S_\mathfrak{q} = A_\mathfrak{q}$. In particular $A_\mathfrak{q}$ is local and we also obtain that $A_\mathfrak{q} = A_\mathfrak{p}$. $\square$

Corollary 6.105. *Let k be an infinite perfect field and A a reduced and equidimensional standard graded k-algebra of dimension d. Write $\mathcal{A}$ for the set of all homogeneous subalgebras $S = k[x_1, \ldots, x_{d+1}]$ of A such that the extension $S \subset A$ is finite and birational, and set $\mathcal{J} = \sum_{S \in \mathcal{A}} A \cdot J(S)$. Then $\sqrt{J(A)} = \sqrt{\mathcal{J}}$.*

Proof. For a fixed prime ideal $\mathfrak{q}$ of A we apply Proposition 6.104 with $g = 1$. The proposition shows that $A_\mathfrak{q}$ is regular if and only if $S_{\mathfrak{q}\cap S}$ is regular for some $S \in \mathcal{A}$. Since S is again equidimensional, one has $J(A) \subset \mathfrak{q}$ if and only if $J(S) \subset \mathfrak{q} \cap S$ if and only if $A \cdot J(S) \subset \mathfrak{q}$ for some $S \in \mathcal{A}$. $\square$

Corollary 6.106. *Let k be an infinite perfect field and A a reduced and equidimensional standard graded k-algebra with $e = \deg(A)$ and $c = \operatorname{codim}(\operatorname{Sing}(A))$. Then the conductor of A contains an ideal of height c generated by forms of degree $e-1$.*

Proof. We claim that the ideal $\mathcal{J}$ of Corollary 6.105 has the desired properties. First, by the corollary, height $\mathcal{J} = c$. Now suppose that $S \in \mathcal{A}$. Since $S = k[x_1, \ldots, x_{d+1}]$ is reduced and equidimensional of dimension d we can write $S = k[X_1, \ldots, X_{d+1}]/(f)$ where f is a form. As the homogeneous extension $S \subset A$ is finite and birational one has $\deg(S) = \deg(A)$, thus $\deg f = e$. Therefore $J(S)$ is generated by forms of degree $e-1$ and hence $\mathcal{J}$ has the same property. Finally by [No50], $J(S)$ is contained in the conductor $\mathfrak{c}(S)$ of S. But $\mathfrak{c}(S) \subset \mathfrak{c}(A)$ since the extension $S \subset A$ is finite and birational. Thus indeed $\mathcal{J} \subset \mathfrak{c}(A)$. $\square$

The proof of the main theorem in this section requires a generic projection lemma that is known in a geometric context, at least for algebras with small singular locus:

Proposition 6.107. *Let k be an infinite perfect field, $A = k[y_1, \ldots, y_n]$ an equidimensional k–algebra of dimension d, and $s \geq 0$ an integer. Let $x_1, \ldots, x_{d+s}$ be general k–linear combinations of $y_1, \ldots, y_n$ and write $S = k[x_1, \ldots, x_{d+s}] \subset A$. If $\operatorname{ecodim}(A_\mathfrak{p}) \leq \dim A_\mathfrak{p}$ for every prime $\mathfrak{p}$ of A with $\dim A_\mathfrak{p} \leq s-1$, then $S_\mathfrak{q} = A_\mathfrak{q}$ for every prime $\mathfrak{q}$ of S with $\dim S_\mathfrak{q} \leq s-1$.*

Proof. Consider the exact sequence

$$0 \to \mathbb{D} \longrightarrow R = A \otimes_k A \stackrel{\text{mult}}{\longrightarrow} A \to 0, \tag{6.10}$$

where $\mathbb{D} = \sum_{i=1}^{n} R(y_i \otimes 1 - 1 \otimes y_i)$. By our assumption and [SUV93, 2.2(b)], $\nu(\mathbb{D}_Q) \leq \dim R_Q$ for every $Q \in V(\mathbb{D})$ with $\dim R_Q \leq d+s-1$, that is, $\mathbb{D}$ satisfies the condition G_{d+s} from [AN72, p. 312]. But then according to [AN72, the proof of 2.3],

$$\text{height } \Big(\sum_{i=1}^{d+s} R(x_i \otimes 1 - 1 \otimes x_i)\Big) : \mathbb{D} \geq d+s. \tag{6.11}$$

Now let $\mathfrak{q}$ be a prime of S with $\dim S_{\mathfrak{q}} \leq s-1$. As $\dim A_{\mathfrak{q}} \otimes_k A_{\mathfrak{q}} \leq \dim A_{\mathfrak{q}} + \dim A \leq d+s-1$, (6.11) implies that

$$(A_{\mathfrak{q}} \otimes_k A_{\mathfrak{q}})\mathbb{D} = \sum_{i=1}^{d+s} (A_{\mathfrak{q}} \otimes_k A_{\mathfrak{q}})(x_i \otimes 1 - 1 \otimes x_i).$$

Thus $A_{\mathfrak{q}} \otimes_{S_{\mathfrak{q}}} A_{\mathfrak{q}} \cong A_{\mathfrak{q}}$ by (6.10). Since $A_{\mathfrak{q}}$ is a finite $S_{\mathfrak{q}}$–module, comparing numbers of generators then yields $\nu_{S_{\mathfrak{q}}}(A_{\mathfrak{q}}) = 1$, hence $S_{\mathfrak{q}} = A_{\mathfrak{q}}$. □

Theorem 6.108. *Let k be a field of characteristic zero and A a reduced and equidimensional standard graded k–algebra of dimension $d \geq 3$ and multiplicity e. Let $A \subset B$ be a finite and birational extension of graded rings and write $t = \min\{d-2, \operatorname{depth}_A B\}$. If A satisfies R_{d-t-1} then*

$$\nu_A(B) \leq 1/2(e(e-1))^{2^{d-t-1}} - 1/2(e(e-1))^{2^{d-t-2}} + 1$$

and

$$\operatorname{embdim}(B) \leq 1/2(e(e-1))^{2^{d-t-1}} - 1/2(e(e-1))^{2^{d-t-2}} + 2d - t.$$

Proof. By Proposition 6.107, we may replace A by a suitable k–subalgebra to assume that $\operatorname{embdim}(A) \leq 2d-t$. Thus it suffices to show that

$$\nu_A(B/A) \leq 1/2(e(e-1))^{2^{d-t-1}} - 1/2(e(e-1))^{2^{d-t-2}}.$$

Let $\mathbf{x} = x_1, \ldots, x_{t-1}$ be a sequence of general linear forms in A. Notice that $\mathbf{x}$ forms a regular sequence on B. Write $B' = B/(\mathbf{x})B$ and A' for the image of A in B'. Since $\dim_A B/A \leq t$ one has $\dim_{A'} B'/A' = \dim_A B'/A' \leq 1$. Furthermore by [Fl77, 4.7], B' satisfies R_{d-t-1}, and is reduced and equidimensional of dimension $d-t+1 \geq 3$. Thus the standard graded k–algebra A' has the same properties. Moreover $\deg(A') = \deg_{A'}(B') = \deg_A(B) = \deg(A) = e$. By the graded Nakayama's Lemma, and since $B'/A' \subset \overline{A'}/A'$, it suffices to show that for every A'–submodule C of $\overline{A'}/A'$,

$$\nu_{A'}(C) \leq 1/2(\deg(A')(\deg(A')-1))^{2^{\dim A'-2}} - 1/2(\deg(A')(\deg(A')-1))^{2^{\dim A'-3}}.$$

Changing notation we prove the following. If A is a reduced and equidimensional standard graded k–algebra of dimension $d \geq 3$ and multiplicity e satisfying R_{d-2}, then for every A–submodule C of $\overline{A}/A$,

$$\nu_A(C) \leq 1/2(e(e-1))^{2^{d-2}} - 1/2(e(e-1))^{2^{d-3}}. \tag{6.12}$$

Now write $B = \overline{A}$ for the integral closure and $\mathfrak{c} = \mathfrak{c}(A)$ for the conductor of A. By Corollary 6.106, $\mathfrak{c}$ contains forms of degree $e-1$ that generate an ideal $\mathcal{J}$ of height at least $d-1$. Let x be a general k–linear combination of these forms. Write $B' = B/xB$ and let A' be the image of A in B'. As height $\mathcal{J} \geq d-1$, it follows that B' satisfies R_{d-3} ([Fl77, 4.7]), and B' is reduced and equidimensional of dimension $d-1 \geq 2$. Since $\dim_{A'} B'/A' \leq 1$, A' then has the same properties. Notice that $\deg(A') = e(e-$

1) because x is a B–regular form of degree $e-1$. On the other hand, as $x \in \mathfrak{c} = \operatorname{ann}_A(B/A)$ we have $B/A = B'/A'$ and therefore $C \subset B/A = B'/A' \subset \overline{A'}/A'$.

We are now ready to prove (6.12) by induction on $d \geq 3$. We may assume that $C \neq 0$. If $d = 3$, let x_1, x_2, x_3 be general linear forms in A, and write $S = k[x_1, x_2, x_3]$ and S' for the image of S in A'. Now $S' \subset A' \subset \overline{A'}$ are finite and birational extensions of two-dimensional rings, $\operatorname{embdim}(S') \leq 3 = \dim S' + 1$, $\deg(S') = \deg(A')$, and S' and $\overline{A'}$ are Cohen–Macaulay. Thus the S'–module $\overline{A'}/S'$ is Cohen–Macaulay of dimension 1 and has multiplicity at most $\binom{\deg(A')}{2}$ by Lemma 6.95(b). Let $D \subset \overline{A'}/S'$ be the preimage of $C \subset \overline{A'}/A'$ under the natural projection from $\overline{A'}/S'$ to $\overline{A'}/A'$. As an S'–submodule of $\overline{A'}/S'$, D is also Cohen–Macaulay of dimension 1. It follows that $\nu_{S'}(D) \leq \deg(D) \leq \deg(\overline{A'}/S')$. Therefore

$$\nu_A(C) \leq \nu_{S'}(C) \leq \nu_{S'}(D) \leq \deg(\overline{A'}/S') \leq \binom{\deg(A')}{2} = 1/2(e(e-1))^2 - 1/2e(e-1),$$

establishing (6.12) for $d = 3$.

If $d \geq 4$ then by the induction hypothesis, since $\nu_A(C) = \nu_{A'}(C)$,

$$\begin{aligned} \nu_A(C) &\leq 1/2(\deg(A')(\deg(A')-1))^{2^{\dim A'-2}} - 1/2(\deg(A')(\deg(A')-1))^{2^{\dim A'-3}} \\ &= 1/2(e(e-1)(e(e-1)-1))^{2^{d-3}} - 1/2(e(e-1)(e(e-1)-1))^{2^{d-4}} \\ &\leq 1/2(e(e-1))^{2^{d-2}} - 1/2(e(e-1))^{2^{d-3}}. \end{aligned}$$

To see the last inequality set $a = e(e-1) \geq 2$ and $\alpha = 2^{d-4}$, and notice that $(a(a-1))^{2\alpha} - (a(a-1))^{\alpha} \leq a^{4\alpha} - a^{2\alpha}$. □

Corollary 6.109. *Let k be a field of characteristic zero and A a reduced and equidimensional standard graded k–algebra of dimension $d \geq 4$ and multiplicity e. If A satisfies R_{d-3} then for $B = \overline{A}$,*

$$\operatorname{embdim}(B) \leq 1/2(e(e-1))^{2^{d-3}} - 1/2(e(e-1))^{2^{d-4}} + 2d - 2.$$

In particular if $d = 4$ and A satisfies R_1 then

$$\operatorname{embdim}(B) \leq 1/2e^4 - e^3 + 1/2e + 6.$$

Bounds on Degrees

Here we give degree bounds for the generators of the integral closure of standard graded integral domains over a field of characteristic zero. Besides being interesting in their own right, such bounds also lead to estimates of the embedding dimension, due to the following observation.

Proposition 6.110. *Let k be a perfect field and A a reduced and equidimensional standard graded k–algebra of dimension $d \geq 1$ and multiplicity e. Let $A \subset B$ be a finite and birational extension of graded rings.*

(a) *If the graded A–module B is generated in degrees at most s then*

$$\nu_A(B) \le \binom{s+e+d-1}{d} - \binom{s+d-1}{d}.$$

(b) *If* embdeg$(B) = s$ *then*

$$\operatorname{embdim}(B) \le \binom{s+e+d-1}{d} - \binom{s+d-1}{d} + d.$$

Proof. We may assume that k is infinite. Replacing A by any $S \in \mathcal{A}$ as in Corollary 6.105, we may further suppose that $\operatorname{embdim}(A) \le d+1$. This means that $A = k[x_1,\ldots,x_{d+1}]/(f)$ with f a homogeneous polynomial of degree e. Let $D \subset B$ be the A–submodule generated by all forms in B of degrees $\le s$. By Corollary 6.106 (or [LS81, Theorem 2]), $A : B$ contains a homogeneous element x of degree $e-1$ that is regular on A, hence on B. Thus $I = xD \cong D[-(e-1)]$ is a homogeneous A–ideal generated by forms of degrees $\le s+e-1$. Let y be a linear form that is regular on A. Multiplying the elements of a homogeneous minimal generating set of I by suitable powers of y one sees that

$$\nu_A(I) \le \dim_k A_{s+e-1} = \binom{s+e+d-1}{d} - \binom{s+d-1}{d}.$$

Now part (a) follows since in this case $B = D \cong I[e-1]$ as A–modules. As to (b), notice that the A–module $D \cong I[e-1]$ is minimally generated by a set of the form $\{1\} \cup W$, where W has $\nu_A(I) - 1$ elements. Since W generates the A–algebra B, the assertion follows in this case as well. □

If A is a standard graded algebra over a field we write $\operatorname{reg}(A)$ for the *Castelnuovo–Mumford regularity* and $\operatorname{ecodim}(A) = \operatorname{embdim}(A) - \dim A$ for the *embedding codimension* of A. Some of our bounds have a conjectural basis, as they depend on the validity of the conjecture from [EG84] as follows:

Conjecture 6.111 (Eisenbud-Goto). Let k be an algebraically closed field and A a standard graded domain over k. Then the Castelnuovo–Mumford regularity of A is bounded by

$$\operatorname{reg}(A) \le \deg(A) - \operatorname{ecodim}(A).$$

This bound has been established in low dimensions (≤ 7), often with additional restrictions.

Theorem 6.112. *Let k be an algebraically closed field of characteristic zero and A a reduced and equidimensional standard graded k–algebra of dimension d and multiplicity* $e \ge 3$. *Assume that A satisfies* R_2. *If* ((6.111)) *holds in dimension* $d-1$ *over the field k, then for* $B = \overline{A}$, *the graded A–module B is generated in degrees at most* $e(e-2)-1$; *in particular*

$$\operatorname{embdeg}(B) \le e(e-2)-1,$$
$$\operatorname{embdim}(B) \le \binom{e(e-1)+d-2}{d} - \binom{e(e-2)+d-2}{d} + d.$$

Proof. We first prove the assertion about the generator degrees. We may assume that A is not Cohen–Macaulay, as otherwise $B = A$. Thus $\operatorname{ecodim}(A) \geq 2$. We may also suppose that A is a domain, as it suffices to prove our claim modulo every minimal prime of B. By Corollary 6.106, the conductor $\mathfrak{c}(A) = A : B$ contains an ideal $\mathcal{J}$ of height ≥ 3 generated by homogeneous elements of degree $e-1$. Let x be a general form of degree $e-1$ in $\mathcal{J}$. Since height $\mathcal{J} \geq 3$ and B satisfies S_2 and has characteristic zero, [Fl77, 4.10] shows that $B' = B/xB$ is a domain. Notice that $xB \subset A$, hence $A' = A/xB$ is a subring of B'. Thus A' is a domain of dimension $d-1$. The finite extension $A' \subset B'$ is birational since $\dim_{A'}(B'/A') = \dim_A(B/A) \leq d-3$. Therefore $\deg(A') = e(e-1)$. As $e-1 \geq 2$ and B is concentrated in nonnegative degrees it follows that $\operatorname{ecodim}(A') = \operatorname{ecodim}(A) + 1$. However, $\operatorname{ecodim}(A) \geq 2$ by the above. According to our assumption, (6.111) holds for the standard graded domain A' of dimension $d-1$. Therefore $\operatorname{reg}(A') \leq e(e-1) - 3$.

Let R be a polynomial ring over k mapping onto A and $\mathfrak{P}$ a prime ideal of R with $A' \cong R/\mathfrak{P}$. The homogeneous R–ideal $\mathfrak{P}$ is generated in degrees at most $\operatorname{reg}(A') + 1 \leq e(e-1) - 2$. On the other hand, since $A' \cong A/xB$ it follows that $\mathfrak{P}$ maps onto xB. Therefore the graded A–module $B \cong xB[e-1]$ is generated in degrees at most $e(e-1) - 2 - (e-1) = e(e-2) - 1$.

The estimates for $\nu_A(B)$ and $\operatorname{embdim}(B)$ now follow from [UV3, Proposition 6.1]. □

Theorem 6.113. *Let k be a perfect field, let A be a reduced and equidimensional standard graded k-algebra of dimension d and multiplicity $e \geq 2$, and let $A \subset B$ be a finite and birational extension of graded rings. If A satisfies R_1 and* $\operatorname{depth}_A B \geq d-1$, *then the A-module B is generated in degrees at most $3e-5$.*

Proof. We may assume that k is infinite and then by Proposition 6.107 we may reduce to the case where $A = k[x_1, \ldots, x_{d+2}]$ where the x_i are linear forms. Take $x_1, \ldots, x_{d+2}$ to be general, map the polynomial ring $R = k[X_1, \ldots, X_{d+2}]$ onto A by sending X_i to x_i, and write $S = k[x_1, \ldots, x_{d+1}] \subset A$. One has $S = k[X_1, \ldots, X_{d+1}]/(f)$ for some form f of degree e. Since S is reduced there exists a form $g \in k[X_1, \ldots, X_{d+1}]$ of degree $e-1$ such that f, g are a regular sequence on $k[X_1, \ldots, X_{d+1}]$ and the image of g in S lies in $J(S)$, hence in the conductor of S ([No50]). Write $k[x_1, \ldots, x_d, x_{d+2}] = k[X_1, \ldots, X_d, X_{d+2}]/(h)$, where h is a form of degree e. As h is monic in X_{d+2}, it follows that f, g, h is an R-regular sequence.

Consider the ring $A' = R/(f, h)$ and its homogeneous Noether normalization

$$S' = k[X_1, \ldots, X_{d-1}, g] \hookrightarrow A'.$$

Since the socle of the complete intersection $A'/(x_1, \ldots, x_{d-1}, g)$ is concentrated in degree $\deg f + \deg h + \deg g - 3 = 3e - 4$, we conclude that the S'-module A' is generated in degrees at most $3e-4$. But A' maps onto A and $gB \subset A$. Thus, as modules over the polynomial ring $T = k[X_1, \ldots, X_{d-1}]$, A/S is generated in degrees at most $3e - 4 - 1 = 3e - 5$ and B/S is finite.

Now consider the exact sequence of graded T-modules,

$$0 \to M = A/S \longrightarrow N = B/S \longrightarrow B/A \to 0.$$

Here N is a maximal Cohen-Macaulay module, hence free, whereas $\dim N/M = \dim B/A \leq d-2 < \dim T$. If the degree of a homogeneous basis element of N exceeds $3e-4$, then N/M has a nontrivial free summand, which is impossible. Thus the T-module N is generated in degrees at most $3e-4$, and then the same holds for the A-module B. □

Homological Degrees of Biduals

Let A be an affine domain over a field and E a finitely generated torsionfree A-module. The construction of the bidual, $E^{**} = \mathrm{Hom}_A(\mathrm{Hom}_A(E,A),A)$, occurs often in our treatment of integral closure and a comparison between the number of generators of E and of E^{**} is required, at least in the case when A is a Gorenstein ring. This is not always possible, but we are going to show how the theory of homological degrees developed in Chapter 2 may help.

Proposition 6.114. *Let $(R,\mathfrak{m})$ be a Gorenstein local ring of dimension d, and E a torsionfree R-module and F its bidual. If $\dim F/E = 2$, then*

$$\mathrm{hdeg}(F) \leq \begin{cases} \mathrm{hdeg}(E) & \text{if } d \geq 6, \\ 2\cdot\mathrm{hdeg}(E) & \text{if } d = 4,5. \end{cases}$$

Proof. Set $C = F/E$ and note that in general $\dim C \leq d-2$. We note also that the case $\dim C \leq 1$ will easily be gleaned in the proof. Consider the exact sequence

$$0 \to E \longrightarrow F \longrightarrow C \to 0,$$

and apply the functor $\mathrm{Hom}_R(\cdot,R)$. We observe that by local duality $\mathrm{Ext}_R^i(C,R) = 0$ for $i \leq d-3$, $\mathrm{Ext}_R^i(F,R) = 0$ for $i \geq d-1$, that $\mathrm{Ext}_R^{d-2}(F,R)$ has finite length because F satisfies condition S_2, and in a similar manner $\mathrm{Ext}_R^d(E,R) = 0$ and $\mathrm{Ext}_R^{d-1}(E,R)$ has finite length because E is torsionfree. With this the long exact sequence of cohomology breaks down into the shorter exact sequences:

$$\mathrm{Ext}_R^i(F,R) = \mathrm{Ext}_R^i(E,R), \quad i \leq d-4,$$

$$\begin{aligned} 0 &\to \mathrm{Ext}_R^{d-3}(F,R) \to \mathrm{Ext}_R^{d-3}(E,R) \to \mathrm{Ext}_R^{d-2}(C,R) \to \mathrm{Ext}_R^{d-2}(F,R) \\ &\to \mathrm{Ext}_R^{d-2}(E,R) \to \mathrm{Ext}_R^{d-1}(C,R) \to 0, \end{aligned}$$

$$\mathrm{Ext}_R^{d-1}(E,R) = \mathrm{Ext}_R^d(C,R).$$

We break down the middle sequence into short exact sequences,

$$0 \to \mathrm{Ext}_R^{d-3}(F,R) \to \mathrm{Ext}_R^{d-3}(E,R) \to L_0 \to 0$$

$$0 \to L_0 \to \mathrm{Ext}_R^{d-2}(C,R) \to L_1 \to 0$$

$$0 \to L_1 \to \mathrm{Ext}_R^{d-2}(F,R) \to L_2 \to 0$$

$$0 \to L_2 \to \mathrm{Ext}_R^{d-2}(E,R) \to \mathrm{Ext}_R^{d-1}(C,R) \to 0.$$

We recall the expression for $\mathrm{hdeg}(E)$:

$$\mathrm{hdeg}(E) = \deg(E) + \sum_{i=1}^{d} \binom{d-1}{i-1} \cdot \mathrm{hdeg}(\mathrm{Ext}_R^i(E,R)),$$

and identify similar values for the expression for $\mathrm{hdeg}(F)$. Note that for $i \leq d-4$, they are the same. We are going to argue that for the higher dimensions the contributions of $\mathrm{Ext}_R^i(E,R)$ are at least as high as those of $\mathrm{Ext}_R^i(F,R)$). We have to concern ourselves with $\mathrm{Ext}_R^{d-3}(F,R)$ and $\mathrm{Ext}_R^{d-2}(F,R)$.

The key module to examine is $H = \mathrm{Ext}_R^{d-2}(C,R)$. Since $\dim C = 2$, its bottom cohomology module is a module of dimension 2 of depth 2 as well: $H = \mathrm{Hom}_{R/I}(C,R/I)$, I being a complete intersection of codimension $d-2$ contained in the annihilator of C. In particular, L_0 is a nonzero module of positive depth and dimension 2, since $\mathrm{Ext}_R^{d-3}(F,R)$ has dimension at most 1, as remarked at the outset; it is the full submodule of $\mathrm{Ext}_R^{d-3}(E,R)$ of dimension ≤ 1.

Since $\mathrm{Ext}_R^{d-2}(C,R)$ has depth 2, we have that $\mathrm{Ext}_R^{d-1}(L_0,R) = \mathrm{Ext}_R^{d}(L_1,R)$, a module of the same length as L_1. On the other hand, the cohomology of the sequence defining L_0 gives the exact sequences

$$\mathrm{Ext}_R^{d-2}(L_0,R) = \mathrm{Ext}_R^{d-2}(\mathrm{Ext}_R^{d-3}(E,R),R)$$

$$0 \to \mathrm{Ext}_R^{d-1}(L_0,R) \to \mathrm{Ext}_R^{d-1}(\mathrm{Ext}_R^{d-3}(E,R),R) \to \mathrm{Ext}_R^{d-1}(\mathrm{Ext}_R^{d-3}(F,R),R) \to 0$$

$$\mathrm{Ext}_R^{d}(\mathrm{Ext}_R^{d-3}(E,R),R) = \mathrm{Ext}_R^{d}(\mathrm{Ext}_R^{d-3}(F,R),R),$$

that establish the equality

$$\mathrm{hdeg}(\mathrm{Ext}_R^{d-3}(E,R)) = \mathrm{hdeg}(\mathrm{Ext}_R^{d-3}(F,R)) + \mathrm{hdeg}(L_0). \tag{6.13}$$

As for $\mathrm{Ext}_R^{d-2}(F,R)$, the modules L_1 and L_2 have finite length and

$$\mathrm{hdeg}(\mathrm{Ext}_R^{d-2}(F,R)) = \lambda(L_1) + \lambda(L_2) \leq \mathrm{hdeg}(L_0) + \mathrm{hdeg}(\mathrm{Ext}_R^{d-2}(E,R)). \tag{6.14}$$

The corresponding terms in the formula for $\mathrm{hdeg}(E)$ and $\mathrm{hdeg}(F)$ are respectively

$$\binom{d-1}{d-4} \cdot \mathrm{hdeg}(\mathrm{Ext}_R^{d-3}(E,R)) + \binom{d-1}{d-3} \cdot \mathrm{hdeg}(\mathrm{Ext}_R^{d-2}(E,R)) \tag{6.15}$$

$$\binom{d-1}{d-4} \cdot \mathrm{hdeg}(\mathrm{Ext}_R^{d-3}(F,R)) + \binom{d-1}{d-3} \cdot \mathrm{hdeg}(\mathrm{Ext}_R^{d-2}(F,R)). \tag{6.16}$$

Replacing $\mathrm{hdeg}(\mathrm{Ext}_R^{d-2}(E,R))$ from (6.13) by $\mathrm{hdeg}(\mathrm{Ext}_R^{d-3}(F,R)) + \mathrm{hdeg}(L_0)$, and moving the term with $\mathrm{hdeg}(L_0)$ over we get

$$\binom{d-1}{d-4}\text{hdeg}(\text{Ext}_R^{d-3}(F,R)) + \binom{d-1}{d-3}\text{hdeg}(\text{Ext}_R^{d-2}(E,R)) + \binom{d-1}{d-4}\text{hdeg}(L_0). \quad (6.17)$$

For $d \geq 6$, $\binom{d-1}{d-4} \geq \binom{d-1}{d-3}$, which shows that taking (6.14) into account gives that the expression (6.17) exceeds (6.16), as desired.

When $d = 4,5$, taking twice $\binom{d-1}{d-4}$ will serve the stated purpose. □

Remark 6.115. Since as they give rise to the inclusion $a/g \in B$. In fact, collecting the coefficient ideal of f in these syzygies, we have

$$B = A[Lf^{-1}].$$

This calculation is realized as follows. Write $A = R/I$, where R is a ring of polynomials in $n = \text{embdim}(A)$ variables. For a set $f_1,\ldots,f_m$ of generators of I, we consider the syzygies $\mathcal{S}$ of $F,G,f_1,\ldots,f_m$, where F and G are lifts of f and g. The desired syzygies over A are the images of $\mathcal{S}$ in A.

Let $\mathcal{G} = \{g_1,\ldots,g_s\}$ be a Gröbner basis of the ideal $J = \{F,G,f_1,\ldots,f_m\}$, and write $\delta = \max\{\deg F, \deg G, \deg f_1,\ldots,\deg f_m\}$. The set $\mathcal{S}$ will arise from division relations amongst the g_i and therefore $\mathcal{S}$ will be generated in degrees bounded by $\Delta = \max\{\deg g_i\}$. According to [BKW93, Appendix], and especially [Gi84], Δ is bounded by a polynomial in δ of degree a^n (with a of the order of $\sqrt{3}$). This is obviously much larger than the bounds of Theorems 6.112 and 6.113.

6.6 Arithmetic Affine Algebras

In this section we shall focus on algebras over $\mathbb{Z}$, or over other Noetherian domains, and extend some of the tools of normalizations from the standard case. Regrettably, the two main tools we have used-Jacobian ideals and Noether normalization-are not fully deployed in general.

Arithmetic Noether Normalization

Let R be a Noetherian domain and $B = R[x_1,\ldots,x_n]$ a finitely generated R-algebra. We will assume that B is R-torsionfree. By abuse of terminology, by *arithmetic Noether normalizations* we shall mean statements asserting the existence of extensions

$$R \subset S = R[y_1,\ldots,y_r] \subset B,$$

B finite over S, where r depends uniformly on the Krull dimensions of B and R. We shall refer to S as a Noether normalization of B.

In one of the forms that the problem appears, B is an integral domain and one seeks a value for r in terms of the Krull dimension of $K \otimes_R B = K[x_1,\ldots,x_n]$, where K is the field of fractions of R. This dimension is a lower bound for r, but it is not always reached–e. g. $B = \mathbb{Z}[1/2]$. We also note that in most cases of interest B is an integral domain admitting a decomposition $B = R + Q$, $R \cap Q = (0)$. By a standard dimension formula (see Lemma 1.21),

$$\dim B = \dim R + \text{height } Q = \dim R + \dim K \otimes_R B.$$

We begin the discussion by recalling a general fact. If R is an integral domain, the usual proof(s) of Noether normalization leads to the following (see [Na62, Theorem 14.4]):

Theorem 6.116. *Let R be an integral domain and $B = R[x_1, \ldots, x_n]$ a finitely generated R-algebra of Krull dimension r that is R-torsionfree. There exists a with $0 \neq a \in R$, and R-algebraically independent elements $y_1, \ldots, y_r \in B$ such that $B[a^{-1}]$ is finite over $R[a^{-1}][y_1, \ldots, y_r]$.*

It is possible to use this result together with induction on the dimension of R and patching on Spec (R) to derive dimension dependent bounds for the number of generators of Noether normalizations for B. Moreover, the bounds below, based on the technique of the above section perform much better.

Theorem 6.117. *Let R be a Noetherian ring (respectively Noetherian domain) and $B = R[x_1, \ldots, x_n]$ a standard graded algebra of Krull dimension d. If R contains an infinite field k, there is a graded subalgebra $S = R[y_1, \ldots, y_r] \subset B$, $r \leq d+1$ (respectively $r \leq d$), over which B is integral.*

Proof. We denote by I the ideal B_+. According to Proposition 1.46, there are $d+1$ k-linear combinations $y_1, \ldots, y_{d+1}$ of the x_i such that $\sqrt{I} = \sqrt{(y_1, \ldots, y_{d+1})}$. Since the y_j are homogeneous of degree 1, this means that

$$B_{m+1} = (y_1, \ldots, y_{d+1})B_m, \quad m \gg 0.$$

This ensures that B is indeed finite over $S = R[y_1, \ldots, y_{d+1}]$. □

Theorem 6.118. *Let R be a universally catenary Noetherian domain and B an integral domain of Krull dimension d finitely generated over R. If R contains an infinite field k, there is a subalgebra $S = R[y_1, \ldots, y_r] \subset B$, with $r \leq d+1$, over which B is integral.*

Proof. Consider a presentation $B = R[x_1, \ldots, x_n] = R[X_1, \ldots, X_n]/I$, where the X_i are independent indeterminates. With $\dim B = d$, I is a prime ideal of codimension $1 + n - d$. Let t be a new indeterminate and define the homogenization I_t of I: this is the ideal of $R[X_1, \ldots, X_n, t]$ obtained by homogenizing a set of generators of I and then saturating it with respect to t (see, for example, [Va98b, p. 32]). Then

$$(I, t-1) = (I_t, t-1), \quad \text{height } I_t = \text{height } I.$$

In particular, the standard graded algebra $B^h = R[Y_1, \ldots, Y_n, t]/I_t$ has Krull dimension $d+1$. It is enough now to apply Theorem 6.117 to B^h, getting a subalgebra A generated by $d+1$ generators, and letting S be its image in $B^h/(t-1) = B$. □

We mention one result relevant to the case when $R = \mathbb{Z}$ ([Sh54, Theorem 1]).

Theorem 6.119 ([Sh54]). *Let R be a ring of algebraic integers of field of fractions K and let B be a homogeneous R-algebra that is a domain. Then there is a graded subalgebra $S = R[y_1,\ldots,y_r]$, $r = \dim K\otimes_R B$, over which B is integral.*

Note that S is actually a ring of polynomials over R, since $\dim K\otimes_R B = \dim B - \dim R$. Clearly this bound is not achieved if $R = \mathbb{C}[t]$.

Corollary 6.120 ([Sh54]). *Let R be a ring of algebraic integers of field of fractions K and let B be a finitely generated R-algebra that is a domain. Then there is a subalgebra $S = R[y_1,\ldots,y_r]$, $r = \dim B$, over which B is integral.*

Proof. As in the proof of Theorem 6.118, we consider a presentation of B and homogenize to get the ideal I_t. Since I_t is a prime ideal, Shimura's result implies our assertion. □

Note that since R is a Jacobson ring, the dimension of the generic fiber $K\otimes_R B$ is at most $\dim B - 1$. Thus in the Corollary, S is an arithmetic hypersurface.

Remark 6.121. Whenever the ring R contains an infinite field k, the Noether normalizations of an algebra $B = R[x_1,\ldots,x_n]$ can be chosen from generic k-linear combinations of the x_i. This depended only on Proposition 1.46 applied to appropriate Rees algebras. We can remove k entirely to obtain a similar finiteness statement in which the y_j are forms of (the same) high degree and in the same number by the following considerations. If B is a finitely generated standard graded algebra over the field k,

$$B = k\oplus B_1\oplus\cdots = k[B_1],$$

the Noether normalization process can be described differently. The argument is that establishing the Hilbert polynomial associated with B. More precisely, if B has Krull dimension 0, then $A = k$ is a Noether normalization. If $\dim B \geq 1$, one begins by choosing a homogeneous element $z\in B_+$ not contained in any of the minimal prime ideals of B. When k is infinite, the usual prime avoidance finds $z\in B_1$. If k is finite, however, one argues differently: If $P_1,\ldots,P_n$ are the minimal primes of B, we must find a form h not contained in any P_i. Suppose the best one can do is to find a form $f\notin \bigcup_{i\leq s} P_i$, for s maximum $< n$. This means that $f\in P_{s+1}$. If $g\in \bigcap_{i\leq P_i} \setminus P_{s+1}$ is a form (easy to show), then $h = f^a + g^b$, where $a = \deg g$, $b = \deg f$ is a form not contained in a larger subset of the associated primes, contradicting the definition of s.

Divisorial Extensions and Descending Chain Conditions

We observed in Theorem 6.57 the fact that divisorial extensions sitting between an affine algebra and its integral closure satisfy a descending chain conditions. A more general observation is:

Proposition 6.122. *Let R be a quasi-unmixed integral domain with a canonical module ω and let A be an integral domain of finite type over R. The set $S_2(A)$ of extensions $A\subset B\subset \overline{A}$ satisfying condition S_2 has the descending chain condition.*

Proof. We may replace R by A and preserve all the hypotheses. We shall view ω as an ideal of A. There is an embedding

$$S_2(A) \hookrightarrow \text{divisorial subideals of } \omega$$

given by

$$B \rightsquigarrow \mathrm{Hom}_A(B, \omega).$$

This map is self-dual on its image. Let

$$\omega = q_1 \cap q_2 \cap \cdots \cap q_n \supset L = \mathrm{Hom}_A(\overline{A}, \omega) = q_1' \cap q_2' \cap \cdots \cap q_n' \cap Q_1 \cap \cdots \cap Q_m$$

be primary decompositions, with $\sqrt{q_i} = \sqrt{q_i'}$, $i = 1, \ldots, n$. Since ω and L satisfy the condition S_2, their primary components all have codimension 1. For any $B \in S_2(A)$, the ideal $\mathrm{Hom}_A(B, \omega)$ contains L and therefore it will correspond to an ideal of the total quotient ring T of A/L. Since T is Artinian, this proves the assertion. □

Examples

Let A be an integral domain that is a finitely generated $\mathbb{Z}$-algebra,

$$A = \mathbb{Z}[x_1, \ldots, x_n]/I.$$

We shall consider the question of 'describing' the integral closure of A as intersection of other integrally closed algebras.

Let us illustrate with an example. Suppose that A is the hypersurface in $\mathbb{Z}[x, y, z]$ given by

$$f = xyz + x^3 + y^3.$$

Its Jacobian ideal (in $\mathbb{Z}[x, y, z]$),

$$L = (f, \frac{\partial f}{\partial x_i}, \quad i = 1, \ldots, n),$$

is generated by $f, yz + 3x^2, xz + 3y^2, xy$. This is an ideal of height ≥ 2 such that whenever a prime integer p is added, (L, p) will have height ≥ 3. Call this property $\mathcal{P}_2$.

Proposition 6.123. *Let A be a hypersurface ring, $A = \mathbb{Z}[x_1, \ldots, x_n]/(f)$, satisfying the condition $\mathcal{P}_2$. Then*

$$\overline{A} = \overline{A \otimes_{\mathbb{Z}} \mathbb{Q}} \bigcap_{i=1}^{n} A_{f_i}, \quad 0 \neq f_i = \frac{\partial f}{\partial x_i}.$$

Proof. The right-hand side is integrally closed since the A_{f_i} are smooth $\mathbb{Z}$-algebras. An element z in the intersection is conducted into the integral closure of A by an ideal L of A of height at least 2, so it is conducted by $L\overline{A}$ as well. Since $\overline{A}$ has the condition S_2, we must have $z \in \overline{A}$. □

This raises two issues: (i) how to carry out the operations, and (ii) what if $\mathcal{P}_2$ is violated? The latter occurs when some minimal prime of L of codimension two contains some prime number. If they occur at all, they will lie in the minimal primes of $(f, \frac{\partial f}{\partial x_i})$, for $\frac{\partial f}{\partial x_i} \neq 0$. Let $p_1, \ldots, p_s$ be the set of these primes, and set δ for their product. The proposition can then be applied to A_δ to obtain its integral closure. What are needed are the integral closures of the localizations $A_{(p_j)}$, $j = 1, \ldots, s$.

Example 6.124. Let A be the hypersurface ring defined by $f = 3x^{10} + 2y^{15} \in \mathbb{Z}[x,y]$. Its Jacobian ideal is $J = (f, 30x^9, 30y^{14})$. There are 4 prime ideals to be examined:

$$\mathfrak{p}_1 = (2,x), \mathfrak{p}_2 = (3,y), \mathfrak{p}_3 = (5, 3x^2 + 2y^3), \mathfrak{q} = (x,y).$$

A calculation shows that $A_{\mathfrak{p}_i}$, $i = 1,2,3$, is a DVR. Thus $\mathfrak{q}$ is the only codimension one singular prime. Set

$$B = \overline{A \otimes_{\mathbb{Z}} \mathbb{Q}} \cap A_{\mathfrak{p}_1} \cap A_{\mathfrak{p}_2} \cap A_{\mathfrak{p}_3}.$$

We claim that $B = \overline{A}$. Suppose that $z \in B$ and set C for the A-conductor of z into $\overline{A}$. We must show that C is not contained in any height one prime ideal of A. The only question is whether C is contained in $\mathfrak{q}$. Since however there exists some nonzero integer in the conductor, we have height $C \geq 2$, and therefore $z \in \overline{A}$.

6.7 Exercises

Exercise 6.125. Let R be a Noetherian ring and $\mathfrak{P}$ a maximal ideal of the polynomial ring $A = R[T_1, \ldots, T_n]$. Show that the ideal $R \cap \mathfrak{P}$ has dimension at most 1.

Exercise 6.126. Let A be an integral domain with a finite integral closure $\overline{A}$. Show that the S_2-ification of A is the ring

$$\bigcap A_{\mathfrak{p}}, \quad \text{height } \mathfrak{p} = 1.$$

Exercise 6.127 (H. Hironaka). Let R be a Noetherian domain with finite integral closure. Suppose that x belongs to the Jacobson radical. If $R/(x)$ is a normal domain, prove that R is normal. (Try your own solution before looking up Nagata's book.)

Exercise 6.128. Let A be a hypersurface ring of dimension d and degree e over a field of characteristic zero. Show that there is an elementary function $c(d,e)$, with $e(e-1)^2$ (or $(e-1)^2$ in the graded case) levels of exponentiation, bounding the degree-complexity for a Gröbner basis computation of the integral closure of the algebra.

Exercise 6.129. Let A be an affine domain over a field of characteristic zero and denote by J the Jacobian ideal. Define

$$\mathcal{P}(A) = \bigcup_{n \geq 1} \operatorname{Hom}_A(J^n, J^n).$$

Show that $\mathcal{P}$ is a proper operation in the sense above, that is, if $A \neq \overline{A}$ then $A \neq \mathcal{P}(A)$.

Exercise 6.130. Let A be an affine domain over a field of characteristic zero, denote by J the Jacobian ideal and by ω its canonical ideal. Define

$$\mathcal{P}(A) = \mathrm{Hom}_A(J\omega^{-1}, J\omega^{-1}).$$

Show that $A = \mathcal{P}(A)$ if and only if A is normal.

Exercise 6.131. Let E be a graded module over the polynomial ring R. Show that

$$\mathrm{tn}(\mathrm{Hom}_R(E,E)) = 0.$$

Exercise 6.132. Let A be a Gorenstein domain of Krull dimension 3 such that $\overline{A} = A[y]$, where y satisfies a monic equation of degree 2. Prove that $\overline{A}$ is a Cohen-Macaulay ring.

Exercise 6.133. Let A be a Noetherian ring and S one of its subrings. If $\mathfrak{N}$ is the nil radical of A and $A/\mathfrak{N}$ is a finitely generated S-module, prove that A is a finitely generated S-module. (In particular, by the Eakin-Nagata theorem, S is a Noetherian ring.)

7

Integral Closure and Normalization of Ideals

This chapter treats in detail several auxiliary constructions and devices to examine the integral closure of ideals, and to study the properties and applications to normal ideals. In Chapter 1, we introduced integrally closed ideals but did not examine the property that all the powers of an ideal I are integrally closed. A *non-direct* construction of the integral closure of an ideal can be sketched as follows. Let R be a normal domain and I an ideal. Its integral closure $\overline{I}$ is the degree 1 component of the integral closure of the Rees algebra of I:

$$I \rightsquigarrow \overline{R[It]} = R + \boxed{\overline{I}t} + \overline{I^2}t^2 + \cdots \rightsquigarrow \overline{I},$$

which we refer to as the *normalization* of I. This begs the question since the construction of $\overline{R[It]}$ takes place in a much larger setting, while in a *direct* construction $I \rightsquigarrow \overline{I}$ the steps of the algorithm would take place entirely in R. These are lacking in the literature. We provide the details of a simple situation, that of monomial ideals of finite colength.

The significant difference between the construction of the integral closure of an affine algebra A and that of $\overline{I}$ lies in the ready existence of *conductors*. Given A by generators and relations (at least in characteristic zero), the Jacobian ideal J of A has the property that

$$J \cdot \overline{A} \subset A,$$

in other words, $\overline{A} \subset A : J$. This fact lies at the root of all current algorithms for building $\overline{A}$. There is no known corresponding *annihilator* for $\overline{I}/I$. In several cases, one can cheat by borrowing part of the Jacobian ideal of $R[It]$ (or of $R[It,t^{-1}]$), proceeding as follows. Let (this will be part of the cheat) $R[It] = R[T_1,\ldots,T_n]/L$ be a presentation of the Rees algebra of I. The Jacobian ideal is a graded ideal

$$J = J_0 + J_1t + J_2t^2 + \cdots,$$

with the components obtained by taking selected minors of the Jacobian matrix. This means that to obtain some of the generators of J_i we do not need to consider all the generators of L. Since J annihilates $\overline{R[It]}/R[It]$, we have that for each i,

$$J_i \cdot \overline{I} \subset I^{i+1},$$

and therefore

$$\overline{I} \subset \bigcap_{i \geq 0} I^{i+1} : J_i.$$

Of course when using subideals $J_i' \subset J_i$, or further when only a few J_i' are used, the comparison gets overstated. Despite these obstacles, in a number of important cases, one is able to understand relatively well the process of integral closure and normalization of ideals. These include monomial ideals and ideals of finite colength in regular local rings. Even here the full panoply of techniques of commutative algebra must be brought into play.

In addition to the lack of direct algorithms for the construction $I \rightsquigarrow \overline{I}$, there are no general numerical benchmarks to tell one ideal from the other. It is fortunate that when the issue of normalization is considered, there are combinatorial indices grounded on Hilbert functions that can be deployed to distinguish subalgebras A such that

$$R[It] \subset A \subset \overline{R[It]},$$

in numerous cases of interest.

Among the themes treated here are:

- Hilbert functions and normal ideals
- Handling ideals of positive dimension
- Integral closure of monomial ideals
- Multiplicities as volumes
- Multiplicities as probabilities
- Normalization of ideals

We begin by focusing on normal ideals. Through the introduction of appropriate Hilbert functions, the filtration defined by the integral closures of the powers of the ideal I may help to measure the relationship between I and some of its reductions or its integral closure $\overline{I}$. In particular for a minimal reduction J it is of interest to examine the corresponding numbers in the equalities

$$\begin{aligned} I^{r+1} &= J \cdot I^r \\ \overline{I}^{s+1} &= I \cdot \overline{I}^s. \end{aligned}$$

The integers such as $r = \mathrm{r}_J(I)$ play a role in the study of the cohomology of the blowup of spec (R) along the closed set $V(I)$, while $s = s(I) = \mathrm{r}_I(\overline{I})$ is a guide for possible algorithmic approaches to the calculation of $\overline{I}$.

Another aspect of normal ideals is their role in the construction of the integral closure of an affine algebra A. Certain constructions of $\overline{A}$ make extensive use of integral closures of ideals. Some are discussed later in this chapter. To obtain numerical information about these issues, at least two tracks are offered: (i) The path of Hilbert functions and the examination of the role of multiplicities (ordinary or extended in the case of higher dimensional ideals) in the stabilization of the processes; (ii) the information that is contained in the syzygies or in the properties of directly-associated objects such as the Koszul homology modules. Since several proposed conjectural bounds will appear, a number of examples will have to be discussed in detail in order to prescribe the appropriate boundary counts.

Monomial ideals are treated next. While it is elementary to give a combinatorial description of the integral closure of a monomial ideal I, it is still not very efficient except when I has finite co-length. In such cases one can use the theory of multiplicities to understand integrality and formulate adequate algorithms. One of these, using a *Monte Carlo* simulation, is offered.

A major theme is opened up with the introduction of normalization indices of an ideal. These are the integers with the property that either $\overline{I^{n+1}} = I \cdot \overline{I^n}$ for $n \geq s(I)$, or that the integral closure of $R[It]$ (for R normal) is generated by elements of degree $\leq s_0(I)$. So $s(I) \leq \dim R - 1$ for a monomial ideal I. This will hold whenever the integral closure of $R[It]$ is Cohen-Macaulay. The general role of the depth of this algebra on these numbers is not yet well understood.

Another topic treated is the examination of the length of chains of graded algebras with the condition S_2 between $R[It]$ and its integral closure B. In general this is connected to some puzzling properties of the Hilbert functions (sometimes the extended Hilbert functions as introduced in Chapter 2) of B. The existence of specific theorems of Briançon-Skoda type often make a treatment possible.

7.1 Hilbert Functions and Integral Closure

We develop first general properties of the integral closure of a Rees algebra. To simplify the exposition and guarantee the existence of several constructions, we shall assume that the base ring R is a quasi-unmixed reduced ring.

Let $(R, \mathfrak{m})$ be a local ring and I an ideal. The integral closure filtration $\{\overline{I^n}, n \geq 0\}$ gives rise to the corresponding Rees algebra

$$\overline{\mathcal{R}} = \sum_{n \geq 0} \overline{I^n} t^n.$$

We shall denote by

$$\overline{\mathcal{G}} = \overline{\mathcal{R}} / \overline{\mathcal{R}}_+[+1] = \bigoplus_{n \geq 0} \overline{I^n} / \overline{I^{n+1}}$$

the corresponding associated graded ring.

Let us introduce some ambiguity in these definitions. The sense of 'integral closure' used above is the usual one. In Chapter 1, we also considered the 'integral closure of I in the ring.' The algebra that results from this relative notion is $\overline{\mathcal{R}} \bigcap R[t]$. Whenever this variant is used, we will make the point clear. The general rule is that the second notion is the default choice when R is not normal.

Ordinary Powers and their Integral Closure

Let R be a normal quasi-unmixed domain and I a nonzero ideal. Set $A = R[It]$ and let B be its integral closure, that is, B is the Rees algebra of the integral closure filtration $\{\overline{I^n}\}$. We examine the equality of some components of A and B.

Proposition 7.1. *Let A and B be as above, and consider the following natural exact sequence of finitely generated A-modules*

$$0 \to A \longrightarrow B \longrightarrow C \to 0.$$

The following assertions about the components of C hold.

(a) *For $n \gg 0$ either all $C_n \neq 0$ or all $C_n = 0$.*
(b) *If A satisfies Serre's condition S_2, then for any $n \geq 0$, if $C_n \neq 0$ then $C_{n+1} \neq 0$.*

Proof. (a) This follows simply from the fact that C is a finitely generated module over the standard graded algebra A, and therefore we have $C_{n+1} = A_1 C_n$ for $n \gg 0$.

(b) Let $P_1, \ldots, P_r$ be the associated prime ideals of C as an A-module, and set $L = R \cap P_1 \cap \cdots \cap P_r$. Note that $L \neq 0$ and some power L^s annihilates C. The ideal of A given by $K = (L, A_1)$ has height at least 2. If $C \neq 0$, since A has the S_2 condition, the K-grade of C is at least 1, which means that K is not contained in any of the prime ideals P_i. But L lies in each P_i so we must have $z \in A_1$ which avoids each associated prime of C. Multiplication by z gives rise to an embedding $C_n \to C_{n+1}$, which proves the claim. □

Hilbert Functions

There are several Hilbert functions attached to the algebra $\overline{R[It]}$. If I is $\mathfrak{m}$-primary, we have its ordinary Hilbert function

$$\overline{H}_I(n) = \lambda(\overline{I^n}/\overline{I^{n+1}}).$$

We shall examine two Hilbert functions, beginning with $\mathfrak{m}$-primary ideals. Let $(R, \mathfrak{m})$ be a normal, Cohen-Macaulay quasi-unmixed local ring and I an $\mathfrak{m}$-primary ideal. Let J be a minimal reduction of I (we may assume that the residue field of R is infinite, so that J is a complete intersection).

We can use the formalism of the Sally module construction of Chapter 2 by defining the module $\overline{S}_I$ via the exact sequence

$$0 \to \overline{I} \cdot R[Jt] \longrightarrow \overline{\mathcal{R}}_+[+1] = \sum_{n \geq 1} \overline{I^n} t^{n-1} \longrightarrow \overline{S}_I = \bigoplus_{n \geq 1} \overline{I^{n+1}}/\overline{I}J^n \to 0. \tag{7.1}$$

An identical calculation to that in Corollary 2.9 will establish the following formula for the Hilbert series:

Theorem 7.2. *Let $(R, \mathfrak{m})$ be a normal Cohen-Macaulay local domain of dimension $d \geq 1$, and I an $\mathfrak{m}$-primary ideal. The Hilbert series of the associated graded ring $\overline{\mathcal{G}}$ is given by*

$$H_{\overline{\mathcal{G}}}(t) = \frac{\lambda(R/J)\cdot t}{(1-t)^d} + \frac{\lambda(R/\overline{I})(1-t)}{(1-t)^d} - (1-t)H_{\overline{S}_I}(t)$$
$$= \frac{\lambda(R/\overline{I}) + \lambda(\overline{I}/J)\cdot t}{(1-t)^d} - (1-t)H_{\overline{S}_I}(t).$$

In particular, the function $H_{\overline{\mathcal{G}}}(t)$ is independent of the choice of the minimal reduction J.

This formula shows that the multiplicity of $\overline{\mathcal{G}}$ is equal to the multiplicity of I, as it is well-known ([Mo84]).

There are significant differences between the ordinary Sally module of the ideal I and the module $\overline{S}_I$. This arises from the fact (view the sequence as composed of finitely generated modules over $R[Jt]$) that (suppose $d \geq 2$), in the exact sequence (7.1) the module $\overline{I}R[Jt]$ is a maximal Cohen-Macaulay module and $\overline{\mathcal{R}}_+$ is a module with property S_2; consequently the Sally module $\overline{S}_I$ either vanishes or has the property that it satisfies S_2 as a module over $R[Jt]/\text{ann}(\overline{S}_I)$.

Corollary 7.3. *For every integer $n \geq 0$,*

$$\lambda(\overline{I^n}/\overline{I^{n+1}}) \leq \lambda(R/J)\binom{n+d-2}{d-1} + \lambda(R/\overline{I})\binom{n+d-2}{d-2}.$$

Proof. This follows when we ignore the contribution of $\overline{S}_I$. Indeed, in degree n the contribution of the term $-(1-t)H_{\overline{S}_I}(t)$ is $\lambda(S_{n-1}) - \lambda(S_n)$. Since $\overline{S}_I$ has depth > 0, its Hilbert function is non-decreasing. □

We are going to sharpen this when we take into account that $\overline{S}_I$ has condition S_2.

If $(R, \mathfrak{m})$ is a regular local ring of dimension two, and I is $\mathfrak{m}$-primary, Lipman and Teissier ([LT81]) show the following.

Theorem 7.4. *For every minimal reduction J of I, $\overline{I^n} = J(\overline{I})^{n-1}$ for all $n \geq 2$.*

Remark 7.5. Applied to the Sally module above, this shows that $\overline{S}_I = 0$, and we obtain the Hilbert function

$$\lambda(\overline{I^n}/\overline{I^{n+1}}) = \lambda(R/J)n + \lambda(R/\overline{I}) = \lambda(R/J)(n+1) - \lambda(\overline{I}/J).$$

In particular the multiplicity of the ideal I is given by

$$e(I) = \lambda(R/J) = \lambda(\overline{I}/\overline{I^2}) - \lambda(R/\overline{I}) = \lambda(R/\overline{I^2}) - 2\lambda(R/\overline{I}) \qquad (7.2)$$

Hilbert Coefficients

As $\overline{\mathcal{G}}$ is a finitely generated module over the standard graded algebra $\mathrm{gr}_J(R)$, it has a Hilbert polynomial

$$P_{\overline{\mathcal{G}}}(t) = \overline{e}_0(I)\binom{t+d-1}{d-1} - \overline{e}_1(I)\binom{t+d-2}{d-2} + \cdots + (-1)^{d-1}\overline{e}_{d-1}(I).$$

Note that $e_0(\overline{I}) = \overline{e}_0(I)$ but that $e_1(\overline{I})$ may be distinct from $\overline{e}_1(I)$. Taking into consideration the Hilbert coefficients $\overline{s}_i(I)$ of the Sally module $\overline{S}$, we obtain that

$$\begin{aligned} \overline{s}_0(I) &= \overline{e}_1(I) - \overline{e}_0(I) + \lambda(R/\overline{I}) \\ \overline{s}_i(I) &= \overline{e}_{i+1}(I), \quad i \geq 1. \end{aligned}$$

If $\overline{S}$ is nonzero it has dimension d, so we must have $\overline{e}_1(I) > \overline{e}_0(I) - \lambda(R/\overline{I})$. In the case of equality we have $\overline{I^p} = J^{p-1}\overline{I}$, which implies that $\overline{I}$ has reduction number at most 1, and $\overline{I^p} = \overline{I}^p$ for all p. In particular $\overline{I}$ will be a normal ideal.

Remark 7.6. One can also show the positivity of $\overline{e}_2(I)$. We might guess that unlike the case of $e_3(I)$, which may be negative, $\overline{e}_3(I)$ might be non-negative due in part by the fact that the Sally module $\overline{S}$ satisfies Serre's condition S_2.

We can already make comparisons between some of the coefficients of the two Hilbert polynomials, but let us consider an alternative approach. We do not assume that R is normal. Consider the exact sequences

$$0 \to \overline{I^n}/I^n \longrightarrow R/I^n \longrightarrow R/\overline{I^n} \to 0.$$

The left-hand module is the degree n component of the finitely generated module $L = \overline{\mathcal{R}}/\mathcal{R}$. Note that L is a module of dimension at most $d = \dim R$. To match the notation, we denote by $(d_1, d_2, \ldots)$ its normalized Hilbert coefficients (with signs attached). Comparing, we have:

Proposition 7.7. *Let $(R, \mathfrak{m})$ be a local ring of dimension $d \geq 1$ and I an $\mathfrak{m}$-primary ideal. Then*

$$\begin{aligned} \overline{e}_0(I) &= e_0(I) \\ \overline{e}_1(I) &= e_1(I) + d_1 \geq e_1(I). \end{aligned}$$

Moreover, if $\overline{e}_1(I) = e_1(I)$ then $\overline{\mathcal{R}}$ is the S_2-ification of $\mathcal{R}$.

Proof. If $d_1 \neq 0$, L must be a module of dimension d, and therefore $d_1 > 0$. Otherwise L is annihilated by an ideal of height at least 2, which proves the claim. □

Example 7.8. If $(R, \mathfrak{m})$ is a two-dimensional regular local ring, for any $\mathfrak{m}$-primary ideal I, Remark 7.5 implies that $\overline{e}_1 \leq \lambda(\overline{I}/J) \leq e_0(I) - \lambda(R/I)$.

Special Fiber

We introduce another Hilbert function attached to an ideal I.

Definition 7.9. Let $(R,\mathfrak{m})$ be a reduced quasi-unmixed local ring and I an $\mathfrak{m}$-primary ideal. Let $\overline{\mathcal{F}}(I)$ be the special fiber of $\overline{R[It]}$. For each positive integer n, set

$$\overline{\mathcal{F}}_I(n) = \nu(\overline{I^n}).$$

This function has some special properties, at least when $(R,\mathfrak{m})$ is a local ring and I is an $\mathfrak{m}$-primary ideal:

Proposition 7.10. *The Hilbert function $\overline{\mathcal{F}}_I(\cdot)$ is monotonic.*

Proof. Since $\overline{I^r}$ is an $\mathfrak{m}$-full ideal for each integer r, by Proposition 1.62 the minimal number of generators of $\overline{I^{r-1}}$ can be at most $\nu(\overline{I^r})$. □

Proposition 7.11. *For every ideal I, $\ell(I) = \ell(\bar{I})$.*

Proof. We recall that J is a reduction of I with $J \subset I$ if and only if the two ideals have the same integral closure. Thus if J is a minimal reduction of I, then $\overline{J} = \bar{I}$, so J is also a reduction of $\bar{I}$. On the other hand, if L is a minimal reduction of $\bar{I}$, with $L \subset J$, then $\overline{L} = \bar{I} \subset \overline{J}$. □

Derivations and Integral Closure

Recall that a derivation of a (commutative) ring R is an additive mapping $D : R \to R$ satisfying Leibniz's rule:

$$D(ab) = D(a) \cdot b + a \cdot d(b), \quad \forall a,b \in R.$$

A derivation D easily 'extends' to a derivation of a ring of fractions of R. To examine the role of derivations on the integral closure of R and on the integral closure of the powers of ideals, we first give a reformulation of integral closure.

Definition 7.12. Let R be a commutative ring. The complete integral closure of R is the set B of elements x of the total ring of fractions of R for which there is a regular element $d \in R$ such that

$$dx^n \in R, \quad \forall n \geq 1.$$

When R is Noetherian, this notion is clearly equivalent to the usual definition of integral closure.

Theorem 7.13. *Let R be a reduced Noetherian ring containing the rationals, with finite integral closure R', and let D be a derivation of R. Then D has an extension to a derivation of R'. Furthermore, if R' is finite over R, the conductor J of R' into R is a differential ideal.*

Proof. If we use the quotient rule for differentiation, D extends to a derivation of the total ring of fractions Q of R. The issue is whether R' is invariant under this extension (still denoted by D), for which purpose we employ a pretty construction of Seidenberg ([Sei66]).

We note that R' is a direct product of integrally closed domains. We may assume that R is a domain. Let z be an element of R'; it is clear that there exists a regular element d of R such that $d \cdot z^n \in R$ for all n; it suffices to consider n bounded by the degree of an integral equation of z with respect to R. Conversely, any element of Q with this property is integral over R.

Denote by $R[[t]]$ the ring of formal power series in the variable t over R. Since R' is Noetherian, $R'[[t]]$ is also Noetherian and is the integral closure of $R[[t]]$. Indeed, since

$$R' = \bigcap V,$$

V running over the discrete valuations of R, we have

$$R'[[t]] = \bigcap V[[t]],$$

where each $V[[t]]$ is a regular local ring and therefore normal. Set

$$\varphi(a) = \sum_{k \geq 0} \frac{D^k(a)}{k!} t^k, \quad \text{for } a \in Q$$
$$\varphi(t^k) = t^k.$$

Then φ extends to an isomorphism of $Q[[t]]$ (still denoted by φ) restricting to an isomorphism of $R[[t]]$. Applying it to an equality $d \cdot z^n = r \in R$, we get

$$\varphi(d) \cdot (\varphi(z))^n = \varphi(r),$$

which shows that $\varphi(z)$ is integral over $R[[t]]$, and therefore has all of its coefficients in R', proving the first assertion.

For the remainder, it suffices to notice that since $J \cdot R' \subset R$, we have $D(J) \cdot R' + J \cdot D(R') \subset R$, so $D(J)$ also conducts R' into R. □

The following application to Rees algebras arose from a discussion with B. Ulrich.

Theorem 7.14. *Let R be a Noetherian ring containing the rationals with finite integral closure, and let D be a derivation of R. For any ideal I and any positive integer r, $D(\overline{I^r}) \subset \overline{I^{r-1}}$.*

Proof. Let A be the extended Rees algebra of I, $A = R[It, t^{-1}]$. The derivation D defines a derivation of the field of fractions of $R[t]$ by setting $\delta = t^{-1}D$. Its restriction to A induces a derivation of A. According to Theorem 7.13, δ extends to a unique derivation Δ of the integral closure $\overline{A}$ of A. In particular, on the elements in $\overline{I^r}t^r$ we have

$$\Delta(\overline{I^r}t^r) \subset D(\overline{I^r})t^{r-1},$$

to prove the assertion. □

Corollary 7.15. *Let $R = k[x_1,\ldots,x_n]$ be a ring of polynomials over a field of characteristic zero. For any homogeneous ideal I and any positive integer r,*

$$\overline{I^r} \subset (x_1,\ldots,x_n)\overline{I^{r-1}}.$$

Proof. Consider the various partial derivatives and use Euler formula. (More generally one may assume that I is quasi-homogeneous.) □

Theorem 7.16. *Let R be a ring of polynomials $k[x_1,\ldots,x_d]$ with $d > 1$, over a field of characteristic zero, and let I be a homogeneous $\mathfrak{m}$-primary ideal, $\mathfrak{m} = (x_1,\ldots,x_d)$. Then the Hilbert function of $\overline{\mathcal{F}}_I$ is strictly monotonic.*

Proof. We use the fact that the ideals $\overline{I^n}$ are $\mathfrak{m}$-full. Let $x \in \mathfrak{m}$ be such that $\overline{I^{n+1}} = \mathfrak{m}\overline{I^{n+1}} : x$. The exact sequence

$$0 \rightarrow \overline{I^{n+1}}/\mathfrak{m}\overline{I^{n+1}} \longrightarrow \overline{I^n}/\mathfrak{m}\overline{I^{n+1}} \xrightarrow{\cdot x} \overline{I^n}/\mathfrak{m}\overline{I^{n+1}} \longrightarrow \overline{I^n}/(x\overline{I^n} + \mathfrak{m}\overline{I^{n+1}}) \rightarrow 0$$

shows that $\nu(\overline{I^{n+1}}) = \lambda(\overline{I^n}/(x\overline{I^n} + \mathfrak{m}\overline{I^{n+1}}))$. Since $\overline{I^{n+1}} \subset \mathfrak{m}\overline{I^n}$ by Corollary 7.15, the equality shows that we cannot have $\nu(\overline{I^n}) \geq \nu(\overline{I^{n+1}})$. In fact, one cannot have $\mathfrak{m}\overline{I^n}$ generated by $f_1\overline{I^n},\ldots,f_r\overline{I^n}$ for $r < d$, as otherwise $\mathfrak{m}$ would be integral over $(f_1,\ldots,f_r)$. In particular we have that $\nu(\overline{I^{n+1}}) \geq \nu(\overline{I^n}) + d - 1$. □

Associated Primes of Filtrations

The associated prime ideals of various filtrations defined on an ideal I were extensively studied by Ratliff and others. A detailed account, with proper attributions, can be found in [Mc83], which is noteworthy for its elegant proofs. We are going to recall two of its results ([Mc83, Propositions 3.9 and 4.1]) that work as a guidebook here. Let R be a Noetherian ring and I an ideal. Denote by $A^*(I)$ the set of associated primes of all the ideals $\overline{I^n}$, for all n.

Theorem 7.17. *$A^*(I)$ is a finite set.*

Theorem 7.18. *Let I be an ideal in a Noetherian ring R and $\mathfrak{p}$ a prime ideal containing I. If* height $\mathfrak{p} = \ell(I_\mathfrak{p})$*, then $\mathfrak{p} \in A^*(I)$. If $R_\mathfrak{p}$ is quasi-unmixed, then the converse is also true.*

As a teaser for the ideas that go into their proofs, we consider a special case.

Proposition 7.19. *Let $(R,\mathfrak{m})$ be a local domain and $I \subset R$ an ideal. Suppose that* $\mathfrak{m}$ *is not an associated prime of any power I^n of I. Then* $\mathfrak{m}$ *is not associated to any $\overline{I^n}$.*

Proof. (*Sketch*) By hypothesis, $\ell(I) < d = \dim R$. Suppose that $\mathfrak{m}$ is associated to some $\overline{I^n}$. Denote by G the associated graded ring corresponding to the filtration $\{\overline{I^n}, n \geq 1\}$. By hypothesis we have $\Gamma_\mathfrak{m}(G) \neq 0$: consider the exact sequence

$$0 \rightarrow \Gamma_\mathfrak{m}(G) \longrightarrow G \longrightarrow G' \rightarrow 0.$$

Since G is unmixed and equidimensional as a module $\Gamma_\mathfrak{m}(G)$ has the same dimension as G. Since

$$\dim \Gamma_\mathfrak{m}(G) \leq \dim G/\mathfrak{m}G = \ell(G) = \ell(I) < d,$$

we have a contradiction. □

Number of Generators of the Integral Closure

Let R be a local ring of dimension n, let $I=(f_1,\ldots,f_m)$ be an ideal and denote by $\overline{I}$ its integral closure. A broad question: Is there an estimate of the number of generators of $\overline{I}$ in terms of m,n and the multiplicities of I? There seems to be very little progress on this issue in general. Even the case of rings of polynomials shows scant advance.

We can use a technique employed in Chapter 2 to obtain overall bounds for the Hilbert function of $\overline{\mathcal{F}}_I$.

Theorem 7.20. *Let $(R,\mathfrak{m})$ be a Cohen-Macaulay local ring of dimension $d\geq 1$.*

(a) *If I is an $\mathfrak{m}$-primary ideal, then for all $n\geq 0$,*

$$\nu(\overline{I^n})\leq e(I)\binom{n+d-2}{d-1}+\binom{n+d-2}{d-2}.$$

(b) *If I is an equimultiple ideal of dimension 1 and $d\geq 2$, then for all $n\geq 0$,*

$$\nu(\overline{I^n})\leq \deg(R/I)\binom{n+d-3}{d-2}+\binom{n+d-3}{d-3}.$$

Proof. (a) The proof here is similar to that of Theorem 2.44. Let $J=(a_1,\ldots,a_{d-1},a_d)$ be a minimal reduction of I and set $J_0=(a_1,\ldots,a_{d-1})$ in order to consider the estimate arising from the embedding

$$\overline{I^n}/J_0^n\hookrightarrow R/J_0^n.$$

We use the multiplicity of R/J_0^n to bound the number of generators of the Cohen-Macaulay ideal $\overline{I^n}/J_0^n$:

$$\nu(\overline{I^n})\leq \deg(R/J_0^n)+\nu(J_0^n),$$

to obtain the desired bound, since $\deg(R/J_0)\leq e(I)$.

(b) There is a minor variation with this case. Suppose that $J=(a_1,\ldots,a_{d-2},a_{d-1})$ is a minimal reduction. By Theorem 7.18, the ideals $\overline{I^n}$ do not have $\mathfrak{m}$ as one of the associated primes. This implies that if now we set $J_0=(a_1,\ldots,a_{d-2})$, then $\overline{I^n}/J_0^n$ are Cohen-Macaulay ideals of the Cohen-Macaulay ring R/J_0^n of dimension 2. We may then proceed as in case (a) to derive the stated bound. □

Reduction Number of a Filtration

The integral closure filtration presents a challenging problem regarding its reduction number. Let us formulate what might be the simplest instance.

Question 7.21. Let $(R,\mathfrak{m})$ be a Cohen-Macaulay local ring of dimension d and I an $\mathfrak{m}$-primary ideal of multiplicity $e=e(I)$. Is there a quadratic polynomial $f(d,e)$ such that the reduction number of the integral closure filtration satisfies the inequality $r\leq f(d,e)$?

7.2 Monomial Ideals

In this section we study the role of monomial ideals in the computation of multiplicities. This is a required undertaking for reasons of efficiency, derived from beautiful formulas of convex geometry and the understanding they provide. Our treatment here is based on [DT2].

We begin our discussion by recalling some general facts about the integral closure of monomial ideals. Let $R = k[x_1,\ldots,x_n]$ be a ring of polynomials over a field k. Its monomials are determined by exponent vectors,

$$\mathbf{x}^v = x_1^{a_1}\cdots x_n^{a_n}, \quad v = (a_1,\ldots,a_n) \in \mathbb{N}^n.$$

Let I be the ideal generated by a set of monomials $\mathbf{x}^{v_1},\ldots,\mathbf{x}^{v_m}$. The exponent vectors $v_i \in \mathbb{N}^n$ and their geometry will determine $\overline{I}$.

Since it is easy to show that $\overline{I}$ is also generated by monomials, our problem is simpler. If $\mathbf{x}^v \in \overline{I}$, it will satisfy an equation

$$(\mathbf{x}^v)^\ell \in I^\ell,$$

and therefore we have the following equation for the exponent vectors:

$$\ell\cdot v = u + \sum_{i=1}^m r_i\cdot v_i, \quad r_i \geq 0, \quad \sum_{i=1}^m r_i = \ell.$$

This means that $v = \dfrac{u}{\ell} + \alpha$, where α belongs to the convex hull $\operatorname{conv}(v_1,\ldots,v_m)$ of $v_1,\ldots,v_m$. The vector v can be written as (set $w = \frac{u}{\ell}$)

$$v = \lfloor w \rfloor + (w - \lfloor w \rfloor) + \alpha,$$

and it is clear that the integral vector

$$v_0 = (w - \lfloor w \rfloor) + \alpha$$

also has the property that $\mathbf{x}^{v_0} \in \overline{I}$.

Proposition 7.22. *Let I be the ideal generated by the monomials $\mathbf{x}^{v_1},\ldots,\mathbf{x}^{v_m}$. Then $\overline{I}$ is generated by $\mathbf{x}^v$, where v lies either on the integral convex hull of the v_i or lies very near the rational convex hull, that is, it differ from a vector in the convex hull by another vector with positive entries that are less than* 1.

In symbols, if C denotes the rational convex hull of $V = \{v_1,\ldots,v_m\}$ and

$$B = [0,1)\times\cdots\times[0,1) = [0,1)^n,$$

then $\overline{I}$ is generated by the monomials x^v with

$$v \in (C+B)\bigcap \mathbb{N}^n = \lceil C \rceil.$$

For simplicity we denote this set by $B(V)$.

We hope to make use of such descriptions of $\overline{I}$ in order to obtain bounds on the number of generators of $\overline{I}$; $B(V)$ itself is not a very straightforward approach to the calculation of the integral closure of monomial ideals.

These observations highlight one of the difficulties of finding $\overline{I}$ from I: the ideal I_c generated by the integral points of C may not be equal to $\overline{I}$; see Example 7.32. A simple example was pointed out to us by R. Villarreal:

$$\begin{aligned} I &= ((4,0),(0,11),(2,6)) \\ I_c &= ((4,0),(0,11),(2,6),(3,3)) \\ \overline{I} &= ((4,0),(0,11),(2,6),(3,3),(1,9)) \end{aligned}$$

Example 7.23. Suppose that $I = J + Lx_n$, in which J and L are monomial ideals in the first $n-1$ variables. In this case,

$$\overline{I} = \overline{J} + \overline{(J+L)x_n}.$$

Indeed, suppose that J is defined by the exponent vectors $V = \{v_1,\ldots,v_p\}$ and L by $U = \{u_1,\ldots,u_q\}$. Let $\mathbf{x}^v$ be a generator of $\overline{I}$:

$$v = \sum r_i v_i + \sum s_j u_j + (\sum_j s_j)e_n + w, \quad w \in B.$$

If $\sum r_i = 1$, then $\mathbf{x}^v \in \overline{J}$. Suppose otherwise, that is, $\sum_j s_j > 0$. In this case, writing $w = w' + (1 - \sum_j s_j)e_n$, where $w' \in [0,1)^{n-1} \times \{0\}$, we clearly have that

$$\sum_i r_i v_i + \sum_j s_j u_j + w' \in B(V \cup U),$$

which shows that $\mathbf{x}^v \in \overline{(J+L)x_n}$, thus proving the assertion.

Example 7.24. To see how this description leads to bounds on the number of generators, consider the following reduction. Let $V = \{v_1,\ldots,v_m\}$ be a set of exponent vectors defining the monomial ideal I of $k[x_1,\ldots,x_n]$. Define $B(V)$ as above, and denote the number of integer vectors in $B(V)$ by $b(V)$. The estimate $\nu(\overline{I}) \leq b(V)$ is unfortunately too large. We want to argue that

$$\nu(\overline{I}) \leq b(V'),$$

where V' is obtained from V as follows. Split the vectors v_s according to the rule: fix one variable, say x_n. Let g be the least nonzero component (we may suppose some exist) of all v_i with respect to e_n (the last base vector). We split the v_s into two groups, v_i, which have 0 as the e_n-coordinate, and $v_j^{\cdot} = u_j' + ge_n$. Now set $V' = v_i$'s, u_j''s.

Suppose that $v \in B(V)$ is an exponent of a minimal generator of I. Say

$$v = \sum r_i v_i + \sum s_j(u_j' + ge_n) + w, w \in B.$$

Consider the vector $(\sum s_j)ge_n + w$. If $\sum s_j + w_n < 1$, then $v \in B(V')$. If not, $v = v' + qe_n$, where $v' \in B(V')$. If v' is the exponent vector of a minimal generator of $\overline{I}$, it is already counted in $B(V')$. Otherwise this says that some unique element $v' + ae_n$ is a minimal generator. We make it correspond to v' in the counting which will complete the assertion. Of course this is just a different kind of bound, not necessarily better!

Another way to describe the integral closure of a monomial ideal is ([Ei95, Exercise 4.23]):

Proposition 7.25. *Suppose that $R = k[x_1, \ldots, x_n]$ and that I is generated by the set of monomials $\mathbf{x}^{v_1}, \ldots, \mathbf{x}^{v_m}$. Let Γ be the set of exponents of monomials in I,*

$$\Gamma = \bigcup_{i=1}^{m} v_i + \mathbb{N}^n.$$

Regarding Γ as a subset of $\mathbb{R}^n_+$, let Λ be the convex hull of $\mathbb{R}^n_+ + \Gamma$, and let Γ^* be the set of integral points in Λ. Then $\overline{I}$ is the ideal generated by the monomials $\mathbf{x}^v$ with $v \in \Gamma^*$.

Remark 7.26. Both general descriptions of the integral closure of a monomial ideal I of $k[x_1, \ldots, x_n]$ highlight the fact that $\overline{I}$ is independent of the field k.

Monomial Subrings and Rees Algebras

The Rees algebra of a monomial ideal I is a special monomial subring of a polynomial ring $k[x_1, \ldots, x_d, t]$. The presence of the x_i makes for a simplifying factor in the discussion of the integral closure. Let us first quote the broad brush description of the integral closure of monomial subrings (see [Vi1, Theorem 7.2.28]).

Theorem 7.27. *Let F be a finite set of monomials of the ring $R = k[x_1, \ldots, x_d]$. The integral closure of the monomial subring $k[F]$ is given by*

$$\overline{k[F]} = k[\{\mathbf{x}^v \mid v \in \mathbb{R}_+\mathcal{A} \cap \mathbb{Z}\mathcal{A}\}],$$

where $\mathcal{A}$ is the set of exponent vectors of F.

If $F = \{v_1, \ldots, v_m\}$ and $I = (\mathbf{x}^{v_1}, \ldots, \mathbf{x}^{v_m}) \subset R = k[x_1, \ldots, x_d]$, then by the theorem,

$$\overline{R[It]} = k[\{\mathbf{x}^v t^b \mid (v, b) \in \mathbb{R}_+\mathcal{A} \cap \mathbb{Z}\mathcal{A}\}],$$

with

$$\mathcal{A} = \{(v_1, 1), \ldots, (v_m, 1), e_1, \ldots, e_d\}, \tag{7.3}$$

where e_i are the exponent vectors of the indeterminates of R. Note that $\mathbb{Z}\mathcal{A} = \mathbb{Z}^{d+1}$.

A simple but useful observation is that if $\mathbf{x}^\alpha t^b \in \overline{R[It]}$, then

$$(\alpha, b) = \sum_i a_i(v_i, 1) + \sum_j b_j e_j, \quad a_i, b_j \in \mathbb{R}_+,$$

and some $a_i > 1$, we can peel away from $\mathbf{x}^{\alpha}t^b$ a factor $\mathbf{x}^{v_i}t$,

$$\mathbf{x}^{\alpha}t^b = \mathbf{x}^{v_i}t \cdot \mathbf{x}^{\beta}t^{b-1}, \quad \mathbf{x}^{\beta}t^{b-1} \in \overline{R[It]}.$$

To obtain a first uniform bound we recall a classical result ([Ew96, Theorem 2.3]).

Theorem 7.28 (Carathéodory). *Let $\alpha_1,\ldots,\alpha_m$ be a set of vectors of $\mathbb{R}^n$, not all zero. If $\alpha \in \mathbb{R}_+\alpha_1+\cdots+\mathbb{R}_+\alpha_m$ there is a linearly independent set $\mathcal{A} \subset \{\alpha_1,\ldots,\alpha_m\}$ such that $\alpha \in \mathbb{R}_+\mathcal{A}$.*

Proposition 7.29. *Let I be a monomial ideal of $R = k[x_1,\ldots,x_d]$. Then*

$$\overline{I^n} = I\overline{I^{n-1}}, \quad n \geq d+1.$$

Proof. Using the notation above, suppose that $\mathbf{x}^{\alpha}t^b$ is a minimal generator of $\overline{R[It]}$. Applying Carathéodory's theorem to the exponent vectors in (7.3), we get

$$(\alpha, b) = \sum_i a_i(v_i, 1) + \sum_j b_j e_j, \quad a_i, b_j \in \mathbb{R}_+,$$

in which the number of nonzero a_i, b_j is at most $d+1$. Furthermore, if any a_i is greater than 1, we could peel away one factor in $\overline{R[It]}$, contrary to the minimality assumption. Thus each $a_i < 1$, and $b = \sum a_i < d+1$. □

A refined analysis by R. Villarreal shows that it suffices to set $n \geq d$, a result that is proved in Theorem 7.58 using another approach.

General Bounds

A first bound arises directly from Proposition 7.22.

Proposition 7.30. *Let I be a monomial ideal of $k[x_1,\ldots,x_n]$ generated by monomials of degree at most d. Then $\overline{I}$ is generated by monomials of degree at most $d+n-1$.*

Grosso modo this solves the complexity problem for the integral closure of a monomial ideal I by putting a bounding box around $\overline{I}$.

In general, for a given set $V = \{v_1,\ldots,v_m\} \subset \mathbb{N}^n$ of exponent vectors of degrees at most d, we would like to have bounds of the kind

$$\nu(\overline{(\mathbf{x}^{v_1},\ldots,\mathbf{x}^{v_m})}) \leq f(n,d,s),$$

for some polynomial $f(\cdot)$, where $s = \dim \operatorname{conv}(V)$. Other bounds might depend on more refined characteristics of $\operatorname{conv}(V)$. Observe that Proposition 7.30 leads to a polynomial bound of degree n when applied to the powers of I, which is not very satisfactory since the corresponding Hilbert polynomial will have degree at most $n-1$.

The bound of Proposition 7.30 might be far from sharp. One would expect a bound dependent upon other properties of the ideal. We are going to consider an observation by P. Gimenez and a contrasting example of R. Villarreal.

Remark 7.31. Suppose that I has dimension zero, and that is generated by monomials of degree at most d. This means that all x_i^d lie in I, so that

$$\overline{(x_1^d,\ldots,x_n^d)} = (x_1,\ldots,x_n)^d \subset \overline{I},$$

and therefore $\overline{I}$ does not require minimal generators of degree $d+1$.

Example 7.32. Let I be the ideal of the ring $R = k[x_1,\ldots,x_8]$ of polynomials defined by the monomials given via exponent vectors $v_1,\ldots,v_8$:

$$\begin{matrix} 1&1&1&1&1&0&0&0\\ 1&1&1&1&0&1&0&0\\ 1&1&1&1&0&0&1&0\\ 1&1&1&1&0&0&0&1\\ 1&0&0&0&1&1&1&1\\ 0&1&0&0&1&1&1&1\\ 0&0&1&0&1&1&1&1\\ 0&0&0&1&1&1&1&1 \end{matrix}$$

Set $L = I^3$ and consider the vector $v = (2,\ldots,2)$. In view of the equality

$$8v = (1,\ldots,1) + 3(v_1 + \cdots + v_8),$$

we have

$$v = (\frac{1}{8}(3v_1) + \cdots + \frac{1}{8}(3v_8)) + (\frac{1}{8},\ldots,\frac{1}{8}),$$

which shows that $\mathbf{x}^v$ lies in the integral closure of L; note that this monomial has degree 16 while L is generated by monomials of degree 15. Since the vectors v_i are linearly independent, one can easily check with *Maple* that decrementing v in any coordinate by 1,

$$v = v_0 + \varepsilon,$$

produces elements that do not lie in the convex hull of $\{3v_1,\ldots,3v_8\}$, that is, $v_0 \notin \overline{L}$. It follows that $\overline{L}$ requires some minimal generators of degree at least 16.

$\mathfrak{m}$-full Monomial Ideals

In practice, the condition for an ideal to be $\mathfrak{m}$-full is hard to set up since x must be a generic element in the maximal ideal. We consider one case where there are automatic choices for x.

Proposition 7.33. *Let I be a monomial ideal of the polynomial ring $R = k[x_1,\ldots,x_d]$. Then I is $\mathfrak{m}$-full if and only if*

$$(x_1,\ldots,x_d)\cdot I : (x_1 + \cdots + x_d) = I.$$

Proof. There is no harm in assuming that k is an infinite field. For I to be $\mathfrak{m}$-full means that there is a generic linear form $f = a_1x_1 + \cdots + a_dx_d$ such that $\mathfrak{m}I : f = I$. We can assume that $a_i \neq 0$ for all i.

Consider the automorphism of R defined by

$$\varphi : R \longrightarrow S, \quad \varphi(x_i) = a_i^{-1}x_i,$$

where we denote by S a copy of R. Using φ, we make S into an R-module. It is actually a faithfully flat R-module. Tensoring the expression $\mathfrak{m}I : f$ by S, we get $\mathfrak{m}I : \Delta$, where $\Delta = x_1 + \cdots + x_d$, since $\mathfrak{m}$ and I are invariant under φ. The faithful flatness condition ensures the rest of the assertion. □

Remark 7.34. Conversations with I. Bermejo and P. Gimenez led to the following approach to finding elements in the integral closure of a monomial ideal I. Denote by $\overline{I}$ the integral closure of I. Since $\overline{I}$ is a monomial ideal (and $\mathfrak{m}$-full), we have

$$L = \mathfrak{m}I : \Delta \subset \mathfrak{m}\overline{I} : \Delta = \overline{I}.$$

This means that all the monomials of the polynomials in L are in $\overline{I}$. Collecting these monomials from a set of generators of L, we get a monomial ideal L_0 such that

$$I \subset L \subset L_0 \subset \overline{I}.$$

The process can be iterated until it stabilizes, which may occur before $\overline{I}$ is reached however. On the other hand, since $\overline{I}$ is J-full for any nonzero ideal J we may use also any prime monomial.

7.3 Multiplicities and Volumes

Let $f_1 = \mathbf{x}^{v_1}, \ldots, f_r = \mathbf{x}^{v_r}$ be a set of monomials generating the ideal I. The integral closure $\overline{I}$ is generated by the monomials whose exponents have the form

$$\sum_{i=1}^{r} r_iv_i + \varepsilon \subset \mathbb{N}^n$$

such that $r_i \geq 0$ and $\sum r_i = 1$ and ε is a positive vector with entries in $[0,1)$.

Consider the integral closure of I^m. According to the valuative criterion, $\overline{I^m}$ is the integral closure of the ideal generated by the mth powers of the f_i. This means that the generators of $\overline{I^m}$ are defined by the exponent vectors of the form

$$\sum r_imv_i + \varepsilon,$$

with r_i and ε as above. We rewrite this as

$$m(\sum r_iv_i + \frac{\varepsilon}{m}),$$

so the vectors enclosed must have denominators dividing m.

This is the first resemblance to Ehrhart polynomials. Suppose that I is of finite co-length, so that $\lambda(R/\overline{I})$ is the number of lattice points 'outside' the ideal. The convex hull $C(V)$ of the v_i's partitions the positive octant into 3 regions: an unbounded connected region, $C(V)$ itself, and the complement $\mathcal{P}$ of the union of other two. It is this bounded region that is most pertinent to our calculation (see also [Sta86, p. 235], [Te88]).

To deal with $\overline{I^m}$ we are going to use the set of vectors v_i, but change the scale by $1/m$. This means that each I^m determines the same $\mathcal{P}$. We get that the number of lattice points is the length ℓ_m of $R/\overline{I^m}$. Note that $\ell_m(1/m)^n$ is the volume of all these hypercubes, and therefore it is a Riemann sum of the volume of $\mathcal{P}$. Thus the limit is just the volume of the region defined by the convex hull of the v_i. This number, multiplied by $n!$, is also the multiplicity of the ideal. This is the second connection with the Ehrhart polynomials. We want to emphasize that the volume of $\mathcal{P}$ may differ from $\lambda(R/I)$, $\lambda(R/\overline{I})$, or even from $\lambda(R/I_c)$, for the ideal generated by the lattice points in the convex hull of V.

Let us sum up some of these relationships between multiplicities and volumes of polyhedra.

Proposition 7.35. *Let I be a monomial ideal of the ring $R = k[x_1,\ldots,x_n]$ generated by $\mathbf{x}^{v_1},\ldots,\mathbf{x}^{v_m}$. Suppose that $\lambda(R/I) < \infty$. If $\mathcal{P}$ is the region of $\mathbb{N}^n$ defined as above by I, then*

$$\boxed{e(I) = n!\cdot \operatorname{Vol}(\mathcal{P}).} \tag{7.4}$$

Example 7.36. We consider an example of an ideal of $k[x,y,z]$. Suppose that

$$I = (x^a, y^b, z^c, x^{\alpha}y^{\beta}z^{\gamma}), \quad \frac{\alpha}{a}+\frac{\beta}{b}+\frac{\gamma}{c} < 1.$$

The inequality ensures that the fourth monomial does not lie in the integral closure of the other three. A direct calculation shows that

$$e(I) = ab\gamma + bc\alpha + ac\beta.$$

It is easy to see that if $a \geq 3\alpha$, $b \geq 3\beta$ and $c \geq 3\gamma$, the ideal $J = (x^a - z^c, y^b - z^c, x^{\alpha}y^{\beta}z^{\gamma})$ is a minimal reduction and that the reduction number satisfies the inequality $\mathrm{r}_J(I) \leq 2$.

A Probabilistic Approach to Multiplicities of Monomial Ideals

We observe that $\mathcal{P}$ is not a polyhedron, but it can be expressed as the difference between two polyhedra directly determined by the set of exponents vectors defining I, namely $V = \{v_1, \ldots, v_m\}$, where $v_i \neq 0$ for all i. Since I has finite co-length, suppose that the first n exponent vectors correspond to the generators of $I \cap k[x_i]$, $i = 1, \ldots, n$. Let Δ be the simplex defined by these vectors, $\Delta = C(v_1, \ldots, v_n)$, and denote by $\mathcal{P}_0$ the convex hull of V. We note that

$$\mathcal{P} = \Delta \setminus \mathcal{P}_0,$$

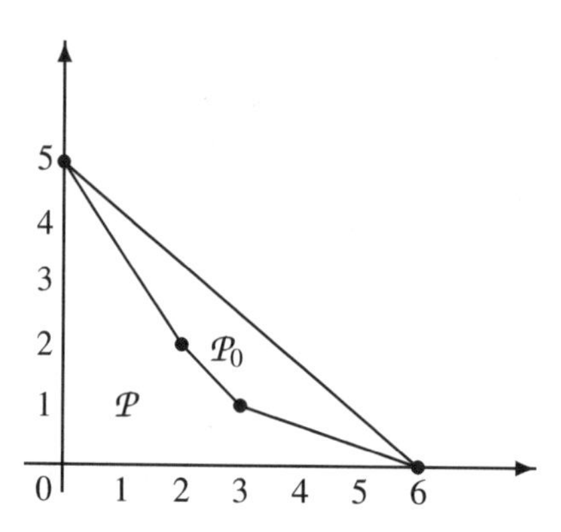

and therefore

$$\mathrm{Vol}(\mathcal{P}) = \mathrm{Vol}(\Delta) - \mathrm{Vol}(\mathcal{P}_0) = \frac{|v_1| \cdots |v_n|}{n!} - \mathrm{Vol}(\mathcal{P}_0).$$

There is an extensive literature on the computation of volumes of polyhedra. [BEK0] has a detailed discussion of several of these methods. The associated costs of the various methods depend on how the convex sets are represented. They often require conversion from one representation to another.

We will now introduce a *Monte Carlo* simulation approach for the calculation of $\mathrm{Vol}(\mathcal{P}_0)$ as a fraction of $\mathrm{Vol}(\Delta)$. We note that Δ is defined by the inequalities

$$\Delta: \quad \frac{x_1}{|v_1|} + \cdots + \frac{x_n}{|v_n|} \leq 1, \quad x_i \geq 0. \tag{7.5}$$

In its turn, $\mathcal{P}_0$ is the convex hull of the vectors v_i, $i = 1, \ldots, m$. According to [Ch0, pp. 284-285], there are linear programming techniques to convert the description of $\mathcal{P}_0$ as $C(v_1, \ldots, v_m)$ into an intersection of halfspaces,

$$\mathcal{P}_0: \quad \mathbb{A} \cdot \mathbf{x} \leq \mathbf{b}. \tag{7.6}$$

Proposition 7.37. *Let I be a monomial ideal of finite colength generated by the monomials $\mathbf{x}^{v_1}, \ldots, \mathbf{x}^{v_m}$. With the notation above, we have*

$$e(I) = (1 - p)|v_1| \cdots |v_n|.$$

Measuring Accuracy

We shall now give a flavor of how accuracies are expressed under these approaches. Our statistical approach is based on classical *Monte Carlo* quadrature methods ([Shr66]). By sampling a very large number of points in Δ and checking when they lie in $\mathcal{P}_0$–both of which are made viable by the rapidity that their descriptions above permit.

Essentially, the proposal consists of a series of N independent trials and the associated count H of hits, and of using the frequency $\frac{H}{N}$ as an approximation for p. According to basic probability theory, these approximations come with an attached probability in the sense that for small $\varepsilon > 0$, the probability

$$\text{Probability}\left\{\left|\frac{H}{N} - p\right| < \varepsilon\right\}$$

is high. The estimation is based on Chebyshev's inequality ([Fe67, p. 233]), which we recall for the convenience of the reader.

Theorem 7.38. *Let X be a random variable with finite second moment $E(X^2)$. For any $t > 0$,*

$$P\{|X| \geq t\} \leq t^{-2}E(X^2).$$

In particular, for a variable X with mean $E(X) = \mu$ and finite variance $Var(X)$,

$$P\{|X - \mu| \geq t\} \leq t^{-2}Var(X) \tag{7.7}$$

for any $t > 0$.

For a set of N independent trials $x_1, \ldots, x_N$ of probability p, the random variable we are interested in is the average number of hits:

$$X = \frac{x_1 + \cdots + x_N}{N} = \frac{H}{N}.$$

We have $E(X) = p$ and $\text{Var}(X) = \sqrt{\frac{p(1-p)}{N}}$. If we set $\varepsilon = t^{-2}\text{Var}(X)$ and substitute into (7.7), we obtain the inequality

$$P\left\{\left|\frac{H}{N} - p\right| < \sqrt{\frac{p(1-p)}{\varepsilon N}}\right\} > 1 - \varepsilon.$$

Since $p(1-p) \leq \frac{1}{4}$, it becomes easy to estimate the required number of trials to achieve a high degree of confidence. Thus, for instance, a crude application shows that in order to obtain a degree of confidence of 0.95, with $\varepsilon = 0.02$, the required number of trials should be at least 12,500. (Actually, a refined analysis, using the law of large numbers, cuts this estimate by $\frac{4}{5}$.) What these considerations do not hide is the remarkable independence of the method from dimension considerations.

Example 7.39. Let us indicate by an example how this will be accomplished. It begins with a conversion that uses PORTA (see [Cr97]), a collection of transformation techniques in linear programming.

```
The points defining the convex hull must be written
in a file with the extension .poi [say mult1.poi]
and the routine ''traf'' is called traf mult1.poi
The content of mult1.poi is:
DIM = 4
CONV_SECTION
4 0 0 0
0 5 0 0
0 0 6 0
0 0 0 7
1 2 1 2
2 0 2 1
0 2 1 2
1 1 1 1
END
The output file is the desired set of linear
inequalities and it is put in the file mult1.poi.ieq:
DIM = 4
VALID
1 1 1 1
INEQUALITIES_SECTION
(  1) -35x1-28x2-57x3-20x4 <= -140
(  2) -35x1-23x2-14x3-12x4 <=  -84
(  3) -25x1-12x2-10x3-13x4 <=  -60
(  4) -15x1-12x2-10x3-23x4 <=  -60
(  5) -11x1-18x2- 7x3- 6x4 <=  -42
(  6) -12x1- 7x2-11x3- 5x4 <=  -35
(  7) - 7x1-12x2- 5x3- 4x4 <=  -28
(  8) - 3x1- 5x2- 2x3- 2x4 <=  -12
(  9) -  x1                <=    0
( 10)      -   x2          <=    0
( 11) +30x1+24x2+20x3+11x4 <=  120
( 12) +35x1+28x2+ 9x3+20x4 <=  140
( 13) +42x1+25x2+28x3+24x4 <=  168
( 14) +31x1+42x2+35x3+30x4 <=  210
END
```

Example 7.40. We shall illustrate an application of the probabilistic method for the calculation of multiplicity. Suppose that $I = (x^3, y^4, z^5, w^6, xyzw)$; Proposition 7.35 gives $e(I) = 342$. To apply the probabilistic method, the exponents are written into a matrix and PORTA is used to obtain the inequalities defining the convex hull. The PORTA input and output are recorded below.

```
DIM = 4
CONV_SECTION
3 0 0 0
0 4 0 0
0 0 5 0
0 0 0 6
1 1 1 1
END
DIM = 4
VALID
1 1 1 1

INEQUALITIES_SECTION
(  1) -23x1-15x2-12x3-10x4 <= -60
(  2) -20x1-15x2-12x3-13x4 <= -60
(  3) -10x1- 9x2- 6x3- 5x4 <= -30
(  4) - 4x1- 3x2- 3x3- 2x4 <= -12
(  5) +20x1+15x2+12x3+10x4 <=  60

END
```

A C program is then used to calculate the probability. Testing with 10 million points gives a probability of 0.05018 and a multiplicity of 341.93520. A test using 500000 points gives a probability of 0.05073 and a multiplicity of 341.73720, whereas a test with 5000 points gives a probability of 0.04839 and a multiplicity of 342.577080.

We observe that the C program itself makes near nil demand on resources, being simply a hit counter. The PORTA routine is also fast, even in much larger examples. It leads to other benefits as well, that are integrated into a package (Polyprob). Polyprob is clearly much better than *Macaulay2* on even medium problems, but *Vinci*-a well known program to compute volumes of polyhedra-appears to be the best (see [BEK0]). However, we note that polyprob will work as accurately if we give it an ideal of the form $(x_1^{\alpha_1},\dots,x_n^{\alpha_n},f_1,\dots f_n)$, where at least one f_i lies in the integral closure of the ideal $(x_1^{\alpha_1},\dots,x_n^{\alpha_n})$, but *Vinci* will fail to give the correct multiplicity in this case. Recently an option for the computation of multiplicities of ideals has been added to *Normaliz*.

Remark 7.41. An effective method to compute the multiplicity of monomial ideals of finite colength is given in [BA04, Theorem 5.1]: It determines an explicit minimal reduction, which is usually not monomial.

POLYPROB Implementation

The fundamental operation of the POLYPROB algorithm is a random trial: that is, generating a random vector within the simplex containing the polytope, and testing whether the vector is in the polytope. Thus, POLYPROB requires an efficient way of getting random vectors uniformly distributed over a simplex. To see how to do this, first consider the general problem of generating a vector $(x_1,\ldots,x_n)$ uniformly distributed over an n-dimensional polytope $\mathcal{P}$. Given a description of the polytope, say as the convex hull of a set of vertices, we can calculate the minimum and maximum values for each coordinate of a vector in the polytope. That is, we can determine that the polytope lies within the hypercube $\prod_{i=1}^{n}[a_i,b_i]$. Our first task, then, is to choose $x_1 \in [a_1,b_1]$ according to an appropriate probability distribution.

Thus, for any $c \in [a_1,b_1]$ we can calculate the quantity $f(c) = \Pr(x_1 \in [a_1,c])$ by calculating the volume of $\mathcal{P} \cap \{(x_1,\ldots,x_n) : a_1 \leq x_1 \leq c\}$ as a percentage of the volume of $\mathcal{P}$. This gives us a monotone increasing distribution function $f : [a_1,b_1] \to [0,1]$. It is from this distribution function that we want to sample x_1. If we can choose a random real number X uniformly distributed over $[0,1]$, then we can just take $x_1 = f^{-1}[X]$. Once x_1 has been sampled, its value determines an $(n-1)$-dimensional cross-section of $\mathcal{P}$, so we have now reduced the problem to choosing a smaller random vector $(x_2,\ldots,x_n)$ uniformly distributed over that cross-section. Thus we can iteratively choose $x_2,\ldots,x_n$ by the same algorithm used to choose x_1.

For general polytopes, there are very large practical problems with this algorithm. However, for simplices all of these problems disappear. Consider a simplex with one vertex at the origin and vertices $v_1,\ldots,v_n$ where $v_i = (0,\ldots,a_i,0,\ldots,0)$. Then

$$\Pr(x_1 \in [c,a_1]) = \left(\frac{a_1 - c}{a_1}\right)^n,$$

so the inverse of the distribution function is just $f^{-1}(X) = 1 - X^{1/n}$ times the scaling factor a_i. And if we sample x_1, the cross-section of the simplex at x_1 is just the $(n-1)$-dimensional simplex with vertices at $(x_1,0,0,\ldots,0)$ and $v_2,\ldots,v_n$ where $v_i = (x_1,0,\ldots,(1-\frac{x_1}{a_1})a_i,\ldots)$.

The source code for our implementation of POLYPROB illustrates our application of this method; it is available at

`http://www.math.rutgers.edu/ nweining/polyprob.tar.gz.`

Tests and Construction

We shall provide now membership and completeness tests and a construction of the integral closure of monomial ideals of finite co-length.

Our treatment is a by-product of the half-spaces description of the convex hull given in equation (7.6). We point out how the following oracle gives a solution to the membership and completeness tests and the construction task in case of an ideal of finite co-length.

Membership Test

There are several tests for membership in the integral closure of a monomial ideal. We begin with the following result.

Proposition 7.42. *Let I be a monomial ideal of finite co-length as above, and let f be a monomial. Denote by $\mathbf{e} = (e_1, \ldots, e_n)$ the exponent vector of f, by $\mathbf{v} = (v_1, \ldots, v_n)$, A and $\mathbf{b}$ the vectors and matrices associated to I as discussed above. Then f is integral over I if one of the following two conditions holds:*

$$A \cdot \mathbf{e} \leq \mathbf{b},$$

$$\sum_{i=1}^{n} \frac{e_i}{v_i} \geq 1.$$

Proof. These conditions simply express the fact that either $\mathbf{e}$ lies in the convex hull of the vectors $\mathbf{v}_1, \ldots, \mathbf{v}_n$ (in which case f would lie in the integral closure of $(\mathbf{x}^{v_1}, \ldots, \mathbf{x}^{v_n})$), or that adding f to I does not affect the volume of $\mathcal{P}$. In the second case, $e(I) = e(I, f)$ and f is integral over I by Rees's theorem. $\square$

Definition 7.43. A *membership oracle* for the integral closure of ideals in a certain class $\mathfrak{A}$ is a boolean function $\mathcal{A}$ on the ring $k[x_1, \ldots, x_n]$ of polynomials such that for every $I \in \mathfrak{A}$, $f \in \overline{I}$ if and only if $\mathcal{A}(f) = \text{true}$.

We illustrate it with the diagram:

$$f \longrightarrow \boxed{\ \mathcal{A}\ } \longrightarrow \{\text{true, false}\}$$

The result above shows that monomial ideals of finite co-length admit such oracles.

Completeness Test

We show now how any membership oracle $\mathcal{A}$ for monomial ideals of finite co-length gives rise to a completeness test.

Proposition 7.44. *Suppose that $\mathcal{A}$ is a membership oracle for the integral closure of monomial ideals of finite co-length. Let I be a monomial ideal of finite co-length and $\{f_1, \ldots, f_s\}$ the monomials in $I : (x_1, \ldots, x_d) \setminus I$. Then*

$$I = \overline{I} \Longleftrightarrow \mathcal{A}(f_i) = \text{false} \textit{ for } i = 1, \ldots, s.$$

Proof. First, we consider the reverse direction. Let $L = I \colon (x_1, \ldots, x_d)$ be the socle ideal of I. Then L is generated by the f_i and monomials in I. Since $\overline{I}$ is a monomial ideal, if f is a monomial in $\overline{I} \setminus I$, then multiplication by another monomial g, gives gf generating a nonzero element in the vector space L/I. This means that gf must be one of the f_i. Since gf is also integral over I, the assertion follows. The other direction is obvious. $\square$

Construction Task

Since I and any monomial ideal between I and $\overline{I}$ have the same oracle $\mathcal{A}$, the construction of $\overline{I}$ follows in a straightforward manner:

Proposition 7.45. *If $I \neq \overline{I}$, suppose that $I : (x_1, \ldots, x_d) = (I, f_1, \ldots, f_s)$ and set*

$$I_1 = (I, f_i \text{ with } \mathcal{A}(f_i) = \text{true},\ i = 1, \ldots, s).$$

Iterating $I \mapsto I_1$ will lead to $\overline{I}$.

General Monomial Ideals

A more comprehensive membership test for the question "$f \in \overline{I}$", valid for any monomial ideal, is the following. It lacks the effectivity discussed above.

Proposition 7.46. *Let $v_1,\dots,v_m$ be a set of vectors in $\mathbb{N}^n$ and A the $n \times m$ matrix whose columns are the vectors $v_1,\dots,v_m$. If $I = (\mathbf{x}^{v_1},\dots,\mathbf{x}^{v_m})$, then a monomial $\mathbf{x}^b$ lies in the integral closure of I if and only if the linear program*

Maximize $x_1+\cdots+x_m$

Subject to $Ax \le b$ and $x \ge 0$

has an optimal value greater or equal than 1 that is attained at a vertex of the rational polytope $\mathcal{P} = \{x \in \mathbb{R}^m \,|\, Ax \le b \text{ and } x \ge 0\}$.

Proof. $\Rightarrow$) Suppose that $\mathbf{x}^b \in \overline{I}$, that is, $\mathbf{x}^{pb} \in I^p$ for some positive integer p. There are non-negative integers r_i satisfying the equality

$$\mathbf{x}^{pb} = \mathbf{x}^{\delta}\,(\mathbf{x}^{v_1})^{r_1}\cdots(\mathbf{x}^{v_m})^{r_m} \text{ and } r_1+\cdots+r_m = p.$$

Hence the column vector c with entries $c_i = r_i/p$ satisfies the following requirement:

$$Ac \le b \text{ and } c_1+\cdots+c_m = 1.$$

This means that the linear program has an optimal value greater or equal than 1.

$\Leftarrow$) Observe that the vertices of $\mathcal{P}$ have rational entries (see [Ch0, Theorem 18.1]) and that the maximum of $x_1+\cdots+x_m$ is attained at a vertex of the polytope $\mathcal{P}$, so that there are non-negative rational numbers $c_1,\dots,c_m$ such that

$$c_1+\cdots+c_m \ge 1 \quad\text{and}\quad c_1v_1+\cdots+c_mv_m \le b.$$

By induction on m it follows rapidly that there are rational numbers $\varepsilon_1,\dots,\varepsilon_m$ such that

$$0 \le \varepsilon_i \le c_i \;\forall i \;\text{ and }\; \sum_{i=1}^{m} \varepsilon_i = 1.$$

Therefore there is a vector $\delta \in \mathbb{Q}^n$ with non-negative entries such that

$$b = \delta + \varepsilon_1 v_1 + \cdots + \varepsilon_m v_m.$$

Thus there is an integer $p > 0$ such that

$$pb = \underbrace{p\delta}_{\in \mathbb{N}^n} + \underbrace{p\varepsilon_1}_{\in \mathbb{N}} v_1 + \cdots + \underbrace{p\varepsilon_m}_{\in \mathbb{N}} v_m,$$

and consequently $\mathbf{x}^b \in \overline{I}$. □

Remark 7.47. According to [Ch0, Theorem 5.1] one can also use the dual problem

Minimize $b_1y_1+\cdots+b_ny_n$

Subject to $yA \geq \mathbf{1}$ and $y \geq 0$

to test whether x^b is in $\overline{I}$. Here $\mathbf{1}$ denotes the vector with all its entries equal to 1. The advantage of considering the dual is that here one has a fixed polyhedron

$$Q = \{y \in \mathbb{R}^n \,|\, yA \geq \mathbf{1} \text{ and } y \geq 0\}$$

that can be used to test membership of any monomial $\mathbf{x}^b$, while in the primal problem the polytope $\mathcal{P}$ depends on b.

Let us illustrate the criterion with a previous example.

Example 7.48. Consider the ideal I of Example 7.32. To verify that $\mathbf{x}^b = x_1^2 \cdots x_8^2$ is in $\overline{I^3}$ one uses the following procedure in *Mathematica*

```
ieq:={
3x1 + 3x2 + 3x3 + 3x4 + 3x5<=2,
3x1 + 3x2 + 3x3 + 3x4 + 3x6<=2,
3x1 + 3x2 + 3x3 + 3x4 + 3x7<=2,
3x1 + 3x2 + 3x3 + 3x4 + 3x8<=2,
3x1 + 3x5 + 3x6 + 3x7 + 3x8<=2,
3x2 + 3x5 + 3x6 + 3x7 + 3x8<=2,
3x3 + 3x5 + 3x6 + 3x7 + 3x8<=2,
3x4 + 3x5 + 3x6 + 3x7 + 3x8<=2}

vars:={x1,x2,x3,x4,x5,x6,x7,x8}

f:=x1+x2+x3+x4+x5+x6+x7+x8

ConstrainedMax[f,ieq,vars]
```

The answer is:

```
{16/15,
{x1 -> 2/15, x2 -> 2/15, x3 -> 2/15, x4 -> 2/15,
 x5 -> 2/15, x6 -> 2/15, x7 -> 2/15, x8 -> 2/15}},
```

where the first entry is the optimal value and the other entries correspond to a vertex of the polytope P. Using the criterion and the procedure above one rapidly verifies that $\mathbf{x}^b$ is a minimal generator of $\overline{I^3}$.

Using *PORTA* one readily obtains that the vertices of the polyhedral set Q described in the remark above are the rows of the matrix

```
M:={{0,0,0,0,1/3,1/3,1/3,1/3},
{0,0,0,1/4,1/12,1/12,1/12,1/12},
{0,0,0,1/3,0,0,0,1/3},
{0,0,0,1/3,0,0,1/3,0},
```

```
{0,0,0,1/3,0,1/3,0,0},
{0,0,0,1/3,1/3,0,0,0},
{0,0,1/4,0,1/12,1/12,1/12,1/12},
{0,0,1/3,0,0,0,0,1/3},
{0,0,1/3,0,0,0,1/3,0},
{0,0,1/3,0,0,1/3,0,0},
{0,0,1/3,0,1/3,0,0,0},
{0,1/4,0,0,1/12,1/12,1/12,1/12},
{0,1/3,0,0,0,0,0,1/3},
{0,1/3,0,0,0,0,1/3,0},
{0,1/3,0,0,0,1/3,0,0},
{0,1/3,0,0,1/3,0,0,0},
{1/15,1/15,1/15,1/15,1/15,1/15,1/15,1/15},
{1/12,1/12,1/12,1/12,0,0,0,1/4},
{1/12,1/12,1/12,1/12,0,0,1/4,0},
{1/12,1/12,1/12,1/12,0,1/4,0,0},
{1/12,1/12,1/12,1/12,1/4,0,0,0},
{1/4,0,0,0,1/12,1/12,1/12,1/12},
{1/3,0,0,0,0,0,0,1/3},
{1/3,0,0,0,0,0,1/3,0},
{1/3,0,0,0,0,1/3,0,0},
{1/3,0,0,0,1/3,0,0,0},
{1/3,1/3,1/3,1/3,0,0,0,0}
}
```

Thus this is a "membership test matrix" in the sense that a monomial $\mathbf{x}^b$ lies in $\overline{I^3}$ iff $Mb \geq \mathbf{1}$. In the case $\mathbf{x}^b = x_1^2 \cdots x_8^2$,

```
Mb={{8/3},{7/6},{4/3},{4/3},{4/3},{4/3},{7/6},{4/3},
{4/3},{4/3},{4/3},{7/6},{4/3},{4/3},{4/3},{4/3},
{16/15},{7/6},{7/6},{7/6},{7/6},{7/6},{4/3},{4/3},
{4/3},{4/3},{8/3}}>=(1,...,1).
```

Computation of Multiplicities

We shall make general observations about the computation of the multiplicity of arbitrary primary ideals. The input data is usually the following. Let $A = k[x_1,\ldots,x_r]/L$ be an affine algebra and I a primary ideal for some maximal ideal $\mathfrak{M}$ of A. The Hilbert-Samuel polynomial is the function given for $n \gg 0$ by the rule

$$n \mapsto \lambda(A/I^n) = \frac{e(I)}{d!} n^d + \text{lower terms}, \quad \dim A_{\mathfrak{M}} = d.$$

In other words, $e(I)$ is the ordinary multiplicity of the standard graded algebra

$$\mathrm{gr}_I(A) = \sum_{n \geq 0} I^n / I^{n+1}.$$

For the actual computation, ordinarily one needs a presentation of this algebra

$$\mathrm{gr}_I(A) = k[T_1, \ldots, T_m]/H,$$

where the right-side is not always a standard graded algebra. In the special case where $I = (x_1, \ldots, x_r)A$ and L is a homogeneous ideal, one has that

$$\mathrm{gr}_M(A) \cong A,$$

and therefore it can be computed in almost all computer algebra systems by making use of the next result.

Theorem 7.49 (Macaulay Theorem). *Given an ideal I and a term ordering >, the mapping*

$$\text{NormalForm}: R/I \longrightarrow R/in_>(I) \tag{7.8}$$

is an isomorphism of k-vector spaces. If I is a homogeneous ideal and > is a degree term ordering, then NormalForm *is an isomorphism of graded k-vector spaces, in particular the two rings have the same Hilbert function.*

For our case, this implies that

$$e(I) = \deg(A) = \deg(k[x_1, \ldots, x_r]/\mathrm{in}_>(L)),$$

where $>$ is any degree term ordering of the ring $k[x_1, \ldots, x_r]$ of polynomials. We can turn the general problem into this case by the following observation (which hides the difficulties of the conversion). Let $(R, \mathfrak{m})$ be a Noetherian local ring and I an $\mathfrak{m}$-primary ideal. To calculate the multiplicity $e(I)$ we need some form of access to a presentation of the associated graded ring $\mathrm{gr}_I(R)$,

$$\mathrm{gr}_I(R) = k[T_1, \ldots, T_s]/(f_1, \ldots, f_m),$$

in order to avail ourselves of the programs that determine Hilbert functions.

Alternatively, one can turn to indirect means. Suppose that $R = k[x_1, \ldots, x_d]$ is a ring of polynomials and I is an $(x_1, \ldots, x_d)$-primary ideal. Let J be a minimal reduction of I, so that

$$e(I) = \lambda(R/J).$$

(A similar approach works whenever R is a Cohen-Macaulay ring.) If $>$ is a term order of R, then

$$\lambda(R/I) = \lambda(R/\mathrm{in}_>(J)).$$

The difficulty is in obtaining J. It usually arises by taking a set of d generic linear combination of a generating system of I. In addition, even when I is homogeneous, J will not be homogeneous (often it is forbidden to be). One positive observation that can be made is:

Proposition 7.50. *Let I be an $(x_1,\ldots,x_d)$-primary ideal. For any term order $>$ of R,*

$$e(I) \leq e(in_>(I)) \leq d!e(I). \tag{7.9}$$

Proof. Set $L = \text{in}_>(I)$. The multiplicities are read from the leading coefficients of the Hilbert polynomials $\lambda(R/I^n)$ and $\lambda(R/L^n)$, $n \gg 0$. We note however that while $\lambda(R/I) = \lambda(R/L)$, for large n we can only guarantee th

$$\lambda(R/I^n) = \lambda(R/\text{in}_>(I^n)) \leq \lambda(R/L^n),$$

since the inclusion

$$(\text{in}_>(I))^n \subset \text{in}_>(I^n)$$

may be proper.

The other inequality will follow from Lech's formula ([Le60]) applied to the ideal L:

$$e(L) \leq d!\lambda(R/L)e(R) \leq d!e(I),$$

since $e(R) = 1$ and $\lambda(R/L) = \lambda(R/I) \leq e(I)$. □

As an illustration, set $I = (xy, x^2+y^2) \subset k[x,y]$. Choosing the deglex ordering with $x > y$ gives $L = \text{in}_>(I) = (xy, x^2, y^3)$. We thus have

$$4 = e(I) < e(L) = 5.$$

We are now going to explain the equality $e(I) = e(L)$. Set $L_n = \text{in}_>(I^n)$. Note that $B = \sum_{n\geq 0} L_n t^n$ is the Rees algebra of the filtration defined by the L_n. Actually, B is the initial algebra $\text{in}_>(R[It])$ of the Rees algebra $R[It]$ for the extended term order of $R[t]$:

$$ft^r > gt^s \Leftrightarrow r > s \quad \text{or} \quad r = s \quad \text{and} \quad f > g.$$

In general, B is not Noetherian (which is the case in the simple example above, according to [El86]).

Theorem 7.51. *Let $R = k[x_1,\ldots,x_d]$ be a ring of polynomials, I an $(x_1,\ldots,x_d)$-primary ideal of R and $>$ a term ordering. The following conditions are equivalent:*

(a) $e(I) = e(in_>(I))$.
(b) *B is integral over $R[Lt]$, in particular B is Noetherian.*

Proof. (a) $\Rightarrow$ (b). To prove that B is contained in the integral closure of $R[Lt]$ it will be enough to show that for each s, the algebra $R[L_s t]$ is integral over $R[L^s t]$, in other words, to prove assertion (b) for the corresponding Veronese subalgebras.

Since, by hypothesis, the functions $\lambda(R/L^n)$ and $\lambda(R/I^n) = \lambda(R/L_n)$, for $n \gg 0$, are polynomials of degree d with the same leading coefficients, and we have

$$\lambda(R/(L^s)^n) \geq \lambda(R/L_s^n) \geq \lambda(R/L_{sn}) = \lambda(R/I^{sn}) = \lambda(R/(I^s)^n),$$

as well as

$$e(L^s) = s^d e(L) = s^d e(I) = e(I^s),$$

it follows that L^s and L_s have the same multiplicities. By Theorem 1.154, ([Re61]), L_s is integral over L^s.

(b) $\Rightarrow$ (a). It is immediate. □

Here is an illustration pointed out by A. Taylor. Suppose that $I = (x^2, y^2, z^2 - xy) \subset k[x,y,z]$, with $z > y > x$. Then $\text{in}_>(I) = (x^2, y^2, z^2)$, so that the multiplicity of both ideals is 8.

Remark 7.52. The assumption $e(I) = e(\text{in}_>(I))$ has other consequences as well, in view of another observation. Since the number of generators of $\text{in}_>(I)$ is the same as that of the Gröbner basis of I for the term order $>$, then by Proposition 2.123 we have $\nu(\text{in}_>(I)) \leq e(I) + d - 1$.

Some of these facts can be extended to more general affine algebras. Suppose that I is a monomial ideal of finite co-length and that $L \subset I$ is a monomial subideal. The multiplicity of I/L arises from the function

$$n \mapsto \lambda(R/(I^n + L)).$$

We shall argue that there is a 'volume formula' similar to Proposition 7.35 that holds in this case. It is an application of the associativity formula for multiplicities. If $\mathfrak{p}_1, \ldots, \mathfrak{p}_r$ are the minimal prime ideals of L of dimension $s =$ height L, then

$$e(I/L) = \sum_{i=1}^{r} \lambda((R/L)_{\mathfrak{p}_i}) \cdot e((I + \mathfrak{p}_i)/\mathfrak{p}_i).$$

Once the $\mathfrak{p}_i$ have been found, we may apply Proposition 7.35 to each monomial ideal $I_i = (I + \mathfrak{p}_i)/\mathfrak{p}_i$. The other terms are co-lengths of monomial ideals. Indeed, the length l_i of the localization $(R/L)_{\mathfrak{p}_i}$ is obtained by setting to 1 in R and in L all the variables which do not belong to $\mathfrak{p}_i$. On the other hand, the ideal $I/\mathfrak{p}_i$ is obtained by setting to 0 the variables that lie in $\mathfrak{p}_i$.

Proposition 7.53. *The multiplicity of the 'monomial' ideal I/L is given by*

$$e(I/L) = \sum_{i=1}^{r} l_i \cdot e(I_i).$$

We can also make comparisons between multiplicities of ideals in general affine rings and the monomial case. Consider an ideal

$$I/L \subset A = R/L = k[x_1, \ldots, x_n]/L$$

of codimension d. For some term order, let L' and I' be the corresponding initial ideals. Denoting by $(\cdot)'$ the initial ideal operation, we have

$$\frac{I'^n + L'}{L'} \subset \frac{(I^n + L)'}{L'}, \; n \geq 0.$$

As in the case when $L = (0)$, we have

$$\lambda(R/(I^n + L)) = \lambda(R/(I^n + L)') \leq \lambda(R/(I'^n + L')),\ n \geq 0,$$

and consequently

$$e(I/L) \leq e(I'/L').$$

On the other hand, by Lech's formula ([Le60]),

$$e(I'/L') \leq d! \cdot \lambda(R/I') \cdot e(R/L') = d! \cdot \lambda(R/I) \cdot e(R/L),$$

the substitution $e(R/L') = e(R/L)$ by an application of Macaulay's theorem.

Remark 7.54. Let us formulate a number of questions about these observations.

(i) Find 'interesting' examples of equality $e(I) = e(\mathrm{in}_>(I))$. Classes of examples arise naturally when I is an ideal generated by forms of degree r that form a Gröbner basis of I. In this case both I (assumed to be an ideal of finite co-length in a ring of polynomials in d variables) and $\mathrm{in}_>(I)$ have multiplicity r^d.
(ii) When is the filtration $\{L_n = \mathrm{in}_>(I^n)\}$ Noetherian? While this is not always the case, which properties of Noetherian filtrations are maintained, for example, is the Hilbert function $\nu(L_n)$, a quasi-polynomial for $n \gg 0$?
(iii) Is there a formulation of Theorem 7.51 for ideals of positive dimension in which a generalized multiplicity replaces the classical multiplicity? What is being asked is what is the numerical meaning of statement (b).
(iv) The subject becomes more delicate when either I or L is not a monomial ideal. We must then consider the issue algorithmically. A place to start is [MR95]. Are there other ways?

7.4 Normalization of an Ideal

Given an ideal I of the Noetherian ring R, the *normalization* of I is the study of the properties of the integral closure of all of its powers. It makes for an obviously difficult task. Nevertheless, in many cases of interest it is possible to establish relations and bounds on these powers. In this section we discuss properties of the normalization of equimultiple ideals and give very sharp bounds for the powers for arbitrary monomial ideals.

The general issue is the following: How to introduce numerical measure for the integral closure of a graded ring? We discuss several issues appropriate to the normalization of ideals ([PUVV4]).

Definition 7.55. Let R be a quasi-unmixed normal domain and let I be an ideal.

(i) The *normalization index* of I is the smallest integer $s = s(I)$ such that

$$\overline{I^{n+1}} = I \cdot \overline{I^n} \quad n \geq s.$$

(ii) The *generation index* of I is the smallest integer $s_0 = s_0(I)$ such that

$$\sum_{n\geq 0} \overline{I^n}t^n = R[\overline{I}t, \ldots, \overline{I^{s_0}}t^{s_0}].$$

For example, if $R = k[x_1, \ldots, x_d]$ and $I = (x_1^d, \ldots, x_d^d)$, then $I_1 = \overline{I} = (x_1, \ldots, x_d)^d$. It follows that $s_0(I) = 1$, while $s(I) = \mathrm{r}_I(I_1) = d - 1$.

Although these integers are well defined–since $\overline{R[It]}$ is finite over $R[It]$–it is not clear, even in case where R is a regular local ring, which invariants of R and of I have a bearing on the determination of $s(I)$. An affirmative case is that of a monomial ideal I of a ring of polynomials in d indeterminates over a field–when $s \leq d - 1$ by Theorem 7.58.

Equimultiple Ideals

For primary ideals and some other equimultiple ideals there are relations between the two indices of normalization.

Proposition 7.56. *Let $(R, \mathfrak{m})$ be an integrally closed, local Cohen-Macaulay domain such that the maximal ideal $\mathfrak{m}$ is normal. Let I be an $\mathfrak{m}$-primary ideal of normalization indices $s(I)$ and $s_0(I)$. Then*

$$s(I) \leq e(I)((s_0(I) + 1)^d - 1) - s_0(I)(2d - 1) + 1,$$

where $e(I)$ is the multiplicity of I.

Proof. Without loss of generality, we may assume that the residue field of R is infinite. Set $S = R[It]$ and denote its integral closure by B. We form the special fiber of B,

$$F = B/(\mathfrak{m}, It)B = \sum_{n\geq 0} F_n,$$

and following Proposition 1.91, we estimate $s(I)$ (the Castelnuovo-Mumford regularity of F) of the algebra F in terms of the indices of nilpotency of the components F_n, for $n \leq s_0(I)$.

Let $J = (z_1, \ldots, z_d)$ be a minimal reduction of I. For each component $I_n = \overline{I^n}$ of B, we collect the following data:

$$\begin{aligned} J_n &= (z_1^n, \ldots, z_d^n), \text{ a minimal reduction of } I_n \\ e(I_n) &= e(I)n^d, \text{ the multiplicity of } I_n \\ r_n = r_{J_n}(I_n) &\leq \frac{e(I_n)}{n}d - 2d + 1, \text{ a bound on the reduction number of } I_n. \end{aligned}$$

The last assertion follows from Theorem 2.45, once it is observed that $I_n \subset \overline{\mathfrak{m}^n} = \mathfrak{m}^n$, by the normality of $\mathfrak{m}$.

We are now ready to estimate the index of nilpotency of the component F_n. With the notation above, we have $I_n{}^{r_n+1} = J_n I_n^{r_n}$. When this relation is read in the special fiber ring F, it means that $r_n + 1$ is at least the index of nilpotency of F_n.

Following Proposition 1.91, we have

$$s(I) \leq 1 + \sum_{n=1}^{s_0(I)} r_n = 1 + \sum_{n=1}^{s_0(I)} e(I)dn^{d-1} - s_0(I)(2d-1) + 1,$$

which we approximate with an elementary integral to get the assertion. □

We can do considerably better when R is a ring of polynomials over a field of characteristic zero.

Theorem 7.57. *Suppose that $R = k[x_1, \ldots, x_d]$, where k is a field of characteristic zero, and let I be a homogeneous ideal that is $(x_1, \ldots, x_d)$-primary. If $s(I)$ and $s_0(I)$ are the normalization indices of I, then*

$$s(I) \leq (e(I) - 1)s_0(I) + 1,$$

where $e(I)$ is the multiplicity of I.

Proof. We begin by localizing R at the maximal homogeneous ideal and choosing a minimal reduction J of I. We denote the associated graded ring of the filtration of integral closures $\{I_n = \overline{I^n}\}$ by G,

$$G = \sum_{n \geq 0} I_n / I_{n+1}.$$

In this affine ring we can take for a Noether normalization a ring $A = k[z_1, \ldots, z_d]$, where the z_i are the images in G_1 of a minimal set of generators of J.

There are two basic algebraic facts about the algebra G. First, its multiplicity as a graded A-module is the same as that of the associated graded ring of I, that is, it is $e(I)$. Second, since the Rees algebra of the integral closure filtration is a normal domain, so is the extended Rees algebra

$$C = \sum_{n \in \mathbb{Z}} I^n t^n,$$

and consequently the algebra $G = C/(t^{-1})$ will satisfy Serre's condition S_1. This means that, as a module over A, C is torsionfree.

We now apply the theory of Cayley-Hamilton equations to the elements of the components of G (see Section 1.5). For $u \in G_n$, we have an equation of integrality over A

$$u^r + a_1 u^{r-1} + \cdots + a_r = 0,$$

where the a_i are homogeneous forms of A, in particular $a_i \in A_{ni}$, and $r \leq e(G) = e(I)$. Since k has characteristic zero, we obtain an equality

$$G_n^r = A_n G_n^{r-1}.$$

At the level of the filtration, this equality means that

$$I_n^r \subset J^n I_n^{r-1} + I_{nr+1},$$

which we weaken by

$$I_n^r \subset I \cdot I_{nr-1} + \mathfrak{m} I_{nr},$$

where we have used Corollary 7.15. Finally, in the special fiber ring $\mathcal{F}(B)$, this equation shows that the indices of nilpotency of the components F_n are bounded by $e(I)$, as desired. Now we apply Proposition 1.91 (and delocalize back to the original homogeneous ideals). □

Normalization of Monomial Ideals

In the case of monomial ideals the picture is very clear, according to the following.

Theorem 7.58. *Let R be a ring of polynomials over a field k, $R = k[x_1, \ldots, x_d]$, and let I be a monomial ideal. Then*

$$\overline{I^n} = I\overline{I^{n-1}}, \quad n \geq d. \tag{7.10}$$

Proof. Let $\overline{\mathcal{R}}$ be the integral closure of the Rees algebra of the ideal I,

$$\overline{\mathcal{R}} = \sum_{n \geq 0} \overline{I^n} t^n.$$

Since the ideals are homogeneous, to prove the asserted equality is enough to localize at the maximal homogeneous ideal M of R. The ring $\overline{\mathcal{R}}$ is Cohen-Macaulay by Hochster's theorem ([BH93, Theorem 6.3.5]).

We first assume that k is an infinite field, and apply Theorem 3.48. The extension to all fields is now immediate: The equality (7.10) of monomial ideals is equivalent to an equality of products of its monomial generators. Since these are independent of the characteristic or any field extension, being completely determined by the convex hull of the corresponding exponent vectors, the assertion is clear. □

The following normality criterion was first proved in [RRV2]:

Theorem 7.59 (Reid-Roberts-Vitulli). *Let $R = k[x_1, \ldots, x_d]$ be a ring of polynomials over a field k and I a monomial ideal. If I^i is integrally closed for $i < d$, then I is normal.*

Corollary 7.60. *Let $R = k[x_1, \ldots, x_d]$ be a ring of polynomials over a field k, and let I be a monomial ideal. Then for $n \geq d-1$, the ideal $\overline{I^n}$ is normal.*

Remark 7.61. Let R be a regular ring of dimension d and I one of its ideals. If $d = 2$, from Zariski's theory ([ZS60]), we have

$$\overline{I^n} = (\overline{I})^n, \quad \forall n \geq 1.$$

Moreover, by [LT81] it will follow that the integral closure of $R[It]$ is Cohen-Macaulay. For $d = 3$, no general bound exists.

Primary Ideals in Regular Rings

We discuss the role of Briançon-Skoda-type theorems in determining some relationships between the coefficients $e_0(I)$ and $e_1(I)$ of the Hilbert polynomial of an ideal. We consider here the case of a normal local ring $(R,\mathfrak{m})$ of dimension d and of an $\mathfrak{m}$-primary ideal I. Set $A = R[It]$ and $B = \overline{R[It]}$; we assume that B is a finite A-module. From the exact sequence

$$0 \to \overline{I^n}/I^n \longrightarrow R/I^n \longrightarrow R/\overline{I^n} \to 0 \tag{7.11}$$

we obtain as above the relationship

$$\overline{e}_1(I) = e_1(I) + \hat{e}_0(I),$$

where $\hat{e}_0(I)$ is the multiplicity of the module of components $\overline{I^n}/I^n$, if $\dim L = d$; otherwise it is set to zero.

Theorem 7.62. *Let $(R,\mathfrak{m})$ be a Cohen-Macaulay local ring of infinite residue field. Suppose the Briançon-Skoda number of R is $c(R)$. Then for any $\mathfrak{m}$-primary ideal I,*

$$\overline{e}_1(I) \leq c(R) \cdot e_0(I).$$

In particular, $e_1(I) \leq c(R) \cdot e_0(I)$.

Proof. We recall the definition of $c = c(I)$. For any ideal L of R,

$$\overline{L^{n+c}} \subset L^n, \quad \forall n.$$

To apply this notion to our setting, let J be a minimal reduction of I. Assume that $\overline{I^{n+c}} \subset J^n$ for all n. To estimate the multiplicity of the module of components $\overline{I^{n+c}}/J^{n+c}$-which is the same as that of the module of components $\overline{I^n}/J^n$–note that $\overline{I^{n+c}} \subset J^n$ and that J^n admits a filtration

$$J^n \supset J^{n+1} \supset \cdots \supset J^{n+c},$$

whose factors all have multiplicity $e_0(J)$. More precisely, for each positive integer k,

$$\lambda(J^{n+k-1}/J^{n+k}) = e_0(J)\binom{n+k-1+d-1}{d-1} = \frac{e_0(J)}{(d-1)!}n^{d-1} + \text{lower terms}.$$

As a consequence we obtain

$$\overline{e}_1(I) \leq e_1(J) + c(R) \cdot e_0(J) = c(R) \cdot e_0(I),$$

since $e_1(J) = 0$. The other inequality, $e_1(I) \leq c(R) \cdot e_0(I)$, follows from (7.11). □

Corollary 7.63. *Let $(R,\mathfrak{m})$ be a Japanese regular local ring of dimension d. Then for every $\mathfrak{m}$-primary ideal I,*

$$\overline{e}_1(I) \leq (d-1)e_0(I), \quad e_1(I) \leq (d-1)e_0(I).$$

Proof. The classical Briançon-Skoda theorem asserts that $c(R) = d-1$. □

Normalization of Rees Algebras

The computation (and its control) of the integral closure of a standard graded algebra over a field benefits greatly from Noether normalizations and of the structures built upon them. If $A = R[It]$ is the Rees algebra of the ideal I of an integral domain R, it does not allow for many such constructions. We would still like to develop some tracking of the complexity of the task required to build $\overline{A}$ (assumed A-finite) through sequences of extensions

$$A = A_0 \to A_1 \to A_2 \to \cdots \to A_n = \overline{A},$$

where A_{i+1} is obtained from an specific procedure $\mathcal{P}$ applied to A_i. At a minimum, we would want to bound the length of such chains. We are going to show how Theorem 7.62 can be used to do just that for a class of Rees algebras.

Let $(R, \mathfrak{m})$ be a Cohen-Macaulay local ring of dimension d that is integrally closed and of Briançon-Skoda number $c(R)$, and let I be an $\mathfrak{m}$-primary ideal of multiplicity $e_0(I)$. Let A and B be distinct algebras satisfying Serre's condition S_2 and such that

$$R[It] \subset A \subset B \subset \overline{R[It]}.$$

For any algebra D such as these, we set $\lambda(R/D_n)$ for its Hilbert function; for $n \gg 0$, one has the Hilbert polynomial

$$\lambda(R/D_n) = e_0(D)\binom{n+d-1}{d} - e_1(D)\binom{n+d-2}{d-1} + \text{lower terms}.$$

The Hilbert coefficients satisfy $e_0(D) = e_0(I)$ and $0 \leq e_1(D) \leq c(R)e_0$ according to Theorem 7.62.

Theorem 7.64. *For any two algebras A and B as above,*

$$c(R)e_0(I) \geq e_1(\overline{R[It]}) \geq e_1(B) > e_1(A) \geq 0;$$

in particular, any chain of such algebras has length bounded by $c(R)e_0(I)$.

Proof. Set $C = B/A$. Since A has Krull dimension $d+1$ and satisfies S_2, it follows easily that C is an A-module of Krull dimension d. From the exact sequence

$$0 \to C_n \longrightarrow R/A_n \longrightarrow R/B_n \to 0,$$

one gets that the multiplicity $e_0(C)$ of C is $e_1(B) - e_1(A)$. As $e_0(C) > 0$, we have all the assertions. □

Corollary 7.65. *If $(R, \mathfrak{m})$ is a regular local ring of dimension d and I an $\mathfrak{m}$-primary ideal, then $(d-1)e_0(I)$ bounds the lengths of the divisorial chains between $R[It]$ and $\overline{R[It]}$.*

Remark 7.66. In sections 3 and 4 of Chapter 6, several bounds for the lengths of divisorial chains of algebras were developed in terms of the multiplicity of the algebras. In the case of a Rees algebra $R[It]$, where $(R,\mathfrak{m})$ is a local ring, the relevant multiplicity would be $\deg(R[It]_{\mathfrak{P}})$, where $\mathfrak{P} = (\mathfrak{m}, R[It]_+)$. This number may however be considerably larger that multiplicities associated with the ideal I in R. For example, if $R = k[x_1,\ldots,x_d]$ is a ring of polynomials over the field k and $I = (x_1,\ldots,x_d)$, then $\deg(R[It]) = d$; see [Jo97] for a discussion of these multiplicities.

We shall now discuss some developments to be found in [PUV4] that considerably extend the previous results. They arise from an examination of Briançon–Skoda-type theorems (see [AH93], [LS81]) in determining some relationships between the coefficients $e_0(I)$ and $\overline{e}_1(I) = e_1(\overline{R[It]})$.

We discuss this role here (see [AH93], [LS81]). We shall use a Briançon-Skoda type theorem that works in non-regular rings. We provide a short proof along the lines of [LS81] for the special case we need, namely $\mathfrak{m}$–primary ideals in a local Cohen-Macaulay ring. The general case is treated by Hochster and Huneke in [HH04, 1.5.5 and 4.1.5]. Let k be a perfect field, R a reduced and equidimensional k-algebra essentially of finite type, and assume that R is affine with $d = \dim R$, or that $(R,\mathfrak{m})$ is local with $d = \dim R + \mathrm{trdeg}_k R/\mathfrak{m}$. Recall that the *Jacobian ideal* $\mathrm{Jac}_k(R)$ of R is defined as the dth Fitting ideal of the module $\Omega_k(R)$ of differentials–it can be computed explicitly from a presentation of the algebra. By a well-known result that can be deduced from [LS81, Theorem 2] by varying Noether normalizations, the Jacobian ideal $\mathrm{Jac}_k(R)$ is contained in the conductor $R{:}\overline{R}$ of R.

Theorem 7.67. *Let k be a perfect field, R a reduced Cohen–Macaulay local k-algebra essentially of finite type, and I an equimultiple ideal of height $g > 0$. Then for every integer n,*

$$\mathrm{Jac}_k(R)\overline{I^{n+g-1}} \subset I^n.$$

Proof. We may assume that k is infinite. Then, passing to a minimal reduction, we may suppose that I is generated by a regular sequence of length g. Let S be a finitely generated k-subalgebra of R such that $R = S_{\mathfrak{p}}$ for some $\mathfrak{p} \in \mathrm{Spec}(R)$, and write $S = k[x_1,\ldots,x_e] = k[X_1,\ldots,X_e]/\mathfrak{a}$ with $\mathfrak{a} = (h_1,\ldots h_t)$ an ideal of height c. Notice that S is reduced and equidimensional. Let $K = (f_1,\ldots,f_g)$ be an S-ideal with $K_{\mathfrak{p}} = I$ and consider the extended Rees ring $B = S[Kt,t^{-1}]$. Now B is a reduced and equidimensional affine k-algebra of dimension $e-c+1$.

Let $\varphi{:}\, k[X_1,\ldots,X_e,T_1,\ldots,T_g,U] \twoheadrightarrow B$ be the k-epimorphism mapping X_i to x_i, T_i to f_it and U to t^{-1}. Its kernel contains the ideal $\mathfrak{b}$ generated by $\{h_i, T_jU - f_j | 1 \leq i \leq t, 1 \leq j \leq g\}$. Consider the Jacobian matrix of these generators,

$$\Theta = \left(\begin{array}{c|ccc} \dfrac{\partial h_i}{\partial X_j} & & 0 & \\ \hline & U & & T_1 \\ & & \ddots & \vdots \\ & & U & T_g \end{array}\right).$$

Notice that $I_{c+g}(\Theta) \supset I_c(\partial h_i/\partial X_j)U^{g-1}(T_1,\ldots,T_g)$. Applying φ, we obtain the inclusion $\mathrm{Jac}_k(B) \supset I_{c+g}(\Theta)B \supset \mathrm{Jac}_k(S)Kt^{-g+2}$. Thus $\mathrm{Jac}_k(S)Kt^{-g+2}$ is contained in the conductor of B. Localizing at $\mathfrak{p}$, we see that $\mathrm{Jac}_k(R)It^{-g+2}$ is in the conductor of the extended Rees ring $R[It,t^{-1}]$. Hence $\mathrm{Jac}_k(R)I\overline{I^{n+g-1}} \subset I^{n+1}$ for every n, which yields the inclusion

$$\mathrm{Jac}_k(R)\overline{I^{n+g-1}} \subset I^{n+1}:I = I^n,$$

as $(\mathrm{gr}_I(R))_+$ has positive grade. □

The next result ([PUV4]), which we simply quote, provides effective bounds for $\overline{e}_1(I)$ in the case of $\mathfrak{m}$–primary ideals.

Theorem 7.68. *Let $(R,\mathfrak{m})$ be an analytically unramified local Cohen-Macaulay ring of dimension $d > 0$ and let I be an $\mathfrak{m}$–primary ideal.*

(a) *If in addition R is an algebra essentially of finite type over a perfect field k with type t, and $\delta \in \mathrm{Jac}_k(R)$ is a non zerodivisor, then*

$$\overline{e}_1(I) \leq \frac{t}{t+1}((d-1)e_0(I) + e_0(I+\delta R/\delta R)).$$

(b) *If the assumptions of (a) hold, then*

$$\overline{e}_1(I) \leq (d-1)(e_0(I) - \lambda(R/\overline{I})) + e_0(I+\delta R/\delta R).$$

(c) *If $R/\mathfrak{m}$ is infinite, then*

$$\overline{e}_1(I) \leq c(I)\,\min\{\frac{t}{t+1}\,e_0(I), e_0(I) - \lambda(R/\overline{I})\}.$$

We extend Theorems 7.64 and 7.68 to arbitrary equimultiple ideals. We place ourselves in the setting of these results with I an equimultiple ideal of codimension g.

The technical change involves Hilbert functions. For $g < d = \dim R$, instead of the length function as in Theorem 7.64, $\deg(R/D_n)$ will denote the multiplicity of the module R/D_n of dimension $d-g$ discussed in Chapter 2 which we recall briefly.

According to the formula for the associativity of multiplicities, this function is easy to express in terms of the multiplicities of the localizations of I at some of its minimal primes. For example, one has

$$\deg(R/I^n) = \sum_{\dim R/I = \dim R/\mathfrak{p}} \lambda(R_\mathfrak{p}/I^n_\mathfrak{p})\deg(R/\mathfrak{p}).$$

It follows that for $n \gg 0$ this function behaves as a polynomial of degree g. Its leading coefficient in the binomial representation is

$$E_0(I) = \sum e_0(I_\mathfrak{p})\deg(R/\mathfrak{p}),$$

the sum extended over such primes. There are similar expressions for the other coefficients $E_i(I)$ in terms of the local Hilbert coefficients $e_i(I_\mathfrak{p})$.

Similarly, for a graded subalgebra D with $R[It] \subset D \subset \overline{R[It]}$, one has Hilbert coefficients $E_0(D)$, $E_1(D)$. We shall use the notation $\overline{E}_i(I)$ when $D = \overline{R[It]}$.

The coefficients E_i treated in Section 2.3 share many properties with the ordinary Hilbert coefficients but differs in subtle ways, particularly in respect to exact sequences. In one case of interest the behavior is the expected one. One point that must be kept in sight is that for each dimension $s \leq \dim R$ there is the function $E_0(\cdot)$, and therefore while considering different modules in an exact sequence one must be explicit about which function is being used on which module.

The version of Theorem 7.64 for equimultiple ideals can now be stated. In its proof we shall discuss only the points that require new justifications.

Theorem 7.69. *Let $(R, \mathfrak{m})$ be a Cohen–Macaulay local ring of dimension d and I an equimultiple ideal of codimension g and of Briançon–Skoda number $c(I)$. Let A and B be distinct graded algebras satisfying Serre's condition S_2 and such that*

$$R[It] \subset A \subset B \subset \overline{R[It]}.$$

Then

$$c(I)E_0(I) \geq \overline{E}_1(I) \geq E_1(B) > E_1(A) \geq 0.$$

Proof. Set $C = B/A$. As in Theorem 7.64, from the exact sequence

$$0 \to C_n \longrightarrow R/A_n \longrightarrow R/B_n \to 0$$

one gets that the multiplicity $E_0(C)$ of C is $E_1(B) - E_1(A)$.

We must argue the positivity of $E_0(C)$ anew. By the associativity formula that we discussed above, it follows that

$$E_0(C) = \sum e_0(C_{\mathfrak{p}}) \deg(R/\mathfrak{p}),$$

with $\mathfrak{p}$ running over the minimal primes of I. The module $C_{\mathfrak{p}}$ is either 0 or has dimension g, so that $E_0(C) = 0$ only when all such localizations are trivial, a condition that is equivalent to saying that the R-annihilator L of C is an ideal of codimension at least $g+1$. Since I is equimultiple of codimension g, it is easy to see that the codimension of the A-ideal LA is at least 2, and as $LAB \subset A$, it follows that $B \subset A$ because A satisfies the condition S_2.

The rest of the proof follows the same tracks as the earlier argument. □

Remark 7.70. The argument shows that when passing from the algebra A to B, one of the values $e_1(A_{\mathfrak{p}})$ is increased. Thus, the integer $\sum_{\mathfrak{p}} e_1(A_{\mathfrak{p}})$ would give tighter control. However, padding the summands with the $\deg(R/\mathfrak{p})$ into an ersatz integral provides a value that becomes 'visible', unlike the $e_1(A_{\mathfrak{p}})$.

It is also possible to derive estimates for equimultiple ideals based on the bounds of Theorem 7.68 and (2.58).

7.5 Algebras of Symbolic Powers

Let I be an ideal of the Noetherian ring R and $D = \{\mathfrak{p}_1, \ldots, \mathfrak{p}_s\}$ a finite set of prime ideals of R. For the integer n, the nth *symbolic power* of I with respect to D is the following ideal of R:

$$I_D^{(n)} = R \cap \bigcap_{\mathfrak{p} \in D} I_{\mathfrak{p}}^n.$$

Sets of primes used are the associated primes of I or of all of its powers, but more commonly the set of minimal primes of I.

Part of the appeal of this notion comes from the fact that under appropriate circumstances it affords a refined version of the Nullstellensatz. This is very clear in the premier characterization of such ideals provided by the theorem of Nagata-Zariski:

Theorem. *If $\mathfrak{p}$ is a prime ideal of $\mathbb{C}[x_1, \ldots, x_d]$, then for any positive integer n,*

$$\mathfrak{p}^{(n)} = \bigcap_{\mathfrak{m}} \mathfrak{m}^n, \quad \mathfrak{p} \subset \mathfrak{m} = \textit{maximal ideal}.$$

The *symbolic power filtration* of I with respect to a set D is the multiplicative filtration $\{I^{(n)}, n \geq 0\}$. The Rees algebra

$$\mathcal{R}_s(I) = \sum_{n \geq 0} I^{(n)}$$

is the *symbolic powers algebra* of I with respect to D.

These algebras present a great diversity of behavior. They may be non-Noetherian, but there exist several criteria to ensure the Noetherian condition. In addition, 'fragments' of the Noetherian condition are more often detected. In this chapter we will examine both aspects, often however with pointers to the literature.

Integral Closure

Let R be a normal domain and I a radical ideal. The Rees valuations of I express both algebras $\overline{R[It]}$ and $\mathcal{R}_s(I)$ in a convenient form whenever I is generically a complete intersection. Indeed, for any minimal prime $\mathfrak{p}$ of I, the prime ideal $\mathfrak{P} = \mathfrak{p}R[t]_{\mathfrak{p}} \cap R[It]$ gives rise to the Rees valuation $V = R[It]_{\mathfrak{P}}$. These valuations are said to be *expected.* The other possible Rees valuations-when they exist–have the form $\overline{R[It]}_{\mathfrak{Q}}$, where $\mathfrak{Q} \cap R$ is not a minimal prime of I. Despite its circularity, the following expressions are useful at times.

Proposition 7.71. *For an ideal I, with $\mathfrak{P}$ and $\mathfrak{Q}$ as above,*

$$\mathcal{R}_s(I) = \bigcap_{\mathfrak{P}} R[It]_{\mathfrak{P}} \bigcap R[t].$$

$$\overline{R[It]} = \bigcap_{\mathfrak{P}} R[It]_{\mathfrak{P}} \bigcap_{\mathfrak{Q}} \overline{R[It]}_{\mathfrak{Q}} \bigcap R[t].$$

Proof. Left as an exercise.

Theorem 7.72. *Let $(R,\mathfrak{m})$ be a Cohen-Macaulay normal domain of dimension d and I a radical height-unmixed ideal. Suppose that I is of linear type on the punctured spectrum and that $\ell(I) < d$. Then $I^{(n)} = \overline{I^n}$ for all n. In particular, if R is quasi-unmixed, then $\mathcal{R}_s(I)$ is a Noetherian ring.*

Proof. We use the criterion of [Va94b, Theorem 5.4.14]. Set $G = \text{gr}_I(R)$ and consider the exact sequence

$$0 \to H(I) \longrightarrow G \longrightarrow G' \to 0.$$

Note that G' is the Rees algebra of the conormal module I/I^2 relative to R/I. Since I is of linear type on the punctured spectrum (in particular it is generically a complete intersection), G' is a reduced ring. In fact, there is an embedding

$$G' \hookrightarrow R[I^{(n)}t^n, t^{-1}, n \geq 0]/(t^{-1}).$$

To prove the assertion it suffices to show that $H(I)$ is the nilradical of G. This follows since some power of $\mathfrak{m}$ annihilates $H(I)$, whereas height $\mathfrak{m}G \geq 1$ as $\ell(I) < d$. Thus the radical ideal $H(I)$ is contained in every minimal prime of G. □

Special Fiber

Although the symbolic Rees algebra of an ideal may not be Noetherian, we have the following result.

Theorem 7.73. *Let $(R,\mathfrak{m})$ be a Cohen-Macaulay local ring such that $\dim R = d > 0$, and I an ideal of dimension 1. Let $J = (a_1, \ldots, a_{d-2}) \subset I$ be an ideal of height $d-2$ generated by a regular sequence. Then*

$$\nu(I^{(n)}) \leq \deg(R/J)\left(\binom{n+d-3}{d-2} + \binom{n+d-3}{d-3}\right)$$

for all $n \geq 0$.

Monomial Ideals

Let $A = k[x_1, \ldots, x_n]$ be the ring of polynomials over a field k and I a monomial ideal generated by square-free monomials. In this case the symbolic powers of I can be described very cleanly. If $P_1, \ldots, P_s$ are the minimal primes of I, then for each m,

$$I^{(m)} = P_1^m \cap \cdots \cap P_s^m.$$

Tagging the generators of all $I^{(m)}$ by another set of integral vectors, G. Lyubeznik ([Ly88]) proved the following result.

Theorem 7.74. *The graded A-algebra*

$$B = A \oplus I \oplus I^{(2)} \oplus \cdots \oplus I^{(m)} \oplus \cdots = A + It + I^{(2)}t^2 + \cdots + I^{(m)}t^m + \cdots$$

is finitely generated.

Proof. For every monomial $z = x_1^{a_1} x_2^{a_2} \cdots x_n^{a_n}$, denote by $\deg_{x_i} y = a_i$ and by $\deg_{P_j} y$ the sum of all a_i such that $x_i \in P_j$. In this notation, $y \in I^{(m)}$ if and only if $\deg_{P_j} y \geq m$ for each P_j. With every monomial of $I^{(m)}$ associate the element

$$v(y) = (\deg_{x_1} y, \ldots, \deg_{x_n} y, \deg_{P_1} y - m, \ldots, \deg_{P_s} y - m) \in \mathbb{Z}^{n+s}.$$

Let $Z(I)$ be the set of all elements $v(y) \in \mathbb{Z}^{n+s}$, for the minimal generators $y \in I^{(m)}$ and all m. Then $Z(I)$ defines a monomial ideal (in a polynomial ring with $n+s$ indeterminates) whose minimal generators have exponent vectors $v(y_1), v(y_2), \ldots, v(y_r)$, with $y_i \in I^{(m_i)}$. The claim is that B is generated by the monomials $y_1, y_2, \ldots, y_r$. Let $y \in I^{(m)}$ be a monomial of degree m in B and assume that y is not one of the y_i. Since $v(y) \geq v(y_i)$, for some $i \leq r$ one has $y = w \cdot y_i$, where $y_i \in I^{(m_i)}$. Since $\deg_{P_j} y - m \geq \deg_{P_j} - m_i$ for each j, it follows that $\deg_{P_j} w \geq m - m_i$ for each P_j, and therefore that $w \in I^{(m-m_i)}$. □

7.6 Exercises

Exercise 7.75. Give an example of a Noetherian ring R and of an ideal$I$ such that the integral closure of $R[It]$ in $R[t]$ is not Noetherian.

Exercise 7.76. Let R be a Noetherian domain and I an ideal such that $\overline{R[It]}$ is Noetherian. Show that I is normal if and only if

$$\overline{I^{n+1}} \cap I^n = I^{n+1}, \quad \forall n.$$

Exercise 7.77. Let $(R, \mathfrak{m})$ be a Cohen-Macaulay local domain and I an $\mathfrak{m}$-primary ideal. Assuming that $e_1(I) = \binom{e_0(I)-1}{2}$ and the ring $G = \mathrm{gr}_I(R)$ has no embedded primes, show that I is integrally closed.

Exercise 7.78. Let $(R, \mathfrak{m})$ be a Cohen–Macaulay local ring of dimension $d > 0$ and I an $\mathfrak{m}$–primary ideal. Compare the normalization indices $s(I)$ and $s_0(I)$ to those of a power I^q.

Exercise 7.79. Let $\mathfrak{p}$ be a prime ideal of the Noetherian ring R. If I is a $\mathfrak{p}$-primary ideal such that $\overline{I} = \mathfrak{p}$, prove that $I = \mathfrak{p}$ if $R_{\mathfrak{p}}$ is a regular local ring.

Exercise 7.80 (D. Rees). Let R be a Noetherian, integrally closed local domain of dimension two. If I is a divisorial ideal, show the equivalence of the following properties: (i) the symbolic power algebra of I is Noetherian; (ii) $\ell(I) = 1$; (iii) the divisor class $[I]$ has torsion.

Exercise 7.81 (R. Villarreal). Let $I = (x^{v_1}, \ldots, x^{v_q}) \subset k[x_1, \ldots, x_n]$ be a square-free monomial ideal of height $g \geq 2$. Show that:

(i) All the minimal primes of $IR[It]$ are of height one and they are in one-to-one correspondence with the Rees valuations of I.
(ii) The minimal primes of $IR[It]$ are of the form

$$(x_{i_1}, \ldots, x_{i_s}, x^{v_{j_1}}t, \ldots, x^{v_{j_r}}t).$$

(iii) The minimal primes of $IR[It]$ are the expected primes if and only if $R_s(I) = \overline{R[It]}$.

Exercise 7.82 (R. Villarreal). Give an example of a non-normal square-free monomial ideal such that $\overline{R[It]} = R_s(I)$.

Exercise 7.83 ([KS3]). Let R be a regular local ring and $x_1, \ldots, x_d$ a regular system of parameters. If I is a monomial ideal in the x_i (i.e. generated by elements of the form $\mathbf{x}^a = x_1^{a_1} \cdots x_n^{a_n}$) is $\overline{I}$ also monomial? (Try the simpler case when R contains a field.)

8

Integral Closure of Modules

Rees algebras of modules exhibit a significant new set of problems compared to ordinary Rees algebras. In this chapter we study the notion of integrality in modules and of algebras built on them. In addition to the general methods of Chapter 6 based on Jacobian criteria, we make use of the properties specific to modules and ways of converting the problems to the setting of ideals via deformation theory or through the intervention of Fitting ideals.

Let R be a Noetherian ring with total ring of fractions K and E a finitely generated R-module. We say that E has a *rank* if $K \otimes_R E \cong K^r$ for some r, in which case r is said to be the rank of E. By $S(E)$ we denote the symmetric algebra of E. As for the *Rees algebra* of E, as discussed in the Introduction, in an embarrassment of riches there are several possible definitions. The path chosen here is that which most resembles the classical case of the blowup algebra of an embedding. For a module E and a mapping $f : E \to R^r$, we take this to mean the subalgebra of $S(R^r)$ generated by the forms in $f(E)$. In other words, $f(E)$ generates a multiplicative filtration of $S(R^r)$, and $\mathcal{R}_f(E)$ is the associated graded algebra. In this definition, one may as well assume that f is an embedding. For modules of rank r we have $f : E \to R^r$, the details of the embedding also get coded in the structure of $\mathcal{R}_f(E)$.

For such modules we may actually adopt the following definition of their Rees algebras. Let R be a Noetherian ring and E a finitely generated R-module having a rank. The *Rees algebra* $\mathcal{R}(E)$ of E is $S(E)$ modulo its R-torsion submodule. If $S(E) = \mathcal{R}(E)$ then $\mathcal{R}(E)$ is said to be of *linear type*. By abuse of terminology, E is said to be of *of linear type*.

There are several algebras that arise in this fashion. First, assume that $\mathbb{A}$ is a finite-dimensional algebra over the field k. For instance, $\mathbb{A}$ may be a Lie algebra or have some other special structure. The *commuting variety* is

$$\mathsf{C}(\mathbb{A}) = \{(u,v) \in \mathbb{A} \times \mathbb{A} \mid uv = vu\}.$$

The defining equations for such an algebraic variety are obtained by choosing a basis $\{e_1, \ldots, e_n\}$ for $\mathbb{A}$ over k and collecting the quadratic polynomials $f_i = f_i(\mathbf{X}, \mathbf{Y})$ that occur as coefficients of the generic commutation relation

$$\sum_{i=1}^{n} f_i e_i = [\sum_{i=1}^{n} X_i e_i, \sum_{i=1}^{n} Y_i e_i].$$

We can write these polynomials as

$$[f_1, \ldots, f_n] = [Y_1, \cdots, Y_n] \cdot \varphi,$$

where φ is a matrix whose entries are linear forms in the variables X_i. It follows that the affine ring of $\mathrm{C}(\mathbb{A})$ is the symmetric algebra $\mathcal{S}_{k[\mathbf{X}]}(\text{coker } \varphi)$. The generic component of this algebra is the Rees algebra $\mathcal{R}(\text{coker } \varphi)$. In several cases of interest (see [Va94b, Chapter 9]) there remains the issue of deciding when these algebras coincide. If $\mathbb{A}$ is a simple Lie algebra over a field of characteristic zero, it is known that coker (φ) is a torsionfree $k[\mathbf{X}]$-module of projective dimension two, and that

$$\mathcal{S}_{k[\mathbf{X}]}(\text{coker } \varphi)_{\text{red}} = \mathcal{R}(\text{coker } \varphi).$$

Another major class of examples are the Rees algebras directly attached to affine algebras. Suppose that A is an algebra essentially of finite type over a field k, and let $\Omega = \Omega_k(A)$ be its module of Kähler differentials. The Rees algebra of A is $\mathcal{R}(\Omega)$. It is the carrier of considerable information about the tangential variety of spec (A) and of its Gauss image (see [SSU2]). In this case, Ω is rarely of linear type, and the construction of $\mathcal{R}(\Omega)$ gives a mechanism for extracting "nonlinear relations" from the module Ω.

A third setting is provided by conormal modules. That is, if $R = S/I$, the Rees algebra of I/I^2, under natural conditions such as I is a prime ideal or is generically a complete intersection, provides the means to study features of the associated graded ring of I though the examination of the natural surjection

$$\mathrm{gr}_I(S) \to \mathcal{R}_{R/I}(I/I^2) \to 0.$$

It is worthwhile pointing out a significant difference between the Rees algebra of an ideal I and the Rees algebra of a module E of rank $r \geq 2$. For example, while a Veronese subring $\mathcal{R}(I)^{(e)}$ of $\mathcal{R}(I)$ is a Rees algebra, the Veronese subring $\mathcal{R}(E)^{(e)}$, is *not* a Rees algebra for $e \geq 2$.

Given the developments that have taken place in the case of ideals, it is natural to pursue a strategy of *idealization*, by which we mean looking for general techniques to reduce the study of the Rees algebra of a module E of rank r to the case of ideals, possibly over another ring. Let us indicate some of these approaches.

- The determinant ideal $\det_0(E)$ associated with the embedding of E into free modules of the same rank,

  $$E \hookrightarrow R^r.$$

  In addition, the Fitting ideals of E and the associated prime ideals of R^r/E play a significant role in the study of the integral closure $\overline{E}$ of E. The *determinant ideal* is a tool in some effective computations with the reductions of E. For instance, if $F \subset E$ are modules of same rank, then one is a reduction of the other if an only if $\det_0(F)$ is a reduction of $\det_0(E)$. As a consequence, a simple formulation of the integral closure of E is terms of the Rees valuations of $\det_0(E)$ is derived.

- The straightforward consideration of the ideal of $R[x_1,\ldots,x_r]$ generated by the linear forms that ;prescribe the inclusion $E \hookrightarrow R^r$.
- The ideals arising from the Bourbaki sequences of E,

$$0 \to R^{r-1} \longrightarrow E \longrightarrow I \to 0.$$

  It permits the rich theory already developed for ideals to be transported here, sometimes in a transparent manner. These ideals (when I is taken in a generic process) are unique up to isomorphisms.

Of these, the most far-reaching is the last approach. The technique of the *generic Bourbaki ideal* gives the study of the Rees algebras of a module $E \subset R^r$ a natural flavor of deformation theory. For simplicity we assume that R is a normal domain with field of quotients K and that E is a torsionfree module of rank $r \geq 2$. A *Bourbaki sequence* for E is a choice of a free submodule F of rank $r-1$ such that E/F is torsionfree, that is, the quotient is isomorphic to an ideal I,

$$0 \to F \longrightarrow E \longrightarrow I \to 0. \tag{8.1}$$

One has that

$$\mathcal{R}(I) = \mathcal{R}(E)/\langle F\rangle, \text{ where } \langle F\rangle = \mathcal{R}(E) \cap (F\mathcal{R}(E))_R \otimes K.$$

If $\mathcal{R}(E)$ is a deformation of $\mathcal{R}(I)$, a condition meaning that

$$\mathcal{R}(I) \cong \mathcal{R}(E)/(\text{modulo a regular sequence}),$$

we have a channel to pass information from $\mathcal{R}(I)$ to $\mathcal{R}(E)$. Meanwhile the exact sequence defining I provides a back-channel to pass information from $\mathcal{R}(E)$ to $\mathcal{R}(I)$. We simplify considerably this approach by choosing I generic (when it is essentially unique). The task becomes to achieve deformation under broad conditions on E and to identify those modules whose Bourbaki ideals have 'good' Rees algebras.

In [SUV3], the reader will find a detailed treatment of several means studying the Cohen-Macaulayness of the Rees algebra of a module. Our aim here, despite some overlap, is to study other aspects of the theory, particularly those dealing with integral closure. Some of these topics will be:

- Numerical invariants of Rees algebras of modules
- Criteria of integrality
- Buchsbaum-Rim multiplicities and reduction number of modules
- Integrality and order determinantal ideals
- Normality criteria for Rees algebras
- Rees algebras of Bourbaki ideals
- Normalization of modules

8.1 Dimensions of Rees Algebras and of their Fibers

There are several measures of size attached to a Rees algebra $\mathcal{R}(E)$, all derived from ordinary Rees algebras. We survey them briefly.

Proposition 8.1. *Let R be a Noetherian ring of dimension d and E a finitely generated R-module of rank r. Then*

$$\dim \mathcal{R}(E) = d + r = d + \text{height } \mathcal{R}(E)_+.$$

Proof. We may assume that E is torsionfree, in which case E can be embedded into a free module $G = R^r$. Now $\mathcal{R}(E)$ is a subalgebra of the polynomial ring $S = \mathcal{R}(G) = R[t_1, \ldots, t_r]$. As in the case of ideals, the minimal primes of $\mathcal{R}(E)$ are precisely those of the form $\mathfrak{P} = \mathfrak{p}S \cap \mathcal{R}(E)$, where $\mathfrak{p}$ ranges over all minimal primes of R. Write $\bar{R} = R/\mathfrak{p}$ and $\bar{E}$ for the image of E in $\bar{R} \otimes_R G$. Since $\mathcal{R}(E)/\mathfrak{P} \cong \mathcal{R}_{\bar{R}}(\bar{E})$, we may replace R and E by $\bar{R}$ and $\bar{E}$ to assume that R is a domain. But then the assertions follow from the dimension formula for graded domains (Lemma 1.21). □

Reductions and Integral Closure of a Module

By analogy with the case of ideals, one introduces a key notion.

Definition 8.2. Let U be a submodule of E. We say that U is a *reduction* of E, or equivalently E is *integral* over U, if $\mathcal{R}(E)$ is integral over the R-subalgebra generated by U.

Alternatively, the integrality condition is expressed by the equations $\mathcal{R}(E)_{s+1} = U \cdot \mathcal{R}(E)_s$, $s \gg 0$. The least integer $s \geq 0$ for which this equality holds is called the *reduction number* of E *with respect to* U and denoted by $r_U(E)$. For any reduction U of E the module E/U is torsion, hence U has the same rank as E. This follows from the fact that a module of linear type, such as a free module, admits no proper reductions.

Let E be a submodule of R^r. The *integral closure* of E in R^r is the largest submodule $\overline{E} \subset R^r$ having E as a reduction. (A broader definition exists, but we shall limit ourselves to this case.)

If R is a local ring with residue field k then the *special fiber* of $\mathcal{R}(E)$ is the ring $\mathcal{F}(E) = k \otimes_R \mathcal{R}(E)$; its Krull dimension is called the *analytic spread* of E and is denoted by $\ell(E)$.

Now assume in addition that k is infinite. A reduction of E is said to be *minimal* if it is minimal with respect to inclusion. For any reduction U of E one has $\nu(U) \geq \ell(E)$ (here $\nu(\cdot)$ denotes the minimal number of generators function), and equality holds if and only if U is minimal. Minimal reductions arise from the following construction. The algebra $\mathcal{F}(E)$ is a standard graded algebra of dimension $\ell = \ell(E)$ over the infinite field k. Thus it admits a Noether normalization $k[y_1, \ldots, y_\ell]$ generated by linear forms; lift these linear forms to elements $x_1, \ldots, x_\ell$ in $\mathcal{R}(E)_1 = E$, and denote by U

the submodule generated by $x_1,\ldots,x_\ell$. By Nakayama's Lemma, for all large r we have $\mathcal{R}(E)_{r+1} = U \cdot \mathcal{R}(E)_r$, making U a minimal reduction of E.

Having established the existence of minimal reductions, we can define the *reduction number* $r(E)$ of E to be the minimum of $r_U(E)$, where U ranges over all minimal reductions of E.

Proposition 8.3. *Let R be a Noetherian local ring of dimension $d \geq 1$ and E a finitely generated R-module having rank r. Then*

$$\boxed{r \leq \ell(E) \leq d + r - 1.} \tag{8.2}$$

Proof. We may assume that the residue field of R is infinite. Let $\mathfrak{m}$ be the maximal ideal of R and U any minimal reduction of E. Now $r = \text{rank } E = \text{rank } U \leq \nu(U) = \ell(E)$. On the other hand, by the proof of Proposition 8.1, $\mathfrak{m}\mathcal{R}(E)$ is not contained in any minimal prime of $\mathcal{R}(E)$. Therefore $\ell(E) = \dim \mathcal{F}(E) \leq \dim \mathcal{R}(E) - 1 = d + r - 1$, where the last equality holds by Proposition 8.1. □

In some cases one can fix completely the analytic spread.

Proposition 8.4. *Let $(R,\mathfrak{m})$ be a Noetherian local ring of dimension $d \geq 1$ with infinite residue field and E a finitely generated R-module having rank r. Suppose that $E \hookrightarrow R^r$ and $0 \neq \lambda(R^r/E) < \infty$. Then $\ell(E) = d + r - 1$.*

Proof. Let $R^s \xrightarrow{\varphi} R^r$ be a homomorphism with $s = \ell(E)$ and such that the image F of φ is a minimal reduction of E. For each prime ideal $\mathfrak{p} \neq \mathfrak{m}$, $F_\mathfrak{p}$ is a reduction of $E_\mathfrak{p} = R^r_\mathfrak{p}$, and therefore $F_\mathfrak{p} = R^r_\mathfrak{p}$. Thus the cokernel of φ is a (nonzero) module of finite length. By the Eagon-Northcott theorem, the height d of the ideal of maximal minors of φ must satisfy the inequality $d \leq s - r + 1$. Together with the estimate of Proposition 8.3, we obtain the desired equality. □

Remark 8.5. In general, if R is a local ring and E is an R-module that is free on the punctured spectrum, it is very difficult to describe its analytic spread in terms of the presentation data of E. We encounter this problem already when E is an ideal; later we shall face it when E is a conormal module I/I^2 of an ideal defining mild singularities.

Example 8.6. Let $\{x_1,\ldots,x_n\}$ and $\{y_1,\ldots,y_m\}$, $m \geq n$, be two sets of indeterminates over a field k, and let $R = k[x_1,\ldots,y_m]$ be the ring of polynomials in the full set of indeterminates. Consider the two ideals $I = (x_1,\ldots,x_n)$ and $J = (y_1,\ldots,y_m)$ and set $E = I \oplus J$.

We claim that $\ell(E) = m + n - 1$ and $\mathrm{r}(E) = n - 1$. Let U be the submodule of E generated by the elements

$$f_r = \sum_{1 \leq i \leq r} (x_i, y_{r-i}), \quad 2 \leq r \leq m + n.$$

Note that the special fiber of the Rees algebra $\mathcal{R}(E)$ is the semigroup ring $k[x_iy_j, 1 \leq i \leq n, 1 \leq j \leq m]$. The image of U in this ring is the minimal reduction used in Theorem 1.114, with the noted reduction number.

Remark 8.7. An advantage of working with a reduction F of a module E is that the corresponding Rees algebra $\mathcal{R}(F)$ has a simpler structure, yet it preserves some of the properties of $\mathcal{R}(E)$, such as its multiplicity. Let us consider two examples. First, suppose R is a Noetherian local domain of dimension 1 and wish infinite residue field, and that E is a torsionfree R-module of rank r. According to Proposition 8.3, $\ell(E) = r$. Thus every minimal reduction F of E is a free module, and so $\mathcal{R}(F)$ is a ring of polynomials over R.

Suppose now that R is a two-dimensional local normal domain (with infinite residue field), and that E is as above. Then $\ell(E) \leq r+1$. When $\ell(E) = r$, E is a free module (and therefore does not admit proper reductions). Suppose then that $\ell(E) = r+1$ and let F be a minimal reduction. Consider a presentation of F,

$$0 \to K \longrightarrow R^{r+1} \longrightarrow F \to 0,$$

in which K has the following description. If $e_1, \ldots, e_{r+1}$ is a basis of R^{r+1} and $z = \sum_{i=1}^{r+1} a_i e_i$ is a nonzero element of K, on setting $I = (a_1, \ldots, a_{r+1})$ we have $K \simeq I^{-1}z$, which can be verified by localizing at the height one primes of R. In the same vein, we obtain for any integer n the following presentation for F^n:

$$0 \to I^{-1}zS_{n-1}(R^{r+1}) \longrightarrow S_n(R^{r+1}) \longrightarrow S_n(F) = F^n \to 0.$$

This presentation shows that F is a module of linear type, and that its Rees algebra

$$\mathcal{R}(F) = \mathcal{R}(R^{r+1})/(I^{-1}z)$$

is a Cohen-Macaulay ring since $(I^{-1}z)$ is a Cohen-Macaulay ideal of $\mathcal{R}(R^{r+1})$ of codimension one.

Integral Closure and Normality

Let E be a module of rank r. Having considered the notion of reductions of E we make the following definition.

Definition 8.8. Let $E \hookrightarrow R^r$ be an embedding of modules of the same rank. Then E is *integrally closed in* R^r if it is not a proper reduction of any submodule of R^r. Further, E is *normal in* R^r if all components $\mathcal{R}(E)_n$ are integrally closed.

Remark 8.9. As defined thus far, the integral closure of a module has a relative character. Thus if E allows two embeddings, $\varphi_1 : E \to R^r$ and $\varphi_2 : E \to R^s$, the integral closures of E may be different. An *absolute integral closure* for E could be defined as follows: let K be the field of fractions of R, and set $T = K \otimes_R E$. For any valuation overring V of R, we can consider the image of $V \otimes_R E$ in $K \otimes_R E = T$, $VE \subset T$. The intersection $\overline{E}$ of these groups, where

$$E \subset \overline{E} = \bigcap_V VE \subset T,$$

is the desired closure. Note that when R is integrally closed it is easy to see that the integral closure of E in any embedding coincides with the absolute integral closure.

Proposition 8.10. *Let R be a normal domain and E a torsionfree finitely generated R-module. The Rees algebra*

$$\mathcal{R}(E) = R \oplus E \oplus E_2 \oplus \cdots$$

is integrally closed if and only if each E_n is an integrally closed module.

Proof. If each component E_n is integrally closed, we have

$$\bigcap_V V\mathcal{R}(E) = \sum_{n\geq 0} \bigcap_V VE_n = \mathcal{R}(E),$$

V running over all the valuation overrings of R. For each V we have $V\mathcal{R}(E) = \mathcal{R}(VE)$, where VE is a finitely generated torsionfree V-module and therefore $\mathcal{R}(VE)$ is a ring of polynomials over V since VE is a free V-module. This gives a representation of $\mathcal{R}(E)$ as an intersection of polynomial rings and it is thus normal. The converse is similar. □

There are scant means of finding the integral closure of modules. We shall return often to this issue under very specific conditions. Nevertheless some of the same general methods used for ideals one can view as steps towards the integral closure of modules. Here is a formulation of Proposition 1.58.

Proposition 8.11. *Let R be a Noetherian domain and $E \hookrightarrow R^r$ an embedding of a module of rank r. For every ideal I of R, $IE :_{R^r} I$ is integral over E.*

Ideal Modules

There is a class of modules that, like ideals, afford a *natural* embedding into a free module of the same rank. They provide the notions of (almost) complete intersection module, equimultiple module, link via a module, in analogy with the case of ideals.

Definition 8.12. Let R be a Noetherian ring and E a nonzero R-module. We say that E is an *ideal module* if E is finitely generated and torsionfree and the double dual E^{**} is free.

Actually the condition requires that E^* be a free R-module. Some of the interest arises from the analysis of the natural mapping $E \to E^{**}$. The following summarizes elementary characterizations of ideal modules ([SUV3, Proposition 5.1]).

Proposition 8.13. *Let R be a Noetherian ring and E a nonzero R-module. The following are equivalent.*

(a) *E is an ideal module.*
(b) *E is finitely generated and torsionfree and E^* is free.*
(c) *$E \subset G$, where G is a free R-module of finite rank and* grade $G/E \geq 2$.
(d) *$E \cong$* image *φ, where φ is a homomorphism of free R-modules of finite rank and* grade coker$(\varphi) \geq 2$.
(e) *$E \cong$* image *ψ^*, where*

$$0 \to G \xrightarrow{\psi} F \longrightarrow M \to 0$$

is a finite free resolution of a torsionfree R-module M.

Let R be a Noetherian local ring and E an ideal module of rank r. The notion of ideal module is useful for enabling the following definitions. As in the case of ideals, we define the *deviation* of E as

$$d(E) = \nu(E) - r + 1 - \text{grade } \det_0(E),$$

and its *analytic deviation* as

$$ad(E) = \ell(E) - r + 1 - \text{grade } \det_0(E).$$

Notice that if E is not free then $d(E) \geq ad(E) \geq 0$ by Proposition 8.25. Accordingly, we say that E is a *complete intersection module*, an *almost complete intersection module*, or an *equimultiple module* if E is an ideal module and if $d(E) \leq 0$, or $d(E) \leq 1$, or $ad(E) \leq 0$, respectively. These definitions coincide with the corresponding notions for ideals provided that the ideals are proper and have grade at least 2.

Example 8.14. Let us discuss a class of examples of such modules. Let R be a Cohen-Macaulay local ring and E a torsionfree R-module of projective dimension 2 with a resolution (not necessarily minimal, with E possibly of projective dimension ≤ 1) of the form

$$0 \to R^r \xrightarrow{\psi} R^n \xrightarrow{\varphi} R^n \longrightarrow E \to 0.$$

If φ is skew-symmetric then E is an ideal module. Indeed, dualizing this resolution, we have that $E^* = \text{Hom}_R(E,R)$ is identified with the kernel of φ^*, which is the same as the kernel of the mapping $-\varphi$. (Modules with such representations arise in the study of *commuting varieties* (see [BPV90], [Va94b, Chapter 9]).)

8.2 Rees Integrality Criteria

We collect here several multiplicity-based criteria and derive them via applications of Section 1.5. Let $B = R(E)$ be the Rees algebra of the module E to which we add the indeterminate t graded by setting deg $t = 0$. Let F be a submodule of E. Denote by C the subalgebra of $R(E)[t]$ generated by the 1-forms in E and Ft. This is the intertwining algebra we defined earlier; it could be denoted by $\mathcal{R}(F,E)$. Further, C is a subalgebra of the Rees algebra of $F \oplus E$ and coincides with it when F and E are ideals. The intertwining module of F and E will be written $T_{E/F}$. In this case one has $\dim C = \dim R + r + 1$ and $\dim T_{E/F} \leq \dim R + r$.

As an immediate consequence of Theorem 1.153, we obtain the following rather general characterization that has also been proved by Kirby and Rees ([KR94]).

Theorem 8.15. *Let R be a Noetherian ring of dimension d, let $F \subset E \subset R^r$ be R-modules, and assume that $\lambda(E/F) < \infty$. Then for $n \gg 0$, $\lambda(E^n/F^n)$ is a polynomial function $f(n)$ of degree at most $d+r-1$. Furthermore, if F is a reduction of E then the degree of $f(n)$ is less than $d+r-1$, and the converse holds when R is a quasi-unmixed local ring and* height ann $R^r/F > 0$.

Proof. We already know, from the general discussion of the modules $T_{E/F}$, at the beginning of Section 1.7, that for $n \gg 0$, $\lambda(E^n/F^n)$ is a polynomial function $f(n)$ of degree $\dim T_{E/F} - 1 \leq d+r-1$.

Now assume that F is a reduction of E. If height ann $E/F > 0$, then indeed $\dim T_{E/F} - 1 \leq d+r-1$ by Proposition 1.153. Otherwise $d = 0$. Now either $E = R^r$ and hence $F = E$ because a free module does not admit a proper reduction; or else some power of E is contained in R^{r-1} and thus $f(n)$ has degree less than $d+r-1$.

Conversely, if R is a quasi-unmixed local ring, height $R^r/F > 0$, and $\dim T_{E/F} - 1 < d+r-1$, then F is a reduction of E by Theorem 1.153. □

Corollary 8.16. *Let R be a quasi-unmixed local ring of dimension d and let $J \subset I$ be ideals such that $\lambda(I/J) < \infty$. Then for $n \gg 0$, $\lambda(I^n/J^n)$ is a polynomial function $f(n)$ of degree at most d. Furthermore, J is a reduction of I if and only if the degree of $f(n)$ is less than d.*

Proof. We replace the local ring $(R, \mathfrak{m})$ by $\widetilde{R} = R[X]_{(\mathfrak{m},X)}$, where X is a variable and the R-ideals $J \subset I$ by the $\widetilde{R}$-ideals $\widetilde{J} = (J, X) \subset \widetilde{I} = (I, X)$. This does not change our assumptions and conclusions because

$$\lambda(I^n/J^n) = \lambda(\widetilde{I}^n/\widetilde{J}^n) - \lambda(\widetilde{I}^{n-1}/\widetilde{J}^{n-1})$$

and J is a reduction of I if and only if $\widetilde{J}$ is a reduction of $\widetilde{I}$, and the corollary follows from Theorem 8.15. □

Proposition 8.17. *Let R be a quasi-unmixed local ring of dimension d, let $F \subset E \subset R^r$ be R-modules with* height (ann $R^r/F) > 0$, *and write $B = \mathcal{R}(E)$, $C = \mathcal{R}_B(FB)$, $G = \mathrm{gr}_{FB}(B)$, $T_{E/F} = (EC)/(FC) = EG$. If F is not a reduction of E, then for $n \gg 0$, $\lambda(E^n/F^n)$ is a polynomial*

$$f(n) = \frac{a}{(d+r-1)!}n^{d+r-1} + \textit{lower terms},$$

where a is the positive integer

$$e(T_{E/F}) = e(C/FC :_C EC) = e(G/0 :_G EG).$$

Proof. In the light of Theorem 8.15 and its proof we only need to show the equality $e(T_{E/F}) = e(B/\text{ann } T_{E/F})$. By the associativity formula for multiplicities, it suffices to check this locally at every prime $\mathfrak{P} \in \text{Supp}_C T_{E/F}$ with $\dim C/\mathfrak{P} = \dim T_{E/F} = d+r$. Let $\mathfrak{m}$ be the maximal ideal of R and set $k = R/\mathfrak{m}$. As $\mathfrak{m} \subset \mathfrak{P}$ we have $E \not\subset \mathfrak{P}$, since otherwise

$$\dim C/\mathfrak{P} \leq \dim C/(\mathfrak{m}, E)C = \dim k \otimes_R \mathcal{R}(F) = \ell(F) \leq d+r-1.$$

But $C_{\mathfrak{P}} = (EC/FC)_{\mathfrak{P}}$ is cyclic, showing that $C_{\mathfrak{P}} \cong (C/\text{ann}_C(T_{E/F}))_{\mathfrak{P}}$. □

Buchsbaum-Rim Multiplicity

Let R be a Noetherian local ring of dimension d. The Buchsbaum-Rim multiplicity ([BR65]) arises in the context of an embedding

$$0 \to E \xrightarrow{\varphi} R^r \longrightarrow C \to 0,$$

where C has finite length.

Denote by

$$\varphi : R^m \longrightarrow R^r$$

a matrix with image $\varphi \cong E$ such that grade coker $\varphi \geq 2$. There is a homomorphism

$$\mathcal{S}(\varphi) : \mathcal{S}(R^m) \longrightarrow \mathcal{S}(R^r)$$

of symmetric algebras whose image is $\mathcal{R}(E)$, and whose cokernel we denote by $C(\varphi)$,

$$0 \to \mathcal{R}(E) \longrightarrow \mathcal{S}(R^r) \longrightarrow C(\varphi) \to 0. \tag{8.3}$$

This exact sequence (with a different notation) is studied in [BR65] in great detail. Of significance for us is the fact that $C(\varphi)$, with the grading induced by the homogeneous homomorphism $S(\varphi)$, has components of finite length which for $n \gg 0$ satisfy the equality

$$\lambda(C(\varphi)_n) = \frac{\text{br}(E)}{(d+r-1)!} n^{d+r-1} + \text{lower degree terms.}$$

Here $\text{br}(E)$ is a non-vanishing positive integer (see Theorem 8.18), the *Buchsbaum-Rim multiplicity* of φ or of E. (In fact, for this one only needs to assume that coker φ is a module of finite length.)

It would be useful to have practical algorithms to find these multiplicities. We simply make the comment that there are raw comparative estimates of them. Let $\mathfrak{a}$ be any $\mathfrak{m}$-primary ideal contained in the annihilator of coker φ, for instance $\mathfrak{a} = I_r(\varphi)$. Then

$$\begin{aligned}\lambda(C(\varphi)_n) &\leq \lambda(S_n(R^r)/\mathfrak{a}^n S_n(R^r)) \\ &= \binom{r+n-1}{r-1} \lambda(R/\mathfrak{a}^n),\end{aligned}$$

since $\mathfrak{a}^n$ annihilates $C(\varphi)_n$. Thus

$$\operatorname{br}(E) \leq \binom{d+r-1}{r-1} \cdot e(\mathfrak{a}),$$

where $e(\mathfrak{a})$ is the Hilbert-Samuel multiplicity of the ideal $\mathfrak{a}$.

One should expect $\operatorname{br}(E)$ to be, in most cases, not larger than the multiplicity of $I_r(\varphi)$. To illustrate the reason, let $(R,\mathfrak{m})$ be a regular local ring of dimension $d \geq 2$ and consider the module $E = \mathfrak{m} \oplus \mathfrak{m}$ with its natural embedding into R^2. Then $I = I_2(\varphi) = \mathfrak{m}^2$, and so I has multiplicity 2^d. The Buchsbaum-Rim polynomial is

$$\lambda(S_n(R^2)/E^n) = \lambda(S_n(R/I^n)) = (n+1)\binom{n+d-1}{d} = \frac{d+1}{(d+1)!}n^{d+1} + \text{lower terms},$$

so that $\operatorname{br}(E) = d+1$.

Theorem 8.18. *If $R^r/E \neq 0$ then $\lambda(S_n(R^r)/E^n)$ is a polynomial in n of degree $d+r-1$ for $n \gg 0$.*

Proof. By Theorem 8.15, $\lambda(S_n(R^r)/E^n)$ is a polynomial of degree at most $d+r-1$ for $n \gg 0$. To show that the degree cannot be less, we may complete R and factor out a minimal prime of maximal dimension to assume that R is a complete domain (by Nakayama's Lemma the assumption $C(\varphi) \neq 0$ is preserved). Now if $d > 0$ then the assertion follows from Theorem 8.15, since the free module R^r does not admit a proper reduction. If on the other hand $d = 0$ then R is a field and the claim is obvious. □

Remark 8.19. The coefficients of the polynomial

$$\lambda(S_n(R^r)/E^n) = \operatorname{br}(E)\binom{n+d+r-2}{d+r-1} - \operatorname{br}_1(E)\binom{n+d+r-3}{d+r-2} + \text{lower terms},$$

called the *Buchsbaum-Rim polynomial* of E, (for $n \gg 0$) are called the *Buchsbaum-Rim coefficients* of E. The leading coefficient $\operatorname{br}(E)$ is the Buchsbaum-Rim multiplicity of φ; if the embedding φ is understood, we shall simply denote it by $\operatorname{br}(E)$. This number is determined by an Euler characteristic of the Buchsbaum-Rim complex ([BR65]).

Corollary 8.20. *Let R be a quasi-unmixed local ring of dimension $d > 0$ and let F and E be R-modules such that $F \subset E \subsetneq R^r$ and $\lambda(R^r/F) < \infty$. Then F is a reduction of E if and only if* $\operatorname{br}(F) = \operatorname{br}(E)$.

Proof. By Remark 8.19, $\operatorname{br}(F) = \operatorname{br}(E)$ if and only if

$$\lambda(E^n/F^n) = \lambda(S_n(R^r)/F^n) - \lambda(S_n(R^r)/E^n)$$

is a polynomial of degree $< d+r-1$ for $n \gg 0$. According to Theorem 8.15, the latter condition is equivalent to the condition that F is a reduction of E. □

The special case $r = 1$ is Theorem 1.154, the classical result of Rees ([Re61]) and the first reduction criterion based on multiplicities.

McAdam Theorem

The next reduction criterion was proved by McAdam for ideals ([Mc83]) and later by Rees for the case of modules ([Re85]). It arises as a consequence of Theorem 1.153.

Theorem 8.21. *Let R be a quasi-unmixed local ring, let F and E be R-modules such that $F \subset E \subset R^r$ and* height ann $R^r/F > 0$, *and write $\ell = \ell(F)$. Then F is a reduction of E if and only if $F_\mathfrak{p}$ is a reduction of $E_\mathfrak{p}$ for every prime $\mathfrak{p}$ of R with* ann $E/F \subset \mathfrak{p}$ *and* height $\mathfrak{p} = \ell(F_\mathfrak{p}) - r + 1 \leq \ell - r + 1$.

Proof. We need to show that the local conditions force F to be a reduction of E. For this we may assume that E/F is cyclic. We use the notation of Theorem 1.153, $B = \mathcal{R}(E)$ and $C = \mathcal{R}_B(FB)$. By that theorem, F is a reduction of E if and only if height $FC :_C EC \geq 2$. Thus, for any minimal prime $\mathfrak{P}$ of $FC :_C EC$, we need to show that height $\mathfrak{P} \geq 2$.

Write $d = \dim R$, let $\mathfrak{m}$ be the maximal ideal of R and set $k = R/\mathfrak{m}$. Localizing at the contraction of $\mathfrak{P}$ (which preserves quasi-unmixedness) we may suppose that $\mathfrak{m}C \subset \mathfrak{P}$. Now if $d = \ell - r + 1$, then height $\mathfrak{P} \geq 2$ by our assumption and Theorem 1.153. If on the other hand $d \neq \ell - r + 1$, then $d \geq \ell - r + 2$. Furthermore, since E/F is cyclic, $C/(\mathfrak{m}, F)C$ is an algebra over $k \otimes_R \mathcal{R}(F)$ generated by one element. Thus $\dim C/(\mathfrak{m}, F)C \leq \ell + 1$. Putting this together and using the quasi-unmixedness of R, we obtain that

$$\begin{aligned}\text{height } \mathfrak{P} &\geq \text{height } (\mathfrak{m}, F)C = \dim C - \dim C/(\mathfrak{m}, F)C \geq d + r + 1 - (\ell + 1) \\ &\geq \ell - r + 2 + r + 1 - (\ell + 1) = 2,\end{aligned}$$

as desired. □

Results of Böger and Kleiman-Thorup

The next result, which has also been proved by Kleiman and Thorup ([KT94]), is a generalization of Theorem 8.15. It follows immediately from Theorems 8.15 and 8.21.

Theorem 8.22. *Let R be a quasi-unmixed local ring, let F and E be R-modules such that $F \subset E \subset R^r$ and* height ann $R^r/F > 0$, *and write $\ell = \ell(F)$. Then F is a reduction of E if and only if for every minimal prime $\mathfrak{p}$ of* ann E/F *with* height $\mathfrak{p} = \ell - r + 1 = \ell(F_\mathfrak{p}) - r + 1$, *the degree of the polynomial $\lambda((E^n/F^n)_\mathfrak{p})$, $n \gg 0$, is less than ℓ.*

If F is an equimultiple module then the above takes a slightly different form. In this form Böger had proved the result for ideals ([Bo69]), as a first generalization of [Re61].

We now explain what we mean by an equimultiple module. Let R be a Noetherian local ring and fix an embedding $F \subsetneq R^r$. Write $\ell = \ell(F)$ and notice that height $\mathfrak{p} \leq \ell - r + 1$ for every minimal prime of ann R^r/F or, equivalently, of the Fitting ideal $\text{Fitt}_0(R^r/F)$. To see this we may assume that the residue field of R is infinite. But then

F has a reduction G generated by ℓ elements. Since a free module does not admit any proper reduction, it follows that

$$\sqrt{\operatorname{ann} R^r/F} = \sqrt{\operatorname{ann} R^r/G}.$$

Thus we may replace F by G to assume that F is generated by ℓ elements. Now by Macaulay's bound, every minimal prime ideal of $\operatorname{Fitt}_0(R^r/F)$ has height at most $\ell - r + 1$.

We say that F is *equimultiple* if height ann $R^r/F \geq \ell - r + 1$. By the above discussion, if F is equimultiple then height $\mathfrak{p} = \ell - r + 1$ for every minimal prime $\mathfrak{p}$ of ann R^r/F. (Our definition of equimultiple module in [SUV3] is more restrictive, but has the advantage of being independent of the embedding $F \subsetneq R^r$.)

Theorem 8.23. *Let R be a quasi-unmixed local ring and let F and E be R-modules such that $F \subsetneq E \subset R^r$. Suppose that F is equimultiple with* height ann $R^r/F > 0$. *Then F is a reduction of E if and only if $\sqrt{\operatorname{ann} R^r/F} = \sqrt{\operatorname{ann} R^r/E}$ and* $\operatorname{br}(F_{\mathfrak{p}}) = \operatorname{br}(E_{\mathfrak{p}})$ *for every minimal prime $\mathfrak{p}$ of the latter ideal.*

Proof. We have already seen that if F is a reduction of E then $\sqrt{\operatorname{ann} R^r/F} = \sqrt{\operatorname{ann} R^r/E}$. Furthermore since F is equimultiple, the prime ideals $\mathfrak{p}$ with $\mathfrak{p} \supset$ ann E/F and height $\mathfrak{p} \leq \ell - r + 1$ are necessarily minimal primes of ann R^r/F. Now the asserted equivalence follows from Theorem 8.21 and Corollary 8.20. □

Multiplicities

Let $A \subset B$ be a homogeneous inclusion of standard graded Noetherian rings with $A_0 = B_0$. It is worthwhile to exploit the situation when $\lambda_R(B_1/A_1)$ is finite by introducing a series of relative multiplicities $e_t(A,B)$, one for each $t \geq 1$, associated with the numerical functions $\lambda_R(B_n/A_{n-t+1}B_{t-1})$.

We follow closely [SUV1], where several other applications are treated. Write $\mathcal{R} = \mathcal{R}_B(A_1B) \subset B[T]$ and $G = \operatorname{gr}_{A_1B}(B)$ for the Rees algebra and the associated graded ring of the B-ideal A_1B, respectively. We endow $\mathcal{R}$ with a bigrading by assigning bidegree $(1,0)$ to the elements of B_1 and bidegree $(0,1)$ to the elements of A_1T. Thus G also becomes a bigraded ring and one has

$$G/B_1G \cong \mathcal{R}/B_1\mathcal{R} \cong \bigoplus_{i\geq 0} \mathcal{R}_{(0,i)} = \bigoplus_{i\geq 0} A_iT^i \cong A,$$

which yields an identification $V(B_1G) = \operatorname{Spec}(A)$.

Lemma 8.24. *In the setting above,*

(a) $\dim A \leq \dim B = \dim G$.
(b) *Suppose that $A_0 = B_0$ is local and B is equidimensional and universally catenary. Then*
 (i) *G is equidimensional;*

(ii) *if* $B_{\mathfrak{q}\cap A}$ *is integral over* $A_{\mathfrak{q}\cap A}$ *for every minimal prime* $\mathfrak{q}$ *of* B*, then every minimal prime of* B *contracts to a minimal prime of* A*, and moreover* A *is equidimensional with* $\dim A = \dim B$.

Proof. (a) Since $A = G/B_1G$, one has $\dim A \leq \dim G$, and as there exists a maximal ideal of B of maximal height containing B_1B it is well known that $\dim G = \dim B$.

(b) To deal with (i), since G is bigraded, we may replace B by the localization at its homogeneous maximal ideal. But then the extended Rees algebra $\mathcal{R}[T^{-1}]$ is an equidimensional and catenary $\mathbb{Z}$-graded ring with a unique maximal homogeneous ideal. As $G \cong \mathcal{R}[T^{-1}]/(T^{-1})$ with T^{-1} a homogeneous non zerodivisor, it follows that G is equidimensional.

As for (ii), notice that every minimal prime $\mathfrak{p}$ of A is a contraction of a minimal prime of B, as can be seen by localizing at $\mathfrak{p}$. The assertion now follows from the dimension formula for graded Noetherian domains (see, e.g., Lemma 1.21). □

Theorem 8.25. *With the notation above, let* s *and* t *be integers with* $t > 0$, $s \geq 0$.

(a) $V((B_1G + 0:_G B_tG)/B_1G) = \operatorname{Supp}_A(B/\sum_{i=0}^{t-1} B_iA)$*; thus* $(B_1G + 0:_G B_tG)/B_1G$ *has height greater than* s *if and only if* $B_\mathfrak{p} = \sum_{i=0}^{t-1} B_iA_\mathfrak{p}$ *for every prime* $\mathfrak{p}$ *of* A *with* $\dim A_\mathfrak{p} \leq s$.

(b) height $0:_G B_tG > s$ *if and only if* B *is integral over* A *and* $B_\mathfrak{p} = \sum_{i=0}^{t-1} B_iA_\mathfrak{p}$ *for every prime* $\mathfrak{p}$ *of* A *with* $\dim A_\mathfrak{p} \leq s$.

Proof. (a) We have

$$V((B_1G + 0:_G B_tG)/B_1G) = \operatorname{Supp}_{G/B_1G}(B_tG/B_{t+1}G) = \operatorname{Supp}_A(B_tG/B_{t+1}G),$$

where the first equality follows from Nakayama's Lemma and the equality $B_{t+1}G = (B_1G)(B_tG)$, while the second uses the identification $G/B_1G \cong A$. Looking at the graded components of G, one sees that there is an isomorphism of A-modules

$$B_tG/B_{t+1}G \cong \sum_{i=0}^{t} B_iA/\sum_{i=0}^{t-1} B_iA,$$

which yields $\operatorname{Supp}_A(B_tG/B_{t+1}G) = \operatorname{Supp}_A(\sum_{i=0}^{t} B_iA/\sum_{i=0}^{t-1} B_iA)$. Finally,

$$\operatorname{Supp}_A(\sum_{i=0}^{t} B_iA/\sum_{i=0}^{t-1} B_iA) = \operatorname{Supp}_A(B/\sum_{i=0}^{t-1} B_iA),$$

since B is standard graded.

(b) If height $0:_G B_tG > s \geq 0$, then $(B_tG)_\mathfrak{q} = 0$ for every minimal prime $\mathfrak{q}$ of G hence $B_tG \subset \mathfrak{q}$, which implies that $B_1G \subset \sqrt{0}$. On the other hand, $B_1G \subset \sqrt{0}$ if and only if $B_1 \subset \sqrt{A_1B}$, which is equivalent to the integrality of B over A.

Thus we may assume in either case that $B_1G \subset \sqrt{0}$ and that B is integral over A. The asserted equivalence now follows from part (a). □

Corollary 8.26. *Let t be a positive integer. Then* height $0:_G B_tG > 0$ *if and only if B is integral over A and $B_\mathfrak{p} = \sum_{i=0}^{t-1} B_iA_\mathfrak{p}$ for every minimal prime $\mathfrak{p}$ of A.*

Notice that, in the terminology of reductions, the equality $B_\mathfrak{p} = \sum_{i=0}^{t-1} B_iA_\mathfrak{p}$ for all minimal primes $\mathfrak{p}$ of A means that the generic reduction number of $A \subset B$ is at most $t-1$.

Corollary 8.27. (a) height $0:_G B_1G > 0$ *if and only if B is integral over A and $B_\mathfrak{p} = A_\mathfrak{p}$ for every minimal prime $\mathfrak{p}$ of A.*
(b) height $0:_G (B_1G)^\infty > 0$ *if and only if B is integral over A.*

Proof. Apply Corollary 8.26, noticing that $0:_G (B_1G)^\infty = 0:_G B_tG$ for $t \gg 0$. □

Let S be a Noetherian ring and M a finitely generated S-module. If $(S, \mathfrak{m})$ is local and $\mathfrak{a}$ is an $\mathfrak{m}$-primary ideal, write $e_\mathfrak{a}(M)$ for the multiplicity of M with respect to $\mathfrak{a}$ and $e(M)$ for $e_\mathfrak{m}(M)$. If on the other hand S is standard graded with S_0 Artinian local and M is graded, $e(M)$ denotes the multiplicity of M. In this case $e(M) = e(M_\mathfrak{m})$, where $\mathfrak{m}$ is the homogeneous maximal ideal of S.

We shall stay within the following setup. Let $A \subset B$ be a homogeneous inclusion of standard graded Noetherian rings with $R = A_0 = B_0$ local and $\lambda(B_1/A_1) < \infty$. Write $d = \dim B$ and $\mathfrak{m}$ for the maximal ideal of R. The bigraded ring $G = \mathrm{gr}_{A_1B}(B)$ introduced also earlier admits a standard grading as an R-algebra given by total degree. If t is a positive integer and $\mathfrak{a} = \mathrm{ann}_R B_1/A_1$, then B_tG is a finitely generated graded module over $G/0:_G B_1G$, the latter being a standard graded ring with $[G/0:_G B_1G]_0 = R/\mathfrak{a}$ Artinian local.

In the next proposition we introduce a family of 'relative' multiplicities $e_t(A,B)$.

Proposition 8.28. *With the assumptions above, the following hold:*

(a) *For every $n \geq t-1$, $\lambda(B_n/A_{n-t+1}B_{t-1}) = \lambda([B_tG]_n)$.*
(b) *For $n \gg 0$, $\lambda(B_n/A_{n-t+1}B_{t-1})$ is a polynomial function $f_t(n)$ of degree*

$$\dim B_tG - 1 = \dim G/0:_G B_tG - 1 \leq \dim G - 1 = d - 1.$$

(c) *The polynomial $f_t(n)$ is of the form*

$$f_t(n) = \frac{e_t(A,B)}{(d-1)!} n^{d-1} + \textit{lower terms},$$

where $e_t(A,B) = 0$ (if $\dim G/0:_G B_tG < d$) or $e_t(A,B) = e(B_tG)$ (if $\dim G/0:_G B_tG = d$).

Proof. Notice that

$$[B_tG]_n = \bigoplus_{j=1}^{n-t+1} A_{j-1}B_{n-j+1}/A_jB_{n-j},$$

which gives (a). The remaining assertions follow from (a) because B_tG is a finitely generated graded module over a ring that is standard graded over an Artinian local ring. □

The sequence of multiplicities $e_t(A,B)$ is obviously non-increasing. According to [SUV1, Proposition 3.4], $e_t(A,B)$ stabilizes for $t \gg 0$.

Remark 8.29. There is room for several other kinds of multiplicities defined on modules. Let us sketch one but leave its development to the reader's imagination. Let $(R,\mathfrak{m})$ be a Noetherian local containing $\mathbb{Z}/(p)$ for some prime integer $p > 0$, and let a $E \hookrightarrow R^r$ be a R–module of rank r and finite colength. We fix a basis $x_1,\ldots,x_r$ for R^r taken as variables in the ring of polynomials $S = R[x_1,\ldots,x_r]$ (like in setting up the Rees algebra of E). For each prime power $q = p^f$, let $E^{[q]}$ be the R-submodule of S generated by the h^q, $h \in E$. This module is actually a R-free module $S_1^{[q]}$ generated by $x_1^q,\ldots,x_r^q$. It is rather tempting to examine the function

$$q \mapsto \lambda(S_1^{[q]}/E^{[q]},$$

an extension of the Hilbert-Kunz function for modules.

8.3 Reduction Numbers of Modules

In this section we develop tools to estimate reduction numbers of modules, making use of the multiplicities of the special fibers. We begin by quoting some results from [SUV3].

Special Fibers and Buchsbaum-Rim Multiplicities

Let $(R,\mathfrak{m})$ be a Noetherian local ring of dimension $d \geq 1$ and E an ideal module of rank r that is a vector bundle (that is, E is free on the punctured spectrum of R) but is not R-free. There is one case where we have an explicit formula for the Buchsbaum-Rim multiplicity $\mathrm{br}(E)$. It arises as follows. Let $(R,\mathfrak{m})$ be a Cohen-Macaulay local ring of dimension d and let

$$E \hookrightarrow R^r$$

be a vector bundle. Assuming that R has an infinite residue field, let U be a minimal reduction of E. According to Proposition 8.4, U is minimally generated by $d+r-1$ elements and is also an ideal module. In the terminology of [BR65], F is a parameter module. We shall refer to such modules as *complete intersection modules*.

Theorem 8.30. *Under the conditions above, let* $\varphi : R^{d+r-1} \to R^r$ *be a mapping whose image is* U. *Then*

$$\mathrm{br}(E) = \mathrm{br}(U) = \lambda(R^r/U) = \lambda(R/\det_0(U)),$$

where $\det_0(U)$ *is the ideal generated by the* r *by* r *minors of* φ.

The first equality follows from [KT94, 5.3(i)], for instance; the second from [BR65, 4.5(2)], and the third from [BV86, 2.8]. We are going to complement this in Theorem 8.37, where the entire Buchsbaum-Rim function of U is given.

Following [SUV3], we shall make use of a modified value of these numbers. We denote by $\mathrm{br}_0(E)$ the infimum of all $\mathrm{br}(\widetilde{E})$, where x is an element of $\mathfrak{m}$ not contained in any of the minimal primes of R, $\overline{R} = R/(x)$, and $\widetilde{E}$ is the image of E in $\overline{R}^r$. We first want to compare this new invariant to $\mathrm{br}(E)$.

We quote some results from [SUV3, Section 4].

Proposition 8.31. *If R is a Cohen-Macaulay local ring with infinite residue field and E an ideal module that is a vector bundle but not free, then* $\mathrm{br}_0(E) \leq \mathrm{br}(E)$.

Theorem 8.32. *Let R be a Noetherian local ring and E an ideal module that is a vector bundle, but not free. Then the special fiber $\mathcal{F}(E)$ has multiplicity at most* $\mathrm{br}_0(E)$.

Corollary 8.33. *Let R and E be as in* Theorem 8.32, *and suppose further that the residue field of R has characteristic zero. If $\mathcal{F}(E)$ satisfies the condition S_1 and is equidimensional then* $\mathrm{r}(E) \leq \mathrm{br}_0(E) - 1$. *If in addition R is Cohen-Macaulay with infinite residue field then $r(E) \leq$* $\mathrm{br}(E) - 1$.

The hypothesis on the condition S_1 and equidimensionality always holds if R is a positively graded domain over a field and E is graded, generated by elements of the same degree. In this case $\mathcal{F}(E)$ embeds into $\mathcal{R}(E)$.

Example 8.34. Suppose that $R = k[x_1, \ldots, x_d]$ and $E = (x_1, \ldots, x_d)^{\oplus r}$, $d \geq 2, r > 0$, with k an infinite field. Applying the argument above we find that $\mathcal{F}(E)$ has multiplicity bounded by $\binom{d+r-2}{r-1}$. This is much too large as a bound for $\mathrm{r}(E)$. In fact, $\mathcal{F}(E) \cong k[x_i y_j \mid 1 \leq i \leq d,\ 1 \leq j \leq r]$. This last ring is studied in [CVV98] and has $\min\{d-1, r-1\}$ for reduction number (in any characteristic).

Maximal Hilbert Function of the Special Fiber

We approach the calculation of the reduction number $\mathrm{r}(E)$ using the strategy employed in the proof of Theorem 2.44. Denote by $\mathcal{F}(E)$ the special fiber of E, $\mathcal{F}(E) = \mathcal{R}(E) \otimes R/\mathfrak{m}$. We seek to estimate the Hilbert function of this standard graded algebra.

Theorem 8.35. *Let $(R, \mathfrak{m})$ be a Cohen-Macaulay local ring of dimension $d > 1$ and E a module as above. For all integers $n \geq 0$,*

$$\lambda(\mathcal{F}(E)_n) \leq \mathrm{br}(E) \cdot \binom{n+d+r-3}{d+r-2} + \binom{n+d+r-3}{d+r-3}. \tag{8.4}$$

Proof. We begin by recalling how generic minimal reductions are formed, according to [SUV3]. Write $E = \sum_{i=1}^n Ra_i$, let z_{ij}, $1 \leq j \leq d+r-1$, $1 \leq i \leq n$, be variables and set

$$S = R(\{z_{ij}\}) = R[\{z_{ij}\}]_{\mathfrak{m}R[\{z_{ij}\}]}, \quad E'' = S \otimes_R E, \quad x_j = \sum_{i=1}^{n} z_{ij} a_i.$$

Consider the S-submodule $U = \sum_{j=1}^{d+r-2} Sx_j$ of E''.

After applying the change of rings, we switch back to the original notation. We note that U is a minimal reduction of E. Since the residue field of the original R is infinite, by evaluation of the variables z_{ij} we would obtain a minimal reduction of the original module. Observe that the change of rings leaves the Hilbert function of $\mathcal{F}(E)$ unaffected.

The method is that of the proof of Theorem 2.44. Let U^n be the component of degree n of the Rees algebra of U. Since

$$\lambda(\mathcal{F}(E)_n) = \nu(E^n) \leq \nu(U^n) + \nu(E^n/U^n)$$

and

$$\nu(U^n) = \binom{n+d+r-3}{d+r-3},$$

it is enough to estimate the number of generators of the submodule

$$E^n/U^n \hookrightarrow S_n(R^r)/U^n.$$

We want to show the following. (i) $S_n(R^r)/U^n$ is a Cohen-Macaulay module of dimension 1, and (ii) it has multiplicity at most

$$\mathrm{br}(E) \cdot \binom{n+d+r-3}{d+r-2}.$$

We shall then use this value to bound $\nu(E^n/U^n)$.

Claim (i) follows from the next result:

Theorem 8.36. *Let R be a Cohen-Macaulay local ring of dimension d and E a complete intersection module of rank r. Then E is of linear type and $\mathcal{R}(E)$ is Cohen-Macaulay. Moreover, if* height $\det_0(E) = c \geq 2$, *then* depth $E^n = d - c + 1$ *for all $n > 0$. In particular, $S_n(R^r)/E^n$ is a Cohen-Macaulay module of depth $d - c$.*

Proof. By [KN95, Theorem 3.2] and [SUV3, Corollary 5.6], E is of linear type and $\mathcal{R}(E)$ is Cohen-Macaulay. To prove the remaining assertion, write $\mathfrak{m}$ for the maximal ideal of R and observe that

$$\mathrm{grade}\ \mathfrak{m}\mathcal{R}(E) = \mathrm{height}\ \mathfrak{m}\mathcal{R}(E) = \dim \mathcal{R}(E) - \dim \mathcal{F}(E) = d + r - \ell(E) = d - c + 1.$$

Thus all the components of $\mathcal{R}(E)$, that is the R-modules E^n have depth at least $d - c + 1$. Therefore $S_n(R^r)/E^n$ are Cohen-Macaulay of depth $d - c$. $\square$

As for the claim (ii), we choose a minimal reduction (x) of the maximal ideal of $R/\mathrm{ann}\,(R^r/U))$ and write $\overline{U}$ for the image of U in $\overline{R}^r = (R/xR)^r$. By [SUV3, Theorem 5.16], $\mathrm{br}(\overline{U}) \leq \mathrm{br}(E)$. Thus it remains to prove that

$$\lambda(S_n(\overline{R}^r)/\overline{U}^n) \leq \mathrm{br}(\overline{U}) \cdot \binom{n+(d-1)+r-2}{(d-1)+r-1}.$$

This is a consequence of the next theorem.

Theorem 8.37. *Let $(R,\mathfrak{m})$ be a Cohen-Macaulay local ring of dimension $d \geq 1$ and let $E \subsetneq R^r$ be a submodule such that $\lambda(R^r/E) = \infty$ and $\nu(E) \leq d+r-1$. For all $n \geq 0$,*

$$\lambda(S_n(R^r)/E^n) = \mathrm{br}(E) \cdot \binom{n+d+r-2}{d+r-1}.$$

Proof. Set $R^r = G$, $s = d+r-1$ and let X be a set of rs indeterminates. Write $\widetilde{R} = R[X]_{(X,\mathfrak{m})}$, $\widetilde{G} = \widetilde{R} \otimes_R G$ and define the module $\widetilde{E}$ as the image of the mapping

$$\widetilde{R}^s \stackrel{\mathbf{X}}{\longrightarrow} \widetilde{G},$$

where $\mathbf{X}$ is the generic $r \times s$ matrix defined by X. Note that $\widetilde{E}$ is a complete intersection module over $\widetilde{R}$. If we denote by φ the mapping

$$R^s \stackrel{\varphi}{\longrightarrow} R^r$$

with image $\varphi = E$, we want to examine the effects of the specialization $\mathbf{X} \rightarrow \varphi$ on the modules $\widetilde{G}^n/\widetilde{E}^n$, since

$$G^n/E^n = (\widetilde{G}^n/\widetilde{E}^n) \otimes_{\widetilde{R}} (\widetilde{R}/I_1(\mathbf{X}-\varphi)).$$

By Theorem 8.36, $\widetilde{G}^n/\widetilde{E}^n$ are Cohen-Macaulay for all $n > 0$, and consequently $\lambda(G^n/E^n) = e(I_1(\mathbf{X}-\varphi); \widetilde{G}^n/\widetilde{E}^n)$.

To obtain this multiplicity, let $\mathfrak{p}$ be a minimal prime of R. Write $\mathfrak{P} = (\mathfrak{p}, I_r(X)) \subset \widetilde{R}$, $S = \widetilde{R}_{\mathfrak{P}}$ and $I = I_r(X)S$. Note that $I_{r-1}(X) \not\subset \mathfrak{P}$. Localizing the mapping

$$\widetilde{R}^s \stackrel{X}{\longrightarrow} \widetilde{G}$$

at $\mathfrak{P}$, we get that the image is given by

$$G_1 \oplus Ie_r \subset G_1 \oplus Se_r = \widetilde{G} \otimes_{\widetilde{R}} S,$$

where G_1 is a free direct summand. Notice that I is generated by a regular sequence of length $s-r+1 = d$. Thus we have

$$\begin{aligned}
\lambda((\widetilde{G}^n/\widetilde{E}^n)_{\mathfrak{P}}) &= \sum_{i=0}^{n} \lambda(G_1^{n-i} \otimes S/I^i) \\
&= \sum_{i=1}^{n} \binom{r-1+n-i-1}{r-2} \binom{d+i-1}{d} \cdot \lambda(S/I) \\
&= \binom{d+r+n-2}{d+r-1} \cdot \lambda((\widetilde{R}/I_r(X))_{\mathfrak{P}}),
\end{aligned}$$

where the last equality holds by a standard summation formula.

Now, applying the associativity formula twice, we obtain the chain of equalities

$$\begin{aligned}
e(I_1(X-\varphi);\widetilde{G}^n/\widetilde{E}^n) &= \sum_{\mathfrak{p}} \lambda((\widetilde{G}^n/\widetilde{E}^n)_{\mathfrak{P}}) e(I_1(X-\varphi);\widetilde{R}/\mathfrak{P}) \\
&= \binom{d+r+n-2}{d+r-1} \sum_{\mathfrak{p}} \lambda(\widetilde{R}/I_r(X))_{\mathfrak{P}}) e(I_1(X-\varphi);\widetilde{R}/\mathfrak{P}) \\
&= \binom{d+r+n-2}{d+r-1} e(I_1(X-\varphi);\widetilde{R}/I_r(X)) \\
&= \binom{d+r+n-2}{d+r-1} \lambda(R/I_r(\varphi)),
\end{aligned}$$

where the last equality holds as $\widetilde{R}/I_r(X)$ is Cohen-Macaulay. Finally, $\lambda(R/I_r(\varphi)) = \text{br}(E)$ by Theorem 8.30. □

8.3.1 Reduction Number of a Module

Let us first clarify, in terms of the coefficients of the Buchsbaum-Rim polynomial, when a module E is a complete intersection.

Let $(R,\mathfrak{m})$ be a Cohen-Macaulay local ring of dimension $d \geq 1$ and E an ideal module which is a vector bundle. Let $E \hookrightarrow R^r$ be the natural embedding and consider the Buchsbaum-Rim polynomial

$$\lambda(S_n(R^r)/E^n) = \text{br}(E)\binom{n+d+r-2}{d+r-1} - \text{br}_1(E)\binom{n+d+r-3}{d+r-2} + \text{lower terms},$$

with $\text{br}(E) = \lambda(R^r/U)$, where U is a minimal reduction of E. From Theorem 8.37, we have that

$$\lambda(S_n(R^r)/U^n) = \text{br}(E)\cdot\binom{n+d+r-2}{d+r-1},$$

and therefore

$$\lambda(E^n/U^n) = \text{br}_1(E)\binom{n+d+r-3}{d+r-2} + \text{lower terms}.$$

In this expression, it follows that $\text{br}_1(E) \geq 0$. Actually one can explain the case when $\text{br}_1(E)$ vanishes in the following manner. Consider the following embedding of Rees algebras

$$0 \to \mathcal{R}(U) \longrightarrow \mathcal{R}(E) \longrightarrow \bigoplus_{n\geq 1} E^n/U^n \to 0.$$

This is an exact sequence of finitely generated modules over $\mathcal{R}(U)$, and since $\mathcal{R}(U)$ is Cohen-Macaulay of dimension $d+r$, the module on the right must have Krull dimension precisely $d+r-1$ unless it vanishes. This means that $\text{br}_1(E) = 0$ only if $U = E$. This is the module extension of the result of [Nor59] for ideals. It would be interesting to have also the inequality $\text{br}_2(E) \geq 0$, as occurs in the case of ideals.

We now turn to the question of the reduction number of the module E as above. We seek submodules U minimally generated by $\ell(E) = d+r-1$ elements such that $\mathcal{R}(E)_{s+1} = U \cdot \mathcal{R}(E)_s$ for some small integer s. For that, we consider the Hilbert function of the special fiber $\mathcal{F}(E)$ in order to apply Theorem 2.36.

Theorem 8.38. *Let R be a Cohen-Macaulay local ring of dimension $d > 1$, with infinite residue field, and let E be a submodule of R^r such that $\lambda(R^r/E) < \infty$. The reduction number of E satisfies the inequality*

$$\mathrm{r}(E) \leq (d+r-1)\cdot \mathrm{br}(E) - 2(d+r-1)+1 = \ell(E)\cdot \mathrm{br}(E) - 2\cdot \ell(E)+1.$$

Proof. From Theorem 8.35, we have

$$\lambda(\mathcal{F}(E)_n) \leq \mathrm{br}(E)\cdot \binom{n+d+r-3}{d+r-2} + \binom{n+d+r-3}{d+r-3}.$$

To apply Theorem 2.36, we need to find n so that

$$\lambda(\mathcal{F}(E)_n) < \binom{n+\ell(E)}{\ell(E)}.$$

Such a value satisfies the inequality $n > \mathrm{r}(E)$. It will be enough to solve for the smallest n in the inequality

$$\mathrm{br}(E)\cdot \binom{n+d+r-3}{d+r-2} + \binom{n+d+r-3}{d+r-3} < \binom{n+\ell(E)}{\ell(E)} = \binom{n+d+r-1}{d+r-1}.$$

Simplifying common factors, we have

$$\mathrm{br}(E)n(d+r-1) + (d+r-1)(d+r-2) < (n+d+r-1)(n+d+r-2),$$

from which we obtain the desired bound for $\mathrm{r}(E)$. □

Remark 8.39. The estimate in Theorem 8.38 still holds, and can even be improved if $d = 1$. Let R be a Cohen-Macaulay local ring of dimension one with infinite residue field k, let $E \subset R^r$ be a nonfree submodule of rank r and write $a = \nu(R^r/E)$. Then

$$\mathrm{r}(E) \leq \mathrm{br}(E) - a \leq \ell(E)\mathrm{br}(E) - 2\ell(E) + 1.$$

If in addition char $k = 0$, then $\mathrm{r}(E) \leq \dfrac{\mathrm{br}(E)}{a}$.

Proof. Write $\mathfrak{m}$ for the maximal ideal of R and let U be a minimal reduction of E. We may assume that $E \subset \mathfrak{m}R^r$, in which case $a = r$. Now $\det(U)$ is a principal ideal contained in $\mathfrak{m}^r$, so that

$$\mathrm{br}(E) = \lambda(R/\det(U)) \geq r\cdot e(R).$$

On the other hand, as the E^n are maximal Cohen-Macaulay R-modules,

$$\nu(E^n) \leq e(R)\binom{n+r-1}{r-1} \leq \frac{\mathrm{br}(E)}{r}\binom{n+r-1}{r-1}.$$

Again Theorem 2.36 gives that $\mathrm{r}(E) \leq \mathrm{br}(E) - r$.

If in addition char $k = 0$, we may assume that R is complete, with coefficient field k. Let x be a system of parameters of R and write

$$A = k[[x]][t_1, \ldots, t_r] \subset \mathcal{R}(U) = R[t_1, \ldots, t_r] \subset \mathcal{R}(E),$$

where the t_i form a free basis of U over R. Note that $\mathcal{R}(E)$ is a torsionfree finite A-module of rank $e(R)$. Therefore [Va94a, Theorem 7 and Proposition 9] shows that $\mathrm{r}(E) < e(R)$, where $e(R) \leq \frac{\mathrm{br}(E)}{r}$ by the calculation above. □

Multiplicity of the Special Fiber of the Rees Algebra of a Module

Let $(R, \mathfrak{m})$ be a Noetherian local ring with infinite residue field, E a finitely generated R-module with Rees algebra $\mathcal{R}(E)$ and U one of its minimal reductions. Set $\mathcal{F}(E)$ and $\mathcal{F}(U)$ for the special fibers of the Rees algebras. The latter is a ring of polynomials in $\ell(E)$ indeterminates.

Proposition 2.142 has the following formulation for the special fiber of the Rees algebra of a module.

Proposition 8.40. *Let $(R, \mathfrak{m})$ be a Noetherian local ring with infinite residue field and E a finitely generated R-module with a minimal reduction U. If $r = \mathrm{r}_U(E)$, then*

$$\deg(\mathcal{F}(E)) \leq 1 + \sum_{j=1}^{r-1} \nu(E^j/UE^{j-1}),$$

and equality holds if and only if $\mathcal{F}(E)$ is Cohen-Macaulay.

A more precise estimate, valid for modules of finite colength, comes from Theorem 8.35 and its proof.

Theorem 8.41. *Let $(R, \mathfrak{m})$ be a Cohen-Macaulay local ring of dimension $d > 1$ and $E \hookrightarrow R^r$ a module of finite colength. Then the multiplicity f_0 of the special fiber $\mathcal{F}(E)$ of $\mathcal{R}(E)$ satisfies inequalities*

$$f_0 \leq \min\{\mathrm{br}(E), b_1(E) + 1\}.$$

Proof. The first bound follows from Theorem 8.35 directly. To prove the other, let U be a minimal reduction of E (we may assume that the residue field of R is infinite). We have

$$\begin{aligned}\nu(E^n) &\leq \nu(U^n) + \nu(E^n/U^n) \leq \nu(U^n) + \lambda(E^n/U^n)\\ &= \nu(U^n) + \lambda(S_n/U^n) - \lambda(S_n/E^n)\\ &= (b_1(E)+1)\binom{n+d+r-3}{d+r-2} + \text{lower terms}.\end{aligned}$$

This result mimics Theorem 2.141, which asserts for the case of an $\mathfrak{m}$-primary ideal I that the multiplicity of $\mathcal{F}(I)$ satisfies the inequality

$$f_0 \leq e_1 - e_0 + \lambda(R/I) + \nu(I) - d + 1.$$

For modules, one would similarly expect to find that

$$f_0 \leq \mathrm{br}_1(E) - \mathrm{br}(E) + \lambda(R^r/E) + \nu(E) - d - r + 2.$$

Integral Closure

We shall now discuss the integral closure of a complete intersection module E. As above, we assume that $(R, \mathfrak{m})$ is a normal, quasi-unmixed Cohen-Macaulay domain of dimension d. Let E be a complete intersection module of rank $r > 1$, defined by a mapping

$$\varphi : R^{r+c-1} \to R^r.$$

We note that $c =$ height $\det_0(E)$.

Proposition 8.42. *Let $(R, \mathfrak{m})$ be a normal, quasi-unmixed Cohen-Macaulay local domain of dimension d, and E a complete intersection module as above. The following hold.*

(i) *For all n, the associated primes of $S_n(R^r)/\overline{E^n}$ are the minimal primes of* $\det_0(E)$.
(ii) *If $c \geq d-1$, the modules $S_n(R^r)/\overline{E^n}$ are Cohen-Macaulay for all n.*

Proof. (i) We may assume that height $\det_0(E) = c < d$. Set $A = \mathcal{R}(E)$ and $B = \overline{A}$. Since $\ell(E) = c + r - 1$, we have that

$$\text{height } \mathfrak{m}A = (d + r) - (c + r - 1) = d + 1 - c \geq 2.$$

It follows that height $\mathfrak{m}B \geq 2$, and as B satisfies Serre's condition S_2, we have that the depth of each component of B must be at least 2.

(ii) Follows directly from (i). □

We shall apply the technique of maximal Hilbert functions to bound the number of generators of $\overline{E^n}$, for some cases of interest. Briefly, if E is a complete intersection module of codimension c with $2 \leq c < d$, we consider a generic submodule U generated by $c + r - 2$ elements. The module U is a complete intersection module of rank r. Consider the exact sequence

$$0 \to U^n \longrightarrow \overline{E^n} \longrightarrow \overline{E^n}/U^n \to 0,$$

from which we write

$$\nu(\overline{E^n}) \leq \nu(\overline{E^n}/U^n) + \nu(U^n) = \nu(\overline{E^n}/U^n) + \binom{n+c+r-3}{c+r-3}.$$

We must now estimate the number of generators of $\overline{E^n}/U^n$ in terms of the underlying data. We observe that this module is a submodule of a Cohen-Macaulay module

$$\overline{E^n}/U^n \hookrightarrow S_n(R^r)/U^n$$

of dimension $d-c+1$, and that $S_n(R^r)/\overline{E^n}$ has depth at least 1, by Proposition 8.42. In particular if $c \geq d-1$, then $\overline{E^n}/U^n$ is Cohen-Macaulay and therefore

$$\nu(\overline{E^n}/U^n) \leq \deg(S_n(R^r)/U^n).$$

We seek now to express $\deg(S_n(R^r)/U^n)$ in terms of some multiplicity associated with $\det_0(U)$. We start with the associativity formula

$$\deg(S_n(R^r)/U^n) = \sum_{\mathfrak{p}} \lambda((S_n(R^r)/U^n)_{\mathfrak{p}}) \deg(R/\mathfrak{p}),$$

where $\mathfrak{p}$ runs over the minimal primes of $\det_0(U)$. From Theorem 8.37, we have

$$\begin{aligned}\lambda((S_n(R^r)/U^n)_{\mathfrak{p}}) &= \lambda((R^r/U)_{\mathfrak{p}}) \cdot \binom{n+c+r-3}{c+r-2} \\ &= \lambda((R/\det_0(U))_{\mathfrak{p}}) \cdot \binom{n+c+r-3}{c+r-2}.\end{aligned}$$

Note that the formula holds even if upon localizing the entries of the matrix do not all lie in the maximal ideal.

Putting all this together, we obtain that

$$\deg(S_n(R^r)/U^n) = \deg(R/\det_0(U)) \cdot \binom{n+c+r-3}{c+r-2},$$

and

$$\nu(\overline{E^n}) \leq \deg(R/\det_0(U)) \cdot \binom{n+c+r-3}{c+r-2} + \binom{n+c+r-3}{c+r-3}.$$

We shall now assemble these calculations into:

Theorem 8.43. *Let $(R, \mathfrak{m})$ be a quasi-unmixed normal Cohen-Macaulay local domain of dimension d and E a complete intersection module of rank r and codimension $c = d-1$. For every integer n,*

$$\nu(\overline{E^n}) \leq \deg(R/\det_0(E)) \cdot \binom{n+d+r-4}{d+r-3} + \binom{n+d+r-4}{d+r-4}.$$

Proof. All that is required is to compare $\deg(R/\det_0(E))$ and $\deg(R/\det_0(U))$. For that, observe that the two ideals $\det_0(E)$ and $\det_0(U)$ are obtained as the ideals of maximal minors of a $r \times (c+r-1)$ matrix ψ with entries in $\mathfrak{m}$ and of the submatrix φ obtained by deleting one column of ψ. The assertion that $\deg(R/I_r(\varphi)) \leq \deg(R/I_r(\psi))$ follows from [SUV3, Lemma 5.16] and [B81, Corollary 1]. □

8.3.2 Extended Degree for the Buchsbaum-Rim Multiplicity

Let R be a Noetherian local ring of dimension d and let E be a submodule of rank r of R^r. Suppose that the ideal $\det_0(E)$ has codimension c. The function

$$n \mapsto \deg(S_n(R^r)/E^n)$$

is the Hilbert function for the *extended degree* associated with the Buchsbaum-Rim multiplicity. If $\mathfrak{p}_1, \ldots, \mathfrak{p}_s$ are the associated prime ideals of R^r/E of maximal dimension, by the associativity formula for multiplicities we have

$$\deg(S_n(R^r)/E^n) = \sum_{i=1}^{s} \lambda((S_n(R^r)/E^n)_{\mathfrak{p}_i}) \deg(R/\mathfrak{p}_i).$$

For $n \gg 0$, this is a polynomial of degree $r + \max\{\dim R_{\mathfrak{p}_i}\} - 1$. When all these primes have the same codimension, say c, it follows that

$$\deg(S_n(R^r)/E^n) = \sum_{j=0}^{c+r-1} (-1)^j b_j(E) \binom{n+c+r-j-2}{c+r-j-1},$$

where the $b_j(E)$ are assembled from the Buchsbaum-Rim coefficients of the modules $E_{\mathfrak{p}_i}$:

$$b_j(E) = \sum_{i=1}^{s} \mathrm{br}_j(E_{\mathfrak{p}_i}) \deg(R/\mathfrak{p}_i).$$

Proposition 8.44. *Let R be a Cohen-Macaulay local ring and let E, $E \subset R^r$, be an equimultiple module of codimension c, and let F be a minimal reduction of E. Then*

$$b_0(E) = \deg(R^r/F).$$

Proof. According to the expression above for $b_0(E)$, we have

$$\begin{aligned} b_0(E) &= \sum_{i=1}^{s} \mathrm{br}(E_{\mathfrak{p}_i}) \deg(R/\mathfrak{p}_i) \\ &= \sum_{i=1}^{s} \lambda((R^r/F)_{\mathfrak{p}_i}) \deg(R/\mathfrak{p}_i) = \deg(R^r/F), \end{aligned}$$

by the rule for computation of multiplicities. □

Let us make a slight extension to subfiltrations of the integral closure filtration $\overline{E^n}$. We are going to assume that E is equimultiple and that the algebra $C = \sum_{n \geq 0} \overline{E^n}$ is Noetherian, that is it is finite over $\mathcal{R}(F)$, for the minimal reduction F of E. We can define the extended degree function $\deg(S_n(R^r)/\overline{E^n})$. It has similar properties to those of $\deg(S_n(R^r)/E^n)$; the coefficients of the polynomial (for $n \gg 0$) will be denoted by $\overline{b}_j(E)$.

Theorem 8.45. *Let $(R,\mathfrak{m})$ be a normal Cohen-Macaulay local ring of dimension d and let E, $E \subset R^r$, be an equimultiple module of codimension c such that $C = \sum_{n\geq 0} \overline{E^n}$ is Noetherian. Let F be a minimal reduction of E and let A and B be distinct graded subalgebras such that*

$$\mathcal{R}(F) \subset A = \sum_{n\geq 0} A_n \subset B = \sum_{n\geq 0} B_n \subset C.$$

If A and B satisfy Serre's condition S_2, then $b_0(A) = b_0(B)$ and $b_1(A) < b_1(B)$. In particular, $b_1(A) < \overline{b}_1(E)$ for all proper subalgebras of C, and all chains of graded subalgebras of C satisfying the condition S_2 have length at most $\overline{b}_1(E)$.

Proof. Consider the following exact sequence of $\mathcal{R}(F)$-modules:

$$0 \to A \longrightarrow B \longrightarrow D \to 0.$$

Since A satisfies S_2, it follows that D has Krull dimension $d+r-1$. We first deal with the case $c = d$. Let us examine $\deg(D_n)$ using the exact sequence

$$0 \to B_n/A_n \longrightarrow S_n(R^r)/A_n \longrightarrow S_n(R^r)/B_n \to 0,$$

in which deg is just ordinary length. Since $\dim D = d+r-1$, $\deg(D_n)$ is just the ordinary Hilbert function and therefore it is a polynomial of degree $d+r-2$ for $n \gg 0$, with leading coefficient $e_0(D)$, the ordinary multiplicity of D. On the other hand, the Buchsbaum-Rim extended degree functions $\deg(S_n(R^r)/A_n)$ and $\deg(S_n(R^r)/B_n)$ are polynomials of degree $d+r-1$ for $n \gg 0$, so their leading coefficients must be equal: $b_0(A) = b_0(B)$, and $b_1(B) = b_1(A) + e_0(D)$. Therefore $b_1(B) > b_1(A)$ since $e_0(D) > 0$. (Observe that comparing A and B individually we get that $b_1(B) > 0$.)

Assume now that $c < d$. Since the special fiber $\mathcal{R}(F) \otimes R/\mathfrak{m}$ has Krull dimension $c+r-1 < d+r-1$, the module $D \otimes R/\mathfrak{m}$ has Krull dimension less than $d+r-1$ as well. This means that the associated primes of D of dimension $d+r-1$ are of the form $\mathfrak{P} = \mathfrak{p} + P \subset \mathcal{R}(F)$, where $\mathfrak{p} \neq \mathfrak{m}$ and P is a homogeneous subideal of $\mathcal{R}(F)_+$. Write $T = \mathcal{R}(F)/\mathfrak{P} = R/\mathfrak{p} + Q_0$. By Lemma 1.21, $\dim T = \dim R/\mathfrak{p} + \text{height } (Q_0)$. As a localization of an equimultiple module is still equimultiple, it follows that the minimal primes that we must consider are still among the associated primes $\mathfrak{p}_i$ of R^r/F. This means that $\dim R/\mathfrak{p} = d - c$ and

$$\text{height } (Q_0) = \dim T - \dim R/\mathfrak{p} = d+r-1-(d-c) = c+r-1.$$

As a consequence, $\lambda((D_n)_\mathfrak{p})$ is a polynomial of degree exactly $c+r-2$ for $n \gg 0$.

Meanwhile the functions $\deg(S_n(R^r)/A_n)$ and $\deg(S_n(R^r)/B_n)$ are both polynomials of degree $c+r-1$ for $n \gg 0$, which again allows us to obtain the equalities $b_0(A) = b_0(B)$ and $b_1(B) = b_1(A) + b_0(D)$, and our assertions follow since $b_0(D) > 0$. □

8.4 Divisors of Modules and Integral Closure

The *divisors* of a module E over a ring R are the ideals of R that carry significant information about the structure and properties of E. The classical Fitting ideals of

E are typical examples. With regard to integral closure, an outright relative notion dependent on a chosen embedding of E into a free module, certain other ideals play significant roles as well. Throughout this section, R will be a Noetherian ring with a total ring of fractions K. Let E be a submodule of R^r of rank r with a fixed embedding

$$E \hookrightarrow R^r.$$

We define ideals to serve as a barometers of this embedding with relation to integral closure. Among the ideals playing the roles, we single out the following.

- Fitting ideals of E
- Fitting ideals of R^r/E
- Associated primes of R^r/E
- Divisor classes

8.4.1 Order Ideal of a Module

Definition 8.46. Let E be a finitely generated torsionfree R-module of rank r. The *order determinant* of the embedding $E \stackrel{\varphi}{\longrightarrow} R^r$ is the ideal defined by the image of the mapping $\wedge^r\varphi$,

$$\text{image } (\wedge^r\varphi) = I \cdot \wedge^r(R^r).$$

When the embedding is clear we shall write $I = \det_0(E)$.

A more appropriate notation would have been $\det_\varphi(E)$. In any event, for any embedding one has

$$\det_0(E) \cong \wedge^r E/(\text{torsion}).$$

For an R-module E, we set $E^* = \text{Hom}_R(E,R)$. We shall have several occasions to use the following notions of determinant (or determinant class) of a module. We shall limit our definition to modules with a *rank*: $K \otimes_R E \cong K^r$, where K is the total ring of fractions of R.

Definition 8.47. Let E be a finitely generated torsionfree R-module of rank r. The *determinant* or *determinant divisor* of E is the fractionary ideal

$$\det(E) = \det_0(E)^{**}.$$

The module E is said to be *orientable*[1] if $\det(E) \cong R$. The isomorphism class of $\det(E)$ as a fractionary ideal will be denoted by $[\det(E)]$ or by $\text{div}(E)$.

Note that if R is factorial, then $\det(E) \cong R$ for every module E of positive rank, while if E is a projective module of positive rank over a ring R, then $\det(E)$ is an invertible ideal.

[1] A more appropriate notation for the bidual $I^{**} = (I^{-1})^{-1}$ of an ideal is I^{--1}, particularly in view of the fact that I^{-2} already stands for $(I^2)^{-1}$.

Example 8.48. Let R be a normal domain and I an ideal such that the Rees algebra $B=R[It]$ is normal. Let $A=R[I^2t^2]$ be the second Veronese subring of B. The algebra B has the decomposition as a module over A,

$$B = A \oplus IAt.$$

Since A is normal (as a direct summand of B) for each ideal $L \subset A$ of height at least 2, the grade of LB is also at least 2. This implies that B is a reflexive A-module and that

$$\det(B) = (IA)^{**} = IA.$$

In particular B is orientable only when I is a principal ideal.

Remark 8.49. Note that for any embedding one has that $\det_0(E) \cong \wedge^r(E)/\text{torsion}$. If E is an orientable module, then $\det_0(E) = aL$, where L is a module of grade at least two.

The following further properties are easy to establish:

(i) The ideal L is independent of the embedding.
(ii) In general, for any embedding the ideal $\det_0(E)\cdot(\det_0(E))^*$ defines the non-free locus of E.
(iii) If F is a submodule of E and ann (E/F) has grade two or larger, then $\det(F) = \det(E)$.

For an ideal I, the analog for an orientable module is a decomposition $I = dJ$, where J is an ideal of grade at least two. In particular, all modules over a factorial domain are orientable.

A general property of orientable module is the following (the converses will not always hold).

Proposition 8.50. *Let E be a finitely generated orientable module of rank r. For every localization $R_{\mathfrak{p}}$ of depth at most one, $E_{\mathfrak{p}} \cong R_{\mathfrak{p}}^r \oplus$ torsion.*

We shall leave the verification to the reader. It has the following consequence.

Corollary 8.51. *Let E be a torsionfree module with a rank and let U be one of its reductions. Then E is orientable if and only if U is orientable.*

Modules with Rank One Relations

A simple but useful application of determinants of modules is the following:

Proposition 8.52. *Let R be a normal domain and E a torsionfree R-module generated by n elements and having rank $n-1$. A presentation of E has the form*

$$0 \rightarrow (\det_0(E))^{-1} \longrightarrow R^n \longrightarrow E \rightarrow 0.$$

Proof. Consider an embedding $E \hookrightarrow R^{n-1}$, and denote by φ the composite map from R^n to R^{n-1} with image E. Then $\det_0(E)$ is the ideal $(\Delta_1, \ldots, \Delta_n)$ generated by the (signed) minors of φ. The corresponding vector in R^n is denoted simply by Δ.

Set $K = \ker(\varphi)$. We have $\Delta \in K$, and for $u \in K$ we must have $u = \alpha \cdot \Delta$ for some α in the field of fractions of R since K has rank one. This means that $K = L\Delta$, where L is a fractionary ideal of R. We are going now to identity L more precisely.

Since the cokernel of the embedding $L\Delta \hookrightarrow R^n$ is torsionfree, the content of all the coordinates of the vectors in $L\Delta$ cannot be contained in an ideal of grade one (this is just the top Fitting ideal of E). Indeed, otherwise $K \subset \mathfrak{p}R^n$ for some height one prime ideal $\mathfrak{p}$. Localizing at this prime, since E is torsionfree, $K_{\mathfrak{p}}$ should split off $R^n_{\mathfrak{p}}$, which would be a contradiction. This implies that the product of the ideals

$$L \cdot (\Delta_1, \ldots, \Delta_n) \subset R$$

is not contained in any height one prime of S. Since E is torsionfree, L is a divisorial ideal of R and the product above means that $L = (\Delta)^{-1}$, as desired. □

Modules with a Finite Free Resolution

The following elementary observation is very useful. It is supple enough to furnish a proof of the factoriality of regular local rings.

Proposition 8.53. *Let R be a Noetherian ring with Serre's condition S_2 and let*

$$0 \to A \longrightarrow B \stackrel{\varphi}{\longrightarrow} C \to 0$$

be an exact sequence of finitely generated torsionfree modules with ranks. If any two of these modules are orientable so is the third.

Proof. Let us set $\text{rank}(A) = s$, $\text{rank}(C) = r$ and $\text{rank}(B) = r + s = n$. We define a pairing of the modules $\wedge^r C$, $\wedge^s A$, into $(\wedge^n B)^{**}$. Suppose that $u_1, \ldots, u_r \in C$ and $v_1, \ldots, v_s \in A$. For each u_i, let $w_i \in B$ be chosen so that $\varphi(w_i) = u_i$. Now define

$$(u_1, \ldots, u_r, v_1, \ldots, v_s) \mapsto (w_1 \wedge \ldots \wedge w_r \wedge v_1 \wedge \ldots \wedge v_s) \in \wedge^n B/\text{modulo torsion} \subset (\wedge^n B)^{**}.$$

Note that the mapping into $\wedge^n B$ itself may not be well-defined.

Note also that if two of the modules are orientable, then at each localization $R_{\mathfrak{p}}$ of depth at most one they are $R_{\mathfrak{p}}$-free, therefore the third module will be $R_{\mathfrak{p}}$-free since it will have projective dimension at most one and is torsionfree. Thus if we set $I = \wedge^s A/\text{modulo torsion}$ and $J = \wedge^r C/\text{modulo torsion}$, we get that the mapping

$$I \otimes J \longrightarrow (\wedge^n B)^{**} = \det(B)$$

is an isomorphism in codimension one and therefore we have an isomorphism

$$(I \otimes J)^{**} \cong \det(B).$$

Since one of the modules, A or C, is orientable, we have that I^* or J^* is an invertible module (actually isomorphic to R). Suppose that I has this property (the choice will be irrelevant). By adjointness, we have natural isomorphisms

$$\mathrm{Hom}(I\otimes J,R)\cong \mathrm{Hom}(J,\mathrm{Hom}(I,R))\cong \mathrm{Hom}(I,R)\otimes \mathrm{Hom}(J,R).$$

Dualizing in the same fashion we obtain that

$$\mathrm{Hom}(I^*\otimes J^*,R)\cong \mathrm{Hom}(J^*,\mathrm{Hom}(I^*,R))\cong \mathrm{Hom}(I^*,R)\otimes \mathrm{Hom}(J^*,R)\cong I^{**}\otimes J^{**},$$

showing that $\det(A)\otimes\det(C)\cong\det(B)$, as desired. □

Corollary 8.54. *Let E be a finitely generated torsionfree R-module with a rank. If E admits a finite free resolution then E is orientable.*

Applying this result to height one prime ideals of regular local rings gives a proof of their factoriality.

Corollary 8.55. *Let R be a Noetherian ring and I an ideal containing regular elements. Then any module E defined by an exact sequence*

$$0\rightarrow R^{r-1}\longrightarrow E\longrightarrow I\rightarrow 0$$

has rank r, and it is orientable if and only if I is orientable.

Normal Domains

Over normal domains there is a great deal of flexibility in the theory of determinantal divisors. At its center are the following results which extend others treated previously.

Proposition 8.56. *Let R be a normal domain and*

$$0\rightarrow A\stackrel{\varphi}{\longrightarrow} B\stackrel{\psi}{\longrightarrow} C\longrightarrow D\rightarrow 0$$

an exact sequence of finitely generated R-modules, with B and C torsionfree. If $\operatorname{codim} A\geq 1$, $\operatorname{codim} D\geq 2$ *and* $\operatorname{rank}(B)>0$, *then* $\det(B)\cong\det(C)$.

Proof. Set $r=\operatorname{rank}(B)=\operatorname{rank}(C)$. The mapping

$$\wedge^r B/\text{torsion}\longrightarrow \wedge^r C/\text{torsion}$$

is an isomorphism at all localizations at primes of codimension at most 1, and therefore at their biduals, that is, $\det(B)\cong\det(C)$. □

Proposition 8.57. *Let R be a normal domain and*

$$0\rightarrow A\stackrel{\varphi}{\longrightarrow} B\stackrel{\psi}{\longrightarrow} C\rightarrow 0$$

a complex of finitely generated R-modules that is an exact sequence of torsionfree $R_{\mathfrak{p}}$-modules for all prime ideals of codimension at most 1. *Then*

$$\det(B)=\det(A)\circ\det(C)=(\det(A)\det(C))^{**}.$$

Proof. We break up the complex into simpler exact complexes:

$$0 \to \ker(\varphi) \longrightarrow A \longrightarrow A' = \text{image}(\varphi) \to 0$$

$$0 \to A' \longrightarrow \ker(\psi) \longrightarrow \ker(\psi)/A' \to 0$$

$$0 \to B' = \text{image}(\psi) \longrightarrow C \longrightarrow C/B' \to 0$$

and

$$0 \to \ker(\psi) \longrightarrow B \longrightarrow B' \to 0.$$

We note that $\operatorname{codim} \ker(\varphi) \geq 1$, $\operatorname{codim} C/B' \geq 2$, $\operatorname{codim} \ker(\psi)/A' \geq 2$ by hypothesis, so that we have the equality of determinantal divisors:

$$\det(A) = \det(A') = \det(\ker(\psi)), \quad \det(C) = \det(B').$$

What this all means is that we may assume that the given complex is exact.

Suppose that $r = \text{rank}(A)$ and $\text{rank}(C) = s$ and set $n = r + s$. Consider the pair $\wedge^r A$, $\wedge^s C$. For $v_1, \ldots, v_r \in A$ and $u_1, \ldots, u_s \in C$, choose w_i in B such that $\psi(w_i) = u_i$ and consider the inclusion

$$v_1 \wedge \cdots \wedge v_r \wedge w_1 \wedge \cdots \wedge w_s \in \wedge^n B.$$

Different choices for w_i would produce elements in $\wedge^n B$ that differ from the above by terms that contain at least $r+1$ factors of the form

$$v_1 \wedge \cdots \wedge v_r \wedge v_{r+1} \wedge \cdots,$$

with $v_i \in A$. Such products are torsion elements in $\wedge^n B$. This implies that modulo torsion we have a well-defined pairing

$$[\wedge^r A/\text{torsion}] \otimes_R [\wedge^s C/\text{torsion}] \longrightarrow [\wedge^n B/\text{torsion}].$$

When localized at primes $\mathfrak{p}$ of codimension at most 1 the complex becomes an exact complex of projective $R_{\mathfrak{p}}$-modules and the pairing is an isomorphism. Upon taking biduals and the $\circ$ composition, we obtain the asserted isomorphism. □

Corollary 8.58. *Let R be a normal domain and E a torsionfree R-module with a presentation*

$$0 \to F \longrightarrow R^n \longrightarrow E \to 0.$$

Then

$$\det(E) \cong (\det(F))^{-1}.$$

A general and useful formulation is:

Proposition 8.59. *Let R be a normal domain and*

$$0 \to E_n \longrightarrow E_{n-1} \longrightarrow \cdots \longrightarrow E_1 \longrightarrow E_0 \to 0$$

a complex of finitely generated R-modules. If the E_i have no associated primes of codimension one and the complex is exact at each localization $R_{\mathfrak{p}}$ of dimension at most 1*, then*

$$\det(E_0) = \prod_{i=1}^{n} \det(E_i)^{(-1)^{i-1}},$$

where the product is multiplication in the divisor class group of R. In particular the divisor classes of their determinants satisfy the equality

$$\sum_{i=0}^{n} (-1)^i [\det(E_i)] = 0.$$

Conormal Modules

A great number of interesting Rees algebras of modules are those corresponding to the conormal modules I/I^2. These modules have remarkable properties, including that given by the following:

Proposition 8.60. *Let R be a Gorenstein local ring and I an ideal of height g. If R/I satisfies Serre's condition S_2 and $I_{\mathfrak{p}}$ is a complete intersection for every prime ideal $I \subset \mathfrak{p}$ of height $\leq g+1$, then* $\det(I/I^2) = \omega^*_{R/I}$. *In particular I/I^2 is orientable if I is a Gorenstein ideal.*

Proof. There is a canonical mapping

$$\operatorname{Ext}^g_R(R/I, R) = \omega_{R/I} \xrightarrow{\varphi} (\wedge^g (I/I^2))^*$$

(cf. [Gr57, supplement]) The hypothesis is that φ is an isomorphism in codimension at most one. Since both modules have Serre's condition S_2, φ must be an isomorphism. □

Suppose that I is a prime ideal. The rank of I/I^2 is the minimal number of generators ν of IR_I. One difficulty in determining the order determinant of some embedding of $E = I/I^2/$(modulo torsion) is that ν may be unrelated to $g =$ height I. They are always equal when I is a smooth point of Spec R, which we assume now. Consider then an embedding

$$E \xrightarrow{\psi} (R/I)^g.$$

We have maps

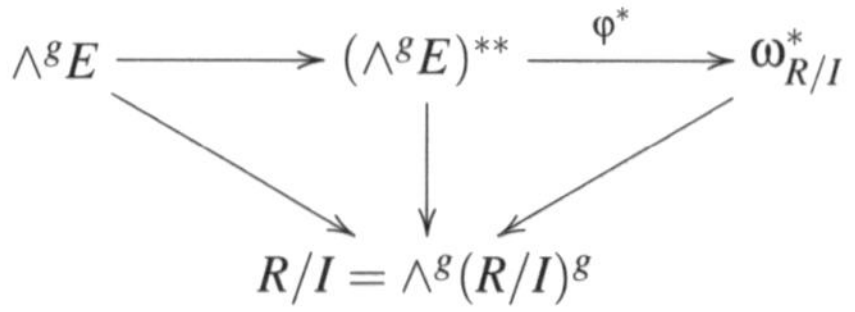

$$\begin{array}{ccccc} \wedge^g E & \longrightarrow & (\wedge^g E)^{**} & \xrightarrow{\varphi^*} & \omega^*_{R/I} \\ & \searrow & \downarrow & \swarrow & \\ & & R/I = \wedge^g (R/I)^g & & \end{array}$$

In the situation of the previous proposition, φ^* is an isomorphism.

We leave to the reader the proof of the following related result:

Proposition 8.61. *Let R be a Gorenstein local ring and I a Cohen-Macaulay ideal of codimension g. If I is a complete intersection in codimension $g+1$, then* $\det(I/I^2) \cong \omega_{R/I}$.

A well-known calculation of determinants occurs in the case of affine domains.

Proposition 8.62. *Let $R = k[x_1, \ldots, x_n]/I$ be a normal affine domain. Let $E = I/I^2$ be the conormal module of R and (when k has characteristic zero) $\Omega_k(R)$ the module of Kähler k-differentials of R. Then*

$$(\det(E))^{-1} = \det(\Omega_k(R)) = \omega_R,$$

where ω_R is the canonical module of R.

Proof. This follows from Proposition 8.59 together with the Jacobian criterion, and the chain complexes

$$0 \to I/I^2 \longrightarrow R^n \longrightarrow \Omega_k(R) \to 0,$$

$$0 \to H_1 \longrightarrow R^m \longrightarrow I/I^2 \to 0,$$

where H_1 is the 1-dimensional Koszul homology module on a set of generators of I with m elements. The multiplication in the Koszul homology algebra readily implies that $\det(H_1) = \omega_R$. □

8.4.2 Determinantal Ideals and Reductions

We give now an effective criterion that checks whether a submodule can be a reduction of a module. It has a very different character from those criteria grounded on multiplicities. We assume here that R is an integral domain with field of fractions K. We shall draw techniques from several sources, ranging from those specific to regular local rings of dimension two treated in [KK97], [Ko95] and [Liu98], to the linkage-theoretic approach of [SUV3].

We begin with a presentation of a result of Rees ([Re87]) in the original introduction of Rees algebras of modules.

Proposition 8.63. *Let F and E be R-modules of the same rank, with $F \subset E \subset R^r$. Then F is a reduction of E if and only if $VF = VE$ for every valuation ring V of R.*

Proof. The notation VF means the V-submodule of V^r generated by F. We may assume that $r = \text{rank}(F)$. Suppose that $S = \mathcal{R}(R^r) = R[T_1, \ldots, T_r]$, and let $\mathcal{R}(F) \subset \mathcal{R}(E)$ be the Rees algebras of F and E. Since the three algebras have the same fields of fractions, they have the same valuation overrings. For a valuation ring $\mathbf{V}$ of S, set $V = \mathbf{V} \cap K$. We note that $V\mathcal{R}(F), V\mathcal{R}(E), VS$ are rings of polynomials over the free modules VF, VE, V^r, where $VF \subset VE \subset V^r$, and therefore they are integrally closed. It follows that F is a reduction of E if and only if $VF = VE$ for each of the valuations such as V. □

Example 8.64. Let R be a local domain and I an ideal minimally generated by $a_1,\ldots,a_n$. Let

$$B_1 \subset Z_1 \subset R^n = \sum_{i=1}^{n} Re_i$$

be the modules of 1-boundaries and 1-cycles of the Koszul complex on the a_i. Suppose that $Z_1 \subset \bar{I}R^n$, that is, all the syzygies of the a_i have coefficients in the integral closure (in R) of I. An example of such is the maximal ideal of R. Then B_1 is a reduction of Z_1. To see this, let V be a valuation ring of R. Suppose that $v(a_1) \leq v(a_i)$, $\forall i$. The elements $a_ie_1 - a_1e_i$, $i = 2,\ldots,n$, lie in VB_1 and clearly span the module of syzygies of VI that have coefficients in I. But that module contains VZ_1, and the assertion follows by the criterion above.

A much deeper result was proved by Rees ([Re87, Theorem 3.1]).

Theorem 8.65. *Let R be a quasi-unmixed domain, and suppose that $a_1,\ldots,a_n$ generate a proper ideal I of height g. Let B_r and Z_r be the modules of r-boundaries and r-cycles of the Koszul complex on the a_i. Then B_r is a reduction of Z_r for $r > n-g$.*

Proposition 8.66. *Let F and E be torsionfree R-modules of rank r with $F \subset E$, and $E \hookrightarrow R^r$ an embedding. Denote by $\det_0(F)$ and $\det_0(E)$ the corresponding order determinants. Then F is a reduction of E if and only if $\det_0(F)$ is a reduction of $\det_0(E)$.*

Proof. We check that the order determinants detect this situation. First, note that for any domain T that is an overring of R, one has that $\det_0(E)T = \det_0(TE)$. This follows because TE is the image of $T \otimes_R E$ in $T \otimes_R R^r = T^r$, and exterior powers commute with base change.

Finally we observe that in the embedding

$$VF \xrightarrow{\alpha} VE \xrightarrow{\beta} V^r$$

of free modules, the two ideals $V\det_0(E)$ and $V\det_0(F)$ are generated by $\det(\beta)$ and $\det(\beta\cdot\alpha)$ respectively. Thus our conditions are equivalent to the requirement that $\det(\alpha)$ be a unit of V. □

Example 8.67. Let I be an equimultiple ideal of the Noetherian ring R, and $J = (x,y,z)$ a minimal reduction. Let $E = I^{\oplus 4} \subset R^4$ be an R-module of rank 4. Consider the submodule F generated by the columns of the matrix

$$\varphi = \begin{bmatrix} x & y & z & 0 & 0 & 0 \\ 0 & x & y & z & 0 & 0 \\ 0 & 0 & x & y & z & 0 \\ 0 & 0 & 0 & x & y & z \end{bmatrix}.$$

Since $\det_0(E) = I^4$ and $\det_0(F) = I_4(\varphi) = J^4$, we see that $\overline{E} = \overline{F}$; F is actually a minimal reduction of E.

Remark 8.68. Observe that, if F and E are submodules of R^r of rank r, and F is a reduction of E, then for the full series of determinantal ideals of the embeddings, we have $\overline{I_t(F)} = \overline{I_t(E)}$.

From this we obtain the following effective criterion.

Corollary 8.69. *Let $(R, \mathfrak{m})$ be a quasi-unmixed local domain, let F and E be torsionfree R-modules of rank r with $F \subset E$ and $E \hookrightarrow R^r$ an embedding. If* $\det_0(F)$ *and* $\det_0(E)$ *are $\mathfrak{m}$-primary ideals then F if a reduction of E if and only if* $\det_0(F)$ *and* $\det_0(E)$ *have the same multiplicity.*

Remark 8.70. We are now able to reduce the membership problem for the integral closure of modules to the case of ideals, as follows. Suppose that E is a submodule of R^r of rank r generated by the elements $v_1, \ldots, v_m$, which we represent as the columns of the matrix φ. Let v be an element of R^r for which we want to decide whether $v \in \overline{E}$. Let φ' be the matrix obtained by enlarging φ by adding v as a new column. Denote by L the ideal generated by all $r \times r$ minors of $\varphi' f$. Then

$$v \in \overline{E} \Leftrightarrow L \subset \overline{\det_0(E)}.$$

As a practical matter, this last result surpasses Corollary 8.20 in usefulness. Note that while these results are helpful in deciding whether or not F is a reduction of E, it says nothing about whether E is integrally closed entirely in terms of its order determinant. Nevertheless, we have the following characterization.

Corollary 8.71. *Let R be a quasi-unmixed integral domain and E a submodule of rank r of the free module R^r. For any integer n, the integral closure of E^n in $S_n(R^r)$ is given by*

$$\overline{E^n} = S_n(R^r) \cap \bigcap_V VE^n,$$

where V runs over the set of Rees valuations of $\det_0(E)$.

Proof. If $n = 1$, the proof follows from the previous remark. For $n > 1$, we make use of the fact (see Theorem 8.104) that $\det_0(E^n)$ has the same integral closure as an appropriate power of $\det_0(E)$, and therefore it will have the same Rees valuations as $\det_0(E)$, according to Corollary 1.42. □

Corollary 8.72. *Let R be a normal domain and E a finitely generated torsionfree R-module. Then*

$$\overline{\mathcal{R}(E)} = \bigcap_V V\mathcal{R}(E),$$

where V runs over the set of valuations of $\det_0(E)$.

Proof. This follows from Proposition 8.10, the preceding corollary and the elementary fact (left as an exercise) that the Rees valuations of E are the same as the Rees valuations of E_n for all $n \geq 1$. □

If R is integrally closed and E is a torsionfree R-module, it is easy to see that the Rees valuations of $\det_0(E)$ are independent of the embedding. One may refer to these valuations as the *Rees valuations of the module*.

We are in position to produce several integrally closed modules from a given module E.

Proposition 8.73. *Let R be an integral domain and E an integrally closed submodule of the free module F. For any finitely generated R-module H, $\mathrm{Hom}_R(H,E)$ is integrally closed. In particular every reflexive R-module is integrally closed.*

Proof. We may assume that H is a torsionfree module. We can view $\mathrm{Hom}_R(H,E)$ as a submodule of the vector space $L = \mathrm{Hom}_K(K\otimes H, K\otimes E)$. The integral closure G of $\mathrm{Hom}_R(H,E)$ is the intersection of the images of $V\otimes_R \mathrm{Hom}_R(H,E)$ in L, where V runs over the valuations of R. In other words,

$$G = \bigcap_V \text{image } (\mathrm{Hom}_V(VH, VE) \to L).$$

But note that this is the same as

$$\bigcap_V \text{image } (\mathrm{Hom}_R(H, VE) \to L) = \mathrm{Hom}_R(H, \bigcap_V VE) = \mathrm{Hom}_R(H,E),$$

since E is integrally closed. □

Ideals and their Determinantal Ideals

Under a number of conditions, the technique above can be applied to decide whether two ideals have the same integral closure.

Let J and I with $J \subset I$ be nonzero ideals of a Noetherian integral domain A and suppose that R is a subring of A over which A is finite. Once the presentation of A over R is given, we also have presentations of J and I as R-modules. If we suppose further that A is R-free (and we have in mind the case of a Cohen-Macaulay, affine integral domain over a field, with R being a Noether normalization), we may define the R-determinant of I and of J (as R-modules of A they both have full rank).

Proposition 8.74. *Under the conditions above, if $\det_0(I)$ and $\det_0(J)$ have the same integral closures as ideals of R, then J is a reduction of I as A-ideals.*

Proof. Let $\{e_1,\ldots,e_n\}$ be an R-basis of A as an R-module, and consider the determinantal ideals $\det_0(J)$ and $\det_0(I)$. Let U be a discrete valuation of A, set $V = \mathrm{Quot}(R)\cap U$; V is a discrete valuation ring of R. The submodules VJ and VI of the free V-module $VA = V\otimes_R A$ have $V\det_0(J)$ and $V\det_0(I)$ for determinantal ideals. By assumption these ideals of V are equal, which implies that the free V-submodules VJ and VI of VA are equal also. In particular we have that $UJ = UI$, and the assertion follows by the valuative test for completeness. □

However, this technique is not sensitive enough to code completeness. Consider the rings

$$R = k[x^2] \subset k[x^2, x^3] = A = R \oplus Rx^3$$

(as subrings of the ring of polynomials $k[x]$ over the field k), and suppose that $J = (x^2) \subset (x^2, x^3) = I$. Then J is obviously a reduction of I, whereas $\det_0(J) = (x^4) \subset (x^2) = \det_0(I)$, as R-ideals.

Integral Closure and Primary Decomposition

We can now approach the primary decomposition of E as a submodule of the free module F. The 'associated' primes of E are by definition the associated prime ideals of F/E. The primes of grade 1 are not greatly significant because they depend on the embedding. The primes of grade two or larger are intrinsic to E, as is easy to show.

The general device we use to relate the integral closure of E and the order determinantal ideal $\det_0(E)$ is the following. Suppose that $E \subset F \subset R^r$, and let L be the annihilator of F/E. Taking the order determinantal ideals of E and F, we obtain the inclusion

$$L^r \cdot \det_0(F) \subset \det_0(E).$$

In particular, if L does not consist of zero divisors mod $\det_0(E)$, we get $\det_0(F) = \det_0(E)$. By Proposition 8.66, $F \subset \overline{E}$.

In general one should not expect equality of the sets of associated primes of R^r/E and $R/\det_0(E)$. There are cases, however, when they coincide.

Proposition 8.75. *Let E be an integrally closed ideal module of rank r. If* $\det_0(E)$ *has no embedded primes then*

$$\mathrm{Ass}(R^r/E) = \mathrm{Ass}(R/\det_0(E)).$$

Proof. Let $\mathfrak{p}$ be an associated prime of R^r/E that does not belong to $\mathrm{Ass}(R/\det_0(E))$. If $\mathfrak{p}$ does not contain any of the minimal primes of $\det_0(E)$, localizing we obtain that $E_{\mathfrak{p}}$ becomes a free module, so there nothing to show. We may assume that $\mathfrak{p}$ contains properly some minimal prime of $\det_0(E)$. If $0 \neq h \in R^r/E$, $\mathfrak{p} \cdot h = 0$, then on lifting Rh to R^r we obtain a proper inclusion $E \subset H$ of submodules of R^r, with $\mathfrak{p} \cdot H \subset E$. This gives rise to the inclusion $\det_0(E) \subset \det_0(H)$, with $\mathfrak{p}^r \cdot \det_0(H) \subseteq \det_0(E)$, which proves that $\det_0(H) = \det_0(E)$. By Proposition 8.66, H is integral over E, which is a contradiction. □

We shall state an element of the proof more generally as:

Corollary 8.76. *Let E be an integrally closed ideal module of rank r. Then*

$$Z(R^r/E) \subset Z(R/\det_0(E)),$$

that is, every zero divisor on R^r/E is a zero divisor on $R/\det_0(E)$.

We shall now describe the integral closure of a restricted but important class of modules.

Proposition 8.77. *Let R be a normal domain and let E be a submodule of R^r of rank r. If $\det_0(E)$ is a divisorial ideal, then the integral closure of E is the bidual of E.*

Proof. Denote the bidual of E by E_0. Since duals are integrally closed, $\overline{E} \subset E_0$. Set $L = \text{ann}\,(E_0/E)$, and note that height $L \geq 2$ since R is normal. If $\det_0(E)$ is divisorial, then $\overline{\det_0(E)} = \det(E)$. From the inclusion $L^r \det_0(E_0) \subset \det_0(E)$ we get $\det_0(E_0) \subset \det_0(E)$, as the associated primes of $\det_0(E)$ are of height 1. We then have that $E_0 \subset \overline{E}$. □

8.4.3 $\mathfrak{m}$-full Modules

The notion of $\mathfrak{m}$-fullness of ideals extends to modules in a very direct manner.

Definition 8.78. Let $(R,\mathfrak{m})$ be a Noetherian local ring and E a submodule of R^r. The R-module E is an $\mathfrak{m}$-*full module relative to* R^r if there is an element $x \in \mathfrak{m}$ such that

$$\mathfrak{m}E :_{R^r} x = E.$$

Observe the reference to the embedding, emphasizing the relative character of the definition. The following is the module analog of Proposition 1.62.

Proposition 8.79. *Suppose that $(R,\mathfrak{m})$ is a local domain and consider submodules E of R^r of rank r. If the residue field of R is infinite, then integrally closed modules are $\mathfrak{m}$-full relative to R^r.*

Proof. Set $L = \det_0(E)$. We make use of the notion of Rees valuation in Proposition 1.41, and its notation. To simplify a little we assume that R is an integral domain. The proof is similar to the ideal case, but takes into account the role of the order determinantal ideal in the description of the integral closure observed above.

Let $(V_1,\mathfrak{p}_1),\ldots,(V_n,\mathfrak{p}_n)$ be the Rees valuations associated with L. Observe that since $\mathfrak{m}V_i \neq 0$ for each i, we have that $\mathfrak{m}$ contains $L_i = \mathfrak{p}_i\mathfrak{m}V_i \cap R$ properly. Since $R/\mathfrak{m}$ is infinite, we have the following comparison of vector spaces:

$$\mathfrak{m}/\mathfrak{m}^2 \neq \sum_{i=1}^{n} L_i + \mathfrak{m}^2/\mathfrak{m}^2.$$

Thus we can choose $x \in \mathfrak{m}$ that is not contained in $\bigcup_{i=1}^{n} \mathfrak{p}_i\mathfrak{m}V_i$. This means that $\mathfrak{m}V_i = xV_i$ for each i. We claim that this element will do. Suppose that $y \in \mathfrak{m}E :_{R^r} x$. For each V_i we have $yx \in \mathfrak{m}V_iE = xV_iE$. Thus $y \in \bigcap_{i=1}^{n} V_iE \cap R^r$, and therefore $y \in \overline{E} = E$. □

As a consequence one has:

Corollary 8.80. *Let E be an $\mathfrak{m}$-full submodule of R^r of rank r. Suppose that G is a module such that $E \subset G \subset R^r$ and $\lambda(G/E) < \infty$. Then $\nu(G) \leq \nu(E)$.*

Proof. Using the notation above, as in the ideal case, consider the exact sequence

$$0 \to E/\mathfrak{m}E \longrightarrow G/\mathfrak{m}E \xrightarrow{\cdot x} G/\mathfrak{m}E \longrightarrow G/(xG, \mathfrak{m}E) \to 0.$$

We have

$$\nu(E) = \lambda(E/\mathfrak{m}E) = \lambda(G/(xG + \mathfrak{m}E)) \geq \nu(G),$$

as desired. □

Determinants and Presentations of Modules

While an embedding of a module into a free module permits a straightforward expression for its determinant, it is convenient to obtain formulas for the determinants from 'external' descriptions of the module, such as are given by free presentations.

Let us calculate the order determinantal ideals of certain families of modules. For simplicity we assume that R is a normal domain. Let E be a torsionfree module of projective dimension one

$$0 \to R^n \xrightarrow{\varphi} R^{n+r} \longrightarrow E \to 0.$$

By assumption the ideal $I_n(\varphi)$ has grade at least 2 and at most $r+1$. If the grade of this ideal is at least 3, then E is reflexive and therefore it is integrally closed, as observed previously, or simply because

$$E = \bigcap_{\text{height } \mathfrak{p}=1} E_{\mathfrak{p}}.$$

Theorem 8.81. *If E is a torsionfree module of rank r with a free presentation as above, then* $\det_0(E) \cong I_n(\varphi)$.

Proof. The presentation of E gives rise to the following presentation of $\wedge^r E$:

$$R^n \otimes \wedge^{r-1} R^{n+r} \xrightarrow{\varphi_r} \wedge^r R^{n+r} \longrightarrow \wedge^r E \to 0.$$

We denote by B the image of φ_r in $F = \wedge^r R^{n+r}$. We compare B with another submodule of F, and consider first the case $n = 1$. Here F is generated by the elements $\varepsilon_i = e_1 \wedge \cdots \wedge \widehat{e_i} \wedge \cdots \wedge e_{r+1}$, for a given basis $\{e_1, \ldots, e_{r+1}\}$ of R^{r+1}, while B is generated by the elements $x_j \varepsilon_i - x_i \varepsilon_j$ with $i < j$, where $\sum x_i e_i$ generates the image of φ. Note that B is actually the submodule of 1-boundaries of a Koszul complex $\mathbb{K}$ on the set of elements $x_1, \ldots, x_{r+1}$. If we denote by Z the corresponding module of 1-cycles, we have

$$0 \to H_1(\mathbb{K}) = Z/B \longrightarrow \wedge^r(E) \cong F/B \longrightarrow F/Z \cong I_1(\varphi) \to 0,$$

which permits us to identify the two torsionfree modules $\wedge^r E/\text{torsion}$ and $I_1(\varphi)$ of rank 1.

The general case is similar, with the identifications done on the Eagon-Northcott complex (see [Ei95] for a detailed discussion of these complexes). It is simpler to work from the complex associated with the dual mapping

$$f = \varphi^* : G = (R^{n+r})^* \longrightarrow (R^n)^* = H.$$

The end of the complex is

$$S_1(H^*) \otimes \wedge^{n+1} G \xrightarrow{\mathrm{d}_f} \wedge^n G \xrightarrow{\wedge^n f} \wedge^n H \to 0.$$

Now we identify $\wedge^i G$ and $\wedge^{m-i} G^*$ in the usual manner. The mapping d_f is then transformed into φ_r, and we finish as in the case $n = 1$ (but without the identification with Koszul homology). □

Corollary 8.82. *Let R be an integral domain and E a torsionfree module defined by single relation,*

$$0 \to R \xrightarrow{\varphi} F = R^{r+1} \longrightarrow E \to 0.$$

Then E is of linear type, and $\det_0(S_n(E)) = \det_0(E)^s$ *for each integer n where $s = \binom{r+n-1}{r}$.*

Proof. We have for each integer n an exact sequence

$$0 \to R \otimes S_{n-1}(F) \longrightarrow S_n(F) \longrightarrow S_n(E) \to 0,$$

since grade $I_1(\varphi) \geq 2$ by the acyclicity lemma. This shows that $S(E)$ is the hypersurface ring defined by $\varphi(1)$. The assertion about the determinants follows from Theorem 8.81. □

In general it is rather cumbersome to obtain formulas for the determinant order ideals of the symmetric powers of modules. Let us look into this by considering a few cases.

First, let E be a module of rank two such that $E \subset R^2$. The elements of E may be seen as linear forms $f = ax + by$ in two variables. Let us calculate $\det_0(E^2)$ under the assumption that 2 is invertible. The module E^2 is generated (as 2 is invertible) by the squares f^2, and for forms $f_i = a_i x + b_i y$, $i = 1, 2, 3$ we have

$$f_1^2 \wedge f_2^2 \wedge f_3^2 = 2(a_1 b_2 - a_2 b_1)(a_2 b_3 - a_3 b_2)(a_3 b_1 - a_1 b_2).$$

By multilinearization, it follows that $\det_0(E^2) = \det_0(E)^3$. Similarly, if R contains the rationals, for every integer n we have $\det_0(E^n) = \det_0(E)^s$, where $s = \binom{n+1}{2}$.

For modules of higher rank we have been unable thus far of deriving such formulas, except indirectly as in the previous corollary.

There are several questions about the relationship between E and $\det_0(E)$ that would be interesting to have clarified:

(i) Let $R = k[x,y]$ be the ring of polynomials in two variables over a field k and let E be the module defined by the exact sequence

$$0 \to R \xrightarrow{\varphi} R^3 \longrightarrow E \to 0,$$

where $\varphi(1) = (x^2, xy, y^2)$. According to the calculation above, $\det_0(E) = (x,y)^2$, which is an integrally closed ideal. Note however that $E \subset (x,y)E^{**} \subset E^{**} \cong R^2$, and that $\lambda((x,y)E^{**}/E) < \infty$. Since $(x,y)E^{**}$ is minimally generated by 4 elements, while E needs 3 generators, it follows from Corollary 8.80 that E is not even (x,y)-full. The reverse statement is also false: it suffices to start with an integrally closed ideal I that is not normal–say I^r is not integrally closed. The module $I^{\oplus r}$ is integrally closed and I^r is its order determinant ideal. See an example (1.52) of a prime ideal P of $k[x,y,z]$ with P^2 not complete.

(ii) If E is a vector bundle, what are the relationships between the reduction numbers of E and of $\det_0(E)$? We pointed out some loose connections in the discussion of Buchsbaum-Rim's multiplicities.

(iii) If E is the Jacobian module of the commuting variety of a simple Lie algebra (see [Va94b, Chapter 8]), then by Theorem 8.81, $\det_0(E)$ is the ideal of the maximal minors of the Jacobian matrix of the fundamental invariants of the adjoint action of the corresponding Lie group. What does this mean?

8.5 Normality of Algebras of Linear Type

In this section we examine the normality of certain classes of modules. More precisely, we want to develop effective normality criteria for algebras of linear type. In particular, given a normal domain R and a torsionfree module E with a free resolution,

$$\cdots \longrightarrow F_2 \xrightarrow{\psi} F_1 \xrightarrow{\varphi} F_0 \longrightarrow E \to 0$$

we study the role of the matrices of syzygies in the normality of the Rees algebra of E. This goal being, in general, not too realistic, we shall restrict ourselves to those algebras which are of linear type, that is, in which $\mathcal{R}(E) = S_R(E)$. Our general reference for this section is [BV3].

An underlying motivation is the following open problem. Let $\mathfrak{g}$ be a simple Lie algebra over an algebraically closed field k of characteristic zero. If $e_1, \ldots, e_n$ is a basis of $\mathfrak{g}$ over k, we consider as in (8.1) the *commuting variety* $\mathbf{V} = C(\mathfrak{g})$ of $\mathfrak{g}$ (see [Va94b, Chapter 9] for details). This is an irreducible variety and an important question is whether it is *normal*. (There is also a more pointed question on whether $\mathbf{V}$ is a rational singularity.)

From the perspective of Rees algebras, there is a torsionfree module E over the ring of polynomials $R = k[x_1, \ldots, x_n]$ such that $\mathcal{R}(E)$ is the affine ring of $\mathbf{V}$. The irreducibility of $\mathbf{V}$ means that

$$\mathcal{R}(E) \cong S(E)_{\text{red}}.$$

The structure of E carries considerable information about $\mathfrak{g}$. Note that E has a projective resolution

$$0 \to R^{\ell} \xrightarrow{\psi} R^{n} \xrightarrow{\varphi} R^{n} \longrightarrow E \to 0$$

in which ℓ is the rank of $\mathfrak{g}$.

The mapping φ is given as a skew-symmetric matrix of linear forms, obtained as a Jacobian matrix of the equations defining commutation in $\mathfrak{g}$. Then ψ is also a Jacobian matrix obtained as follows. Let $p_1, \ldots, p_\ell$ be the fundamental invariants of the adjoint action of the corresponding Lie group G. The mapping ψ is the Jacobian matrix of these polynomials. A fundamental fact about the ideal $I = I_\ell(\psi)$ is that it has codimension 3, and its radical has one or two components, according to the root length of $\mathfrak{g}$. Elementary calculations show that I is a prime ideal in the cases of the algebras A_2 and A_3.

This wealth of information makes an appealing case for a program to examine normality in Rees algebras of modules. In this section we develop some techniques targeted at simpler classes of modules.

The main result of this section (Theorem 8.98) characterizes normality in terms of the ideal $I_c(\psi)S(E)$ and of the completeness of the first s symmetric powers of E, where $c = \text{rank}\psi$ and $s = \text{rank}F_0 - \text{rank}E$. It requires that R be a regular domain. Some of the other characterizations, although valid only for restricted classes of modules, do not assume regularity in R. Furthermore, they are more accessible for computation.

- Complete intersection modules
- Complete intersection algebras
- Almost complete intersection algebras
- Algebras of linear type

8.5.1 Complete Intersection Modules

Let R be a Cohen-Macaulay ring and let E be a complete intersection module. Suppose that E is given by a mapping

$$\varphi : R^{r+c-1} \longrightarrow R^{r}.$$

This means that $E = \text{image}\ \varphi$ has rank r, and that the ideal $I = \det_0(E)$ has codimension $c \geq 2$.

Our purpose is to describe, entirely in terms of the ideal I, when E is integrally closed. It turns out that more is achieved.

Theorem 8.83. *Let R be a Cohen-Macaulay integrally closed domain and E a complete intersection module. The following conditions are equivalent.*

(a) *E is integrally closed.*
(b) *E is normal.*
(c) $\det_0(E)$ *is an integrally closed generic complete intersection.*

Moreover if $\det_0(E)$ *is a prime ideal, then the above conditions hold.*

Before we start the proof, we refer the reader to Chapter 9, where we discuss some tests of normality for ideals which are generic complete intersection.

The proof of the theorem for modules will follow the arguments in Proposition 1.69, with some technicalities stripped away due to the assumption that R is a Cohen-Macaulay domain.

Proof. Let $\mathcal{R} = \mathcal{R}(E)$ be the Rees algebra of the module E. According to Theorem 8.36, E is a module of linear type and $\mathcal{R}$ is a Cohen-Macaulay ring.

To check the normality of $\mathcal{R}$, we need to understand the height 1 prime ideals P of $\mathcal{R}$. Set $\mathfrak{p} = P \cap R$ and localize at $R_{\mathfrak{p}}$. If $\mathfrak{p} = 0$ then $\mathcal{R}(E)_{\mathfrak{p}}$ is a ring of polynomials over the field of fractions of R, and we have nothing to be concerned about. We may assume that $(R, \mathfrak{m})$ is a local ring and that $P = \mathfrak{m}\mathcal{R}(E)$. If $\dim R = 1$ then R is a discrete valuation ring and E is a free R-module.

We may thus assume thatdim$R \geq 2$. If E as a submodule of R^r contains a free summand of R^r, $E = E_0 \oplus R^s$, then $\mathcal{R}(E)$ is a Rees algebra of a complete intersection module of rank $r-s$ over the ring of polynomials $R[T_1, \ldots, T_s]$. Thus all conditions would be preserved in E_0. We may finally assume that $E \subset \mathfrak{m}R^r$.

We claim that $\dim R = d = c$, that is, $\mathfrak{m}$ is a minimal prime of $\det_0(E)$, in particular the module R^r/E has finite length. This follows simply from the equality

$$\begin{aligned} 1 = \text{height } \mathfrak{m}\mathcal{R} &= \dim \mathcal{R} - \ell(E) \\ &= (d+r) - (c+r-1) = (d-c)+1, \end{aligned}$$

since E is of linear type and therefore its analytic spread $\ell(E)$ is equal to its minimal number of generators.

Let us assume that condition (a) holds, that is, E is integrally closed. We denote by d_0 the embedding dimension of R. Among other things, we must show that $d = d_0$, which means that R is a regular local ring. According to Proposition 8.79, E is $\mathfrak{m}$-full. Since $\lambda(\mathfrak{m}R^r/E) < \infty$ by Corollary 8.80, we must have

$$\nu(E) \geq \nu(\mathfrak{m}R^r).$$

Because the module $\mathfrak{m}R^r$ is minimally generated by rd_0 elements, while E itself is generated by $r+d-1$ elements, we have

$$r+d-1 \geq rd_0,$$

which means that

$$(r-1)(d-1) \leq r(d-d_0) \leq 0.$$

Since we have assumed that $d \geq 2$, this means that $d = d_0$ and $r = 1$. In other words, E is isomorphic to an ideal generated by system of parameters of a regular ring. The assertion thus follows from the ideal case ([Go87]).

The other implications are either immediate, or in the case of (c) $\Rightarrow$ (a) it follows from the equalities of associated primes:

$$\forall n, \quad \text{Ass}\,(S_n(R^r)/E^n) = \text{Ass}\,(R/\text{det}_0(E)).$$

□

Example 8.84. Let $\mathfrak{p}$ be a perfect ideal of codimension 2 generated by maximal minors of the $n \times (n+1)$ matrix φ. According to the assertion above, the columns of φ generate a normal submodule of R^n. In contrast, $\mathfrak{p}$ is often a non-normal ideal.

Let R be a Cohen-Macaulay normal domain and E a torsionfree R-module. The criteria above can be used to test the completeness of other modules that are not complete intersection modules. Let E be an ideal module of projective dimension $c-1 \geq 1$. It will follow that the associated primes of the cokernel of the embedding $E \hookrightarrow E^{**} = R^r$ have codimension at most c. This means that in comparing E to $\overline{E}$, namely

$$\overline{E}/E \hookrightarrow R^r/E,$$

we may consider only those localizations at the associated primes of R^r/E.

We shall assume one additional requirement on E, namely that it satisfies the condition $\mathcal{F}_1$ (or G_∞) on the local number of generators:

$$\forall \mathfrak{p} \in \text{Spec}\,(R), \mathfrak{p} \neq (0), \quad \nu(E_\mathfrak{p}) \leq r + \dim R_\mathfrak{p} - 1.$$

This condition is always present when the module is of linear type. If the codimension of R^r/E is c, localizing at a prime ideal associated to R^r/E, $E_\mathfrak{p}$ is a complete intersection module. As a consequence of these observations we obtain:

Corollary 8.85. *Let E be an ideal module for which E^{**}/E is equidimensional. If E satisfies the condition $\mathcal{F}_1$, then E is integrally closed if and only if* $\text{det}_0(E)$ *is an integrally closed generic complete intersection.*

8.5.2 Symbolic Powers and Normal Modules

Let R be a normal domain and E a torsionfree R-module of rank r with an embedding $E \hookrightarrow R^r$. We are going to give a normality criterion for the Rees algebra of E that makes strong demands on the order determinant ideal $\text{det}_0(E)$, but which will not require that E be of linear type. We shall then give an example of a module where conjecturally the condition is met.

The condition we seek involves the normality of the powers of $\text{det}_0(E)$. It requires formulas for the order determinantal ideal of the powers E^n, $n \geq 1$.

Theorem 8.86. *Suppose that R contains the rationals. Then*

$$\text{det}_0(E^n) = \text{det}_0(E)^s, \quad s = \binom{r+n-1}{r}.$$

Proof. This is a consequence of Theorem 8.104; see the discussion on it. □

A Determinantal Criterion of Normality

A previous result allows for several applications to establishing normality in some Rees algebras of modules.

Theorem 8.87. *Let R be a normal domain containing the rationals and E a torsionfree R-module of rank r. Suppose that $I = \det_0(E)$ is a prime (more generally a radical) ideal that is generically a complete intersection and whose symbolic and ordinary powers coincide. If the Rees algebra $\mathcal{R}(E)$ satisfies Serre's condition S_2, then it is normal.*

Proof. The ideal $I = \det_0(E)$ is defined from a fixed embedding $E \hookrightarrow R^r$. Denote by E^n the component of degree n of $\mathcal{R}(E)$, and let $\overline{E^n}$ be its integral closure. According to Theorem 8.86, the order determinantal ideal of E^n is a power of I, which by assumption is integrally closed and height unmixed. This means that $\det_0(\overline{E^n}) = I^s$ and therefore by Proposition 8.75, $\overline{E^n}$ is an unmixed module. Thus for every integer n, the module $S_n(R^r)/\overline{E^n}$ has just $\mathfrak{p}$ for associated prime and therefore for every prime ideal $\mathfrak{q}$ properly containing $\mathfrak{p}$, the $\mathfrak{q}$-depth of $\overline{E^n}$ is at least 2.

Suppose that $\mathcal{R}(E)$ has S_2 but it is not normal. This means that there is a prime $\mathfrak{q}$ associated to some module $\overline{E^n}/E^n$. We note that $\mathfrak{q} \neq \mathfrak{p}$. Indeed, localizing at $\mathfrak{p}$ we obtain $\det_0(E_{\mathfrak{p}}) = \mathfrak{p}R_{\mathfrak{p}}$, which implies that $E_{\mathfrak{p}} \cong R_{\mathfrak{p}}^{r-1} \oplus (\text{ideal})$. Therefore $E_{\mathfrak{p}}$ is a complete intersection module that meets the conditions of Theorem 8.83 (which is really not required in this case), and therefore $\mathcal{R}(E_{\mathfrak{p}})$ is normal. This shows that $\mathfrak{q} \neq \mathfrak{p}$. Localizing at a minimal such prime, we may assume that $\mathfrak{q}$ is the unique maximal ideal of R. If we now set $A = \mathcal{R}(E)$, $B = \overline{\mathcal{R}(E)}$ and $C = B/A$, we obtain that height $\mathfrak{q}B \geq 2$, since the $\mathfrak{q}$-depth of each component of B is at least two by the previous argument. This implies that height $\mathfrak{q}A \geq 2$ also. Since some power of $\mathfrak{q}$ annihilates C by construction, $A = B$ if A has condition S_2. □

Complete Intersection Algebras

Let R be an integrally closed Cohen-Macaulay domain and E a finitely generated torsionfree R-module. It is clear that if the Rees algebra $\mathcal{R}(E)$ of E is a complete intersection, then E has a projective resolution

$$0 \to R^m \xrightarrow{\varphi} R^n \longrightarrow E \to 0, \tag{8.5}$$

and it is of linear type. According to [Va94b, Theorem 3.1.6], for $S(E)$ to be a complete intersection we must have

$$\text{grade } I_t(\varphi) \geq \text{rank}(\varphi) - t + 1 = m - t + 1, \quad 1 \leq t \leq m.$$

Moreover, since $S(E) = \mathcal{R}(E)$ is an integral domain, this requirement will hold modulo any nonzero element of R, so it will be strengthened to

$$\text{grade } I_t(\varphi) \geq \text{rank}(\varphi) - t + 2 = m - t + 2, \quad 1 \leq t \leq m.$$

This can be rephrased in terms of the local number of generators as

$$\nu(E_{\mathfrak{p}}) \leq n - m + \text{height } \mathfrak{p} - 1, \quad \mathfrak{p} \neq 0.$$

For these modules, the graded components of the Koszul complex of the forms of $A = R[T_1, \ldots, T_n]$ (see [Av81]), namely

$$[f_1, \ldots, f_m] = [T_1, \ldots, T_n] \cdot \varphi,$$

give R-projective resolutions of the symmetric powers of E:

$$0 \to \wedge^s R^m \to \wedge^{s-1} R^m \otimes R^n \to \cdots \to R^m \otimes S_{s-1}(R^n) \to S_s(R^n) \to S_s(E) \to 0.$$

In particular we have:

Proposition 8.88. *Let R be a Cohen-Macaulay normal domain and E a finitely generated torsionfree R-module such that the Rees algebra $\mathcal{R}(E)$ is a complete intersection defined by m equations. The non-normal R-locus of $\mathcal{R}(E)$ has codimension at most $m+1$.*

Proof. It suffices to consider the following observation. For any torsionfree R-module G contained in a free module F, the embedding

$$\overline{G}/G \hookrightarrow F/G$$

shows that if G has projective dimension at most r, the associated primes of $\overline{G}/G$ have codimension at most $r+1$. In the case of the symmetric powers $S_s(E)$ of E, the projective dimensions are bounded by m, as follows from the comments above. □

The next aim is to discuss the normality of the algebra $\mathcal{R}(E)$ versus the completeness of the module E and of a few other symmetric powers.

Hypersurfaces

We shall begin our analysis with the special case of a hypersurface.

Assume that the module E is torsionfree with a single defining relation:

$$0 \to R\mathbf{a} \longrightarrow R^n \longrightarrow E \to 0, \quad \mathbf{a} = (a_1, \ldots, a_n).$$

The ideal generated by the a_i has grade at least 2, and

$$S(E) = R[T_1, \ldots, T_n]/(f), \quad f = a_1T_1 + \cdots + a_nT_n.$$

Proposition 8.89. *In the setting above, $S(E)$ is normal if and only if for every prime ideal $\mathfrak{p}$ of codimension two that contains $(a_1, \ldots, a_n)$, the following conditions hold:*

(i) *$R_{\mathfrak{p}}$ is a regular local ring.*
(ii) $(a_1, \ldots, a_n) \not\subset \mathfrak{p}^{(2)}$.

Proof. Since $S(E)$ is Cohen-Macaulay, to test for normality if suffices to check the localizations $S(E)_P$, where P is a prime ideal of $A = R[T_1, \ldots, T_n]$ that contains f. Set $\mathfrak{p} = P \cap R$. If height $\mathfrak{p} \leq 1$, then $E_{\mathfrak{p}}$ is a free $R_{\mathfrak{p}}$-module and $S(E_{\mathfrak{p}})$ is a ring of polynomials over a DVR. Otherwise $P = \mathfrak{p}A$, with height $\mathfrak{p} = 2$.

The listed conditions just express the fact that $(\mathfrak{p}A/(f))_P$ is a cyclic module. □

Corollary 8.90. *Suppose further that $R = k[x_1, \ldots, x_d]$ is a ring of polynomials over a field of characteristic zero. Then $S(E)$ is normal if and only if*

$$\text{height }(a_1, \ldots, a_n, \frac{\partial a_i}{\partial x_j},\ i = 1 \ldots n,\ j = 1 \ldots d) \geq 3.$$

Proof. We must show that this condition is equivalent to saying that the Jacobian ideal J of $S(E)$ has codimension at least 2. Note that

$$J = (a_1, \ldots, a_n, \frac{\partial a_i}{\partial x_j} T_j,\ i = 1 \ldots n,\ j = 1 \ldots d)/(f).$$

We consider the height of the lift J' of J in the ring $R[T_1, \ldots, T_n]$. Let P be a minimal prime of J'; if some variable T_j lies in P, $(a_1, \ldots, a_n, T_j) \subset P$ and so height $P \geq 3$. On the other hand, if no variable T_j is contained in P, then P must contain the ideal of the assertion. □

While the normality is straightforward, the completeness of E requires a different kind of analysis.

Theorem 8.91. *Suppose that R is a Cohen-Macaulay integral domain that is regular in codimension two. If E is a module as above, then E is complete if and only if E is normal.*

Proof. According to the previous discussion, E is not normal precisely when there exists a prime ideal $\mathfrak{p}$ of codimension two containing $(a_1, \ldots, a_n)$ such that $(\mathfrak{p}A/(f))_{\mathfrak{p}A}$ is not principal. We shall argue that in such case there exists an element h in the integral closure of $S(E)$ of degree 1 but not lying in E. In other words, E is not complete.

We replace R by $R_{\mathfrak{p}}$, and denote by x, y a set of generators of its maximal ideal. In the one-dimensional local ring $B = S(E)_{\mathfrak{p}S(E)}$, we have $P = \mathfrak{p}B$ and $\text{Hom}_B(P, P) = P^{-1}$, since P is not principal. It will suffice to find elements of degree 1 in $(\mathfrak{p}S(E))^{-1}$ that do not belong to $E = S_1$.

Suppose that $z \in \mathfrak{p}$, so that x, f is a regular sequence in $\mathfrak{p}A$. Writing

$$\begin{aligned} z &= ax + by \\ f &= cx + dy, \quad \text{where } c \text{ and } d \text{ are elements of } E \end{aligned}$$

we obtain that

$$(z, f) : \mathfrak{p}A = (z, f, ad - bc).$$

The fraction $h = (ad - bc)z^{-1}$ does not lie in E and has the desired property. □

General Complete Intersections

When $S(E)$ is defined by more than one hypersurface,

$$S(E) = R[T_1,\ldots,T_n]/(f_1,\ldots,f_m), \quad [f_1,\ldots,f_m] = [T_1,\ldots,T_n]\cdot\varphi,$$

it may be more convenient to use an algorithmic formulation; this is carried out in Chapter 9.

Theorem 8.92. *Let E be a module as above. Then $S(E)$ is normal if and only if the following conditions hold.*

(i) *For every prime ideal $\mathfrak{p}$ of height $m-t+2$ containing $I_t(\varphi)$, $R_\mathfrak{p}$ is a regular local ring.*
(ii) *The modules $S_s(E)$ are complete for $s = 1\ldots m$.*

Proof. Let us assume that $S(E)$ is normal, and establish (i). Let $\mathfrak{p}$ be a prime ideal as in (i). Since height $I_{t-1}(\varphi) \geq m-t+3$, localizing at $\mathfrak{p}$ we obtain a presentation of the module in the following form (we set still $R = R_\mathfrak{p}$):

$$0 \to R^{m-t+1} \xrightarrow{\varphi'} R^{n-t+1} \longrightarrow E \to 0,$$

where the entries of φ' lie in $\mathfrak{p}$. Changing notation, this means that we can assume that all entries of φ lie in the maximal ideal $\mathfrak{p}$. Setting $A = R[T_1,\ldots,T_n]$ and $P = \mathfrak{p}A$, we get that

$$S(E)_P = (A/(f_1,\ldots,f_m))_P,$$

and therefore if S_P is a DVR, A_P must be a regular local ring, and therefore R will also be a regular local ring.

For the converse, let P be a prime ideal of the ring A of polynomials, of height $m+1$, containing the forms f_i and set $\mathfrak{p} = P\cap R$. The normality of $S(E)$ means that for all such P, $S(E)_P$ is a DVR. The claim is that failure of this to hold is controlled by either (i) or (ii).

We may localize at $\mathfrak{p}$. Suppose that height $\mathfrak{p} = m+1$. In this case, there exists $z \in \mathfrak{p}$ so that $z, f_1,\ldots,f_m$ is a regular sequence in $\mathfrak{p}A$. If $x_1,\ldots,x_{m+1}$ is a regular system of parameters of the local ring R, from a representation

$$[z, f_1,\ldots,f_m] = [x_1,\ldots,x_{m+1}]\cdot\psi,$$

as in Theorem 8.91, we obtain the socle equality

$$(z, f_1,\ldots,f_m) : \mathfrak{p}A = (z, f_1,\ldots,f_m, \det\psi).$$

The image u of $z^{-1}\det\psi$, provides us with a nonzero form of degree m in the field of fractions of $S(E)$. If $S(E)_P$ is not a DVR, then

$$\mathrm{Hom}_{S(E)}(PS(E)_P, PS(E)_P) = (PS(E)_P)^{-1},$$

and, given that $u \in (PS(E)_P)^{-1}$, we have obtained a fresh element in the integral closure of $S_m(E)$.

On the other hand, if height $\mathfrak{p} \leq m$ we may assume that for some t with $1 < t \leq m$, we have $I_t(\varphi) \subset \mathfrak{p}$ but $I_{t-1}(\varphi) \not\subset \mathfrak{p}$. This implies that height $\mathfrak{p} = m-t+2$. We can localize at $\mathfrak{p}$ and argue as in the previous case. □

Almost Complete Intersection Algebras

An *almost complete intersection* Rees algebra arises from a module E with a projective resolution

$$0 \to R \stackrel{\psi}{\longrightarrow} R^m \stackrel{\varphi}{\longrightarrow} R^n \longrightarrow E \to 0.$$

For the module to be of linear type, $S(E) = \mathcal{R}(E)$, the roles of the determinantal ideals $I_t(\varphi)$ and of $I_1(\psi)$ are less well behaved than in the case of complete intersections. They are nevertheless determined by any minimal resolution of E, the $I_t(\varphi)$ giving the Fitting ideals of E, while $I_1(\psi)$ is the annihilator of $\text{Ext}^2_R(E,R)$. Let us give a summary of some of the known results, according to [Va94b, Section 3.4]:

Theorem 8.93. *Let R be a Cohen-Macaulay integral domain and E a module with a resolution as above.The following hold.*

(a) *If* $S(E) = \mathcal{R}(E)$ *then* height $I_1(\psi)$ *is odd.*
(b) *If* $I_1(\psi)$ *is a strongly Cohen-Macaulay ideal of codimension* 3 *satisfying the condition* $\mathcal{F}_1$, *and* E^* *is a third syzygy module, then* $S(E)$ *is a Cohen-Macaulay integral domain.*

This shows the kind of requirement that must be present when one wants to construct integrally closed Rees algebras in this class.

Example 8.94. Let $R = k[x_1, \ldots, x_d]$ be a ring of polynomials. In [Ve73], for each $d \geq 4$ there is described an indecomposable vector bundle on the punctured spectrum of R of rank $d-2$. Its module E of global sections has a resolution

$$0 \to R \stackrel{\psi}{\longrightarrow} R^d \stackrel{\varphi}{\longrightarrow} R^{2d-3} \longrightarrow E \to 0,$$

with φ having linear forms as entries, and $\psi(1) = [x_1, \ldots, x_d]$. If d is odd, according to [SUV93, Corollary 3.10], E is of linear type and normal.

The analysis of the normality of the two previous classes of Rees algebras was made simpler because they were naturally Cohen-Macaulay. This is no longer the case with almost complete intersections, requiring that the S_2 condition be imposed in some fashion. We choose one such imposition closely related to normality.

Theorem 8.95. *Let R be a regular integral domain and E a torsionfree module whose second Betti number is* 1. *Suppose that E is of linear type with a resolution as above. If the ideal* $I_1(\psi)\mathcal{R}(E)$ *is principal at all localizations of* $\mathcal{R}(E)$ *of depth* 1, *then* $\mathcal{R}(E)$ *satisfies Serre's condition* S_2.

Proof. The condition on $L = I_1(\psi)\mathcal{R}(E)$ can be recast in several global ways, such as the requirement that $(L \cdot L^{-1})^{-1} = S$.

We may assume that R is a local ring and that the resolution of E is minimal. Choose in $A = R[T_1, \ldots, T_n]$ a prime ideal P for which depth $S_P = 1$. We must show that $\dim S_P = 1$. As in the other cases, we may assume that $P \cap R$ is the maximal ideal of R.

We derive now a presentation of the ideal $J = J(\varphi) = (f_1, \ldots, f_m)A$, modulo J:

$$0 \to K \to (A/J)^m = S^m \longrightarrow J/J^2 \to 0,$$

and analyze the element of S^m induced by $v = \psi(1)$. This is a nonzero 'vector' whose entries in S_P^m generate L_P, which by assumption is a principal ideal. This means that $v = \alpha v_0$ for some nonzero $\alpha \in S_P$, where v_0 is an unimodular element S_P^m. Since $v \in K$, this means that the image u of v_0 in $(J/J^2)_P$ is a torsion element of the module. Two cases arise. If $u = 0$, then $(J/J^2)_P$ is a free S_P-module since it also has rank $m-1$, and therefore J_P is a complete intersection in the regular local ring A_P, according to [Va67]. This implies that the Cohen-Macaulay local ring S_P has dimension 1. On the other hand, if $u \neq 0$, then u is a torsion element of $(J/J^2)_P$ which is also a minimal generator of the module. We thus have that in the exact sequence

$$0 \to \text{torsion } (J/J^2)_P \to (J/J^2)_P \to C \to 0,$$

C is a torsionfree S_P-module of rank $m-1$ generated by $m-1$ elements. We thus have that the ideal J_P of the regular local ring A_P has the property that

$$J_P/J_P^2 \cong (A_P/J_P)^{m-1} \oplus (\text{torsion}).$$

According to [Va67] again, J_P is a complete intersection since it has codimension $m-1$. This shows that S_P must have dimension 1. □

In studying the normality in an algebra $S = S(E)$ (E torsionfree) of linear type, the presence of S_2 is extremely useful. We recall briefly some technical facts from [Va94b, p. 138]. For a prime ideal $\mathfrak{p} \subset R$ there is an associated prime ideal in $S(E)$ defined by the rule

$$T(\mathfrak{p}) = \ker(S_R(E) \longrightarrow S_{R/\mathfrak{p}}(E/\mathfrak{p}E)_\mathfrak{p}).$$

In our case, $T(\mathfrak{p})$ is the contraction $\mathfrak{p}S(E)_\mathfrak{p} \cap S(E)$, so height $T(\mathfrak{p}) =$ height $\mathfrak{p}S(E)$.

The prime ideals we are interested in are those of height 1. If R is equidimensional, we see that

$$\text{height } T(\mathfrak{p}) = 1 \text{ if and only if } \nu(E_\mathfrak{p}) = \text{height } \mathfrak{p} + \text{rank}(E) - 1.$$

Let us quote the following result from [Va94b, Proposition 5.6.2]:

Proposition 8.96. *Let R be a universally catenary Noetherian ring and E a finitely generated module such that $S(E)$ is a domain. Then the set*

$$\{T(\mathfrak{p}) \mid \text{height } \mathfrak{p} \geq 2 \textit{ and } \text{height } T(\mathfrak{p}) = 1\}$$

is finite. More precisely, for any presentation $R^m \xrightarrow{\varphi} R^n \longrightarrow E \to 0$, this set is in bijection with

$$\{\mathfrak{p} \subset R \mid E_\mathfrak{p} \textit{ not free, } \mathfrak{p} \in \text{Min}(R/I_t(\varphi)) \textit{ and } \text{height } \mathfrak{p} = \text{rank}(\varphi) - t + 2\},$$

where $1 \leq t \leq \text{rank}(\varphi)$.

Theorem 8.97. *Let R be a regular integral domain and E a torsionfree module with a free resolution*

$$0 \to R \xrightarrow{\psi} R^m \xrightarrow{\varphi} R^n \longrightarrow E \to 0.$$

Suppose that E is of linear type. Then E is normal if and only if the following conditions hold.

(i) *The ideal $I_1(\psi)S(E)$ is principal at all localizations of $S(E)$ of depth* 1.
(ii) *The modules $S_s(E)$ are complete for $s = 1 \ldots m-1$.*

Proof. We only have to show that (i) and (ii) imply that E is normal. In view of Theorem 8.95, it suffices to verify Serre's condition R_1.

Let P be a prime ideal of A such that its image in $S = A/J$ has height 1. We may assume that $P \cap R$ is the maximal ideal of R and that E has projective dimension 2, that is, $I_1(\psi) \subset \mathfrak{m}$ as otherwise we could apply Theorem 8.92.

From Proposition 8.96 and the paragraph preceding it, $\dim R = \operatorname{rank}(\varphi) - t + 2$ on the one hand and $\dim R = n - r + 1$ on the other. Thus, $t = 1$ and $m = \dim R$. Choose a such that $0 \neq a \in \mathfrak{m}$, and consider the ideal $I = (a, f_1, \ldots, f_m)$ of A. With $P = \mathfrak{m}A$, set $L = I : P$. For a set $x_1, \ldots, x_m$ of minimal generators of $\mathfrak{m}$, we have

$$[a, f_1, \ldots, f_m] = [x_1, \ldots, x_m] \cdot B(\Phi), \tag{8.6}$$

where $B(\Phi)$ is a $m \times (m+1)$ matrix whose first column has entries in R, and the other columns are linear forms in the T_i. We denote by L_0 the ideal of A generated by the minors of order m that fix the first column of $B(\Phi)$. These are all forms of degree $m-1$, and $L_0 \subset L$. When we localize at P, however, we get that $L_0A_P = LA_P$, since by condition (i) and the proof of Theorem 8.95, J_P is a complete intersection and the assertion follows from Theorem 1.113. This means that the image C of $a^{-1}L_0$ in the field of fractions of S is not contained in S and has the property that $C \cdot PS_P \subset S_P$, giving rise to two possible outcomes:

$$C \cdot PS_P = \begin{cases} PS_P, \\ S_P. \end{cases}$$

In the first case, C would consist of elements in the integral closure of S_P, but it is not contained in S_P. This cannot occur, since by condition (ii) all the symmetric powers of E up to order $m-1$ are complete. This means that the second possibility occurs, and S_P is a DVR. □

Effective Criteria

We are going to make a series of observations leading to the proof of a generalization of Theorem 8.97 to all algebras of linear type over regular domains. The arguments show great similarity, for the following technical reason. If E is a torsionfree R-module of linear type, a prominent role is played by the prime ideals of $S(E)$ of codimension one, which we denoted by $T(\mathfrak{p})$. If $\mathfrak{p}$ is the unique maximal ideal-as a reduction will lead to–then height $T(\mathfrak{p}) = 1$ if and only if $\dim R + \operatorname{rank}(E) = \nu(E) + 1$, a condition equivalent to (when E has finite projective dimension) the second Betti number of E being 1.

Theorem 8.98. *Let R be a regular integral domain and E a torsionfree module of rank r with a free presentation*

$$R^p \xrightarrow{\psi} R^m \xrightarrow{\varphi} R^n \longrightarrow E \to 0.$$

Suppose that E is of linear type. Then $\mathcal{R}(E)$ is normal if and only if the following conditions hold.

(i) *The ideal $I_c(\psi)S(E)$, $c = m+r-n$, is principal at all localizations of $S(E)$ of depth* 1.
(ii) *The modules $S_s(E)$ are complete for $s = 1\dots n-r$.*

Proof. We note that $n-r$ is the height of the defining ideal $J(\varphi)$ of S, while $c = m+r-n$ is the rank of the second syzygy module.

The necessity of these conditions follows *ipso literis* from the discussion of Theorem 8.97, except for a clarification of the role of condition (i). We consider the complex induced by tensoring the tail of the presentation by S, that is,

$$S^p \xrightarrow{\overline{\psi}} S^m \longrightarrow J/J^2 \to 0.$$

Note that $\overline{\psi}$ has rank c, while J/J^2 has rank $n-r$. This means that the kernel of the natural surjection $C = \text{coker}\,(\overline{\psi}) \to J/J^2$ is a torsion S-module.

Lemma 8.99. *Let E be a module as above, set $A = R[T_1,\dots,T_n]$, and $J = (f_1,\dots,f_m) = [T_1,\dots,T_n]\cdot\varphi$ for the defining ideal of $S(E) = A/J$. Let P be a prime ideal of A containing J. If $I_c(\overline{\psi})S_P$ is a principal ideal then J_P is a complete intersection.*

Proof. According to [Va98b, Proposition 2.4.5], since the Fitting ideal $I_c(\overline{\psi})_P$ is principal, the module C_P decomposes as

$$C_P \cong S_P^{n-r} \oplus (\text{torsion}).$$

This means that we have a surjection

$$S_P^{n-r} \to (J_P/J_P^2)/(\text{torsion})$$

of torsionfree S_P-modules of the same rank. Therefore

$$J_P/J_P^2 \cong S_P^{n-r} \oplus (\text{torsion}).$$

At this point, we invoke [Va67] to conclude that J_P is a complete intersection. □

The rest of the proof of Theorem 8.98 would proceed as in the proof of Theorem 8.97. Choosing $P = \mathfrak{m}A$ so that S_P has dimension 1, J_P is a complete intersection of codimension $n-r = \dim R - 1 = d-1$. As in setting up the equation (8.6), we choose a such that $0 \neq a \in \mathfrak{m}$, choose a minimal set $\{x_1,\dots,x_d\}$ of generators for $\mathfrak{m}$, and define the matrix $B(\Phi)$

$$[a, f_1,\dots,f_m] = [x_1,\dots,x_d]\cdot B(\Phi). \tag{8.7}$$

Note that when localizing at P the ideals $(a, f_1,\dots,f_m)_P$ and $(x_1,\dots,x_d)_P$ are generated by regular sequences and we can use the same argument as that employed in the proof of Theorem 8.97. □

8.5.3 Complete Modules and Finite Projective Dimension

Let E be a complete submodule of R^r. The associated primes of R^r/E are particularly restricted when E is a module of finite projective dimension. This phenomenon was described by Burch ([Bu68, Corollary 3] in the case of ideals, and more recently using a different approach by Goto and Hayasaka ([GH1]). The latter technique extends easily to modules as pointed out to us by J. Hong. It serves to show the naturality of the regularity hypothesis on some of the normality criteria described.

Proposition 8.100. *Let $(R,\mathfrak{m})$ be a Noetherian local ring and E a submodule of R^r such that $\mathfrak{m}E :_{R^r} x = E$ for some $x \in \mathfrak{m}$. Set $\ell = \lambda((E :_{R^r} \mathfrak{m})/E)$ and $E :_{R^r} \mathfrak{m} = (y_1,\ldots,y_\ell) + E$, where $y_i \in (E :_{R^r} \mathfrak{m}) \setminus E$ for every $i = 1,\ldots,\ell$. Then the following hold.*

(a) $E :_{R^r} \mathfrak{m} = E :_{R^r} x$.
(b) *If* $\mathfrak{m} \in \text{Ass}\,(R^r/E)$, *then* $x \notin \mathfrak{m}^2$.
(c) $\{xy_i\}_1^\ell$ *is a part of a minimal system of generators of* E.
(d) *Suppose that* $E = (xy_1,\ldots,xy_\ell) + (z_1,\ldots,z_m)$ *where* $z_i \in E$ *and* $\ell + m = \nu(E)$. *Then*

$$E/xE = \sum_1^\ell R\overline{xy_i} \oplus \sum_1^m R\overline{z_j},$$

where $^-$ *denotes reduction modulo* xE.

Proof. (a) For $\alpha \in E : x$ and $a \in \mathfrak{m}$ we have $(a\alpha)x = a(\alpha x) \in \mathfrak{m}E$, so that $a\alpha \in \mathfrak{m}E : x = E$.

(b) Suppose that $x \in \mathfrak{m}^2$. Then

$$E = \mathfrak{m}E : x \supseteq \mathfrak{m}E : \mathfrak{m}^2 = (\mathfrak{m}E : \mathfrak{m}) : \mathfrak{m} \supseteq E : \mathfrak{m}.$$

Let $\mathfrak{m} = (E :_R \alpha)$ for some $\alpha \in R^r \setminus E$. Then $\alpha \in E :_{R^r} \mathfrak{m} \subseteq E$, which is a contradiction.

(c) Let $\{a_i\}_1^\ell$ be elements in R such that $\sum_1^\ell a_i(xy_i) = x(\sum_1^\ell a_iy_i) \in \mathfrak{m}E$. Then $\sum_1^\ell a_iy_i \in \mathfrak{m}E :_{R^r} x = E$ so that $a_i \in \mathfrak{m}$ because $y_i \in E : \mathfrak{m}$.

(d) Let $\{a_i\}_1^l$ and $\{b_j\}_1^n$ be elements in R such that $\sum_1^l a_i(xy_i) + \sum_1^n b_jz_j \in xE$. Since $a_i \in \mathfrak{m}$ for each i and $y_i \in E : \mathfrak{m}$, we have $a_iy_i \in E$ for each i and hence $\sum a_i(xy_i) = x(\sum a_iy_i) \in xE$. It follows that $\sum b_jz_j \in xE$. □

Corollary 8.101. *Let $(R,\mathfrak{m})$ be a Noetherian local ring and E a submodule of R^r such that $\mathfrak{m}E :_{R^r} x = E$ for some $x \in \mathfrak{m}$. Then*

$$E/xE = (E :_{R^r} \mathfrak{m})/E \oplus (E + xR^r)/xR^r = (E :_{R^r} \mathfrak{m})/E \oplus E/x(E :_{R^r} \mathfrak{m}).$$

Proof. Consider the exact sequence

$$\mathbf{C}_\bullet : \quad 0 \to E \to R^r \to R^r/E \to 0$$

and the mapping induced by multiplication by x on $\mathbf{C}_\bullet$. By the snake lemma, we have the exact sequence

$$0 \to 0 :_E x \to 0 :_{R^r} x \to E :_{R^r} x \to E/xE \to R^r/xR^r \to R^r/(xR^r + E) \to 0.$$

Let γ be the map $E :_{R^r} x \to E/xE$ in this exact sequence. Note that $\ker \gamma = \{\alpha \in R^r | \alpha x \in xE\} \supseteq E$. For any $\alpha \in \ker\ \gamma$, we have $\alpha x = x\beta$ for some $\beta \in E$. Thus $(\alpha - \beta)x = 0$ so that $\alpha - \beta \in 0 :_{R^r} x \subseteq \mathfrak{m}E : x = E$. Therefore $\ker \gamma = E$. Since $E :_{R^r} \mathfrak{m} = E :_{R^r} x$, we get an exact sequence

$$0 \to (E : \mathfrak{m})/E \xrightarrow{f} E/xE \to R^r/xR^r \to R^r/(E + xR^r) \to 0.$$

If we assume that $E : \mathfrak{m} = (y_1, \ldots y_l) + E$ where $y_i \in E : \mathfrak{m}$, then $f(y \bmod E) = xy \bmod xE$ for all $y \in E : \mathfrak{m}$, so that $\text{image}(f) = \sum_1^l R\overline{xy_i}$, where $^-$ denotes reduction modulo xE. By the proposition above, the sequence

$$0 \to (E : \mathfrak{m})/E \to E/xE \to E/(x(E : \mathfrak{m})) \to 0$$

is split exact. □

Theorem 8.102. *Let $(R, \mathfrak{m})$ be a Noetherian local ring and E a submodule of R^r such that $\mathfrak{m}E :_{R^r} x = E$ for some regular element $x \in \mathfrak{m}$. Suppose that* proj dim $(E) < \infty$ *and* $\mathfrak{m} \in$ Ass (R^r/E). *Then R is a regular local ring.*

Proof. It suffices to show that $R/\mathfrak{m}$ has finite projective dimension as an R-module. Since proj dim $(E) < \infty$, proj dim $(E/xE) < \infty$. By the decomposition in the Corollary, proj dim $((E : \mathfrak{m})/E) < \infty$. Since $\mathfrak{m} \in$ Ass (R^r/E), proj dim $(R/\mathfrak{m}) < \infty$. □

Kähler Differentials

An application to the completeness of some modules of Kähler differentials arise as follows.

Theorem 8.103. *Let $R = k[x_1, \ldots, x_n]/I$ be an affine domain over a field of characteristic zero. If R is a local complete intersection then the module $\Omega_k(R)$ of Kähler differentials is torsionfree and complete if and only if R is regular in codimension two. Equivalently, if and only if $\Omega_k(R)$ is a reflexive module that is free in codimension ≤ 2.*

Proof. The module of differentials has a presentation

$$0 \to I/I^2 \longrightarrow R^n \longrightarrow \Omega_k(R) \to 0,$$

which in codimension 1 is an exact free complex since R is a domain and $\Omega_k(R)$ is torsionfree. Since R is a local complete intersection I/I^2 is a projective R-module and the sequence gives a projective R-resolution of $\Omega_k(R)$. We can then embed $\Omega_k(R)$ into a free R-module and the associated primes of the cokernel have codimension at most two. According to Theorem 8.102, the localization $R_\mathfrak{p}$ is a regular local ring; but then $\Omega_k(R)_\mathfrak{p}$ is a free module by the Jacobian criterion. □

8.5.4 Determinantal Ideals of Symmetric Powers

Let E be a submodule of rank r of the module R^r defined by the columns of an $m \times r$ matrix φ. The determinantal ideals of E^n should be coded already in the matrix φ. That this is indeed the case, at least if R contains the rationals, is proved in [BVa3].

Let R be an integral domain and $\varphi : R^m \to R^n$ an R-linear map of rank r. It is an easy exercise to show that the d-th symmetric power $S^d(\varphi) : S^d(R^m) \to S^d(R^n)$ has rank $\binom{r+d-1}{d}$. Let $I_t(\varphi)$ denote the ideal generated by the minors (of a matrix representing φ). Since

$$\operatorname{rank} \varphi = \max\{r : I_r(\varphi) \neq 0\},$$

one can immediately determine the radicals of the ideals $I_t(S^d(\varphi))$, namely

$$\sqrt{I_t(S^d(\varphi))} = \sqrt{I_r(\varphi)} \quad \text{if} \quad \binom{r+d-1}{d} \le t < \binom{r+d}{d}.$$

Here we want to derive a more precise description of the $I_t(S^d(\varphi))$, the ideals of minors of the exterior powers of φ, and the ideals of minors of a tensor product.

At least for the ideals associated with $\operatorname{rank}\varphi = n$, one has a very satisfactory result. To simplify notation we set $I(\varphi) = I_n(\varphi)$.

Suppose that $m = n$. Then the ideals $I(\varphi)$ and $I(S^d(\varphi))$ are principal, generated by the determinants of square matrices, and

$$\det(S^d(\varphi)) = \det(\varphi)^s, \qquad s = \binom{n+d-1}{d-1}, \tag{8.8}$$

as follows immediately by transformation to a triangular matrix (it is enough to consider a generic matrix over $\mathbb{Z}$, which is contained in a field). For non-square matrices we can replace $\det(\varphi)$ by $I(\varphi)$:

Theorem 8.104. *Let R be a commutative ring and $\varphi : R^m \to R^n$, $\psi : R^p \to R^q$ R-linear maps. Set $s = \binom{n+d-1}{d-1}$ and $e = \binom{n-1}{d-1}$.*

(1) *Then*

$$I(\varphi \otimes \psi) \subset I(\varphi)^q I(\psi)^n, \qquad I(S^d(\varphi)) \subset I(\varphi)^s, \qquad I(\bigwedge^d(\varphi)) \subset I(\varphi)^e,$$

with equality up to integral closure.

(2) *Suppose that R contains the field of rational numbers. Then equality holds in* (1).

Proof. For simplicity of notation we treat only the case of the symmetric powers in detail. That of the exterior powers is completely analogous. The tensor product requires slight modifications, which we shall indicate below.

(1) The formation of $I(X)$ and $I(S^d(\varphi))$ commutes with ring extensions, for obvious reasons. Thus it is enough to prove the inclusion for $R = \mathbb{Z}[X] = \mathbb{Z}[X_{ij} : i = 1,\dots,n,\ j = 1,\dots,r]$ and the linear map $\varphi : R^m \to R^n$ given by the matrix $X = (X_{ij})$. It is well-known ([Tr79]; see also [BV88, (9.18)]) that the ideals $I(X)^k$ are primary with radical $I(X)$. It follows that $I(X)^k$ is integrally closed, and therefore

$$\overline{I(X)} = \bigcap I(X)V,$$

where V runs through the discrete valuation rings extending R. To sum up, we may assume that R is a discrete valuation ring.

Note that all ideals under consideration are invariant under base change in R^n and R^m. In fact, they are Fitting invariants of φ and $S^d(\varphi)$. By the elementary divisor theorem we can therefore assume that φ is given by a matrix with non-zero entries only in the diagonal positions (i,i), $i = 1,\dots,n$. Then $S^d(\varphi)$ is also given by a diagonal matrix, and it is an easy exercise to show that indeed $I(S^d(\varphi)) = I(\varphi)^s$ for such matrices φ.

As we have observed, equality holds for the ideals under consideration if R is a discrete valuation ring. This implies equality up to integral closure.

(2) We have to establish the inclusion opposite to that in (1), and it is enough to prove it for $R = \mathbb{Q}[X]$. Since $\operatorname{rank}\varphi = n$, the linear map $S^d(\varphi)$ has rank $\binom{n+d-1}{d}$. Therefore the ideal under consideration is non-zero, and its generators have the same degree as those of $I(\varphi)^s$. The group $G = \mathrm{Gl}_m(\mathbb{Q}) \times \mathrm{Gl}_r(\mathbb{Q})$ acts on R via the linear substitution sending each entry of X to the corresponding entry of

$$AXB^{-1}, \qquad A \in \mathrm{Gl}_m(\mathbb{Q}),\ B \in \mathrm{Gl}_r(\mathbb{Q}). \tag{8.9}$$

It is well-known that the $\mathbb{Q}$-vector space W generated by the degree rs elements of $I(X)^s$ is the irreducible G-representation associated with Young bitableaux of rectangular shape $s \times n$ (see [DEP80] or [BV88, Section 11]). So the desired inclusion follows if the ideal $I(S^d(\varphi))$ is G-stable. This is not difficult to see. In fact let, σ be the automorphism induced by the substitution (8.9). We write $\sigma(\varphi)$ for the linear map which we obtain from φ by replacing each entry with its image under φ, i. e. $\sigma(\varphi) = \varphi \otimes \sigma$ for the ring extension $\sigma : R \to R$. Then

$$\begin{aligned}\sigma\left(I(S^d(\varphi))\right) &= I\left(\sigma(S^d(\varphi))\right) = I\left(S^d(\sigma(\varphi))\right)\\ &= I\left(S^d(\beta^{-1}\varphi\alpha)\right) = I\left(S^d(\beta)^{-1} \circ S^d(\varphi) \circ S^d(\alpha)\right) = I(S^d(\varphi))\end{aligned}$$

where α is the automorphism of R^m induced by the matrix A, and β the automorphism of R^n induced by B. The very last equation uses again the fact that Fitting ideals are invariant under base change in the free modules.

For the tensor product the arguments above have to be modified at two places. We can of course assume that ψ is also given by a matrix Y of indeterminates. The first critical point is whether $I(X)^q I(Y)^n$ is again integrally closed as an ideal of the ring $R = \mathbb{Z}[X,Y]$.

We can argue as follows: $\mathbb{Z}[X,Y]$ is a free $\mathbb{Z}$-module whose basis is given by the products $\mu\nu$, where μ is a standard monomial in the variables X_{ij} and ν is a standard monomial in the Y_{kl} (see [BV88, Section 4]). An element f of $\mathbb{Z}[X,Y]$ belongs to $I(X)^q$ if and only if $\mu \in I(X)^q$ for all the factors μ appearing in the representation of f as a linear combination of standard monomials, and a similar assertion holds with respect to $I(Y)^n$ and the factors ν. It follows immediately that $I(X)^q I(Y)^n =$

$I(X)^q \cap I(Y)^n$. Since both $I(X)^q$ and $I(Y)^n$ are integrally closed, their intersection is also integrally closed.

The other crucial question is whether the $\mathbb{Q}$-vector space generated by the degree rq elements of $I(X)^q I(Y)^n$ is an irreducible $G \times G'$-representation, where $G' = \mathsf{Gl}_p(\mathbb{Q}) \times \mathsf{Gl}_q(\mathbb{Q})$ acts on $\mathbb{Q}[Y]$ in the same way as G acts on $\mathbb{Q}[X]$, and the action of $G \times G'$ on $K[X,Y] = K[X] \otimes K[Y]$ is the induced one. The answer is positive since $W \otimes W'$ is an irreducible $G \times G'$-representation if W and W' are irreducible for G and G' respectively. It is enough to test this after extending $\mathbb{Q}$ by $\mathbb{C}$, and then one can apply a classical theorem (see [Hup67, II.10.3]).

8.6 Bourbaki Ideals and Rees Algebras

Let E be a torsionfree R-module of rank r. A *Bourbaki sequence* for E is an exact sequence

$$0 \to R^{r-1} \longrightarrow E \longrightarrow I \to 0,$$

where I is torsionfree. Then I may be identified with an ideal. The ideal I is referred to as a *Bourbaki ideal* of E (see [B6183, Chapter 7, §4, Théorème 6]). One of the most elementary properties of such sequences is that they lead to a surjection of the Rees algebras $\mathcal{R}(E) \to \mathcal{R}(I)$, that permit some transfer of properties back and forth, particularly when the kernel is a complete intersection.

The construction of these sequences requires elements of the theory of basic elements. We outline a simplified version of this. Suppose that E is an ideal module, that is, $E \hookrightarrow R^r$ has a cokernel whose annihilator has grade at least 2. Let $y_1, \ldots, y_{r-1}$ be 'general' elements of E (which requires some conditions on R to which we shall return to later); these elements will generate a free submodule F of rank $r-1$. The issue is whether the maximal minors of the matrix representing the embedding $F \hookrightarrow R^r$ generate an ideal of grade 2. When this is the case, one has an acyclic complex (the Hilbert-Burch complex), viz

$$0 \to F = R^{r-1} \xrightarrow{\varphi} R^r \longrightarrow R \to 0,$$

and the Bourbaki sequence is its restriction to E.

In [SUV3] there is developed a theory of such sequences particularly tailored for the examination of Cohen-Macaulayness. We refer the reader to it, as we shall not reproduce it here. We shall only recall, without proofs, the basic construction.

Let R be a Noetherian ring and E a finitely generated R-module having a rank $r > 0$. We say that E satisfies G_s, s an integer, if $\nu(E_\mathfrak{p}) \leq \dim R_\mathfrak{p} + r - 1$ for every $\mathfrak{p} \in \operatorname{Spec}(R)$ with $1 \leq \dim R_\mathfrak{p} \leq s-1$. In terms of Fitting ideals this condition is equivalent to the requirement that height $F_i(E) \geq i - r + 2$ for $r \leq i \leq r+s-2$. We say that E satisfies $\widetilde{G}_s$ provided the conditions above hold with "depth" and "grade" in place of "dim" and "height" respectively. If G_s is satisfied for every s, then E is said to satisfy G_∞ (or $\mathcal{F}_1$).

Lemma 8.105. *Let R be a Noetherian ring and $U = \sum_{i=1}^n Ra_i$ an R-module having a rank $r \geq 2$. Let $R' = R[z_1, \ldots, z_n]$ be a polynomial ring, set $U' = R' \otimes_R U$, and suppose that $x = \sum_{i=1}^n z_i a_i \in U'$. Then $U'/(x)$ has rank $r-1$. If U satisfies G_s or $\widetilde{G}_s$, then so does $U'/(x)$.*

Proposition 8.106. *Let R be a Noetherian ring and E a finitely generated torsionfree R-module having a rank $r > 0$. Let $U = \sum_{i=1}^n Ra_i$ be a submodule of E. Assume that E satisfies $\widetilde{G}_2$ and that* grade $E/U \geq 2$. *Further, let $R' = R[\{z_{ij}, 1 \leq i \leq n,\ 1 \leq j \leq r-1\}]$ be a polynomial ring. Set*

$$U' = R' \otimes_R U, \quad E' = R' \otimes_R E, \quad x_j = \sum_{i=1}^n z_{ij} a_i \in U', \quad F = \sum_{j=1}^{r-1} R' x_j.$$

(a) *The module F is a free R'-module of rank $r-1$ and $E'/F \cong I$ for some R'-ideal I. Let J denote the image of U'/F in I. Then* grade $J > 0$ *and $J_{\mathfrak{p}} = I_{\mathfrak{p}} \cong R'_{\mathfrak{p}}$ for every prime ideal $\mathfrak{p}$ with* depth $R'_{\mathfrak{p}} \leq 1$.
(b) *Assume that* height ann $(E/U) \geq s$. *Then I satisfies G_s if and only if J does if and only if E does. The same holds for $\widetilde{G}_s$, assuming that grade $E/U \geq s$.*
(c) *The ideal I can be chosen to have grade ≥ 2 if and only if E is orientable. In this case,* grade $J \geq 2$.

Let $(R, \mathfrak{m})$ be a Noetherian local ring. If $\mathbf{Z}$ is a set of indeterminates over R, we denote by $R(\mathbf{Z})$ the localization $R[\mathbf{Z}]_{\mathfrak{m}R[\mathbf{Z}]}$. Let E be a finitely generated R-module having a rank $r > 0$, and $U = \sum_{i=1}^n Ra_i$ a submodule of E such that E/U is a torsion module. Further let z_{ij}, $1 \leq i \leq n$, $1 \leq j \leq r-1$, be indeterminates, set

$$R'' = R(\{z_{ij}\}), \quad U'' = R'' \otimes_R U, \quad E'' = R'' \otimes_R E, \quad x_j = \sum_{i=1}^n z_{ij} a_i \in E'',$$

and $F = \sum_{j=1}^{r-1} R'' x_j$ (which, by Lemma 8.105, is a free R''-module of rank $r-1$). Assume that E''/F is torsionfree over R'' (which, by Proposition 8.106(a), holds if E is a torsionfree module satisfying $\widetilde{G}_2$ and the assumption that grade $E/U \geq 2$). In this case $E''/F \cong I$ for some R''-ideal I with grade $I > 0$. We shall denote the image of U''/F in I by J.

Definition 8.107. We call an R''-ideal I with $I \cong E''/F$ a *generic Bourbaki ideal of E with respect to U*. If $U = E$, we simply talk about a *generic Bourbaki ideal of E* and write $I = I(E)$.

Remark 8.108. (a) A generic Bourbaki ideal of E with respect to U is essentially unique. Indeed, suppose that $I \subset R(Z)$ and $K \subset R(Y)$ are two such ideals defined using generating sequences $a_1, \ldots, a_n$ and $b_1, \ldots, b_m$ of U, and sets of variables $Z = \{z_{ij} \mid 1 \leq i \leq n,\ 1 \leq j \leq r-1\}$ and $Y = \{y_{ij} \mid 1 \leq i \leq m,\ 1 \leq j \leq r-1\}$, respectively; then there exists an automorphism φ of the R-algebra $S = R(Y, Z)$ and a unit u of $\mathrm{Quot}(S)$ so that

$$\varphi(IS) = uKS.$$

Furthermore, $u = 1$ if I and K have grade ≥ 2.

(b) With the notation preceding Definition 8.107, $J = I(U)$.

(c) If $E \cong R^{r-1} \oplus L$ for some R-ideal L, then $LR'' = I(E)$.

(d) (See also Proposition 8.106(c). Conversely, if $I = I(E)$ has grade ≥ 3, then $E \cong R^{r-1} \oplus L$ for some R-ideal L.

Having defined the Bourbaki ideal I of a module E, we now state the main result of [SUV3] on the comparison of their Rees algebras.

Theorem 8.109. *Let R be a Noetherian local ring and E a finitely generated R-module having a rank $r > 0$. Let U be a reduction of E. Further let $I \cong E''/F$ be a generic Bourbaki ideal of E with respect to U, and write J for the image of U''/F in I.*

(a) (i) *$\mathcal{R}(E)$ is Cohen-Macaulay if and only if $\mathcal{R}(I)$ is Cohen-Macaulay.*
(ii) (*In When R is universally catenary*) *$\mathcal{R}(E)$ is normal with* depth $\mathcal{R}(E) \otimes_R R_{\mathfrak{p}} \geq r+1$ *for every nonzero prime $\mathfrak{p}$ of R if and only if $\mathcal{R}(I)$ is normal.*
(iii) *E is of linear type with* grade $\mathcal{R}(E)_+ \geq r$ *if and only if I is of linear type.*

(b) *If any of the conditions* (i), (ii), (iii) *of* (a) *hold, then $\mathcal{R}(E'')/(F) \cong \mathcal{R}(I)$ and the generators $x_1, \ldots, x_{r-1}$ of F form a regular sequence on $\mathcal{R}(E'')$.*

(c) *If $\mathcal{R}(E'')/(F) \cong \mathcal{R}(I)$, then*
(i) *J is a reduction of I and $\mathrm{r}_U(E) = \mathrm{r}_J(I)$;*
(ii) (*When the residue field of R is infinite and $U = E$*) $\mathrm{r}(E) = \mathrm{r}(I)$.

We are now going to display some of the results of [SUV3] that provide instances of Rees algebras with good arithmetic properties. In some manner or another they are linked to Theorem 8.109.

In analogy with Theorem 3.39, one has ([SUV3, Theorem 4.2]):

Theorem 8.110. *Let R be a Cohen-Macaulay local ring of dimension $d > 0$ with infinite residue field and E a finitely generated R-module having a rank r. If $\mathcal{R}(E)$ is Cohen-Macaulay then*

$$\mathrm{r}(E) \leq \ell(E) - r \leq d - 1.$$

In the case of modules of projective dimension one ([SUV3, Theorem 4.7]) we have the following result.

Theorem 8.111. *Let R be a Gorenstein local ring with infinite residue field and E a finitely generated R-module with* proj dim $E = 1$. *Write $e =$* rank(E), *let s be an integer not less than r, and assume that E satisfies G_{s-r+1} and is torsionfree locally in codimension* 1. *Further let φ be the matrix presenting E with respect to a generating sequence $a_1, \ldots, a_n$ where $n \geq s$.*

(a) *The following are equivalent:*
(i) *$\mathcal{R}(E)$ is Cohen-Macaulay and $\ell(E) \leq s$;*
(ii) $\mathrm{r}(E) \leq \ell(E) - e \leq s - r$;
(iii) $\mathrm{r}(E_{\mathfrak{p}}) \leq s - r$ *for every prime $\mathfrak{p}$ with* $\dim R_{\mathfrak{p}} = \ell(E_{\mathfrak{p}}) - r + 1 = s - r + 1$, *and* $\ell(E) \leq s$;

(iv) *after elementary row operations,* $I_{n-s}(\varphi)$ *is generated by the maximal minors of the last* $n-s$ *rows of* φ*;*
(v) *after changing the generating sequence, one has an equality* $F_s(E) = F_0(E/U)$ *of Fitting ideals for* $U = \sum_{i=1}^{s} Ra_i$.

(b) *If the equivalent conditions of* (a) *hold then* U *is a reduction of* E *with* $\mathrm{r}_U(E) = \mathrm{r}(E)$*. Furthermore,* E *is of linear type or else* $\ell(E) = s$ *and* $\mathrm{r}(E) = s - r$.

The following result describes a very large class of Cohen-Macaulay Rees algebras ([SUV3, Theorem 5.4]):

Theorem 8.112. *Let* R *be a Cohen-Macaulay local ring of dimension* d *and* E *a finitely generated torsionfree* R*-module that has a rank* r *and is free locally in codimension one. Assume that* E *is integral over an almost complete intersection module with* $\nu(U) = s$*, and that* E *satisfies* G_{s-r+1}.

(a) *If* $E = U$ *then* E *is of linear type.*
(b) *If* $\mathrm{r}_{U_\mathfrak{p}}(E_\mathfrak{p}) \leq 1$ *for every prime* $\mathfrak{p}$ *with* $\dim R_\mathfrak{p} = \ell(E_\mathfrak{p}) - r + 1 = s - r + 1$*, then* $\mathcal{R}(E)$ *is Cohen-Macaulay if and only if* depth $E \geq d - s + r$*. In this case either* E *is of linear type or else* U *is a minimal reduction of* E *and* $\mathrm{r}_U(E) = \mathrm{r}(E) = 1$ (*if* R *has an infinite residue field*).

We list some consequences but refer to [SUV3] for their proofs.

Corollary 8.113. *Let* R *be a Cohen-Macaulay local ring of dimension* d *and* E *an almost complete intersection module with* $n = \nu(E)$ *and* $r = \mathrm{rank}(E)$ *satisfying* G_{n-r+1}*. Then*

(a) *E is of linear type,*
(b) $\mathcal{R}(E)$ *is Cohen-Macaulay if and only* depth $E \geq d - n + r$.

Proof. One applies Theorem 8.112 with $U = E$ and $s = n$. ☐

Corollary 8.114. (see also [KN95, Theorem 3.2]) *Let* R *be a Cohen-Macaulay local ring of dimension* d *and* E *a complete intersection module. Then* E *is of linear type and* $\mathcal{R}(E)$ *is Cohen-Macaulay.*

Corollary 8.115. *Let* R *be a Cohen-Macaulay local ring of dimension* d *with infinite residue field and* E *an ideal module with* $\ell = \ell(E)$ *and* $e = \mathrm{rank} E$ *satisfying* $G_{\ell-r+1}$*. Assume that* $\nu(E) - \ell(E) \leq 1$ *and* $\mathrm{r}(E_\mathfrak{p}) \leq 1$ *for every prime* $\mathfrak{p}$ *with* $\dim R_\mathfrak{p} = \ell(E_\mathfrak{p}) - r + 1 = \ell - e + 1$*. Then* $\mathcal{R}(E)$ *is Cohen–Macaulay if and only if* depth $E \geq d - \ell + r$*. In this case either* E *is of linear type or else* $\mathrm{r}(E) = 1$.

Corollary 8.116. *Let* R *be a Cohen-Macaulay local ring of dimension* d *with infinite residue field and* E *an ideal module of rank* r*. Assume that* E *is free locally in codimension* $d-2$*, satisfies* G_d*, and that* $\mathrm{r}(E) \leq 1$*. Then* $\mathcal{R}(E)$ *is Cohen-Macaulay. Furthermore either* E *is of linear type or else* $\ell(E) = d + r - 1$ *and* $\mathrm{r}(E) = 1$.

8.7 Normalization of Modules

Let R be a normal domain and consider a module with an embedding $E \hookrightarrow R^r$ (where $r = \text{rank}(E)$). In this section, following some results of [HUV4], we shall show how the issues of normalization of ideals and modules resemble one another. We shall denote by E_n the components of the Rees algebra $\mathcal{R}(E)$.

The *normalization* problem for a module of E is understood here in several senses. A strong one as either the description of the integral closure of $\mathcal{R}(E)$ or its construction. In more general terms, it is seen as the study of the relationships (expressed in the comparison between their invariants) between the two algebras $\mathcal{R}(E)$ and $\overline{\mathcal{R}(E)}$.

The normalization problem for modules over two-dimensional regular local rings is well clarified in ([Ko95]; see also [KK97]):

Theorem 8.117. *Let R be a two-dimensional regular local ring and E a finitely generated torsionfree R-module. Then $\overline{\mathcal{R}(E)} = \mathcal{R}(\overline{E})$, and it is a Cohen-Macaulay algebra.*

In higher dimensions, or for nonregular two-dimensional normal domains, the picture is much more diffuse. To exhibit some of this diversity, first we make several observations on how the transition between Rees algebras of ideals and of modules can take place besides the technique of Bourbaki sequences. Let E be a submodule of the free module R^r, and denote by $\mathcal{R}(E)$ and $\mathcal{R}(R^r) = S$ the corresponding Rees algebras. We can consider the ideal (E) of S generated by the 1-forms in E. In other words, (E) is the ideal of relations of $S/(E) = S_R(R^r/E)$. In particular some properties of (E) can be examined through the general tools developed for symmetric algebras. Thus for example to examine the codimension of (E) we may replace E by one of its reductions F, and focus on $S_R(R^r/F)$.

Our aim in this section being normalization, we start by providing a source of complete modules derived from ideals.

Proposition 8.118. *Let M be a graded ideal of the polynomial ring $S = R[x_1, \ldots, x_d]$, $M = \oplus_{n \geq 0} M_n$. If M is an integrally closed S-ideal then every component M_n is an integrally closed R-module.*

Proof. M_n is a R-submodule of the module S_n freely R-generated by the monomials T_α of degree n in the x_i. Denote by $\mathcal{R}(M_n)$ and $\mathcal{R}(S_n)$ the corresponding Rees algebras. Let $u \in S_n$ be integral over M_n; there is an equation in $\mathcal{R}(S_n)$ of the form

$$u^m + a_1 u^{m-1} + \cdots + a_m = 0, \quad a_i \in M_n^i.$$

Map this equation using the natural homomorphism $\mathcal{R}(S_n) \to S$ that sends the variable T_α into the corresponding monomial of S. The equation converts into an equation of integrality of $u \in S$ over the ideal M. Since $M = \overline{M}$, it follows that $\mathrm{x}u \in M_n$. $\square$

Proposition 8.119. *Set $S = \mathcal{R}(R^r) = R[x_1,\ldots,x_r]$ and denote by (E) the ideal of S generated by the forms in E. For every positive integer n,*

$$\overline{E_n} = \overline{(E^n)}_n.$$

Proof. It suffices to verify the equality of these two integrally closed modules at the valuations of R. For some valuation V, $VS = V[z_1,\ldots,z_r]$ and VE is generated by forms $a_1z_1,\ldots,a_rz_r$, with $a_i \in V$. It suffices to show that the ideal (VE) is normal. We may assume that a_1 divides all a_i, $a_i = a_1b_i$, so that the ideal (VE) is $a_1(z_1,b_2z_2,\ldots,b_rz_r)$, and it will be normal if and only if that is the case for $(z_1,b_2z_2,\ldots,b_rz_r)$. Obviously we can drop the indeterminate z_1 and iterate. □

We shall apply the Briançon-Skoda theory to the ideal (E) and to its powers. To illustrate, suppose that R is a regular domain of dimension d so that the localizations of S are regular of dimension at most $d+r$; its Briançon-Skoda number is therefore $c(S) \leq d+r-1$. Since the analytic spread of E is $d+r-1$, according to Theorem 1.161, the Briançon-Skoda number of E is $c \leq \ell(E)-1 = d+r-2$.

Let $E \hookrightarrow R^r$ be a module of finite colength and F a minimal reduction of E (we may assume that R has infinite residue field). Then F is a complete intersection module generated by $d+r-1$ elements ($\dim R = d$). From the definition of $c = c(E)$, we have

$$\overline{(E^{n+c})} \subset (F^n), \quad n \geq 1,$$

which, by the Proposition above, can be written in degree $n+c$ in the form

$$\overline{E^{n+c}} \subset (F^n)_{n+c} = F^n S_c, \quad n \geq 1.$$

Consider now the corresponding filtrations

$$0 \to \overline{E^{n+c}}/F^{n+c} \longrightarrow S_{n+c}/F^{n+c} \longrightarrow S_{n+c}/\overline{E^{n+c}} \to 0. \tag{8.10}$$

As in the case of ideals,

$$\lambda(\overline{E^{n+c}}/F^{n+c}) = \lambda(S_{n+c}/F^{n+c}) - \lambda(S_{n+c}/\overline{E^{n+c}}) - \leq \lambda(S_{n+c}/F^{n+c}) - \lambda(S_{n+c}/F^nS_c),$$

since $\overline{E^{n+c}} \subset F^nS_c$, by the definition of $c(E)$.

We now identify these numerical functions. The simplest to do is $\lambda(S_{n+c}/F^{n+c})$. This, according to Theorem 8.37, is given by the Buchsbaum-Rim polynomial of F,

$$\lambda(S_{n+c}/F^{n+c}) = \text{br}(F) \cdot \binom{n+c+d+r-2}{d+r-1}.$$

Here $\text{br}(F) = \text{br}(E) = \lambda(R^r/F)$ is the Buchsbaum-Rim multiplicity of E. More generally, we use the following notation for the Buchsbaum-Rim polynomials of E and of the integral closure of $\mathcal{R}(E)$:

$$P(n) = \lambda(S_n/E^n) = \mathrm{br}(E)\binom{n+d+r-2}{d+r-1} - \mathrm{br}_1(E)\binom{n+d+r-3}{d+r-2} + \text{lower terms}$$

$$\overline{P}(n) = \lambda(S_n/\overline{E^n}) = \overline{\mathrm{br}}(E)\binom{n+d+r-2}{d+r-1} - \overline{\mathrm{br}}_1(E)\binom{n+d+r-3}{d+r-2} + \text{lower terms}$$

For any graded subalgebra A such that $\mathcal{R}(E) \subset A \subset \overline{\mathcal{R}(E)}$ one can define $\lambda(S_n/A_n)$, and associate with it the Buchsbaum-Rim polynomial. In such case, the coefficients will be denoted $\mathrm{br}(A)$, $\mathrm{br}_1(A)$, and so on.

Theorem 8.120. *Let $(R,\mathfrak{m})$ be a Cohen-Macaulay normal local domain of infinite residue field. Let E be a module of rank $r \geq 2$ and let S be as above such, that the Rees algebra $\mathcal{R}(E)$ has finite integral closure. If the Briançon-Skoda number of S is $c(S)$, then*

$$\overline{\mathrm{br}}_1(E) \leq \mathrm{br}(E) \cdot \binom{r+c-1}{c-1}.$$

In addition,

$$\mathrm{br}_1(E) \leq \mathrm{br}(E) \cdot \binom{r+c-1}{c-1}.$$

Proof. According to the definition of $c = c(S)$ and the discussion above, $\overline{E^{n+c}} \subset F^n S_c$ for $n > 0$. We set up the exact sequences

$$0 \to \overline{E^{n+c}}/F^{n+c} \longrightarrow S_{n+c}/F^{n+c} \longrightarrow S_{n+c}/\overline{E^{n+c}} \to 0.$$

For $n \gg 0$, we use it to make comparisons between the coefficients of the polynomials. First note the embedding $\overline{E^{n+c}}/F^{n+c} \hookrightarrow F^n S_c/F^{n+c}$, which means that the Hilbert polynomial $p(n)$ of the $\mathcal{R}(F)$-module

$$C = \bigoplus_{n \geq 0} F^n S_c/F^{n+c}$$

dominates that of $\bigoplus_{n\geq 1} \overline{E^{n+c}}/F^{n+c}$, so that we have

$$p(n+c) \geq P(n+c) - \overline{P}(n+c), \quad n \gg 0.$$

Since C is annihilated by a power of $\mathfrak{m}$, its Krull dimension is at most $\dim \mathcal{R}(F) - 1 = d+r-1$, and consequently the polynomial $p(n)$ has degree at most $d+r-2$. Comparing leading coefficients for $P(n+c)$ and $\overline{P}(n+c)$, this implies that $\overline{\mathrm{br}}(E) = \mathrm{br}(E)$. One also has the inequality

$$0 \leq \overline{\mathrm{br}}_1(E) \leq e_0(C),$$

where $e_0(C)$ is set to the multiplicity of C if $\dim C = d+r-1$, or to 0 otherwise. We are going to see that $\dim C = d+r-1$, and calculate its multiplicity of C. First note that C is defined by the exact sequence

$$0 \to \mathcal{R}(F) \longrightarrow R \oplus F \oplus \cdots \oplus F^{c-1} \oplus \mathcal{R}(F)S_c \longrightarrow C \to 0.$$

Since $\mathcal{R}(F)$ is Cohen-Macaulay, it follows that C has dimension precisely $\dim \mathcal{R}(F) - 1$.

The determination of $e_0(C)$ is more delicate but we offer an estimation that looks natural. We begin by filtering C itself. From the chain

$$F^{n+c} \subset F^{n+c-1}S_1 \subset \cdots \subset F^n S_c$$

we define the $\mathcal{R}(F)$-modules

$$D_i = \bigoplus_{n \geq 0} F^n S_i / F^{n+1} S_{i-1}, \quad i = 1, \ldots, c$$

that give the factors of a filtration of C.

Lemma 8.121. $\deg(D_i) \leq \operatorname{br}(E) \cdot \binom{r+i-2}{i-1}$.

Proof. We set $S_i = S_1 \cdot S_{i-1}$, $s = \lambda(S_1/F) = \operatorname{br}(E)$ and write

$$S_1 = (F, a_1, \ldots, a_s),$$

where the submodules

$$F_j = (F, a_1, \ldots, a_j), \quad F_0 = F, \quad j \leq s,$$

of S_1 define a composition series for S_1/F, that is, $\lambda(F_{j+1}/F_j) = 1$. We use these s modules for a final filtering of the D_i obtaining factors of the form

$$\bigoplus_{n \geq 0} F^n \cdot F_{j+1} S_{i-1} / F^n \cdot F_j S_{i-1}.$$

Note that this module is annihilated by the maximal ideal of R and is generated as an $\mathcal{R}(F)$-module by the elements $a_{j+1} \cdot m_\alpha$, where the m_α are the monomials of degree $i-1$ of the polynomial ring $S = R[x_1, \ldots, x_r]$. As a consequence it is generated by $\binom{r+i-2}{i-1}$ elements over the special fiber of $\mathcal{R}(F)$. Since F is of linear type, this ring is a ring of polynomials and the multiplicity of the module is at most $\binom{r+i-2}{i-1}$. Adding these contributions, we get the asserted bound. $\square$

To complete the proof of Theorem 8.120, collecting the partial multiplicities gives

$$e_0(C) \leq \operatorname{br}(E) \sum_{i=1}^{c} \binom{r+i-2}{i-1} = \operatorname{br}(E) \cdot \binom{r+c-1}{c-1},$$

establishing the desired formula. $\square$

Corollary 8.122. *If R is a regular local ring of dimension d and E is a submodule of R^r of finite colength so that $\mathcal{R}(E)$ has finite integral closure, then for any distinct graded subalgebras A and B such that*

$$\mathcal{R}(E) \subset A \subset B \subset \overline{\mathcal{R}(E)},$$

and satisfying Serre's condition S_2,

$$0 \leq \operatorname{br}_1(A) < \operatorname{br}_1(B) \leq \operatorname{br}(E) \cdot \binom{2r+d-3}{r+d-3}.$$

In particular, any chain of such subalgebras has length at most $\operatorname{br}(E) \cdot \binom{2r+d-3}{r+d-3}$.

Proof. As in the proof above, the module $C = B/A$ has dimension $d+r-1$, and we would obtain $b_1(B) = \operatorname{br}_1(A) + e_0(C)$. □

Cohen-Macaulay Algebras

Let $(R, \mathfrak{m})$ be a Cohen-Macaulay local domain of dimension d and E a torsionfree R-module of rank r. Denote by A the Rees algebra $\mathcal{R}(E)$ of E, and let B be a graded, finite A-subalgebra of $\overline{A}$. The main result of this section is that if B is Cohen-Macaulay then its reduction number (in a sense to be made explicit) is at most $d-1$. In particular, B is generated by its components of degree at most $d-1$.

We fix an embedding $E \hookrightarrow R^r$ and identify $\mathcal{R}(E)$ with the subalgebra of $S = R[T_1, \ldots, T_r]$ generated by the 1-forms in E. The integral closure of $\mathcal{R}(E)$ and the subalgebra B will be written

$$A = \mathcal{R}(E) \subset B = \sum_{n \geq 0} E_n \subset \overline{A} = \sum_{n \geq 0} \overline{E^n}.$$

We can refer to B as an algebra of the multiplicative filtration $\{E_n\}$ of R-submodules of S.

The following gives an extension of Theorem 8.110 to these more general filtrations.

Theorem 8.123. *Let $(R, \mathfrak{m})$ be a Cohen-Macaulay local ring with infinite residue field and let $\{E_n, \neq 0, E_0 = R\}$ be a multiplicative filtration such that the Rees algebra $B = \sum_{n \geq 0} E_n$ is Cohen-Macaulay and finite over A. Let G be a minimal reduction of E. Then*

$$E_{n+1} = GE_n = EE_n, \quad n \geq d-1,$$

namely and in particular, B is generated over A by forms of degrees at most $d-1$,

$$\sum_{n \geq 0} E_n = R[E_1, \ldots, E_{d-1}].$$

For $r = 1$, that is, the case of ideals, is proved in Theorem 3.48. That result was based on the characterization of the Cohen-Macaulayness of the Rees algebra of an ideal I in terms of its associated graded ring and of its reduction number. It turns out that Theorem 8.123 is a direct consequence of the ideal case and of the technique of Bourbaki sequences.

Proof. We assume that $r \geq 2$. In the algebra A, the prime ideal $A_+ = EA$ has codimension r and therefore, by the going-up theorem, height $(EB) = r$; more precisely, grade $(EB) = r$ since B is Cohen-Macaulay by assumption.

Now we carry out a generic extension of $R \to R'$ to construct a prime ideal of $B \otimes_R R'$ generated by a regular sequence of $r-1$ elements of $R' \otimes E$. For example, if $e_1, \ldots, e_n$ is a set of R-generators of E and $\mathbf{X}$ is a $n \times (r-1)$ matrix of distinct indeterminates over R, taking the $r-1$ linear combinations

$$F = (f_1, \ldots, f_{r-1}) = (e_1, \ldots, e_n) \cdot \mathbf{X}$$

and setting $R' = R[\mathbf{X}]_{\mathfrak{m}[\mathbf{X}]}$, gives rise to the prime ideal (F) of B with the asserted properties. In particular, this construction gives rise to the following exact sequence of R-modules (after we replace, innocuously, R by R')

$$0 \to F \longrightarrow E_1 \longrightarrow I_1 \to 0,$$

where I_1 is torsionfree R-module of rank 1, which we identity with an ideal I.

Consider the R-algebra homomorphism

$$B = \sum_{n \geq 0} E_n \longrightarrow \sum_{n \geq 0} E_n / F E_{n-1} \longrightarrow \sum_{n \geq 0} I_n = C,$$

where $E_n / F E_{n-1} = I_n$ is also torsionfree of rank 1. Observe that C is finite over the Rees algebra of I and thus C is contained in $\overline{R[It]}$. Since (F) is a complete intersection ideal of the Cohen-Macaulay ring B, C is Cohen-Macaulay. We may thus apply Theorem 3.48 to obtain a minimal reduction J for C of reduction number at most $d-1$. Lifting J to E_1 and adding F gives rise to a reduction L for B generated by at most $d+r-1$ elements and having reduction number at most $d-1$. It will easily be seen to have the property that $E_{n+1} = LE_n = EE_n$ for $n \geq d-1$. Note that if G is any minimal reduction of E, we can change A by the algebra $\mathcal{R}(G)$ and obtain in addition that $E_{n+1} = GE_n$, $n \geq d-1$. □

8.8 Exercises

Exercise 8.124. Let R be a Noetherian integral domain of dimension d containing an infinite field, and let E be a torsionfree R-module of rank r. Prove that E has a reduction generated by $d+r$ elements.

Exercise 8.125. Let $(R, \mathfrak{m})$ be a regular local ring and A a commutative, finite, free R-algebra that is a singular ring. Show that if I is an integrally closed $\mathfrak{m}$-primary ideal then IA is an integrally closed R-submodule of A but it is not an integrally closed A-ideal.

Exercise 8.126. Let R be a normal domain. A complex of finitely generated R-modules

$$0 \to A \longrightarrow B \longrightarrow C \to 0$$

is said to be quasi-exact if for every prime $\mathfrak{p}$ of codimension at most 1 the localization is an exact complex of free modules. If $\varphi : R \to S$ is a homomorphism of normal domains such that $\varphi^{-1}(\mathfrak{P})$ has codimension at most 1 for every prime $\mathfrak{P}$ of S of codimension 1, show that

$$0 \to S \otimes_R A \longrightarrow S \otimes_R B \longrightarrow S \otimes_R C \to 0$$

is a quasi-exact complex of S-modules.

Exercise 8.127. Let E be a finitely generated torsionfree module over a regular local ring R of dimension 2. Suppose that E is defined by a single relation and that the order determinantal ideal of the natural embedding $E \hookrightarrow E^{**}$ is integrally closed. Show that E is normal. Can this be extended to modules of higher Betti numbers?

Exercise 8.128. Let B be a finitely generated torsionfree module over an integral domain R. If B is complete, prove that for any finitely generated R-module A the module $\mathrm{Hom}_R(A,B)$ is complete.

Exercise 8.129. Let R be a regular ring and I a prime ideal of height g such that R/I is a normal domain. If I is generated by n elements and $H_\bullet(I)$ is the homology of the corresponding Koszul complex, show that $\det(H_1(I)) = H_{n-g}(I)$.

Exercise 8.130. Let R be a Cohen-Macaulay local ring of dimension $d \geq 1$ and E a submodule of a free module R^r of finite colength. Show that the Buchsbaum-Rim coefficient $b_1(E)$ is zero if and only if E is a complete intersection module.

Exercise 8.131. Let R be a regular local ring and let E, $E \subset R^r$, be an equimultiple module of rank r and codimension c. Show that

$$\mathrm{br}_1(E) \leq \mathrm{br}(E) \cdot \binom{2r+c-2}{r+c-2}.$$

9

HowTo

Most algorithms used for exact computation in Algebraic Geometry and Commutative Algebra have a non-elementary complexity–while most problems they are applied to have highly structured input data. It is from the synergy of these pulls that many problems draw their interest. Ultimately most of the problems will have to be solved 'almost' entirely by mathematical means.

In this chapter we discuss a bare minimum set of methods and approaches dealing with integral closure. They are intended to supplement those treated in Chapters 6, 7, and 8.

9.1 Module Operations

Several constructions of modules appear in our discussion: dualization (and bidualization), modules of endomorphisms, etc. Some of these constructions are now direct procedures in symbolic computation packages when the base ring is a ring of polynomials. We shall treat some constructions that involve other rings by reducing to constructions relative to a Noether normalization.

Reflexive Modules

Let us indicate a method to approach the problem of how to construct and decide reflexivity. We start with the problem of deciding when the canonical mapping $E \longrightarrow E^{**}$ is an isomorphism.

Let R be a normal domain (see the comment below on using more general rings) and let E be a module of rank r with a presentation

$$R^m \stackrel{\varphi}{\longrightarrow} R^n \longrightarrow E \rightarrow 0.$$

We use the following notation. If $x \in R$, we set $0 :_E x = {}_xE$.

Proposition 9.1. *Let I be the Fitting ideal $I_{n-r}(\varphi) = \mathrm{Fitt}_r(E)$. The following hold:*

(a) *E is a torsionfree R-module if and only if* ${}_xE = 0$ *for some nonzero element* $x \in I$.
(b) *E is a reflexive R-module if and only* height $I \geq 2$ *and* ${}_xE = 0$ *and* ${}_y(E/xE) = 0$ *for some regular sequence* x, y *in* I.

Proof. (a) If ${}_xE = 0$, we have an embedding $E \hookrightarrow E_x$. But the condition $I_x = R_x$ implies that the localization of the presentation splits, so that E_x is a projective R_x-module. The converse is clear.

(b) The two conditions mean that E, by part (a), is a torsionfree R-module, E_x and E_y are projective modules over R_x and R_y respectively, and x, y is a regular sequence on E. The last condition readily implies that

$$E = E_x \cap E_y.$$

On the other hand, for normal domains, the bidual of a torsionfree R-module F is given as

$$F^{**} = \bigcap_{\mathfrak{p}} F_{\mathfrak{p}},$$

where $\mathfrak{p}$ runs over the height 1 primes of R. Given the representation of E as the intersection of two modules E_x and E_y, each of which are finitely generated reflexive modules over the normal domains R_x and R_y respectively, we obtain that $E = E^{**}$. The converse is given by the Krull-Serre criterion for reflexive modules. □

Remark 9.2. These observations do not necessarily require that R be a normal domain. A broader class for which it works equally well is that of domains with Serre's condition S_2 such that the codimension one localizations $R_{\mathfrak{p}}$ are Gorenstein.

Let us recast part of Proposition 9.1(a) in a method to determine the torsion submodule of a module.

Proposition 9.3. *Let R be a domain and E a module of rank r with a presentation*

$$R^m \xrightarrow{\varphi} R^n \longrightarrow E \to 0.$$

Let I be the Fitting ideal $I_{n-r}(\varphi) = \mathrm{Fitt}_r(E)$, *and denote by L the image of* φ. *For any nonzero* $x \in I$, *the torsion submodule of E is given by*

$$\bigcup_{s \geq 1} (L :_{R^n} x^s)/L.$$

In particular, E is torsionfree if and only if $L :_{R^n} x = L$.

Biduals

Let E be a torsionfree module as above, of rank r, and denote by I its Fitting ideal $\mathrm{Fitt}_r(E)$. The bidual E^{**} can be obtained as follows:

Proposition 9.4. *Let* $E \hookrightarrow F$ *be an embedding of E into a free module, let* x, y *be a regular sequence on R contained in I, and set* $L = (x, y)$. *Then*

$$E^{**} = \bigcup_{n \geq 1} E :_F L^n.$$

Rings of Endomorphisms

Let A be a reduced equidimensional affine ring, equipped with a Noether normalization $R \hookrightarrow A$. Let E be a finitely generated torsionfree A-module; and we seek an approach to the calculation of $B = \mathrm{Hom}_A(E,E)$. A first observation, enabled by Noether normalization, is that if S is a hypersurface ring over which A is rational, that is,

$$R \hookrightarrow S = R[\tau] = R[t]/(f(t)) \hookrightarrow A,$$

then

$$B = \mathrm{Hom}_A(E,E) = \mathrm{Hom}_S(E,E).$$

Proposition 9.5. *Denote by τ the image of t in A. One has an exact sequence*

$$0 \to B \longrightarrow \mathrm{Hom}_R(E,E) \stackrel{\varphi}{\longrightarrow} \mathrm{Hom}_R(E,E),$$

where for $\alpha \in \mathrm{Hom}_R(E,E)$, we have $\varphi(\alpha) = [\tau, \alpha] = \tau \cdot \alpha - \alpha \cdot \tau$. That is, B is the centralizer in $\mathrm{Hom}_R(E,E)$ of the element τ.

9.2 Integral Closure of an Algebra

Rees Algebras Method

We discuss the steps concerned with the method studied in Proposition 6.65. It is briefly described as follows:

Algorithm 9.6 *Let S be a hypersurface ring defined by a square-free polynomial of degree e. There is an integer $r <$ (quadratic/cubic as main counts) such that for any finite, rational extension $S \subset B$,*

- *Set $I = \gamma(B) = \mathrm{Hom}_S(B,S)$*
- *Set $C = \mathrm{Hom}_S(I^r, I^r)$*
- *Set $L = \gamma(C) = \mathrm{Hom}_S(C,S)$*
- *If $I \not\subset B$ then*
$$\lambda(I \cdot \mathrm{Quot}(S/\gamma(\overline{S}))/L \cdot \mathrm{Quot}(S/\gamma(\overline{S}))) \geq 2.$$

The Computation

There are several approaches to the actual computation. Here are some comments.

- To begin, what is a nice way to carry out the calculation of $\mathfrak{c}(A/S)$,

$$S = k[x_1, \ldots, x_d, x_{d+1}]/(f) \hookrightarrow k[x_1, \ldots, x_n]/I = A?$$

  One approach is the following. Let g be a nonzero element in the Jacobian ideal of S and set $L = gA \subset S$. Note that

$$\mathfrak{c}(A/S) = g \cdot \mathrm{Hom}_S(gA, S) = Sg :_S L.$$

For this we need gA to be expressed as an ideal of S. This is an elimination question, for example gA is the image in S of the ideal

$$(g, I) \bigcap k[x_1, \ldots, x_{d+1}].$$

If we want to track closely the computation, choose an elimination term order for the x_i such that $x_i > x_{i+1}$ (such ordering as might have been required earlier to achieve the Noether normalization). Let $\mathcal{G}_>(I)$ be a Gröbner basis of I and let $NF(\cdot)$ be the corresponding normal form function. Then $gA \subset S$ is generated by all the elements

$$NF(g\mathbf{x}^\alpha), \quad \mathbf{x}^\alpha = x_{d+2}^{\alpha_{d+2}} \cdots x_n^{\alpha_n}, \quad \text{where } \alpha_i < e.$$

- From the point we have the ideal L, the computations are always with ideals of S, until we reach the integral closure. That is, given an extension $S \subset A$, we assume that A is represented as a fractionary ideal, $A = Lx^{-1}$, where $L \subset S$ and x is a regular element of S. The closure of A is

$$C(A) = (L^{-1})^{-1}x^{-1}$$

while its conductor is

$$\mathfrak{c} = Sx :_S L.$$

Note that the conductors of A and of $C(A)$ are the same. This means that the closure operation is being used to control the chain of extensions but it is only required at the last step of the computation.
- The hard part is the calculation of $B = H^0(X, O_X)$, as indicated above. We don't know an elegant way of doing it.
- Note that the test of termination is given by the equality $A = B$. If reached, say $\overline{S} = Lx^{-1} = (a_1, \ldots, a_n)x^{-1}$, a setup for a presentation of $\overline{S}$ can go as follows. Consider the homomorphism

$$k[x_1, \ldots, x_{d+1}, T_1, \ldots, T_n] \xrightarrow{\varphi} Lx^{-1} \subset Sx^{-1} \subset S[x^{-1}],$$

where T_i is mapped to $a_i x^{-1}$, $i = 1, \ldots, n$. If we use the presentation

$$S[x^{-1}] = k[x_1, \ldots, x_d, x_{d+1}, u]/(f, u\tilde{x} - 1),$$

where u is a new indeterminate and $\tilde{x}$ is a lift of x, the defining ideal of $\overline{S}$ is then

$$\ker(\varphi) = (f, u\tilde{x} - 1, T_i - \widetilde{a}_i u, i = 1 \ldots n) \bigcap k[x_1, \ldots, x_{d+1}, T_1, \ldots, T_n].$$

9.3 Integral Closure of an Ideal

Integrally Closed Generic Complete Intersections

We give here a full treatment (with pointers to the literature) of the problem of deciding when a given ideal has the property of the title.

Throughout, R will be a Noetherian ring which is locally Gorenstein (it would not be difficult to recast the tests if R is Cohen-Macaulay with a canonical module). On those occasions when Jacobian ideals are used, R will be a ring of polynomials over a field of characteristic zero (mostly). We shall need the construction of several regular sequences; we refer to [ESt94], and [Va98b, Section 5.5] for convenient methods. Finally, we emphasize that an ideal I of codimension c is *height-unmixed* if all of its associated primes have codimension c. A height unmixed ideal I is said to be a *generic complete intersection*, or simply a *g.c.i*, if $I_{\mathfrak{p}}$ is a complete intersection for each associated prime $\mathfrak{p}$. There are more general notions of *g.c.i.* but in this section we adhere to this terminology. Finally, if I is an ideal of a ring of polynomials $R = k[x_1,\ldots,x_n]$, say $I = (f_1,\ldots,f_m)$, then the *Jacobian ideal* of I is the ideal J generated by the $c \times c$ minors, $c =$ height I, of the Jacobian matrix of the f_i with respect to the x_j. Clearly, only the image of J in R/I is unambiguously defined; this will not hamper us here since we shall only use J in quotient operations of the kind $I : J$.

Height-Unmixed Ideals

There are several ways to test for such properties. A simple one is:

Proposition 9.7. *Let I be an ideal of codimension c and let I_0 be a subideal of I generated by a regular sequence of c elements. Then I is height-unmixed if and only if*

$$I = I_0 : (I_0 : I).$$

Proof. See [Va98b, Corollary 3.2.2].

Radicals

We recall the method of [EHV92] (see [Va98b, Theorem 5.4.2] also).

Theorem 9.8. *Let I be a height-unmixed ideal and let J be its Jacobian ideal. Then*

$$\sqrt{I} = I : J$$

if and only if I is a generic complete intersection.

To be used as a test for *g.c.i.*, we need another path to $\sqrt{I}$.

Theorem 9.9. *Let I be a height-unmixed ideal of codimension c, and let I_0, $I_0 \subset I$, be a complete intersection of codimension c. If J_0 is the Jacobian ideal of I_0, set $L = I_0 : J_0$. Then*

$$\sqrt{I} = L : (L : I).$$

Proof. See [EHV92], [Va98b].

Generic Complete Intersections

When the number of generators of the ideal I is not large relative to its codimension c, ideal-theoretic means are preferable in testing for the *g.c.i.* property.

Proposition 9.10. *Let R be a Cohen-Macaulay integral domain and I a height-unmixed ideal of codimension c. Then I is a generic complete intersection if one of the following two conditions hold.*

(a) $\operatorname{codim} \wedge^{c+1} I \geq c+1$.
(b) *If $\varphi : R^m \to R^n$ is a presentation of I, then* $\operatorname{height} I_{n-c}(\varphi) \geq c+1$.

Proof. Both conditions express the fact that for each prime ideal $\mathfrak{p}$ of codimension c, $I_{\mathfrak{p}}$ is generated by at most c elements. □

Integrally Closed Complete Intersections

The first study of when a complete intersection ideal is integrally closed was made by Goto ([Go87]). In the original proof the ring was not assumed to be Cohen-Macaulay.

Theorem 9.11. *If $I = (a_1, \ldots, a_c)$ is a complete intersection (of codimension c) then the following are equivalent.*

(a) *I is integrally closed.*
(b) *For each minimal prime $\mathfrak{p}$ of I, $R_{\mathfrak{p}}$ is a regular local ring that has a system of parameters $(x_1, \ldots, x_c)$ such that*

$$I_{\mathfrak{p}} = (x_1, \ldots, x_{c-1}, x_c^s).$$

Integrally Closed Generic Complete Intersection

We shall now formulate the promised tests.

Theorem 9.12. *Let R be a Cohen-Macaulay integral domain and I a generic complete intersection ideal of codimension c. Then I is integrally closed if and only if the following conditions hold.*

(i) height ann $\wedge^{c+1} \sqrt{I} \geq c+1$;
(ii) height ann $\wedge^{2} (\sqrt{I}/I) \geq c+1$.

Proof. The meaning of these numerical conditions is simply the following. (i) For each minimal prime $\mathfrak{p}$ of I, $R_{\mathfrak{p}}$ is a regular local ring. The condition (ii) means that at each such prime $\mathfrak{p}$, $I_{\mathfrak{p}}$ has at least $c-1$ of the elements in a minimal generating set of $\mathfrak{p}R_{\mathfrak{p}}$. Together they imply that the primary components of I are integrally closed by Theorem 9.11. □

There are other tests of completeness for these ideals. In [CHV98], the following test was developed for ideals in rings of polynomials over a field of characteristic zero.

Theorem 9.13. *Let I be a height-unmixed ideal in a Cohen-Macaulay ring R. Suppose that I is generically a complete intersection. Then I is integrally closed if and only if*

$$\sqrt{I} = IL : L^2, \tag{9.1}$$

where $L = I : \sqrt{I}$.

These issues (for the type of ideals here) lie in a class of computational problems for which there are oracles:

$$I \longrightarrow \boxed{\mathcal{A}} \longrightarrow \begin{cases} I = \overline{I}\ ? \\ \text{build } \sqrt{I} \\ I \text{ generically c.i.?} \end{cases}$$

Norms of Ideals

Noether normalization plays a role in the examination of the integral closure of ideals. Despite the burden of achieving a Noether normalization from a given affine domain A, the payoff may be significant. To pave the way we describe a technique of [HNV4].

Let A be an affine domain over a field k and let $R \hookrightarrow A$ be a Noether normalization. We introduce the following notion.

Definition 9.14. With $b \in A$ associate the R-endomorphism of A given by

$$f_b : A \mapsto A, \quad a \mapsto ba.$$

The *norm* of b is the element $\text{norm}_R(b) = \det f_b \in R$. If there is no ambiguity we denote it by $\text{N}(b)$.

$\text{N}(\cdot)$ is a non-additive mapping. Despite this drawback we make the following

Definition 9.15. Let A be an affine domain and $R \hookrightarrow A$ a Noether normalization. For an ideal I of A, its *norm* is the ideal $\text{N}(I)$ of R generated by all $\text{N}(b)$, $b \in I$.

Remark 9.16. Let I, L be ideals of A. Then

(i) $\text{N}(I) \subset I \cap R$.
(ii) $\text{N}(a)\text{N}(b) = \text{N}(ab)$ for $a, b \in A$.
(iii) $\text{N}((a)) = (\text{N}(a))$ for $a \in A$.
(iv) $b \in A$ is a unit if and only if $\text{N}(b) \in R$ is a unit.
(v) $\text{N}(I)\text{N}(L) \subseteq \text{N}(IL)$.

Noether normalization already provides a setting to which we can apply the integral closure techniques of modules. For example, suppose that I is an ideal of A; we can think of A as embedded in a free R-module R^r, where r is the rank of A over R (this is just the degree of the field of fractions of A over the field of fractions of R). Note that if $I \neq 0$, then I has also torsionfree rank r over R. We then have two notions of integral closures associated to an ideal I of A:

$$\overline{I}_1 = \bigcap_U IU, \quad U \text{ valuation of } A$$

$$\overline{I}_2 = \bigcap_V IV, \quad V \text{ valuation of } R,$$

i.e., $\overline{I}_1$ denotes the integral closure of I as an A-ideal and $\overline{I}_2$ denotes that of I as an R-module. The following collects some elementary observations. For simplicity we assume that A is integrally closed.

Proposition 9.17. *For ideals J and I of A:*

(i) $\overline{I}_2 = \bigcap_V IV$, *$V$ Rees valuations of* $\det(I)$.
(ii) $\overline{\det(J)} = \overline{\det(I)}$ *if and only if* $\overline{J}_2 = \overline{I}_2$ *as R-modules.*
(iii) *If* $\overline{\det(I)} = \overline{\det(J)}$, *then* $\overline{J}_1 = \overline{I}_1$ *as A-ideals.*

Proof. For (i) – (ii), this follows from the earlier discussion.

(iii) We just emphasize the fact that the integral closure of an ideal I is not always equal to the integral closure of I as an R-module. Indeed, for the two closures

$$\overline{I}_1 = \bigcap_U IU, \quad U \text{ valuation of } A$$

$$\overline{I}_2 = \bigcap_V IV, \quad V \text{ valuation of } R$$

of I, $\overline{I}_2$ could be properly contained in $\overline{I}_1$, since for each valuation U its restriction to R (rather to its field of fractions) yields a valuation V of R. Above any such V there are only finitely many such U, say $U_1, \ldots, U_s$, and

$$IV = \bigcap_{i=1}^{s} IU_i.$$

□

For ideals J, I, $J \subset I$, of A, the integral closures of $\det_0(I)$ and $\det_0(J)$ detect whether $\overline{J} = \overline{I}$ as R-modules while they cannot capture the whole relationship of I over J as A-ideals. There is a companion observation that uses norms but requires modifications. Note its relative strength.

Proposition 9.18. *Let A be an affine domain over a field of characteristic zero and let J, I, $J \subset I$, be ideals of A. Then, $\overline{I} = \overline{J}$ as A-ideals if and only if* $\overline{\mathrm{N}(I)} = \overline{\mathrm{N}(J)}$.

Proof. Let F be the quotient field of A and K be the quotient field of R. We first assume $\overline{\mathrm{N}(I)} = \overline{\mathrm{N}(J)}$. Let U be a valuation of A and V its contraction to R. Set $B = AV$, a V-submodule of V^r, and let $\overline{B}$ be the integral closure of B in F, where r is the rank of A over R. Since V is a DVR and F is finite over K, $\overline{B}$ is Noetherian of dimension 1 by Krull-Akizuki. Further, $\overline{B}$ is a Dedekind domain with finitely many maximal ideals since $\overline{B}$ is normal, i.e., $\overline{B}$ is a PID and hence there exist $h \in J$, $g \in I$ such that $J\overline{B} = h\overline{B}$ and $I\overline{B} = g\overline{B}$. Since $J\overline{B} \subseteq I\overline{B}$, we have $h = gb \in J\overline{B}$ for some $b \in \overline{B}$. Since

$\overline{B}$ is a finitely generated free V-submodule of V^r, we can apply the norm function on $\overline{B}$ over V. Since $\overline{\mathrm{N}(I)} = \overline{\mathrm{N}(J)}$ in R, we have $\mathrm{N}(I)V = \mathrm{N}(J)V$, which also implies that $\mathrm{N}(I)\overline{B} = \mathrm{N}(J)\overline{B}$. Since $J\overline{B} = (h)$ and $I\overline{B} = (g)$, we have $\mathrm{N}(J)V = (\mathrm{N}(h))$ and $\mathrm{N}(I)V = (\mathrm{N}(g))$ by Remark 9.16. Therefore, $v(\mathrm{N}(h)) = v(\mathrm{N}(g))$ and hence $h = gb \in \overline{B}$ which implies that $\mathrm{N}(h) = \mathrm{N}(g)\mathrm{N}(b) \in V$ and hence $v(\mathrm{N}(h)) = v(\mathrm{N}(g)) + v(\mathrm{N}(b))$; that is, $v(\mathrm{N}(b)) = 0$ and hence $\mathrm{N}(b)$ is a unit and therefore b is a unit in $\overline{B}$ by Remark 9.16. Therefore $h\overline{B} = g\overline{B}$, i.e., $I\overline{B} = J\overline{B}$ which implies that $IU = JU$. Therefore $\overline{I} = \overline{J}$ as A-ideals.

Conversely, let us assume that $\overline{I} = \overline{J}$ as A-ideals. Let V be a discrete valuation of R. Set $B = VA$ and $\overline{B}$ as above. Then $\overline{B}$ is finitely generated over V and therefore it is a Dedekind domain with finitely many maximal ideals $P_1, P_2, \ldots, P_s$, in particular $\overline{B}$ is a PID. Denote $\overline{B}_{P_i}$ by U_i for each i. Then U_i is a 1-dimensional Noetherian normal domain, and hence a DVR of A for each i. Since both $V \subseteq U_i \cap R$ are DVRs of K, we have $U_i \cap K = V$ for all i. Since $\overline{I} = \overline{J}$, we have $IU_i = JU_i$ for all such i. It further implies that $I\overline{B} = J\overline{B}$ since $\overline{B} = \bigcap_i \overline{B}_{P_i}$ is a Krull domain. Let N' denote the norm of $\overline{B}$ over V. Then $\mathrm{N}'|_A = \mathrm{N}$. Since $\overline{B}$ is a PID, there exists some $g \in I$ such that $I\overline{B} = g\overline{B}$. Then

$$\mathrm{N}(g)V \subseteq \mathrm{N}(I)V \subseteq \mathrm{N}(I\overline{B} \cap A)V = \mathrm{N}'(I\overline{B} \cap A) = \mathrm{N}'(I\overline{B}) = \mathrm{N}'(g\overline{B}) = \mathrm{N}'(g)V = \mathrm{N}(g)V,$$

which implies that $\mathrm{N}(I)V = \mathrm{N}(g)V$. Since $g\overline{B} = I\overline{B} = J\overline{B}$, we have $\mathrm{N}(g) \in \overline{\mathrm{N}(J)}$, i.e., $\mathrm{N}(I)V = \mathrm{N}(g)V \subseteq \mathrm{N}(J)V$, which implies that $\mathrm{N}(I)V = \mathrm{N}(J)V$; and this completes the proof. □

Corollary 9.19. *Let J, I, $J \subset I$, be ideals of A. Then Let $J \subset I$ be ideals of A. Then J is a reduction of I if and only if $\mathrm{N}(J)$ is a reduction of $\mathrm{N}(I)$.*

There is a real puzzle here on how to determine $\mathrm{N}(I)$: it is not clear to see which $\mathrm{N}(b_i)$, for $b_i \in I$, will generate $\mathrm{N}(I)$. It will not be enough to take a set of generators of I, either as an A-ideal or even as an R-module. The following are examples of generating sets of $I = (b_1, b_2, \ldots, b_n)$ such that (1) $\mathrm{N}(I) = (\mathrm{N}(b_i))_{i=1}^n$ and (2) $N(I) \neq (\mathrm{N}(b_i))_{i=1}^n$, respectively.

We are going to discuss where a finite set of generators for I can be found so that the integral closure $\overline{\mathrm{N}(I)}$ is the closure of the ideal of their norms. We are going to assume that R contains a field k of characteristic zero. Let $I = (a_1, \ldots, a_n)$ be an ideal of A. For a discrete valuation V of R, suppose that $B = VA$ and that $\overline{B}$ is the integral closure of B in the quotient field of A. Then $\overline{B}$ is finitely generated over V (this seems to be the place where we need a field of characteristic zero) and therefore it is a Dedekind domain with finitely many maximal ideals, in particular $\overline{B}$ is a PID. By the Chinese Remainder Theorem, since the ideal $I\overline{B}$ is principal, and in any localization of $\overline{B}$ one of the a_i generates $I\overline{B}$, there exists a linear combination

$$b = \sum_{i=1}^{n} r_i a_i,$$

with $r_i \in A$ such that $I\overline{B} = b\overline{B}$. By standard prime avoidance, we can choose r_i in the field k. More concretely, we can even assume that there is $t \in k$ for which the choices $r_i = t^i$ work. (This is essentially Vandermonde plus prime and subspace avoidance.)

This argument shows what kind of elements are needed to define $\overline{\mathrm{N}(I)}$. The difficulty is that we don't know how many such combinations to take. Therefore we are going to provide another set of generators.

Form the extension $R \to R[x]$, where x is an indeterminate, and consider the Noether normalization $R[x] \hookrightarrow A[x]$. We observe that the corresponding *norm* mapping is the extension of the norm mapping of $R \hookrightarrow A$. One of its properties is that if

$$b = f_1(x)a_1 + \cdots + f_n(x)a_n,\ f_i(x) \in R[x],$$

and $u \in k$, then evaluating $\mathrm{N}(b)$ by setting $x = u$ gives the same answer as

$$\mathrm{N}(b') = \mathrm{N}(f_1(u)a_1 + \cdots + f_n(u)a_n) \in \mathrm{N}(I),$$

that is, norm and evaluation at an element of R commute. In particular, we get that $\overline{\mathrm{N}(I)}$ is the integral closure of the ideal generated by all such $\mathrm{N}(b')$; here it suffices to consider $f_i(x) = x^i$.

Proposition 9.20. *Let $I = (a_1, \ldots, a_n)$ be an ideal of A and let*

$$\alpha = \sum_{i=1}^{n} x^i a_i$$

be the 'generic' element of $I[x]$. Denote by L the ideal of R generated by the coefficients of the polynomial $N(\alpha) \in R[x]$. Then

$$\overline{N(I)} = \overline{L}.$$

Proof. We first show that $L \subset N(I)$, which will consequently imply that $\overline{L} \subset \overline{N(I)}$. Write

$$N(\alpha) = \sum_{j=1}^{m} b_j x^j$$

and $L = (b_1, \ldots, b_m)$. We note that for each evaluation $x \to u$ for $u \in k$, we have $N(\alpha') = \sum_{j=1}^{m} b_j u^j \in N(I)$. It follows that the polynomial $N(\alpha)$ has coefficients in $N(I)$ for each such evaluation. Since k is infinite, by Vandermonde it follows that each b_j lies in $N(I)$, and therefore $L \subset N(I)$.

To prove the converse, $\overline{L} \supset \overline{N(I)}$, we use the early set of observations. □

9.4 Integral Closure of a Module

Filtering the Integral Closure

Let R be a Cohen-Macaulay, equidimensional, normal domain of dimension d and E a torsionfree R-module of rank r. We shall assume as given an embedding $E \hookrightarrow R^r = F$, and denote by I the corresponding order determinantal ideal $\det_0(E)$. We want to control the integral closure $\overline{E}$ using I. This is not entirely clear how to do it in general.

We are going to outline all the steps in the construction of $\overline{E}$ under the following restrictions. (i) I is a height unmixed ideal; (ii) for each minimal prime $\mathfrak{p}$ of I, $E_{\mathfrak{p}}$ is integrally closed; and (iii) R has a canonical module.

Set $g = \text{height } I$. Consider the embedding $\overline{E}/E \hookrightarrow F/E$. By assumption (ii), the associated primes of $\overline{E}/E$ have codimension at least $g+1$. We recall from Proposition 8.75 the following description of $\overline{E}$.

Proposition 9.21. *If* $\det_0(E)$ *is height-unmixed and* $E_{\mathfrak{p}}$ *is complete for each minimal prime of* $\det_0(E)$*, then*

$$\overline{E} = \{f \in F \mid \text{height } (E :_F f) > \text{height } \det_0(E)\}.$$

In particular, if $\text{Ass}(F/E) = \text{Ass}(R/\det_0(E))$ *then E is integrally closed.*

Proof. One direction is part of the hypotheses, while the other uses that $\det_0(E)$ is unmixed and the argument of Proposition 8.75 that these elements of F are indeed integral over E. □

We shall now describe a filtration

$$E = E_0 \subset E_1 \subset E_2 \subset \cdots \subset E_{d-g} = \overline{E}$$

of submodules of $\overline{E}$ with the property that the associated primes of F/E_r have dimension at least r. This would mean, by the proposition above, that $E_{d-g} = \overline{E}$.

Set $E_0 = E$. If $E \neq \overline{E}$, let us construct E_1. Define it as

$$E_1 = \{f \in F \mid \text{height } (E_0 :_F f) \geq d\}.$$

Note that from the sequence

$$0 \to E_1/E_0 \longrightarrow E_0 \longrightarrow F/E_1 \to 0$$

and the definition of E_1, F/E_1 has no associated prime of dimension 0. The associated primes of F/E_0 of dimension 0 are now associated to E_1/E_0 alone. The passage from E_r to E_{r+1} is similar, with

$$E_{r+1} = \{f \in F \mid \text{height } (E_r :_F f) \geq d - r\};$$

from the sequence

$$0 \to E_{r+1}/E_r \longrightarrow F/E_r \longrightarrow F/E_{r+1} \to 0$$

we decompose the associated primes of F/E_r into the disjoint union

$$\text{Ass}(F/E_r) = \text{Ass}(F/E_{r+1}) \biguplus \text{Ass}(E_{r+1}/E_r).$$

We shall now describe the process of obtaining E_{r+1} from E_r. It is reminiscent of the method developed in [EHV92] for deriving equidimensional decomposition of ideals. One of the properties that we shall use, derived from local duality theory, is the fact that the associated primes of F/E of dimension $d-r$ are precisely the associated primes of the 'dual'

$$\mathrm{Ext}_R^{d-r}(\mathrm{Ext}_R^{d-r}(F/E,C),C).$$

Making using of the cohomology long exact sequence, we obtain that

$$\mathrm{Ext}_R^i(F/E_1,C) \cong \mathrm{Ext}_R^i(F/E_1,C), \quad i \leq d-1,$$

and similar isomorphisms relating $\mathrm{Ext}_R^j(F/E_r,C)$ to $\mathrm{Ext}_R^j(F/E_{r+1},C)$ in appropriate ranges. The key point is that

$$\mathrm{Ext}_R^{d-r}(\mathrm{Ext}_R^{d-r}(F/E,C),C) \cong \mathrm{Ext}_R^{d-r}(\mathrm{Ext}_R^{d-r}(F/E_r,C),C).$$

Let C denote the canonical ideal of R. The associated primes of F/E of codimension d show up as the associated primes of the module $\mathrm{Ext}_R^d(F/E,C)$ (which are the same as those for $\mathrm{Ext}_R^d(\mathrm{Ext}_R^d(F/E,C),C)$ at this dimension). Denote by I_1 the annihilator of this module.

Lemma 9.22. *We have* $E_1 = \{f \in F \mid \bigcup_{n\geq 1}(E_0 :_F I_1^n)\}$.

In general we set I_r for the annihilator of $\mathrm{Ext}_R^{d-r}(\mathrm{Ext}_R^{d-r}(F/E,C),C)$, and construct E_{r+1} in the same manner as done for E_1,

$$E_{r+1} = \{f \in F \mid \bigcup_{n\geq 1}(E_r :_F I_{r+1}^n)\} = \mathcal{P}(E_r; I_{r+1}).$$

If at any step we have that $I_r = R$, it follows that $E_{r+1} = E_r$. This simply means that F/E has no associated prime of dimension $d-r$. The process terminates when $r = d-g$.

Of course there is no need to filter, except for the purpose of making the computation simpler. We can do all in one step:

Algorithm 9.23 *For a module E as above, set* $L = I_1 \cdots I_{d-g}$. *The integral closure of E is given by*

$$\overline{E} = \bigcup_{n\geq 0}(E :_F L^n).$$

Alternatively, let $z \in L$ *be a regular element modulo* $\mathrm{det}_0(E)$, *then*

$$\overline{E} = E :_F z^\infty.$$

Normality of a Module

We shall assume that R is a normal domain, more often a ring of polynomials over a field. Although in many cases one can use the Jacobian criterion as a test for normality, we shall find that in many cases a direct approach is useful. In this manner one will handle cases of positive characteristics as well.

Complete Intersections

Let E be a torsionfree R-module whose Rees algebra $\mathcal{R}(E)$ is a complete intersection. It is easy to see that this implies that E is a module of linear type with a presentation

$$0 \to R^m \xrightarrow{\varphi} R^n \longrightarrow E \to 0.$$

The hypothesis is expressed (see [Va94b, Theorem 3.1.6] and its references) entirely by the determinantal ideals of φ:

$$S(E) \text{ is a complete intersection } \Leftrightarrow \text{grade } I_t(\varphi) \geq m-t+2, \quad 1 \leq t \leq m.$$

Let us further assume that R is a ring of polynomials over a field, and seek to follow the prescriptions of [Va94b, p. 123]. Since $S(E)$ is Cohen-Macaulay, we shall examine conditions under which it satisfies Serre's condition R_1. Often it is simple enough to use the Jacobian ideal, but it is preferable to do all the calculation in the base ring R.

Write $S(E) = R[T_1,\ldots,T_n]/J$, where

$$J = (f_1,\ldots,f_m) = (T_1,\ldots,T_n)\cdot\varphi.$$

Let P be a prime ideal of the ring of polynomials containing J, of height $m+1$. We must check whether $S(E)_P$ is a discrete valuation domain. If some entry a_{ij} of φ does not lie in P, we consider $\mathfrak{p} = P\cap R$, and write the equations over $R_{\mathfrak{p}}$, say $J = (T_1,\ldots,T_n)\cdot\varphi$. Let $(x_1,\ldots,x_s)$ is a set of generators of $\mathfrak{p}R_{\mathfrak{p}}$, we can also write

$$J = (x_1,\ldots,x_s)\cdot B(\varphi),$$

where $B(\varphi)$ is a $s\times m$ matrix with entries in $R[T_1,\ldots,T_n]$. If s is the minimal number of generators of the maximal ideal of $R_{\mathfrak{p}}$, the condition we must have is then $I_m(B(\varphi)) \notin P$. This condition does not depend on the chosen set of generators, much less on whether it is a minimal set (verify!).

Unfortunately it is not clear how to deal quickly with all these prime ideals, so let us address some of the issues.

(i) Let L be the component (if it exists) of height $m+1$ of the radical of $I_1(\varphi)$. It can be determined from $I_1(\varphi)$. If $\mathbf{z} = (z_1,\ldots,z_s)$ is a set of generators of L_1 we write $J = \mathbf{z}\cdot B(\varphi)$. Note that if $\mathfrak{p}$ is the prime ideal discussed above then $L_{\mathfrak{p}}$ is the maximal ideal of $R_{\mathfrak{p}}$.

(ii) The condition we must have is that $I_m(B(\varphi))$ is not contained in any prime of the radical ideal L, which can be expressed by

$$L : I_m(B(\varphi)) = L.$$

We note that in this quotient operation, we may replace $I_m(B(\varphi))$ by the ideal generated by the coefficients of the polynomials in the T_i variables, so the colon operation is computed in R.

(iii) Thus far we do not need localization and have dealt with all prime ideals of height $m+1$ containing $I_1(\varphi)$. Suppose that height $P\cap R = r < m+1$. If we localize at $\mathfrak{p} = P\cap R$, we obtain a presentation

$$0 \to R_{\mathfrak{p}}^{u} \xrightarrow{\psi} R_{\mathfrak{p}}^{v} \longrightarrow E_{\mathfrak{p}} \to 0$$

of $E_{\mathfrak{p}}$, where we have the following equality of determinantal ideals:

$$I_t(\psi) = (I_{t+m-u}(\varphi))_{\mathfrak{p}},$$

corresponding to the presentation

$$\varphi = \left[\begin{array}{ccc|ccc} 1 & \dots & 0 & 0 & \dots & 0 \\ \vdots & \ddots & \vdots & \vdots & \ddots & \vdots \\ 0 & \dots & 1 & 0 & \dots & 0 \\ \hline 0 & \dots & 0 & & & \\ \vdots & \ddots & \vdots & & \psi & \\ 0 & \dots & 0 & & & \end{array}\right].$$

The $m-u$ generators of J (corresponding to the identity block on the upper left corner) that are dropped are indeterminates of $R_{\mathfrak{p}}[T_1,\ldots,T_n]$, so we can actually assume that P is an ideal of height $u+1$, namely

$$S(E_{\mathfrak{p}}) = R_{\mathfrak{p}}[T_1,\ldots,T_n]/(T_1,\ldots,T_{m-u},J_t).$$

(iv) If L is the codimension $u+1$ component of the radical of $I_{m-u+1}(\varphi)$ then $I_{m-u}(\varphi)$ contains regular elements modulo L. If we can find a $(m-u)\times(m-u)$ minor Δ that is regular modulo L then we can carry out elementary row and column operations in the ring $R[\Delta^{-1}]$ enough to reduce φ to the form above. It would be interesting to find such an element, short of appealing to generic linear combinations. That done we could write the entries of the remaining columns with coefficients in L and form the corresponding matrix $B(\varphi)$.

We can formulate the following method.

Algorithm 9.24 *Let φ be the matrix defining E.*

(i) *For each integer t with $1 \leq t \leq m$, let L_t be the component of codimension $m-t+2$ of the radical of $I_t(\varphi)$.*
(ii) *For each t such that L_t exists, find a minor $\Delta_t \in I_{t-1}(\varphi)$ that is regular modulo L_t.*
(iii) *With Δ_t as above, carry out elementary row operations as outlined above and write the equations of J_t with entries in L_t,*

$$J_t = (x_1,\ldots,x_s)\cdot B(\varphi_t), \quad (x_1,\ldots,x_s) = L_t.$$

(iv) *Finally we check for*

$$L_t : I_u(B(\varphi_t)) = L_t.$$

The ideal $I_u(B(\varphi_t))$ is taken in $R[T_1,\ldots,T_n]$ after clearing powers of Δ_t.

General Modules

We shall rephrase quickly the criterion of normality of Theorem 8.98. (It is very algorithmic already.)

Algorithm 9.25 *Let R be a regular integral domain and E a torsionfree module of rank r with a free presentation*

$$R^p \xrightarrow{\psi} R^m \xrightarrow{\varphi} R^n \longrightarrow E \to 0.$$

Suppose that $\mathcal{R}(E)$ is of linear type. Then E is normal if and only if the following conditions hold:

(i) *Choose h with $0 \neq b \in I = I_c(\psi)$, set $L = (b,J) : I$. Then*

$$\text{grade}\,(b^{-1}L(J,I)\mathcal{R}(E)) \geq 2.$$

(ii) *Choose a with $0 \neq a \in I_r(\varphi)$. For each integer t with $1 \leq t \leq r$, let L_t be the component of codimension $r-t+2$ of the radical of $I_t(\varphi)$. Then*

$$\text{height}\,(a^{-1}((a,J) : L_t)L_t, J) \geq r+2.$$

The conversion, from Theorem 8.98 to this formulation, makes use of the fact that while in the proof of Theorem 8.98 we always used minimal presentations–a fact that would change the ideal J–we would still have $J_{\mathfrak{p}} = (J_1, T_1', \ldots, T_s')$, and the colon ideals would localize properly.

This should be contrasted (in characteristic zero) with the Jacobian criterion:

$$K = \text{Jac}(\mathcal{R}(E)), \text{ then grade } K\mathcal{R}(E) \geq 2.$$

Its computation could be rather cumbersome. Another point: Theorem 8.98 does not require characteristic zero or that R be a ring of polynomials, just that it be regular up to codimension $r+1$.

9.5 Exercises

Exercise 9.26. Let $R \hookrightarrow A$ be a Noether normalization as above and let I, J be two ideals of A. Show that $\overline{\text{N}(IJ)} = \overline{\text{N}(I)\text{N}(\text{J})}$.

References

[AHu95] I. M. Aberbach and S. Huckaba, Reduction number bounds on analytic deviation two ideals and the Cohen-Macaulayness of associated graded rings, *Comm. Algebra* **23** (1995), 2003-2026.

[AHH95] I. M. Aberbach, S. Huckaba and C. Huneke, Reduction numbers, Rees algebras, and Pfaffian ideals, *J. Pure & Applied Algebra* **102** (1995), 1–16.

[AH93] I. M. Aberbach and C. Huneke, An improved Briançon-Skoda theorem with applications to the Cohen-Macaulayness of Rees algebras, *Math. Annalen* **297** (1993), 343–369.

[AH96] I. M. Aberbach and C. Huneke, A theorem of Briançon-Skoda type for regular local rings containing a field, *Proc. Amer. Math. Soc.* **124** (1996), 707–713.

[AHT95] I. M. Aberbach, C. Huneke and N. V. Trung, Reduction numbers, Briançon-Skoda theorems and depth of Rees algebras, *Compositio Math.* **97** (1995), 403–434.

[Ab67] S. Abhyankar, Local rings of high embedding dimension, *American J. Math.* **89** (1967), 1073–1077.

[AMa93] R. Achilles and M. Manaresi, Multiplicity for ideals of maximal analytic spread and intersection theory, *J. Math. Kyoito Univ.* **33** (1993), 1029–1046.

[Al78] G. Almkvist, K-theory of endomorphisms, *J. Algebra* **55** (1978), 308–340; *Erratum*, *J. Algebra* **68** (1981), 520–521.

[AN72] M. Artin and M. Nagata, Residual intersections in Cohen-Macaulay rings, *J. Math. Kyoto Univ.* **12** (1972), 307–323.

[Abr69] M. Auslander and M. Bridger, Stable module theory, *Mem. Amer. Math. Soc.* **94** (1969).

[Av81] L. Avramov, Complete intersections and symmetric algebras, *J. Algebra* **73** (1981), 248–263.

[AH80] L. Avramov and J. Herzog, The Koszul algebra of a codimension 2 embedding, *Math. Z.* **175** (1980), 249–280.

[AH94] L. Avramov and J. Herzog, Jacobian criteria for complete intersections. The graded case, *Inventiones Math.* **117** (1994), 75–88.

[BF85] J. Backelin and R. Fröberg, Koszul algebras, Veronese subrings and rings with linear resolutions, *Rev. Roumaine Math. Pures Appl.* **30** (1985), 85–97.

[BM91] D. Bayer and D. Mumford, What can be computed in Algebraic Geometry? *Computational Algebraic Geometry and Commutative Algebra*, Proceedings, Cortona 1991 (D. Eisenbud and L. Robbiano, Eds.), Cambridge University Press, 1993, 1–48.

[BS92] D. Bayer and M. Stillman, *Macaulay*: A system for computation in algebraic geometry and commutative algebra, 1992. Available via anonymous ftp from `math.harvard.edu`.

[BKW93] T. Becker, H. Kredel and V. Weispfenning, *Gröbner Bases*, Springer, Heidelberg, 1993.

[BA04] C. Bivià-Ausina, Non-degenerate ideals in formal power seris, *Rocky Mountain J. Math.* **34** (2004), 495–511.

[Bo69] E. Böger, Eine Verallgemeinerung eines Multiplizitätensatzes von D. Rees, *J. Algebra* **12** (1969), 207–215.

[B6183] N. Bourbaki, *Algèbre Commutative*, Chap. I–IX, Hermann, Masson, Paris, 1961–1983.

[BER79] M. Boratynski, D. Eisenbud and D. Rees, On the number of generators of ideals in local Cohen-Macaulay rings, *J. Algebra* **57** (1979), 77–81.

[BPV90] J. Brennan, M. Vaz Pinto and W. V. Vasconcelos, The Jacobian module of a Lie algebra, *Trans. Amer. Math. Soc.* **321** (1990), 183–196.

[BUV1] J. Brennan, B. Ulrich and W. V. Vasconcelos, The Buchsbaum-Rim polynomial of a module, *J. Algebra* **341** (2001), 379–392.

[BV93] J. Brennan and W. V. Vasconcelos, Effective computation of the integral closure of a morphism, *J. Pure & Applied Algebra* **86** (1993), 125–134.

[BV1] J. Brennan and W. V. Vasconcelos, The structure of closed ideals, *Math. Scand.* **88** (2001), 3–16.

[BV3] J. Brennan and W. V. Vasconcelos, Effective normality criteria for algebras of linear type, *J. Algebra* **273** (2004), 640–656.

[BS74] J. Briançon and H. Skoda, Sur la clôture intégrale d'un idéal de germes de fonctions holomorphes en un point de $\mathbb{C}^n$, *C. R. Acad. Sci. Paris* **278** (1974), 949–951.

[Br74] M. Brodmann, Asymptotic stability of $\text{Ass}(M/I^nM)$, *Proc. Amer. Math. Soc.* **74** (1979), 16–18.

[Br82] M. Brodmann, Rees rings and form rings of almost complete intersections, *Nagoya Math. J.* **88** (1982) 1–16.

[Br83] M. Brodmann, Einige Ergebnisse der lokalen Kohomologie Theorie und ihre Anwendung, Osnabrücker Schriften zur Mathematik, Reihe M, Heft 5, (1983).

[Br86] M. Brodmann, A few remarks on blowing-up and connectedness, *J. reine angew. Math.* **370** (1986), 52–60.

[BrS98] M. Brodmann and R. Y. Sharp, *Local Cohomology*, Cambridge University Press, 1998.

[BSV88] P. Brumatti, A. Simis and W. V. Vasconcelos, Normal Rees algebras, *J. Algebra* **112** (1988), 26–48.

[B81] W. Bruns, The Eisenbud-Evans generalized principal ideal theorem and determinantal ideals, *Proc. Amer. Math. Soc.* **83** (1981), 19–24.

[B82] W. Bruns, The canonical module of a determinantal ring, in *Commutative Algebra*: Durham 1981 (R. Sharp, Ed.), London Math. Soc., Lecture Note Series **72**, Cambridge University Press, 1982, 109–120.

[B91] W. Bruns, Algebras defined by determinantal ideals, *J. Algebra* **142** (1991), 150–163.

[BC98] W. Bruns and A. Conca, KRS and powers of determinantal ideals, *Compositio Math.* **111** (1998), 111-122.

[BC1] W. Bruns and A. Conca, Algebras of minors, *J. Algebra* **246** (2001), 311–330.

[BH92] W. Bruns and J. Herzog, On the computation of a-invariants, *Manuscripta Math.* **77** (1992), 201–213.

[BH93] W. Bruns and J. Herzog, *Cohen-Macaulay Rings*, Cambridge University Press, 1993.

[BK98] W. Bruns and R. Koch, *Normaliz*: A program to compute normalizations of semigroups. Available by anonymous ftp from ftp.mathematik.Uni-Osnabrueck.DE/pub/osm/kommalg/software.

[BR2] W. Bruns and G. Restuccia, Canonical modules of Rees algebras, Preprint, 2002.

[BVa3] W. Bruns and W. V. Vasconcelos, Minors of symmetric and exterior powers, *J. Pure & Applied Algebra* **179** (2003), 235–240.

[BVV97] W. Bruns, W. V. Vasconcelos and R. Villarreal, Degree bounds in monomial subrings, *Illinois J. Math.* **41** (1997), 341–353.

[BV86] W. Bruns and U. Vetter, Length formulas for the local cohomology of exterior powers, *Math. Z.* **191** (1986), 145–158.

[BV88] W. Bruns and U. Vetter, *Determinantal Rings*, Lecture Notes in Mathematics **1327**, Springer, Berlin-Heidelberg-New York, 1988.

[BR65] D. Buchsbaum and D. S. Rim, A generalized Koszul complex II. Depth and multiplicity, *Trans. Amer. Math. Soc.* **111** (1965), 197–224.

[BEK0] B. Büeler, A. Enge and K. Fukuda, Exact volume computations for convex polytopes: a practical study. In G. Kalai and G. Ziegler, editors, *Polytopes-Combinatorics and Computation*, DMV-Seminar **29**, Birkhäuser Verlag, 2000.

[Bu68] L. Burch, On ideals of finite homological dimension in local rings, *Math. Proc. Camb. Phil. Soc.* **74** (1968), 941–948.

[Bu72] L. Burch, Codimension and analytic spread, *Math. Proc. Camb. Phil. Soc.* **72** (1972), 369–373.

[Ca84] F. Catanese, Commutative algebra methods and equations of regular surfaces, *in Algebraic Geometry*, Bucharest, 1982, Lecture Notes in Math. **1056**, Springer, Berlin, 1984, 68–111.

[CGW91] B. Char, K. Geddes and S. Watt, *Maple V Library Reference Manual*, Springer, Berlin Heidelberg New York, 1991.

[Cr97] T. Christof, revised by A. Löbel and M. Stoer, *PORTA*, 1997. Available by anonymous ftp from www.zib.de/Optimization/Software/Porta.

[Ch0] V. Chvátal, *Linear Programming*, W. H. Freeman and Company, New York, 2000.

[Ci1] C. Ciuperca, A numerical characterization of the S_2-ification of a Rees algebra, *J. Algebra* **202** (2001), 782–794.

[Co3] A. Conca, Reduction numbers and initial ideals, *Proc. Amer. Math. Soc.* **130** (2003), 1015–1020.

[C3] A. Corso, Sally modules of $\mathfrak{m}$-primary ideals in local rings, Preprint, 2003.

[CGPU3] A. Corso, L. Ghezzi, C. Polini and B. Ulrich, Cohen-Macaulayness of special fiber rings, *Comm. Algebra* **31** (2003), 3713–3734.

[CHKV3] A. Corso, C. Huneke, D. Katz and W. V. Vasconcelos, Integral closure of ideals and annihilators of homology, to appear in Lecture Notes in Pure and Applied Math, Marcel Dekker.

[CHV98] A. Corso, C. Huneke and W. V. Vasconcelos, On the integral closure of ideals, *Manuscripta Math.* **95** (1998), 331–347.

[CP95] A. Corso and C. Polini, Links of prime ideals and their Rees algebras, *J. Algebra* **178** (1995), 224–238.

[CP1] A. Corso and C. Polini, A class of strongly Cohen-Macaulay ideals, *Bull. London Math. Soc.* **33** (2001), 662–668.

[CPR3] A. Corso, C. Polini and M. E. Rossi, Depth of associated graded rings via Hilbert coefficients of ideals, *J. Pure & Applied Algebra*, to appear.

[CPU1] A. Corso, C. Polini and B. Ulrich, The structure of the core of ideals, *Math. Annalen* **321** (2001), 89–105.

[CPU2] A. Corso, C. Polini and B. Ulrich, Core and residual intersections of ideals, *Trans. Amer. Math. Soc.* **354** (2002), 2579–2594.

[CPV94] A. Corso, C. Polini and W. V. Vasconcelos, Links of prime ideals, *Math. Proc. Camb. Phil. Soc.* **115** (1994), 431–436.

[CPV3] A. Corso, C. Polini and W. V. Vasconcelos, The multiplicity of the special fiber of blowups, *Math. Proc. Camb. Phil. Soc.*, to appear.

[CPVz98] A. Corso, C. Polini and M. Vaz Pinto, Sally modules and associated graded rings, *Comm. Algebra* **26** (1998), 2689–2708.

[CVV98] A. Corso, W. V. Vasconcelos and R. Villarreal, Generic gaussian ideals, *J. Pure & Applied Algebra* **125** (1998), 117–127.

[CZ97] T. Cortadellas and S. Zarzuela, On the depth of the fiber cone of filtrations, *J. Algebra* **198** (1997), 428–445.

[CZ1] T. Cortadellas and S. Zarzuela, Burch's inequality and the depth of the blow up rings of an ideal, *J. Pure & Applied Algebra* **157** (2001), 183–204.

[Co84] R. C. Cowsik, Symbolic powers and number of defining equations, in *Algebra and its Applications*, Lecture Notes in Pure & Applied Mathematics **91**, Marcel Dekker, New York, 1984, 13–14.

[CN76] R. C. Cowsik and M. V. Nori, On the fibers of blowing up, *J. Indian Math. Soc.* **40** (1976), 217–222.

[Cu90] S. D. Cutkosky, A new characterization of rational surface singularities, *Inventiones Math.* **102** (1990), 157–177.

[CH2] S. D. Cutkosky and H. T. Hà, Arithmetic Macaulayfication of projective schemes, Preprint, 2002.

[DV3] K. Dalili and W. V. Vasconcelos, The tracking number of an algebra, *American J. Math.*, to appear.

[DEP80] C. De Concini, D. Eisenbud and C. Procesi, Young diagrams and determinantal varieties, *Inventiones Math.* **56** (1980), 129–165.

[Jo98] T. de Jong, An algorithm for computing the integral closure, *J. Symbolic Computation* **26** (1998), 273–277.

[DT2] D. Delfino, A. Taylor, W. V. Vasconcelos, R. Villarreal and N. Weininger, Monomial ideals and the computation of multiplicities, in *Commutative Ring Theory and Applications* (M. Fontana, S.-E. Kabbaj and S. Wiegand, Eds.), Lecture Notes in Pure and Applied Mathematics **231**, Marcel Dekker, New York, 2002, 87–107.

[Do97] L. R. Doering, *Multiplicities, cohomological degrees and generalized Hilbert functions*, Ph.D. Thesis, Rutgers University, 1997.

[DGV98] L. R. Doering, T. Gunston and W. V. Vasconcelos, Cohomological degrees and Hilbert functions of graded modules, *American J. Math.* **120** (1998), 493–504.

[DK84] L. van den Dries and K. Schmidt, Bounds in the theory of polynomial rings over fields. A nonstandard approach, *Inventiones Math.* **76** (1984), 77–91.

[DVz0] L. R. Doering and M. Vaz Pinto, On the monotonicity of the Hilbert function of the Sally module, *Comm. Algebra* **28** (2000), 1861–1866.

[ES76] P. Eakin and A. Sathaye, Prestable ideals, *J. Algebra* **41** (1976), 439–454.

[Ei95] D. Eisenbud, *Commutative Algebra with a View Toward Algebraic Geometry*, Springer, Berlin Heidelberg New York, 1995.

[EG84] D. Eisenbud and S. Goto, Linear free resolutions and minimal multiplicity, *J. Algebra* **88** (1984), 89–133.

[EH99] D. Eisenbud and J. Harris, *The Geometry of Schemes*, Springer, Berlin Heidelberg New York, 1999.

[EH83] D. Eisenbud and C. Huneke, Cohen-Macaulay Rees algebras and their specializations, *J. Algebra* **81** (1983), 202–224.

[EHU3] D. Eisenbud, C. Huneke and B. Ulrich, What is the Rees algebra of a module?, *Proc. Amer. Math. Soc.* **131** (2003), 701–708.

[EHV92] D. Eisenbud, C. Huneke and W. V. Vasconcelos, Direct methods for primary decomposition, *Inventiones Math.* **110** (1992), 207–235.

[ERT94] D. Eisenbud, A. Reeves and B. Totaro, Initial ideals, Veronese subrings, and rates of algebras, *Adv. Math.* **109** (1994), 168–187.

[ESt94] D. Eisenbud and B. Sturmfels, Finding sparse systems of parameters, *J. Pure & Applied Algebra* **94** (1994), 143–157.

[El86] S. Eliahou, A problem about polynomial ideals, *Contemporary Math.* **58** (1986), 107–120.

[E99] J. Elias, On the depth of the tangent cone and the growth of the Hilbert function, *Trans. Amer. Math. Soc.* **351** (1999), 4027–4042.

[EVY3] C. Escobar, R. Villarreal and Y. Yoshino, Torsion freeness and normality of associated graded rings and Rees algebras of monomial ideals, Preprint, 2003.

[EG81] E. G. Evans and P. Griffith, The syzygy problem, *Annals of Math.* **114** (1981), 323–333.

[EG85] E. G. Evans and P. Griffith, *Syzygies*, London Math. Soc. Lecture Notes **106**, Cambridge University Press, Cambridge, 1985.

[Ew96] G. Ewald, *Combinatorial Convexity and Algebraic Geometry*, Graduate Texts in Mathematics **168**, Springer, New York, 1996.

[Fe67] W. Feller, *An Introduction to Probability Theory and its Applications*, 3rd Edition, Wiley, New York, 1967.

[Fi71] M. Fiorentini, On relative regular sequences, *J. Algebra* **18** (1971), 384–389.

[Fl77] H. Flenner, Die Sätze von Bertini für lokale Ringe, *Math. Ann.* **229** (1977), 253–294.

[FOV99] H. Flenner, L. O'Carroll and W. Vogel, *Joins and Intersections*, Monographs in Mathematics, Springer, Berlin Heidelberg New York, 1999.

[FM01] H. Flenner and M. Manaresi, A numerical characterization of reduction ideals, *Math. Z.* **238** (2001), 205–214.

[Fo73] R. Fossum, *The Divisor Class Group of a Krull Domain*, Springer, Berlin Heidelberg New York, 1973.

[Ga92] T. Gaffney, Integral closure of modules and Whitney equisingularity, *Inventiones Math.* **107** (1992), 301–322.

[Gh2] L. Ghezzi, On the depth of the associated graded ring of an ideal, *J. Algebra* **248** (2002), 688–707.

[GMS92] P. Gimenez, M. Morales and A. Simis, The analytic spread of the ideal of a monomial curve in projective 3-space, *Progress in Math.* **109** (1992), 1082–1095.

[GMS99] P. Gimenez, M. Morales and A. Simis, The analytic spread of toric ideals of codimension 2, *Results in Mathematics* **35** (1999), 252–259.

[Gi84] M. Giusti, Some effectivity problems in polynomial ideal theory, *EUROSAM* 1984, Lecture Notes in Computer Science **174**, Springer, Berlin Heidelberg New York, 1984, 159–171.

[Go87] S. Goto, Integral closedness of complete intersection ideals, *J. Algebra* **108** (1987), 151–160.

[GH1] S. Goto and F. Hayasaka, Finite homological dimension and primes associated to integrally closed ideals, *Proc. Amer. Math. Soc.* **130** (2002), 3159–3164.

[GN94] S. Goto and K. Nishida, The Cohen-Macaulay and Gorenstein Rees algebras associated to filtrations, *Mem. Amer. Math. Soc.* **526** (1994).

[GH94] S. Goto and S. Huckaba, On graded rings associated to analytic deviation one ideals, *American J. Math.* **116** (1994), 905–919.

[GN94a] S. Goto and Y. Nakamura, Cohen-Macaulay Rees algebras of ideals having analytic deviation two, *Tohuku Math. J.* **46** (1994), 573–586.

[GN94b] S. Goto and Y. Nakamura, On the Gorensteinness of graded rings associated to ideals of analytic deviation one, *Contemporary Math.* **159** (1994), 51–72.

[GNN96] S. Goto, Y. Nakamura and K. Nishida, Cohen-Macaulay graded rings associated to ideals, *American J. Math.* **118** (1996), 1197–1213.

[GS79] S. Goto and Y. Shimoda, On the Rees algebras of Cohen-Macaulay local rings, Lecture Notes in Pure and Applied Mathematics **68**, Marcel Dekker, New York, 1979, 201–231.

[GW78] S. Goto and K. Watanabe, On graded rings I, *J. Math. Soc. Japan* **30** (1978), 179–213.

[GRe84] H. Grauert and R. Remmert, *Coherent Analytic Sheaves*, Grundlehrender Mathematischen Wissenschaften **265**, Springer, Berlin, 1984.

[GRi70] H. Grauert and O. Riemenschneider, Verschwindungssätze für analytische Kohomologiegruppen auf komplexen Räumen, *Inventiones Math.* **11** (1970), 263–292.

[GS96] D. Grayson and M. Stillman, *Macaulay2*, 1996. Available via anonymous ftp from `math.uiuc.edu`.

[GPS95] G.-M. Greuel, G. Pfister and H. Schoenemann, *Singular*: A system for computation in algebraic geometry and singularity theory. 1995. Available for download from `http://www.singular.uni-kl.de`.

[GP2] G.-M. Greuel and G. Pfister, *A* **Singular** *Introduction to Commutative Algebra*, Springer, Heidelberg, 2002.

[GHO84] U. Grothe, M. Herrmann and U. Orbanz, Graded rings associated to equimultiple ideals, *Math. Z.* **186** (1984), 531–556.

[Gr57] A. Grothendieck, Théorèmes de dualité pour les faisceaux algébriques cohérents, Séminaire Bourbaki, **147**, Paris, 1957.

[Gr67] A. Grothendieck, *Local Cohomology*, (Notes by R. Hartshorne) Lecture Notes in Mathematics **41**, Springer, Berlin Heidelberg New York, 1967.

[Gu94] A. Guerrieri, On the depth of the associated graded ring of an $\mathfrak{m}$-primary ideal, *J. Algebra* **167** (1994), 745–757.

[Gu95] A. Guerrieri, On the depth of the associated graded ring, *Proc. Amer. Math. Soc.* **123** (1995), 11–20.

[GL67] T. H. Gulliksen and G. Levin, *Homology of Local Rings*, Queen's Papers in Pure and Applied Math. **20**, Queen's Universiy, Kingston, 1967.

[Gu98] T. Gunston, *Cohomological degrees, Dilworth numbers and linear resolution*, Ph.D. Thesis, Rutgers University, 1998.

[Ha2] D. Hanes, Bounds on multiplicities of local rings, *Comm. Algebra* **30** (2002), 3789–3812.

[Har77] R. Hartshorne, *Algebraic Geometry*, Springer, Berlin Heidelberg New York, 1977.

[HRT99] H. Hauser, J.-J. Risler and B. Teissier, The reduced Bautin index of planar vector fields, *Duke Math. J.* **100** (1999), 425–445.

[HHK96] M Herrmann, E. Hyry and T. Korb, Filter-regularity and Cohen-Macaulay multigraded Rees algebras, *Comm. Algebra* **24** (1996), 2177–2200.

[HHK98] M Herrmann, E. Hyry and T. Korb, On Rees algebras with a Gorenstein Veronese subring, *J. Algebra* **200** (1998), 279–311.

[HHK0] M Herrmann, E. Hyry and T. Korb, On a-invariant formulas, *J. Algebra* **227** (2000), 254–267.

[Hal95] M. Herrmann, C. Huneke and J. Ribbe, On reduction exponents of ideals with Gorenstein formrings, *Proc. Edinburgh Math. Soc.* **38** (1995), 449–463.

[HHR93] M. Herrmann, E. Hyry and J. Ribbe, On the Cohen-Macaulay and Gorenstein property of multi-Rees algebras, *Manuscripta Math.* **79** (1993), 343–377.

[HHR95] M. Herrmann, E. Hyry and J. Ribbe, On multi-Rees algebras (with an appendix by N. V. Trung), *Math. Annalen* **301** (1995), 249–279.

[HIO88] M. Herrmann, S. Ikeda and U. Orbanz, *Equimultiplicity and Blowing-up*, Springer, Berlin, 1988.

[HRZ94] M. Herrmann, J. Ribbe and S. Zarzuela, On the Gorenstein property of Rees and form rings of powers of ideals, *Trans. Amer. Math. Soc.* **342** (1994), 631–643.

[He70] J. Herzog, Generators and relations of abelian semigroups and semigroup rings, *Manuscripta Math.* **3** (1970), 153–193.

[He81] J. Herzog, Homological properties of the module of differentials, Atas VI Escola de Algebra, Soc. Brasileira de Matemática, 1981, 33–64.

[HPT2] J. Herzog, D. Popescu and N. V. Trung, Regularity of Rees algebras, *J. London Math. Soc.* **65** (2002), 320–338.

[HSV81] J. Herzog, A. Simis and W. V. Vasconcelos, Koszul homology and blowing-up rings, in *Commutative Algebra*, Proceedings: Trento 1981 (S. Greco and G. Valla, Eds.), Lecture Notes in Pure and Applied Mathematics **84**, Marcel Dekker, New York, 1983, 79–169.

[HSV82] J. Herzog, A. Simis and W. V. Vasconcelos, Approximation complexes of blowing-up rings, *J. Algebra* **74** (1982), 466–493.

[HSV83] J. Herzog, A. Simis and W. V. Vasconcelos, Approximation complexes of blowing-up rings.II, *J. Algebra* **82** (1983), 53–83.

[HSV84] J. Herzog, A. Simis and W. V. Vasconcelos, On the arithmetic and homology of algebras of linear type, *Trans. Amer. Math. Soc.* **283** (1984), 661–683.

[HSV87] J. Herzog, A. Simis and W. V. Vasconcelos, On the canonical module of the Rees algebra and the associated graded ring of an ideal, *J. Algebra* **105** (1987), 285–302.

[HSV91] J. Herzog, A. Simis and W. V. Vasconcelos, Arithmetic of normal Rees algebras, *J. Algebra* **143** (1991), 269–294.

[HV85] J. Herzog and W. V. Vasconcelos, On the divisor class group of Rees algebras, *J. Algebra* **93** (1985), 182–188.

[HVV85] J. Herzog, W. V. Vasconcelos and R. Villarreal, Ideals with sliding depth, *Nagoya Math. J.* **99** (1985), 159–172.

[HO98] T. Hibi and H. Ohsugi, Normal polytopes arising from finite graphs, *J. Algebra* **207** (1998), 409–426.

[Hoa93] L. T. Hoa, Reduction numbers and Rees algebras of powers of an ideal, *Proc. Amer. Math. Soc.* **119** (1993), 415–422.

[Hoa96] L. T. Hoa, Reduction numbers of equimultiple ideals, *J. Pure & Applied Algebra* **109** (1996), 111–126.

[HT2] L. T. Hoa and N. V. Trung, Borel-fixed ideals and reduction numbers, *J. Algebra* **270** (2003), 335–346.

[HZ94] L. T. Hoa and S. Zarzuela, Reduction number and a-invariant of good filtrations, *Comm. Algebra* **22** (1994), 5635–5656.

[Ho72] M. Hochster, Rings of invariants of tori, Cohen-Macaulay rings generated by monomials, and polytopes, *Annals of Math.* **96** (1972), 318–337.

[Ho73a] M. Hochster, Properties of Noetherian rings stable under general grade reduction, *Arch. Math.* **24** (1973), 393–396.

[Ho73b] M. Hochster, Criteria for equality of ordinary and symbolic powers of primes, *Math. Z.* **133** (1973), 53–65.

[HH04] M. Hochster and C. Huneke, Tight closure in equal characteristic zero, in preparation.

[Hon3] J. Hong, *Rees algebras of conormal modules*, Ph.D. Thesis, Rutgers University, 2003.

[Hon3a] J. Hong, Rees algebras of conormal modules, *J. Pure & Applied Algebra* **193** (2004), 231–249.

[HNV4] J. Hong, S.-S. Noh and W. V. Vasconcelos, Divisors of integrally closed modules, *Comm. Algebra*, to appear.

[HUV4] J. Hong, B. Ulrich and W. V. Vasconcelos, Normalization of modules, Preprint, 2004.

[HuH1] R. Hübl and C. Huneke, Fiber cones and the integral closure of ideals, *Collect. Math.* **52** (2001), 85–100.

[Huc87] S. Huckaba, Reduction numbers for ideals of higher analytic spread, *Math. Proc. Camb. Phil. Soc.* **102** (1987), 49–57.

[Huc96] S. Huckaba, A d-dimensional extension of a lemma of Huneke's and formulas for the Hilbert coefficients, *Proc. Amer. Math. Soc.* **124** (1996), 1393–1401.

[Huc3] S. Huckaba, Length one ideal extensions and their associated graded rings, Preprint, 2003.

[HH92] S. Huckaba and C. Huneke, Powers of ideals having small analytic deviation, *American J. Math.* **114** (1992), 367–403.

[HH93] S. Huckaba and C. Huneke, Rees algebras of ideals having small analytic deviation, *Trans. Amer. Math. Soc.* **339** (1993), 373–402.

[HM99] S. Huckaba and C. Huneke, Normal ideals in regular rings, *J. reine angew. Math.* **510** (1999), 63–82.

[HM94] S. Huckaba and T. Marley, Depth formulas for certain graded rings associated to an ideal, *Nagoya Math. J.* **133** (1994), 57–69.

[HM97] S. Huckaba and T. Marley, Hilbert coefficients and the depths of associated graded rings, *J. London Math. Soc.* **56** (1997), 64–76.

[HH99] S. Huckaba and T. Marley, On associated graded rings of normal ideals, *J. Algebra* **222** (1999), 146–163.

[Hu80] C. Huneke, On the symmetric and Rees algebras of an ideal generated by a d-sequence, *J. Algebra* **62** (1980), 268–275.

[Hu81] C. Huneke, Symbolic powers of prime ideals and special graded algebras, *Comm. Algebra* **9** (1981), 339–366.

[Hu82a] C. Huneke, On the associated graded ring of an ideal, *Illinois J. Math.* **26** (1982), 121-137.

[Hu82b] C. Huneke, Linkage and Koszul homology of ideals, *American J. Math.* **104** (1982), 1043–1062.

[Hu83] C. Huneke, Strongly Cohen-Macaulay schemes and residual intersections, *Trans. Amer. Math. Soc.* **277** (1983), 739–763.

[Hu86a] C. Huneke, Determinantal ideals of linear type, *Arch. Math.* **47** (1986), 324–329.

[Hu86b] C. Huneke, The primary components of and integral closures of ideals in 3-dimensional regular local rings, *Math. Annalen* **275** (1986), 617–635.

[Hu87] C. Huneke, Hilbert functions and symbolic powers, *Michigan Math. J.* **34** (1987), 293–318.

[Hu96] C. Huneke, *Tight Closure and its Applications*, CBMS **88**, American Mathematical Society, Providence, RI, 1996.

[HuR86] C. Huneke and M. E. Rossi, The dimension and components of symmetric algebras, *J. Algebra* **98** (1986), 200–210.

[HSV89] C. Huneke, A. Simis and W. V. Vasconcelos, Reduced normal cones are domains, *Contemporary Math.* **88** (1989), 95–101.

[HS95] C. Huneke and I. Swanson, Cores of ideals in 2-dimensional regular local rings, *Michigan J. Math.* **42** (1995), 193–208.

[HU88] C. Huneke and B. Ulrich, Residual intersections, *J. reine angew. Math.* **390** (1988), 1–20.

[HUV92] C. Huneke, B. Ulrich and W. V. Vasconcelos, On the structure of certain normal ideals, *Compositio Math.* **84** (1992), 25–42.

[Hup67] B. Huppert, *Endliche Gruppen I*, Springer 1967.

[Hy99] E. Hyry, Necessary and sufficient conditions for the Cohen-Macaulayness of form rings, *J. Algebra* **212** (1999), 17–27.

[Hy1] E. Hyry, Coefficient ideals and the Cohen-Macaulay property of Rees algebras, *Proc. Amer. Math. Soc.* **129** (2001), 1299–1308.

[Hy2] E. Hyry, Cohen-Macaulay multi-Rees algebras, *Compositio Math.* **130** (2002), 319–343.

[HyJ2] E. Hyry and T. Järvilehto, Hilbert coefficients and the Gorenstein property of the associated graded ring, Preprint, 2002.

[HyV98] E. Hyry and O. Villamayor, A Briançon-Skoda theorem for isolated singularities, *J. Algebra* **204** (1998), 656–665.

[Ik86] S. Ikeda, On the Gorensteinness of Rees algebras over local rings, *Nagoya Math. J.* **102** (1986), 135–154.

[IT89] S. Ikeda and N. V. Trung, When is the Rees algebra Cohen-Macaulay?, *Comm. Algebra* **17** (1989), 2893–2922.

[It88] S. Itoh, Integral closures of ideals generated by regular sequences, *J. Algebra* **117** (1988), 390–401.

[It92] S. Itoh, Coefficients of normal Hilbert polynomials, *J. Algebra* **150** (1992), 101–117.

[It95] S. Itoh, Hilbert coefficients of integrally closed ideals, *J. Algebra* **176** (1995), 638–652.

[Ja2] A. S. Jarrah, Integral closures of Cohen-Macaulay monomial ideals, *Comm. Algebra* **30** (2002), 5473–5478.

[Jo97] M. Johnson, Note on multiplicities of blow-ups and residual intersections, *Comm. Algebra* **25** (1997), 3797–3801.

[JM1] M. Johnson and S. Morey, Normal blowups and their expected defining equations, *J. Pure & Applied Algebra* **162** (2001), 303–313.

[JU96] M. Johnson and B. Ulrich, Artin-Nagata properties and Cohen-Macaulay associated graded rings, *Compositio Math.* **103** (1996), 7-29.

[JU99] M. Johnson and B. Ulrich, Serre's condition R_k for associated graded rings, *Proc. Amer. Math. Soc.* **127** (1999), 2619–2624.

[JK94] B. Johnston and D. Katz, On the relation type of large powers of an ideal, *Mathematika* **41** (1994), 209–214.

[JK95] B. Johnston and D. Katz, Castelnuovo regularity and graded rings associated to an ideal, *Proc. Amer. Math. Soc.* **123** (1995), 727–734.

[Kap74] I. Kaplansky, *Commutative Rings*, University of Chicago Press, Chicago, 1974.

[Ka94] D. Katz, Generating ideals up to projective equivalence, *Proc. Amer. Math. Soc.* **120** (1994), 77–83.

[Ka95] D. Katz, Reduction criteria for modules, *Comm. Algebra* **23** (1995), 4543–4548.

[KK97] D. Katz and V. Kodiyalam, Symmetric powers of complete modules over a two-dimensional regular local ring, *Trans. Amer. Math. Soc.* **349** (1997), 481–500.

[KN95] D. Katz and C. Naudé, Prime ideals associated to symmetric powers of a module, *Comm. Algebra* **23** (1995), 4549–4555.

[Kaw2] T. Kawasaki, On arithmetic Macaulayfication of local rings, *Trans. Amer. Math. Soc.* **354** (2002), 123–149.

[KM82] D. Kirby and H. A. Mehran, A note on the coefficients of the Hilbert-Samuel polynomial for a Cohen-Macaulay module, *J. London Math. Soc.* **25** (1982), 449–457.

[KR94] D. Kirby and D. Rees, Multiplicities in graded rings I: The general theory, *Contemporary Math.* **159** (1994), 209–267.

[KS3] K. Kiyek and J. Stückrad, Integral closure of monomial ideals on regular sequences, *Rev. Mat. Iberoamericana* **19** (2003), 483–508.

[KT94] S. Kleiman and A. Thorup, A geometric theory of the Buchsbaum-Rim multiplicity, *J. Algebra* **167** (1994), 168–231.

[Ko95] V. Kodiyalam, Integrally closed modules over two-dimensional regular local rings, *Trans. Amer. Math. Soc.* **347** (1995), 3551–3573.

[Kot91] B. V. Kotsev, Determinantal ideals of linear type of a generic symmetric matrix, *J. Algebra* **139** (1991), 488–504.

[Ku74] E. Kunz, Almost complete intersections are not Gorenstein rings, *J. Algebra* **28** (1974), 111–115.

[Ku86] E. Kunz, *Kähler Differentials*, Vieweg, Wiesbaden, 1986.

[Kus95] A. Kustin, Ideals associated to two sequences and a matrix, *Comm. Algebra* **23** (1995), 1047–1083.

[KMU92] A. Kustin, M. Miller and B. Ulrich, Generating a residual intersection, *J. Algebra* **146** (1992), 335–384.

[Le60] C. Lech, Note on multiplicities of ideals, *Arkiv för Matematik* **4** (1960), 63–86.

[Le64] C. Lech, Inequalities related to certain couples of local rings, *Acta Math.* **112** (1964), 69–89.

[Li69a] J. Lipman, On the Jacobian ideal of the module of differentials, *Proc. Amer. Math. Soc.* **21** (1969), 422–426.

[Li69b] J. Lipman, Rational singularities with applications to algebraic surfaces and unique factorization, *Publ. Math. I.H.E.S.* **36** (1969), 195–279.

[Li88] J. Lipman, On complete ideals in regular local rings, in *Algebraic Geometry and Commutative Algebra in Honor of Masayoshi Nagata* (1988), 203–231.

[Li94a] J. Lipman, Cohen-Macaulayness in graded algebras, *Math. Research Letters* **1** (1994), 149–159.

[Li94b] J. Lipman, Adjoints of ideals in regular local rings, *Math. Research Letters* **1** (1994), 739–755.

[LS81] J. Lipman and A. Sathaye, Jacobian ideals and a theorem of Briançon-Skoda, *Michigan Math. J.* **28** (1981), 199–222.

[LT81] J. Lipman and B. Teissier, Pseudo-rational local rings and a theorem of Briançon-Skoda about integral closures of ideals, *Michigan Math. J.* **28** (1981), 97–116.

[Liu98] J.-C. Liu, Rees algebras of finitely generated torsion-free modules over a two-dimensional regular local ring, *Comm. Algebra* **26** (1998), 4015–4039.

[Ly86] G. Lyubeznik, A property of ideals in polynomial rings, *Proc. Amer. Math. Soc.* **98** (1986), 399–400.

[Ly88] G. Lyubeznik, On the arithmetical rank of monomial ideals, *J. Algebra* **112** (1988), 87–89.

[Mar89] T. Marley, The coefficients of the Hilbert polynomials and the reduction number of an ideal, *J. London Math. Soc.* **40** (1989), 1–8.

[Mat0] R. Matsumoto, On computing the integral closure, *Comm. Algebra* **28** (2000), 401–405.

[Ma80] H. Matsumura, *Commutative Algebra*, Benjamin/Cummings, Reading, 1980.

[Ma86] H. Matsumura, *Commutative Ring Theory*, Cambridge University Press, Cambridge, 1986.

[Mc80] S. McAdam, Asymptotic prime divisors and analytic spread, *Proc. Amer. Math. Soc.* **80** (1980), 555–559.

[Mc83] S. McAdam, *Asymptotic Prime Divisors*, Lecture Notes in Mathematics **1023**, Springer, Berlin, 1983.

[Mi64] A. Micali, Sur les algèbres universalles, *Annales Inst. Fourier* **14** (1964), 33–88.

[MSS65] A. Micali, P. Salmon and P. Samuel, Integrité et factorialité des algèbres symétriques, Atas do IV Colóquio Brasileiro de Matemática, SBM, (1965), 61–76.

[Mo83] P. Monsky, The Hilbert-Kunz function, *Math. Ann.* **263** (1983), 43–49.

[MM84] H. M. Möller and T. Mora, Upper and lower bounds for the degree of Gröbner bases, *EUROSAM* 1984, Lecture Notes in Computer Science **174**, Springer, Berlin Heidelberg New York, 1984, 172–183.

[MR95] T. Mora and M. E. Rossi, An algorithm for the Hilbert function of a primary ideal, *Comm. Algebra* **23** (1995), 1899–1911.

[Mo84] M. Morales, Polynôme de Hilbert-Samuel des clôtures intégrales des puissances d'un idéal $\mathfrak{m}$-primaire, *Bull. Soc. Math. France* **112** (1984), 343–358.

[MTV90] M. Morales, N. V. Trung and O. Villamayor, Sur la fonction de Hilbert-Samuel des clôtures intégrales des puissances d'idéaux engendrés par un système de paramètres, *J. Algebra* **129** (1990), 96–102.

[Mor96] S. Morey, Equations of blowups of ideals of codimension two and three, *J. Pure & Applied Algebra* **109** (1996), 197–211.

[MNV95] S. Morey, S. Noh and W. V. Vasconcelos, Symbolic powers, Serre conditions and Cohen-Macaulay Rees algebras, *Manuscripta Math.* **86** (1995), 113–124.

[MU96] S. Morey and B. Ulrich, Rees algebras of ideals with low codimension, *Proc. Amer. Math. Soc.* **124** (1996), 3653–3661.

[MV1] S. Morey and W. V. Vasconcelos, Special divisors of blowup algebras, in *Ring Theory and Algebraic Geometry*, Proceedings SAGA V (A. Granja, J. A. Hermida Alonso and A. Verschoren, Eds.), Lecture Notes in Pure and Applied Mathematics **221**, Marcel Dekker, New York, 2001, 257–288.

[Mu66] D. Mumford, *Lectures on Curves on an Algebraic Surface*, Annals of Mathematics Studies **59**, Princeton University Press, Princeton, New Jersey, 1966.

[Na62] M. Nagata, *Local Rings*, Interscience, New York, 1962.

[Nag3] U. Nagel, Comparing Castelnuovo-Mumford regularity and extended degree: the borderline cases, *Trans. Amer. Math. Soc.*, to appear.

[Nar63] M. Narita, A note on the coefficients of Hilbert characteristic functions in semi-regular local rings, *Math. Proc. Camb. Phil. Soc.* **59** (1963), 269–275.

[No50] E. Noether, Idealdifferentiation und Differente, *J. reine angew. Math.* **188** (1950), 1–21.

[NV93] S. Noh and W. V. Vasconcelos, The S_2-closure of a Rees algebra, *Results in Math.* **23** (1993), 149–162.

[Nor59] D. G. Northcott, A generalization of a theorem on the contents of polynomials, *Math. Proc. Camb. Phil. Soc.* **55** (1959), 282–288.

[Nor60] D. G. Northcott, A note on the coefficients of the abstract Hilbert function, *J. London Math. Soc.* **35** (1960), 209–214.

[Nor63] D. G. Northcott, A homological investigation of a certain residual ideal, *Math. Annalen* **150** (1963), 99–110.

[NR54] D. G. Northcott and D. Rees, Reductions of ideals in local rings, *Math. Proc. Camb. Phil. Soc.* **50** (1954), 145–158.

[OR90] J. S. Okon and L. J. Ratliff, Jr., Reductions of filtrations, *Pacific J. Math.* **144** (1990), 137–154.

[Oo82] A. Ooishi, Castelnuovo's regularity of graded rings and modules, *Hiroshima Math. J.* **12** (1982), 627–644.

[Oo87] A. Ooishi, Genera and arithmetic genera of commutative rings, *Hiroshima Math. J.* **17** (1987), 361–372.

[Oo91] A. Ooishi, On the associated graded modules of canonical modules, *J. Algebra* **141** (1991), 143–157.

[Oo93] A. Ooishi, On the Gorenstein property of the associated graded ring and the Rees algebra of an ideal, *J. Algebra* **155** (1993), 397–414.

[Pol0] C. Polini, A filtration of the Sally module and the associated graded ring of an ideal, *Comm. Algebra* **28** (2000), 1335–1341.

[PU98] C. Polini and B. Ulrich, Linkage and reduction numbers, *Math. Annalen* **310** (1998), 631–651.

[PU99] C. Polini and B. Ulrich, Necessary and sufficient conditions for the Cohen-Macaulayness of blowup algebras, *Compositio Math.* **119** (1999), 185–207.

[PU2] C. Polini and B. Ulrich, A formula for the core of an ideal, *Math. Annalen*, to appear.

[PUV4] C. Polini, B. Ulrich and W. V. Vasconcelos, Normalization of ideals and Briançon-Skoda numbers, *Math. Research Letters*, to appear.

[PUVV4] C. Polini, B. Ulrich, W. V. Vasconcelos and R. Villarreal, Indices of the normalization of ideals, Preprint, 2004.

[Rag94] K. R. Raghavan, Powers of ideals generated by quadratic sequences, *Trans. Amer. Math. Soc.* **343** (1994), 727–747.

[Ra74] L. J. Ratliff, Jr., Locally quasi-unmixed Noetherian rings and ideals of the principal class, *Pacific J. Math.* **52** (1974), 185–205.

[Ra78] L. J. Ratliff, Jr., *Chain Conjectures in Ring Theory*, Lecture Notes in Mathematics **647**, Springer, Berlin, 1978.

[RR78] L. J. Ratliff, Jr. and D. E. Rush, Two notes on reductions of ideals, *Indiana Univ. Math. J.* **27** (1978), 929–934.

[Re56] D. Rees, Valuations associated with ideals. II, *J. London Math. Soc.* **31** (1956), 221–228.

[Re61] D. Rees, α-transforms of local rings and a theorem on multiplicities of ideals, *Math. Proc. Camb. Phil. Soc.* **57** (1961), 8–17.

[Re85] D. Rees, Amao's theorem and reduction criteria, *J. London Math. Soc.* **32** (1985), 404–410.

[Re87] D. Rees, Reductions of modules, *Math. Proc. Camb. Phil. Soc.* **101** (1987), 431–449.

[RS88] D. Rees and J. D. Sally, General elements and joint reductions, *Michigan J. Math.* **35** (1988), 241–254.

[RRV2] L. Reid, L. G. Roberts and M. A. Vitulli, Some results on normal homogeneous ideals, *Comm. Algebra* **31** (2003), 4485–4506.

[RoS90] L. Robbiano and M. Sweedler, Subalgebra bases, *in Commutative Algebra*, Proceedings, Salvador 1988 (W. Bruns and A. Simis, Eds.), Lecture Notes in Mathematics **1430**, Springer, Berlin Heidelberg New York, 1990, 61–87.

[Ros0a] M. E. Rossi, Primary ideals with good associated graded ring, *J. Pure & Applied Algebra* **145** (2000) 75–90.

[Ros0b] M. E. Rossi, A bound on the reduction number of a primary ideal, *Proc. Amer. Math. Soc.* **128** (2000), 1325–1332.

[RTV3] M. E. Rossi, N. V. Trung and G. Valla, Castelnuovo-Mumford regularity and extended degree, *Trans. Amer. Math. Soc.* **355** (2003), 1773–1786.

[RV96] M. E. Rossi and G. Valla, A conjecture of Sally, *Comm. Algebra* **24** (1996), 4249–4261.

[RV3] M. E. Rossi and G. Valla, The Hilbert function of the Ratliff-Rush filtration, *J. Pure & Applied Algebra*, to appear.

[RVV1] M. E. Rossi, G. Valla and W. V. Vasconcelos, Maximal Hilbert functions, *Results in Math.* **39** (2001), 99–114.

[Sa76] K. Saito, On a generalization of De Rham lemma, *Ann. Inst. Fourier* **26** (1976), 165–170.

[SS87] J. B. Sancho de Salas, Blowing-up morphisms with Cohen-Macaulay associated graded rings, Géométrie Algébrique et Applications I, Géométrie et calcul algèbrique, Deuxiéme conférence internationale de la Rabida, Travaux en cours no. 22, Hermann, Paris, 1987, 201–209.

[Sal76] J. D. Sally, Bounds for numbers of generators of Cohen-Macaulay ideals, *Pacific J. Math.* **63** (1976), 517–520.

[Sal78] J. D. Sally, *Numbers of Generators of Ideals in Local Rings*, Lecture Notes in Pure and Applied Mathematics **36**, Marcel Dekker, New York, 1978.

[Sal79] J. D. Sally, Cohen-Macaulay local rings of maximal embedding dimension, *J. Algebra* **56** (1979), 168–183.

[Sal80] J. D. Sally, Tangent cones at Gorenstein singularities, *Compositio Math.* **40** (1980), 167–175.

[Sal92] J. D. Sally, Hilbert coefficients and reduction number 2, *J. Algebraic Geometry* **1** (1992), 325–333.

[Sal93] J. D. Sally, Ideals whose Hilbert function and Hilbert polynomial agree at $n = 1$, *J. Algebra* **157** (1993), 534–547.

[Sc69] G. Scheja, Multiplizitätenvergleich unter Verwendung von Testkurven, *Comment. Math. Helv.* **44** (1969), 438–445.

[ScS70] G. Scheja and U. Storch, Über differentielle Abhängigkeit bei Idealen analytischer Algebren, *Math. Z.* **114** (1970), 101–112.

[Sch90] P. Schenzel, Castelnuovo's index of regularity and reduction numbers, Banach Center Publications **26**, PWN-Polish Scientific Publishers, Warsaw, 1990, 201–208.

[Sch98] P. Schenzel, On the use of local cohomology in algebra and geometry, in *Six Lectures on Commutative Algebra*, Progress in Mathematics **166**, Birkhäuser, Boston, 1998, 241–292.

[Sho3] H. Schoutens, Number of generators of a Cohen-Macaulay ideal, *J. Algebra* **259** (2003), 235–242.

[Shr87] A. Schrijver, *Theory of Linear and Integer Programming*, John Wiley & Sons, New York, 1987.

[Sei66] A. Seidenberg, Derivations and integral closure, *Pacific J. Math.* **16** (1966), 167–173.

[Sei75] A. Seidenberg, Construction of the integral closure of a finite integral domain II, *Proc. Amer. Math. Soc.* **52** (1975), 368–372.

[Ser65] J.-P. Serre, *Algèbre Locale. Multiplicités*, Lecture Notes in Mathematics **11**, Springer, Berlin, 1965.

[Sha91] K. Shah, Coefficient ideals, *Trans. Amer. Math. Soc.* **327** (1991), 373–384.

[Sh54] G. Shimura, A note on the normalization theorem of an integral domain, *Sci. Papers Coll. Gen. Ed. Univ. Tokyo* **4** (1954), 1–8.

[Shr66] Yu. A. Shreider, *The Monte Carlo Method*, Pergamon Press, Oxford, 1966.

[SSU2] A. Simis, K. Smith and B. Ulrich, An algebraic proof of Zak's inequality for the dimension of the Gauss image, *Math. Z.* **241** (2002), 871–881.

[ST88] A. Simis and N. V. Trung, The divisor class group of ordinary and symbolic blowups, *Math. Z.* **198** (1988), 479–491.

[SUV93] A. Simis, B. Ulrich and W. V. Vasconcelos, Jacobian dual fibrations, *American J. Math.* **115** (1993), 47–75.

[SUV95] A. Simis, B. Ulrich and W. V. Vasconcelos, Cohen-Macaulay Rees algebras and degrees of polynomial relations, *Math. Annalen* **301** (1995), 421–444.

[SUV1] A. Simis, B. Ulrich and W. V. Vasconcelos, Codimension, multiplicity and integral extensions, *Math. Proc. Camb. Phil. Soc.* **130** (2001), 237–257.

[SUV3] A. Simis, B. Ulrich and W. V. Vasconcelos, Rees algebras of modules, *Proceedings London Math. Soc.* **87** (2003), 610–646.

[SV81a] A. Simis and W. V. Vasconcelos, On the dimension and integrality of symmetric algebras, *Math. Z.* **177** (1981), 341–358.

[SV81b] A. Simis and W. V. Vasconcelos, The syzygies of the conormal module, *American J. Math.* **103** (1981), 203–224.

[SV88] A. Simis and W. V. Vasconcelos, The Krull dimension and integrality of symmetric algebras, *Manuscripta Math.* **61** (1988), 63–78.

[SVV94] A. Simis, W. V. Vasconcelos and R. Villarreal, On the ideal theory of graphs, *J. Algebra* **167** (1994), 389–416.

[SVV98] A. Simis, W. V. Vasconcelos and R. Villarreal, The integral closure of subrings associated to graphs, *J. Algebra* **199** (1998), 281–289.

[ST97] V. Srinivas and V. Trivedi, On the Hilbert function of a Cohen-Macaulay local ring, *J. Algebraic Geometry* **6** (1997), 733–751.

[Sta86] R. Stanley, *Enumerative Combinatorics*, Vol. I, Wadsworth & Brooks/Cole, Monterey, California, 1986.

[Sta91] R. Stanley, On the Hilbert function of a graded Cohen-Macaulay domain, *J. Pure & Applied Algebra* **73** (1991), 307–314.

[Sto68] G. Stolzenberg, Constructive normalization of an algebraic variety, *Bull. Amer. Math. Soc.* **74** (1968), 595–599.

[StV86] J. Stückrad and W. Vogel, *Buchsbaum Rings and Applications*, Springer, New York, 1986.

[Sw67] R. Swan, The number of generators of a module, *Math. Z.* **102** (1967), 318–322.

[TZ3] S.-L. Tan and D.-Q. Zhang, The determination of integral closures and geometric applications, *Adv. Math.*, to appear.

[Ta94] Z. Tang, Rees rings and associated graded rings of ideals having higher analytic deviation, *Comm. Algebra* **22** (1994), 4855–4898.

[Te82] B. Teissier, Variétés polaires, in *Algebraic Geometry*, La Rabida, Lecture Notes in Mathematics **961**, Springer, Berlin, 1982, 314–491.

[Te88] B. Teissier, Monômes, volumes et multiplicités, in *Introduction à la théorie des Singularités*, Vol. II (Lê Dung Trang, Ed.), Travaux en Cours **37**, Hermann, Paris, 1988.

[Tri97] V. Trivedi, Hilbert functions, Castelnuovo-Mumford regularity and uniform Artin-Rees numbers, *Manuscripta Math.* **94** (1997), 485–499.

[Tr79] N. V. Trung, On the symbolic powers of determinantal ideals, *J. Algebra* **58** (1979), 361–369.

[Tr87] N. V. Trung, Reduction exponent and degree bound for the defining equations of graded rings, *Proc. Amer. Math. Soc.* **101** (1987), 229–236.

[Tr94] N. V. Trung, Reduction numbers, a-invariants and Rees algebras of ideals having small analytic deviation, in *Commutative Algebra*, Proceedings, Trieste 1992 (G. Valla, N. V. Trung and A. Simis, Eds.), World Scientific, Singapore, 1994, 245–262.

[Tr98] N. V. Trung, The Castelnuovo regularity of the Rees algebra and the associated graded ring, *Trans. Amer. Math. Soc.* **350** (1998), 2813–2832.

[Tr99] N. V. Trung, The largest non-vanishing degree of local cohomology modules, *J. Algebra* **215** (1999), 481–499.

[Tr2] N. V. Trung, Constructive characterization of the reduction numbers, *Compositio Math.* **137** (2003), 99–113.

[Ul92] B. Ulrich, Remarks on residual intersections, in *Free Resolutions in Commutative Algebra and Algebraic Geometry*, Proceedings, Sundance 1990 (D. Eisenbud and C. Huneke, Eds.), Research Notes in Mathematics **2**, Jones and Bartlett Publishers, Boston-London, 1992, 133–138.

[Ul94a] B. Ulrich, Artin-Nagata properties and reductions of ideals, *Contemporary Math.* **159** (1994), 373–400.

[Ul94b] B. Ulrich, Cohen-Macaulayness of associated graded rings and reduction numbers of ideals, Lecture Notes, ICTP, Trieste, 1994.

[Ul96] B. Ulrich, Ideals having the expected reduction number, *American J. Math.* **118** (1996), 17–28.

[UV93] B. Ulrich and W. V. Vasconcelos, The equations of Rees algebras of ideals with linear presentation, *Math. Z.* **214** (1993), 79–92.

[UV3] B. Ulrich and W. V. Vasconcelos, On the complexity of the integral closure, *Trans. Amer. Math. Soc.* **357** (2005), 425–442.

[VV78] P. Valabrega and G. Valla, Form rings and regular sequences, *Nagoya Math. J.* **72** (1978), 91–101.

[Val76] G. Valla, Certain graded algebras are always Cohen-Macaulay, *J. Algebra* **42** (1976), 537–548.

[Val80] G. Valla, On the symmetric and Rees algebras of an ideal, *Manuscripta Math.* **30** (1980), 239–255.

[Val81] G. Valla, Generators of ideals and multiplicities, *Comm. Algebra* **9** (1981), 1541–1549.

[Val98] G. Valla, Problems and results on Hilbert functions of graded algebras, in *Six Lectures on Commutative Algebra*, Progress in Mathematics **166**, Birkhäuser, Boston, 1998, 293–344.

[Va67] W. V. Vasconcelos, Ideals generated by R-sequences, *J. Algebra* **6** (1967), 309–316.

[Va78] W. V. Vasconcelos, On the homology of I/I^2, *Comm. Algebra* **6** (1978), 1801–1809.

[Va87] W. V. Vasconcelos, On linear complete intersections, *J. Algebra* **111** (1987), 306–315.

[Va91a] W. V. Vasconcelos, On the equations of Rees algebras, *J. reine angew. Math.* **418** (1991), 189–218.

[Va91b] W. V. Vasconcelos, Computing the integral closure of an affine domain, *Proc. Amer. Math. Soc.* **113** (1991), 633–638.

[Va94a] W. V. Vasconcelos, Hilbert functions, analytic spread and Koszul homology, *Contemporary Math.* **159** (1994), 401–422.

[Va94b] W. V. Vasconcelos, *Arithmetic of Blowup Algebras*, London Math. Soc., Lecture Note Series **195**, Cambridge University Press, 1994.

[Va96] W. V. Vasconcelos, The reduction number of an algebra, *Compositio Math.* **104** (1996), 189–197.

[Va98a] W. V. Vasconcelos, The homological degree of a module, *Trans. Amer. Math. Soc.* **350** (1998), 1167–1179.

[Va98b] W. V. Vasconcelos, *Computational Methods in Commutative Algebra and Algebraic Geometry*, Springer, Heidelberg, 1998.

[Va98c] W. V. Vasconcelos, Cohomological degrees of graded modules, in *Six Lectures on Commutative Algebra*, Progress in Mathematics **166**, Birkhäuser, Boston, 1998, 345–392.

[Va99] W. V. Vasconcelos, The reduction numbers of an ideal, *J. Algebra* **216** (1999), 652–664.

[Va0] W. V. Vasconcelos, Divisorial extensions and the computation of integral closures, *J. Symbolic Computation* **30** (2000), 595–604.

[Va3a] W. V. Vasconcelos, Multiplicities and reduction numbers, *Compositio Math.* **139** (2003), 361–379.

[Va3b] W. V. Vasconcelos, Multiplicities and the number of generators of Cohen-Macaulay ideals, *Contemporary Math.* **331** (2003), 343–352.

[Ve73] U. Vetter, Zu einem Satz von G. Trautmann über den Rang gewisser kohärenter analytischer Moduln, *Arch. Math.* **24** (1973), 158–161.

[Vil89] O. Villamayor, On class groups and normality of Rees rings, *Comm. Algebra* **17** (1989), 1607–1625.

[Vi88] R. Villarreal, Rees algebras and Koszul homology, *J. Algebra* **119** (1988), 83–104.

[Vi90] R. Villarreal, Cohen-Macaulay graphs, *Manuscripta Math.* **66** (1990), 277–293.

[Vi96] R. Villarreal, Normality of subrings generated by square free monomials, *J. Pure & Applied Algebra* **113** (1996), 91–106.

[Vi1] R. Villarreal, *Monomial Algebras*, Monographs and Textbooks in Pure and Applied Mathematics **238**, Marcel Dekker, New York, 2001.

[Vz95] M. Vaz Pinto, *Structure of Sally modules and Hilbert functions*, Ph.D. Thesis, Rutgers University, 1995.

[Vz97] M. Vaz Pinto, Hilbert functions and Sally modules, *J. Algebra* **192** (1997), 504–523.

[Wal62] R. J. Walker, *Algebraic Curves*, Dover, New York, 1962.

[Wa97] H.-J. Wang, On Cohen-Macaulay local rings with embedding dimension $e+d-2$, *J. Algebra* **190** (1997), 226–240.

[Wa98] H.-J. Wang, A note on powers of ideals, *J. Pure & Applied Algebra* **123** (1998), 301–312.

[We94] C. Weibel, *An Introduction to Homological Algebra*, Cambridge University Press, Cambridge, 1994.

[ZS60] O. Zariski and P. Samuel, *Commutative Algebra*, Vol. II, Van Nostrand, Princeton, 1960.

[Za92] S. Zarzuela, On the structure of the canonical module of the Rees algebra and the associated graded ring of an ideal, *Publ. Mat.* **36** (1992), no. 2B, 1075–1084.

[Zi95] G. M. Ziegler, *Lectures on Polytopes*, Graduate Texts in Mathematics **152**, Springer, Heidelberg, 1995.

Notation and Terminology

$\overline{A}$	integral closure of the affine algebra A
ann (E)	annihilator of the module E
$a(G)$	a-invariant of the graded ring G
$\mathrm{bdeg}(E)$	a cohomological degree of the module E, 159
$\mathrm{bigrank}(E)$	bigrank of the module E, 73
$\mathrm{bigr}(I)$	big reduction number of the ideal I, 75
$\mathrm{br}(E)$	Buchsbaum-Rim multiplicity of the module E, 422
$\mathfrak{c}(A)$	conductor of the algebra A, 318
C.I.	complete intersection
C-M	Cohen-Macaulay
$\mathrm{Deg}(E)$	a cohomological degree of the module E, 140
$\det(E)$	determinant of the module E, 71, 439
$\det_0(E)$	order determinant ideal of the module E, 439
$\mathrm{codim} E$	codimension of the module E
$\deg(E)$	multiplicity of a graded module
depth E	depth of the module M relative to the maximal ideal
$\dim E$	Krull dimension of a module
E^*	dual of the R-module E, $E^* = \mathrm{Hom}_R(E,R)$
$\overline{E}$	integral closure of the R-module E
$E[-a]$	shift of a graded module, $E[-a]_n = E_{n-a}$
$e(\mathrm{gr}_I(R))$	multiplicity of the ring $\mathrm{gr}_I(R)$
$e(I)$	multiplicity of a primary ideal of a local ring
$\mathrm{embdim}(A)$	embedding dimension of the affine algebra A, 348
$\mathrm{embdeg}(A)$	embedding degree of the graded algebra A, 349
$F_s(E)$	Fitting ideal of the module E
$\mathrm{Fitt}_s(E)$	Fitting ideal of the module E
$\Gamma_I(E)$	0th local cohomology module of E relative to I
$\mathrm{gr}_I(R)$	associated graded ring of the ideal I, 7
$H_A(t)$	Hilbert series of the graded algebra A
$h(A)$	h-vector of the graded algebra A, 100

$\mathrm{hdeg}(A)$	homological degree of the module A, 141
$\mathrm{hdeg}_I(A)$	homological degree of the module A relative to I, 142
height I	height of the ideal I
$\overline{I}$	integral closure of the ideal I, 33
$I^{(n)}$	nth symbolic power of the ideal I
$I_t(\varphi)$	ideal generated by the $t \times t$ minors of the matrix φ
$\lambda(E)$	length of a module with a composition series
$\ell(E)$	analytic spread of a module over a local ring, 45
$\mathcal{M}(I;R)$	$\mathcal{M}$-complex of the ideal I, 294
$\mathrm{nil}(I)$	index of nilpotency of the ideal I, 121
$\nu(E)$	minimal number of generators of the module E
proj dim E	projective dimension of the module E
$\mathcal{Z}(I;R)$	$\mathcal{Z}$-complex of the ideal I, 294
ω_R	canonical module of the ring R
$\mathcal{R}(E)$	Rees algebra of the module E, 9
$\mathrm{r}(I)$	reduction number of the ideal I, 54
$\mathrm{reg}(E)$	Castelnuovo-Mumford regularity of the module E, 130
$\mathrm{r}_J(I)$	reduction number of the ideal I relative to J, 32
$\mathrm{tn}(E)$	tracking number of a module or algebra, 337
$V(I)$	closed set of $\mathrm{Spec}(R)$ defined by I

Index

Springer Monographs in Mathematics

This series publishes advanced monographs giving well-written presentations of the "state-of-the-art" in fields of mathematical research that have acquired the maturity needed for such a treatment. They are sufficiently self-contained to be accessible to more than just the intimate specialists of the subject, and sufficiently comprehensive to remain valuable references for many years. Besides the current state of knowledge in its field, an SMM volume should also describe its relevance to and interaction with neighbouring fields of mathematics, and give pointers to future directions of research.

Abhyankar, S.S. **Resolution of Singularities of Embedded Algebraic Surfaces** 2nd enlarged ed. 1998
Alexandrov, A.D. **Convex Polyhedra** 2005
Andrievskii, V.V.; Blatt, H.-P. **Discrepancy of Signed Measures and Polynomial Approximation** 2002
Angell, T. S.; Kirsch, A. **Optimization Methods in Electromagnetic Radiation** 2004
Ara, P.; Mathieu, M. **Local Multipliers of C*-Algebras** 2003
Armitage, D.H.; Gardiner, S.J. **Classical Potential Theory** 2001
Arnold, L. **Random Dynamical Systems** corr. 2nd printing 2003 (1st ed. 1998)
Arveson, W. **Noncommutative Dynamics and E-Semigroups** 2003
Aubin, T. **Some Nonlinear Problems in Riemannian Geometry** 1998
Auslender, A.; Teboulle M. **Asymptotic Cones and Functions in Optimization and Variational Inequalities** 2003
Bang-Jensen, J.; Gutin, G. **Digraphs** 2001
Baues, H.-J. **Combinatorial Foundation of Homology and Homotopy** 1999
Brown, K.S. **Buildings** 3rd printing 2000 (1st ed. 1998)
Cherry, W.; Ye, Z. **Nevanlinna's Theory of Value Distribution** 2001
Ching, W.K. **Iterative Methods for Queuing and Manufacturing Systems** 2001
Crabb, M.C.; James, I.M. **Fibrewise Homotopy Theory** 1998
Chudinovich, I. **Variational and Potential Methods for a Class of Linear Hyperbolic Evolutionary Processes** 2005
Dineen, S. **Complex Analysis on Infinite Dimensional Spaces** 1999
Dugundji, J.; Granas, A. **Fixed Point Theory** 2003
Elstrodt, J.; Grunewald, F. Mennicke, J. **Groups Acting on Hyperbolic Space** 1998
Edmunds, D.E.; Evans, W.D. **Hardy Operators, Function Spaces and Embeddings** 2004
Fadell, E.R.; Husseini, S.Y. **Geometry and Topology of Configuration Spaces** 2001
Fedorov, Y.N.; Kozlov, V.V. **A Memoir on Integrable Systems** 2001
Flenner, H.; O'Carroll, L. Vogel, W. **Joins and Intersections** 1999
Gelfand, S.I.; Manin, Y.I. **Methods of Homological Algebra** 2nd ed. 2003
Griess, R.L. Jr. **Twelve Sporadic Groups** 1998
Gras, G. **Class Field Theory** corr. 2nd printing 2005
Hida, H. **p-Adic Automorphic Forms on Shimura Varieties** 2004
Ischebeck, F.; Rao, R.A. **Ideals and Reality** 2005
Ivrii, V. **Microlocal Analysis and Precise Spectral Asymptotics** 1998
Jech, T. **Set Theory** (3rd revised edition 2002)
Jorgenson, J.; Lang, S. **Spherical Inversion on SLn (R)** 2001
Kanamori, A. **The Higher Infinite** corr. 2nd printing 2005 (2nd ed. 2003)
Kanovei, V. **Nonstandard Analysis, Axiomatically** 2005
Khoshnevisan, D. **Multiparameter Processes** 2002
Koch, H. **Galois Theory of p-Extensions** 2002
Komornik, V. **Fourier Series in Control Theory** 2005
Kozlov, V.; Maz'ya, V. **Differential Equations with Operator Coefficients** 1999
Landsman, N.P. **Mathematical Topics between Classical & Quantum Mechanics** 1998
Leach, J.A.; Needham, D.J. **Matched Asymptotic Expansions in Reaction-Diffusion Theory** 2004
Lebedev, L.P.; Vorovich, I.I. **Functional Analysis in Mechanics** 2002
Lemmermeyer, F. **Reciprocity Laws: From Euler to Eisenstein** 2000
Malle, G.; Matzat, B.H. **Inverse Galois Theory** 1999

Mardesic, S. **Strong Shape and Homology** 2000
Margulis, G.A. **On Some Aspects of the Theory of Anosov Systems** 2004
Murdock, J. **Normal Forms and Unfoldings for Local Dynamical Systems** 2002
Narkiewicz, W. **Elementary and Analytic Theory of Algebraic Numbers** 3rd ed. 2004
Narkiewicz, W. **The Development of Prime Number Theory** 2000
Parker, C.; Rowley, P. **Symplectic Amalgams** 2002
Peller, V. (Ed.) **Hankel Operators and Their Applications** 2003
Prestel, A.; Delzell, C.N. **Positive Polynomials** 2001
Puig, L. **Blocks of Finite Groups** 2002
Ranicki, A. **High-dimensional Knot Theory** 1998
Ribenboim, P. **The Theory of Classical Valuations** 1999
Rowe, E.G.P. **Geometrical Physics in Minkowski Spacetime** 2001
Rudyak, Y.B. **On Thom Spectra, Orientability and Cobordism** 1998
Ryan, R.A. **Introduction to Tensor Products of Banach Spaces** 2002
Saranen, J.; Vainikko, G. **Periodic Integral and Pseudodifferential Equations with Numerical Approximation** 2002
Schneider, P. **Nonarchimedean Functional Analysis** 2002
Serre, J-P. **Complex Semisimple Lie Algebras** 2001 (reprint of first ed. 1987)
Serre, J-P. **Galois Cohomology** corr. 2nd printing 2002 (1st ed. 1997)
Serre, J-P. **Local Algebra** 2000
Serre, J-P. **Trees** corr. 2nd printing 2003 (1st ed. 1980)
Smirnov, E. **Hausdorff Spectra in Functional Analysis** 2002
Springer, T.A. Veldkamp, F.D. **Octonions, Jordan Algebras, and Exceptional Groups** 2000
Sznitman, A.-S. **Brownian Motion, Obstacles and Random Media** 1998
Taira, K. **Semigroups, Boundary Value Problems and Markov Processes** 2003
Talagrand, M. **The Generic Chaining** 2005
Tauvel, P.; Yu, R.W.T. **Lie Algebras and Algebraic Groups** 2005
Tits, J.; Weiss, R.M. **Moufang Polygons** 2002
Uchiyama, A. **Hardy Spaces on the Euclidean Space** 2001
Üstünel, A.-S.; Zakai, M. **Transformation of Measure on Wiener Space** 2000
Vasconcelos, W. **Integral Closure. Rees Algebras, Multiplicities, Algorithms** 2005
Yang, Y. **Solitons in Field Theory and Nonlinear Analysis** 2001